Series: Environmental and Agricultural Nanotechnology

Series Editor: Kamel A. Abd-Elsalam

Volume

Biopolymer Nanoparticles

A Novel Approach for Eco-conscious Environmental Applications

Books in the Series

- Biopolymer Nanoparticles: A Novel Approach for Eco-conscious Environmental Applications
  Ayat F. Hashim, Rabab M.A. El-Sherif, Farah K. Ahmed and Sabu Thomas (eds.), 2026
- Green Nanotechnology for Stress-Resilient Crops and Sustainable Agroecosystems
  Jen-Tsung Chen (ed.), 2026
- Soil Nanotechnology
  Akhtar Hameed and Muhammad Zahid (eds.), 2026

Series: Environmental and Agricultural Nanotechnology

Series Editor: Kamel A. Abd-Elsalam

Volume

Biopolymer Nanoparticles

A Novel Approach for Eco-conscious Environmental Applications

Editors

Ayat F. Hashim
Food Industries and Nutrition Research Institute
National Research Centre (NRC)
Giza, Egypt

Rabab M.A. El-Sherif
Faculty of Nanotechnology
Cairo University
Giza, Egypt

Farah K. Ahmed
Biotechnology program, Faculty of Agriculture
Cairo University
Giza, Egypt

Sabu Thomas
School of Energy Materials
Mahatma Gandhi University
Kottayam, Kerala, India

CRC Press
Taylor & Francis Group
Boca Raton London New York

CRC Press is an imprint of the
Taylor & Francis Group, an **informa** business
A SCIENCE PUBLISHERS BOOK

First edition published 2026
by CRC Press
2385 NW Executive Center Drive, Suite 320, Boca Raton FL 33431

and by CRC Press
4 Park Square, Milton Park, Abingdon, Oxon, OX14 4RN

CRC Press is an imprint of Taylor & Francis Group, LLC

Library of Congress Cataloging-in-Publication Data (applied for)

ISBN: 978-1-032-87488-3 (hbk)
ISBN: 978-1-032-90499-3 (pbk)
ISBN: 978-1-003-55833-0 (ebk)

DOI: 10.1201/9781003558330

Typeset in Times New Roman
by Prime Publishing Services

Preface to the Series

Welcome to the "Environmental and Agricultural Nanotechnology" book series, a specialty forum revealing the breakthrough promise of nanotechnology in two of the most vital areas of human endeavor: environment and agriculture. As Kamel A. Abd-Elsalam, Series Editor from the Plant Pathology Research Institute, Agricultural Research Center (ARC), I'm pleased to present this series of publications by leading figures in the field.

Nanotechnology is a promise technology, with creative solutions to the majority of the globe's most contentious problems. It presents innovative alternatives in environmental applications to prevent pollution, protect resources, and effectively address environmental effects. Consider new nanomaterials with previously unavailable pollutant absorption properties from air, water, and soil, or new sensors with specific environmental detection and cleansing. These are no longer concepts on paper anymore; they are becoming things and hard facts in a rush to redefine our way of environmental stewardship.

Parallel to the concerns, nanotechnology is also going to revolutionize agriculture, ushering in new avenues of food production and sustainable land management. This includes higher yields from crops through improved delivery of nutrients, optimization of nutrient use efficiency, enhanced fertility of the soil, and development of climate change-resilient agro-systems. The combination of the nanotechnological advances is the door to a future where farming is not only productive but considerably more sustainable.

This book series is to be a comprehensive treatise, taking the reader through the fundamentals of and concepts in nanoscience, nanomaterials, and their unique properties in relation to environmental and agronomic applications. We cover the specifics of the application of nanotechnology in controlling environmental pollution, e.g., its worth in contaminant decontamination and environmental monitoring. Moreover, we detail how nanotechnology can be applied in sustainable agricultural use and resource management, such as the application of targeted agrochemical delivery as well as crop adaptation to climate change.

Through these volumes, we hope to impart a deep appreciation of recent advances and envision future breakthroughs. This volume is a must-read for researchers, academics, policymakers, and entrepreneurs interested in leveraging nanotechnology for a sustainable and green future. In our opinion, by bridging the gap between frontier nanoscience and practical use in the environment and agriculture, we can contribute importantly to global efforts for sustainable development and higher living standards.

Preface

Pollution, climate change, resource depletion, and trash accumulation are some of the 21st-century environmental problems that require meaningful and innovative solutions. Biopolymers have emerged as a crucial resource in the search for environmentally friendly alternatives because they provide renewable, biodegradable, and adaptable materials for a variety of uses. When paired with nanotechnology, biopolymer nanoparticles could assist in identifying and solving some of the most urgent problems that society and the environment are currently facing.

The book "Biopolymer Nanoparticles: A Novel Approach for Eco-conscious Environmental Applications" investigates the synthesis, properties, and uses of biopolymer-based nanoparticles as sustainable materials in environmental science. The combination of nanotechnology and biopolymer research has resulted in advances in water purification, soil remediation, biodegradable packaging, controlled delivery, and other critical disciplines. This multidisciplinary research area combines environmental engineering, chemistry, biotechnology, and material science to provide new and long-term solutions to global concerns. The chapters in this book provide an overview of the most important features of eco-friendly materials and advancements in various methods of nanoparticle synthesis and their applications for addressing environmental challenges. This book balances theoretical and practical approaches, serving as both an academic reference and a practical guide for researchers, students, and practitioners interested in sustainable materials development.

The goal of this book is to provide a thorough analysis of recent developments and research in the field of biopolymeric nanoparticles for eco-conscious environmental applications. The authors and contributors to this book are leading experts whose research work is related to the advancement of sustainable technologies. Their combined expertise underlines the transformative potential of biopolymer nanoparticles in driving innovation while preserving the health of our planet. This book is envisioned to inspire researchers, scientists, policymakers, and the industry into pursuing a myriad of exciting opportunities availed by biopolymer nanoparticles. Our own hope is that this work shall contribute to nurturing sustainable practices and innovations that shall be in concert with the core principles of environmental stewardship and a circular economy.

The chapters discuss the use of biopolymeric nanocomposites for water purification, removal, and degradation of environmental and organic pollutants, biosensing applications, environmental monitoring, soil remediation, and adsorbent material. The book also discusses the environmental fate and toxicity of biopolymer-based metal nanoparticles and their effect on plant growth and health. It also goes into the use of biopolymer-based silver nanoparticles for food packaging or edible coatings.

The book chapters were written by well-known experts in the field and provide important new information about how biopolymeric nanoparticles positively impact the environment. Inspiring further research and innovation in this fascinating field, we believe this book will be a valuable tool for academics, students, and practitioners in environmental science, agriculture, and nanotechnology. We thank the authors who wrote it and the reviewers who provided insightful comments and suggestions.

Ayat F. Hashim National Research Centre, Egypt
Rabab M.A. El-Sherif Cairo University, Egypt
Farah K. Ahmed Cairo University, Egypt
Sabu Thomas Mahatma Gandhi University, India

Contents

CHAPTER-1

Metal-biopolymer Based Nanocomposite

A Potential Antibiotics Material

Rishabh Anand Omar,[1] *Neetu Talreja,*[2,*] *Divya Chauhan*[3] and *Mohammad Ashfaq*[4,*]

1. Introduction

An increasing worldwide health concern is antibiotic resistance, which makes treating bacterial infections extremely difficult. Many standard antibiotics are now useless due to the overuse and abuse of antibiotics, which has expedited the creation of resistant strains (Aguirre-Becerra et al., 2022). This alarming trend necessitates the development of new and emerging materials to control bacterial infections, ensuring the effectiveness of medical interventions and the safety of patients. Among the promising solutions emerging in the field of antimicrobial materials are Metal-biopolymer (MBP)-based nanocomposites (MBP-NC) (Elsakhawy et al., 2018). MBP-NC are a new class of materials that combine the special qualities of biopolymers with metal nanoparticles (M-NPs). Ag, Cu, and Zn-based M-NPs are widely known for having strong antibacterial qualities (Nasrollahzadeh, 2021; Ashfaq et al., 2016; Omar et al., 2019a; Sasidharan et al., 2021; Omar et al., 2019b; Chauhan et al., 2023). These M-NPs can rapture cell membranes of bacteria, produce Reactive Oxygen Species (ROS), and interfere with essential bacterial enzymes and DNA, leading to the destruction of bacterial cells. However, the direct use of M-NPs can pose toxicity issues and stability challenges, limiting their practical applications (Zhao et al., 2023).

Biopolymers, nevertheless, offer a range of beneficial properties, including biodegradability, non-toxicity, biocompatibility, and the ability to form hydrogels, films, and scaffolds. These natural or synthetic polymers, derived from renewable resources, provide a versatile platform for the incorporation of M-NPs. The combination of M-NPs with biopolymers not only enhances the antimicrobial efficacy but also mitigates the potential toxicity of the metals, providing a safer and more effective material for medical applications (Kučuk et al., 2023; Nathanael et al., 2020). The synthesis of MBP-NC involves various techniques, such as *in situ* reduction, electrostatic assembly,

[1] Centre for Environmental Science and Engineering, Indian Institute of Technology Kanpur, Kanpur, 208016, India.
[2] Faculty of Science and Technology, Department of Science, Alliance University, Bengaluru, Karnataka 562106, India.
[3] Department of Drinking Water and Sanitation, Ministry of Jal Shakti, 1208-A, Pandit Deendayal Antyodaya Bhawan, CGO Complex, Lodhi Road, New Delhi 110003.
[4] School of Sciences, Woxsen University, Telangana, Hyderabad 502345, India.
* Corresponding authors: neetutalreja99@gmail.com; mohdashfaqbiotech@gmail.com
DOI: 10.1201/9781003558330-1

and solvent casting. These methods allow for precise control over the dimension, distribution, and loading of M-NPs in between the biopolymer matrix, tailoring the nanocomposite characteristics for specific applications. Characterization techniques, including electron microscopy, spectroscopy, and thermal analysis, are employed to understand the structural, morphological, and functional attributes of these nanocomposites (Shaikh et al., 2024). The antimicrobial mechanisms of MBP-NC are multifaceted. The metal ions that are released from the M-NPs exert a direct bactericidal effect, while the biopolymer matrix can improve these ions' stability and regulated release. Additionally, the nanocomposites can disrupt biofilms, which are protective layers formed by bacterial communities, making them more susceptible to treatment (Ramisetti et al., 2024). The synergy between the M-NPs and biopolymers results in enhanced antimicrobial activity, making these nanocomposites highly effective against a number of bacteria, including the strains having antibiotic-resistance (Swarupa et al., 2024).

The potential applications of MBP-NC are vast and diverse. In the medical field, these nanocomposites can be used in the development of advanced wound dressings, coatings for medical devices, and drug delivery systems. Wound dressings incorporating MBP-NC can provide sustained antimicrobial activity, promoting faster healing and reducing the risk of infections (Al-Mosawi and Abdulsada, 2024). Medical device coatings, such as those for implants and catheters, can stop bacteria from colonizing and forming biofilms, which lowers the risk of illnesses linked to the devices. Furthermore, the controlled release characteristics of these nanocomposites enhance targeted drug delivery, increasing the effectiveness and reducing the side effects of antimicrobial therapies (Al-Kaisy et al., 2024). Figure 1 shows the schematic illustration of the MBC-NC and their applications.

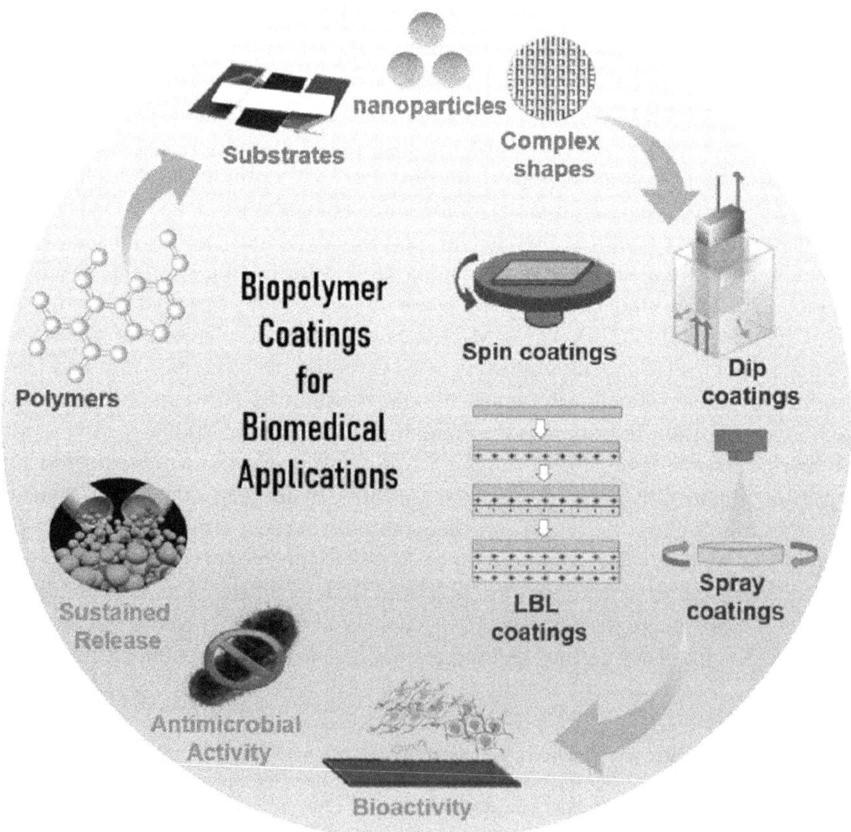

Figure 1. A schematic illustration of the MBC-NC and their applications. The image was reproduced with the permission, under common creative license (Nathanael et al., 2020).

Despite their promising potential, several challenges must be addressed to optimize MBP-NC for clinical use. These include ensuring the consistency and reproducibility of nanocomposite synthesis, understanding the long-term biocompatibility and environmental impact, and overcoming regulatory hurdles for clinical approval. The goal of the research is to solve these issues, with advancements in nanotechnology and materials science paving the way for the development of next-generation antibiotic materials.

The emergence of antibiotic resistance has highlighted the urgent need for new antimicrobial strategies. MBP-NC offer a promising solution, leveraging the antimicrobial properties of M-NPs and the biocompatibility of biopolymers (Borah et al., 2024). This chapter explores the synthesis, characterization, and antimicrobial mechanisms of these nanocomposites, highlighting their potential applications and addressing the challenges in their development. The findings underscore the potential of MBP-NC as next-generation antibiotic materials, offering a sustainable and efficient solution to the global antibiotic resistance crisis.

2. Metal-Biopolymer Based Nanocomposite

The development of MBP-NC represents a noteworthy development in the antibacterial materials sector. These nanocomposites integrate the beneficial properties of M-NPs and biopolymers to create materials with enhanced antimicrobial efficacy, biocompatibility, and environmental sustainability (Borah et al., 2024). The synergy between metals like Ag, Cu, and Zn, which are known for their potent antibacterial properties, and biopolymers, which provide a biocompatible and biodegradable matrix, results in a versatile platform for combating bacterial infections. The incorporation of M-NPs into biopolymers can be achieved through various synthesis techniques, such as *in situ* reduction, which involves the reduction of metal salts within the polymer matrix, or electrostatic assembly, where M-NPs are electrostatically bound to the biopolymer. These methods ensure the uniform distribution of NPs and control over their size and morphology, which are critical for optimizing antimicrobial activity and minimizing cytotoxicity (Tadge et al., 2023).

The antimicrobial properties of MBP-NC are primarily attributed to the M-NPs. These NPs exert their bactericidal effects through multiple mechanisms, including the generation of reactive oxygen species (ROS), disruption of bacterial cell membranes, and interference with vital cellular processes. For instance, AgNPs are known to interact with thiol groups in proteins and enzymes, leading to the inactivation of essential bacterial functions. Cu and ZnNPs also exhibit similar antimicrobial mechanisms, disrupting cell membrane integrity and generating ROS that cause oxidative stress and cellular damage (Zhao et al., 2023). The biopolymer matrix in these nanocomposites plays a crucial role in enhancing the stability and controlled release of metal ions, providing a sustained antimicrobial effect. Additionally, the biopolymer can prevent the aggregation of NPs, ensuring their effective dispersion and interaction with bacterial cells (Matar et al., 2023). MBP-NC have a wide range of potential uses in the biomedical industry. Wound dressings represent one of the most promising uses, where these nanocomposites can provide sustained antimicrobial activity, promote the healing of wounds, and lower the risk of infection. The incorporation of M-NPs into biopolymer-based wound dressings can create an environment that is hostile to bacteria while being conducive to tissue regeneration (Yan et al., 2020). For example, AgNPs embedded in a chitosan biopolymer matrix have been shown to significantly enhance the antimicrobial properties of wound dressings, accelerating the healing process, and preventing bacterial colonization. Similarly, metal-biopolymer nanocomposites can be used as coatings for medical devices, such as catheters and implants, to prevent biofilm formation and reduce the incidence of device-associated infections (Borah et al., 2024). The biopolymer coating provides a biocompatible interface, while the M-NPs inhibit bacterial adhesion and proliferation, ensuring the long-term functionality and safety of medical devices (Matar et al., 2023).

Drug delivery is another critical application area for MBP-NC. These nanocomposites can be engineered to provide targeted and controlled release of antimicrobial agents, enhancing the efficacy of treatment, and reducing side effects. The biopolymer matrix can be designed to degrade in response to specific physiological conditions, releasing the encapsulated M-NPs or drugs at the site of infection (Warkara and Meena, 2022). This targeted delivery approach not only improves the therapeutic outcomes but also minimizes the risk of developing antibiotic resistance by reducing the overall dosage required. For instance, M-NPs encapsulated in a biodegradable polymer can be used to deliver antibiotics directly to infected tissues, bypassing systemic circulation and reducing the potential for adverse reactions (Tabasi et al., 2021).

Despite the promising potential of MBP-NC, several challenges remain to be addressed to ensure their successful clinical translation. One of the primary challenges is the need for consistent and reproducible synthesis methods that can produce nanocomposites with uniform properties. Variations in nanoparticle size, distribution, and loading can significantly impact the antimicrobial efficacy and biocompatibility of the final product. Additionally, the long-term biocompatibility and environmental impact of these nanocomposites need to be thoroughly investigated to ensure their safety for clinical use. The potential for M-NPs to cause cytotoxic effects and the degradation products of biopolymers to elicit immune responses are areas of concern that require comprehensive evaluation. Furthermore, regulatory approval processes for new antimicrobial materials are stringent, necessitating extensive preclinical and clinical studies to demonstrate safety and efficacy (Yan et al., 2020). MBP-NC offer a promising solution to the growing problem of antibiotic resistance. By combining the antimicrobial properties of M-NPs with the biocompatibility and biodegradability of biopolymers, these nanocomposites provide a versatile and effective platform for a wide range of biomedical applications. The ongoing research and development efforts in this field are focused on optimizing the synthesis, characterization, and functionalization of these materials to maximize their antimicrobial efficacy and minimize potential risks. The successful integration of MBP-NC into clinical practice could provide a sustainable and efficient solution to the global antibiotic resistance crisis, offering new hope for the prevention and treatment of bacterial infections (Liz et al., 2018).

2.1 Types of MBC-NC

MBP-NC encompass a diverse range of materials, each tailored to leverage the unique properties of specific metals and biopolymers for various applications. These nanocomposites can be classified based on the type of M-NPs used, the biopolymer matrix employed, and the intended application. Here are some of the key types of MBP-NC:

2.2 Ag-based MBC-NC

Ag-NPs, are well known for having strong antibacterial qualities, making them a popular choice for developing nanocomposites. When integrated into biopolymer matrices, such as chitosan, alginate, or cellulose, Ag-NPs enhance the antimicrobial efficacy of the composite while maintaining biocompatibility and biodegradability. These nanocomposites are widely used in bandages for wounds, coverings for medical equipment, and water filtration systems. The Ag ions released from the NPs can disrupt bacterial cell membranes, inhibit enzyme function, and generate reactive oxygen species, leading to the effective killing of a broad spectrum of bacteria.

2.3 Cu-based MCB-NC

Cu-NPs offer strong antimicrobial activity, second only to Ag, but at a lower cost. Cu-based MCB-NC are typically synthesized using biopolymers like chitosan, gelatin, or Poly (Lactic Acid) (PLA). These nanocomposites are particularly effective against Gram-negative and Gram-positive bacteria, as well as fungi. The release of Cu ions from the NPs induces oxidative stress and damages cellular components, leading to microbial death. Cu-biopolymer nanocomposites find applications in antimicrobial coatings, wound dressings, and food packaging materials, providing long-lasting protection against microbial contamination (Pathak et al., 2021).

2.4 ZnO-based MCB-NC

ZnO-NPs are another widely studied antimicrobial agent. ZnO-based MCB-NC combine the antimicrobial properties of ZnO with the biocompatibility of biopolymers like starch, chitosan, and polycaprolactone (PCL). ZnO-NPs exhibit photocatalytic activity and generate reactive oxygen species under UV light, which can effectively kill bacteria. These nanocomposites are used in sunscreen formulations, antimicrobial coatings, and biomedical devices. The ability of ZnO-NPs to induce apoptosis in bacterial cells without harming human cells makes them an attractive option for various medical applications (Ramesan et al., 2018).

2.5 Au-based MCB-NC

Au-NPs are valued for their stability, biocompatibility, and unique optical properties. Au-based MCB-NC are typically synthesized using biopolymers such as collagen, hyaluronic acid, or Poly(Vinyl Alcohol) (PVA). Although Au itself has limited antimicrobial properties, the high surface area of Au-NPs allows for the functionalization with various antimicrobial agents, enhancing the overall efficacy of the nanocomposite. These composites are used in biosensors, drug delivery systems, and imaging applications due to their excellent stability and functionalization capabilities (Suresh et al., 2018).

2.6 TiO_2-based MBC-NC

TiO_2-NPs are known for their photocatalytic activity and strong antimicrobial properties. TiO_2-based MCB-NC are typically synthesized using biopolymers like cellulose, chitosan, and poly (vinyl alcohol) (PVA). Under UV light, TiO_2-NPs generate reactive oxygen species that can degrade organic pollutants and kill bacteria. These nanocomposites are widely used in environmental applications, such as water and air purification, as well as in antimicrobial coatings for medical devices and surfaces (Gumus et al., 2020).

2.7 Fe_3O_4-based MBC-NC

Fe_3O_4-NPs offer unique magnetic properties, in addition to their antimicrobial activity. Iron oxide-biopolymer nanocomposites are typically synthesized using biopolymers such as dextran, chitosan, and alginate. These composites are utilized in targeted medication administration, Magnetic Resonance Imaging (MRI) contrast agents, and cancer therapy heat. The magnetic properties of Fe_3O_4-NPs allow for the controlled manipulation of the nanocomposites using external magnetic fields, enabling precise targeting and delivery of therapeutic agents (Dhivya et al., 2016).

3. Synthesis Methods of MBC-NC

The synthesis of MBP-NC involves the integration of M-NPs into biopolymer matrices, which requires precise control over the size, distribution, and loading of the NPs. This process is critical for optimizing the antimicrobial efficacy, biocompatibility, and stability of the resulting nanocomposites. Several synthesis methods have been developed to achieve these objectives, each with its unique advantages and challenges. Here, we delve into the detailed synthesis methods commonly used to fabricate MBP-NC.

3.1 In situ Reduction

In situ reduction is one of the most widely used methods for synthesizing metal-biopolymer nanocomposites. This technique involves the reduction of metal ions within the biopolymer matrix, leading to the formation of M-NPs. The biopolymer serves as both a reducing and stabilizing agent, facilitating the formation and stabilization of the NPs. A metal salt, like $AgNO_3$/$CuSO_4$, is dissolved in a solution containing the biopolymer (e.g., chitosan, gelatin, or alginate). The biopolymer solution acts as a reducing agent, converting metal ions into M-NPs. This can be achieved by adjusting the pH, and temperature, or by adding additional reducing agents like sodium borohydride ($NaBH_4$). The reaction mixture is stirred continuously to ensure uniform distribution of NPs within the biopolymer matrix. After the reduction process is complete, the nanocomposite is purified by removing unreacted metal ions and other by-products through filtration, centrifugation, or dialysis (Sharma et al., 2018). The final product can be processed into films, fibres, or hydrogels, depending on the desired application. It provides a uniform distribution of M-NPs within the biopolymer matrix with a simple and cost-effective process. The biopolymer matrix acts as a capping agent, preventing the agglomeration of NPs. In this process control over nanoparticle size and shape is still challenging. Also, reaction conditions need to be carefully optimized to avoid unwanted side reactions (Zienkiewicz-Strzałka and Deryło-Marczewska, 2020).

3.2 Electrostatic Assembly

Electrostatic assembly leverages the electrostatic interactions between charged M-NPs and the functional groups of biopolymers to create nanocomposites. This method allows for precise control over the loading and distribution of NPs within the biopolymer matrix. M-NPs with surface charges are synthesized separately using methods such as chemical reduction or green synthesis. The biopolymer solution is prepared, ensuring that it has functional groups (e.g., amino, carboxyl) that can interact electrostatically with the charged NPs.The M-NPs are mixed with the biopolymer solution under controlled conditions, allowing the NPs to bind to the biopolymer through electrostatic interactions. The resulting nanocomposite is purified and processed into the desired form, such as films, coatings, or scaffolds (Tadi et al., 2022). The process provided precise control over nanoparticle loading, distribution, and mild reaction conditions, preserving the functional integrity of both NPs and biopolymers. The method could be applied to a wide range of biopolymers and NPs. Challenges in this method are the stability of the electrostatic interactions can be influenced by pH and ionic strength and it requires pre-synthesized M-NPs, adding a step to the process (Del et al., 2020).

3.3 Solvent Casting

Solvent casting is a straightforward method for fabricating metal-biopolymer nanocomposites, where M-NPs are dispersed in a biopolymer solution and then cast into films or other desired shapes. M-NPs are synthesized and stabilized using appropriate agents to prevent aggregation. The biopolymer is dissolved in a suitable solvent (e.g., water, ethanol) to form a solution. M-NPs are dispersed into the biopolymer solution using techniques such as ultrasonication or stirring to ensure uniform distribution. The mixture is cast onto a flat surface or mould and allowed to dry, forming a solid film or shaped nanocomposite. The solvent is evaporated under controlled conditions (e.g., room temperature, vacuum), resulting in the formation of a nanocomposite with well-dispersed M-NPs (Ramisetti et al., 2024). Solvent casting is a simple and scalable process that can produce thin films and coatings with uniform nanoparticle distribution and is suitable for a wide range of biopolymers and M-NPs. However, solvent removal can be time-consuming and may require specific conditions. Also, control over nanoparticle dispersion can be challenging without adequate mixing techniques (Meera and Ramesan, 2023).

3.4 Green Synthesis

Green synthesis methods employ natural reducing agents, such as plant extracts, to synthesize M-NPs within the biopolymer matrix. This approach is environmentally friendly and sustainable, reducing the reliance on toxic chemicals. Plant extracts containing natural reducing agents (e.g., phenolic compounds, flavonoids) are prepared by boiling plant materials in water or ethanol. A solution of the metal salt is mixed with the plant extract, leading to the reduction of metal ions and the formation of M-NPs. The biopolymer solution is added to the mixture, allowing the NPs to be encapsulated within the biopolymer matrix. The resulting nanocomposite is purified and processed into the desired form, such as films, gels, or fibres (Jafarzadeh et al., 2024). Green synthesis is an environmentally friendly and sustainable process, that reduces the use of hazardous chemicals, and solvents and utilizes renewable resources for nanoparticle synthesis. However, variability in the composition of plant extracts can lead to inconsistent results. Also, optimization of reaction conditions is required to achieve the desired nanoparticle properties (Mondal et al., 2020).

3.5 Emulsion-Based Methods

Emulsion-based methods involve the formation of emulsions (mixtures of two immiscible liquids) to create nanocomposites with well-dispersed M-NPs. A biopolymer solution (aqueous phase) is prepared, and metal salts are dissolved in this solution. An organic solvent (oil phase) containing a surfactant is added to the aqueous phase to form an emulsion. The emulsion is stabilized by continuous stirring or ultrasonication. A reducing agent is added to the emulsion to convert metal ions into NPs within the biopolymer matrix. The emulsion is broken by adding a demulsifier or by solvent evaporation, resulting in the formation of a nanocomposite. The nanocomposite is purified and processed as needed (Zewde et al., 2020). This process enables the formation of NPs with a controlled size and shape. It is suitable for producing nanocomposites with high nanoparticle loading that can be used to create a variety of nanocomposite forms, such as films, fibres, and beads. However, the process requires careful control of emulsion stability and droplet size. Also, the presence of surfactants and solvents may necessitate additional purification steps (Kedir et al., 2022).

4. Antibacterial Applications of Metal-Doped Biopolymers

4.1 Ag-based MBC-NC

Ag-based MBC-NC are extensively utilized in various industries due to their strong antimicrobial properties against bacteria, fungi, and viruses. These composites are utilized in targeted medication administration, Magnetic Resonance Imaging (MRI) contrast agents, and cancer therapy heat (Singh et al., 2017). Additionally, they are employed in the production of textiles, food packaging, and drinking water treatment. The applications of these nanocomposites are broad, ranging from bone regeneration to wound dressings. For instance, chitosan/nanohydroxyapatite scaffolds containing Ag-NPs have shown antimicrobial activity against bacteria without harming human osteoprogenitor cells. Similarly, antibacterial bioactive glass-substituted AgO-NPs are used in bone graft nanocomposites, promoting the formation of hydroxyapatite. Porous antibacterial sponges made from cellulose composite hydrogels containing Ag-NPs demonstrate excellent antibacterial properties and accelerated wound healing (Makvandi et al., 2019). Other applications include soft tissue regeneration, where Ag oxide NPs loaded in silk fibroin nanocomposites exhibit potent antibacterial activity and enhanced healing ability. Various polymers are also used in combination with Ag-NPs for antimicrobial scaffold fabrication, further expanding their applications in tissue engineering and wound healing. The antimicrobial mechanisms of Ag nanostructures have been extensively studied (Babu et al., 2018). Different bacteria' cell surfaces physically interact with Ag-NPs. It is thought that interactions between Ag ions and the peptidoglycan cell wall of bacteria result in structural alterations that raise membrane permeability and ultimately kill the cells. Additionally, bacterial proteins' exposed sulphhydryl groups interact with Ag-NPs to prevent DNA replication. Similarly, AgO-NPs interact with bacterial DNA, rendering them unable to replicate. Because of this interaction, DNA damage stops the cell cycle in the G2/M phase, which eventually results in oxidative stress-induced cell death (Dizaj et al., 2014). Viruses without an exterior lipid membrane are more stable than those with an envelope. Compared to Au-NPs (6 to 20%), Ag-NPs have a greater antiviral efficacy against HIV-1 (98%) (Castro-Mayorga et al., 2017).

4.2 TiO_2-based MBC-NC

The antibacterial and antifungal characteristics of TiO_2-NPs have been brought to light in recent research. To improve the quality of food, titanium compounds and NPs are frequently added to oral medications and food-related goods. As a result, research has concentrated on the cytotoxicity of TiO_2-NPs in the digestive system. While some studies show low entrance into the bloodstream, others imply that these NPs can pass the intestines and disperse in multiple organs. Cellular-level cytotoxicity of TiO_2-NPs has yielded conflicting results, with both toxic and non-toxic outcomes reported. Oral administration studies in rats have shown uptake of TiO_2-NPs in the 100–300 nm range, and liver tissue examinations in mice exposed to TiO_2-NPs orally demonstrated mitochondrial edema in hepatocytes, disrupting liver metabolism. TiO_2-NPs have been mixed with dental resins in dentistry to create tooth-filling materials with antimicrobial properties (Tassinari et al., 2014). TiO_2-NPs have also been added to antibacterial paints or coatings for touch surfaces in hospitals, where they may be activated by visible light to provide antibacterial effects. TiO_2 nanotubes have been loaded with Ag-NPs, which have an average diameter of 5 nm, to improve their antibacterial capabilities. This hybrid material generates hydroxyl radicals and superoxide anions when exposed to ultraviolet or visible light, demonstrating increased antimicrobial activity against bacteria like *S. aureus* under sunlight exposure (Kubacka et al., 2014).

4.3 ZnO-MBC-NC

It is known that ZnO-NPs have broad-spectrum antibacterial activity. Research has revealed that ZnO-NPs exhibit significantly higher bactericidal activity against *S. mutans* compared to Ag-NPs at the same concentration. ZnO-NPs have an antibacterial mechanism that includes producing reactive oxygen species (ROS) inside cells, which ultimately results in the death of bacteria. Studies utilizing confocal microscopy indicate that the generation of reactive oxygen species (ROS) and the build-up of NPs in the cytoplasm or on the outer membrane of the cell, which physically damages the bacterial cell wall, are both responsible for this action (Makvandi et al., 2019). Particle size affects ZnO-NPs' antibacterial and antifungal properties, with smaller particles generally exhibiting greater antimicrobial effects. Under physiological settings, ZnO-NPs have significant positive surface charges that facilitate their contact with bacteria through electrostatic interactions. This promotes phagocytosis, cellular absorption, and bacterial toxicity (Haddaoui and Raouafi, 2015). Both ZnO and TiO_2-NPs are commonly used in cosmetic nonprescription drug products due to their ability to absorb ultraviolet light without the white chalky appearance of traditional sunscreens. Modifications to the ultraviolet protection system of these NPs have been reported to enhance sun protection. ZnO-NPs in sunscreens demonstrate antibacterial activity against *E. coli*, with smaller NPs producing more reactive oxygen species (ROS) than larger ones. The food sector also uses ZnO-NPs as additives for active or smart packaging and as antibacterial agents against foodborne microorganisms. According to Valencia et al. (2019), they transmit antibacterial activity and improve mechanical qualities, which have uses in tissue regeneration and restorative dentistry. Because of its antibacterial qualities, ZnO nanosheets, nanorods, and nanowires are used in biomedicine in addition to spherical materials. Through both the release of Ag ions and the direct killing of bacteria by the ZnO nanorods, a hybrid nanomaterial made of Ag-NPs placed on ZnO nanorods demonstrates dual antibacterial activity. With regards to *E. coli* and *Bacillus subtilis*, this hybrid material has strong antibacterial activity without causing cytotoxicity in human cells (Agnihotri et al., 2015). *In vitro* studies have shown that Zn has a proliferative effect on osteoblastic cells and inhibits osteoclastic bone resorption. For bone tissue engineering, porous and nanostructured ZnO-incorporated TiO_2 coatings on titanium implants have been developed to enhance the growth of bone marrow stem cells, exhibiting antimicrobial activity against *S. aureus* and *E. coli* while promoting bone formation (Feng et al., 2014).

4.4 Ni-based MBC-NC

When applied to strains of *K. pneumoniae* and *E. coli* that are resistant to many drugs, nickel, and Ni $(OH)_2$-NPs show antibacterial properties. These NPs are used in food packaging, pharmaceuticals, textiles, and water disinfection. While the antimicrobial properties of Ni-based NPs are not as strong as those of Ag and Si-NPs, they surpass those of Au-NPs. The antibacterial capabilities of Ni NPs are becoming closer to those of Ag due to advancements in surface modification and production procedures (Zhang et al., 2013). The manufacture of Ni-NPs by the use of plant and bacterial extracts has replaced conventional physical and chemical processes, as the focus on environmentally friendly manufacturing grows. Polymer-mediated surface modification of Ni-NPs has been explored to enhance their antibacterial properties and cytocompatibility. For instance, chitin nanogels loaded with Ni-NPs (120–150 nm) show higher toxicity than pure Ni-NPs to bacterial strains at lower doses, while being completely non-toxic to A549 and L929 cells *in vitro*. The features of M-NPs that are antibacterial and cytocompatible can be affected by the selection of bacterium, metal supply, and polymer (Kumar et al., 2013). The changes to the size and form of Ni-NPs can lead to further improvements in their cytocompatibility and antibacterial qualities. Shape-determining crystallographic growth rates may be regulated by varying experimental parameters such pH, metal-to-reducing agent ratio, irradiation duration, and microwave synthesis energy. These qualities can also be improved by employing nickel hydroxide NPs or nickel oxide nanoplates, which provide more morphological and size variations. To increase their antibacterial action, for instance, new platinum-on-flower-like Ni NPs have been created to cause ROS buildup when exposed to UV light (Zhu et al., 2017). Figure 2 Shows the schematic representation of the mode of action of M-NPs against bacterial strains.

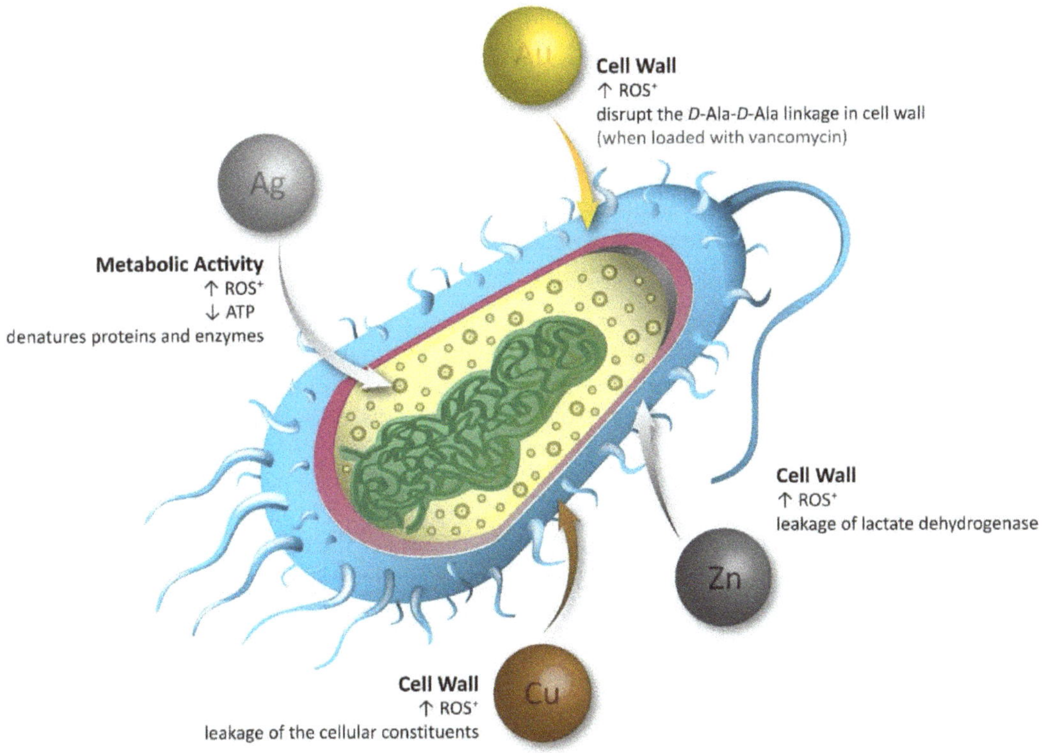

Figure 2. A schematic representation of the mode of action of M-NPs against bacterial strains. The image was reproduced with the permission (Ghazzy et al., 2023).

4.5 Cu-MBC-NC

The small size and high surface-to-volume ratio of Cu-NPs allow them to interact closely with microbial membranes, which has an antibacterial effect. The purpose of synthesizing these NPs was to demonstrate their antibacterial effectiveness against a variety of bacteria and fungus, such as MRSA, *Micrococcus luteus*, *S. aureus*, *E. coli*, *K. pneumoniae*: *aeruginosa*, and several strains of fungal (Kruk et al., 2015). While Cu-NPs exhibit potent antimicrobial activity, especially against MRSA, their bactericidal efficacy is reported to be slightly lower than that of Ag-NPs and some antibiotics. Similar to how Ag works, the mechanism of action of Cu-NPs involves the interaction of released Cu ions with DNA. However, compared to Cu-NPs, Ag-NPs have been shown to be more effective against Gram-negative *E. coli*. This might be because Gram-positive and Gram-negative bacteria have different cell wall structures. To completely comprehend the bactericidal action of Cu-NPs against a larger variety of bacterial strains, more investigation is required (Kailasa et al., 2019). Using herbal extracts, bimetallic core-shell Cu-platinum NPs have been created, and they exhibit strong antimicrobial action against *E. Coli*, *S. aureus*, and *P. aeruginosa*. Because of their antibacterial qualities, Cu-NPs are applied in medicine for many purposes, such as lowering the risk of hospital infections, treating burns, and stopping the growth of germs and fungi on dental materials, catheters, vascular grafts, and prosthesis. To cleanse water, Cu-NPs coated on polyurethane foam and enclosed in alginate beads have been developed as antibacterial water filters against *Escherichia coli* (Dobrucka and Dlugaszewska, 2018). Researchers are investigating the use of Cu-NPs embedded in mussel-inspired dendritic polyglycerol (MI-dPG) as an antibacterial coating for implants and other medical equipment. MI-dPG uses a combination of crosslinking and catechol anchoring to firmly bind to surfaces in a manner akin to that of mussel foot protein-5. After the coating has formed, the Cu-NPs can be incorporated in the MI-dPG coating via *in situ* reduction of $CuCl_2$ solution, diffusion of Cu ions, or during the coating process. These coatings exhibit long-term antibacterial properties against various bacteria through an "attract-kill-release" strategy, involving the production of reactive oxygen species (ROS) and the release of Cu^{2+}. The coatings inhibit biofilm formation and are biocompatible with eukaryotic cells. Compared to Ag and Au-NPs, CuO-based NPs are less expensive (Li et al., 2017). Recently, eco-friendly CuO-NPs synthesized using green approaches have shown lower cytotoxicity compared to those prepared by chemical methods. These NPs exhibit potent antibacterial activity against multidrug-resistant bacteria and demonstrate both antimicrobial and anticancer activities, broadening their potential applications in biomedicine (Dadi et al., 2019).

4.6 Au-based MBC-NC

Au nanomaterials are highly favoured for biomedical applications due to their ease of preparation and biological inertness, which make them ideal for bioconjugation. Incorporating AuNPs into biomedical devices enhances their antimicrobial properties, as seen in the example of Au NPs improving the mechanical properties of poly(methyl) methacrylate (PMMA)-based bone cement and reducing the formation of *S. aureus* biofilms (De Santis, 2017). In spite of its benefits, evaluating AuNPs' long-term impact on human health requires a knowledge of how they interact with biological systems, notably their cytotoxicity. Research from the 1990s has demonstrated that red blood cells and other specific cells may be specifically targeted by Au colloid-based monolayers. The toxicity of these compounds depends on the size, content, and surface characteristics of the Au colloid films (Makvandi et al., 2019). Recent studies have identified Au-NPs in organs like the spleen and liver, indicating their potential uptake by macrophages (Semmler-Behnke et al., 2008). Au-NPs and nanoclusters have been shown to be harmful *in vivo* by altering the number of red blood cells and causing harm to organs including the kidneys, liver, and spleen. To guarantee that goods based on nanotechnology are accepted by the general public, extensive studies into the biosafety of these NPs are required (Zheng et al., 2017). Due to their special optical characteristics, Au-NPs may amplify, absorb, scatter, and fluoresce the light's electromagnetic field. This feature allows for photothermal treatment of bacteria by irradiating Au-based nanostructures with near-infrared radiation, which can be further enhanced by altering the shape of the NPs. These nanostructures show promise in water sterilization and photothermal/photodynamic therapy for bacteria elimination (Ma et al., 2018). In addition to their optical properties, Au-NPs can be used for bacterial identification and antimicrobial therapy. They can be coated with ligands to form hybrid NPs, enhancing their stability and antibacterial activity. For instance, peptide-coated Au-NPs have shown improved stability against trypsin digestion and enhanced antibacterial activities. Moreover, hybrid assemblies of AuNPs have been created for near-infrared photodynamic antimicrobial treatment and Surface-Enhanced Raman Scattering (SERS) imaging, showing great promise in the fight against bacterial strains resistant to antibiotics while protecting eukaryotic cells (Zhou et al., 2018).

4.7 Pd-based MBC-NC

Pd-NPs are highly valued in various fields due to their cost-effectiveness, recyclability, and thermal and mechanical stability compared to the other group of 10 elements. Pd-NPs with size-dependent antibacterial activity against *S. aureus* and *E. coli* have been synthesized by chemical techniques. Green chemistry approaches have also been developed to produce antimicrobial Pd-NPs using substrates like fruit extracts, leaf extracts, microbes, or marine life, which are relatively nontoxic to human cells (Anasdass et al., 2018). Compared to other metals, Pd-NPs has a greater tendency to interchange ligands, which results in the formation of active metal species that are easily able to interact with other substances. This elevated reactivity makes Pd-NPs highly toxic to bacteria, especially when they are reduced to smaller sizes. However, this reactivity also introduces a degree of cytotoxicity that can potentially impact organ function (Gnanasekar et al., 2018). Hybrid organic-Pd-NPs exhibit antibacterial property are gaining popularity. For instance, it has been demonstrated that produced Pd (II) complexes with halo substituted benzylamine ligands have antibacterial activity against both Gram-positive (*S. aureus* and *B. subtilis*) and Gram-negative (*E. coli* and *P. aeruginosa*) bacteria. Another example is an oxime-based Pd (II) complex that demonstrates strong antibacterial action against both Gram-positive and Gram-negative pathogenic microorganisms (/Sharma et al., 2018). Additionally, Pd-loaded nanostructured silicon carbide NPs have been created, and they have shown antibacterial activity against *S. aureus* and *E. coli* (Baig et al., 2018). Furthermore, Pd-N-heterocyclic carbene complexes functionalized with bidentate pyridines that have different wingtip substituents have been produced and assessed for their medicinal uses as antibacterial and antiproliferative therapeutic candidates. Recent research has also examined the antibacterial activity of Pd (II) complexes incorporating thiosemicarbazone ligands. Compared to the free thiosemicarbazone ligand, these complexes have greater antibacterial action against both Gram-positive and Gram-negative bacteria because they can chelate metal ions (El-Sawaf et al., 2008).

4.8 Se-based MBC-NC

Due of their ability to inhibit biofilm formation, se-NPs are frequently used to prevent medical or foodborne infections. Unlike ZnO or Ag-based NPs, which can be toxic to healthy cells and the environment due to the generation of reactive oxygen species (ROS). Se is considered safer and less harmful to healthy cells. Smaller Se NPs exhibit greater inhibitory effects against *S. aureus* infections compared to larger NPs. Additionally, Se-NPs exhibit anti-fungal characteristics, as evidenced by a more pronounced decrease in the viability of Candida albicans in probiotics containing Se-NPs compared to those without Se enrichment (Kheradmand et al., 2014). In the biomedical field, the green synthesis of Se-NPs from non-pathogenic, ecofriendly actinobacteria was first reported in 2015. These NPs demonstrate potent antibacterial activities against a range of bacteria including *P. aeruginosa*, *S. aureus*, *E. coli*, and *Streptococcus pyogenes*, as well as antifungal properties against *Aspergillus clavatus*; their antibacterial efficacy is comparable to the antibiotic ampicillin. Se-NPs have also been found to possess antibiofilm and antioxidant activities and have been experimentally applied to accelerate wound healing in rats with a high success rate.

Food industry: Se-NPs are used to fight major foodborne pathogens including Listeria monocytogenes, *E. coli*, *S. aureus*, and other Salmonella species. Comprehensive reviews on the biomedical applications of Se-NPs are available for interested readers seeking more detailed information (Khiralla et al., 2015; Nguyen et al., 2017).

4.9 Fe-based MBC-NC

Nanoscale Zero-Valent Iron (nZVI)-NPs gained attention initially for their environmental remediation capabilities, detoxifying contaminants through redox reactions (Pandey et al., 2024). Recent studies have focused on assessing the *in vitro* toxicity of nZVI towards terrestrial and aquatic biota. Most studies have used uncoated nZVI-NPs, lacking polymeric coatings. The toxicity mechanisms likely involve oxidative stress and other mechanisms such as cell membrane disruption via Fe^{2+} generation, chlorosis, and hypoxia. The toxicity of naked nZVI-NPs and those coated with cellulose, tetrapolyphosphate, or bismuth to different species, such as bacteria, plants, water fleas, and earthworms, was studied in 2018. By altering the physicochemical characteristics of the NPs, surface modification of nZVI dramatically decreases environmental toxicity, according to particle characterizations (Yoon et al., 2018). A vancomycin-modified, Ag-coated Fe_3O_4 magnetic nanoparticle hybrid has been developed for rapid bacteria detection using surface-enhanced Raman scattering (SERS). Fe_3O_4@Ag is created when 200 nm superparamagnetic Fe_3O_4-NPs are coated with Ag. Pathogens are subsequently separated and concentrated from the solution by conjugating these magnetic NPs with vancomycin (Van). Through its interaction with D-alanine-d-alanine on the bacterial cell walls, conjugated vancomycin aids in the NPs' attachment to the bacterium. In addition, plasmonic core-shell Au@Ag-NPs are produced with the purpose of enhancing secondary SERS. All these biosensors work together to efficiently detect and distinguish between a variety of bacteria, such as MRSA, *E. coli*, and *S. aureus* (Wang et al., 2018). By precisely delivering 3-azido-d-alanine, a chemical that labels the metabolic activity of bacteria, using MIL-100 (Fe)-NPs as nanocarriers, a new approach for simultaneous *in vivo* bacterial metabolic labelling and removal has been created. The NPs accumulate within H_2O_2-rich inflamed tissues and degrade rapidly upon exposure to H_2O_2, releasing 3-azido-d-alanine. This molecule interacts with a fluorescent dye to get specifically integrated inside the bacterial cell wall, where it is then detectable. For the purpose of eliminating germs using photodynamic treatment, the attached molecules are then reacted with a photosensitizer using *in vivo* click chemistry. This therapeutic strategy significantly reduces the number of bacteria within infected tissues. Because of their surface engineering potential, colloidal stability, and biocompatibility, Fe-NPs are highly prized in biomedicine. They can also be readily degraded, which makes them suitable for *in vivo* applications (Mao et al., 2018). Iron oxide NPs' antibacterial and antifungal qualities have been thoroughly investigated. They are useful as catalysts for H_2O_2 activation in root canal treatment and show antibacterial ability against various bacteria strains. Green synthesis methods have been developed for producing iron oxide NPs, and surface modifications with chitosan have enhanced their antimicrobial activity. Cobalt ferrite NPs and bismuth ferrite nanorods have also been synthesized with improved antibacterial properties. These iron oxide-based nanomaterials have dual antimicrobial and magnetic properties, making them valuable for diverse applications in biomedicine and bioanalysis (Bukhari et al., 2018).

4.10 Al-based MBC-NC

Numerous sectors, including alloys, explosives, rocket fuel, wear-resistant coatings, sensors, personal hygiene products, and medicine delivery systems, use Al-NPs extensively. They exhibit a broad inhibitory effect on the growth of *E. coli*, comparable to other metal oxide nanomaterials, throughout a broad concentration range (10–1000 $\mu g\ mL^{-1}$). One of the primary bactericidal mechanisms of Al_2O_3-NPs is the increase in reactive oxygen species (ROS) generation (Ansari et al., 2014). Al_2O_3-NPs have remarkable antibacterial and barrier qualities when mixed with other materials.

For example, adding organic ingredients to the surface of Al_2O_3 to coat board paper with polylactic acid-Al_2O_3 or double-coating polylactic acid films with layers of Al_2O_3 and alginate-chitosan greatly increases the films' water retention, oxygen content, and scent (Valerini et al., 2018). These antibacterial components are frequently used in films used for food packaging. Strong antibacterial activity against *E. coli* is demonstrated by alumina-doped ZnOcoatings on polylactic acid films, which qualifies them for use in food packaging. Chitosan/silica nanocomposite membranes doped with varying concentrations of Al_2O_3-NPs demonstrate excellent spectroscopic and antimicrobial properties, making them ideal for optical and biological applications (El Nahrawy et al., 2019). Nano-Ag-TiO_2 hybrids implanted in anodic Al_2O_3 templates have demonstrated potential in the field of inorganic Al_2O_3 materials; the high porosity of these templates amplifies the contact- and release-killing activities of the NPs against biofilms (Sabry et al., 2018).

4.11 Mg-based MBC-NC

Mg is not commonly used in its pure form for antimicrobial purposes. Instead, MgO and hybrid Mg materials are more frequently employed to combat infections. The effect of MgO-NPs on the survival of mesenchymal stem cells produced from bone marrow and different types of infectious bacteria (such as Gram-positive *Staphylococcus epidermidis* and Gram-negative *E. coli*) depends on the concentration, as was demonstrated *in vitro*. Consequently, whether incorporating MgO-NPs into coating materials or using them as a composite filler for possible medical applications, it is imperative to choose the concentration with great care (He et al., 2016). Bacteria and MgO-NPs combine to cause oxidative stress, leaking of the cell membrane, and ultimately, cell death. MgO-NPs applied thinly and quickly corrode titanium surfaces, exhibiting antibacterial and biocompatible qualities. Moreover, doping silicate hydroxyapatite with Mg^{2+} ions, which are liberated by MgO-NPs, and Ni^{2+} ions produce materials that have significant antimicrobial activity against *E. coli* and *P. aeruginosa* (Alshemary et al., 2015).

4.12 Other MBC-NC

In addition to the more commonly discussed M-NPs mentioned earlier, novel M-NPslately been developed. Cerium (IV) oxide (CeO_2) serves as a notable example of a redox-active nanoparticle capable of transitioning between 4^+ and 3^+ oxidation states. These NPs high nonstoichiometric condition in the monocrystalline form allows them to actively participate in redox reactions. They demonstrate potential for use in antibacterial, antioxidant, and anticancer applications in addition to their efficacy in neutralizing free radicals such as nitroxyl, superoxide, and hydrogen peroxide radicals (Thakur et al., 2019). CeO_2-NPs mimic the actions of catalase and superoxide dismutase to decrease excess reactive oxygen species (ROS), hence acting as antioxidants in healthy human cells. Through direct and indirect interactions, they demonstrate significant antimicrobial activity against Gram-positive and Gram-negative bacteria (Rehana et al., 2017). When CeO_2-NPs meet bacteria directly, they injure the outer cell wall and cause intracellular ROS to be produced. Through indirect interactions, the bacterial microenvironment, including the extracellular polysaccharide matrix, is impacted by the NPs. As a result, ROS are produced that enter bacterial cells through the cell membrane, inflicting damage at different intracellular levels, and ultimately causing cell death. CeO_2-NPs have a redox activity that increases the permeability of bacterial outer membranes to β-lactam antibiotics, which makes them an effective supplement against multidrug-resistant infections like *K. pneumonia* (Bellio et al., 2018). Fungal enzyme activity is irreversibly altered when CeO_2-NPs interact with the components of their cell walls, such as those of *Candida albicans*, producing $Ce3^+$ species that raise the pH of the surrounding milieu and cause oxidative stress (Farias et al., 2018). Novel water-soluble nanocomplexes of platinum (II) with superior antibacterial capabilities have also been created, in addition to CeO_2-NPs. Furthermore, the 2-((E)-(pyridine-2-ylimino) methyl) napthalen-1-ol ligand has been incorporated into new nanoscale Cr (III), Fe (II),

Co (II), and Ni (II) complexes that exhibit enhanced antiviral and antibacterial capabilities over the free ligand (Abdel-Rahman et al., 2016). Table 1 describes various M-NPs and composites and their applications in the biomedical sector.

Table 1. Different MBC-NC in biomedical and antibacterial applications.

S. No.	Metal nanoparticle	Size of NPs (nm)	Application	Microorganism/Cell line	Reference
1.	Ag	20 – 40	Antimicrobial activity in hospital infection	*S. aureus*	Ravindran et al., 2018
2.	Zein-coated Au-NPs	50 – 80	Antimicrobial activity against bacteria and virus	*Bacillus pumilus, Bacillus subtilis, Zika virus*	Suganya et al., 2017
3.	MgO-NPs	50 – 70	Antimicrobial activity against multidrug-resistant microorganisms	*S. aureus, E. coli*	Hayat et al., 2018
4.	Fe-CuO-NPs	21	Antimicrobial activity against Methicillin-resistant bacterial and fungi	*S. aureus, epidermidis, C. albicans*	Pugazhendhi et al., 2018
5.	Ag NPs	15 – 20	Conventional antibiotic therapy with blue light	*P. aeruginosa*	Nour El Din et al., 2016
6.	Ag NPs	10 – 15	Antimicrobial activity against multidrug-resistant bacteria	*A. baumanii*	Neethu et al., 2018
7.	Small-sized Au NPs	0.8 and 1.2 mm	Antimicrobial activity against multidrug-resistant bacteria	*S. aureus, epidermidis*	Boda et al., 2015
8.	Fe-NPs	18	Cancer therapy	Leukemia (Jurkat cells), breast cancer (MCF-7 cells) cell lines	Namvar et al., 2014
9.	MgO-NPs	100 – 200	Coatings for medical implants	Human breast cancer, cells (MCF-7) cell lines	Katsumiti et al., 2016
10.	Cu NPs	2	Antimicrobial activity against drug-resistant superbugs	*S. aureus, E. coli*	Khurana et al., 2018
11.	Palladium NPs	1.5	Antimicrobial activity against multidrug-resistant microorganism	*Cronobactersakazakii*	Hazarika et al., 2017

5. Conclusion

In conclusion, the advent of microorganisms resistant to antibiotics has led to a critical need for new materials to combat infections the development of antibiotic-resistant microbes due to their special qualities, MBP-NC have become attractive options. Combining the non-toxicity and biocompatibility of biopolymers with the antibacterial capabilities of M-NPs, these nanocomposites offer a potential solution to the antibiotic resistance crisis. This review has discussed the synthesis, characterization, and antimicrobial mechanisms of these nanocomposites, highlighting their applications in medical devices, wound dressings, and drug delivery systems. Despite their potential, challenges remain in optimizing these materials for clinical use. Future research in this area will focus on addressing these challenges and further advancing the field. Overall, MBP-NC represent a sustainable and efficient approach to combating antibiotic resistance.

References

Abdel-Rahman, L. H., Abu-Dief, A. M., Newair, E. F. and Hamdan, S. K. (2016). Some new nano-sized Cr (III), Fe (II), Co (II), and Ni (II) complexes incorporating 2-((E)-(pyridine-2-ylimino) methyl) napthalen-1-ol ligand: Structural characterization, electrochemical, antioxidant, antimicrobial, antiviral assessment and DNA interaction. Journal of Photochemistry and Photobiology B: Biology, 160: 18–31.

Agnihotri, S., Bajaj, G., Mukherji, S. and Mukherji, S. (2015). Arginine-assisted immobilization of silver NPs on ZnO nanorods: an enhanced and reusable antibacterial substrate without human cell cytotoxicity. Nanoscale, 7(16): 7415–7429.

Aguirre-Becerra, H., Feregrino-Pérez, A. A., Esquivel, K., Perez-Garcia, C. E., Vazquez-Hernandez, M. C. and Mariana-Alvarado, A. (2022). Nanomaterials as an alternative to increase plant resistance to abiotic stresses. Frontiers in Plant Science, 13: 1023636.

Al-Kaisy, H. A., Al-Tamimi, B. H., Hamad, Q. A. and Abed, M. S. (2024). Investigation of roughness, morphology, and wettability characteristics of biopolymer composite coating on SS 316L for biomedical applications. International Journal of Biomaterials, 2024(1): 5568047.

Al-Mosawi, A. I. and Abdulsada, S. A. 2024. Biopolymer-based coatings for anti-corrosion of Ti-alloys used in biomedical applications: A review. Polymer Engineering & Science, 64(5): 1905–1920.

Alshemary, A. Z., Akram, M., Goh, Y. F., Tariq, U., Butt, F. K., Abdolahi, A. et al. 2015. Synthesis, characterization, *in vitro* bioactivity and antimicrobial activity of magnesium and nickel doped silicate hydroxyapatite. Ceramics International, 41(9): 11886–11898.

Anasdass, J. R., Kannaiyan, P., Raghavachary, R., Gopinath, S. C. and Chen, Y. (2018). Palladium nanoparticle-decorated reduced graphene oxide sheets synthesized using *Ficus carica* fruit extract: A catalyst for Suzuki cross-coupling reactions. PLoS One, 13(2): e0193281.

Ansari, M. A., Khan, H. M., Khan, A. A., Cameotra, S. S., Saquib, Q. and Musarrat, J. (2014). Interaction of Al2O3 nanoparticles with *Escherichia coli* and their cell envelope biomolecules. Journal of Applied Microbiology, 116(4): 772–783.

Ashfaq, M., Verma, N., Khan, S. (2016). Copper/zinc bimetal nanoparticles-dispersed carbon nanofibers: A novel potential antibiotic material. Materials Science and Engineering: C, 59: 938–947.

Babu: J., Doble, M. and Raichur, A. M. (2018). Silver oxide nanoparticles embedded silk fibroin spuns: Microwave mediated preparation, characterization and their synergistic wound healing and anti-bacterial activity. Journal of Colloid and Interface Science, 513: 62–71.

Baig, U., Gondal, M.A., Ansari, M.A. and Akhtar, S. 2018. Facile synthesis, characterization and antibacterial activity of nanostructured palladium loaded silicon carbide. Ceramics International, 44(14): 16908–16914.

Bellio, P., Luzi, C., Mancini, A., Cracchiolo, S., Passacantando, M., Di Pietro, L. et al. 2018. Cerium oxide nanoparticles as potential antibiotic adjuvant. Effects of CeO_2 nanoparticles on bacterial outer membrane permeability. Biochimica et Biophysica Acta (BBA)-Biomembranes, 1860(11): 2428–2435.

Boda, S. K., Broda, J., Schiefer, F., Weber-Heynemann, J., Hoss, M., Simon, U. et al. (2015). Cytotoxicity of ultrasmall gold nanoparticles on planktonic and biofilm encapsulated gram-positive staphylococci. Small, 11(26): 3183–3193.

Borah, N., Kar, A. and Karak, N. (2024). Biocomposites of biopolymers with metals and their derivatives. In Advances in Biocomposites and their Applications (pp. 167–200). Woodhead Publishing.

Bukhari, S., Kim, D., Liu, Y., Karabucak, B. and Koo, H. (2018). Novel endodontic disinfection approach using catalytic nanoparticles. Journal of Endodontics, 44(5): 806–812.

Chauhan, D., Ashfaq, M., Mangalaraja, R. V. and Talreja, N. (2023). 2D-Nanosheets based hybrid nanomaterials interaction with plants. *In*: Al-Khayri, J. M., Alnaddaf, L. M. and Jain, S. M. (eds.). Nanomaterial Interactions with Plant Cellular Mechanisms and Macromolecules and Agricultural Implications. Springer, Cham. https://doi.org/10.1007/978-3-031-20878-2_11.

Castro-Mayorga, J. L., Randazzo, W., Fabra, M. J., Lagaron, J. M., Aznar, R. and Sánchez, G. (2017). Antiviral properties of silver nanoparticles against norovirus surrogates and their efficacy in coated polyhydroxyalkanoates systems. LWT-Food Science and Technology, 79: 503–510.

Dadi, R., Azouani, R., Traore, M., Mielcarek, C. and Kanaev, A. (2019). Antibacterial activity of ZnO and CuO nanoparticles against gram positive and gram negative strains. Materials Science and Engineering: C, 104: 109968.

De Santis, R.T.G.A. (2017). R. D'Amora U. Balato G. Vollaro A. Oliviero O. Improta G. Triassi M. Ambrosio L. Preliminary focus on the mechanical and antibacterial activity of a PMMA-based bone cement loaded with gold nanoparticles. Bioact. Mater, 2: 156–161.

del Mar Orta, M., Martín, J., Santos, J. L., Aparicio, I., Medina-Carrasco, S. and Alonso, E. (2020). Biopolymer-clay nanocomposites as novel and ecofriendly adsorbents for environmental remediation. Applied Clay Science, 198: 105838.

Dhivya, S. M., Sathiya, S. M., Manivannan, G. and Rajan, M. J. (2016). A comparative study on the biopolymer functionalized iron oxide nanocomposite for antimicrobial activity. Materials Today: Proceedings, 3(10): 3866–3871.

Dizaj, S. M., Lotfipour, F., Barzegar-Jalali, M., Zarrintan, M. H. and Adibkia, K. (2014). Antimicrobial activity of the metals and metal oxide nanoparticles. Materials Science and Engineering: C, 44: 278–284.

Dobrucka, R. and Dlugaszewska, J. 2018. Antimicrobial activity of the biogenically synthesized core-shell Cu@ Pt nanoparticles. Saudi Pharmaceutical Journal, 26(5): 643–650.

El Nahrawy, A. M., Abou Hammad, A. B., Abdel-Aziz, M. S. and Wassel, A. R. (2019). Spectroscopic and antimicrobial activity of hybrid chitosan/silica membranes doped with Al2 O 3 nanoparticles. Silicon, 11(3): 1677–1685.

Elsakhawy, T., Omara, A. E. D., Alshaal, T. and El-Ramady, H. (2018). Nanomaterials and plant abiotic stress in agroecosystems. Environment, Biodiversity and Soil Security, 2(2018): 73–94.

El-Sawaf, A. K., El-Essawy, F., Nassar, A. A. and El-Samanody, E. S. A. (2018). Synthesis, spectral, thermal and antimicrobial studies on cobalt (II), nickel (II), copper (II), zinc (II) and palladium (II) complexes containing thiosemicarbazone ligand. Journal of Molecular Structure, 1157: 381–394.

Farias, I. A. P., Santos, C. C. L. D. and Sampaio, F. C. (2018). Antimicrobial activity of cerium oxide nanoparticles on opportunistic microorganisms: a systematic review. BioMed Research International, 2018(1): 1923606.

Feng, P., Wei, P., Shuai, C. and Peng, S. (2014). Characterization of mechanical and biological properties of 3-D scaffolds reinforced with zinc oxide for bone tissue engineering. PloS one, 9(1): e87755.

Ghazzy, A., Naik, R. R. and Shakya, A. K. (2023). Metal–Polymer Nanocomposites: A Promising Approach to Antibacterial Materials. Polymers, 15(9): 2167. https://doi.org/10.3390/polym15092167.

Gnanasekar, S., Murugaraj, J., Dhivyabharathi, B., Krishnamoorthy, V., Jha, P.K., Seetharaman, P. et al. (2018). Antibacterial and cytotoxicity effects of biogenic palladium nanoparticles synthesized using fruit extract of Couroupitaguianensis. Aubl. Journal of Applied Biomedicine, 16(1): 59–65.

Gumus, H., Aydemir, D., Altuntas, E., Kurt, R. and Imren, E. (2020). Cellulose nanofibrils and nano-scaled titanium dioxide-reinforced biopolymer nanocomposites: Selecting the best nanocomposites with multicriteria decision-making methods. Journal of Composite Materials, 54(7): 923–935.

Haddaoui, M. and Raouafi, N. (2015). Chlortoluron-induced enzymatic activity inhibition in tyrosinase/ZnO NPs/SPCE biosensor for the detection of ppb levels of herbicide. Sensors and Actuators B: Chemical, 219: 171–178.

Hayat, S., Muzammil, S., Rasool, M. H., Nisar, Z., Hussain, S. Z., Sabri, A. N. et al. 2018. *In vitro* antibiofilm and anti-adhesion effects of magnesium oxide nanoparticles against antibiotic resistant bacteria. Microbiology and Immunology, 62(4): 211–220.

Hazarika, M., Borah, D., Bora, P., Silva, A. R. and Das, P. (2017). Biogenic synthesis of palladium nanoparticles and their applications as catalyst and antimicrobial agent. PLoS One, 12(9): e0184936.

He, Y., Ingudam, S., Reed, S., Gehring, A., Strobaugh, T. P. and Irwin, P. (2016). Study on the mechanism of antibacterial action of magnesium oxide nanoparticles against foodborne pathogens. Journal of Nanobiotechnology, 14: 1–9.

Jafarzadeh, S., Nooshkam, M., Zargar, M., Garavand, F., Ghosh, S., Hadidi, M. et al. (2024). Green synthesis of nanomaterials for smart biopolymer packaging: challenges and outlooks. Journal of Nanostructure in Chemistry, 14(2): 113–136.

Kailasa, S. K., Park, T. J., Rohit, J. V. and Koduru, J. R. (2019). Nanoparticles in Pharmacotherapy.

Kailasa, S. K., Park, T. -J., Rohit, J. V. and Koduru, J. R. (2019). Chapter 14 - Antimicrobial activity of silver nanoparticles. *In*: Grumezescu, A. M. (Ed.). Nanoparticles in Pharmacotherapy, William Andrew Publishing, pp. 461–484. https://doi.org/https://doi.org/10.1016/B978-0-12-816504-1.00009-0.

Katsumiti, A., Arostegui, I., Oron, M., Gilliland, D., Valsami-Jones, E. and Cajaraville, M. P. (2016). Cytotoxicity of Au, ZnO and SiO2 NPs using *in vitro* assays with mussel hemocytes and gill cells: relevance of size, shape and additives. Nanotoxicology, 10(2): 185–193.

Kedir, W. M., Abdi, G. F., Goro, M. M. and Tolesa, L. D. (2022). Pharmaceutical and drug delivery applications of chitosan biopolymer and its modified nanocomposite: A review. Heliyon, 8(8).

Kheradmand, E., Rafii, F., Yazdi, M. H., Sepahi, A. A., Shahverdi, A. R. and Oveisi, M. R. (2014). The antimicrobial effects of selenium nanoparticle-enriched probiotics and their fermented broth against *Candida albicans*. DARU Journal of Pharmaceutical Sciences, 22: 1–6.

Khiralla, G. M. and El-Deeb, B. A. (2015). Antimicrobial and antibiofilm effects of selenium nanoparticles on some foodborne pathogens. LWT-Food Science and Technology, 63(2): 1001–1007.

Khurana, C. and Chudasama, B. (2018). Nanoantibiotics: strategic assets in the fight against drug-resistant superbugs. International Journal of Nanomedicine, 13(sup1): 3–6.

Kruk, T., Szczepanowicz, K., Stefańska, J., Socha, R. P. and Warszyński, P. (2015). Synthesis and antimicrobial activity of monodisperse copper nanoparticles. Colloids and surfaces B: Biointerfaces, 128: 17–22.

Kubacka, A., Diez, M. S., Rojo, D., Bargiela, R., Ciordia, S., Zapico, I. et al. (2014). Understanding the antimicrobial mechanism of TiO 2-based nanocomposite films in a pathogenic bacterium. Sci. Rep. 4: 4134.

Kučuk, N., Primožič, M., Knez, Ž. and Leitgeb, M. (2023). Sustainable biodegradable biopolymer-based nanoparticles for healthcare applications. International Journal of Molecular Sciences, 24(4): 3188.

Kumar, N. A., Rejinold, N. S., Anjali, P., Balakrishnan, A., Biswas, R. and Jayakumar, R. (2013). Preparation of chitin nanogels containing nickel nanoparticles. Carbohydrate Polymers, 97(2): 469–474.

Li, M., Gao, L., Schlaich, C., Zhang, J., Donskyi, I. S., Yu, G. et al. (2017). Construction of functional coatings with durable and broad-spectrum antibacterial potential based on mussel-inspired dendritic polyglycerol and *in situ*-formed copper nanoparticles. ACS Applied Materials & Interfaces, 9(40): 35411–35418.

Luz, E. P. C. G., Borges, M. D. F., Andrade, F. K., Rosa, M. D. F., Infantes-Molina, A., Rodríguez-Castellón, E. et al. (2018). Strontium delivery systems based on bacterial cellulose and hydroxyapatite for guided bone regeneration. Cellulose, 25: 6661–6679.

Ma, H., Li, Z., Xue, N., Cheng, Z. and Miao, X. (2018). A gold nanoparticle based fluorescent probe for simultaneous recognition of single-stranded DNA and double-stranded DNA. Microchimica Acta, 185: 1–6.

Makvandi, P., Ali, G. W., Della Sala, F., Abdel-Fattah, W. I. and Borzacchiello, A. (2019). Biosynthesis and characterization of antibacterial thermosensitive hydrogels based on corn silk extract, hyaluronic acid and nanosilver for potential wound healing. Carbohydrate Polymers, 223: 115023.

Mao, D., Hu, F., Kenry, Ji, S., Wu, W., Ding, D., Kong, D. et al. (2018). Metal–organic-framework-assisted *in vivo* bacterial metabolic labeling and precise antibacterial therapy. Advanced Materials, 30(18): 1706831.

Matar, G. H., Kaymazlar, E., Andac, M. and Andac, O. (2023). Novel binary blended hydrogel films (Chitosan-Vanillin Schiff Base/Locust Bean Gum and Fe (III), Cu (II) & Zn (II) Complexes): Synthesis, Characterization, Conductivity, and Antibacterial Activity. Journal of Polymers and the Environment, 31(8): 3509–3521.

Meera, K. and Ramesan, M. T. (2023). Development of high-performance biopolymer nanocomposites derived from carboxymethyl chitosan/boehmite via green synthesis. Polymer Composites, 44(2): 1135–1148.

Mondal, P., Anweshan, A. and Purkait, M. K. (2020). Green synthesis and environmental application of iron-based nanomaterials and nanocomposite: a review. Chemosphere, 259: 127509.

Nathanael, A. J. and Oh, T. H. (2020). Biopolymer coatings for biomedical applications. Polymers, 12(12): 3061. https://doi.org/10.3390/polym12123061.

Namvar, F., Rahman, H. S., Mohamad, R., Baharara, J., Mahdavi, M., Amini, E. et al. (2014). Cytotoxic effect of magnetic iron oxide nanoparticles synthesized via seaweed aqueous extract. International Journal of Nanomedicine: 2479–2488.

Nasrollahzadeh, M. (2021). Biopolymer-Based Metal Nanoparticle Chemistry for Sustainable Applications: Volume 2: Applications. Elsevier.

Neethu, S., Midhun, S. J., Radhakrishnan, E. K. and Jyothis, M. (2018). Green synthesized silver nanoparticles by marine endophytic fungus *Penicillium polonicum* and its antibacterial efficacy against biofilm forming, multidrug-resistant *Acinetobacter baumanii*. Microbial Pathogenesis, 116: 263–272.

Nguyen, T. H., Vardhanabhuti, B., Lin, M. and Mustapha, A. (2017). Antibacterial properties of selenium nanoparticles and their toxicity to Caco-2 cells. Food Control, 77: 17–24.

Nour El Din, S., El-Tayeb, T. A., Abou-Aisha, K. and El-Azizi, M. (2016). *In vitro* and *in vivo* antimicrobial activity of combined therapy of silver nanoparticles and visible blue light against *Pseudomonas aeruginosa*. International Journal of Nanomedicine, pp. 1749–1758.

Omar, R. A., Afreen, S., Talreja, N., Chauhan, D. and Ashfaq, M. (2019a). Impact of nanomaterials in plant systems. *In*: Prasad, R. (eds.). Plant Nanobionics. Nanotechnology in the Life Sciences. Springer, Cham. https://doi.org/10.1007/978-3-030-12496-0_6.

Omar, R. A., Afreen, S., Talreja, N., Chauhan, D., Ashfaq, M. and Srituravanich, W. (2019b). Impact of nanomaterials on the microbial system. *In*: Prasad, R. (eds.). Microbial Nanobionics. Nanotechnology in the Life Sciences. Springer, Cham. https://doi.org/10.1007/978-3-030-16383-9_6.

Pandey, K., Omar, R. A., Verma, N. and Gupta, G. (2024). Fe-carbon nanofiber-modified Mo-MOF for the controlled release and translocation of micronutrients in plants. Environmental Science: Nano.

Pathak, P., Hwang, J. H., Li, R. H., Rodriguez, K. L., Rex, M. M., Lee, W. H. et al. (2021). Flexible copper-biopolymer nanocomposite sensors for trace level lead detection in water. Sensors and Actuators B: Chemical, 344: 130263.

Pugazhendhi, A., Kumar, S. S., Manikandan, M. and Saravanan, M. (2018). Photocatalytic properties and antimicrobial efficacy of Fe doped CuO nanoparticles against the pathogenic bacteria and fungi. Microbial Pathogenesis, 122: 84–89.

Ramesan, M. T., Siji, C., Kalaprasad, G., Bahuleyan, B. K. and Al-Maghrabi, M. A. (2018). Effect of silver doped zinc oxide as nanofiller for the development of biopolymer nanocomposites from chitin and cashew gum. Journal of Polymers and the Environment, 26: 2983–2991.

Ramisetti, P., Dumpala, M., Danikonda, S. K., Suramoni, R., Nampally, B. and Katakam, M. (2024). Synthesis of biopolymer blends nanocomposites embedded with mono-(Ag, Fe) and bi-(Ag-Fe) metallic nanoparticles using an eco-friendly approach for antimicrobial activities. Bioprocess and Biosystems Engineering: 1–14.

Ramisetti, P., Dumpala, M., Danikonda, S. K., Suramoni, R., Nampally, B. and Katakam, M. (2024). Synthesis of biopolymer blends nanocomposites embedded with mono-(Ag, Fe) and bi-(Ag-Fe) metallic nanoparticles using an eco-friendly approach for antimicrobial activities. Bioprocess and Biosystems Engineering: 1–14.

Ravindran, D., Ramanathan, S., Arunachalam, K., Jeyaraj, G. P., Shunmugiah, K. P. and Arumugam, V. R. (2018). Phytosynthesized silver nanoparticles as antiquorum sensing and antibiofilm agent against the nosocomial pathogen *Serratia marcescens*: an *in vitro* study. Journal of Applied Microbiology, 124(6): 1425–1440.

Rehana, D., Mahendiran, D., Kumar, R. S. and Rahiman, A. K. (2017). Evaluation of antioxidant and anticancer activity of copper oxide nanoparticles synthesized using medicinally important plant extracts. Biomedicine & Pharmacotherapy, 89: 1067–1077.

Sabry, R. S., Ali Al-fouadi, A. H. and Habool, H. K. (2018). Enhanced antibacterial activity of anodic aluminum oxide membranes embedded with nano-silver-titanium dioxide. Journal of Adhesion Science and Technology, 32(8): 874–888.

Sasidharan, V., Sachan, D., Chauhan, D. et al. (2021). Three-dimensional (3D) polymer—metal–carbon framework for efficient removal of chemical and biological contaminants. Sci. Rep. 11: 7708. https://doi.org/10.1038/s41598-021-86661-w.

Semmler-Behnke, M., Kreyling, W. G., Lipka, J., Fertsch, S., Wenk, A., Takenaka, S. et al. (2008). Biodistribution of 1.4-and 18-nm gold particles in rats. Small, 4(12): 2108–2111.

Shaikh, A. A., Datta, P., Dastidar, P., Majumder, A., Das, M. D., Manna, P. et al. (2024). Biopolymer-based nanocomposites for application in biomedicine: a review. Journal of Polymer Engineering, 44(2): 83–116.

Sharma, B., Malik, P. and Jain, P. (2018). Biopolymer reinforced nanocomposites: A comprehensive review. Materials Today Communications, 16: 353–363.

Sharma, N. K., Ameta, R. K. and Singh, M. (2016). From synthesis to biological impact of Pd (II) complexes: synthesis, characterization, and antimicrobial and scavenging activity. Biochemistry Research International, 2016(1): 4359375.

Singh, S. P., Bhargava, C. S., Dubey, V., Mishra, A. and Singh, Y. (2017). Silver nanoparticles: Biomedical applications, toxicity, and safety issues. Int. J. Res. Pharm. Pharm. Sci., 4(2): 1–10.

Suganya, P., Vaseeharan, B., Vijayakumar, S., Balan, B., Govindarajan, M., Alharbi, N. S. et al. (2017). Biopolymer zein-coated gold nanoparticles: Synthesis, antibacterial potential, toxicity and histopathological effects against the Zika virus vector Aedes aegypti. Journal of Photochemistry and Photobiology B: Biology, 173: 404–411.

Suresh, L., Brahman, P. K., Reddy, K. R. and Bondili, J. S. (2018). Development of an electrochemical immunosensor based on gold nanoparticles incorporated chitosan biopolymer nanocomposite film for the detection of prostate cancer using PSA as biomarker. Enzyme and Microbial Technology, 112: 43–51.

Swarupa, S. and Thareja, P. (2024). Techniques, applications and prospects of polysaccharide and protein based biopolymer coatings: A review. International Journal of Biological Macromolecules, pp. 131104.

Tabasi, H., Babaei, M., Abnous, K., Taghdisi, S. M., Saljooghi, A. S., Ramezani, M. et al. (2021). Metal–polymer-coordinated complexes as potential nanovehicles for drug delivery. Journal of Nanostructure in Chemistry, 11: 501–526.

Tadge, T., Garje, S., Saxena, V. and Raichur, A. M. (2023). Application of shape memory and self-healable polymers/composites in the biomedical field: a review. ACS omega, 8(36): 32294–32310.

Tadi, K. K., Reddy, N. M., Chandaluri, C. G., Sakala, G. P. and Ramesh, G. V. (2022). Functionalized biopolymer nanocomposites for the degradation of textile dyes. In Functional polymer nanocomposites for wastewater treatment (pp. 175–200). Cham: Springer International Publishing.

Tassinari, R., Cubadda, F., Moracci, G., Aureli, F., D'Amato, M., Valeri, M. et al. (2014). Oral, short-term exposure to titanium dioxide nanoparticles in Sprague-Dawley rat: focus on reproductive and endocrine systems and spleen. Nanotoxicology, 8(6): 654–662.

Thakur, N., Manna, P. and Das, J. (2019). Synthesis and biomedical applications of nanoceria, a redox active nanoparticle. Journal of Nanobiotechnology, 17: 1–27.

Valencia, G. A., Zare, E. N., Makvandi, P. and Gutiérrez, T. J. (2019). Self-assembled carbohydrate polymers for food applications: A review. Comprehensive Reviews in Food Science and Food Safety, 18(6): 2009–2024.

Valerini, D., Tammaro, L., Di Benedetto, F., Vigliotta, G., Capodieci, L., Terzi, R. et al. (2018). Aluminum-doped zinc oxide coatings on polylactic acid films for antimicrobial food packaging. Thin Solid Films, 645: 187–192.

Wang, C., Gu, B., Liu, Q., Pang, Y., Xiao, R. and Wang, S. (2018). Combined use of vancomycin-modified Ag-coated magnetic nanoparticles and secondary enhanced nanoparticles for rapid surface-enhanced Raman scattering detection of bacteria. International Journal of Nanomedicine, pp. 1159–1178.

Warkara, S. G. and Meena, J. (2022). Synthesis and applications of biopolymer/FeO nanocomposites: A review. Journal of New Materials for Electrochemical Systems, 25(1).

Yan, K., Xu, F., Wang, C., Li, Y., Chen, Y., Li, X. et al. (2020). A multifunctional metal-biopolymer coordinated double network hydrogel combined with multi-stimulus responsiveness, self-healing, shape memory and antibacterial properties. Biomaterials Science, 8(11): 3193–3201.

Yoon, H., Pangging, M., Jang, M. H., Hwang, Y. S. and Chang, Y. S. (2018). Impact of surface modification on the toxicity of zerovalent iron nanoparticles in aquatic and terrestrial organisms. Ecotoxicology and Environmental Safety, 163: 436–443.

Zewde, B., Atoyebi, O., Zvonkina, I. J., Thomspon, T., Hargrove, A., Scott, A. et al. (2020). Biopolymer nanocomposites and their applications. In Soft Matter and Biomaterials on the Nanoscale: The WSPC Reference on Functional Nanomaterials—Part I Volume 2: Polymers on the Nanoscale: Nano-structured Polymers and Their Applications (pp. 45–108).

Zhang, W., Li, Y., Niu, J. and Chen, Y. (2013). Photogeneration of reactive oxygen species on uncoated silver, gold, nickel, and silicon nanoparticles and their antibacterial effects. Langmuir, 29(15): 4647–4651.

Zhao, C., Liu, G., Tan, Q., Gao, M., Chen, G., Huang, X. et al. (2023). Polysaccharide-based biopolymer hydrogels for heavy metal detection and adsorption. Journal of Advanced Research, 44: 53–70.

Zheng, K., Setyawati, M. I., Leong, D. T. and Xie, J. (2017). Antimicrobial gold nanoclusters. ACS Nano, 11(7): 6904–6910.

Zhou, Z., Peng, S., Sui, M., Chen, S., Huang, L., Xu, H. et al. (2018). Multifunctional nanocomplex for surface-enhanced Raman scattering imaging and near-infrared photodynamic antimicrobial therapy of vancomycin-resistant bacteria. Colloids and Surfaces B: Biointerfaces, 161: 394–402.

Zhu, L., Zheng, T., Zheng, J., Yu, C., Zhou, Q., Hua, J. et al. (2017). Synthesis of novel platinum-on-flower-like nickel catalysts and their applications in hydrogenation reaction. Applied Surface Science, 423: 836–844.

Zienkiewicz-Strzałka, M. and Deryło-Marczewska, A. (2020). Small AgNP in the biopolymer nanocomposite system. International Journal of Molecular Sciences, 21(24): 9388.

CHAPTER-2

Nano-based Biopolymer as Adsorbent Material

Thandapani Gomathi, Sivalingam Revathi, Arumugam Mahalakshmi, Balaraman Revathi* and *Ramachandran Sasirekha*

1. Introduction

In the realm of polymer technology, there has been continuous innovation and progress over time. The growing emphasis on environmental sustainability in today's world has stimulated numerous researchers to explore, characterize, design, and produce novel polymer, biopolymer, and nano biopolymer materials. These materials are increasingly favored for their wide availability, sustainability, biocompatibility, and biodegradability, which have garnered significant scientific and engineering interest in current years (Azizi Samir et al., 2005; Balakrishnan et al., 2017; Bamba et al., 2017; Moon et al., 2011).

Biopolymers, such as nanocellulose, nano starch, nano chitin, and nano silk, derived from living organisms, offer promising advantages of high petroleum-based polymers due to their enhanced biodegradability and sustainability (Iftekhar et al., 2021). It is crucial to develop comprehensive green strategies to manage environmental pollution effectively while conserving natural resources (Chincholikar et al., 2023). Nano biopolymers, produced through chemical, mechanical, or microbiological techniques, have emerged as a viable solution. These bio nanoparticles exhibit affordability, low toxicity, and robust mechanical properties, making them well-suited for environmental remediation efforts (Lavoine et al., 2012; Ogawa et al., 2011; Omenetto and Kaplan, 2010). By integrating principles of green chemistry with nanotechnology, a holistic approach can be adopted towards achieving long-term sustainability (Chincholikar et al., 2023). This synergy between green practices and advanced nanomaterials holds promise for addressing contemporary environmental challenges.

Water, being essential for all life forms on Earth, faces severe degradation due to human activities such as industrialization, urban sprawl, and agricultural practices. The contamination of water formed by several dye complexes, originating primarily from industries like textiles, paper, printing, and others, poses a significant environmental threat. These industries discharge substantial volumes of colored wastewater, necessitating urgent measures to mitigate these pollutants (Arshadi et al., 2015, 2016). While numerous water treatment methods have been employed, including

PG and Research Department of Chemistry, D.K.M. College for Women (Autonomous), Vellore, Tamil Nadu, India.
* Corresponding author: drgoms1@gmail.com
DOI: 10.1201/9781003558330-2

advanced oxidation (Amiri et al., 2021), ion-exchange, membrane filtration (Cheng et al., 2020), ultrafiltration, and adsorption, escalating pollution levels have rendered many of these methods inefficient (Mangla et al., 2022; Singh et al., 2019; Singh and Purkait, 2017). Nanotechnology offers promising avenues for treating contaminated water causes, such as underground water, surface water, and wastewater. Nanoparticles, with their unique characteristics, production of a vital role in adsorption and desorption processes, enabling effective detoxification (Iravani, 2021; Lu and Astruc, 2020).

Emerging technologies like nanofiber filters, nanofiltration and CNTs are revolutionizing remedy efforts, enhancing water treatment efficiency, and safeguarding environmental health (Annu et al., 2024; Yang et al., 2021). The present chapter reports the promising advancement of nano biopolymers as efficient adsorbents. With enhanced surface area, tunable properties, and eco-friendly characteristics, nano biopolymers show significant potential for mitigating surfactant pollution and promoting sustainable water management practices. Their renewability and cost-effectiveness further highlight their importance in addressing environmental challenges. The objective of this chapter is to discover nano biopolymers as effective adsorbents for justifying surfactant pollution and helping sustainable water management practices. These bio nanoparticles show low toxicity, affordability, and strong mechanical properties, making them well-suited for environmental remediation efforts. The synergy between green practices and advanced nanomaterials holds promise for addressing current environmental challenges.

2. Biopolymer

For centuries, humans have relied on biopolymers consequential from agricultural sources like leather, cotton, silk and wool (Flórez et al., 2023). These bio-based materials are composed of building blocks obtained from living organisms. In recent years, there has been increasing attention in developing biopolymers outstanding to their probable environmental benefits (Du et al., 2016). These materials help reduce carbon dioxide emissions during production and contribute to waste reduction by being biodegradable, thus lessening the burden on landfills (Patra et al., 2018; Valdés and Garrigós, 2016).

Biopolymers encompass a wide range of living macromolecules designed by the polymerization of monomers such as nucleic acids (formed from nucleotides), proteins (composed of amino acids) and polysaccharides (comprising sugars) (Pattanashetti et al., 2017). They are incredibly diverse and versatile, sourced from biological systems or synthesized from biological feedstocks (Singh et al., 2021). Like conventional polymers, biopolymers consist of repeating monomeric units linked together (Wróblewska-Krepsztul et al., 2019). Samples of biopolymers include DNA, RNA, gelatine, fatty acids, keratin, starch, and cellulose. In the pursuit of a sustainable lifestyle, biopolymers provide a pathway towards achieving eco-friendly practices (Leja and Lewandowicz, 2010). Their use aligns with the goal of creating a more environmentally conscious and sustainable world.

3. Classification of Biopolymers

Biopolymers are categorized in three main sets based on the nature of their repetition monomer units: polysaccharides, polypeptides and polynucleotides (Fig. 1) (Thekkila-Veedu et al., 2023). Polysaccharides contain monosaccharide components connected together by glycosidic bonds. Examples contain starch, cellulose, glycogen, chitin/chitosan, pectin, and alginate. Polysaccharide biopolymers can be further classified into four categories: sugar-based, starch-based, cellulose-based, and lignin-based biopolymers. Polynucleotides, such as DNA and RNA, are collected from nucleotide units and serve as biopolymers. They store and transmit genetic information within living organisms. Another class of biopolymers comprises amino acid monomers linked together by peptide bonds, forming polypeptides. Examples include silk, collagen, and keratin, which exhibit diverse structural and functional properties in biological systems (Mallik, 2022).

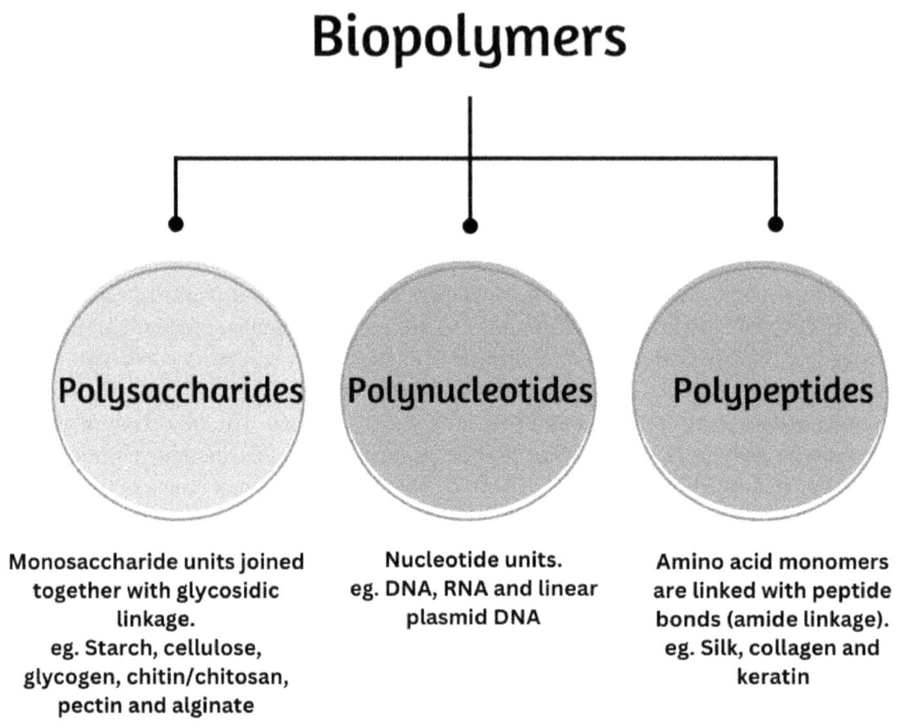

Figure 1. Classification of Biopolymers.

Biopolymers, naturally sourced from alive organisms, are increasingly known as ecological, biocompatible, and supportable resources for environmental remediation (Biswas and Pal, 2021). They find wide applications in wastewater treatment, serving as coagulants for clarification and adsorbents for dyes and heavy metals (Pal and Pal, 2017, 2019; Sánchez-Martín et al., 2013). Various forms like beads, composites, and hydrogels are utilized, with hydrogels particularly beneficial due to their high water absorption and permeable structure (Qureshi et al., 2020; Sakib et al., 2021). Despite their efficacy, challenges like poor mechanical strength and stability hinder practical use, prompting exploration into chemical modification and reinforcement. Integration of nanomaterials offers promising solutions, leveraging their large surface area and reactivity for efficient pollutant removal (Dhariwal and Mittal, 2023; Mittal and Jana, 2023; Sharma et al., 2023; Singh, Garg, et al., 2023). Additionally, research focuses on synthesizing biodegradable polymers tailored for wastewater remediation.

4. Nanotechnology in Wastewater Treatment

Nanotechnology is attracting considerable interest for pioneering wastewater treatment process approaches (Marimuthu et al., 2023). Propelled by the unique physicochemical properties of nanomaterials, nanotechnology is rapidly expanding. Various nanomaterials are under investigation to reduce absorptions of contaminated pollutants in wastewater, capitalizing on their more properties at the nanoscale. These materials demonstrate improved acuteness, outstanding adsorption capabilities, and a maximum surface-to-volume ratio (Nasrollahzadeh et al., 2021). Initially applied across diverse industries, nanotechnology now plays a crucial role in wastewater treatment, providing cost-effective, efficient, and versatile solutions through methods such as adsorption and catalysis based on nanomaterials (Humplik et al., 2011; Jain, 2019; Khan et al., 2019).

Nanomaterials, well-defined as materials with at minimum one dimension smaller than 100 nm, offer promising solutions for wastewater treatment process outstanding to their similar properties such as large surface area, nano size and high reactivity (Khan and Malik, 2019). These properties enable nanomaterials to effectively eliminate various pollutants from wastewater, including heavy metals, inorganic and organic solvents, colorants, living contaminants, and disease-causing pathogens like cholera and typhoid (Kumar et al., 2014). Incorporating nanomaterials into water and wastewater treatment processes capitalizes on their size-dependent effects, enhancing treatment efficiency and water quality. Many studies have demonstrated the considerable capabilities of nanomaterials in membrane processes, adsorption, catalytic oxidation, disinfection, and sensing applications (Wang, 2019). Despite their potential, challenges remain in deploying nanotechnology for wastewater treatment.

One key challenge is the tendency of nanoparticles to agglomerate in fluidized or rigid bed systems, leading to decreased activity and pressure loss. Additionally, separating exhausted nanoparticles, except magnetic ones, from treated water poses difficulties, impacting reusability (Ambashta and Sillanpää, 2010). Concerns also persist regarding the environmental and health implications of nanomaterial use, necessitating further research into their ?behavior and consequences (Lu et al., 2014).

5. Polymeric Nanocomposites

Nanocomposites emerge as an encouraging approach to address these tests, combining nanoparticles with supporting materials like membranes or polymers. These nanocomposites demonstrate enhanced performance, cost-effectiveness, compatibility with existing infrastructure, and recyclability. Research efforts focus on optimizing nanocomposite properties and manufacturing methods to improve water treatment efficiency (Li et al., 2008; Santhosh et al., 2016; Zhu et al., 2019). Nanocomposites represent a novel class of materials characterized by ultrafine stage dimensions, naturally on the scale of a rare ?nanometers. They differ from atomic and bulk counterparts and are defined by the presence of at least one phase with nanoscale dimensions (Roy et al., 1986). The size of nanoparticles within nanocomposites is pivotal, imparting unique properties not typically found in micro composites and offering novel technological and business opportunities (Giannelis, 1996).

Nanocomposites are categorized based on the matrix materials used in their preparation into three classes: Polymer Medium Nanocomposites, Ceramic Medium Nanocomposites, and Metal Medium Nanocomposites (Chung, 2003). In the current economic and environmental slandscape, there is a developing imperative to grow supportable composites leveraging nanotechnology. Polymers, known for their favorable attributes such as low cost, lightweight nature, ease of dispensation, corrosion resistance, ductility, and major durability, present a promising avenue for the manufacture of sustainable nanocomposites (Gomathi et al., 2019; Thostenson et al., 2005).

Polymer nanocomposites consistently outperform individual materials due to factors beyond just chemical structure (Kumar and Krishnamoorti, 2010). Nanoparticle parameters, including shape, size, optimal pore size, large surface area, loading, dispersion, interaction, mobility, temperature, ease of regeneration, cost-effectiveness, polymer entanglement, elasticity, surface chemistry, high adsorption capacity, robust mechanical strength, and bio persistence, influence polymer matrix reinforcement and nanocomposite properties (Fig. 2). Understanding these relationships aids optimization (Kinnear et al., 2017; Pandey et al., 2017).

5.1 Structure

The structure of composite materials, especially in the context of nanocomposites, is of paramount importance. Achieving a homogeneous distribution of nanoparticles inside the polymer matrix is vital for enhancing mechanical properties, such as strength and stiffness (Keledi et al., 2012). The specific arrangement of these nanoparticles, whether in zero-dimensional clusters, one-dimensional nanotubes, or other configurations, greatly influences the material's overall performance and

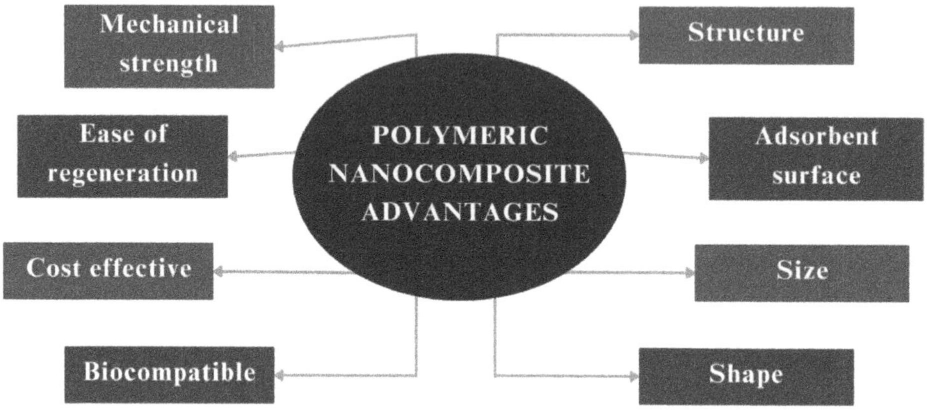

Figure 2. Advantages of polymeric nanocomposites.

functionality. Understanding the structure enables the customization of properties to fulfil particular application demands, such as ensuring crashworthiness in aerospace vehicles or providing protective armor (Dhiman et al., 2017; Singh et al., 2023). Hence, attention to structure is indispensable for optimizing the performance and functionality of composite materials.

5.2 Shape

In polymer matrices, distributing spherical nanoparticles is crucial, in battling aggregation. Nanomaterials, classified by crystallite shape, include 3 dimensions (zeolites), 2 dimensions (thin films), 1 dimension (quantum wires), and 0D (quantum dots). The shape significantly impacts material behavior: 3D materials feature equiaxed grains allowing electron movement in all directions, while 2D restricts to two dimensions. In 1D, electrons move in one direction, confined in two. Zero-dimensional materials confine electrons in all three spatial directions without delocalization. Particle shapes (spherical, cubic, platelet) diversely affect properties and costs, often enhancing isotropic materials. Manipulating the shape is vital for optimizing composite performance (Bednarcyk, 2003; Suryanarayana, 1994).

5.3 Size

Nanoparticle size plays a pivotal role in water treatment as an adsorbent. Recent efforts focus on tailoring the nanoparticle size to enhance adsorption onto cell surfaces. Optimal sizes, typically ranging between 30–50 nm, demonstrate superior uptake efficiency for pollutants. Larger nanoparticles, exceeding 100 nm, show distinct trends with rods exhibiting higher adsorption rates compared to spheres. Sub-100-nm spheres, however, outperform rods, indicating size-dependent adsorption efficiency. This highlights the significance of nanoparticle size in water treatment applications, particularly for achieving effective pollutant removal through tailored adsorbents.

5.4 Adsorbent Surface

Surface area is critical in enhancing nanocomposite applications, but high surface area can lead to agglomeration. Effective mixing, including dispersion and distribution, breaks down filler agglomerates, ensuring uniform distribution without altering particle size. Improved properties result from better nanoparticle dispersion, facilitating enhanced polymer-filler interactions. While nanoparticle incorporation improves several properties, the polymer's interaction with the nanofiller chiefly determines final properties. High specific area promotes macromolecular chain adsorption, influencing structural development and phase separation. Achieving active filler dispersal and enhancing polymer-filler interactions remain key challenges in polymer nanocomposite research.

6. Adsorbent material

6.1 Cellulose

Cellulose, a recyclable and renewable polymer, remains insoluble in maximum solvents due to hydrogen bonding and crystallinity (Suhas et al., 2016). Its natural form exhibits lower adsorption capacity, but chemical modification enhances its properties. Abundant in nature, cellulose supports plant, oomycete, and algae cell walls. It possesses hydrophilicity, sorbent potential, and facile chemical modification, making it environmentally friendly (Klemm et al., 2005). Cellulose's active hydroxyl groups allow for easy functionalization. Cotton fiber, a cellulose derivative, offers renewability, degradability, affordability (O'Connell et al., 2008), and a great exact surface area ideal for adsorption. Together, cellulose and its derivatives provide versatile, sustainable solutions with minimal environmental impact.

Cellulose, a linear syndiotactic homopolymer, consists of glucose components connected by β-(1-4)-glycosidic bonds, creating cellobiose dimmers (Demirbas, 2008; Figueiredo et al., 2010; Huber et al., 2012; Kocherbitov et al., 2008). Complete hydrolysis yields glucose, confirming its composition. Cellulose's properties include biodegradability, hydrophilicity, chirality, and versatile biochemical modification measurements. Its hydroxyl groups exhibit distinct behaviors; the C-1 end shows dipping properties, while the C-4 end does not. Cellulose's polar oxygen and hydrogen atoms form inter- and intramolecular hydrogen bonds, providing stiffness (Figueiredo et al., 2010; Klemm et al., 2005). It exists in four major polymorphs, notably cellulose I and II. Cellulose's crystallinity impedes chemical processing but aids in its isolation, often through environmentally friendly methods (Huber et al., 2012).

Cellulose serves as a natural adsorbent for several substances, including water, organic compounds, metal ions, and dyes (Acemioglu and Alma, 2001; Dridi-Dhaouadi et al., 2011; Gupta and Suhas, 2009). Sorption of water vapor is a significant application, particularly in textile characterization, where it undergoes chemisorption at low relative humidity, multilayer sorption at medium humidity, and capillary condensation at high humidity. Kocherbitov et al. (Kocherbitov et al., 2008) studied water and nitrogen sorption on different types of cellulose, revealing that amorphous cellulose exhibited the highest water sorption due to its more accessible sorption sites. Metal ion adsorption by cellulose has also been investigated. Acemioglu and Alma (Acemioglu and Alma, 2001) considered Cu(II) adsorption on cellulose, finding it to be exothermic and spontaneous, with Cu(II) forming a monolayer coverage on the adsorbent. Júnior et al. (Júnior et al., 2009) observed higher Cu(II) removal on mercerized cellulose due to increased surface area and enhanced accessibility of hydroxy groups after transformation to cellulose II. These studies underscore cellulose's potential as an effective adsorbent for diverse applications. Its abundance, renewability, and biodegradability make it an attractive option for environmental remediation and wastewater treatment.

6.2 Chitin/chitosan

Polysaccharides like chitosan and its predecessor chitin have found favor as natural polymer biosorbents (Fig. 3) (Sarode et al., 2019). Chitin, the second most abundant biopolymer after cellulose, primarily sourced from crab and shrimp shells (El Knidri et al., 2018), suffers from poor solubility, limiting its widespread application. To address this, soluble chitosan boasts abundant hydrophilic hydroxyl and polycationic amine groups in its structure is derived from chitin through alkaline deacetylation (Hamed et al., 2016; Muxika et al., 2017). Chitosan's structure, enriched with -NH_2 and -OH groups, makes it effective for biosorption of various contaminants (Sharififard et al., 2018). Despite its biodegradability, non-toxicity, biocompatibility, and bioactivity, chitosan suffers from weaknesses in mechanical properties, thermal stability, porosity, and surface area (Vakili et al., 2014).

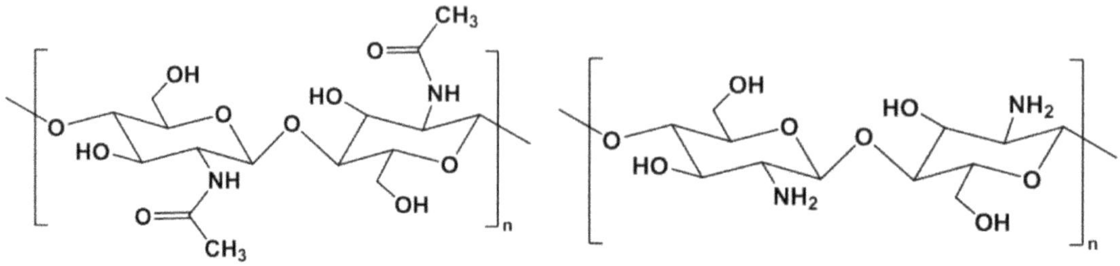

Figure 3. Structure of Chitin and Chitosan.

Chitosan, derived from the alkaline deacetylation of chitin, extracted from shellfish, is a partially deacetylated polymer. It exhibits a linear hydrophilic amino polysaccharide structure containing glucosamine and acetylglucosamine units (poly-β-(1 → 4)-2-amino-2-deoxy-d-glucose). Chitin, characterized by poly-β-(1 → 4)-N-acetyl-d-glucosamine units, stands as one of the most abundant normal biopolymers (Anastopoulos et al., 2017; Renault et al., 2009; Tanhaei et al., 2020). Chitosan distinguishes itself from chitin through the inclusion of amino groups, endowing it with solubility in dilute acids (pH < 6) and the capability to produce complexes with metal ions. This renders chitosan a valuable tool in wastewater treatment and purification efforts (Croisier and Jérôme, 2013; Rinaudo, 2006). Conversely, the practical utility of chitin remains severely restricted due to its inherent insolubility (Pavinatto et al., 2010). Intriguingly, chitosan's aqueous solubility is contingent upon pH, facilitating its processing under mild conditions (Madihally and Matthew, 1999).

Chitosan boasts a multitude of desirable traits, including biodegradability, hydrophobicity, biocompatibility and non-toxicity, complemented by highly reactive amino ($-NH_2$) and hydroxyl (–OH) groups within its structure (Samadi et al., 2016). These characteristics equip chitosan to serve as an effective adsorbent for purging contaminants from wastewater. Its unique chemical composition allows for precise modifications, tailoring polymers for specific applications. Leveraging its reactive groups, researchers develop composites with diverse compounds, augmenting pollutant adsorption and resilience in acidic conditions (Anastopoulos and Kyzas, 2015; Karimi-Maleh et al., 2021; Kyzas et al., 2013; Terzopoulou et al., 2016). Examples encompass kaolinite, zeolites, montmorillonite, bentonite, oil palm ash, polyurethane, and magnetite. Addressing limitations, chitosan undergoes various modifications—beads, membranes, films, semi-IPN hydrogels, nano-magnetic particles, grafting polymers, and composites—to bolster adsorption capacity and physical properties (Liu and Bai, 2014; Purnama Sari et al., 2019). Further, its singular cationic charge, distinct among polysaccharides, facilitates the neutralization and efficient flocculation of anionic suspended colloidal particles, thereby curbing chlorides, COD, and turbidity in wastewater (You et al., 2017).

6.3 Alginate

Alginate, specifically, is an anionic polymer characteristically derived from brown marine algae (Fig. 4). It is composed of unbranched polysaccharide copolymers, consisting of alternating blocks of d-Mannonate (M) and l-Guluronic (G) linked by 1,4-glycosidic linkages (Mohan et al., 2007). A significant feature of alginate is its capacity to form hydrogels through a gelation phenomenon involving divalent cations like Ba^{2+} and Ca^{2+}. These cations interact with the G blocks to form so-called egg-box complexes, creating a hydrogel matrix. This hydro gelling capability significantly broadens alginate's applications, particularly in biomedical and pharmaceutical research.

Also, sodium alginate, has garnered significant attention for its impressive physicochemical properties, particularly its efficacy in adsorbing heavy metal ions such as Cu^{2+}, Cd^{2+}, and Pb^{2+}. These capabilities stem from its molecular structure and composition, which enable strong interactions with various metal ions, making it a valuable tool for environmental remediation. However, sodium

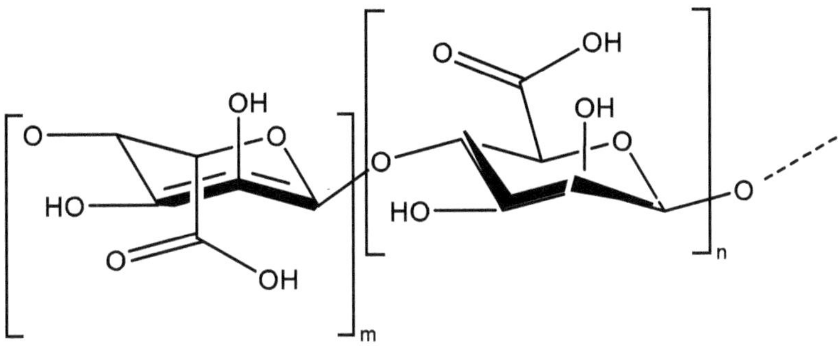

Figure 4. Structure of Alginate.

alginate is not without its drawbacks. One major limitation is its susceptibility to thermal degradation, which can limit its stability and efficacy in applications requiring sustained performance under varying temperature conditions. Additionally, sodium alginate's inherent poor stability can further restrict its utility, necessitating modifications to enhance its functional properties and broaden its application scope.

To overcome these limitations, researchers have explored various strategies to modify and enhance sodium alginate. For instance, the incorporation of other materials or the chemical change of alginate can improve its thermal stability and mechanical strength. Such modifications are crucial for expanding the use of sodium alginate in different industrial applications, particularly those involving the removal of hazardous pollutants.

The environmental utility of alginate is partially attributed to its abundant surface functional groups, such as carboxyl and hydroxyl, which facilitate the capture of metallic or cationic ions through ion exchange between the cross-linking cations and target pollutants like heavy metals or dyes. Nonetheless, alginate gel suffers from drawbacks including high rigidity, fragility, poor elasticity, and mechanical properties, as noted by (Thakur et al., 2016). To address these limitations, both organic and inorganic alginate-based composites have been developed to enhance the mechanical and thermal stability as well as the swelling properties of pure alginate gels. These composites exhibit distinctive physicochemical properties and excellent biocompatibility. In recent years, researchers have extensively explored alginate-based composites that combine alginate gels with other polymers, natural and engineered nanoparticles, and microorganisms for the purpose of removing pollutants from aqueous solutions (Prof. Ali, 2016; Allouss et al., 2019).

6.4 Starch

Starch, a widely available and affordable polysaccharide, contains of homopolymers of glucose components with the molecular formula $(C_6H_{10}O_5)_n$ (Fig. 5). Amylose and amylopectin are the two main types of polymer chains that constitute starch. Amylopectin is a branched polymer composed of α-1,4-glycosidic bonds with branch points linked by α-1,6-glycosidic bonds, while amylose is a linear chain with only α-1,4-glycosidic links. These structural differences impart distinct characteristics to each polymer. For example, the presence of short branches in amylopectin, formed by α-1,6-glycosidic linkages, contributes to the crystalline regions within starch granules. In its natural state, starch typically comprises 20–30% amylose and 70–80% amylopectin.

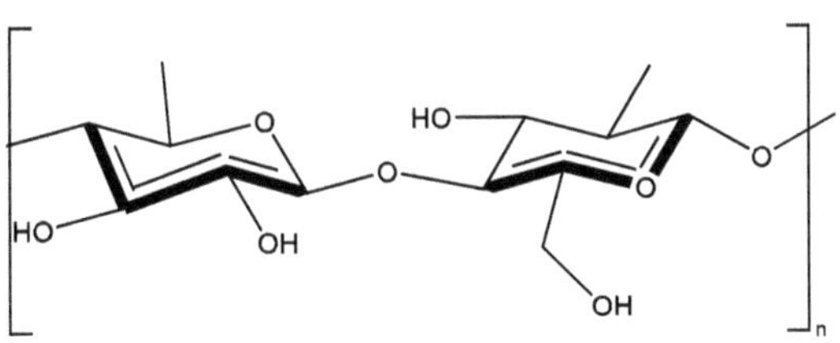

Figure 5. Structure of Starch.

Starch synthesis mostly occurs in the chloroplasts of plant leaves or in the amyloplasts of plant storage organs, and it may contain lipids and phosphate groups. Botanical sources of starch include rice, potato, cassava, and corn.

It is characterized by numerous -OH groups on its surface, rendering it an ionic hydrophilic linear polymer with remarkable attributes such as non-toxicity, biodegradability, biocompatibility, and cost-effectiveness. However, native starch also presents significant limitations, including a low surface area, inadequate water solubility, low molecular weight, restricted thermal stability, and a deficiency of functional reaction groups. These shortcomings pose challenges to its direct application as an adsorbent for wastewater treatment (Fang et al., 2022).

Replacing starch with xanthate esters allows for the creation of chelates with different heavy metal ions in water. This substitution is cost-effective because xanthation is a straightforward procedure, and starch is an inexpensive biopolymer. Nonetheless, xanthate undergoes gradual degradation over time, leading to the production of an undesirable sulfide odor during storage and use (Kim and Lim, 1999).

Many researchers employ starch by also modifying its properties or chemically altering it (Aziz et al., 2004). This allows comparison with the original starch, even in applications such as heavy metal removal combined with filtration processes. In the PEUF process, the performance of starch is evaluated against that of Polyethylene Glycol (PEG), a commonly used water-soluble polymer (Kim and Lim, 1999). Starch has recently gathered attention due to its diverse functions and forms, serving as a water binder, emulsion stabilizer, and gelling agent. However, there is limited research on the binding of metal cations by unmodified starch, especially in membrane process applications.

In this study, unmodified starch derived from natural plant sources, readily available in tropical Asian countries, is utilized. No additional chemicals are introduced to enhance the binding mechanism of starch, yet it still effectively binds with metal ions. The selection of unmodified starch as a water-soluble polymer is based on its availability, relatively low cost, and ease of handling. Furthermore, the lower feed concentration and final concentration of the sample render the treated solutions safe for discharge into the environment (Copeland et al., 2009).

6.5 Pectin

Pectin, a prevalent macromolecule in the main cell walls of plants, is detected both in the matrix and middle lamellae (Willats et al., 2001) (Fig. 6). It is predominantly rich in galacturonic acid (GalA), which serves as the backbone for three distinct domains: homogalacturonan (HGA), rhamnogalacturonan-I (RG-I), and rhamnogalacturonan-II (RG-II). Approximately 70% of pectin consists of Galacturonic Acid (GA) (Mohnen, 2008). Comprising three polysaccharides interconnected through covalent bonds, pectin forms networks within the cell wall matrix and middle lamellae.

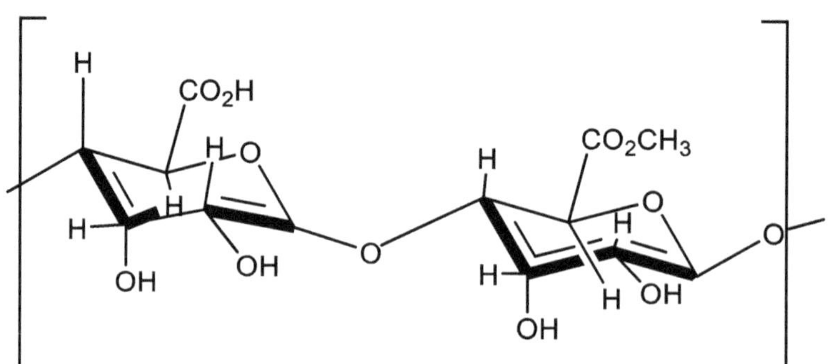

Figure 6. Structure of Pectin.

Pectin, a high molecular weight polysaccharide found in nearly all plants, plays an energetic role in continuing the reliability of cell structure . In the food industry, pectin is utilized to rise the viscosity of crops such as jams, beverages, and jellies. Its applications extend to the pharmaceutical industry, particularly in drug preparations, where it serves as an excipient, influencing release kinetics. With the growing request for pectin across food, pharmaceutical, and therapeutic sectors, efficient extraction processes are imperative. Various extraction methods have been developed to enhance pectin yield, including heating in acidic conditions to convert insoluble pectin into soluble forms. Fruit ripening also facilitates this conversion. While it can be extracted from many fruits, the most common profitable source is the peels of citrus fruits, typically achieved through alcohol precipitation. Citrus fruits contain pectin in concentrations ranging from 0.5 to 3.5%, predominantly concentrated in the fruit peels (Renard and Jarvis, 1999).

In a recent study, composite beads collected of Pectin and Cellulose Microfibers (CF) were developed, as detailed by (Lessa et al., 2016). The integration of cellulose microfibers into the Pec-Ca(II) network was determined by their reinforcing properties and their ability to effectively capture various solutes through adsorption, as noted by (Urruzola et al., 2014). These attributes stem from the wide surface area of CF and the abundance of hydroxyl groups along CF molecules. Both Pec and CF were extracted from orange waste using straightforward methods, yielding renewable, low-cost, and biodegradable raw materials. Previously, Pec-CF beads had been tested for the removal of methylene blue, an organic dye (Lessa et al., 2016). In this study, we explore the adsorption capacity of pectin/cellulose microfibers (Pec-CF) beads for removing multi-metal ions from water. To achieve this, a series of experiments were conducted to determine the optimum conditions for adsorbing copper (Cu), cadmium (Cd), and iron (Fe) onto Pec-CF beads. The impact of various experimental parameters on the adsorption capacities of these metals was assessed using a factorial design approach.

7. Applications

7.1 Nanosorbent for Heavy Metal Removal

Heavy metals, also known as suggestion elements or metallic elements, are characterized by having an atomic density of 6 g/cm^3 or greater (Suyambulingam et al., 2023). Rapid industrial growth in sectors like metal plating, mining, and battery manufacturing has escalated heavy metal contamination, especially in developing countries. Heavy metals such as arsenic, lead, cadmium, copper, and chromium do not degrade and bioaccumulate, posing severe health risks due to their toxicity and carcinogenicity (Akpor, 2014; Ali et al., 2023; Khalef et al., 2022; Tchounwou et al., 2012; P. Zhang et al., 2023). Arsenic contaminates water and is highly toxic (Gupta et al., 2022). Lead affects multiple body systems, causing neurological and cardiovascular issues. Cadmium, found in industrial waste, harms soil and water. Excessive copper leads to conditions like Wilson's disease. Chromium's toxicity varies; hexavalent chromium is highly carcinogenic, posing significant risks to industrial workers and the environment (Bansal, 2020; Mitra et al., 2022).

Various methodologies remove heavy metals from wastewater, including chemical precipitation, Flocculation, electrodialysis, ion exchange, coagulation, ultrafiltration, reverse osmosis, and adsorption (Fig. 7) (Bisht et al., 2017). Adsorption, especially using nanomaterials, has gained widespread acceptance due to its high effectiveness in taking heavy metal ions (Sadegh et al., 2017). Unlike methods like chemical precipitation and ion exchange, which produce substantial sludge and require precise conditions, adsorption with nanomaterials is more effective and versatile. It offers design flexibility, high-quality treated effluent, and reversibility with adsorbent regeneration, making it a promising alternative for heavy metal removal in wastewater treatment (grĕgorio et al., 2018; Sridevi et al., 2021).

Chitosan nanoparticles exhibit notable efficacy in eliminating heavy metals from wastewater by adsorption (Zhang et al., 2021). Their diminutive size and expansive surface area enhance their capacity to adsorb heavy metals significantly (Garcia-Aguilar et al., 2022). The adsorption

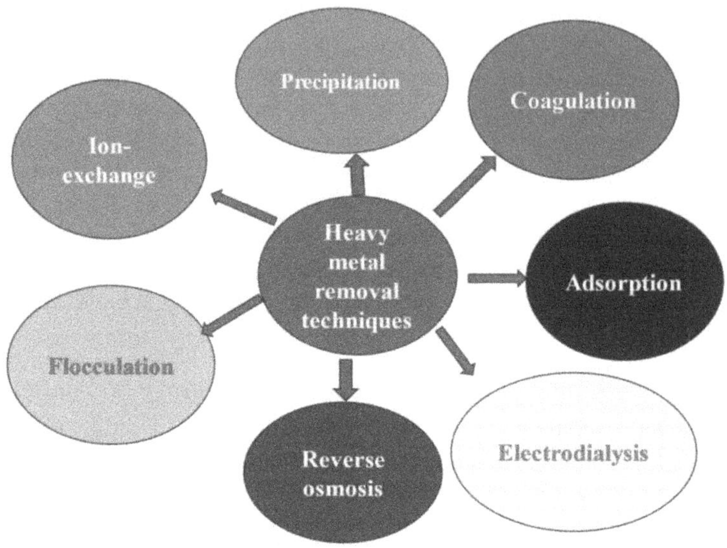

Figure 7. Heavy metal removal techniques.

mechanism primarily relies on the electrostatic attraction between the positively charged amino groups on chitosan's surface and the negatively charged pollutant ions (Pang et al., 2021). This interaction results in the conjugate adsorption of positively and negatively charged ions onto the nanoparticles' surface, facilitating their extraction from the water. Moreover, the nitrogen lone pair electrons present in the amino and N-acetylamino groups can form dative bonds with transition metal ions. Once the heavy metal ions adhere to the chitosan nanoparticles' surface, they can be separated from wastewater through filtration or centrifugation (Kim et al., 2005). Subsequently, the heavy metal-laden chitosan nanoparticles can either be regenerated for reuse or appropriately disposed of, depending on the specific requirements. Notably, chitosan nanoparticles demonstrate effectiveness in adsorbing lead, cadmium, and copper ions from aqueous solutions. Recent advancements highlight the use of biopolymeric nanocomposites for water purification. Rawat et al., created chitosan-based beads with iron oxyhydroxide nanocomposites for arsenic removal, achieving levels below 10 μg/L (Rawat et al., 2023). Ehsan et al., developed an alginate-based nanocomposite with graphene oxide for oil-water separation. These innovations demonstrate the potential of biopolymeric nanocomposites in tackling environmental pollutants (Ehsan et al., 2023).

Utilizing many methods such as films, beads, and fibers, alginate serves as an effective biosorbent for purifying wastewater (Table 1) (Cataldo et al., 2013; Gomez-Estaca et al., 2012; Lahiri et al., 2006; Torres and Dangl, 2005). For instance, encapsulated pillared clay microbeads with alginate have been employed in fixed bed columns to eliminate a wide range of pollutants like pentachlorophenol. Moreover, the emergence of nano alginate has garnered attention for its rapid interaction with toxic contaminants in wastewater, showcasing promising potential in wastewater treatment processes.

The use of Ca-alginate hydrogel beads generally enhances metal adsorption capability through surface grafting and cross-linking. However, these beads often suffer from instability and low mechanical strength, as noted by (Cai et al., 2017; Zhang et al., 2021). To address these limitations, carbonaceous materials like biochar, known for their robust mechanical strength and large specific surface area, have become popular for enhancing metal adsorption capacities in biochar-alginate bead formulations (Lee 2013; Gao et al., 2020; Liu et al., 2020).

A novel nanocomposite comprising graphene oxide, chitosan and 1,2-cyclohexylenedinitrilotetraacetic acid (CDTA) was synthesized to enhance the efficiency of hexavalent Cr (VI) ion removal and adsorption. By incorporating graphene, which possesses abundant hydroxyl, carboxyl, and epoxy groups, the nanocomposite provides improved adsorption

Table 1. Performance comparison of recent nano biocomposite for removal of heavy metal.

Adsorbent	Metal	pH	Cmax (mg/g)	References
Nanochitosan/sodium alginate/microcrystalline cellulose	Cu (II)	5	-	Vijayalakshmi et al., 2016
Sawdust-Chitosan nanocomposite beads (SDNCB)	Cu (II)	5	7.32	K et al., 2022
Nanochitosan/Carboxymethyl cellulose/ Graphene oxide composite	Cu (II)	4	145.26	Sugashini et al., 2022
Thiourea-modified magnetic Chitosan Microsphere Composite (TMCS)	Cu (II)	5	66.7	L et al., 2009
Nanochitosan /Bentonite/alginate	Cu (II)	6	12.21	Rahmani et al., 2017
Salicylaldehyde derivative of nano-chitosan	Cu (II)	5	72.77	Ms et al., 2020
Chitosan NPs	Cu (II)	7	95.93	Abd-Elhakeem et al., 2015
Cross-linked chitosan magnetic modified with methionine glutaraldehyde	Cu (II)	5.5	172.4	Salehi and Moghimi, 2022
Nanochitosan/sodium alginate/CMC/HAP	Cu (II)	6	512.79	Sasirekha et al., 2023
Chitosan nanoparticles (CNPs)	Fe (II) Mn (II) Cr (VI)	2.5-7 3-9	116.2 74.1 1395.32	El Knidri et al., 2018
Magnetic CS	Zn (II)	5	32.2	Gelsefidi and Mamaghanian, 2013
Pure chitosan nanoparcles	Ni (II)	5.2	384	Kong et al., 2018
CS/Fe_3O_4 NPs	Cr (VI)	3.37	162	Pourmortazavi et al., 2019
Fe_3O_4/CS NPs	(Hadi Najafabadi et al., 2015) Pb (II), Cd (II)	6	79.24, 36.42	Fan et al., 2017
Chitosan/Graphene oxide	Pb (II), Cu (II), Cr (VI)	-	461.3, 423.8,310.4	Hadi Najafabadi et al., 2015
Chitosan/Graphene oxide-SH	Cu (II), Pb (II), Cd (II)	-	425, 447,177	Li et al., 2015
MWCNT-PDA-CS-GO	Gd (III)	-	150.9 mg/g	Zhang et al., 2021
Cross-linked chitosan magnetic modified with methionine-glutaraldehyde	Cu (II) Pb (II) Cd (II)	5.5	172.4 mg/g 175.4 mg/g 163.9 mg/g	Salehi et al., 2020
Magnetic chitosan NPS	RR-141 RY-14	-	98.8 mg/g 89.7 mg/g	J et al., 2020; Jaafari et al., 2020

capabilities for heavy metals in aqueous solutions. Under optimized conditions of pH, temperature, and duration, the nanocomposite achieved a great adsorption capacity of 166.98 mg/g for hexavalent chromium ions (Ali, 2016). This surpasses the previously reported highest adsorption capacity of 153.85 mg/g for chitosan composites in studies focusing on Cr (VI) ion adsorption (Boddu et al., 2003).

Cellulose derivatives have been extensively studied for their adsorption capacities towards various metal ions including Cd (II), Ag (I), Pb (II), Cu (II), Hg (II), and Ni (II), yielding a diverse range of adsorption measurements values. Moreover, modifications of carboxymethyl cellulose (CMC) have resulted in the production of hydrogels through crosslinking with agents such as epichlorohydrin, polyacrylamide, and polyvinyl alcohol. Additionally (Liu et al., 2020) conducted a comprehensive investigation to obtain cellulose sponges from CMC. This was achieved by treating the cellulose derivative with varying concentrations of sulfuric acid followed by a freeze-thawing process, eliminating the necessity for an organic crosslinker (Fakhri et al., 2023).

Two variabilities of altered cassava starch specimens, namely MCS and MWS, resulting from commercially sourced Native Cassava Starch (NCS) and native cassava starch extracted through the wet process (NWS), experienced scrutiny for their efficacy in eliminating Pb(II) ions from aqueous solutions (Gunawardene et al., 2021). Synthesized under acidic circumstances, MCS and MWS samples were crafted utilizing Pluronic 123 as the structure-directing component and tetra ethyl orthosilicate (TEOS) as the chemical adjusting element. The altered samples confirmed a desorption effectiveness exceeding 97%. This investigation showcases the potential utility of modified starch in the removal of Pb (II) from industrial wastewater (Gunawardene et al., 2021).

Corn starch experienced alteration using obviously occurring compounds, Maleic Acid (MA) and Itaconic Acid (IA), through esterification in an aqueous medium catalyzed by NaOH, resulted in the creation of adsorbents for the removal of heavy metals from water. Furthermore, the efficacy of the adapted starches containing itaconic acid in adsorbing aqueous metal cations such as Cd^{2+}, Ni^{2+}, Zn^{2+} and Pb^{2+} was investigated, with results related against native corn starch. Both the modified and native starches displayed reasonable adsorption capacities for major cations. The study exposed that the absence of native starch led to selectivity for any cation, whereas itaconate starches exhibited a maximum capacity for removing Pb^{2+} with the smallest hydration radius compared to other ions (Soto et al., 2015).

Utilizing citrus peels for the biosorption of lead (Pb) from aqueous solutions provides a promising and economical method for lead removal from industrial wastewaters. These peels cover pectin, a biopolymer with a high attraction for metal ions (Balaria and Schiewer, 2008). A deeper type of the chemistry underlying these interactions could enable the development of profitable biosorbents using waste citrus peels. Research has shown that citrus fruit waste material helps as a definite lead biosorbent. Surveys into the impact of methylation and acetylation degrees on metal ion binding by pectin have indicated that the required mechanism relies not only on the interaction of Pb with carboxylic oxygen but also on the degree of acetylation, which effects the conformation of the system (Renard and Jarvis, 1999).

7.2 Nanosorbent for Dye Removal

Saha et al., directed a study examining the influence of chitosan powder dosage on the elimination of Reactive Black 5 dye from aqueous solutions (Saha et al., 2011). Their results confirmed that augmenting the chitosan powder dosage led to greater effectiveness of dye removal. This phenomenon is explained by the better obtainability of adsorption sites subsequent from increased chitosan dosage, thereby fostering improved interactions between the dye molecules and the adsorbent. Furthermore, the results showed that the adsorption capacity rose as the pH decreased.

Bekci et al., 2008, examined how chitosan beads were limited to the adsorption of the cationic dye Malachite Green. Same as chitosan powder, the effectiveness of adsorption with chitosan beads improved with higher doses of the adsorbent, as this provided more binding sites for dye molecules. The study also inspected the impact of solution pH, revealing that chitosan beads could efficiently adsorb cationic dye at high pH levels of 8 or above. At pH 8 and over a period of 300 minutes, the extreme adsorption capacity reached 93.55 mg/g. Additionally, the research showed that chitosan beads reached a maximum saturation point compared to chitosan powder, suggesting that the optimal dose for chitosan beads might be higher before attainment its adsorption measurements limit.

Karimifard et al., led a study where coir fiber was hydrolyzed with sulfuric acid to yield rod-like nanocrystals with diameters ranging from 10 to 30 nm. Also, cellulose nano whiskers with lengths varying from 80 to 500 nm and widths of 6 nm were mined from coconut husk fibers using nitric acid (Karimifard and Alavi Moghaddam, 2018). For high-scale production, nanocellulose can be professionally synthesized through an economical alkali-acid hydrolysis method (Tavakolian et al., 2020). However, post-adsorption processes such as centrifugation or coagulation are naturally required to recover the nanocellulose, preventing their widespread application. To address this limitation, nanocellulose can be cross-linked with other polymers to form hydrogel beads. The consumption of hydrogel beads in wastewater treatment has garnered significant attention from researchers due to their swelling capacity, stable hydrophobic-hydrophilic characteristics, and reusability (Allouss et al., 2019). Therefore, the primary aim of this study was to fabricate and enhance an actual nanocrystalline cellulose-chitosan (NCC-CH) hydrogel bead for the removal of CV and MB dyes from textile wastewater.

A cellulose-based adsorbent resulting from cotton fibers was developed. Originally, cotton fibers led to activation using a NaOH aqueous solution, following which the activated cotton was subjected to chemical alteration using β-CD and NH_2-HBP, respectively (Moon et al., 2011). These altered cotton fibers were employed for the removal of azo dyes from aqueous solutions. Congo Red (CR) and Methylene Blue (MB) were selected as illustrative examples of anionic and cationic dyes for the experiments. The study explored in investigating the effects of solution pH, adsorption time, and initial concentration of CR and MB, all of which are crucial parameters for assessing the adsorption properties of the adsorbent.

The synthesis of nanocomposites (Rebar et al., 2015) compounding calcium alginate polysaccharides with iron oxide yields materials with desirable properties for effective adsorption (Zirpe and Thakur, 2023). In this search, beads composed of calcium alginate and magnetic composites were invented and utilized as adsorbents for the extraction of MV from aqueous solutions. The adsorption of MV was examined from both equilibrium and kinetics perspectives. Moreover, the study explored the effect of various limits, such as temperature, adsorbent mass, and initial concentration of MV, on the adsorption process.

According to (Rebar et al., 2015) ZnO nanoparticles can be synthesized using several methods. Nanostructured ZnO, known for its improved surface area and photocatalytic activity as highlighted by (Bibi et al., 2020) and (Karimi-Maleh et al., 2021), holds promise as an additive to recover dye removal from wastewater. Thermodynamic studies specify that adsorption by ZnO is non-spontaneous and exothermic (Pholnak et al., 2022), whereas in the case of zeolite incorporating ZnO, the adsorption process is spontaneous and endothermic (Elfeky et al., 2019). In this study, alginate beads containing ZnO were manufactured for the purpose of eliminating dye molecules from wastewater, which is critical for environmental conservation. The dye removal efficacies of fresh and dry beads with varying ZnO loadings were compared by treating dye solutions usually found in the production of Thale Noi grey sedge handicrafts.

Al-Gorair et al positively synthesized a pectin/acrylic acid/CNC nanocomposite through γ-irradiation for the purpose of remediating Methylene Blue (MB) dye from a typical solution. Their findings indicated that the adsorption capacity increased with the rise in MB concentration (Al-Gorair et al., 2022), reaching a peak at 70 mg/L at pH 9, regardless of the temperature variations. On the other hand, (Ar et al., 2014) documented the development of composite films using Amidated Pectin (AP) with varying montmorillonite (MMT) content of 10, 30, and 50% for the adsorption of Undeveloped Yellow 28. They observed a direct correlation between the adsorption capacity and the MMT content, with values increasing from 85 mg/g (AP/MMT 10%) to 103 mg/g (AP/MMT 30%) and further to 126 mg/g (AP/MMT 50%), when the initial dye concentration was maintained at 30 mg/L.

7.3 Nanosorbents for Effluent Treatment

Frequent research efforts have showcased chitosan's notable affinity for various contaminants (Guibal et al., 2006). These studies have emphasized its effectiveness in eliminating dyes from solution or textile wastewater (Lu et al., 2011; Rodrigues et al., 2008), organic substances like chlorinated compounds and lignin found in pulp and paper mill effluents, heavy metals (Bai, 2002;

Renault et al., 2009), and phenolic compounds present in cardboard-mill wastewater, as well as inorganic suspensions such as kaolinite suspension (Ariffin et al., 2009; Li et al., 2020).

In their current study, Abu Hassan et al., underlined that utilizing chitosan at dosages ranging from 12 mg/l to 30 mg/l resulted in substantial reductions in chemical oxygen demand and turbidity. They qualified this effect to the high charge density of chitosan, which facilitates the rapid destabilization of colloidal equilibrium (Nechita, 2015).

Chitosan extracted from Indian prawn and employed it for treating wastewater from the ghee industry (Ali, 2013). They studied various factors influencing the flocculation process, including chitosan dosage, optimal pH, and contact times. The results verified that chitosan effectively flocculated the negatively charged suspended particles, leading to significant reductions in chemical oxygen demand (by 80.1%), turbidity (by 91.8%), total dissolved solids (by 72.5%), and conductivity (by 73.7%) of the evaluated wastewater.

Nechita et al. studied the adsorption capacity of chitosan for heavy metal ions present in textile wastewaters, specifically Cu (II) and Pb (II), as well as in aqueous solutions containing Zn (II) and Fe (III). We observed a removal efficiency of 91.67% for Pb ions and 54.15% for Cu ions. The pH value played a crucial role, influencing the surface charge of the wastewater solutions and thereby affecting the stability of suspensions. In our studies, we found that the maximum adsorption of Cu (II) and Pb (II) ions from textile wastewaters occurred at a pH value of 8 (Nechita, 2015).

Moradi Dehaghi et al., worked at a sonication process to develop chitosan-ZnO nanoparticle composites (Dehaghi et al., 2014). They led dissolution and swelling studies on these composites and characterized their crystallinity and surface morphology via techniques such as X-ray diffraction, Fourier transform infrared spectroscopy (FT-IR), and scanning electron microscopy. Besides, they tested the sorbent performance of these materials for pesticide adsorption from aqueous solutions containing 25% permethrin, employing a column packed with chitosan-ZnO nanocomposite beads (Hritcu et al., 2012).

Singh et al., examined raw cotton, an economical and readily existing cellulose-based material, as a sorbent for oil (Singh et al., 2013). They placed raw cotton within a circular steel mesh and flooded it in a glass dish filled with raw crude oil, allowing the mesh to float freely in the oil. Their findings revealed that lower microlayer cotton exhibited superior sorption capacities. Specifically, they stated that immature cotton with a micronaire of 3.1 had a crude oil sorption capacity of 30.5 g g^{-1}. Micronaire serves as an indicator for fiber quality and maturity. In comparison to synthetic adsorbents, raw cotton demonstrated higher crude oil adsorption capacity and a favorable environmental footprint, making it an eco-friendly choice for oil spill cleanup. Dicarboxyl cellulose (DCC) samples were characterized by varying levels of carboxylation. The synthesis of these DCCs involved giving bleached birch chemical wood pulp with sodium periodate followed by sodium chlorite. Through this method, cellulose nanocrystals (CNCs) with carboxyl contents ranging from 0.38 to 1.75 mmol g^{-1} were obtained. Their findings revealed that CNCs with a carboxyl content of 1.75 mmol g^{-1} showed the ability to eliminate up to 80% of turbidity and 60% of (COD) from water. They noted that increasing the flocculant dosage enhanced flocculation up to doses of 5 mg L^{-1}, beyond which further dosage increments did not influence the flocculation ability significantly. Moreover, they reported that pH exerted minimal impact on the cellulose's flocculating efficacy.

Addressing the extensive concern of environmental pollution caused by heavy metals (Xu et al., 2021) introduced a novel approach for justifying this issue. They established a bio adsorbent utilizing calcium-crosslinked alginate-encapsulated bacteria to successfully remove cadmium from contaminated water. Their study demonstrated that 10 grams of the synthesized biomaterial could achieve significant removal of Cd^{2+} ions when tested with a loading amount of 1 mM Cd^{2+} in a 50 mL of a water sample.

Fang et al explored the efficacy of starch-based magnetic adsorbents in eliminating dye, heavy metals, pharmaceuticals, and oil. Meanwhile, Haq et al., explored water pollution by different pollutants, including new micropollutants and traditional pollutants like organics and heavy metals, which is a major source of concern (e.g., consumer care products and interfering endocrine-related substances) (Fang et al., 2022; Haq et al., 1988). Scholars are currently investigating sustainable materials and methods with the purpose of eliminating pollutants from wastewater. The benefits of both starch and magnetic nanoparticles can be combined in Magnetic Starch-based Adsorbents (MSAs), which have exceptional surface properties and unique critical features like size, shape, crystallinity, stability, magnetic properties, availability, and cost-effectiveness. However, there may only be a few reviews accessible presently on the characterizations, applications, and adsorption mechanisms of MSAs. As a result, this review covers both their physical (such as mechanical activation treatment, microwave radiation treatment, sonication, and extrusion), chemical (such as grafting, cross-linking, oxidation, and esterification), and enzymatic modifications to improve their adsorption properties, as well as their preparation and activation methods. Additionally, it provides an all-encompassing, state-of-the-art analysis of the entire spectrum of their characterization techniques, the adsorption of different contaminants, and the underlying adsorption mechanisms. In the end, this analysis concentrates on the performance of reclamation and recycling and identifies the primary gaps in the areas that require additional research. Our goal for this review is to initiate an interdisciplinary discussion and revolutionize the use of recent developments in starch-based magnetic adsorbents for the removal of contaminants from wastewater. (Fang et al., 2022) discussed the utilization of starch-based hydrogels, composites, and nanoparticles in removing dyes and heavy metals. Additionally, Gupta et al., directed their attention towards starch modification employing various techniques such as oxidation, irradiation, grafting, crosslinking, and esterification for the elimination of heavy metals (Gupta and Rastogi, 2008).

8. Multifunctional Material of the Future—Biopolymer Nanomaterial

Water serves as a vital element not just for sustaining life on our planet but also for driving numerous industrial processes. Various sectors like paper manufacture, textile industries, agriculture, leather processing, among others, create considerable volumes of wastewater laden with potential contaminants (Iftekhar et al., 2021). The swift pace of industrial growth and urban development aimed at enhancing human standards of living has inflicted substantial harm on the environment.

Natural biopolymers possess the advantage of being biodegradable, having expansive surface areas, and being easily prepared, broadly accessible, highly active in sorption, and cost-effective. Examples like peat, Rhizopus, yeast, and algae are well-suited as sustainable substitutes for conventional sorbents that are challenging to prepare (Iftekhar et al., 2021). These biopolymers are under scrutiny for their possible application as sorbents in wastewater treatment. Especially, coir pith has demonstrated remarkable efficiency in removing heavy metal cations from polluted water sources.

Alginate biopolymers are frequently utilized in bead form, characteristically formed through crosslinking with calcium ions, for wastewater treatment. Norris et al., (2014) developed magnetic alginate beads tricking iron particles for adsorbing Cetyl Pyridinium Chloride (CPC). Besides surfactant removal, alginate biopolymers have gained significant traction in recent years for wastewater decontamination purposes. Kube et al., (2018) highlighted the effectiveness of algae-loaded alginate beads in removing nutrients from urban wastewater. Additionally (Kausar et al., 2020), utilized alginate-clay composite material for treating dye wastewater (Blue FBN and Rose FRN), achieving maximum removal efficiencies of 76.39% for Blue FBN and 59.85% for Rose FRN at pH 2 and a temperature of 30°C. Additionally, (Allouss et al., 2020) synthesized graphene oxide encapsulated carboxymethylcellulose-alginate microspheres through ionotropic gelation for adsorbing Cu(II) from real wastewater samples.

9. Conclusion

In summary, the paper highlights the consequence of sustainable materials, such as biopolymers and nano biopolymers, in an environmental remedy, notably in water treatment. It investigates how nanotechnology develops the efficiency of these materials as adsorbents for eliminating pollutants like heavy metals and dyes from effluent water sources. Moreover, it vitrines the development of polymeric nanocomposites and nano sorbents with sensitive surface area and reactivity to tackle pollutant purification more efficiently. The conclusion emphasizes the potential of biopolymer nanomaterials as versatile, eco-friendly solutions to address current environmental challenges. Overall, the paper offers a complete outlook on advancements in biopolymer technology for environmental sustainability, signifying that combining green chemistry with nanotechnology could cover the way for stable solutions in water purification and pollution management.

References

Abd-Elhakeem, M. A., Ramadan, M. M. and Basaad, F. S. (2015). Removing of heavymetals from water by chitosan nanoparticles. Journal of Advances In Chemistry, 11(7), Article 7. https://doi.org/10.24297/jac.v11i7.2200.

Acemioglu, B. and Alma, M. H. (2001). Equilibrium studies on adsorption of Cu(II) from aqueous solution onto cellulose. Journal of Colloid and Interface Science, 243(1): 81–84. https://doi.org/10.1006/jcis.2001.7873.

Akpor, O. (2014). Heavy metal pollutants in wastewater effluents: sources, effects and remediation. Advances in Bioscience and Bioengineering, 2: 37. https://doi.org/10.11648/j.abb.20140204.11.

Al-Gorair, A. S., Sayed, A. and Mahmoud, G. A. (2022). Engineered superabsorbent nanocomposite reinforced with cellulose nanocrystals for remediation of basic dyes: isotherm, kinetic, and thermodynamic studies. Polymers, 14(3), Article 3. https://doi.org/10.3390/polym14030567.

Ali, M. (2016). Synthesis and adsorption properties of chitosanCDTA-GO nanocomposite for removal of hexavalent chromium from aqueous solutions. Arabian Journal of Chemistry, 11. https://doi.org/10.1016/j.arabjc.2016.09.010.

Ali, Prof. I., Imran Ali, Zeid A. AL-Othman, Abdulrahman Alwarthan and Hassan Y. Aboul-Enein. (2016). Capillary electrophoresis: An versatile technique in pharmaceutical analysis. *In*: Suvardhan Kanchi, S., Bisetty, K. and Sabela, M. I. (eds.). Capillary electrophoresis: Recent developments and trends in Pharmaceutical research. The Pan Stanford Publications, Singapore (2016). Hard copy ISBN: 10: 981477412X and Electronic copy ISBN: 13: 978-9814774123.

Ali, Q., Zia, M. A., Kamran, M., Shabaan, M., Zulfiqar, U., Ahmad, M. et al. (2023). Nanoremediation for heavy metal contamination: A review. Hybrid Advances, 4: 100091. https://doi.org/10.1016/j.hybadv.2023.100091.

Ali, Z. (2013). Extraction and Characterization of Chitosan from Indian Prawn (Fenneropenaeus Indicus) and its Applications on Waste Water Treatment of Local Ghee Industry. IOSR Journal of Engineering, 3: 28–37. https://doi.org/10.9790/3021-031022837.

Allouss, D., Essamlali, Y., Amadine, O., Chakir, A. and Zahouily, M. (2019). Response surface methodology for optimization of methylene blue adsorption onto carboxymethyl cellulose-based hydrogel beads: Adsorption kinetics, isotherm, thermodynamics and reusability studies. RSC Advances, 9(65): 37858–37869. https://doi.org/10.1039/C9RA06450H.

Allouss, D., Essamlali, Y., Chakir, A., Khadhar, S. and Zahouily, M. (2020). Effective removal of Cu(II) from aqueous solution over graphene oxide encapsulated carboxymethylcellulose-alginate hydrogel microspheres: Towards real wastewater treatment plants. Environmental Science and Pollution Research International, 27(7): 7476–7492. https://doi.org/10.1007/s11356-019-06950-w.

Ambashta, R. D. and Sillanpää, M. (2010). Water purification using magnetic assistance: A review. Journal of Hazardous Materials, 180(1): 38–49. https://doi.org/10.1016/j.jhazmat.2010.04.105.

Amiri, M. J., Faraji, A., Azizi, M., Nejad, B. G. and Arshadi, M. (2021). Recycling bone waste and cobalt-wastewater into a highly stable and efficient activator of peroxymonosulfate for dye and HEPES degradation. Process Safety and Environmental Protection, 147: 626–641. https://doi.org/10.1016/j.psep.2020.12.039.

Anastopoulos, I., Bhatnagar, A., Bikiaris, D. N. and Kyzas, G. Z. (2017). Chitin Adsorbents for Toxic Metals: A Review. International Journal of Molecular Sciences, 18(1): 114. https://doi.org/10.3390/ijms18010114.

Anastopoulos, I. and Kyzas, G. Z. (2015). Composts as biosorbents for decontamination of various pollutants: a review. Water, Air and Soil Pollution, 226(3): 61. https://doi.org/10.1007/s11270-015-2345-2.

Jin, L. and Bai, R. (2002). Mechanisms of lead adsorption on chitosan/PVA hydrogel beads. Langmuir, 18(25): 9765–9770.

Annu, Mittal, M., Tripathi, S. and Shin, D. K. (2024). Biopolymeric nanocomposites for wastewater remediation: an overview on recent progress and challenges. Polymers, 16(2), Article 2. https://doi.org/10.3390/polym16020294.

Ar, N., Sj, V. and Dg, A. (2014). Novel composite films based on amidated pectin for cationic dye adsorption. Colloids and Surfaces. B, Biointerfaces, 116. https://doi.org/10.1016/j.colsurfb.2013.10.031.

Ariffin, M., Hassan, M., Li, T. and Zainon Noor, Z. (2009). Coagulation and flocculation treatment of wastewater in textile industry using chitosan. Journal of Chemical and Natural Resources Engineering, 4: 43–53.

Arshadi, M., Mehravar, M., Amiri, M. J. and Faraji, A. R. (2015). Synthesis and adsorption characteristics of an heterogenized manganese nanoadsorbent towards methyl orange. Journal of Colloid and Interface Science, 440: 189–197. https://doi.org/10.1016/j.jcis.2014.10.053.

Arshadi, M., Mousavinia, F., Amiri, M. J. and Faraji, A. R. (2016). Adsorption of methyl orange and salicylic acid on a nano-transition metal composite: Kinetics, thermodynamic and electrochemical studies. Journal of Colloid and Interface Science, 483: 118–131. https://doi.org/10.1016/j.jcis.2016.08.032.

Aziz, A., Daik, R., Daud, N. and Yamin, B. (2004). Hydroxypropylation and Acetylation of Sago Starch. Malaysian J. Chem, 6.

Azizi Samir, M. A. S., Alloin, F. and Dufresne, A. (2005). Review of recent research into cellulosic whiskers, their properties and their application in nanocomposite field. Biomacromolecules, 6(2): 612–626. https://doi.org/10.1021/bm0493685.

Bai, R. and Leow, H. F. (2002). Microfiltration of activated sludge wastewater—the effect of system operation parameters. Separation and Purification Technology, 29(2): 189–198.

Balakrishnan, P., Sreekala, M. S., Kunaver, M., Huskić, M. and Thomas, S. (2017). Morphology, transport characteristics and viscoelastic polymer chain confinement in nanocomposites based on thermoplastic potato starch and cellulose nanofibers from pineapple leaf. Carbohydrate Polymers, 169: 176–188. https://doi.org/10.1016/j.carbpol.2017.04.017.

Balaria, A. and Schiewer, S. (2008). Assessment of biosorption mechanism for Pb binding by citrus pectin. Separation and Purification Technology, 63(3): 577–581. https://doi.org/10.1016/j.seppur.2008.06.023.

Bamba, Y., Ogawa, Y., Saito, T., Berglund, L. A. and Isogai, A. (2017). Estimating the Strength of Single Chitin Nanofibrils via Sonication-Induced Fragmentation. Biomacromolecules, 18(12): 4405–4410. https://doi.org/10.1021/acs.biomac.7b01467.

Bansal, O. P. (2020). Health risks of potentially toxic metals contaminated water. In heavy metal toxicity in public health. IntechOpen. https://doi.org/10.5772/intechopen.92141.

Bednarcyk, B. A. (2003). An inelastic micro/macro theory for hybrid smart/metal composites. Composites Part B: Engineering, 34(2): 175–197. https://doi.org/10.1016/S1359-8368(02)00067-7.

Bekçi, Z., Özveri, C., Seki, Y. and Yurdakoç, K. (2008). Sorption of malachite green on chitosan bead. Journal of Hazardous Materials, 154(1-3): 254–261.

Bibi, A., Naseer, N. and Habib, Z. (2020). Study habits of students and academic achievement: A correlational study. Global Educational Studies Review, V: 114–122. https://doi.org/10.31703/gesr.2020(V-III).12.

Bisht, R., Agarwal, M. and Singh, K. (2017). Methodologies for removal of heavy metal ions from wastewater: An overview. Interdisciplinary Enviornmental Review, 18: 124–142. https://doi.org/10.1504/IER.2017.10008828.

Biswas, S. and Pal, A. (2021). Application of biopolymers as a new age sustainable material for surfactant adsorption: A brief review. Carbohydrate Polymer Technologies and Applications, 2: 100145. https://doi.org/10.1016/j.carpta.2021.100145.

Boddu, V., Abburi, K., Talbott, J. and Smith, E. (2003). Removal of hexavalent chromium from wastewater using a new composite chitosan biosorbent. Environmental Science & Technology, 37: 4449–4456. https://doi.org/10.1021/es021013a.

Cai, Y., Wu, C., Liu, Z., Zhang, L., Chen, L., Wang, J. et al. (2017). Fabrication of a phosphorylated graphene oxide–chitosan composite for highly effective and selective capture of U(VI). Environmental Science: Nano, 4(9): 1876–1886. https://doi.org/10.1039/C7EN00412E.

Cataldo, R., John, J., Chandran, L., Pati, S. and Shroyer, A. L. W. (2013). Impact of physical activity intervention programs on self-efficacy in youths: A systematic review. ISRN Obesity, 586497. https://doi.org/10.1155/2013/586497.

Cheng, J., Zhan, C., Wu, J., Cui, Z., Si, J., Wang, Q. et al. (2020). Highly efficient removal of methylene blue dye from an aqueous solution using cellulose acetate nanofibrous membranes modified by polydopamine. ACS Omega, 5(10): 5389–5400. https://doi.org/10.1021/acsomega.9b04425.

Chincholikar, P., Singh, K. R., Natarajan, A., Kerry, R. G., Singh, J., Malviya, J. et al. (2023). Green nanobiopolymers for ecological applications: A step towards a sustainable environment. RSC Advances, 13(18): 12411–12429. https://doi.org/10.1039/d2ra07707h.

Chung, D. D. L. (2003). Composite Materials: Functional Materials for Modern Technologies. Springer Science & Business Media.

Copeland, L., Blazek, J., Salman, H. and Tang, M. C. (2009). Form and functionality of starch. Food Hydrocolloids, 23(6): 1527–1534. https://doi.org/10.1016/j.foodhyd.2008.09.016.

Croisier, F. and Jérôme, C. (2013). Chitosan-based biomaterials for tissue engineering. European Polymer Journal, 49(4): 780–792. https://doi.org/10.1016/j.eurpolymj.2012.12.009.

Dehaghi, S., Rahmanifar, B., Mashinchian Moradi, A. and Aberoomand azar, P. (2014). Removal of Permethrin pesticide from water by Chitosan-Zinc oxide nanoparticles composite as an adsorbent. Journal of Saudi Chemical Society, 18. https://doi.org/10.1016/j.jscs.2014.01.004.

Demirbas, A. (2008). Heavy metal adsorption onto agro-based waste materials: A review. Journal of Hazardous Materials, 157(2): 220–229. https://doi.org/10.1016/j.jhazmat.2008.01.024.

Dhariwal, S. and Mittal, M. (2023). Wastewater treatment with perovskite-based photocatalysts: Environmental sustainability from a green perspective. Materials Today: Proceedings. https://doi.org/10.1016/j.matpr.2023.03.048.

Dhiman, B., Divtrannum, null, Dhiman, A. and Saini, S. (2017). An appraisal on various methods of nano particulate formulations. Pharmaceutical Nanotechnology, 5(4): 255–262. https://doi.org/10.2174/2211738505666171108143805.

Dridi-Dhaouadi, S., Douissa-Lazreg, N. B. and M'Henni, M. F. (2011). Removal of lead and Yellow 44 acid dye in single and binary component systems by raw *Posidonia oceanica* and the cellulose extracted from the raw biomass. Environmental Technology, 32(3): 325–340. https://doi.org/10.1080/09593330.2010.499545.

Du, Y., Li, S., Zhang, Y., Rempel, C. and Liu, Q. (2016). Treatments of protein for biopolymer production in view of processability and physical properties: A review. Journal of Applied Polymer Science, 133(17). https://doi.org/10.1002/app.43351.

Ehsan, M., Razzaq, H., Razzaque, S., Kanwal, M. and Hussain, I. (2023). Engineering nanocomposite membranes of sodium alginate-graphene oxide for efficient separation of oil-water and antifouling performance. Journal of Environmental Chemical Engineering, 11(1): 109185. https://doi.org/10.1016/j.jece.2022.109185.

El Knidri, H., Belaabed, R., Addaou, A., Laajeb, A. and Lahsini, A. (2018). Extraction, chemical modification and characterization of chitin and chitosan. International Journal of Biological Macromolecules, 120: 1181–1189. https://doi.org/10.1016/j.ijbiomac.2018.08.139.

Elfeky, A., Youssef, H. and Elzaref, A. (2019). Adsorption of dye from wastewater onto ZnO nanoparticles-loaded zeolite: Kinetic, thermodynamic and isotherm studies. Zeitschrift Für Physikalische Chemie, 234. https://doi.org/10.1515/zpch-2018-1342.

Fakhri, V., Jafari, A., Layaei Vahed, F., Su, C.-H. and Pirouzfar, V. (2023). Polysaccharides as eco-friendly bio-adsorbents for wastewater remediation: Current state and future perspective. Journal of Water Process Engineering, 54: 103980. https://doi.org/10.1016/j.jwpe.2023.103980.

Fan, H.-L., Zhou, S.-F., Jiao, W.-Z., Qi, G.-S. and Liu, Y.-Z. (2017). Removal of heavy metal ions by magnetic chitosan nanoparticles prepared continuously via high-gravity reactive precipitation method. Carbohydrate Polymers, 174: 1192–1200. https://doi.org/10.1016/j.carbpol.2017.07.050.

Fang, K., Deng, L., Yin, J., Yang, T., Li, J. and He, W. (2022). Recent advances in starch-based magnetic adsorbents for the removal of contaminants from wastewater: A review. International Journal of Biological Macromolecules, 218: 909–929. https://doi.org/10.1016/j.ijbiomac.2022.07.175.

Figueiredo, J. A., Ismael, M. I., Anjo, C. M. S. and Duarte, A. P. (2010). Cellulose and derivatives from wood and fibers as renewable sources of raw-materials. pp. 117–128. *In*: Rauter, A. P., Vogel, P. and Queneau, Y. (eds.). Carbohydrates in Sustainable Development I. Springer. https://doi.org/10.1007/128_2010_88.

Flórez, M., Cazón, P. and Vázquez, M. (2023). Selected biopolymers' processing and their applications: A review. Polymers, 15(3): 641. https://doi.org/10.3390/polym15030641.

Gao, J., Zheng, P., Jia, Y., Chen, H., Mao, Y., Chen, S. et al. (2020). Mental health problems and social media exposure during COVID-19 outbreak. PloS One, 15(4): e0231924. https://doi.org/10.1371/journal.pone.0231924.

Garcia-Aguilar, J., Patil, S., Gollub, M. J., Kim, J. K., Yuval, J. B., Thompson, H. M. et al. (2022). Organ preservation in patients with rectal adenocarcinoma treated with total neoadjuvant therapy. Journal of Clinical Oncology, 40(23): 2546–2556. https://doi.org/10.1200/JCO.22.00032.

Gelsefidi, S. and Mamaghanian, J. (2013). Stabilization of a Weak Low Plasticity Clay Soil Using Nanomaterial. 134–137. https://doi.org/10.3233/978-1-61499-297-4-134.

Giannelis, E. P. (1996). Polymer layered silicate nanocomposites. Advanced Materials, 8(1): 29–35. https://onlinelibrary.wiley.com/doi/abs/10.1002/adma.19960080104.

Gomathi, T., Rajeshwari, K., Kanchana, V., Sudha, P. N. and Parthasarathy, K. (2019). Impact of nanoparticle shape, size, and properties of the sustainable nanocomposites. pp. 313–336. *In*: Inamuddin, S. Thomas, R. Kumar Mishra and Asiri, A. M. (eds.). Sustainable Polymer Composites and Nanocomposites. Springer International Publishing. https://doi.org/10.1007/978-3-030-05399-4_11.

Gomez-Estaca, J., Balaguer, M. P., Gavara, R. and Hernandez-Munoz, P. (2012). Formation of zein nanoparticles by electrohydrodynamic atomization: Effect of the main processing variables and suitability for encapsulating the food coloring and active ingredient curcumin. Food Hydrocolloids, 28(1): 82–91. https://doi.org/10.1016/j.foodhyd.2011.11.013.

grégorio, C., Lichtfouse, E., Wilson, L. and Crini, N. (2018). Adsorption-Oriented Processes Using Conventional and Non-conventional Adsorbents for Wastewater Treatment (pp. 23–71). https://doi.org/10.1007/978-3-319-92111-2_2.

Guibal, E., Vooren, M., Dempsey, B. and Roussy, J. (2006). A review of the use of chitosan for the removal of particulate and dissolved contaminants. Separation Science and Technology, 41: 2487–2514. https://doi.org/10.1080/01496390600742807.

Gunawardene, O. H. P., Gunathilake, C. A., Amaraweera, A. P. S. M., Fernando, N. M. L., Manipura, A., Manamperi, W. A. et al. (2021). Removal of Pb(II) Ions from aqueous solution using modified starch. Journal of Composites Science, 5(2): 46. https://doi.org/10.3390/jcs5020046.

Gupta, A., Dubey, P., Kumar, M., Roy, A., Sharma, D., Khan, M. M. et al. (2022). Consequences of arsenic contamination on plants and mycoremediation-mediated arsenic stress tolerance for sustainable agriculture. Plants, 11(23): 3220. https://doi.org/10.3390/plants11233220.

Gupta, V. K. and Rastogi, A. (2008). Biosorption of lead(II) from aqueous solutions by non-living algal biomass *Oedogonium* sp. And *Nostoc* sp.—A comparative study. Colloids and Surfaces. B, Biointerfaces, 64(2): 170–178. https://doi.org/10.1016/j.colsurfb.2008.01.019.

Gupta, V. K. and Suhas. (2009). Application of low-cost adsorbents for dye removal—A review. Journal of Environmental Management, 90(8): 2313–2342. https://doi.org/10.1016/j.jenvman.2008.11.017.

Hadi Najafabadi, H., Irani, M., Roshanfekr Rad, L., Sojoudi, A. and Haririan, I. (2015). Removal of Cu2+, Pb2+ and Cr6+ from aqueous solutions using a chitosan/graphene oxide composite nanofibrous adsorbent. RSC Advances, 5: 16532–16539. https://doi.org/10.1039/c5ra01500f.

Hamed, I., Özogul, F. and Regenstein, J. M. (2016). Industrial applications of crustacean by-products (chitin, chitosan, and chitooligosaccharides): A review. Trends in Food Science & Technology, 48: 40–50. https://doi.org/10.1016/j.tifs.2015.11.007.

Haq, B., Hardenbol, J. and VAIL, P. (1988). Mesozoic and cenozoic chronostratigraphy and cycles of sea-level change. In Sea-level changes: An integrated approach (Vol. 42, pp. 7–108). https://doi.org/10.2110/pec.88.01.0071.

Hritcu, D., Dodi, G. and Popa, M. I. (2012). Heavy metal ions adsorption on chitosan-magnetite microspheres. International Review of Chemical Engineering, 4(3): 364–368.

Huber, T., Müssig, J., Curnow, O., Pang, S., Bickerton, S. and Staiger, M. P. (2012). A critical review of all-cellulose composites. Journal of Materials Science, 47(3): 1171–1186. https://doi.org/10.1007/s10853-011-5774-3.

Humplik, T., Lee, J., O'Hern, S. C., Fellman, B. A., Baig, M. A., Hassan, S. F. et al. (2011). Nanostructured materials for water desalination. Nanotechnology, 22(29): 292001. https://doi.org/10.1088/0957-4484/22/29/292001.

Iftekhar, S., Wasayh, M. A., Asif, M. B., Lehto, V.-P. and Sillanpää, M. (2021). 2—Methods for synthesis of nanobiopolymers. pp. 13–35. *In*: Kanwar, S., Kumar, A., Nguyen, T. A., Sharma, S. and Slimani, Y. (eds.). Biopolymeric Nanomaterials. Elsevier. https://doi.org/10.1016/B978-0-12-824364-0.00009-5.

Iravani, S. (2021). Nanomaterials and nanotechnology for water treatment: Recent advances. Inorganic and Nano-Metal Chemistry, 51(12): 1615–1645. https://doi.org/10.1080/24701556.2020.1852253.

J, J., H, B., S, M., N, A. A. F., K, S., P, S. and Ga, H. (2020). Effective adsorptive removal of reactive dyes by magnetic chitosan nanoparticles: Kinetic, isothermal studies and response surface methodology. International Journal of Biological Macromolecules, 164. https://doi.org/10.1016/j.ijbiomac.2020.07.042.

Jafari, H., Lista, A., Siekapen, M. M., Ghaffari-Bohlouli, P., Nie, L., Alimoradi, H. et al. (2020). Fish collagen: Extraction, characterization, and applications for biomaterials engineering. Polymers, 12(10): 2230.

Jain, K. (2019). Nanohybrids of dendrimers and carbon nanotubes: A benefaction or forfeit in drug delivery? Nanoscience & Nanotechnology-Asia, 9(1): 21–29. https://doi.org/10.2174/2210681208666171204163622.

Júnior, O. K., Gurgel, L. V. A., de Freitas, R. P. and Gil, L. F. (2009). Adsorption of Cu(II), Cd(II), and Pb(II) from aqueous single metal solutions by mercerized cellulose and mercerized sugarcane bagasse chemically modified with EDTA dianhydride (EDTAD). Carbohydrate Polymers, 77(3): 643–650. https://doi.org/10.1016/j.carbpol.2009.02.016.

K, K., Nmi, A., D, S., Sb, M., K, K., A, A., Am, A.-M., Mr, A. and Ms, E. (2022). Adsorption of copper and nickel by using sawdust chitosan nanocomposite beads—A kinetic and thermodynamic study. Environmental Research, 203. https://doi.org/10.1016/j.envres.2021.111814.

Karimifard, S. and Alavi Moghaddam, M. R. (2018). Application of response surface methodology in physicochemical removal of dyes from wastewater: A critical review. The Science of the Total Environment, 640–641: 772–797. https://doi.org/10.1016/j.scitotenv.2018.05.355.

Karimi-Maleh, H., Ayati, A., Davoodi, R., Tanhaei, B., Karimi, F., Malekmohammadi, S. et al. (2021). Recent advances in using of chitosan-based adsorbents for removal of pharmaceutical contaminants: A review. Journal of Cleaner Production, 291: 125880. https://doi.org/10.1016/j.jclepro.2021.125880.

Kausar, S., Jammu, A., Kashmir, K., Kashmir, P., Solomon, S., Oyelere, S. et al. (2020). Mining Smart Learning Analytics Data Using Ensemble Classifiers. 15: 81–102.

Keledi, G., Hári, J. and Pukánszky, B. (2012). Polymer nanocomposites: Structure, interaction, and functionality. Nanoscale, 4(6): 1919–1938. https://doi.org/10.1039/C2NR11442A.

Khalef, R. N., Hassan, A. I., Saleh, H. M., Khalef, R. N., Hassan, A. I. and Saleh, H. M. (2022). Heavy metal's environmental impact. In Environmental Impact and Remediation of Heavy Metals. IntechOpen. https://doi.org/10.5772/intechopen.103907.

Khan, I., Saeed, K. and Khan, I. (2019). Nanoparticles: Properties, applications and toxicities. Arabian Journal of Chemistry, 12(7): 908–931. https://doi.org/10.1016/j.arabjc.2017.05.011.

Khan, S. T. and Malik, A. (2019). Engineered nanomaterials for water decontamination and purification: From lab to products. Journal of Hazardous Materials, 363: 295–308. https://doi.org/10.1016/j.jhazmat.2018.09.091.

Kim, B. S. and Lim, S.-T. (1999). Removal of heavy metal ions from water by cross-linked carboxymethyl corn starch. Carbohydrate Polymers, 39(3): 217–223. https://doi.org/10.1016/S0144-8617(99)00011-9.

Kim, D., Petrisor, I. G. and Yen, T. F. (2005). Evaluation of biopolymer-modified concrete systems for disposal of cathode ray tube glass. Journal of the Air & Waste Management Association, 55(7): 961.

Kinnear, C., Moore, T. L., Rodriguez-Lorenzo, L., Rothen-Rutishauser, B. and Petri-Fink, A. (2017). Form follows function: Nanoparticle shape and its implications for nanomedicine. Chemical Reviews, 117(17): 11476–11521. https://doi.org/10.1021/acs.chemrev.7b00194.

Klemm, D., Heublein, B., Fink, H.-P. and Bohn, A. (2005). Cellulose: Fascinating biopolymer and sustainable raw material. Angewandte Chemie International Edition, 44(22): 3358–3393. https://doi.org/10.1002/anie.200460587.

Kocherbitov, V., Ulvenlund, S., Kober, M., Jarring, K. and Arnebrant, T. (2008). Hydration of microcrystalline cellulose and milled cellulose studied by sorption calorimetry. The Journal of Physical Chemistry B, 112(12): 3728–3734. https://doi.org/10.1021/jp711554c.

Kong, A., Ji, Y., Ma, H., Song, Y., He, B. and Li, J. (2018). A novel route for the removal of Cu(II) and Ni(II) ions via homogeneous adsorption by chitosan solution. Journal of Cleaner Production, 192: 801–808. https://doi.org/10.1016/j.jclepro.2018.04.271.

Kube, M., Mohseni, A., Fan, L. and Roddick, F. (2018). Impact of alginate selection for wastewater treatment by immobilised *Chlorella vulgaris*. Chemical Engineering Journal, 358. https://doi.org/10.1016/j.cej.2018.10.065.

Kumar, S., Ahlawat, W., Bhanjana, G., Heydarifard, S., Nazhad, M. M. and Dilbaghi, N. (2014). Nanotechnology-Based water treatment strategies. Journal of Nanoscience and Nanotechnology, 14(2): 1838–1858. https://doi.org/10.1166/jnn.2014.9050.

Kumar, S. K. and Krishnamoorti, R. (2010). Nanocomposites: Structure, phase behavior, and properties | annual reviews. Annual Review of Chemical And Biomolecular Engineering, 1: 37–58. https://www.annualreviews.org/content/journals/10.1146/annurev-chembioeng-073009-100856.

Kyzas, G. Z., Lazaridis, N. K. and Kostoglou, M. (2013). On the simultaneous adsorption of a reactive dye and hexavalent chromium from aqueous solutions onto grafted chitosan. Journal of Colloid and Interface Science, 407: 432–441. https://doi.org/10.1016/j.jcis.2013.05.052.

L, Z., Y, W., Z, L. and Q, H. (2009). Characteristics of equilibrium, kinetics studies for adsorption of Hg(II), Cu(II), and Ni(II) ions by thiourea-modified magnetic chitosan microspheres. Journal of Hazardous Materials, 161(2-3). https://doi.org/10.1016/j.jhazmat.2008.04.078.

Lahiri, S., Nayak, D. and Korschinek, G. (2006). Separation of no-carrier-added 52Mn from bulk chromium: A simulation study for accelerator mass spectrometry measurement of 53Mn. Analytical Chemistry, 78(21): 7517–7521. https://doi.org/10.1021/ac0607459.

Lavoine, N., Desloges, I., Dufresne, A. and Bras, J. (2012). Microfibrillated cellulose - its barrier properties and applications in cellulosic materials: A review. Carbohydrate Polymers, 90(2): 735–764. https://doi.org/10.1016/j.carbpol.2012.05.026.

Lee, H. and Park, H.J. (2013). Testing the impact of message interactivity on relationship management and organizational reputation. Journal of Public Relations Research, 25(2): 188–206. Journal of Public Relations Research, 25, 188–206.

Leja, K. and Lewandowicz, G. (2010). Polymer biodegradation and biodegradable polymers—A review. Polish Journal of Environmental Studies, 19(2): 255–266. https://www.pjoes.com/Polymer-Biodegradation-and-Biodegradable-Polymers-a-Review,88379,0,2.html.

Lessa, I., Corrêa Seabra Guimarães, T., de Godoy Bergallo, H., Cunha, A. and M. Vieira, E. (2016). Domestic dogs in protected areas: A threat to Brazilian mammals? Natureza & Conservação, 14(2): 46–56. https://doi.org/10.1016/j.ncon.2016.05.001.

Li, Q., Mahendra, S., Lyon, D. Y., Brunet, L., Liga, M. V., Li, D. et al. (2008). Antimicrobial nanomaterials for water disinfection and microbial control: Potential applications and implications. Water Research, 42(18): 4591–4602. https://doi.org/10.1016/j.watres.2008.08.015.

Li, X., Zhou, H., Wu, W., Wei, S., Xu, Y. and Kuang, Y. (2015). Studies of heavy metal ion adsorption on chitosan/sulfydryl-functionalized graphene oxide composites. Journal of Colloid and Interface Science, 448: 389–397. https://doi.org/10.1016/j.jcis.2015.02.039.

Li, Y., Yang, C., Wang, S., Yang, D., Zhang, Y., Xu, L. et al. (2020). Copper and iron ions accelerate the prion-like propagation of α-synuclein: A vicious cycle in Parkinson's disease. International Journal of Biological Macromolecules, 163: 562–573. https://doi.org/10.1016/j.ijbiomac.2020.06.274.

Liu, C. and Bai, R. (2014). Recent advances in chitosan and its derivatives as adsorbents for removal of pollutants from water and wastewater. Current Opinion in Chemical Engineering, 4: 62–70. https://doi.org/10.1016/j.coche.2014.01.004.

Liu, H., Gong, P., Wang, J., Clinton, N., Bai, Y. and Liang, S. (2020). Annual dynamics of global land cover and its long-term changes from 1982 to 2015, link to GeoTIFF files [Dataset]. PANGAEA. https://doi.org/10.1594/PANGAEA.913496.

Lu, F. and Astruc, D. (2020). Nanocatalysts and other nanomaterials for water remediation from organic pollutants. Coordination Chemistry Reviews, 408: 213180. https://doi.org/10.1016/j.ccr.2020.213180.

Lu, L., Wang, C. and Sye, W. (2011). Applications of chitosan beads and porous crab shell powder for the removal of 17 organochlorine pesticides (OCPs) in water solution. Carbohydrate Polymers, 83: 1984–1989. https://doi.org/10.1016/j.carbpol.2010.11.003.

Lu, X., Wang, X., Jin, J., Zhang, Q. and Chen, J. (2014). Electrochemical biosensing platform based on amino acid ionic liquid functionalized graphene for ultrasensitive biosensing applications. Biosensors and Bioelectronics, 62: 134–139. https://doi.org/10.1016/j.bios.2014.06.036.

Madihally, S. V. and Matthew, H. W. (1999). Porous chitosan scaffolds for tissue engineering. Biomaterials, 20(12): 1133–1142. https://doi.org/10.1016/s0142-9612(99)00011-3.

Mallik, D. T. (2022). Biopolymers: Classification and Properties, 9(1).

Mangla, D., Annu, Sharma, A. and Ikram, S. (2022). Critical review on adsorptive removal of antibiotics: Present situation, challenges and future perspective. Journal of Hazardous Materials, 425: 127946. https://doi.org/10.1016/j.jhazmat.2021.127946.

Marimuthu, T., Chee, C. Y. and Sulaiman, N. M. N. (2023). A review on the use of cellulose nanomaterials for wastewater remediation of heavy metal ions. International Journal of Environmental Science and Technology, 20(3): 3421–3436. https://doi.org/10.1007/s13762-022-04209-5.

Mitra, S., Chakraborty, A. J., Tareq, A. M., Emran, T. B., Nainu, F., Khusro, A. et al. (2022). Impact of heavy metals on the environment and human health: Novel therapeutic insights to counter the toxicity. Journal of King Saud University - Science, 34(3): 101865. https://doi.org/10.1016/j.jksus.2022.101865.

Mittal, M. and Jana, A. (2023). Role of Nanophotocatalysts in Water Remediation. In Nanotechnology for Sustainable Agriculture, Food and Environment. CRC Press.

Mohan, V., Sandeep, S., Deepa, R., Shah, B. and Varghese, C. (2007). Epidemiology of type 2 diabetes: Indian scenario. The Indian Journal of Medical Research, 125(3): 217–230.

Mohnen, D. (2008). Pectin structure and biosynthesis. Current Opinion in Plant Biology, 11(3): 266–277. https://doi.org/10.1016/j.pbi.2008.03.006.

Moon, R. J., Martini, A., Nairn, J., Simonsen, J. and Youngblood, J. (2011). Cellulose nanomaterials review: Structure, properties and nanocomposites. Chemical Society Reviews, 40(7): 3941–3994. https://doi.org/10.1039/c0cs00108b.

Ms, H., Sg, M., Mi, B. and Mi, M. (2020). Salicylaldehyde derivative of nano-chitosan as an efficient adsorbent for lead(II), copper(II), and cadmium(II) ions. International Journal of Biological Macromolecules, 147. https://doi.org/10.1016/j.ijbiomac.2020.01.091.

Muxika, A., Etxabide, A., Uranga, J., Guerrero, P. and de la Caba, K. (2017). Chitosan as a bioactive polymer: Processing, properties and applications. International Journal of Biological Macromolecules, 105: 1358–1368. https://doi.org/10.1016/j.ijbiomac.2017.07.087.

Nasrollahzadeh, M., Sajjadi, M., Iravani, S. and Varma, R. S. (2021). Carbon-based sustainable nanomaterials for water treatment: State-of-art and future perspectives. Chemosphere, 263: 128005. https://doi.org/10.1016/j.chemosphere.2020.128005.

Nechita, P. (2015). Natural additives used in adsorption of pollutants from textile wastewaters. American Journal of Environmental Engineering, 5(2): 39–42.

Norris, M. L., Robinson, A., Obeid, N., Harrison, M., Spettigue, W. and Henderson, K. (2014). Exploring avoidant/restrictive food intake disorder in eating disordered patients: A descriptive study. The International Journal of Eating Disorders, 47(5): 495–499. https://doi.org/10.1002/eat.22217.

O'Connell, D. W., Birkinshaw, C. and O'Dwyer, T. F. (2008). Heavy metal adsorbents prepared from the modification of cellulose: A review. Bioresource Technology, 99(15): 6709–6724. https://doi.org/10.1016/j.biortech.2008.01.036.

Ogawa, Y., Kimura, S. and Wada, M. (2011). Electron diffraction and high-resolution imaging on highly-crystalline β-chitin microfibril. Journal of Structural Biology, 176(1): 83–90. https://doi.org/10.1016/j.jsb.2011.07.001.

Omenetto, F. G. and Kaplan, D. L. (2010). New opportunities for an ancient material. Science (New York, N.Y.), 329(5991): 528–531. https://doi.org/10.1126/science.1188936.

Pal, P. and Pal, A. (2017). Surfactant-modified chitosan beads for cadmium ion adsorption. International Journal of Biological Macromolecules, 104: 1548–1555. https://doi.org/10.1016/j.ijbiomac.2017.02.042.

Pal, P. and Pal, A. (2019). Treatment of real wastewater: Kinetic and thermodynamic aspects of cadmium adsorption onto surfactant-modified chitosan beads. International Journal of Biological Macromolecules, 131, 1092–1100. https://doi.org/10.1016/j.ijbiomac.2019.03.121.

Pandey, N., Shukla, S. K. and Singh, N. B. (2017). Water purification by polymer nanocomposites: An overview. Nanocomposites, 3(2): 47–66. https://doi.org/10.1080/20550324.2017.1329983.

Pang, Y. L., Tan, J. H., Lim, S. and Chong, W. C. (2021). A State-of-the-Art Review on Biowaste Derived Chitosan Biomaterials for Biosorption of Organic Dyes: Parameter Studies, Kinetics, Isotherms and Thermodynamics. Polymers, 13(17), Article 17. https://doi.org/10.3390/polym13173009.

Patra, J. K., Das, G., Fraceto, L. F., Campos, E. V. R., Rodriguez-Torres, M. del P., Acosta-Torres, L. S. et al. (2018). Nano based drug delivery systems: Recent developments and future prospects. Journal of Nanobiotechnology, 16(1): 71. https://doi.org/10.1186/s12951-018-0392-8.

Pattanashetti, N. A., Heggannavar, G. B. and Kariduraganavar, M. Y. (2017). Smart Biopolymers and their Biomedical Applications. Procedia Manufacturing, 12: 263–279. https://doi.org/10.1016/j.promfg.2017.08.030.

Pavinatto, F. J., Caseli, L. and Oliveira, O. N. (2010). Chitosan in nanostructured thin films. Biomacromolecules, 11(8): 1897–1908. https://doi.org/10.1021/bm1004838.

Pholnak, P., Sittiyothee, S., Sirisathitkul, C. and Sirisathitkul, Y. (2022). Dye removal efficiency of fresh and dry alginate beads incorporating zinc oxide. Arab Journal of Basic and Applied Sciences, 29: 53–64. https://doi.org/10.1080/25765299.2022.2040173.

Pourmortazavi, S. M., Sahebi, H., Zandavar, H. and Mirsadeghi, S. (2019). Fabrication of Fe3O4 nanoparticles coated by extracted shrimp peels chitosan as sustainable adsorbents for removal of chromium contaminates from wastewater: The design of experiment. Composites Part B: Engineering, 175: 107130. https://doi.org/10.1016/j.compositesb.2019.107130.

Purnama Sari, F., Agusnar, H. and Taufik, M. (2019). Preparation and Characterization of Chitosan with Activated Carbon as Adsorbent to Reduce Level Metal Cadmium (Cd) and Nickel (Ni): Proceedings of the 1st International Conference on Chemical Science and Technology Innovation, 280–286. https://doi.org/10.5220/0008921802800286.

Qureshi, M. A., Nishat, N., Jadoun, S. and Ansari, M. Z. (2020). Polysaccharide based superabsorbent hydrogels and their methods of synthesis: A review. Carbohydrate Polymer Technologies and Applications, 1: 100014. https://doi.org/10.1016/j.carpta.2020.100014.

Rahmani, O., Bouzid, B. and Guibadj, A. (2017). Extraction and characterization of chitin and chitosan: Applications of chitosan nanoparticles in the adsorption of copper in an aqueous environment. E-Polymers, 17(5): 383–397. https://doi.org/10.1515/epoly-2016-0318.

Rawat, S., Chaudhary, M. and Maiti, A. (2023). Synergistic arsenic removal using chitosan-based nanocomposite beads and cross-flow ultrafiltration: A significant reduction of membrane fouling. Journal of Environmental Chemical Engineering, 11(2): 109431. https://doi.org/10.1016/j.jece.2023.109431.

Rebar, A. L., Stanton, R., Geard, D., Short, C., Duncan, M. J. and Vandelanotte, C. (2015). A meta-meta-analysis of the effect of physical activity on depression and anxiety in non-clinical adult populations. Health Psychology Review, 9(3): 366–378. https://doi.org/10.1080/17437199.2015.1022901.

Renard, C. M. G. C. and Jarvis, M. C. (1999). Acetylation and methylation of homogalacturonans 2: Effect on ion-binding properties and conformations. Carbohydrate Polymers, 39(3): 209–216. https://doi.org/10.1016/S0144-8617(99)00015-6.

Renault, F., Sancey, B., Badot, P.-M. and Crini, G. (2009). Chitosan for coagulation/flocculation processes–An eco-friendly approach. European Polymer Journal, 45(5): 1337–1348. https://doi.org/10.1016/j.eurpolymj.2008.12.027.

Rinaudo, M. (2006). Chitin and chitosan: Properties and applications. Progress in Polymer Science, 31(7): 603–632. https://doi.org/10.1016/j.progpolymsci.2006.06.001.

Rodrigues, A. C., Boroski, M., Shimada, N. S., Garcia, J. C., Nozaki, J. and Hioka, N. (2008). Treatment of paper pulp and paper mill wastewater by coagulation–flocculation followed by heterogeneous photocatalysis. Journal of Photochemistry and Photobiology A: Chemistry, 194(1): 1–10. https://doi.org/10.1016/j.jphotochem.2007.07.007.

Roy, R., Roy, R. A. and Roy, D. M. (1986). Alternative perspectives on "quasi-crystallinity": Non-uniformity and nanocomposites. Materials Letters, 4(8): 323–328. https://doi.org/10.1016/0167-577X(86)90063-7.

Sadegh, H., Ali, G. A. M., Gupta, V. K., Makhlouf, A. S. H., Shahryari-ghoshekandi, R., Nadagouda, M. N. et al. (2017). The role of nanomaterials as effective adsorbents and their applications in wastewater treatment. Journal of Nanostructure in Chemistry, 7(1): 1–14. https://doi.org/10.1007/s40097-017-0219-4.

Saha, T. K., Bhoumik, N. C., Karmaker, S., Ahmed, M. G., Ichikawa, H. and Fukumori, Y. (2011). Adsorption Characteristics of Reactive Black 5 from Aqueous Solution onto Chitosan. CLEAN – Soil, Air, Water, 39(10): 984–993. https://doi.org/10.1002/clen.201000315.

Sakib, M. N., Mallik, A. K. and Rahman, M. M. (2021). Update on chitosan-based electrospun nanofibers for wastewater treatment: A review. Carbohydrate Polymer Technologies and Applications, 2: 100064. https://doi.org/10.1016/j.carpta.2021.100064.

Salehi, N. and Moghimi, A. (2022). Synthesis of crosslinked magnetic chitosan to removal of trace amounts of some heavy metals from aqueous solutions. Journal of Color Science and Technology, 16(2): 135–145.

Salehi, N., Moghimi, A. and Shahbaazi, H. (2020). Preparation of cross-linked magnetic chitosan with methionine-glutaraldehyde for removal of heavy metals from aqueous solutions. International Journal of Environmental Analytical Chemistry, 102: 1–17. https://doi.org/10.1080/03067319.2020.1753718.

Samadi, F. Y., Mohammadi, Z., Yousefi, M. and Majdejabbari, S. (2016). Synthesis of raloxifene–chitosan conjugate: A novel chitosan derivative as a potential targeting vehicle. International Journal of Biological Macromolecules, 82: 599–606. https://doi.org/10.1016/j.ijbiomac.2015.10.041.

Sánchez-Martín, J., Beltrán-Heredia, J., Delgado-Regaña, A., Rodríguez-González, M. A. and Rubio-Alonso, F. (2013). Adsorbent tannin foams: New and complementary applications in wastewater treatment. Chemical Engineering Journal, 228: 575–582. https://doi.org/10.1016/j.cej.2013.05.009.

Santhosh, C., Velmurugan, V., Jacob, G., Jeong, S. K., Grace, A. N. and Bhatnagar, A. (2016). Role of nanomaterials in water treatment applications: A review. Chemical Engineering Journal, 306: 1116–1137. https://doi.org/10.1016/j.cej.2016.08.053.

Sarode, S., Upadhyay, P., Khosa, M. A., Mak, T., Shakir, A., Song, S. et al. (2019). Overview of wastewater treatment methods with special focus on biopolymer chitin-chitosan. International Journal of Biological Macromolecules, 121: 1086–1100. https://doi.org/10.1016/j.ijbiomac.2018.10.089.

Sasirekha, R., Vijayakumar, S., Pazhanisamy, P., Joseph, J. J. and Gomathi, T. (2023). Biosorption of copper (II) ion using chitosan nanoparticles/alginate/carboxymethyl cellulose/novel hydroxyapatite biocomposite from aqueous solution. Materials Today Communications, 35: 106339. https://doi.org/10.1016/j.mtcomm.2023.106339.

Sharififard, H., shahraki, Z. H., Rezvanpanah, E. and Rad, S. H. (2018). A novel natural chitosan/activated carbon/iron bio-nanocomposite: Sonochemical synthesis, characterization, and application for cadmium removal in batch and continuous adsorption process. Bioresource Technology, 270: 562–569. https://doi.org/10.1016/j.biortech.2018.09.094.

Sharma, T., Yadav, S. and Mittal, M. (2023). Recent advancement in the sustainable synthesis of BiOX (X = I, Br, and Cl) nanomaterials and their applications in the photocatalysis. Materials Today: Proceedings. https://doi.org/10.1016/j.matpr.2023.02.209.

Singh, R., Gautam, S., Sharma, B., Jain, P. and Chauhan, K. D. (2021). Chapter 2—Biopolymers and their classifications. pp. 21–44. *In*: Thomas, S., Gopi, S. and Amalraj, A. (eds.). Biopolymers and their Industrial Applications. Elsevier. https://doi.org/10.1016/B978-0-12-819240-5.00002-X.

Singh, R. and Purkait, M. K. (2017). Role of poly(2-acrylamido-2-methyl-1-propanesulfonic acid) in the modification of polysulfone membranes for ultrafiltration. Journal of Applied Polymer Science, 134(37): 45290. https://doi.org/10.1002/app.45290.

Singh, R., Yadav, V. S. K. and Purkait, M. K. (2019). Cu2O photocatalyst modified antifouling polysulfone mixed matrix membrane for ultrafiltration of protein and visible light driven photocatalytic pharmaceutical removal. Separation and Purification Technology, 212: 191–204. https://doi.org/10.1016/j.seppur.2018.11.029.

Singh, S., Garg, R., Jana, A., Bathula, C., Naik, S. and Mittal, M. (2023). Current developments in nanostructurally engineered metal oxide for removal of contaminants in water. Ceramics International, 49(5): 7308–7321. https://doi.org/10.1016/j.ceramint.2022.10.183.

Singh, S., Jaiswal, V., Singh, J. K., Semwal, R. and Raina, D. (2023). Nanoparticle formulations: A smart era of advanced treatment with nanotoxicological imprints on the human body. Chemico-Biological Interactions, 373: 110355. https://doi.org/10.1016/j.cbi.2023.110355.

Singh, V., Kendall, R. J., Hake, K. and Ramkumar, S. (2013). Crude oil sorption by raw cotton. Industrial & Engineering Chemistry Research, 52(18): 6277–6281. https://doi.org/10.1021/ie4005942.

Soto, D., Urdaneta, J., Pernia, K., León, O., Muñoz-Bonilla, A. and Fernández-García, M. (2015). Removal of heavy metal ions in water by starch esters. Starch-Starke. https://www.semanticscholar.org/paper/Removal-of-heavy-metal-ions-in-water-by-starch-Soto-Urdaneta/6a9e21f019a87ae65a2a77fd13e604ae01196aa3.

Sridevi, M., Nirmala, C., Jawahar, N., Arthi, G., Vallinayagam, S. and Sharma, V. K. (2021). Role of nanomaterial's as adsorbent for heterogeneous reaction in waste water treatment. Journal of Molecular Structure, 1241: 130596. https://doi.org/10.1016/j.molstruc.2021.130596.

Sugashini, S., Gomathi, T., Pavithra, S., Sudha, P. N., Florence, J. A. K. and Mubashirunnisa, A. (2022). Removal of Copper(II) Ion using nanochitosan/carboxymethyl cellulose/grapheme oxide composite biosorbent. Asian Journal of Chemistry, 34(6): 1465–1471. https://doi.org/10.14233/ajchem.2022.23631.

Suhas, null, Gupta, V. K., Carrott, P. J. M., Singh, R., Chaudhary, M. and Kushwaha, S. (2016). Cellulose: A review as natural, modified and activated carbon adsorbent. Bioresource Technology, 216: 1066–1076. https://doi.org/10.1016/j.biortech.2016.05.106.

Suryanarayana, C. (1994). Structure and properties of nanocrystalline materials. Bulletin of Materials Science, 17(4): 307–346. https://doi.org/10.1007/BF02745220.

Suyambulingam, I., Gangadhar, L., Sana, S. S., Divakaran, D., Siengchin, S., Kurup, L. A. etal. (2023). Chitosan Biopolymer and Its Nanocomposites: Emerging Material as Adsorbent in Wastewater Treatment. Advances in Materials Science and Engineering, 2023, 1–20. https://doi.org/10.1155/2023/9387016.

Tanhaei, B., Ayati, A., Iakovleva, E. and Sillanpää, M. (2020). Efficient carbon interlayed magnetic chitosan adsorbent for anionic dye removal: Synthesis, characterization and adsorption study. International Journal of Biological Macromolecules, 164: 3621–3631. https://doi.org/10.1016/j.ijbiomac.2020.08.207.

Tavakolian, M., Jafari, S. M. and Ven, T. G. M. van de. (2020). A review on surface-functionalized cellulosic nanostructures as biocompatible antibacterial materials. Nano-Micro Letters, 12. https://doi.org/10.1007/s40820-020-0408-4.

Tchounwou, P. B., Yedjou, C. G., Patlolla, A. K. and Sutton, D. J. (2012). Heavy Metals Toxicity and the Environment. EXS, 101: 133–164. https://doi.org/10.1007/978-3-7643-8340-4_6.

Terzopoulou, Z., Papageorgiou, M., Kyzas, G. Z., Bikiaris, D. N. and Lambropoulou, D. A. (2016). Preparation of molecularly imprinted solid-phase microextraction fiber for the selective removal and extraction of the antiviral drug abacavir in environmental and biological matrices. Analytica Chimica Acta, 913: 63–75. https://doi.org/10.1016/j.aca.2016.01.059.

Thakur, S., Pandey, S. and Arotiba, O. A. (2016). Development of a sodium alginate-based organic/inorganic superabsorbent composite hydrogel for adsorption of methylene blue. Carbohydrate Polymers, 153: 34–46. https://doi.org/10.1016/j.carbpol.2016.06.104.

Thekkila-Veedu, S., Mohanan, D. P., Banerjee, S., Ravichandiran, V. and Natesan, S. (2023). 14—Natural biopolymers in ophthalmology. pp. 369–405. *In*: Jayakumar, R. and Murali, V. P. (eds.). Natural Biopolymers in Drug Delivery and Tissue Engineering. Woodhead Publishing. https://doi.org/10.1016/B978-0-323-98827-8.00002-3.

Thostenson, E. T., Li, C. and Chou, T.-W. (2005). Nanocomposites in context. Composites Science and Technology, 65(3): 491–516. https://doi.org/10.1016/j.compscitech.2004.11.003.

Torres, M. A. and Dangl, J. L. (2005). Functions of the respiratory burst oxidase in biotic interactions, abiotic stress and development. Current Opinion in Plant Biology, 8(4): 397–403. https://doi.org/10.1016/j.pbi.2005.05.014.

Urruzola, I., Robles, E., Serrano, L. and Labidi, J. (2014). Nanopaper from almond (*Prunus dulcis*) shell. Cellulose, 21. https://doi.org/10.1007/s10570-014-0238-y.

Vakili, M., Rafatullah, M., Salamatinia, B., Abdullah, A. Z., Ibrahim, M. H., Tan, K. B. et al. (2014). Application of chitosan and its derivatives as adsorbents for dye removal from water and wastewater: A review. Carbohydrate Polymers, 113: 115–130. https://doi.org/10.1016/j.carbpol.2014.07.007.

Valdés, A. and Garrigós, M. C. (2016). Carbohydrate-based advanced biomaterials for food sustainability: A review. Materials Science Forum, 842: 182–195. https://doi.org/10.4028/www.scientific.net/MSF.842.182.

Vijayalakshmi, K., Gomathi, T., Latha, S., Hajeeth, T. and Sudha, P. N. (2016). Removal of copper(II) from aqueous solution using nanochitosan/sodium alginate/microcrystalline cellulose beads. International Journal of Biological Macromolecules, 82: 440–452. https://doi.org/10.1016/j.ijbiomac.2015.09.070.

Wang, D. (2019). A critical review of cellulose-based nanomaterials for water purification in industrial processes. Cellulose, 26. https://doi.org/10.1007/s10570-018-2143-2.

Wang, K., Zhang, F., Xu, K., Che, Y., Qi, M. and Song, C. (n.d.). Modified magnetic chitosan materials for heavy metal adsorption: A review. RSC Advances, 13(10): 6713–6736. https://doi.org/10.1039/d2ra07112f.

Willats, W. G., McCartney, L., Mackie, W. and Knox, J. P. (2001). Pectin: Cell biology and prospects for functional analysis. Plant Molecular Biology, 47(1-2): 9–27.

Wróblewska-Krepsztul, J., Rydzkowski, T., Michalska-Pożoga, I. and Thakur, V. K. (2019). Biopolymers for biomedical and pharmaceutical applications: Recent advances and overview of alginate electrospinning. Nanomaterials, 9(3), Article 3. https://doi.org/10.3390/nano9030404.

Xu, X., Wang, D., Li, Y., Deyrup, S. and Zhang, H.-J. (2021). Xu et al.-2021-Phytochemistry Reviews. https://doi.org/10.1007/s11101-021-09758-0(0123456789().,-volV).

Yang, Y., Li, Y., Goh, K., Tan, C. H. and Wang, R. (2021). Liposomes-assisted fabrication of high performance thin film composite nanofiltration membrane. Journal of Membrane Science, 620: 118833. https://doi.org/10.1016/j.memsci.2020.118833.

You, L., Song, L., Wang, A., Lu, F. and Zhang, Q. (2017). Fabrication of a Cationic Polysaccharide for High Performance Flocculation. Colloids and Surfaces A: Physicochemical and Engineering Aspects, 520. https://doi.org/10.1016/j.colsurfa.2017.02.078.

Zhang, P., Yang, M., Lan, J., Huang, Y., Zhang, J., Huang, S. et al. (2023). Water quality degradation due to heavy metal contamination: Health impacts and eco-friendly approaches for heavy metal remediation. Toxics, 11(10), 828. https://doi.org/10.3390/toxics11100828.

Zhang, Y., Zhao, M., Cheng, Q., Wang, C., Li, H., Han, X. et al. (2021). Research progress of adsorption and removal of heavy metals by chitosan and its derivatives: A review. Chemosphere, 279: 130927. https://doi.org/10.1016/j.chemosphere.2021.130927.

Zhu, Y., Liu, X., Hu, Y., Wang, R., Chen, M., Wu, J. e tal. (2019). Behavior, remediation effect and toxicity of nanomaterials in water environments. Environmental Research, 174: 54–60. https://doi.org/10.1016/j.envres.2019.04.014.

Zirpe, M. and Thakur, J. (2023). Magnetically Recoverable CMC/Fe3O4 Nanocomposite for efficient removal of Hg(II) and Cd(II) ions from the Aqueous Solution. Separation Science and Technology. https://www.tandfonline.com/doi/abs/10.1080/01496395.2023.2225733.

CHAPTER- 3

Environmental Fate and Toxicity of Biopolymer-Based Metal Nanoparticles

Nandini Teji,[1,2] *Rahul Bhardwaj,*[1,2] *Anjali Nisha,*[1,2] *Harpreet Kaur,*[1,2]
Rubina Dhiman[1] and *Amitabha Acharya*[1,2,*]

1. Introduction

With the dawn of the 21st century, science has been revolutionized with the consumption of nanoparticles instead of their bulk counterparts. The world has drenched itself with chemicals and toxins produced for betterment, but the repercussions are unfathomable. The trendy terms such as sustainability and ecofriendly are superficial if examined cautiously. Out of two sides of the coin, the advantageous side is highly appreciated, whereas the negative side needs multiple scrutinizing eyes. Every scientific and technological activity has seen a complete alteration in the situation. Material science is finding suitable alternatives to replace the hazardous synthesized material usage in different sectors such as agriculture, diagnostics, pharmaceuticals, cosmeceuticals, etc. Out of many synthesized chemicals, with the advent of nanotechnology, different nanoparticles are synthesized as they offer multiple advantages due to their small size, large surface area, high stability, and biocompatibility. Various nanoparticles have manifold applications, such as in drug delivery systems, nanopesticide formulations, point-of-care diagnostics, biosensors, and electrical equipment (Nasimi and Haidari, 2013; Hou et al., 2023).

Multiple nanoparticles are being synthesized, such as 0D (quantum dots), 1D (nanowires), 2D (nanotubes), and 3D (nanoflowers), according to their specific functions (Joudeh and Linke, 2022). A distinct class of nanoparticles known as nanocomposites employ biopolymers as a matrix and metal nanoparticles as fillers. Biopolymer-based metal nanoparticles (BMNPs) have surged in recent years due to their numerous advantages, such as the controlled release of metal ions, improved stability, biocompatibility and biodegradability (Kumar et al., 2021). Biopolymers, on the other hand, are frequently synthesized from non-toxic monomers and are carbon-neutral. They are readily deposited at the nanoscale, facilitating the synthesis of nanomaterials. Biopolymers can be isolated from various natural sources, such as plants, algae, fungi, bacteria, and animals. Examples include macromolecules like starch, alginate, chitosan, carrageenan, silk, keratin, chitin, etc., and others

[1] Biotechnology Division, CSIR-Institute of Himalayan Bioresource Technology, Palampur (H.P.) 176061, India.
[2] Academy of Scientific and Innovative Research (AcSIR), Ghaziabad- 201002, India, 140306, India.
* Corresponding author: amitabhachem@gmail.com; amitabha@ihbt.res.in
DOI: 10.1201/9781003558330-3

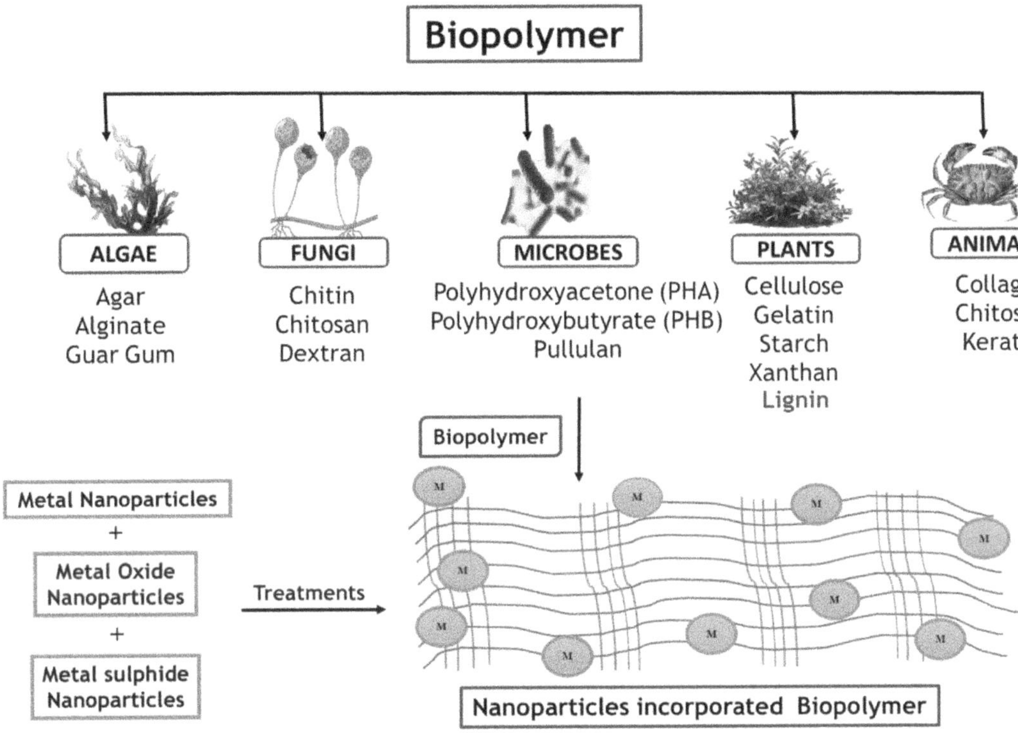

Figure 1. Isolation of biopolymers from various natural sources and their incorporation with metal nanoparticles to fabricate biopolymer-based metal nanoparticles (BMNPs).

have been depicted in Fig. 1. Compounds like pectin, cellulose, lignin, and hyaluronic acid can be used to synthesize nanocomposite materials (Carrion et al., 2021). Biodegradable materials like chitosan nanoparticles are being employed for targeted delivery of anticancer drugs. Biopolymers offer a diverse range of components with varied physiological properties. Furthermore, biopolymeric nanomaterials can be synthesized by attaching metals to biopolymers. Compared to bare metal nanoparticles, biopolymer attachment with metal nanoparticles has increased their efficiency and performance. There are certain examples, such as chitosan-silver nanocomposites used in wound dressings as an effective anti-bacterial agent targeting an array of bacteria. Another example includes, palladium nanoparticles embedded within starch matrix useful for the degradation of organic pollutants in water and soil (You et al., 2017). Similarly, gold and cellulose nanocomposites have been used for biosensor application in disease detection (Sarkar et al., 2020). The better and safer picture of BMNPs are their advantages, but the ecotoxic nature of high concentrations of metal ions released into the environment during synthesis, usage, and disposal, which harms the environment, is incomparable.

There are certain environmental factors like salinity, pH, and the presence of organic matter that majorly influence the mobility of released metal nanoparticles from BMNPs. There are multiple incidents where accidental spills, product degradation, and discharge from manufacturing units occur. The release of such BMNPs can reach varied sectors of the environment, such as water bodies, agricultural land, the atmosphere, etc. There is a chance that when BMNPs come in contact with the prior present chemicals in the environment, they will lose their stability and function. This will pave the way for the production of more toxins harming the environment and creatures. Certain reports suggest that these BMNPs can pose a health risk by entering the body through many mechanistic pathways, such as ingestion, inhalation, or skin penetration (Prajitha et al., 2019). Furthermore, it is quite evident that the ecosystem functions in a balanced way, and to sustain the balance, various cycling processes occur in nature. Nature has its processes for nutrient cycling and cleanup. The decomposition of organic matter aids in releasing nutrients back into the environment, ensuring it acts as a long-term reservoir of required elements (Hussain et al., 2021). Additionally, this supports the healthy bacteria around the plants. However, BMNPs may interfere with the cycling of nutrients when they enter the picture. They can accumulate in the vicinity of plant roots, hampering the rhizosphere, vigor, and vitality of the growing plant.

Nature was never the dump yard, nor will it ever be. The persistence of BMNPs in the atmosphere for a longer duration has a notable detrimental impact. The bioaccumulation of such BMNPs entering the body system of living organisms might not be readily eliminated, thus disrupting the physiology while making the body more vulnerable to diseases. Amplified toxicity at higher trophic levels can also be asserted through biomagnification once the BMNPs enter the food web. The biodegradable nature of biopolymers enhances the efficacy of metal NPs, but it is observed that the majority of ecotoxicity arises due to the unwanted release of metal NPs from the metal-polymer complex into the ecosystem (Kumar et al., 2021). Prior and thorough knowledge is required while designing and synthesizing BMNPs. Certain alternative metals should be used for similar applications, which may have better polymer-metal binding capacities, minimal metal release when not required, and less toxic effects (Yu et al., 2021). Compared to others, biopolymer-based gold nanocomposites might be used in cellular heavy metal detection, including Pb^{2+} (Bharti et al., 2022). Sincere efforts are required to reduce the toxic effects of BMNPs as there is a large space for understanding the challenges in managing the waste generated.

2. Properties Influencing the Toxicity of BMNPs

Biopolymer-based metal nanoparticles (BMNPs) represent a burgeoning class of nanomaterials poised to revolutionize the fields of medicine, food processing, agriculture, bioengineering, bioremediation, etc. (Kaliaraj et al., 2017). This surge in interest stems from the unique ability of biopolymers to address key limitations of bare metallic NPs, such as potential toxicity and uncontrolled reactivity. Biopolymers, with their inherent biocompatibility and biodegradability, serve as ideal stabilizing matrices. BMNPs offer promising applications but raise concerns about human health and environmental effects upon exposure. By precisely manipulating the physicochemical properties of the encapsulated NPs, researchers can achieve targeted delivery, enhanced functionality, and minimized environmental impact. This targeted approach represents a paradigm shift, paving the way for a new generation of nanomaterials that are not only efficacious but also sustainable and biocompatible. However, a comprehensive understanding of the interplay between BMNPs and biological response remains crucial. Unraveling the complex relationship between NP's physicochemical properties (size, shape, surface chemistry) and it's *in vivo* fate (uptake, distribution, clearance, and potential toxicity) is essential for the responsible development and utilization of these BMNPs (Remédios et al., 2012). Additionally, continued research is needed to address the environmental challenges associated with traditional metal NP synthesis, often reliant on harmful organic solvents and reducing agents (Rashid, 2023). Biopolymer-based synthesis provides a promising avenue for a more sustainable, safe, and environmentally friendly approach. The biological response and nanotoxicological profile of BMNPs are significantly influenced by their physicochemical determinants. Notably, the incorporation of metal nanoparticles within the polymer matrix plays a major role in determining their overall toxicity. Here is a detailed breakdown of the key determinants impacting the biological response and nanotoxicological profile of BMNPs:

2.1 Size

The dimensions of BMNPs, particularly their inherent small size and high surface area to volume ratio, emerge as a critical determinant of cytotoxicity within the cellular milieu. Smaller particles exhibit enhanced reactivity due to a greater surface area for interaction with biological components. This potentiates their uptake through various endocytosis pathways, ultimately directing their cellular fate. The increased surface area also promotes adsorption and binding to biomolecules, potentially leading to a cascade of cytotoxic events. The high surface area to volume ratio of BMNPs can be exploited for targeted drug delivery and specific cellular interactions (Fazal et al., 2023). It can also lead to unintended consequences such as increased cellular uptake and its accumulation (Kamkaew et al., 2016), potentially causing damage to cellular organelles and inducing cytotoxicity. However, the relationship between size and toxicity is not always straightforward. Studies have shown that differently sized nanomaterials can exhibit diverse toxicity profiles and may follow distinct cellular uptake pathways, leading to a lack of direct correlation between size and adverse effects (Sharifi et al., 2012; Dhawan and Sharma, 2010). The size of nanomaterials significantly influences their interactions with the cellular milieu, impacting cellular fate and potentially leading to long-term environmental toxicity. Smaller nanoparticles exhibit enhanced environmental mobility and stability compared to larger ones (Navarro et al., 2008). This translates to increased persistence in air, water, and soil, facilitating their transport over longer distances and potentially affecting a wider range of organisms across diverse ecosystems. Furthermore, the diminutive size of nanoparticles significantly increases their bioavailability (Li et al., 2015). This signifies a greater propensity for uptake by organisms through ingestion, inhalation, or dermal contact. This enhanced uptake can lead to bioaccumulation within these organisms, potentially culminating in adverse toxicological effects.

2.2 *Shape*

Considering that the toxicity of metallic nanoparticles often stems from their elemental composition and size being key determinants, it is noteworthy that BMNPs can be engineered into various shapes, including spheres, rods, wires, tubes, and even more complex geometries. Compared to isotropic spheres, anisotropic shapes (rods, wires) can exhibit different cellular uptake, circulation, or retention times and interact with various biomolecules inside the cell. Shape is one of the primary factors affecting cytotoxicity in the cellular environment. For instance, spherical NPs are less toxic than other shapes like nanotubes or nanofibers (Abbasi et al., 2023). Sharp or irregular shapes might cause more physical damage to membranes or disrupt cellular processes to a greater extent than spherical NPs. It has been reported that the extent and kinetics of phagocytosis are influenced by the curvature at the particle-cell interface (Yang et al., 2019) and anisotropic shapes may be taken up by cells more readily due to their enhanced interaction with the cell membrane. For example, studies have shown that rod-like nanoparticles can induce more pronounced inflammatory responses than cubic or octahedral NPs with similar compositions (Han et al., 2023). This difference is attributed to the larger surface area and potentially higher propensity of rod-like shapes to interact with and activate immune cells. Similarly, shape plays a critical role in the environmental fate of BMNPs. Nanofibers, for instance, exhibit superior aerodynamic properties compared to spherical particles. This translates to longer airborne persistence and greater travel distances, ultimately influencing their potential for inhalation by organisms, thus leading to toxicity.

2.3 Surface Chemistry/Functionalization

The surface chemistry or functionalization of BMNPs is crucial in determining their environmental and cellular toxicity. Functional groups on the surface can alter cellular interaction and internalization. Studies have shown that cationic surfaces promote uptake via clathrin-mediated endocytosis, while neutral or anionic surfaces favor other pathways (Marcellus et al., 2024). Increased uptake often leads to higher intracellular accumulation and a higher potential for cytotoxicity (Schaeublin et al., 2014). Surface chemistry also determines the interaction of BMNPs with biomolecules such as proteins and lipids within the cell. Certain functionalities can lead to the formation of a protein corona, a layer of adsorbed biomolecules that changes the identity of the nanoparticle and influences its interaction with cellular components (Corbo et al., 2016). These interactions can disrupt vital processes and lead to oxidative stress, inflammation, and cell death. In addition to the biological interactions, surface chemistry also influences the environmental behavior of BMNPs. Functional groups can improve the ability of nanoparticles to adsorb pollutants or leach toxic metals, depending on their chemical properties (Zhao et al., 2021). In addition, they also influence the stability and dispersion of nanoparticles in the environmental matrix, which affects their bioavailability and potential uptake by organisms. Depending on the surface properties, BMNPs can interact with and disrupt important environmental processes. For example, they can interfere with nutrient cycling or microbial activity in the soil, potentially affecting the ecosystem's health (Ameen et al., 2021).

2.4 Aggregation Ability

The aggregation of BMNPs, i.e., the formation of larger clusters of individual nanoparticles, significantly impacts their environmental and cellular toxicity. This phenomenon alters the interaction of nanoparticles with biological systems and leads to a complex interplay between reduced cellular uptake and potential environmental consequences. Larger aggregates may have difficulty penetrating cells through established pathways, which could attenuate cellular toxicity if the individual particles are inherently toxic (Stark, 2011). However, this reduced uptake may also interfere with the body's natural elimination mechanisms, resulting in prolonged exposure and potentially delayed toxicity. In addition, aggregation directly affects the surface area available for interaction with biomolecules. It influences the formation of the protein corona, a crucial layer that affects the BMNPs' interaction with cells. Changes in the corona due to aggregation could lead to incidental consequences, such as increased interaction with certain cell types or altered cellular responses. In some cases, aggregation may even increase reactivity, leading to increased cellular uptake despite the larger size (Liu et al., 2013). This can lead to an accumulation of nanoparticles in cells, disrupting cellular processes and potentially causing disease. The effects of aggregation go beyond the cellular level. Aggregated or agglomerated nanoparticles tend to settle faster compared to their counterparts in water or air (Shrestha et al., 2020). This limited mobility and transport distance can reduce their exposure to a larger number of organisms and thus mitigate environmental toxicity. However, this sedimentation may also lead to the accumulation of nanoparticles in certain environmental compartments, such as bottom sediments, potentially affecting local organisms in a concentrated form. They can also be detrimental to benthic organisms (those living at the bottom of water bodies) by disrupting the food chain. These BMNPs can adsorb pollutants and other toxins present in the water. As benthic organisms ingest these contaminated nanoparticles, they can accumulate toxins in their bodies, impacting their health and potentially affecting predators higher up the food chain. The various BMNPs may not be readily broken down or excreted by benthic organisms. This can lead to bioaccumulation, where the concentration of these NPs increases in their bodies over time, potentially causing chronic health problems. In addition, aggregation can reduce the bioavailability of BMNPs to organisms. Larger aggregates may be less readily taken up by organisms compared to individual NPs, which could reduce their overall environmental toxicity. However, some organisms may still ingest aggregated BMNPs, and the potential consequences of such ingestion are not fully understood. Further research is needed to explore the long-term effects of these interactions on different organisms in different environmental niches. Thus, it is crucial to understand these complex interactions between BMNPs, their aggregation behavior, and the biological environment.

2.5 Fabrication Methods

Different manufacturing techniques can lead to variations in surface chemistry and the presence of functional groups on the BMNPs. These variations can significantly impact how the nanocomposite interacts with cells. For example, some manufacturing processes may leave chemical residues on the surface that could be cytotoxic (Srivastava et al., 2020). Conversely, certain manufacturing processes may intentionally incorporate biocompatible functional groups to minimize cellular toxicity. Manufacturing methods can influence the size distribution and morphology (shape) of the final BMNPs. As mentioned above, smaller sizes and elongated shapes can lead to increased cellular uptake and potentially higher cytotoxicity. It is easy to say that all factors affecting the toxicity of BMNPs are somehow interrelated. Techniques with better control over size and morphology can help minimize unintended cellular interactions and toxicity. Certain manufacturing processes may use harsh chemicals or high temperatures, which can release impurities or unwanted by-products (Rane et al., 2018). These impurities can also contribute to toxicity if they are released into the environment during the manufacture or disposal of BMNPs. The synthesis methods can influence the stability and degradation profile of BMNPs in the environment. For example, some methods may result in composites that are more persistent, which can lead to long-term environmental exposure and toxicity. Traditional chemical synthesis methods for Fe_2O_3 NPs often use hazardous substances like sodium borohydride, hypophosphite, and hydrazine hydrate (Jafarzadeh et al., 2024). Green chemistry provides a more sustainable approach by utilizing natural extracts and biopolymers as reducing and stabilizing agents. Conversely, methods that produce biodegradable nanocomposites can minimize long-term environmental risks (Vilaplana et al., 2010). Therefore, researchers and manufacturers need to carefully consider the fabrication method when designing nanocomposites for specific applications. Choosing methods that minimize unwanted byproducts, promote biocompatibility, and ensure controlled size and morphology can help create safer and more environment-friendly nanomaterials.

2.6 Elemental Makeup

The elemental makeup of BMNPs plays a complex role in both environmental and cellular toxicity. Depending on the specific metal used in the BMNPs, there is potential for metal ion release within the cellular environment. Certain metals are inherently toxic, and their free ions can disrupt cellular processes, leading to oxidative stress, DNA damage, and ultimately, cell death (Sengul and Asmatulu, 2020). The rate and extent of metal ion release depend on factors like the type of metal, its interaction with the biopolymer matrix, and the cellular environment's pH (Geetha et al., 2021). The biocompatibility of the biopolymer used in the BMNPs significantly impacts cellular toxicity. Some biopolymers might be biodegradable and non-toxic, while others might trigger an immune response or exhibit inherent cytotoxicity. Biopolymer-based nanocomposites containing a synergistic mixture of metal NPs can exhibit significantly improved antibacterial properties compared to single metal NPs. Studies have shown that the combination of metal oxides, such as copper oxide (CuO) and zinc oxide (ZnO), in a biopolymer matrix can achieve this effect. For example, nanocomposite films of Sodium Alginate (SA) and chitosan (CS) loaded with both CuO and ZnO nanoparticles showed excellent antibacterial activity against *E. coli* and *S. aureus*. The incorporation of ZnO nanoparticles not only enhanced the photocatalytic activity of the nanocomposite by generating Reactive Oxygen Species (ROS) under light irradiation but also improved the overall mechanical, and optical properties. This synergistic effect resulted in superior antibacterial performance, with the SA-CS@CuO/ZnO nanocomposite inhibiting bacterial growth by over 60% in the dark and over 90% when exposed to light. The enhanced efficacy is attributed to a combination of factors, including the destruction of the bacterial cell wall, impairment of the cellular redox balance, and DNA damage caused by the nanocomposite (Guan et al., 2021). Choosing biocompatible biopolymers minimizes the risk of cellular damage beyond the potential effects of the metal component. The combined properties of the metal and biopolymer determine the BMNP's persistence and degradation profile in the environment (Pathak et al., 2021). Ideally, the biopolymer component should be biodegradable, allowing for natural breakdown of the composite over time. However, the presence of the metal might hinder biodegradation, potentially leading to long-term environmental persistence and exposure risks. Depending on the specific elements involved, BMNPs can disrupt the delicate balance of local ecosystems. Some metals can accumulate in plants and animals, potentially affecting their growth and survival. Additionally, the biodegradation products of the biopolymer component might also have indirect consequences for the environment if not carefully considered. In conclusion, the shape, size, surface functionalization, fabrication method, and elemental makeup of BMNPs are critical parameters that determine their environmental fate and toxicity. Understanding how these factors influence the behavior of NPs is essential for assessing their risks and developing strategies to mitigate their potential adverse effects on human health and the environment.

3. Unveiling the Dark Side: Toxicity Concerns of Biopolymer-Based Metal Nanoparticles

Biopolymer-based metal nanoparticles (BMNPs) offer numerous advantages in various applications. This is because they hybridize two components, i.e., biopolymer and metal. Biopolymers are mainly responsible for increased stability, biocompatibility, and non-immunogenic response, while the metals are responsible for various biological activities (Kučuk et al., 2023). However, these BMNPs have a disadvantage that needs to be addressed, as the metals they contain can be potentially toxic.

3.1 Cellular Uptake and Toxicity Mechanisms of BMNPs

BMNPs can enter cells and cause toxicity through various methods, affecting the cells' state and functionality. Evaluating the toxicity of these BMNPs requires understanding the cellular uptake mechanisms resulting in harmful effects. Here, we discuss how BMNPs interact with cells and cause toxicity. BMNPs can be internalized by cells through various mechanisms based on their size, shape, charge, functional moiety, and type of cell involved. Endocytosis is the primary mechanism of cellular uptake in which phagocytosis takes up larger particles, and smaller particles enter into cells via pinocytosis and passive diffusion. Another mechanism is receptor-mediated uptake, in which BMNPs bind to the cell surface and trigger endocytosis. Once the BMNPs are inside the cell, they can cause toxicity through several mechanisms, i.e., oxidative stress, DNA damage, protein dysfunction, and membrane dysfunction, as shown in Fig. 2. Oxidative stress is caused by the generation of ROS and the depletion of antioxidants. Biopolymers such as polyethylene, polyether, and polyurethanes have a structural tendency to create free radicals, allowing them to undergo oxidative degradation (Sairam et al., 2023). DNA damage occurs by direct interaction of BMNPs with the DNA, causing strand breaks or indirectly interfering with the DNA repair mechanism. They can also induce protein aggregation, impair their function, inhibit cellular enzymes, disrupt metabolic processes,

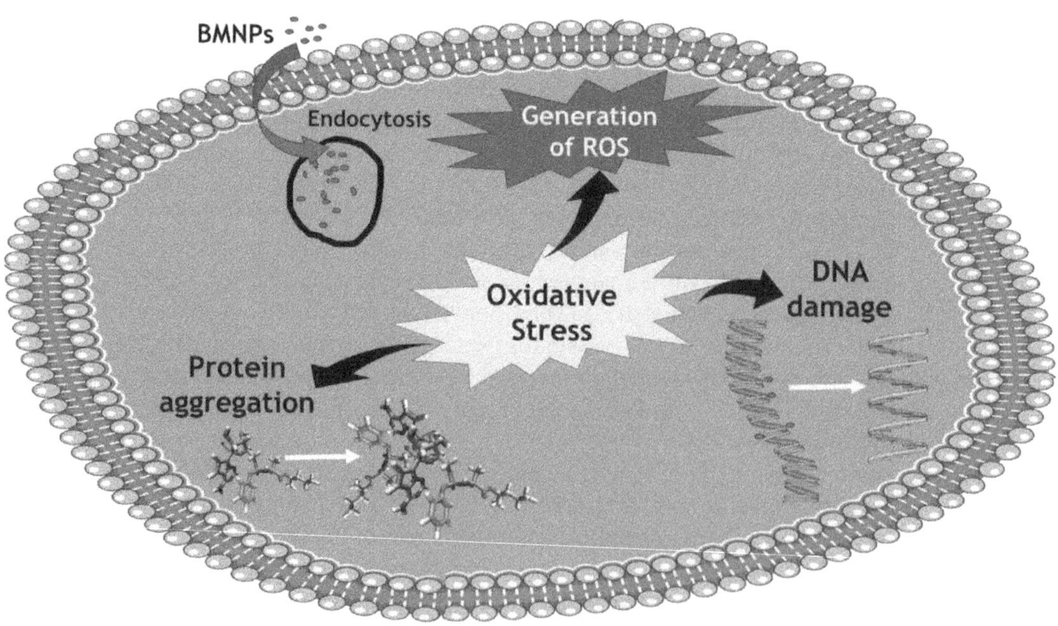

Figure 2. Schematic representation of cellular toxicity induced by the biopolymer-based metal nanoparticles (BMNPs).

and lead to cellular toxicity (Rai et al., 2015). BMNPs physically interact with the cell membrane, causing membrane damage and potential cell lysis by changing membrane permeability and ion channel function. Understanding BMNPs' cellular absorption and toxicity pathways is critical for determining their possible health effects. By discovering these pathways, researchers can build safer nanoparticle formulations and decrease the hazards associated with their use in various applications.

3.2 Potential Impact of BMNPs on Organisms

BMNPs are showing great promise as tools in environmental cleanup, agriculture, and medicine. However, the characteristics that make BMNPs useful also raise fears regarding possible detrimental effects on living organisms. These BMNPs may interact with different species in ways that have unforeseen harmful consequences on individual creatures and broader ecological groups. In the hybrid of biopolymer and metal NPs, biopolymers, due to their biodegradable nature, undergo degradation, whereas metal NPs, once released, can persist in the atmosphere, harming different lives. There is a potential chance if such NPs have longer persistence in the air, they can be inhaled by humans, leading to accumulation in the respiratory tract and causing respiratory disorders (Iyer et al., 2015). Extensive research has investigated the potential toxicity of various metal NPs using *in vitro* and *in vivo* models. Studies have shown that the cytotoxicity of BMNPs can vary depending on factors like the type of metal nanoparticle, its surface coating, and its dose. For instance, a study by Kai et al., observed varying cytotoxic effects of oleic acid-coated Fe_3O_4 NPs on human hepatoma BEL-7402 cells. They found that NPs with a larger surface area exhibited increased cytotoxicity along with induction of apoptosis and cell cycle arrest (Kai et al., 2011). Similarly, Lee et al., reported that Fe_2O_3 NPs caused elevated levels of lactate dehydrogenase (LDH) and Tumor Necrosis Factor-α (TNF-α) in mouse macrophage cells, indicating cellular stress and potential inflammation (Lee et al., 2014). Furthermore, Martínez-Torres et al., demonstrated that chitosan-coated gold NPs exhibited dose-dependent cytotoxicity toward MCF-7 and HeLa cancer cell lines. They proposed that these BMNPs induced cell death by overproduction of ROS, which can damage cellular components (Martínez-Torres et al., 2018). In addition to potential human health risks, BMNP persistence in the environment threatens aquatic ecosystems due to their widespread use and release into waterways. These BMNPs can have acute and chronic effects on aquatic life based on their exposure. For instance, nanocomposite having silver NPs is highly toxic to fish, causing damage to gills and other vital organs (Kakakhel et al., 2021). In terrestrial ecosystems, they can affect the soil quality and interfere with the uptake of essential elements, induce toxicity, and ultimately lead to reduced plant growth. As the antimicrobial properties of metal NPs are well explored, and if the soil is contaminated with these BMNPs, their antimicrobial attribute can also disrupt the microbial community that plays a significant role in the nutrient cycle and soil fertility (Ameen et al., 2021). In animals, they accumulate in the tissues and cause toxicity, reducing survival and reproduction rates. The potential impact of BMNPs on organisms is diverse and concerning. These BMNPs offer significant benefits, but it is crucial to maintain a balance between their possible use and health risks. Developing safer NP formulations and implementing robust environmental monitoring are necessary to mitigate the risks associated with these BMNPs.

4. Biocompatibility and Biodegradability of BMNPs

Biocompatibility is the key element for evaluating the safety and applicability of BMNPs. It refers to the potential ability of BMNPs to perform desired functions without causing undesirable local or systemic changes in the host organism. The factors (size, shape, surface chemistry, chemical composition) that influence the toxicity of BMNPs are found to be responsible for their biocompatibility as well. For instance, BMNPs composed of naturally occurring polymers like chitosan or alginate tend to be more biocompatible due to their biodegradability and minimal toxicity (Dodero et al., 2021). The surface properties of BMNPs can be modified to improve interaction with biological systems. For example, surface coating of Polyethylene glycol (PEG) improves biocompatibility by lowering protein adsorption and suppressing immune recognition (Pucci et al., 2022). There are many studies that report, that NPs like gold and iron oxide are generally considered less toxic. In contrast, metals like copper and silver may exhibit higher cytotoxicity due to their ability to generate ROS and interaction with cellular components (Ameh et al., 2022). The surface coating may reduce the cytotoxicity by providing a biocompatible surface that reduces direct metal exposure to cells. For example, chitosan and dextran-coated AuNPs used for diabetic foot ulcers have better biocompatibility and promisingly controlled release, reduce cytotoxicity, and enhance the healing process (Vijayakumar et al., 2019). Clinical studies are the most definitive test of biocompatibility in medical applications. Trials using BMNPs for targeted drug administration have shown promising results in decreasing side effects and improving therapeutic efficacy, highlighting the potential of these materials when appropriately engineered. Preclinical investigations with animal models provide critical information about the systemic impact of BMNPs. A study on mice showed that CuNPs stabilized in a biopolymer matrix could accumulate in the liver and spleen, emphasizing the necessity of evaluating potential organ-specific toxicity (Antonio-Pérez et al., 2023). Biodegradation is another important aspect to look into while evaluating the toxicity of BMNPs. Biodegradability refers to a material's ability to degrade into more straightforward, safer substances via biological processes and microorganisms. This feature is crucial for mitigating the long-term environmental and health concerns associated with NP disposal and unintentional leakage. When it comes to BMNPs, biodegradation refers to the process by which they gradually change into simpler substances over time with the intervention of biological factors. Biodegradable BMNPs offer several advantages including sustainability, reduced toxicity, and environmental compatibility. Types of biopolymers such as polylactic acid (PLA), polyglycolic acid (PGA), and polyhydroxyalkanoates (PHA) can significantly influence the biodegradability of BMNPs. Therefore, choosing such a type of polymer can dramatically reduce the environmental burden of NP accumulation in the ecosystem. Microbial intervention plays a major role in the biodegradation of BMNPs. Biological agents such as bacteria like *Bacillus subtilis* and *Pseudomonas aeruginosa* possess enzymes that degrade chitosan (Li et al., 2015). Different enzymes that take part in their breakdown include proteases, lipases, cellulases, and amylases (Vijayakumar et al., 2019). These enzymatic breakdowns occur either by surface attack or intracellular degradation. Because of the unique physicochemical characteristics of BMNPs, microbes stick to their surface and secrete enzymes. These enzymes break down BMNPs and release metal ions. The successful incorporation of biocompatibility and biodegradability into the design of BMNPs improves the functionality of these materials and is also in line with the ideas of green chemistry and sustainability. Researchers and developers are working on novel synthesis methods that can tailor the surface properties and degradation dynamics of BMNPs for specific applications, reducing potential adverse impacts while increasing performance. This comprehensive strategy is crucial to promote the use of BMNPs in a safe and environmentally sustainable way.

5. Environmental Fate of BMNPs

The environmental fate of BMNPs is fundamental to understanding their effects on the ecosystem. From synthesis to disposal, BMNPs take several routes to enter the environment. Each environmental domain interacts specifically with BMNPs which leads them to various fates.

5.1 Release Pathways of BMNPs into the Environment

The growing field of nanotechnology is leading to a significant rise in the use of BMNPs. Although these nanomaterials offer exciting possibilities such as biocompatibility, enhanced properties in terms of mechanical strength, protection against physicochemical factors, and antimicrobial activity (Jafarzadeh et al., 2024). However, their increasing presence in the environment raises concern due to their potential impact on human health and the ecosystem. To ensure responsible development, it is crucial to understand not only their function but also their potential environmental impact. BMNPs, like many other nanomaterials, have a unique ability to enter the environment at various

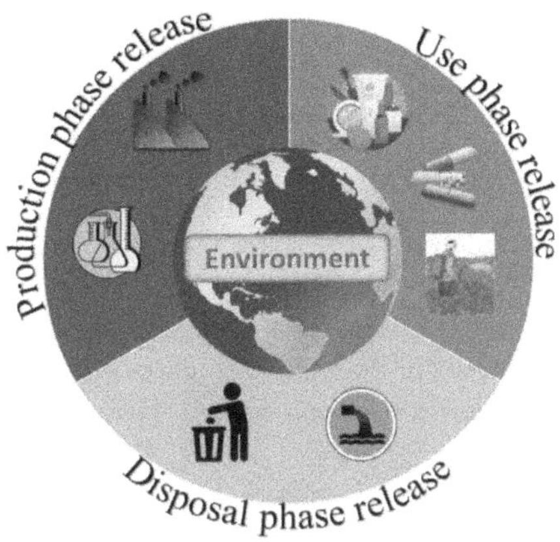

Figure 3. An illustration depicting the three primary release phases of biopolymer-based metal nanoparticles (BMNPs) in the environment.

stages of their life cycle. Three of these primary release phases are; (i) production phase release, (ii) use phase release, and (iii) end-of-life release (Bundschuh et al., 2018) (Fig. 3). The production phase release is the unintentional escape of BMNPs during the extraction and processing of raw materials used in their synthesis. In addition to these accidental releases, the unethical disposal of waste produced during the synthesis of BMNPs also plays a role in the leaking of these NPs into the environment (Zahra et al., 2022). In-use phase release, wear and tear, abrasion, and even everyday use can all lead to the unintentional release of BMNPs. For example, chitosan-AgNPs, alginate-AgNPs, and starch-AgNPs, among others, are being used extensively in textile and biomedical industries (Fernandes et al., 2022). The incorporated silver NPs in textiles disperse gradually as they are washed, worn, and disposed off. In their final stage of release following product disposal, BMNPs can escape from landfills or waste streams. Processes such as biodegradation, incineration, and leaching from landfills all play a role in this final act.

5.2 Transport and Distribution of BMNPs in Different Environmental Domains

Having established the three primary release pathways for BMNPs throughout their life cycle (production, use, and end-of-life), our focus now shifts to the transport and distribution of BMNPs in different environmental domains such as air, land, and water. BMNPs can become airborne via several methods; for example, if suitable ventilation and filtration measures are not in place, BMNPs can be released into the air during manufacturing activities. This is the principal source of aerosolized BMNPs. Spills or disruptions during BMNP research or industrial applications could cause them to become airborne. Improper incineration of BMNP waste might release them as part of flue gas (Walser et al., 2012). From the air, there is a possibility that BMNPs can enter into soil/land and water bodies through rain and irrigation water. As mentioned earlier, improper disposal of BMNP containing wastewater from research laboratories and industrial facilities can also introduce the former into water systems (Auffan et al., 2009). From these water bodies BMNPs can be directly absorbed by aquatic plants and algae situated at the lower tropic level of the food chain. Filter feeders, such as mussels, can consume BMNPs, thereby introducing them into the food web (Oberdorster et al., 2005). Airborne BMNPs can settle on land surfaces under the effect of gravity as well as rainwater and become incorporated into the soil. BMNPs could be intentionally applied in soil for purposes like bioremediation or plant growth enhancement (Santos et al., 2022). But once the biopolymer coating of BMNPs undergoes degradation by soil microorganisms, it will potentially release the metal core into the soil which ultimately harms the plants by disrupting nutrient uptake or causing other physiological stress. Improper disposal of BMNP-containing waste (e.g., sludge) can contaminate soil. Inside the soil, BMNPs will possibly interact with both abiotic (soil texture, pH, organic matter content) and biotic (microbiota, earthworms) components of soil and ultimately affect the quality of soil. Accumulation of BMNPs in soil can also have long-term effects on soil fertility. If plants take up BMNPs from the soil, they could enter the food chain and pose risks to consumers as well (Rajput et al., 2020).

5.3 Disruption of Ecosystem Services

BMNPs can disturb the fundamental ecosystem services that impact ecological balance and human well-being. The release of BMNPs into the environment, whether intentional through their use or accidental via spills, can interfere with ecological services through various pathways, affecting the core system that supports life on the planet. Ecological services like clean water, fertile soil, food, and a favorable climate are vital for human beings. Here, the potential effects of BMNPs on ecological services are discussed.

Healthy soil is crucial for plant growth, maintaining ecosystem productivity, and regulating water quality. Microorganisms play a vital role in decomposing organic matter, cycling nutrition, and maintaining soil microbiome. BMNPs, those having NPs like silver and copper with antimicrobial properties, can disturb the microbiome by inhibiting microbial activity. If microbial growth is altered, it disrupts the nutrient cycle, and ultimately, plant growth is affected. This can impair plant growth and agricultural productivity, affecting food production and the ecosystem (Machado et al., 2022). Aquatic plants and algae form the foundation of aquatic food webs and ecological health. BMNPs can accumulate on water plants and algae surfaces, obscuring sunlight and preventing photosynthesis. This can diminish the primary productivity of aquatic habitats, affecting food supply for higher trophic levels. NPs also interfere with the nutrient uptake by the aquatic plant, resulting in reduced growth and vitality. BMNPs can accumulate in the tissue of aquatic organisms and biomagnify through the food web, as illustrated in Fig. 4. Aquatic species may consume BMNPs directly from the water or through their feed. This accumulation can have hazardous consequences, particularly for top predators who consume tainted prey. BMNP concentrations can increase at higher trophic levels, leading to greater exposure and possible toxicity for predators that feed on fish and other aquatic creatures, including humans (Tamayo-Belda et al., 2023).

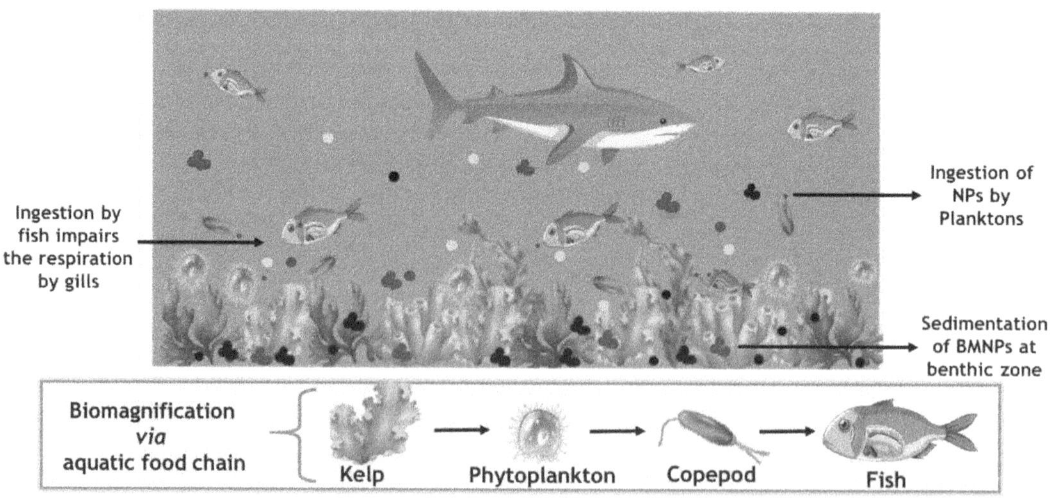

Figure 4. An illustration depicting biopolymer-based metal nanoparticles (BMNPs) affecting aquatic life through various pathways.

5.4 Transformation and Degradation of BMNPs

In the environment, BMNPs may undergo various transformations. Understanding these transformations and the degradation of BMNPs in the environment is critical for determining their potential impact and predicting their persistence. The transformations and degradation processes

Table 1. Various environmental transformation mechanisms of biopolymer-based metal nanoparticles (BMNPs).

S. No.	Types of transformation of BMNPs	Mechanism	Description	References
1.	Chemical transformation	Oxidative dissolution	Release of metal ions by oxygen	Lead et al., 2018
		Reductive dissolution	Electron-driven breakdown of BMNPs	Lead et al., 2018
		Sulfidation	Formation of sulfide complexes with metal ion	Wang and Liu, 2022
2.	Physical transformation	Light-induced chain incision	Degradation of conjugated polymers	Louis et al., 2018
		Aggregation	Formation of tightly bonded cluster due to short-range thermodynamic interaction	Paula et al., 2012
		Agglomeration	Formation of reversible clusters due to electrostatic interaction, van der Waal forces, and Brownian motion	Goswami et al., 2017
3.	Biological transformation	Enzymatic breakdown by micro-organisms	*P. aeruginosa* degrades PLGA-coated NPs via esterase secretion	Noor et al., 2020
		Insect mediated transformation	BMNPs transform inside the gut of insects like beetles, caterpillars	Gwenzi et al., 2024
		Plant mediated transformation	Inside the plant, BMNPs undergo enzymatic transformation, -compartmentalization, etc.	Lv et al., 2019

that BMNPs undergo can be classified into three main categories namely chemical, physical, and biological (Wang and Liu, 2022) (Table 1). Chemical transformations involve changes in the molecular structure or composition of the BMNPs, such as dissolution, sulfidation, and oxidation, among others. Physical transformations, on the other hand, involve changes in the physical state or arrangement of the BMNPs, that result in their agglomeration, aggregation, sedimentation, and deposition (Goswami et al., 2017). Lastly, the biological transformations refer to the transformation processes mediated by living organisms, such as biomodification, where the BMNPs undergo structural or functional modifications due to biological interactions, and biodegradation, where the BMNPs are broken down by biological entities.

5.4.1 Chemical transformations

When BMNPs are subjected to different environmental conditions, they can undergo chemical transformations. Numerous processes, including oxidation, reduction, and interactions with other substances, might result in these changes (Lead et al., 2018). The characteristics, reactivity, and environmental fate of NPs can all be strongly impacted by their changed chemical state. The environmental mobility of the metal part of BMNPs is significantly influenced by oxidative and reductive dissolution reactions. Also, the behavior of BMNPs in soil, water, or air depends upon their size, content, and environmental factors. For example, cellulose-based metal NPs, such as cellulose platinum NPs, eventually dissociate when they encounter water. Water molecules penetrate the nanoparticle structure during the dissolution process, breaking the hydrogen bonds that hold the cellulose chains together and causing the particles to disintegrate (Azeez et al., 2022). Sulfidation is another crucial chemical transformation of BMNPs in the environment. Sulfide, produced by microbial sulfate reduction, is frequently found in hypoxic settings such as riverine sediments, lakes, and wastewater treatment facilities (Liu et al., 2018). Different biopolymers such as cellulose, alginate, and chitosan-based AgNPs are being extensively produced for biomedical purposes (Alavi and Rai, 2019). These AgNPs undergo a quick, direct NPs–fluid reaction at high sulfide concentrations, producing Ag_2S. However, at low concentrations, AgNPs first transform into silver ions, which then react with sulfide ions to produce Ag_2S (Liu et al., 2018). On the whole, these chemical transformations make BMNPs prone to various other degradation processes such as physical and biological.

5.4.2 Physical transformations

The physical transformation of BMNPs refers to the changes in their structure, size, shape, and properties that can occur due to either external stimuli or inherent instabilities. Factors such as temperature, UV radiations, pH, solvent conditions, and the presence of varied environments can induce structural rearrangements, leading to changes in the morphology and chemistry of BMNPs. These modifications cause BMNPs to assemble into larger reversible aggregates or clusters called agglomerates (Goswami et al., 2017). Exposure to ultraviolet (UV) light can degrade the biopolymer coating of BMNPs. This can happen through several mechanisms, including chain scission (breaking of polymer chains) and photo-oxidation (degradation by reactive oxygen species generated by UV) (Louis et al., 2018). A study by Hu et al., investigated the degradation of chitosan-coated gold NPs under UV light, demonstrating a significant decrease in particle size due to biopolymer breakdown (Sun et al., 2008). Aggregation is another transformation in which irreversible, tightly bonded clusters are formed by BMNPs. In aggregation, short-range thermodynamic interactions are essential. Large aggregates may affect the movement and behavior of BMNPs. Since particle cores need to come into contact for particles to aggregate, the rate of aggregation is thought to be proportional to the chance that two particles will collide (Paula et al., 2012). The size of the NPs also affects the aggregation process. The smaller size NPs have a greater number of atoms on the surface thereby resulting in a change in surface charge and promoting aggregation. Ionic strength, pH and the nature of surface coating are some other key factors affecting the aggregation of BMNPs in environmental systems. Lastly, the sedimentation which is another important process in which BMNPs settle out of a liquid suspension and accumulate at the bottom of the surface (Lead et al., 2018). In natural stagnant water bodies like lakes, ponds, etc., BMNPs may settle to the bottom and might be degraded by the local flora and fauna which will be discussed in the next section. Whereas, BMNPs dumped in fresh running water may experience shear stress which can disrupt the biopolymer coating or even fragment the entire BMNP. Overall, these processes affect their fate, distribution, and transport (Wang and Liu, 2022).

5.4.3 Biological transformation

Biological transformation is a phenomenon where various biological systems, encompassing microbes, plants, and insects, interact with and modify BMNPs. This interaction can potentially mitigate possible ecological imbalances. Studies have shown bacteria like *Pseudomonas* sp. and *Shewanella* sp. can degrade biopolymer coatings on NPs through enzymatic breakdown. These enzymes are highly specific to the type of the biopolymer. For instance, research by Xu et al., demonstrated that *Pseudomonas aeruginosa* could effectively degrade Poly (Lactic-co-Glycolic Acid) (PLGA) coated NPs due to its secretion of specific esterases (Noor et al., 2020; Iravani and Varma, 2020). Fungal species like *Aspergillus fumigatus* and *Phanerochaete chrysosporium* have been shown to degrade BMNPs through a combination of mechanisms. They secrete organic acids (e.g., citric acid) that can break down the biopolymer coating or directly degrade the metal nanoparticle itself, depending on the metal type (Burford et al., 2003). Plants can take up BMNPs, particularly smaller NPs, through their root systems. Once inside the plant, various mechanisms such as compartmentalization, enzymatic modification, etc., could potentially transform the BMNPs (Lv et al., 2019). For instance, plant enzymes can degrade the biopolymer or interact with the metal nanoparticle, altering its properties. Research by Hu et al., investigated the uptake and degradation of PLGA NPs by maize (*Zea mays*) roots, demonstrating enzymatic degradation of the biopolymer within the plant tissues (Hu et al., 2017). The root zone (rhizosphere) is a hotspot for microbial activity. Plant root exudates stimulate the growth of specific microbes adept at BMNP degradation. These microbes can utilize the biopolymer for energy and potentially disintegrate the metal nanoparticle (Wang et al., 2023). A study by Cheng et al., explored the role of rhizosphere microbes in degrading polystyrene NPs, highlighting their contribution to BMNP transformations in the soil environment (Awet et al., 2018). Insects like beetles and caterpillars possess complex gut microbiomes containing bacteria that can degrade the biopolymer coating of BMNPs. Insects can excrete BMNPs after this degradation. This excretion process can alter the chemical form of the NPs, potentially making them more susceptible to further degradation by other organisms in the environment (Gwenzi et al., 2024). Zhu et al., explored the excretion of silver NPs by collembola (springtails), highlighting potential transformations of the NPs on the passage through the insect gut (Zhu et al., 2018). To sum up, numerous paths of release, transport, distribution, degradation, and transformation control the fate and behavior of BMNPs in the environment. Comprehending these processes is essential for evaluating the possible ecological consequences and hazards linked to BMNPs. Ultimately, a thorough understanding of these pathways will guide us in developing sustainable and responsible usage of BMNPs in many applications.

6. Risk Assessment and Environmental Safety

As technology in material science advances, BMNPs are being used as a unique solution for various issues, including drug resistance, medication delivery, alternative pesticide usage, improving soil quality, etc. Metal NPs used in BMNPs pose several environmental risks. Thus, it is critical to have a thorough framework for risk assessment to guarantee their safety during synthesis, use, and disposal (Babaremu et al., 2023). With the inevitable upsurge in applications of BMNPs, it is our primary concern to look into the dimensions of its negative implications on human health and different environment niches. The undermentioned aspects should be considered for careful evaluation while managing to prevent the pernicious effects.

6.1 Potential Hazards and Exposure Assessment

The initial phase of fabrication of BMNPs requires the usage of certain chemicals and harsh conditions that have potentially toxic effects on humans and the environmental health. The use of certain chemicals and reaction conditions is requisite, leading to occupational exposures to workers, which can reach unexpected bodily routes. Administration, dosage, and exposure duration may cause a cumulative build-up in particular organs. These nano-sized BMNPs are directly related to elevated toxicity due to their high propensity to cross biological barriers (Gatoo et al., 2014). Various reports state the exposure hazards while evaluating the *in vitro* toxicity. During *in vitro* testing of NPs, these can undergo various biological transformations, causing the formation of bio-corona, agglomeration, and generation of new nano-sized nanoparticles (Sharma et al., 2016). These transformations digress the exposure assessment of the nanoparticles, hampering the authenticated evaluation.

It is not forbidden that the usage of biopolymer will not pose any threat. It is of utmost significance that toxicity evaluation of biopolymer and metal NPs should be done separately for the ease of understanding its impact in different ecosystems. The desired application might be affected if the combined toxicity of biopolymer and metal NPs is witnessed. In understanding the nanotoxicological profile of certain biopolymers along with metal NPs, various studies have been performed. Wang et al reported that the HeLa cell line was used to examine the toxicity assessment of chitosan nanoparticles in the production of reactive oxygen species, and the study revealed dose-dependent toxicity. Maximum ROS production was obtained as chitosan NPs concentration increased (Wang et al., 2018). Further, literature reports are suggesting that bare copper nanoparticles when coated with chitosan resulted in a higher inflammatory response in the lungs and increased

Table 2. Application and toxic hazards of various biopolymer- based metal nanoparticles.

S. No.	Biopolymer Metal Nanoparticles	Application	Function	Potential Toxic hazard	References
1.	Chitosan/Silver NPs, Alginate/Iron oxide NPs	Biomedicine	Antibacterial activity, Magnetic Resonance Imaging (MRI) contrasting agent	Cellular toxicity	Argentiere et al., 2016; Hidayat et al., 2023; Manuja et al., 2021; Bar-Shir et al., 2014
2.	Gelatin/Gold NPs	Drug delivery	Targeted drug delivery	Biomagnification in the food chain	Cascone et al., 2002; Maharramov et al., 2019
3.	Pectin/Copper oxide NPs, Cellulose/Zero-valent Iron NPs	Environmental Remediation	Removal of heavy metal ions from wastewater	ROS generation and oxidative stress	Mahmoud et al., 2021; Lizundia et al., 2021; Naz et al., 2020
4.	Starch/Silver NPs, Pectin/Zinc oxide NPs	Food Packaging	For extension of the shelf life of food, Free radical scavenging to improve food quality	High mortality rate of embryos of Zebrafish	Sharma et al., 2023; Sharaby et al., 2022; Zavitri et al., 2023
5.	Chitosan/Palladium NPs, Dextran/Gold NPs	Biosensors	Detection of environmental pollutants	Protein misfolding	Chen and Ostrom, 2015; Huang et al., 2013; Bhatti et al., 2022
6.	Cellulose/Copper NPs, Alginate/ Magnesium Oxide NPs	Agriculture	Enhancement of plant growth and disease resistance	Genotoxic in nature	Naz et al., 2020; Fernandes et al., 2020; Mangalampalli et al., 2017
7.	Starch/Iron oxide Nps, Chitosan/Gold Nps	Drug delivery and Environmental remediation	Removal of magnetic pollutants and ease of drug targeting	Causes cellular toxicity, Impaired embryonic development of Aquatic organisms	Malhotra et al., 2020
8.	Gelatin/Silver Nps, Collagen/Titanium dioxide Nps	Biomedical	AntibacterialCoating material on medical devices	Cytotoxic in nature	Das et al., 2021
9.	Silk fibroin/Gold Nps	Imaging	Better contrasting agents used in imaging	Excessive ROS generation	Özcan et al., 2023
10.	Alginate/Platinum Nps	Industrial use	Used extensively in fuel cells	Cytotoxic in nature	Buchtelova et al., 2017

inflammatory cytokines in blood levels (Worthington et al., 2013). Various nanocomposites have been exploited for their application in fields like biomedicine, environmental bioremediation, food packaging, etc., and their potential toxicity has been discussed in Table 2.

Looking at the broader picture of the ecosystem, aquatic life is largely affected when nanotechnology intervenes in the wastewater management regime. The industries in the name of biopolymer composite formation, use some polymers that are hazardous organic pollutants requiring efficient waste management. These comprise styrene, epoxy resins, polyurethanes, etc. The disposal of such polymers into marine bodies imposes several issues that require their replacement and usage of biodegradable biopolymers (Liang et al., 2021). Hence it is important to use biodegradable polymers in composite formation such as cellulose, starch, gelatin, guar gum, etc. There are some instances where even biopolymers are not bereft of the toxicity profile. To acknowledge the fact that silver nanoparticles are well known for the removal of organic dyes present in wastewater, they are coupled with chitosan biopolymer to boost the efficacy in dye removal; nevertheless, aquatic life is negatively impacted by such nanocomposite. Various studies observed a great impact of Ag-CS nanocomposite on the gills of various fishes. This is likely because gills function by filtering water, making them the first point of contact for any contaminants suspended in the water (Mohamed et al., 2021).

6.2 Risk Mitigation Strategies and Environmental Monitoring

The scenario is not restricted to regulating the degradation of BMNPs rather the whole process of its life cycle should be monitored. There is a variety of parameters that should be checked to have a holistic view of the synthesis, usage, and degradation process. To proceed, the persistence of harmful compounds released into the environment and their discharge teach us indirectly how to reduce their poisonous character. Hazardous byproducts are frequently released during the biopolymer manufacturing process. Such hazardous byproducts are mobile and affect living things as well as soil, water, and air. For instance, the synthesis process of polylactic acid (PLA) involves the expulsion of ammonia from the reactor; the toxicity arising from it can impact the fishes in aquatic life, as reported that 0.02 mg/l in water bodies can harm freshwater invertebrates such as Daphnia. Not only the aquatic life but also the plants are affected intensely as ammonia vapors can cause the burning of leaves (Zhang et al., 2022). In this way, the three ecosystems are intertwined to support the notion that if one gets affected, the other one will ultimately be under certain pernicious consequences. Lastly, it should be mentioned that some inorganic and organic substances are being released into the environment with the coexistence of pollutants like polycyclic aromatic hydrocarbons. They can adsorb on certain low molecular weight biopolymer nanoparticles that have undergone degradation, which might result in coupled poisoning. Studies highlight the need for more investigation to comprehend the origins, spread, and possible risks related to BMNPs to create efficient mitigation plans and safeguard the environment and public health.

7. Conclusion and Future Perspective

Biopolymer-based metal nanoparticles (BMNPs) present a fascinating avenue in nanotechnology, offering distinct advantages over traditional metallic nanoparticles. Biopolymers enhance biocompatibility, biodegradability, and stability, and enable targeted delivery, making BMNPs attractive for various applications. However, thoroughly understanding their environmental fate and toxicity is crucial for their safe development and use. This chapter has highlighted the critical factors influencing environmental behavior and potential risks associated with BMNPs. These factors encompass the size, shape, surface chemistry, aggregation ability, fabrication methods, and elemental makeup, significantly impacting cellular and environmental toxicity. We explored the potential impacts of BMNPs on organisms, including cellular uptake mechanisms, biocompatibility concerns, and potential disruption of ecosystem services. Additionally, the chapter delved into the environmental fate of BMNPs, encompassing their release pathways, transport and distribution in various environmental compartments, and their transformation and degradation processes. Moving forward, a multi-pronged approach is necessary to ensure the safe development and utilization of BMNPs. Building upon the knowledge of BMNPs' factors influencing environmental fate and toxicity, future research should prioritize deciphering the intricate relationship between these variables. This can be achieved by the use of robust characterization and comprehensive toxicity studies and by designing BMNPs with minimal environmental impact through proper biopolymer selection, surface chemistry optimization, and aggregation control. Conducting life cycle assessments is also important to evaluate the environmental footprint along with the introduction of stricter regulations for production and disposal. Another approach to mitigate risks and environmental threats is to optimize manufacturing methods and research fully biodegradable biopolymers for the sustainable development of BMNPs. By addressing these challenges, BMNPs can become a truly green and sustainable nanotechnology platform with immense potential for diverse applications.

Acknowledgments

The authors would like to thank the Director, CSIR-IHBT, for his constant support and encouragement. AA acknowledges financial assistance in the form of project grant MLP-0201 and GAP-0295 from CSIR, Government of India. Nandini Teji, Rahul Bhardwaj, Anjali Nisha and Harpreet Kaur acknowledge the Academy of Scientific and Innovative Research (AcSIR), DST-INSPIRE-JRF (IF220319), UGC-JRF (211610162757), DST-INSPIRE-JRF (IF210287) and UGC-JRF (231610127265) for their respective fellowships. Rubina Dhiman acknowledges project fellowship in the form of GAP-0295. The CSIR-IHBT communication number for this manuscript is 5586.

References

Abbasi, R., Shineh, G., Mobaraki, M., Doughty, S. and Tayebi, L. (2023). Structural parameters of nanoparticles affecting their toxicity for biomedical applications: A review. Journal of Nanoparticle Research, 25(3): 43.

Alavi, M. and Rai, M. (2019). Recent progress in nanoformulations of silver nanoparticles with cellulose, chitosan, and alginic acid biopolymers for antibacterial applications. Applied Microbiology and Biotechnology, 103(21): 8669–8676.

Ameen, F., Alsamhary, K., Alabdullatif, J. A. and ALNadhari, S. (2021). A review on metal-based nanoparticles and their toxicity to beneficial soil bacteria and fungi. Ecotoxicology and Environmental Safety, 213: 112027.

Ameh, T., Gibb, M., Stevens, D., Pradhan, S. H., Braswell, E. and Sayes, C. M. (2022). Silver and copper nanoparticles induce oxidative stress in bacteria and mammalian cells. Nanomaterials, 12(14): 2402.

Antonio-Pérez, A., Durán-Armenta, L. F., Pérez-Loredo, M. G. and Torres-Huerta, A. L. (2023). Biosynthesis of copper nanoparticles with Medicinal plants extracts: from extraction methods to applications. Micromachines, 14(10): 1882.

Argentiere, S., Cella, C., Cesaria, M., Milani, P. and Lenardi, C. (2016). Silver nanoparticles in complex biological media: Assessment of colloidal stability and protein corona formation. Journal of Nanoparticle Research, 18: 1–13.

Auffan, M., Rose, J., Bottero, J. Y., Lowry, G. V., Jolivet, J. P. and Wiesner, M. R. (2009). Towards a definition of inorganic nanoparticles from an environmental, health and safety perspective. Nature Nanotechnology, 4(10): 634–641.

Awet, T. T., Kohl, Y., Meier, F., Straskraba, S., Grün, A. L., Ruf, T. et al. (2018). Effects of polystyrene nanoparticles on the microbiota and functional diversity of enzymes in soil. Environmental Sciences Europe, 30: 1–10.

Azeez, L. A., Lateef, A. and Olabode, O. (2022). Biogenic metallic nanoparticles for water treatment and purification: The state of the art–A review. Available at SSRN 4092472.

Buchtelova, H., Dostalova, S., Michalek, P., Krizkova, S., Strmiska, V., Kopel, P. et al. (2017). Size-related cytotoxicological aspects of polyvinylpyrrolidone-capped platinum nanoparticles. Food and Chemical Toxicology, 105: 337–346.

Babaremu, K., Oladijo, O. P. and Akinlabi, E. (2023). Biopolymers: A suitable replacement for plastics in product packaging. Advanced Industrial and Engineering Polymer Research, 6(4): 333–340.

Bar-Shir, A., Avram, L., Yariv-Shoushan, S., Anaby, D., Cohen, S., Segev-Amzaleg, N. et al. (2014). Alginate-coated magnetic nanoparticles for noninvasive MRI of extracellular calcium. NMR in Biomedicine, 27(7): 774–783.

Bharti, K., Sahu, J. K. and Sadhu, K. K. (2022). Origin of luminescence properties and synthetic methods for gold-and bimetallic gold-based nanomaterials. Materials Advances, 3(14): 5698–5724.

Bhatti, R., Shakeel, H., Malik, K., Qasim, M., Khan, M. A., Ahmed, N. et al. (2022). Inorganic nanoparticles: toxic effects, mechanisms of cytotoxicity and phytochemical interactions. Advanced Pharmaceutical Bulletin, 12(4): 757.

Bin Rashid, A. (2023). Utilization of nanotechnology and nanomaterials in biodiesel production and property enhancement. Journal of Nanomaterials, 2023, 1–14.

Bundschuh, M., Filser, J., Lüderwald, S., McKee, M. S., Metreveli, G., Schaumann, G. E. et al. (2018). Nanoparticles in the environment: where do we come from, where do we go to? Environmental Sciences Europe, 30: 1–17.

Burford, E. P., Fomina, M. and Gadd, G. M. (2003). Fungal involvement in bioweathering and biotransformation of rocks and minerals. Mineralogical Magazine, 67(6): 1127–1155.

Carrion, C. C., Nasrollahzadeh, M., Sajjadi, M., Jaleh, B., Soufi, G. J. and Iravani, S. (2021). Lignin, lipid, protein, hyaluronic acid, starch, cellulose, gum, pectin, alginate and chitosan-based nanomaterials for cancer nanotherapy: Challenges and opportunities. International Journal of Biological Macromolecules, 178: 193–228.

Cascone, M. G., Lazzeri, L., Carmignani, C. and Zhu, Z. (2002). Gelatin nanoparticles produced by a simple W/O emulsion as delivery system for methotrexate. Journal of Materials Science: Materials in Medicine, 13: 523–526.

Chen, A. and Ostrom, C. (2015). Palladium-based nanomaterials: synthesis and electrochemical applications. Chemical Reviews, 115(21): 11999–12044.

Corbo, C., Molinaro, R., Parodi, A., Toledano Furman, N. E., Salvatore, F. and Tasciotti, E. (2016). The impact of nanoparticle protein corona on cytotoxicity, immunotoxicity and target drug delivery. Nanomedicine, 11(1): 81–100.

Das, N. C., Roy, B., Patra, R., Choudhury, A., Ghosh, M. and Mukherjee, S. (2021). Surface-modified noble metal nanoparticles as antimicrobial agents: biochemical, molecular and therapeutic perspectives. Nanotechnology for Advances in Medical Microbiology, 165–205.

Dhawan, A. and Sharma, V. (2010). Toxicity assessment of nanomaterials: methods and challenges. Analytical and Bioanalytical Chemistry, 398: 589–605.

Dodero, A., Alberti, S., Gaggero, G., Ferretti, M., Botter, R., Vicini, S. et al. (2021). An up-to-date review on alginate nanoparticles and nanofibers for biomedical and pharmaceutical applications. Advanced Materials Interfaces, 8(22): 2100809.

Dos Santos, O. A. L., Pizzorno Backx, B., Abumousa, R. A. and Bououdina, M. (2022). Environmental implications associated with the development of nanotechnology: From synthesis to disposal. Nanomaterials, 12(23): 4319.

Fazal, T., Murtaza, B. N., Shah, M., Iqbal, S., Rehman, M. U., Jaber, F. et al. (2023). Recent developments in natural biopolymer based drug delivery systems. RSC Advances, 13(33): 23087–23121.

Fernandes, M., Padrão, J., Ribeiro, A. I., Fernandes, R. D., Melro, L., Nicolau, T. et al. (2022). Polysaccharides and metal nanoparticles for functional textiles: A review Nanomaterials, 12(6): 1006.

Fernandes, M., RB Singh, K., Sarkar, T., Singh, P. and Pratap Singh, R. (2020). Recent applications of magnesium oxide (MgO) nanoparticles in various domains. Advanced Materials Letters, 11(8): 1–10.

Gatoo, M. A., Naseem, S., Arfat, M. Y., Mahmood Dar, A., Qasim, K. and Zubair, S. (2014). Physicochemical properties of nanomaterials: Implication in associated toxic manifestations. BioMed research international, 2014.

Geetha, N., Bhavya, G., Abhijith, P., Shekhar, R., Dayananda, K. and Jogaiah, S. (2021). Insights into nanomycoremediation: Secretomics and mycogenic biopolymer nanocomposites for heavy metal detoxification. Journal of Hazardous Materials, 409: 124541.

Goswami, L., Kim, K. H., Deep, A., Das, P., Bhattacharya, S. S., Kumar, S. et al. (2017). Engineered nano particles: nature, behavior, and effect on the environment. Journal of Environmental Management, 196: 297–315.

Guan, G., Zhang, L., Zhu, J., Wu, H., Li, W. and Sun, Q. (2021). Antibacterial properties and mechanism of biopolymer-based films functionalized by CuO/ZnO nanoparticles against *Escherichia coli* and *Staphylococcus aureus*. Journal of Hazardous Materials, 402: 123542.

Gwenzi, W., Gufe, C., Alufasi, R., Makuvara, Z., Marumure, J., Shanmugam, S. R. et al. (2024). Insects to the rescue? Insights into applications, mechanisms, and prospects of insect-driven remediation of organic contaminants. Science of The Total Environment, 925: 171116.

Han, R., Xiao, Y., Bai, Q. and Choi, C. H. J. (2023). Self-therapeutic metal-based nanoparticles for treating inflammatory diseases. Acta Pharmaceutica Sinica B, 13(5): 1847–1865.

Hidayat, M. I., Adlim, M., Suhartono, S., Hayati, Z. and Bakar, N. H. H. A. (2023). Comparison of antibacterial properties between chitosan stabilized silver and zinc oxide nanoparticles immobilized on white silica beads. South African Journal of Chemical Engineering, 45: 111–119.

Hou, F., Sun, S., Abdullaha, S. W., Tang, Y., Li, X. and Guo, H. (2023). The application of nanoparticles in point-of-care testing (POCT) immunoassays. Analytical Methods.

Hu, Y., Xu, M., Liu, Y., Xie, X., Bao, W., Song, A. et al. (2017). Chitosan gel incorporated peptide-modified AuNPs for sustained drug delivery with smart pH responsiveness. Journal of Materials Chemistry B, 5(6): 1174–1181.

Huang, C. F., Yao, G. H., Liang, R. P. and Qiu, J. D. (2013). Graphene oxide and dextran capped gold nanoparticles based surface plasmon resonance sensor for sensitive detection of concanavalin A. Biosensors and Bioelectronics, 50: 305–310.

Hussain, I., Ali, I., Ullah, S., Iqbal, A., Tawaha, A. A., Al-Tawaha, A. R. et al. (2021, June). Agricultural soil reclamation and restoration of soil organic matter and nutrients via application of organic, inorganic and bio fertilization (Mini review). In IOP Conference Series: Earth and Environmental Science (Vol. 788, No. 1, p. 012165). IOP Publishing.

Iravani, S. and Varma, R. S. (2020). Bacteria in heavy metal remediation and nanoparticle biosynthesis. ACS Sustainable Chemistry & Engineering, 8(14): 5395–5409.

Iyer, R., CW Hsia, C. and T Nguyen, K. (2015). Nano-therapeutics for the lung: state-of-the-art and future perspectives. Current Pharmaceutical Design, 21(36): 5233–5244.

Jafarzadeh, S., Nooshkam, M., Zargar, M., Garavand, F., Ghosh, S., Hadidi, M. et al. (2024). Green synthesis of nanomaterials for smart biopolymer packaging: challenges and outlooks. Journal of Nanostructure in Chemistry, 14(2): 113–136.

Jafarzadeh, S., Nooshkam, M., Zargar, M., Garavand, F., Ghosh, S., Hadidi, M. et al. (2024). Green synthesis of nanomaterials for smart biopolymer packaging: challenges and outlooks. Journal of Nanostructure in Chemistry, 14(2): 113–136.

Joudeh, N. and Linke, D. (2022). Nanoparticle classification, physicochemical properties, characterization, and applications: a comprehensive review for biologists. Journal of Nanobiotechnology, 20(1): 262.

Kai, W., Xiaojun, X., Ximing, P., Zhenqing, H. and Qiqing, Z. (2011). Cytotoxic effects and the mechanism of three types of magnetic nanoparticles on human hepatoma BEL-7402 cells. Nanoscale Research Letters, 6: 1–10.

Kakakhel, M. A., Wu, F., Sajjad, W., Zhang, Q., Khan, I., Ullah, K. et al. (2021). Long-term exposure to high-concentration silver nanoparticles induced toxicity, fatality, bioaccumulation, and histological alteration in fish (*Cyprinus carpio*). Environmental Sciences Europe, 33: 1–11.

Kaliaraj, G. S., Subramaniyan, B., Manivasagan, P. and Kim, S. K. (2017). Green synthesis of metal nanoparticles using seaweed polysaccharides. Seaweed Polysaccharides, 101–109.

Kamkaew, A., Chen, F., Zhan, Y., Majewski, R. L. and Cai, W. (2016). Scintillating nanoparticles as energy mediators for enhanced photodynamic therapy. ACS nano, 10(4): 3918–3935.

Kučuk, N., Primožič, M., Knez, Ž. and Leitgeb, M. (2023). Sustainable biodegradable biopolymer-based nanoparticles for healthcare applications. International Journal of Molecular Sciences, 24(4): 3188.

Kumar, A., Choudhary, A., Kaur, H., Mehta, S. and Husen, A. (2021). Smart nanomaterial and nanocomposite with advanced agrochemical activities. Nanoscale Research Letters, 16: 1–26.

Lead, J. R., Batley, G. E., Alvarez, P. J., Croteau, M. N., Handy, R. D., McLaughlin, M. J. et al. (2018). Nanomaterials in the environment: Behavior, fate, bioavailability, and effects—an updated review. Environmental Toxicology and Chemistry, 37(8): 2029–2063.

Lee, J. H., Ju, J. E., Kim, B. I., Pak, P. J., Choi, E. K., Lee, H. S. et al. (2014). Rod-shaped iron oxide nanoparticles are more toxic than sphere-shaped nanoparticles to murine macrophage cells. Environmental Toxicology and Chemistry, 33(12): 2759–2766.

Leopold, K., Philippe, A., Wörle, K. and Schaumann, G. E. (2016). Analytical strategies to the determination of metal-containing nanoparticles in environmental waters. TrAC Trends in Analytical Chemistry, 84: 107–120.

Li, S., Li, Z., Ke, X., Wu, L. and Christie, P. (2022). Biological transfer of silver under silver nanoparticle exposure and nitrogen transfer via a collembolan-predatory mite food-chain and ecotoxicity of silver sulfide. Soil Ecology Letters, 4(4): 435–443.

Li, H., Fei, Z., Gong, J., Yang, T., Xu, Z. and Shi, J. (2015). Screening and characterization of a highly active chitosanase based on metagenomic technology. Journal of Molecular Catalysis B: Enzymatic, 111: 29–35.

Li, Z., Jiang, H., Xu, C. and Gu, L. (2015). A review: Using nanoparticles to enhance absorption and bioavailability of phenolic phytochemicals. Food Hydrocolloids, 43: 153–164.

Liang, Y., He, J. and Guo, B. (2021). Functional hydrogels as wound dressing to enhance wound healing. ACS nano, 15(8): 12687–12722.

Liu, J., Zhang, F., Allen, A. J., Johnston-Peck, A. C. and Pettibone, J. M. (2018). Comparing sulfidation kinetics of silver nanoparticles in simulated media using direct and indirect measurement methods. Nanoscale, 10: 22270–22279.

Liu, X., Chen, Y., Li, H., Huang, N., Jin, Q., Ren, K. et al. (2013). Enhanced retention and cellular uptake of nanoparticles in tumors by controlling their aggregation behavior. ACS Nano, 7(7): 6244–6257.

Lizundia, E., Sipponen, M. H., Greca, L. G., Balakshin, M., Tardy, B. L., Rojas, O. J. et al. (2021). Multifunctional lignin-based nanocomposites and nanohybrids. Green Chemistry, 23(18): 6698–6760.

Louis, B., Caubergh, S., Larsson, P. O., Tian, Y. and Scheblykin, I. G. (2018). Light and oxygen induce chain scission of conjugated polymers in solution. Physical Chemistry Chemical Physics, 20(3): 1829–1837.

Lv, J., Christie, P. and Zhang, S. (2019). Uptake, translocation, and transformation of metal-based nanoparticles in plants: recent advances and methodological challenges. Environmental Science: Nano, 6(1): 41–59.

Malhotra, N., Lee, J. S., Liman, R. A. D., Ruallo, J. M. S., Villaflores, O. B., Ger, T. R. et al. (2020). Potential toxicity of iron oxide magnetic nanoparticles: A review. Molecules, 25(14): 3159.

Machado, T. O., Grabow, J., Sayer, C., de Araújo, P. H., Ehrenhard, M. L. and Wurm, F. R. (2022). Biopolymer-based nanocarriers for sustained release of agrochemicals: A review on materials and social science perspectives for a sustainable future of agri-and horticulture. Advances in Colloid and Interface Science, 303: 102645.

Maharramov, A. M., Hasanova, U. A., Suleymanova, I. A., Osmanova, G. E. and Hajiyeva, N. E. (2019). The engineered nanoparticles in food chain: Potential toxicity and effects. SN Applied Sciences, 1: 1–25.

Mohamed, A. S., Dajem, S. B., Al-Kahtani, M., Ali, S. B., Ibrahim, E., Morsy, K. et al. (2021). Silver/chitosan nanocomposites induce physiological and histological changes in freshwater bivalve. Journal of Trace Elements in Medicine and Biology, 65: 126719.

Mangalampalli, B., Dumala, N. and Grover, P. (2017). Acute oral toxicity study of magnesium oxide nanoparticles and microparticles in female albino Wistar rats. Regulatory Toxicology and Pharmacology, 90: 170–184.

Manuja, A., Kumar, B., Kumar, R., Chhabra, D., Ghosh, M., Manuja, M. et al. (2021). Metal/metal oxide nanoparticles: Toxicity concerns associated with their physical state and remediation for biomedical applications. Toxicology Reports, 8: 1970–1978.

Marcellus, K. A., Bugiel, S., Nunnikhoven, A., Curran, I. and Gill, S. S. (2024). Polystyrene nano-and microplastic particles induce an inflammatory gene expression profile in rat neural stem cell-derived astrocytes *in vitro*. Nanomaterials, 14(5): 429.

Martínez-Torres, A. C., Zarate-Triviño, D. G., Lorenzo-Anota, H. Y., Ávila-Ávila, A., Rodríguez-Abrego, C. and Rodríguez-Padilla, C. (2018). Chitosan gold nanoparticles induce cell death in HeLa and MCF-7 cells through reactive oxygen species production. International Journal of Nanomedicine, 3235–3250.

Mohamed, A. S., Dajem, S. B., Al-Kahtani, M., Ali, S. B., Ibrahim, E., Morsy, K. et al. (2021). Silver/chitosan nanocomposites induce physiological and histological changes in freshwater bivalve. Journal of Trace Elements in Medicine and Biology, 65: 126719.

Nasimi, P. and Haidari, M. (2013). Medical use of nanoparticles: Drug delivery and diagnosis diseases. International Journal of Green Nanotechnology, 1: 1943089213506978.

Navarro, E., Baun, A., Behra, R., Hartmann, N. B., Filser, J., Miao, A. J. et al. (2008). Environmental behavior and ecotoxicity of engineered nanoparticles to algae, plants, and fungi. Ecotoxicology, 17: 372–386.

Naz, S., Gul, A. and Zia, M. (2020). Toxicity of copper oxide nanoparticles: a review study. IET Nanobiotechnology, 14(1): 1–13.

Noor, H., Satti, S. M., ud Din, S., Farman, M., Hasan, F., Khan, S. et al. (2020). Insight on esterase from *Pseudomonas aeruginosa* strain S3 that depolymerize poly (lactic acid) (PLA) at ambient temperature. Polymer Degradation and Stability, 174: 109096.

Oberdörster, G., Oberdörster, E. and Oberdörster, J. (2005). Nanotoxicology: an emerging discipline evolving from studies of ultrafine particles. Environmental Health Perspectives, 113(7): 823–839.

Özcan, O., Ünal, İ., Tufan, E., Alturfan, E. E. and Tunalı-akbay, T. (2023). Investigation of the toxicologic and biochemical effects of silk fibroin/gold nanoparticles-based nanofiber using zebrafish embryos. Experimed, 13(2): 109–114.

Pathak, P., Hwang, J. H., Li, R. H., Rodriguez, K. L., Rex, M. M., Lee, W. H. et al. (2021). Flexible copper-biopolymer nanocomposite sensors for trace level lead detection in water. Sensors and Actuators B: Chemical, 344: 130263.

Prajitha, N., Athira, S. S. and Mohanan, P. V. (2019). Bio-interactions and risks of engineered nanoparticles. Environmental Research, 172: 98–108.

Pucci, C., Degl'Innocenti, A., Gümüş, M. B. and Ciofani, G. (2022). Superparamagnetic iron oxide nanoparticles for magnetic hyperthermia: Recent advancements, molecular effects, and future directions in the omics era. Biomaterials Science, 10(9): 2103–2121

Rai, M., Ingle, A. P., Gupta, I. and Brandelli, A. (2015). Bioactivity of noble metal nanoparticles decorated with biopolymers and their application in drug delivery. International Journal of Pharmaceutics, 496(2): 159–172.

Rajput, V., Minkina, T., Mazarji, M., Shende, S., Sushkova, S., Mandzhieva, S. et al. (2020). Accumulation of nanoparticles in the soil-plant systems and their effects on human health. Annals of Agricultural Sciences, 65(2): 137–143.

Rane, A. V., Kanny, K., Abitha, V. K. and Thomas, S. (2018). Methods for synthesis of nanoparticles and fabrication of nanocomposites. In Synthesis of inorganic nanomaterials (pp. 121–139). Woodhead Publishing.

Rashid, B. A. (2023). Utilization of nanotechnology and nanomaterials in biodiesel production and property enhancement. Journal of Nanomaterials, 2023, 1–14.

Remédios, C., Rosário, F. and Bastos, V. (2012). Environmental nanoparticles interactions with plants: Morphological, physiological, and genotoxic aspects. Journal of Botany.

Sairam, A. B., Sanmugam, A., Pushparaj, A., Mahesh Kumar, G., Sundarapandian, N., Balaji, S. et al. (2023). Toxicity of Polymeric Nanodrugs as Drug Carriers. ACS Chemical Health & Safety, 30(5): 236–250.

Sarkar, S., Ponce, N. T., Banerjee, A., Bandopadhyay, R., Rajendran, S. and Lichtfouse, E. (2020). Green polymeric nanomaterials for the photocatalytic degradation of dyes: A review. Environmental Chemistry Letters, 18(5): 1569–1580.

Schaeublin, N. M., Comfort, K. K., Schlager, J. J., Singh, S. B. and Hussain, S. M. (2014). Nanotoxicity and cellular stress response: Physical and chemical properties and their link to translational research. Translational Research in Environmental and Occupational Stress, 69–80.

Sengul, A. B. and Asmatulu, E. (2020). Toxicity of metal and metal oxide nanoparticles: A review. Environmental Chemistry Letters, 18(5): 1659–1683.

Sharaby, M. R., Soliman, E. A., Abdel-Rahman, A. B., Osman, A. and Khalil, R. (2022). Novel pectin-based nanocomposite film for active food packaging applications. Scientific Reports, 12(1): 20673.

Sharifi, S., Behzadi, S., Laurent, S., Forrest, M. L., Stroeve, P. and Mahmoudi, M. (2012). Toxicity of nanomaterials. Chemical Society Reviews, 41(6): 2323–2343.

Sharma, M., Shatkin, J. A., Cairns, C., Canady, R. and Clippinger, A. J. (2016). Framework to evaluate exposure relevance and data needs for risk assessment of nanomaterials using in vitro testing strategies. Risk Analysis, 36(8): 1551–1563.

Sharma, S., Sharma, N. and Kaushal, N. (2023). Utilization of novel bacteriocin synthesized silver nanoparticles (AgNPs) for their application in antimicrobial packaging for preservation of tomato fruit. Frontiers in Sustainable Food Systems, 7: 1072738.

Shrestha, S., Wang, B. and Dutta, P. (2020). Nanoparticle processing: Understanding and controlling aggregation. Advances in Colloid and Interface Science, 279: 102162.

Srivastava, V., Zare, E. N., Makvandi, P., Zheng, X. Q., Iftekhar, S., Wu, A. et al. (2020). Cytotoxic aquatic pollutants and their removal by nanocomposite-based sorbents. Chemosphere, 258: 127324.

Stark, W. J. (2011). Nanoparticles in biological systems. Angewandte Chemie International Edition, 50(6): 1242–1258.

Sun, C., Qu, R., Chen, H., Ji, C., Wang, C., Sun, Y and Wang, B. (2008). Degradation behavior of chitosan chains in the 'green'synthesis of gold nanoparticles. Carbohydrate Research, 343(15): 2595–2599.

Tamayo-Belda, M., Venâncio, C., Fernandez-Piñas, F., Rosal, R., Lopes, I. and Oliveira, M. (2023). Effects of petroleum-based and biopolymer-based nanoplastics on aquatic organisms: A case study with mechanically degraded pristine polymers. Science of The Total Environment, 883: 163447.

Tourinho, P. S., Cornelis, A. M., Gestel, V., Lofts, S., Svendsen, C., Amadeu, M. V. M. et al. (2012). Metal-based nanoparticles in soil: Fate, behavior, and effects on soil invertebrates. Environmental Toxicology and Chemistry, 31(8): 1679–1692.

Vijayakumar, A. S., Malaikozhundan, B., Thangaraj, M. P., Ekambaram, P., Murugan, T. et al. (2020). Chitosan-coated silver nanoparticles promoted antibacterial, antibiofilm, wound-healing of murine macrophages and antiproliferation of human breast cancer MCF 7 cells. Polymer Testing, 90: 106675.

Vijayakumar, V., Samal, S. K., Mohanty, S. and Nayak, S. K. (2019). Recent advancements in biopolymer and metal nanoparticle-based materials in diabetic wound healing management. International Journal of Biological Macromolecules, 122: 137–148.

Vilaplana, F., Strömberg, E. and Karlsson, S. (2010). Environmental and resource aspects of sustainable biocomposites. Polymer Degradation and Stability, 95(11): 2147–2161.

Walser, T., Limbach, L. K., Brogioli, R., Erismann, E., Flamigni, L., Hattendorf, B. et al. (2012). Persistence of engineered nanoparticles in a municipal solid-waste incineration plant. Nature Nanotechnology, 7(8): 520–524.

Wang, H., Yu, X., Su, C., Shi, Y. and Zhao, L. (2018). Chitosan nanoparticles triggered the induction of ROS-mediated cytoprotective autophagy in cancer cells. Artificial cells, Nanomedicine, and Biotechnology, 46(sup1): 293–301.

Wang, Q., Zhang, P., Zhao, W., Li, Y., Jiang, Y., Rui, Y. et al. (2023). Interplay of metal-based nanoparticles with plant rhizosphere microenvironment: Implications for nanosafety and nano-enabled sustainable agriculture. Environmental Science: Nano, 10(2): 372–392.

Wang, T. and Liu, W. (2022). Emerging investigator series: Metal nanoparticles in freshwater: Transformation, bioavailability and effects on invertebrates. Environmental Science: Parthasarathy Nano, 9(7): 2237–2263.

Wang, T., Bai, J., Jiang, X. and Nienhaus, G. U. (2012). Cellular uptake of nanoparticles by membrane penetration: a study combining confocal microscopy with FTIR Spectro electrochemistry. ACS Nano, 6(2): 1251–1259.

Worthington, K. L., Adamcakova-Dodd, A., Wongrakpanich, A., Mudunkotuwa, I. A., Mapuskar, K. A., Joshi, V. B. et al. (2013). Chitosan coating of copper nanoparticles reduces *in vitro* toxicity and increases inflammation in the lung. Nanotechnology, 24(39): 395101.

Yang, L., Zhou, Z., Song, J. and Chen, X. (2019). Anisotropic nanomaterials for shape-dependent physicochemical and biomedical applications. Chemical Society Reviews, 48(19): 5140–5176.

You, C., Li, Q., Wang, X., Wu, P., Ho, J. K., Jin, R. et al. (2017). Silver nanoparticle loaded collagen/chitosan scaffolds promote wound healing via regulating fibroblast migration and macrophage activation. Scientific Reports, 7(1): 10489.

Yu, J., Wang, D., Geetha, N., Khawar, K. M., Jogaiah, S. and Mujtaba, M. (2021). Current trends and challenges in the synthesis and applications of chitosan-based nanocomposites for plants: A review. Carbohydrate Polymers, 261: 117904.

Zahra, Z., Habib, Z., Hyun, S. and Sajid, M. (2022). Nanowaste: Another future waste, its sources, release mechanism, and removal strategies in the environment. Sustainability, 14(4): 2041.

Zavitri, N. G., Syahbaniati, A. P., Primastuti, R. K., Putri, R. M., Damayanti, S. and Wibowo, I. (2023). Toxicity evaluation of zinc oxide nanoparticles green synthesized using papaya extract in zebrafish. Biomedical Reports, 19(6): 1–11.

Zhang, N., Xiong, G. and Liu, Z. (2022). Toxicity of metal-based nanoparticles: Challenges in the nano era. Frontiers in Bioengineering and Biotechnology, 10: 1001572.

Zhao, J., Lin, M., Wang, Z., Cao, X. and Xing, B. (2021). Engineered nanomaterials in the environment: Are they safe? Critical Reviews in Environmental Science and Technology, 51(14): 1443–1478.

Zhu, D., Chen, Q. L., An, X. L., Yang, X. R., Christie, P., Ke, X. et al. (2018). Exposure of soil collembolans to microplastics perturbs their gut microbiota and alters their isotopic composition. Soil Biology and Biochemistry, 116: 302–310.

CHAPTER-4

Effect of Biopolymer Based Nanoparticles on Plant Growth and Health

M. K. Mohammad Ziaul Hyder, Syed Julkar Nine, Md. Akkas Uddin* and *Arup Kumer Roy*

1. Introduction

Agricultural sustainability faces numerous challenges, including soil degradation, nutrient deficiency, pest infestations, and the overuse of synthetic agrochemicals. These challenges not only hinder crop productivity but also pose environmental and health risks. Widespread plant diseases and soil degradation have led to a significant reliance on agrochemicals to ensure global food production. The controlled release of fertilizers and pesticides presents a promising approach to reducing environmental toxicity while enhancing the efficiency and reducing the quantity of agrochemical applications in agriculture (Machado et al., 2022). Nanotechnology, particularly through the use of nanoparticles, provides innovative solutions to these challenges. Nanoparticles can enhance the delivery of nutrients, pesticides, and growth regulators due to their high surface area, reactivity, and ability to penetrate plant tissues more effectively than conventional formulations (Riseh et al., 2024). Among the various types of nanoparticles, biopolymer-based nanoparticles are increasingly favored due to their biodegradability, biocompatibility, and reduced toxicity compared to synthetic counterparts.

Biopolymers, which are naturally derived polymers from renewable sources like plants, animals, and microorganisms, provide several advantages over synthetic polymers, including environmental safety, versatility in functionalization, and the ability to form nanostructures that can be customized for specific agricultural applications (An et al., 2022). Biopolymer-based nanoparticles can serve as carriers for nutrients, pesticides, and bioactive compounds, enabling targeted delivery and controlled release, thereby enhancing the efficiency of agrochemicals, and reducing their environmental impact (Fig. 1) (Machado et al., 2022). This approach aligns with sustainable agriculture principles, which seek to minimize negative farming impacts while maintaining or increasing crop yields.

Department of Chemistry, Chittagong University of Engineering & Technology, Chattogram-4349, Bangladesh.
* Corresponding author: ziaulhyder@cuet.ac.bd
DOI: 10.1201/9781003558330-4

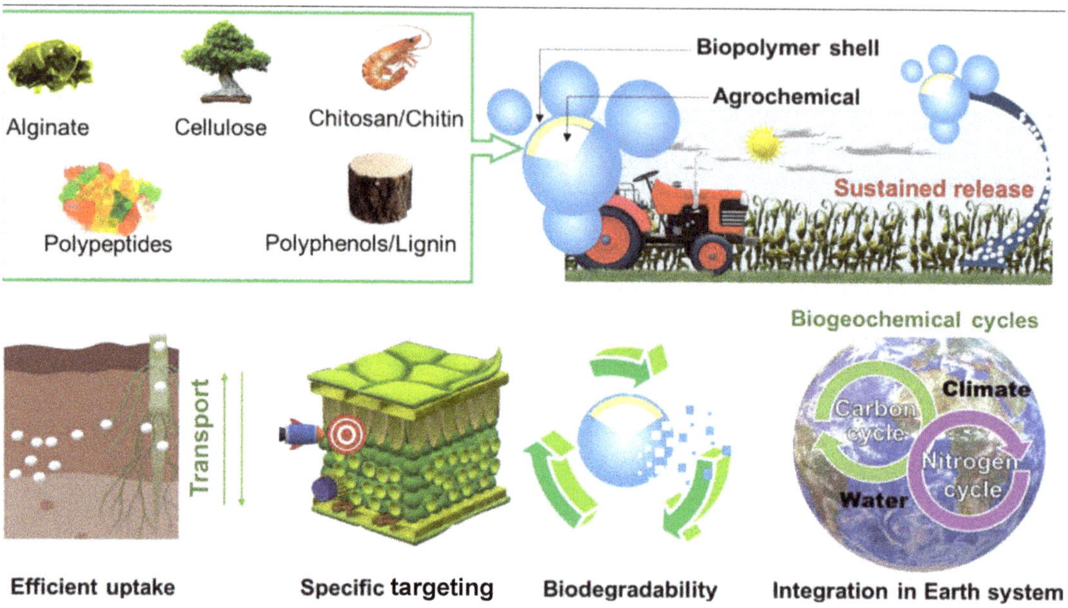

Figure 1. Biopolymer based nanoparticles in agriculture enabling targeted delivery and controlled release, thereby enhancing the efficiency of agrochemicals, and reducing their environmental impact. Permission with creative common from Machado et al., 2022.

1.1 Role of Biopolymers in Agriculture

Biopolymers, which include chitosan, alginate, starch, and other natural polymers, are gaining popularity in agriculture due to their biodegradable nature, low toxicity, and multifunctional benefits for plant growth and health. These natural polymers can significantly enhance plant growth by improving nutrient uptake, stimulating the production of growth hormones, and promoting seed germination. For instance, chitosan, a derivative of chitin, has been widely studied for its role in stimulating plant growth. It can activate the production of growth hormones such as auxins, gibberellins, and cytokinins, which are crucial for root and shoot development. Chitosan has been shown to enhance the growth of various crops, including tomatoes, beans, and wheat, by improving root growth and increasing biomass accumulation (Ingle et al., 2022). Additionally, biopolymers can modify the surface properties of seeds, thereby improving water absorption and protecting them from pathogens, which results in higher germination rates and more vigorous seedlings (Xu et al., 2020).

Biopolymers also play a significant role in enhancing plant health by inducing resistance to diseases and pests. Certain biopolymers, such as chitosan, can trigger Induced Systemic Resistance (ISR) in plants, which enhance their defensive responses against a wide range of pathogens. This resistance is mediated through the activation of plant defense pathways, such as the salicylic acid, jasmonic acid, and ethylene signaling pathways. These pathways lead to the production of pathogenesis-related proteins, phytoalexins, and other defensive compounds that help the plant resist infection (Kamle et al., 2020). The antimicrobial properties of chitosan further contribute to its protective effects, as it can directly inhibit the growth of harmful microorganisms, including bacteria, fungi, and viruses. Studies have shown that chitosan can effectively suppress the growth of Fusarium oxysporum in tomatoes and Pythium aphanidermatum in cucumbers, both of which are major plant pathogens (Mosa and El-Abeid, 2023).

In addition to improving plant growth and health, biopolymers positively impact soil health and microbial activity. When applied as soil amendments, biopolymers can improve soil structure by enhancing soil aggregation and stability. This results in better aeration, water retention, and root penetration, which are essential for optimal plant growth. Alginate, a biopolymer derived from

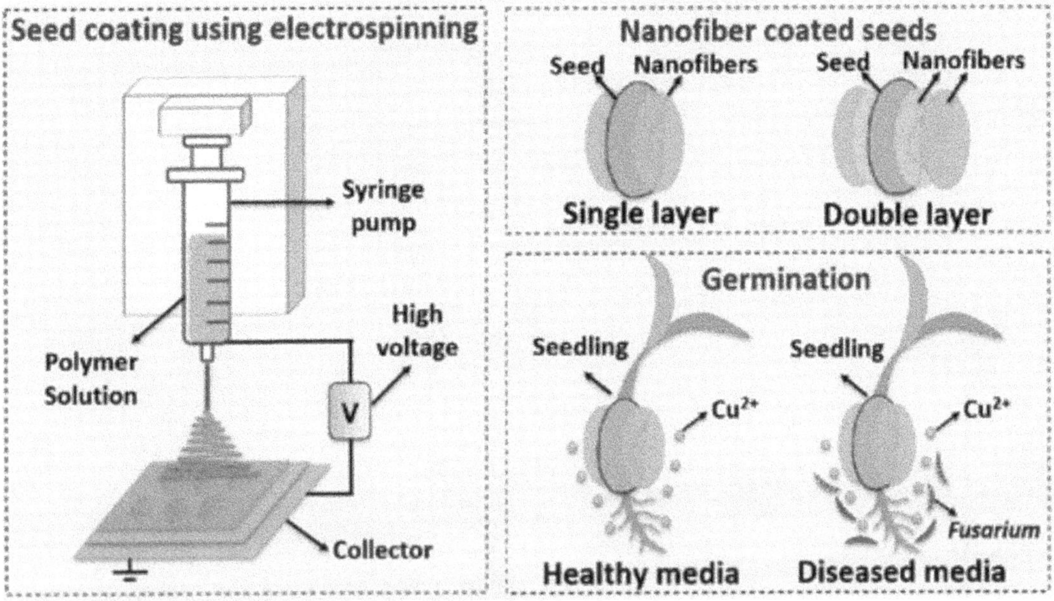

Figure 2. Enhanced seedling development with biodegradable, tunable, biopolymer-based nanofiber seed coatings (Xu et al., 2020). Permission from American Chemical Society 2020.

brown seaweed, is known for its ability to form hydrogels that improve soil water retention. This property is particularly beneficial in arid and semi-arid regions, where water availability is a limiting factor for crop production (El-Diehy et al., 2024).

The sustainable nature of biopolymers further amplifies their value in agriculture. Being biodegradable and environmentally friendly, biopolymers offer a sustainable alternative to synthetic chemicals and polymers, which often have adverse environmental impacts. By enhancing plant growth and health naturally, biopolymers can reduce the reliance on chemical fertilizers and pesticides, thereby lowering the environmental footprint of agricultural practices (Gamage et al., 2022). For example, the use of chitosan as a biostimulant in organic farming systems has been shown to improve crop yields and quality while reducing the need for synthetic inputs. The environmental benefits of biopolymers extend beyond their direct effects on plants; their biodegradability ensures that they do not accumulate in the environment, thus preventing soil and water pollution.

Overall, biopolymers represent a promising approach to enhancing plant growth and health while promoting sustainable agricultural practices. Their ability to improve nutrient uptake, enhance disease resistance, support soil health, and mitigate abiotic stress makes them versatile tools for modern agriculture. As research continues to uncover new applications and mechanisms of action, biopolymers are likely to play an increasingly important role in the development of sustainable and resilient agricultural systems.

1.2 Role of Nanoparticles in Agriculture

On the other hand, nanoparticles have emerged as a promising tool in agriculture, offering potential benefits for enhancing plant growth and health. These tiny particles, typically ranging from 1 to 100 nanometers in size, have unique physical, chemical, and biological properties that make them highly effective in various agricultural applications. Their high surface area-to-volume ratio, reactivity, and ability to penetrate plant tissues enable them to influence plant physiology in ways that conventional fertilizers and pesticides cannot. Nanoparticles, including metal and metal oxide nanoparticles such as silver, zinc oxide, titanium dioxide, and iron oxide, have been extensively studied for their potential to improve nutrient delivery, enhance stress tolerance, and protect plants against diseases (Fig. 3) (Raza et al., 2024). However, while nanoparticles offer numerous advantages, their impact on plant

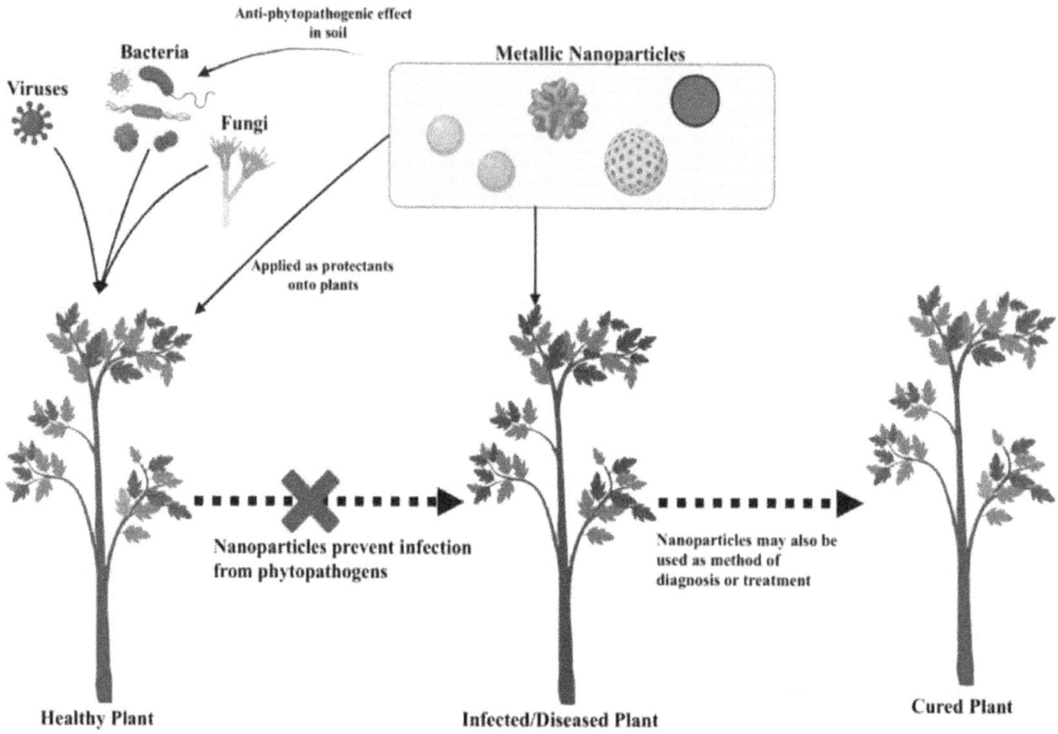

Figure 3. The potential of metallic and metal oxide nanoparticles for reinforced disease management in agricultural systems (Raza et al., 2024). Permission from Elsevier.

health can vary significantly depending on their type, concentration, and method of application, as well as the specific plant species being treated.

One of the most significant positive effects of nanoparticles on plants is their ability to enhance nutrient delivery and uptake. Nanoparticles can be engineered to act as carriers for essential nutrients, improving their bioavailability and absorption by plant roots and leaves. For instance, zinc oxide nanoparticles have been shown to enhance zinc uptake in plants, leading to improved growth and development (Faizan et al., 2021). Zinc is a critical micronutrient involved in various plant metabolic processes, including enzyme activation, protein synthesis, and hormone regulation. By increasing zinc availability, nanoparticles can help alleviate deficiencies that would otherwise limit plant growth. Similarly, iron oxide nanoparticles have been found to enhance chlorophyll synthesis and photosynthetic efficiency in plants, resulting in greater biomass production (Rizwan et al., 2019). The high solubility and reactivity of nanoparticles allow them to release nutrients more efficiently than conventional fertilizers, providing a steady supply of essential elements over time.

In addition to enhancing nutrient uptake, nanoparticles can also improve plant resilience to abiotic stresses such as drought, salinity, and extreme temperatures. Under drought conditions, nanoparticles like silicon dioxide and titanium dioxide have been shown to improve water use efficiency and reduce transpiration rates, helping plants conserve water (Naseer et al., 2022). Silicon dioxide nanoparticles, for example, can enhance the expression of aquaporin proteins that regulate water transport in plant cells, thus improving the plant's ability to retain water during periods of low moisture availability. Similarly, titanium dioxide nanoparticles can increase the antioxidant capacity of plants, protecting them from oxidative stress caused by environmental challenges. By enhancing the activity of antioxidant enzymes like superoxide dismutase, catalase, and peroxidase, these nanoparticles help mitigate the damaging effects of Reactive Oxygen Species (ROS), thereby improving plant survival and growth under stress conditions (Wang et al., 2023).

Nanoparticles also exhibit strong antimicrobial properties that can protect plants from a variety of pathogens, including bacteria, fungi, and viruses. Silver nanoparticles, in particular, are well-known for their broad-spectrum antimicrobial activity and have been widely used in agriculture to control plant diseases (Hernández-Díaz et al., 2021). These nanoparticles can disrupt the cell membranes of pathogens, inhibit their metabolism, and ultimately lead to cell death. This protective effect not only reduces the incidence of disease but also minimizes the need for chemical pesticides, which can have harmful environmental impacts. Furthermore, the small size of nanoparticles allows them to penetrate plant tissues more effectively, providing a more thorough and long-lasting defense against infections. Copper nanoparticles have similarly demonstrated efficacy in controlling fungal pathogens, such as those responsible for downy mildew and powdery mildew, by generating ROS that inhibit fungal growth (El-Beltagi et al., 2022).

However, the use of nanoparticles in agriculture is not without challenges, and their impact on plant health can be complex. While low to moderate concentrations of nanoparticles generally promote plant growth and stress tolerance, high concentrations can lead to toxicity and adverse effects. For example, excessive levels of silver nanoparticles can generate ROS within plant tissues, leading to oxidative stress, membrane damage, and impaired photosynthesis (Rastogi et al., 2019). This toxicity is often dose-dependent, with the severity of the impact increasing with higher nanoparticle concentrations. Additionally, the size, shape, and surface charge of nanoparticles play critical roles in determining their phytotoxicity. Smaller nanoparticles with higher surface reactivity are more likely to interact with plant cell components, potentially causing greater damage. Moreover, the accumulation of nanoparticles in plant tissues can interfere with essential physiological processes, such as nutrient uptake and hormone signaling, leading to stunted growth and reduced crop yields.

The environmental impact of nanoparticles also raises concerns, particularly regarding their persistence and potential accumulation in soil and water ecosystems. Although some nanoparticles are designed to be biodegradable, many metal and metal oxide nanoparticles are not, and their long-term environmental fate remains poorly understood. Accumulation of nanoparticles in the soil can disrupt soil microbiota, which are essential for nutrient cycling and maintaining soil health (Sun et al., 2021). For instance, high concentrations of zinc oxide nanoparticles have been shown to inhibit the growth of beneficial soil bacteria, thereby affecting nitrogen fixation and other vital soil processes (Sheteiwy et al., 2021). Furthermore, nanoparticles can be taken up by non-target organisms, including beneficial insects and aquatic species, leading to unintended ecological consequences. This highlights the need for careful management of nanoparticle applications in agriculture to avoid environmental contamination and ensure the safety of non-target organisms.

In conclusion, nanoparticles provide significant potential for enhancing plant growth, nutrient uptake, stress tolerance, and disease resistance, making them a valuable tool in sustainable agriculture. However, their use must be carefully managed to balance the benefits with potential risks to plant health and the environment. Ongoing research is needed to optimize nanoparticle formulations, determine safe and effective application rates, and understand the long-term effects of nanoparticles on plants and ecosystems. By addressing these challenges, it may be possible to fully harness the potential of nanoparticles to improve agricultural productivity in a sustainable and environmentally friendly manner.

1.3 Classification of Biopolymer Based Nanoparticles

Biopolymer-based nanoparticles can be classified into several categories based on the type of biopolymer used, the nature of the incorporated nanomaterial, and their specific agricultural applications. These primary categories include biopolymeric nanoparticles, biopolymer nanocomposites, biopolymer-based metal nanoparticles, biopolymer-based metal oxide nanoparticles.

1.3.1 Biopolymeric nanoparticles

Biopolymeric nanoparticles are nanoparticles entirely composed of biopolymers without the addition of other materials like metals or oxides. They are synthesized from biopolymers such as chitosan, alginate, cellulose, and proteins, providing a sustainable, biodegradable, and biocompatible platform for agricultural applications. These nanoparticles are used for the controlled release of nutrients, pesticides, and plant growth regulators, improving their efficacy while minimizing environmental impacts. For example, nanoparticles made from starch or cellulose can be engineered to encapsulate fertilizers or pesticides, releasing them slowly in response to environmental conditions like pH or moisture levels (Fig. 4) (Salimi et al., 2023). Biopolymeric nanoparticles are typically synthesized through methods such as ionic gelation, emulsion cross-linking, and nanoprecipitation, which allow for the fine-tuning of particle size, shape, and surface properties to suit specific agricultural needs.

Chitosan, a polysaccharide derived from chitin, is one of the most extensively studied biopolymers for nanoparticle synthesis. Chitosan-based nanoparticles exhibit excellent biocompatibility, biodegradability, and antimicrobial properties, making them suitable for protecting plants against various pathogens (Yu et al., 2021). These nanoparticles can enhance seed germination, stimulate plant growth, and increase crop yield by improving nutrient uptake and promoting plant defense mechanisms against biotic stresses (Bandara et al., 2020). Chitosan nanoparticles can be synthesized using methods such as ionic gelation, emulsion cross-linking, and nanoprecipitation, allowing for customization of particle size and surface properties to suit specific agricultural needs.

Alginate, a polysaccharide extracted from brown seaweed, is known for its gel-forming ability, and is widely used to encapsulate active ingredients, making it ideal for controlled release formulations. Alginate-based nanoparticles are used to deliver nutrients, fertilizers, and bioactive compounds in a slow-release manner, reducing the frequency of applications and minimizing nutrient leaching into the environment (Haydar et al., 2024). This controlled release mechanism ensures that plants receive a steady supply of nutrients, enhancing growth and reducing environmental impact.

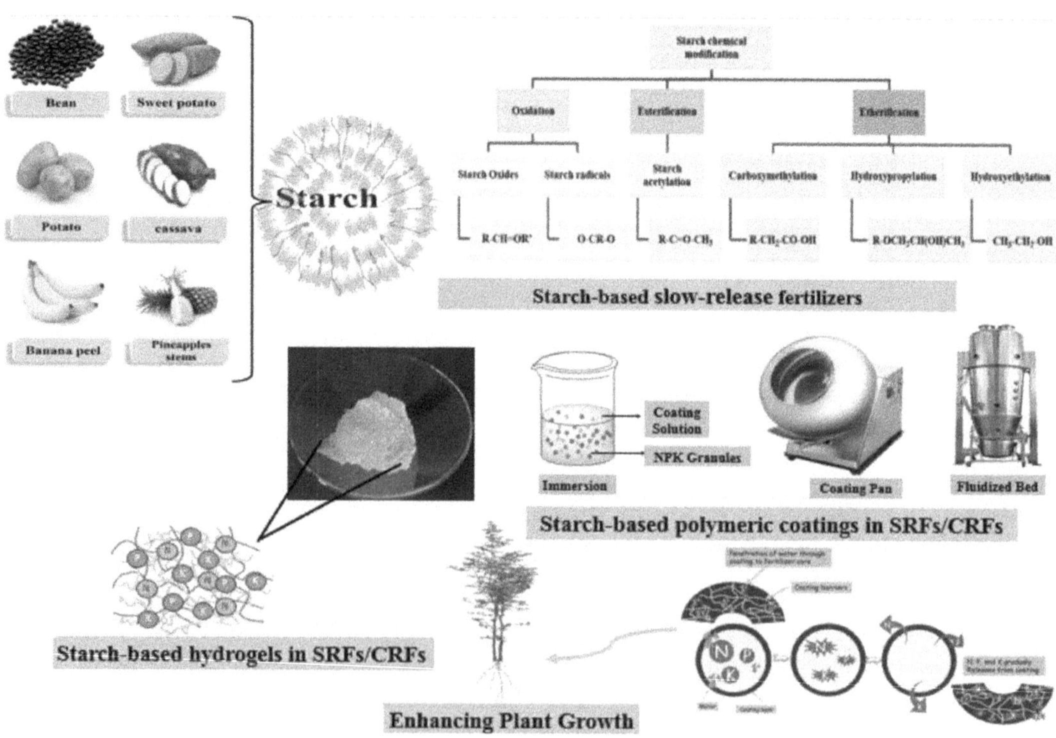

Figure 4. Starch based polymeric nanoparticles in slow/controlled-release fertilizers in agriculture (Salimi et al., 2023). With the permission from Elsevier.

Alginate nanoparticles are typically prepared using ionic cross-linking with calcium ions, which stabilizes the nanoparticles and enhances their functionality as delivery vehicle.

Starch, a naturally occurring polysaccharide derived from crops like corn, potatoes, and rice, is a promising material for nanoparticle synthesis due to its biodegradability, biocompatibility, and ease of modification. Starch-based nanoparticles are used in agriculture for the controlled delivery of micronutrients and pesticides (Salimi et al., 2023). Their surface can be modified to enhance loading capacity and release rates, making them versatile carriers for various agrochemicals. Starch nanoparticles improve nutrient use efficiency, enhance plant growth, and provide sustained release of active ingredients, thereby reducing the overall input of chemicals in agriculture.

Cellulose, the most abundant biopolymer on earth, derived from plant cell walls, is used for production of cellulose nanocrystals (CNCs) and cellulose nanofibers (CNFs). These cellulose-based nanoparticles are highly crystalline, have a large surface area, and exhibit excellent mechanical properties, making them suitable for reinforcing materials and as carriers for agrochemicals. In agriculture, cellulose-based nanoparticles can be used to improve soil mechanical stability, enhance water retention, and deliver nutrients and pesticides in a controlled manner (Skrzypczak et al., 2023). The high surface area and reactive hydroxyl groups allow for functionalization with various bioactive compounds, making cellulose nanoparticles effective tools for targeted delivery in precision agriculture. For example, Cellulose-based Nanocarriers (NCs) offer a promising solution for sustainable agriculture, serving as effective vehicles for the stimuli-responsive release of agrochemicals. These NCs can specifically target cellulase-secreting fungi, which are responsible for serious plant diseases like Apple Canker, making them a valuable tool in combating such diseases. Their ability to respond to specific environmental triggers enhances the precision of agrochemical delivery, reducing unnecessary chemical usage and promoting more sustainable farming practices (Fig. 5) (Machado et al., 2021).

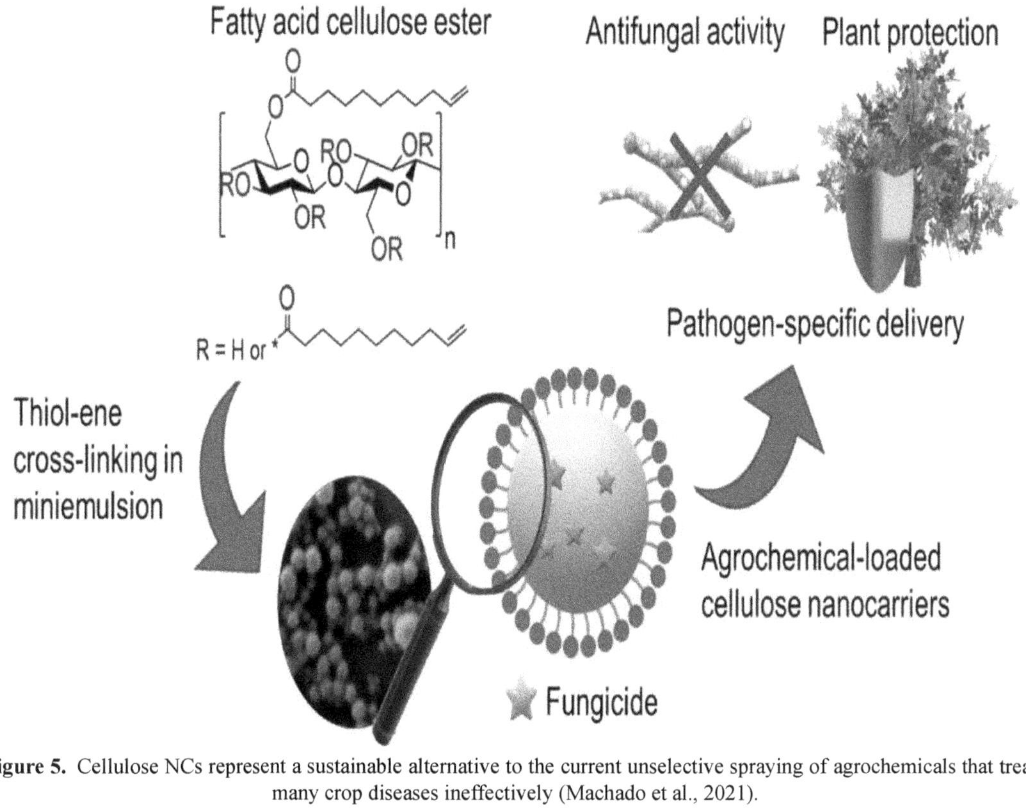

Figure 5. Cellulose NCs represent a sustainable alternative to the current unselective spraying of agrochemicals that treats many crop diseases ineffectively (Machado et al., 2021).

Proteins such as zein, casein, and silk fibroin are increasingly explored for nanoparticle synthesis due to their biocompatibility, biodegradability, and ability to interact with a wide range of bioactive molecules. Protein-based nanoparticles can encapsulate and protect sensitive agrochemicals from degradation, ensuring their stability and effectiveness in the field (Jiang et al., 2024). For example, zein nanoparticles, derived from corn, have been used to encapsulate essential oils and pesticides, providing a slow and sustained release that enhances pest control efficacy and reduces the need for frequent applications (Banerjee et al., 2022). The ability of protein nanoparticles to form stable colloids in aqueous environments further enhances their applicability in agricultural formulations.

1.3.2 Biopolymer nanocomposites

Biopolymers are natural polymer-based macromolecules composed of regularly repeating subunits that are found in living organisms and natural sources. Biopolymers contain essential biomolecules, such as proteins, nucleic acids, lipids, and carbohydrates, which offer versatility across numerous applications, particularly in biomedicine, agricultural, and food sectors. Moreover, due to being environmentally friendly, biopolymers have gained attention to mitigate sustainability concerns by minimizing dependence on synthetic petrochemical-based materials. Biopolymers can also play a pivotal role by reducing greenhouse gas emissions, particularly carbon dioxide, by enhancing agricultural productivity.

Biopolymer nanocomposites consist of a biopolymer matrix reinforced with nanoparticles or nanofillers, such as clay, carbon nanotubes, or other nanoparticles, to enhance their mechanical, thermal, and barrier properties. Biopolymer-based nanocomposites have the potential to serve as slow-release carriers for agrochemicals like fertilizers or pesticides, enhancing crop yields. Likewise, biopolymer-based nanofilms and hydrogels can be used as coatings to prolong the product shelf life, enhance seed germination, or protect against pests and pathogens. In food packaging, biopolymers offer significant promise, but their properties, including mechanical strength and resistance to gas, water, or microbes, can be significantly enhanced by incorporating nanofillers such as nanoparticles (Wypij et al., 2023). These nanocomposites are utilized in agriculture for applications such as slow-release fertilizers, encapsulation of pesticides, and soil conditioning agents (Fig. 6) (Feba Mohan and Praseetha, 2023).

Incorporating nanoparticles into the biopolymer matrix enhances the composite's strength, durability, and functionality, making them ideal for controlled release systems. For example,

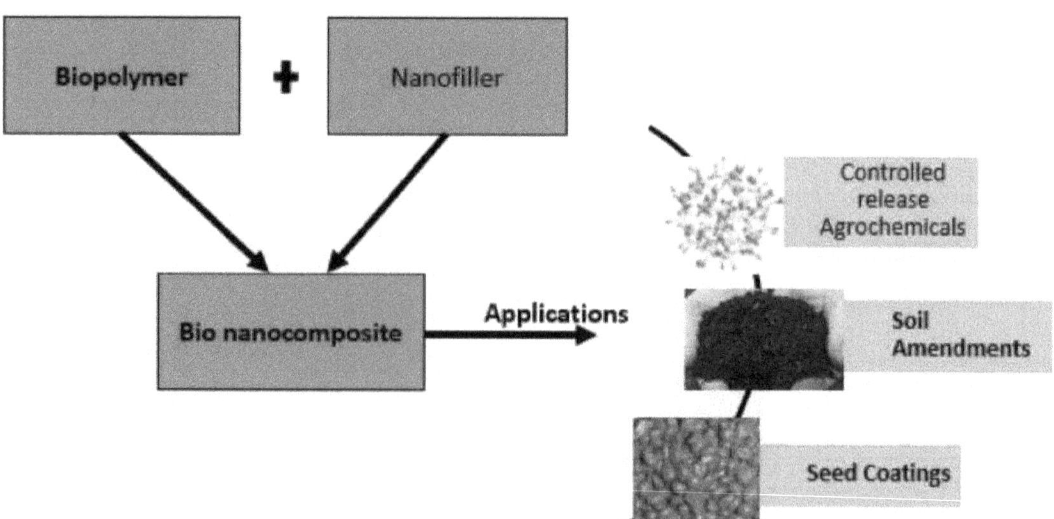

Figure 6. Prospects of biopolymers-based nanocomposites for the slow and controlled release of agrochemicals formulations (Feba Mohan and Praseetha, 2023). Perrmission from Springer.

chitosan-clay nanocomposites have been developed for the controlled release of pesticides, reducing application frequency and environmental contamination (Mishra et al., 2023). The development of biopolymer nanocomposites offers multifunctional materials that address plant nutrition, pest control, and soil health simultaneously.

1.3.3 Biopolymer-based metal nanoparticles

In recent years, the fusion of biotechnology and nanotechnology has gained growing attention in sustainable agriculture to explore innovative strategies for enhancing plant growth and health (Prasad et al., 2017). The integration of natural-derived biopolymers with metal-based nanoparticles (BMPs) can potentially revolutionize crop management strategies (Dhlamini et al., 2022). They also offer potential benefits ranging from increased yield and nutrient absorption to enhanced resistance against environmental stressors. The biopolymer-metal based nanoparticles (BPMNs) have drawn significant interest and as a promising candidate in agricultural research, owing to their biodegradability, environmentally friendly, and abundance in nature. Metal-based nanoparticles (MNPs), on the other hand, exhibit unique properties compared to their bulk counterparts due to their tiny size and high surface area (Ehsan et al., 2023). MNPs are common to use in catalysis, electronics, optics, healthcare, and agricultural industries. MNPs have also garnered particular interest in medical diagnostics, drug delivery, and environmental remediation. Currently, many studies suggest that MNPs also help to enhance plant growth and development with a minimal dose which leads to substantial benefits both economically and environmentally (Chen, 2018). Various metal nanoparticles, namely gold (Au), silver (Ag), manganese (Mn), iron (Fe) are commonly used for the development of plant growth and their health conditions (Rico et al., 2015). MNPs promote nutrient intake by sequestration of nutrients to plant roots and the root junctions through plant tissue (Rasheed et al., 2022; Rastogi et al., 2017). For instance, due to photocatalytic activity, at low concentration of TiO_2, increase the photosynthesis rate and enhance the plant immunity, fruit yield as well as chlorophyll concentration by penetrating plant tissues. Otherwise, by intaking lower concentration of AgNPs from the soil, plants exhibit positive responses in the growth of both roots and shoots across various plant species (Khan et al., 2023). AgNPs as well as CuNPs also possess antiviral, antibacterial and antifungal activities, which protect plants from detrimental microorganisms (Vatankhah et al., 2023). Although, MNPs are most preferable for plant germination and development but very few studies have highlighted the adverse effects of metal nanoparticles on plants in the form of decrease in plant growth, productivity, and pigments levels (Da Costa and Sharma, 2016; Gupta and Xie, 2018; Rastogi et al., 2017).

Among the BPMNs, natural polysaccharide-based metal nanoparticles have been extensively investigated in plant and agriculture research (Banerjee et al., 2017; Taher et al., 2022; Wang et al., 2017). Polysaccharides, such as chitosan, cellulose, starch, alginate, etc., contribute to the reduction and stabilization of metal nanoparticles through their unique functional groups. In this instance, polysaccharide act as both reducing and stabilizing agents to host the MNPs through electrostatic interactions (Fernandes et al., 2022). These fusion metallic nanoparticles help to improve the efficiency of nutrient absorption by plant roots and serving as essential nutrients carriers to develop overall plant health. For example, the incorporation of metallic nanoparticles into the chitosan matrix plays an essential role in combating pesticide in the agricultural sector. Additionally, the incorporation of various MNPs, such as Ca, Cu and Zn into chitosan act as nanocarriers and enhance the plant growth by inhibiting antifungal agents (Choudhary et al., 2019; Saharan et al., 2015; Yáñez et al., 2023).

Biopolymer-based metal nanoparticles are synthesized by reducing metal ions (such as silver, gold, or copper) in the presence of a biopolymer matrix, which stabilizes and prevents nanoparticle agglomeration. These nanoparticles are notable for their unique properties, including antimicrobial activity, catalytic efficiency, and enhanced delivery capabilities. In agriculture, biopolymer-based metal nanoparticles like chitosan-silver and alginate-gold nanoparticles are used to combat plant pathogens, promote seed germination, and enhance plant growth. For instance, chitosan-silver

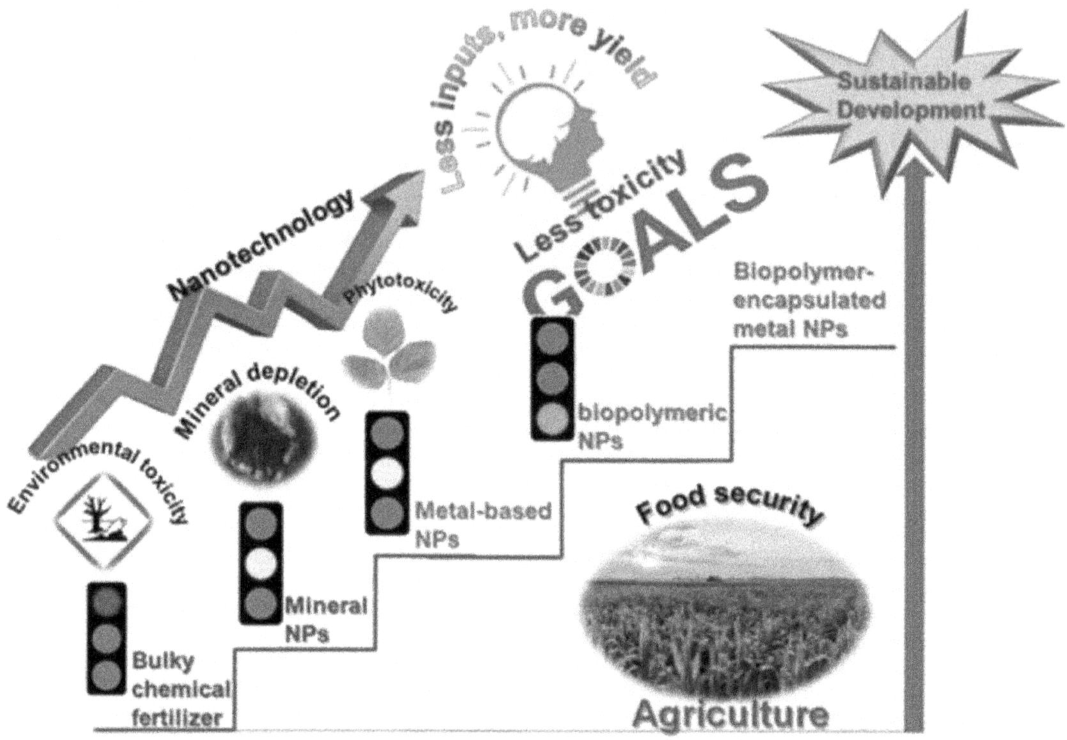

Figure 7. Nano-engineering metal-based fertilizers using biopolymers: An innovative strategy for a more sustainable agriculture (Dhlamini et al., 2022).

nanoparticles exhibit potent antifungal activity against common plant pathogens, reducing the reliance on synthetic fungicides (Alghuthaymi, et al., 2020). The biopolymer matrix not only stabilizes the metal nanoparticles but also enhances their biocompatibility, making them safer for agricultural use.

In conclusion, BPMNs represent a paradigm shift in agricultural technology and offer a potent combination of biocompatibility, functionality, and sustainability. As we strive to meet the growing global demand for food security and environmental stewardship, these BPMNs hold immense promise as transformative tools for revolutionizing modern agriculture (Fig. 7) (dhlamini et al., 2022). With ongoing research and innovation, their full potential in unlocking the full productivity and resilience of agricultural systems is yet to be fully realized.

1.3.4 Biopolymer-based metal oxide nanoparticles

The global population has increased fourfold in the past century, and this surge is projected to lead to a 70% rise in worldwide food demand by the year 2050 (Ryder, 2017). Therefore, the growing demand for food production necessitates sustainable agriculture practices. The synergistic utilization of bioplomer based metal oxide NPs represents a revolutionary approach to realizing sustainable and resource-efficient agricultural systems. Bioplomer-based metal oxide NPs have been designed for applications such as antimicrobial activity, pest control, and weed inhibition, offering potential for addressing biotic stresses, disease detection and management, stress tolerance, weed control, and improving crop yields. Metallic NPs can repel pests, exhibit larvicidal and ovicidal effects, fight plant pathogens, deliver agrochemicals with precision, and inhibit weed growth, thereby enhancing agricultural productivity (Rajpal et al., 2024).

In this regard, incorporating metal oxide nanoparticles into the biopolymer matrices may improve the plant's health and nutrition as well as the environment. The biopolymer-based metal oxide nanoparticles also showed exceptional adaptability for specific agricultural applications with tuning their desirable properties including improved mechanical strength, stability, and functionalities of plants (Paul and Sharma, 2023). Moreover, this hybrid nanocomposite may also increase crop productivity, manage pests as well as disease management by enhancing soil health, and stress tolerance to plants.

Biopolymer-based metal oxide nanoparticles, such as zinc oxide (ZnO), titanium dioxide (TiO_2), and iron oxide (Fe_3O_4) nanoparticles, are synthesized using biopolymers as stabilizing agents. These nanoparticles are highly valued in agriculture for their photocatalytic properties, ability to enhance nutrient uptake, and potential to protect plants from abiotic stressors such as UV radiation and drought. For example, alginate-stabilized ZnO nanoparticles have been shown to improve wheat and maize growth by enhancing photosynthesis and nutrient absorption (Singh et al., 2024). The biopolymer matrix stabilizes the metal oxide nanoparticles, controls their release, and reduces their potential toxicity to plants and soil microorganisms. Additionally, biopolymer-based metal oxide nanoparticles can function as fertilizers, providing essential micronutrients like zinc and iron directly to plants in a controlled manner, thus enhancing plant growth and reducing the need for traditional fertilizers. For example, ZnO nanoparticles (ZnO NPs) encapsulated in alginate beads, cross-linked with CaCl2, exhibited the slowest zinc release kinetics. When tested on maize plants grown in zinc-deficient acidic soil (LUFA 2.1, pH 5.2), the consistent zinc release from these ZnO NPs/alginate beads led to a more stable zinc concentration in soil pore water over time (Fig. 8). The findings showed that ZnO NPs/alginate beads effectively met the zinc requirements of maize without causing early-stage zinc toxicity, which is often associated with traditional zinc supplements. This suggests that engineered nanomaterials (ENMs) offer a sustainable method for delivering zinc in a controlled manner in acidic soils. Additionally, the study explored the influence of maize root exudates on zinc bioavailability in the rhizosphere, emphasizing the importance of understanding micronutrient dynamics in the rhizosphere to better predict long-term zinc availability in bulk soils (Martin et al., 2020).

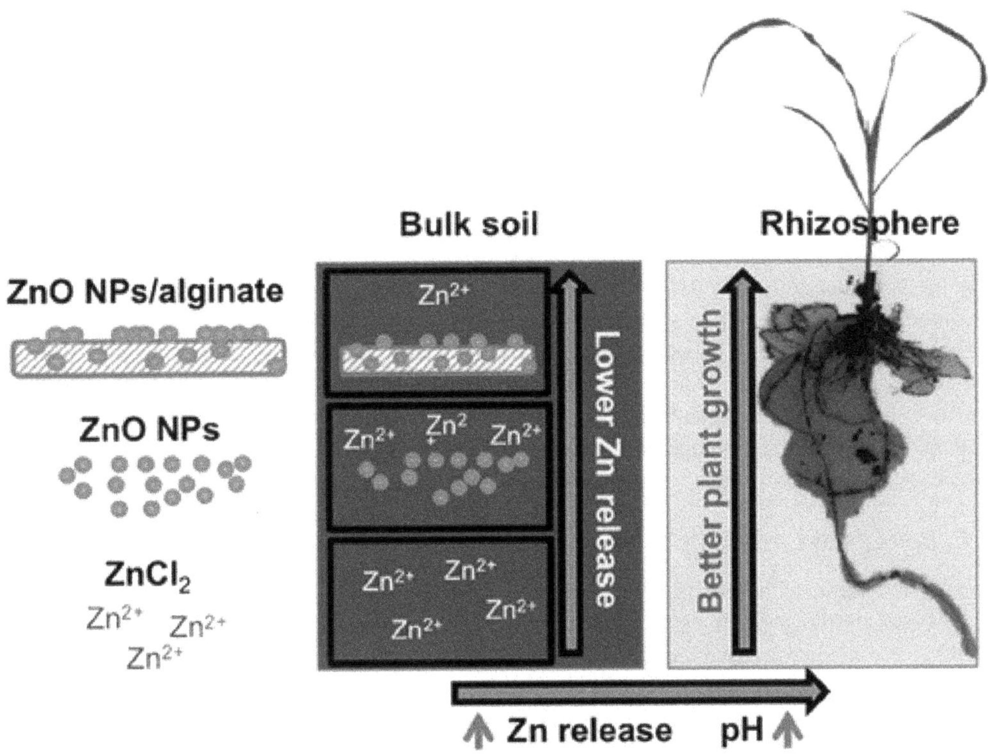

Figure 8. Composites of biopolymers and ZnO NPs for controlled release of zinc in agricultural soils and timed delivery for maize (Martin et al., 2020)

In conclusion, biopolymeric-metal and metal oxide nanoparticles hold great promise for enhancing plant growth, nutrient uptake, stress tolerance, and disease resistance, providing a sustainable approach to modern agriculture. However, their application requires careful management to balance the advantages with potential risks to plant health and the environment. Future research should focus on optimizing formulations, understanding species-specific responses, and evaluating the long-term environmental impacts to fully harness the potential of these innovative nanomaterials in agriculture.

2. Mechanism of Nanoparticle-Plant Interaction

The interaction between nanoparticles (NPs) and plants is a complex process that involves several mechanisms of uptake, translocation, and accumulation. Understanding these mechanisms is crucial for optimizing the use of nanoparticles in agriculture, particularly for enhancing nutrient delivery, improving stress tolerance, and providing protection against pathogens. The uptake of nanoparticles by plants is influenced by several factors, including the size, shape, surface charge, and chemical composition of the nanoparticles, as well as the type of plant species and the environmental conditions. Generally, nanoparticles can enter plants through roots, leaves, and even seeds, depending on the application method, such as soil amendment, foliar spray, or seed priming.

2.1 Root Uptake of Nanoparticles

Root uptake is one of the primary pathways through which nanoparticles enter plants. The process begins with the adsorption of nanoparticles onto the root surface, facilitated by the high surface area and reactivity of the nanoparticles, which can interact with root exudates and cell wall components. Nanoparticles can penetrate root tissues primarily through three routes: apoplastic, symplastic, and transcellular pathways. The apoplastic pathway involves the movement of nanoparticles along the cell walls and intercellular spaces without crossing cell membranes. This route allows nanoparticles to reach the root cortex but is generally restricted by the Casparian strip in the endodermis, which acts as a selective barrier (Schwab et al., 2016). In contrast, the symplastic pathway involves the entry of nanoparticles into root cells through the plasma membrane, followed by their transport from cell to cell via plasmodesmata, which are intercellular channels that connect the cytoplasm of adjacent cells. The transcellular pathway, although less common, involves the endocytosis of nanoparticles by root cells, followed by their transport within vesicles through the cell cytoplasm.

The size and surface properties of nanoparticles significantly influence their uptake efficiency. Smaller nanoparticles, typically less than 20 nm, are more readily taken up by roots due to their ability to navigate the apoplastic and symplastic pathways more efficiently. For instance, zinc oxide nanoparticles of smaller size have been shown to penetrate root tissues more effectively than their larger counterparts, enhancing zinc uptake and translocation in plants Fig. 9 (Yusefi-Tanha et al., 2023). Surface charge also plays a critical role, with positively charged nanoparticles generally exhibiting higher uptake rates due to their electrostatic attraction to the negatively charged components of the root cell walls. However, excessive positive charge can also lead to nanoparticle aggregation, which may reduce uptake efficiency.

2.2 Foliar Uptake of Nanoparticles

Nanoparticles can also be absorbed through the leaves when applied as foliar sprays. The foliar uptake of nanoparticles involves their adhesion to the leaf surface, penetration through the cuticle, and entry into the epidermal cells. The cuticle, a hydrophobic layer covering the epidermis, serves as the first barrier to foliar uptake. Nanoparticles can penetrate the cuticle through three main mechanisms: diffusion through the cuticular waxes, entry through stomatal openings, and penetration via trichomes or leaf hairs. Stomata, the pores on the leaf surface that facilitate gas exchange, are particularly significant entry points for nanoparticles due to their size and relatively

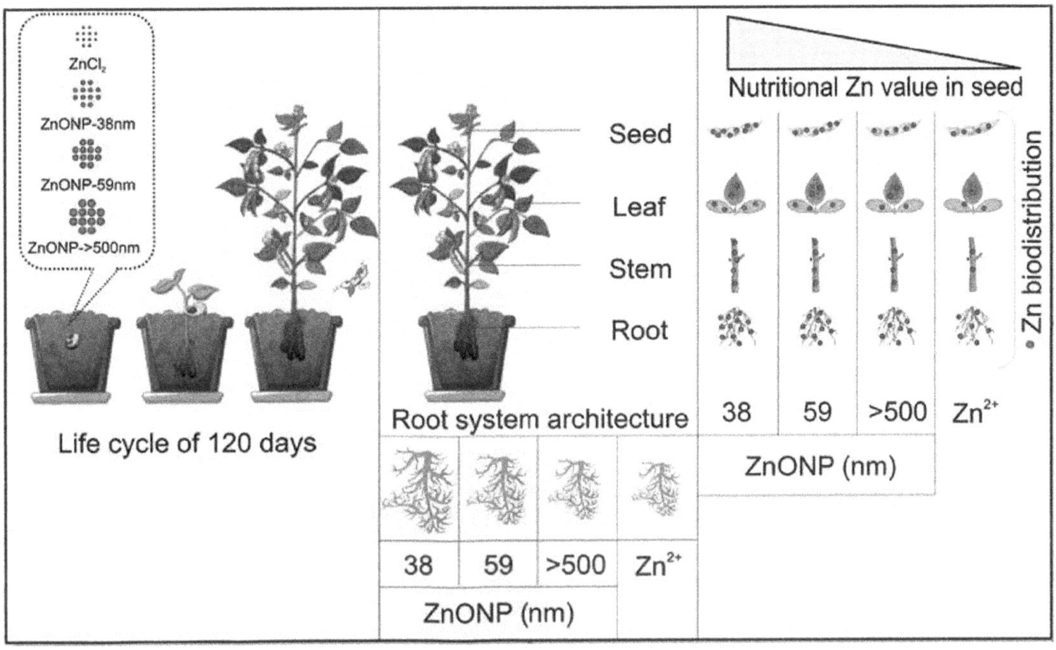

Figure 9. Responses of soybean (*Glycine max* [L.] Merr.) to zinc oxide nanoparticles: Understanding changes in root system architecture, zinc tissue partitioning and soil characteristics (Yusefi-Tanha et al., 2022).

open structure. Studies have shown that nanoparticles can enter leaves through the stomata and then translocate into the mesophyll tissues, where they can be distributed throughout the plant via the vascular system (Wang et al., 2023).

Once inside the leaf tissue, nanoparticles can move apoplastically through cell walls or symplastically through the cytoplasm, depending on their size and surface properties. The presence of surfactants or other additives in nanoparticle formulations can enhance foliar uptake by altering the surface tension and wettability of the leaf surface, facilitating better adhesion and penetration (Pimentel et al., 2023). Moreover, the shape of nanoparticles can influence their interaction with leaf tissues; for example, rod-shaped nanoparticles may penetrate more efficiently than spherical ones due to their ability to align along the cell walls and navigate the extracellular spaces more effectively.

Chitosan nanocarriers are often applied as foliar sprays, which allows for direct uptake through the leaves. This method is particularly effective for delivering nutrients and PGRs to support plant growth (Kumaraswamy et al., 2021; Sangwan et al., 2023) showed that foliar application of chitosan nanoparticles on rice plants resulted in enhanced growth parameters, including plant height, leaf area, and grain yield.

2.3 Translocation and Distribution within Plants

After uptake, the translocation of nanoparticles within the plant occurs primarily through the vascular system, which includes the xylem and phloem. The xylem is responsible for the upward transport of water and solutes from the roots to the shoots and leaves, while the phloem distributes photosynthates and other nutrients from the leaves to other parts of the plant. Nanoparticles taken up by roots can be loaded into the xylem vessels through the root endodermis, where they are translocated to aerial parts of the plant via the transpiration stream. The efficiency of xylem loading and transport depends on the size and aggregation state of the nanoparticles; smaller, well-dispersed nanoparticles are more readily translocated compared to larger aggregates, which may be filtered out or immobilized within the root tissues (Huang et al., 2022).

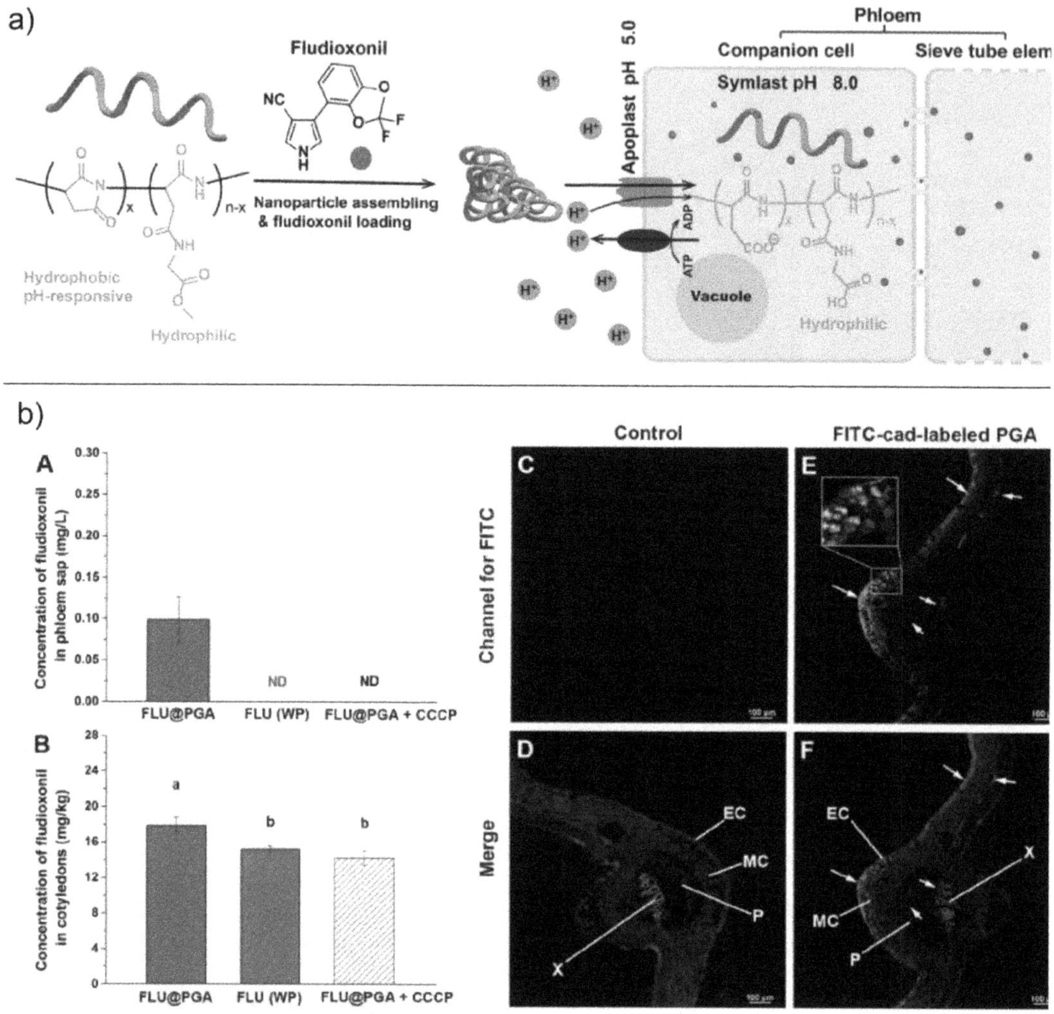

Figure 10. (a) Schematic Illustration of the PGA Nanoparticles as Targeted Nanocarriers of the Fungicide FLU and the Proposed Carrier-Mediated Transport to Plant Phloem. (b) Uptake and phloem transport of the PGA nanoparticles in castor beans. (A) Concentration of FLU in phloem sap. The data are presented as the mean ± SE with 8 plants for each treatment (n = 8). (B) Concentration of FLU in cotyledon tissues at the end of the experiment represented as mean ± SE (n = 8). The Kruskal–Wallis test was used to assess significant differences at the P = 0.05 level, and different letters indicated significant differences. ND, not detected. (C and D) cross sections of midveins were taken from cotyledons that were incubated without (control) 100 mg/L FITC-cad-labeled PGA nanoparticles. E and F: Cotyledons that were incubated with 100 mg/L of FITC-cad-labeled PGA nanoparticles. The arrowheads indicate the presence of FITC- cad-labeled PGA nanoparticles. EC: epidermal cells; MC: mesophyll cells; P: phloem; X: xylem. Scale bar = 100 μm (Fig. 10) (Machado et al., 2022).

Nanoparticles can also move through the phloem, although this process is generally less efficient compared to xylem transport. Phloem translocation is more selective, as it involves the active loading of solutes into the sieve elements, which are the main conduits of the phloem (Fig. 11) (Wu et al., 2021). This process can facilitate the downward movement of nanoparticles from leaves to roots and other growing tissues, enabling a more systemic distribution of the nanoparticles throughout the plant. For example, cerium oxide nanoparticles have been observed to translocate through both the xylem and phloem, reaching different parts of the plant and affecting physiological processes such as photosynthesis and antioxidant activity (Fox et al., 2020).

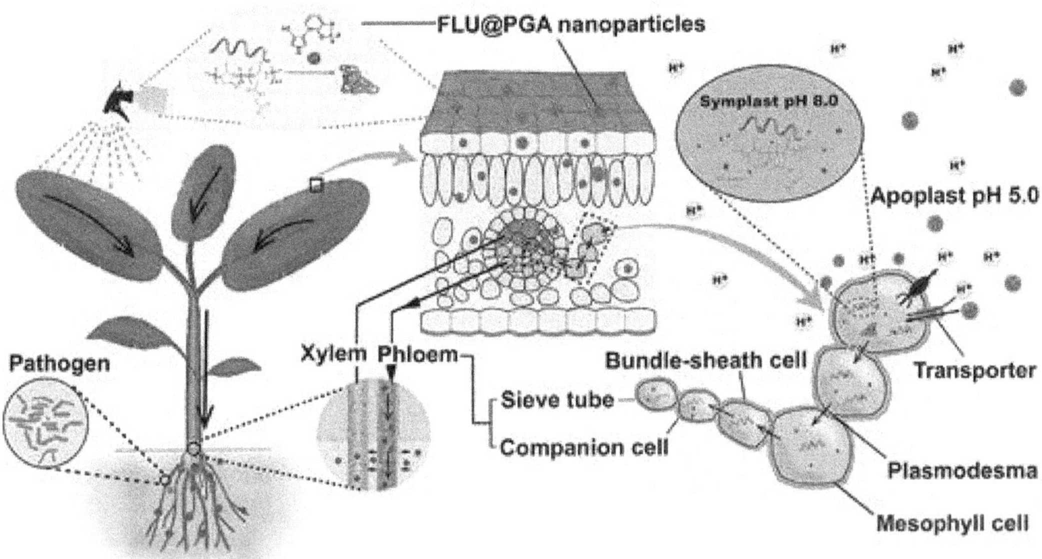

Figure 11. Phloem delivery of fludioxonil by plant amino acid transporter-mediated poly succinimide nanocarriers for controlling fusarium wilt in banana (Wu et al., 2021).

2.4 Cellular and Subcellular Localization

At the cellular level, nanoparticles can be localized in various compartments, including the cell wall, cytoplasm, vacuoles, and even organelles such as chloroplasts and mitochondria. The precise localization depends on the uptake pathway and the interactions of the nanoparticles with cellular components. For instance, nanoparticles that enter cells via endocytosis can be trafficked to vacuoles for storage or detoxification, particularly if they are recognized as foreign particles. Nanoparticles that enter the cytoplasm can interact with cellular organelles, potentially affecting their function. For example, titanium dioxide nanoparticles have been shown to accumulate in chloroplasts, where they can enhance light absorption and photosynthetic efficiency, although excessive accumulation can lead to oxidative stress and cellular damage (Kumar et al., 2023).

The interaction of nanoparticles with plant cells also involves various molecular and biochemical responses. Up uptake, nanoparticles can induce the generation of reactive oxygen species (ROS), which serve as signaling molecules that activate stress response pathways. While low levels of ROS can stimulate defense mechanisms and promote growth, excessive ROS production can cause oxidative stress, leading to cell damage and impaired physiological functions (Xie et al., 2019). Plants may respond to nanoparticle exposure by altering the expression of genes involved in stress response, antioxidant defense, and metal detoxification, which helps mitigate the potential toxic effects of nanoparticles.

In summary, the uptake of nanoparticles by plants involves complex interactions that depend on the properties of the nanoparticles, the plant species, and environmental conditions. Nanoparticles can enter plants through roots, leaves, and other tissues, and are subsequently translocated throughout the plant via the vascular system. Their cellular and subcellular localization, as well as their impact on plant physiology, are influenced by the specific uptake pathways and the interactions of the nanoparticles with cellular components. Understanding these mechanisms is essential for optimizing the use of nanoparticles in agriculture and minimizing potential risks to plant health and the environment.

3. Applications of biopolymer-based nanoparticles in the plant's growth

Environmental factors, such as low temperature, pests, and weeds, are among the primary causes of low crop productivity. Therefore, it is crucial to identify growth stimulants and biological sensors that can effectively evaluate the biochemical and physiological changes in plant growth. Nanomaterials, including biopolymeric nanoparticles, can serve as delivery systems for plant growth stimulants to enhance their effectiveness. In a study, two systems were developed using alginate/chitosan (ALG/CS) or chitosan/tripolyphosphate (CS/TPP) to deliver the plant hormone Gibberellic Acid (GA). These systems demonstrated effects on both morphological and biochemical parameters (Santo Pereira et al., 2017). At the biochemical level, chitosan nanoparticle systems had varying impacts on pigment contents. The ALG/CS–GA nanoparticles increased the levels of chlorophyll "a" relative to the control, while the CS/TPP–GA nanoparticles increased the levels of chlorophyll "b". Only the ALG/CS–GA nanoparticles were found to increase the content of carotenoids (Santo Pereira et al., 2017). Morphological parameters, such as leaf area and root length, also played a significant role. Roots play a critical role in plant growth as they are responsible for water and nutrient uptake (Lynch and Brown, 2012; Zinta et al., 2022). The CS/TPP–GA nanoparticles were observed to increase root length by 5.5 cm at concentrations of 0.012 and 0.025%, compared to the control. In comparison, the free hormone resulted in an increase in root length of 4 cm at concentrations of 0.025% (Santo Pereira et al., 2017). A larger leaf area enhances the photosynthetic activity of the plant, thereby increasing the energy available for plant development. The experiment revealed that using ALG/CS–GA nanoparticles at concentrations of 0.012 led to a larger leaf area compared to other variants. Additionally, the use of chitosan-based nanosystems offers the added benefit of improved antifungal activity of chitosan nanoparticles (Chakraborty et al., 2020; Chowdappa et al., 2014), which can aid in enhancing plant growth while protecting the crop.

In addition to delivering plant growth stimulants, biopolymeric nanoparticles have also demonstrated growth-promoting properties. Multiple studies have indicated that chitosan nanoparticles promote seed germination in crops and can therefore be utilized to produce robust seedlings (Gomes et al., 2021; Li et al., 2019). Kadam et al. demonstrated the impact of chitosan-based nanoparticles on plant growth by creating salicylic acid–chitosan nanoparticles (SA–CS NPs). They found that the application of these nanoparticles increased the activities of seed reserve food-remobilizing enzymes during seedling growth, such as α-amylase and proteases. Higher enzyme activities were observed from the 3rd to the 7th days. Additionally, they noticed that the shoot–root length and fresh weight were considerably higher in the nanoparticle treatments compared to the control, bulk chitosan, and salicylic acid treatments (Kadam et al., 2021).

Chitosan nanoparticles also influence the germination and seedling growth of wheat (*Triticum aestivum* L.), as evidenced by the research of Li et al. They observed a positive impact of chitosan nanoparticles on the seed germination and seedling growth of wheat at a lower concentration compared to chitosan treatment. This efficacy was suggested to be due to the high adsorption of chitosan nanoparticles on the surface of wheat seeds, as observed via energy-dispersive spectroscopy and confocal laser scanning microscopy (Li et al., 2019). In addition to laboratory experiments on seedling growth, studies in greenhouses also demonstrated the effects of chitosan nanoparticles. Their application increased the chlorophyll content and the photosynthesis intensity in coffee leaves compared to controls. They also enhanced the nutrient uptake and had a significant impact on the growth of the coffee seedlings, with the growth parameters of the treated coffee being higher than those of the controls (Van et al., 2013).

Foliar application of biopolymeric nanoparticles loaded with macronutrients such as nitrogen, phosphorus, and potassium (NPK) has been shown to improve plant growth. These nanoparticles were easily applied to the leaf surfaces of wheat plants and entered the stomata via gas uptake. Transmission electron microscopy revealed that the nanoparticles were taken up and transported through phloem tissues. The foliar application of the nanoparticles with NPK increased the wheat yield (Abdel-Aziz et al., 2016). Loading or encapsulating nutrients in biopolymeric nanoparticles can also help minimize economic losses and control the release of nutrients. In addition to chitosan, other carbohydrate- and protein-based polymeric nanoparticles have been utilized in the preparation of such systems for the controlled release of nutrients (Melaj and Daraio, 2014).

3.1 Biopolymer Based Nanoparticles as Fertilizer

Nanofertilizers (NFs) are plant nutrients that consist of nanostructured formulations. These formulations are designed to release active substances into the soil gradually and in a regulated manner. This helps to reduce nutrient loss, eutrophication, and the contamination of water and air. Due to their elevated surface area to volume ratio, NFs exhibit superior efficacy, performance, availability, and utilization compared to CFs. Consequently, NFs provide a promising foundation for the creation of innovative and sustainable nutrition delivery systems (Elemike et al., 2019). Their composition can also facilitate efficient absorption by crops, restoration of soil fertility, extremely high absorption, enhanced photosynthesis, increased production, decreased soil toxicity, reduced application frequency, improved plant health, and minimized environmental pollution (Naderi and Danesh-Shahraki, 2013). The structure of nano fertilizer covers is designed to be porous in order to slowly release the nutrient content (Kubavat et al., 2020). The timing and amount of nutrient release will depend on the specific needs of the plant (Lawrencia et al., 2021).

In a study, alginate chains were cross-linked with Zn (II) and Cu (II), and Scanning Electron Microscopy (SEM) analysis revealed that ZnO and CuO nanoparticles were embedded within the alginate matrix. Experiments on the release of cations in soil and water using the tea bag method showed that the release rate of cations gradually increased over time. Micronutrient uptake in tomato plants was monitored through leaf analyses over 30 consecutive days. To assess micronutrient release behavior in the presence of compost, the nano fertilizer was applied to tomato plants along with pre-determined amounts of compost. Flame Atomic Absorption Spectroscopy (FAAS) results demonstrated that copper concentrations in the plants increased steadily, while zinc remained undetectable (Fig. 12) (Ekanayake and Godakumbura, 2021).

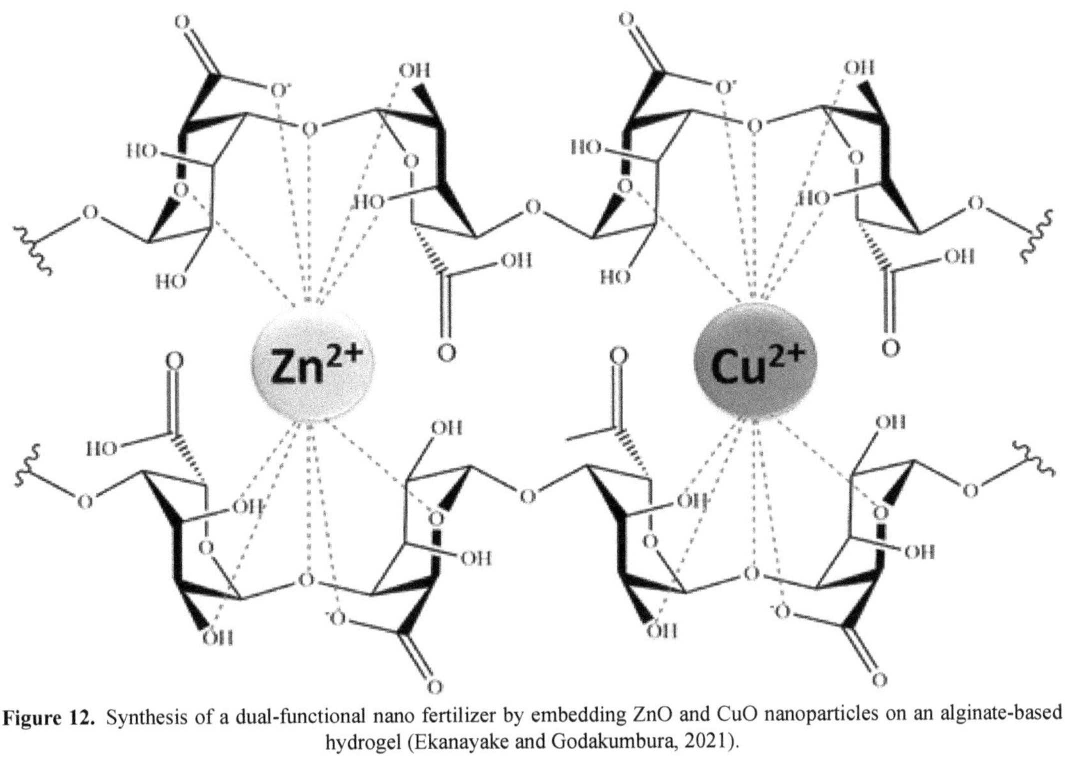

Figure 12. Synthesis of a dual-functional nano fertilizer by embedding ZnO and CuO nanoparticles on an alginate-based hydrogel (Ekanayake and Godakumbura, 2021).

The effectiveness of employing nanoparticles as nano fertilizers for enhancing plant growth mostly relies on the type of crops and the size, concentration, composition, and chemical characteristics of the nanomaterials (Thakur et al., 2018). When NP suspensions containing NFs come into contact with water, nutrients are released into the soil, which are essential for the growth of crops. The application of a polymer coating on NFs or the use of a thin coating to encapsulate NPs can effectively hinder early exposure to water and soil, hence preventing the undesired depletion of nutrients (Shang et al., 2019). Although commercial slow-release fertilizers coated with petrochemical polymers have significantly advanced agricultural productivity, there is a growing need for the development of superhydrophobic biopolymer coatings with exceptional slow-release capabilities from sustainable and eco-friendly materials. In a study, Superhydrophobic Biopolymer-coated Slow-release Fertilizers (SBSF) with outstanding slow-release performance, water resistance, and scalability was fabricated using a simple, solvent-free self-assembly process. The superhydrophobic surfaces of SBSF were achieved by incorporating uniformly dispersed Fe_3O_4 Superhydrophobic Magnetic-Sensitive Nanoparticles (SMNs). The results showed that SBSF exhibited a prolonged slow-release duration (exceeding 100 days), outperforming unmodified biopolymer-coated slow-release fertilizers, with excellent durability across different environmental conditions (Xie et al., 2019).

Commonly used nano fertilizers include urea, ammonium nitrate, ammonium sulfate, sulfuret, calcium nitrate, calcium phosphate, mono and diammonium phosphate, triple superphosphate, potash, silicon, zeolite, zinc, and others (Alimohammadi et al., 2020; Rajonee et al., 2017; Rameshaiah et al., 2015). Agricultural waste products such as banana peels, which are rich in minerals like manganese, magnesium, and potassium, have the potential to be used in the production of NFs. NFs derived from banana peels, as studied by Hussein et al. (2019), contained chelated potassium, chelated iron, tryptophan, urea, amino acids, proteins, and citric acid, and showed significant effects when tested for tomato and fenugreek seed germination (Hussein et al., 2019). Commercial nanonutrients provide advantages such as controlled release due to the cover materials used for these fertilizers. Controlled release refers to the gradual delivery of the nutrient over several months (Vejan et al., 2021). Some available products coated with patented biopolymers are Agrocote (United States), ESN Smart nitrogen (United States), Meister (Australia), Multicote (Israel), Nutricote of Florikan CRF (United States), Osmocote (United States), and zeolites, which are commonly used as fertilizers for fruit trees, coffee, bananas, sugar cane, vegetables, potatoes, rice, corn, and wheat, among others. Additionally, nano fertilizers such as Nano-Gro™ (United States), Nano-green (India), Nano-Ag Answer (United States), Biozar nanofertilizer (Iran), Nano max NPK with microelements and microorganisms (India), Master Nano chitosan organic fertilizer (Thailand), NanoMax (PowerMax) (Taiwan), TAG Nano fertilizer (India), Nanofertilizer LITHOVIT (Germany) and more are available in the market. They are commonly used in various crops including vegetables, fruit trees, wheat, rice, cotton, tea, and others. Ultimately, biopolymers as nanocarriers of NFs or micronutrients have advantages over conventional fertilizers such as the low amount applied and the controlled release of the nutrient, which is more profitable for increasing crop production and fruit quality (Elshamy et al., 2019; Lawrencia et al., 2021). The future of NFs looks promising due to their ecological approach.

Chitosan nanoparticles can encapsulate essential nutrients, such as nitrogen, phosphorus, or micronutrients, and release them gradually, improving nutrient use efficiency. For example, (Colman et al., 2024; Malerba and Cerana, 2016; Sangwan et al., 2023) studied the effect of chitosan nanoparticles loaded with nitrogen on tomato plants and found significant improvements in plant growth and yield. The controlled release of nitrogen helped in better root and shoot development, leading to increased fruit production. Chitosan nanoparticles can also be used to deliver Plant Growth Regulators (PGRs) like auxins, gibberellins, or cytokinins. In a study by (Mahmoud et al., 2024; Pereira et al., 2019), chitosan nanoparticles were loaded with the plant hormone Gibberellic Acid (GA3) and applied to wheat plants. The results showed enhanced germination, improved root and shoot growth, and increased biomass accumulation.

The combination of copper and chitosan has been found to enhance seedling growth and plant biomass (Saharan et al., 2016). Polymer hydrogels composed of hydroxyapatite nanoparticles and urea are used as soil remediation fertilizers to enhance crop production of maize, kale, and capsicum (Rop et al., 2018). Combining silica nanoparticles (NPs) and NPK fertilizers with carboxymethyl cellulose and acrylic acid results in desirable characteristics like suitable pH levels, sensitivity to salt, and water retention in loamy sand soil (Olad et al., 2018). Nanocomposite fertilizers derived from corncob biochar are also viable options for promoting sustainable agriculture (Lateef et al., 2019).

3.2 Biopolymer Based Nanoparticles as Agrochemicals

Agrochemicals, also known as agricultural chemicals or agrichemicals, encompass a variety of chemical products used in farming. Typically, the term refers to a wide range of pesticides, such as insecticides, herbicides, fungicides, and nematicides (which target roundworms). It can also include synthetic fertilizers, growth hormones, and other chemical growth agents, along with concentrated forms of raw animal manure. Numerous studies highlight the potential of biopolymer-based nanotechnologies as valuable tools for safeguarding the environment against the detrimental impacts of agrochemicals. In addition to plant pathogens, abiotic stress factors such as salinity and drought contribute to reduced crop productivity.

The inefficient delivery of agrichemicals in agrifood systems is a major contributor to negative environmental and public health impacts, primarily due to the failure to deliver agrichemicals at the right target, time, and dose. This study addresses the issue by developing scalable, biodegradable, and sustainable biopolymer-based multi stimuli-responsive core–shell nanostructures for smart agrichemical delivery. Three types of responsive core/shell nanostructures, loaded with model agrichemicals ($CuSO_4$ and NPK fertilizer), were synthesized using coaxial electrospray, resulting in spherical nanostructures with an average diameter of 160 nm. Agrichemical release kinetics were tunable by modifying the surface hydrophobicity of the nanostructures, with pH and enzyme responsiveness demonstrated through controlled release of analytes for up to 7 days in an aqueous solution. The efficacy of these stimuli-responsive nanostructures was evaluated in soil-based greenhouse experiments with soybean and wheat, focusing on photosynthesis efficiency and Linear Electron Flow (LEF), key indicators of seedling development and health. Results showed plant-specific benefits; for soybean, nanostructures led to a 34.3% increase in relative chlorophyll content and a 41.2% improvement in PS1 center activity in photosystem I, compared to ionic controls at equivalent agrichemical concentrations. In wheat, LEF was 37.6% higher with nanostructures than with ionic agrichemicals applied at four times the concentration, highlighting the precision and effectiveness of the core–shell nanostructures in reducing agrichemical inputs. Additionally, the Zn and Na content in the leaves of 4-week-old soybean seedlings increased significantly with nanostructure application, suggesting potential for these nanostructures to enhance micronutrient accumulation as part of a biofortification strategy (Fig. 13) (Xu et al., 2022).

Several studies have indicated that chitosan nanoparticles can enhance the response of plants to salinity stress (Balusamy et al., 2022; Mosavikia et al., 2020; Sen et al., 2020; Sheikhalipour et al., 2021). Balusamy et al. discussed the positive impact of chitosans, chitosan nanoparticles (CsNPs), and modified chitosan biomaterials (CsBMs) in combating salt stress to enhance crop performance by inducing physiological changes in plants. These changes include alterations in primary metabolite production, the jasmonic acid signaling pathway, antioxidant activity, secondary metabolite production, and membrane permeability, ultimately leading to improved salt resistance (Balusamy et al., 2022).

In addition to enhancing salt tolerance, chitosan nanoparticles have also demonstrated beneficial effects on plants under drought conditions (Ahmad et al., 2020). The findings of Behboudi et al. suggested that the application of chitosan NPs, particularly at 90 ppm, can alleviate the adverse effects of drought on wheat (Behboudi et al., 2019). Ali et al. reported that the application of

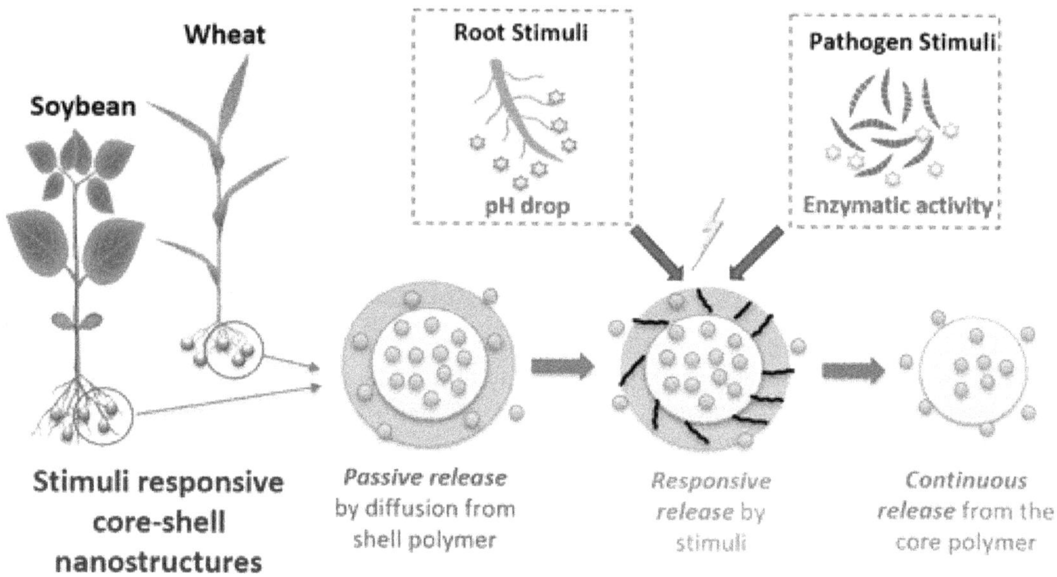

Figure 13. Potential applications of nanostructures to enhance micronutrient accumulation as part of a biofortification strategy (Xu et al., 2022).

chitosan NPs mitigated the negative effects of drought stress, such as the reduction in relative water content, stomatal conductance, and total chlorophyll. NPs increased the activity of enzymes catalase and ascorbate peroxidase, as well as the accumulation of proline. They also reduced oxidative stress and the accumulation of H_2O_2 and malondialdehyde, thereby preserving membrane integrity. Their results also demonstrated an increase in the accumulation of alkaloids, which was associated with the expression of genes involved in the alkaloid biosynthetic pathway, including strictosidine synthase, deacetylvindoline-4-O-acetyltransferase, peroxidase 1, and geissoschizine synthase (Ali et al., 2021). These works illustrate that chitosan nanoparticles aid in mitigating the adverse effects of abiotic stress through biochemical changes and modulation of gene expression.

Pesticides are one of the forms of agrochemicals. The term "nano pesticides" denotes any pesticide formulation involving very small particles of an active pesticide ingredient or other small engineered structures with useful pesticidal properties (Bergeson, 2010; Kookana et al., 2014). Nano-formulations for pesticide products provide several benefits, such as increased solubility, mobility, and durability, as well as a significantly reduced amount of active ingredients. This results in less harm to non-target organisms, thereby reducing the development of resistance and providing protection against premature degradation (Kah et al., 2013; Kah and Hofmann, 2014). Currently, polymers are the main components of NP and nano capsule synthesis. They are bound together by covalent bonds, making the polymer sturdier and more robust than liposomes (Holister et al., 2003). By adjusting their concentration and pH, the sizes, morphology, and shapes of polymers can be more effectively controlled, making them more efficient against chemical compounds or metals. In addition, polymeric nano capsules can be easily combined with different substances, increasing their value for achieving specific objectives. Natural biopolymers such as chitosan and chitin derivatives are popular subjects of research in the field of nano capsules containing natural polymers (Prashanth and Tharanathan, 2007). These polysaccharides are biodegradable and consist of linear β-1, 4 linkages of glucose residues, making them the second most common in nature after cellulose. Chitosan can convert drugs into thin membranous films and can be utilized in the biomedical or pharmaceutical industry. The mucoadhesive properties of chitosan support its potential use in the agrochemical industry, where it is well-recognized for its antimicrobial properties against various microorganisms. Additionally, chitosan induces defense responses in plants and exhibits antibacterial and insecticidal activity (Scrinis and Lyons, 2007; Torney, 2009). Several researchers have conducted experiments on NPs as antifungal agents (Sridhar, 2012). Natural polymers have been employed to encase traditional fungicides to produce nanofungicides (Maluin et al., 2019), nano insecticides (Wang et al., 2019), or nano herbicides (Rychter, 2019) for the gradual release of active ingredients and for the improvement of their disease-prevention efficacy. Figure 1 summarizes the application of biopolymeric NPs as carriers of fungicides, insecticides, and herbicides.

Insects are the most populous animals and can be found in various environments worldwide. Their success can be attributed to several evolutionary factors such as diversification of habits, high reproductive potential, resistant eggs, ability to undergo metamorphosis, and the development of wings and exoskeletons. There are about 500,000 identified insect species that primarily feed on green leaves, with many of them specializing in a small number of plant species. These insects, in large numbers, can cause significant crop losses, especially during the flowering period, by feeding on crop leaves, flowers, and fruits, rendering them unfit for use. Some insects also cause damage to fruits after they have been harvested.

In a study, Kumar et al., 2014 involved the creation of sodium alginate nano capsules loaded with the insecticide imidacloprid (IMI) using the Water-in-Oil-in-Water (W/O/W) double-emulsion method. The researchers supplemented an aqueous solution of sodium alginate with solid IMI, then subjected it to sonication and emulsification using methylene chloride and dioctyl sodium sulfosuccinate. The resulting NPs were collected through ultracentrifugation and tested on leafhoppers, insects that feed on plant sap, to assess the effectiveness of the IMI nano formulations. Leaf infestation was evaluated by counting the leafhopper population on three leaves in a scientific manner. The findings indicated that blank NPs had no positive effect, but the nano-encapsulated IMI was more effective than the free insecticide or the commercial form of IMI. Toxicity studies also revealed that the nano-formulated IMI was significantly less toxic than the conventional pesticide.

Similarly, carboxymethyl chitosan and sodium alginate nano-formulations containing methomyl and pyridalyl insecticides were used to manage Helicoverpa armigera larvae (Saini et al., 2015). Moreover, a controlled-release nanoformulation was developed for the low-water-solubility insecticide acetamiprid (Faizan et al., 2021). The creation of acetamiprid-loaded alginate–chitosan nanocapsules involved the use of ionic pregelation and polyelectrolyte complexation. A polyionic complex was formed with a loading efficiency of 62% due to ionic interactions between the positively charged ammonium groups of chitosan and the negatively charged carboxylate groups of alginates. Studies on its controlled release in various soil mediums demonstrated its superiority over commercial acetamiprid, with the best results achieved in acidic soils.

Essential oils have also been encapsulated in chitosan and other natural biopolymers to enhance their insecticidal activity. For instance, chitosan/gum arabic, chitosan/zinc oxide, and chitosan NPs encapsulating geraniol, azadirachtin, carvacrol, and linalool were synthesized to control whitefly (*B. tabaei*), groundnut bruchid (*C. serratus*), and mite (*T. urticae*), respectively, by the authors (Campos et al., 2018; Song et al., 2019). Additionally, spinosad- and permethrin-loaded chitosan NPs were developed to control Drosophila melanogaster by the authors (Sharma et al., 2019). Emamectin benzoate and thiamethoxam were encapsulated into carboxymethyl chitosan and cellulose NPs to control the insect pests Mythimna separate and Phenacoccus solenopsis by the authors (Song et al., 2019). In a comprehensive review, the authors (Sun et al., 2020) discussed the use of 15 polymers as nanocarriers of insecticides.

Unwanted plants that grow alongside cultivated crops are known as herbs or weeds. These plants present a new challenge for modern agriculture. The use of modern agrochemicals has led to increased agricultural productivity and disease resistance, but has also resulted in the emergence of new types of herbs that need to be eliminated to enhance productivity and maintain soil health. Paraquat, a widely used contact herbicide, is fast-acting and non-selective. However, its water solubility and soil sorption can lead to toxicity issues for non-target organisms and humans through groundwater and rivers. To address this toxicity problem, researchers prepared alginate/chitosan conjugated NPs as a carrier system for paraquat, with a zeta potential of -22.8 ± 2.3 mV and an association efficiency of 74.2%. The encapsulated paraquat within the conjugated NPs improved the release profile of the herbicide and its interaction with the soil (dos Santos Silva et al., 2011), showing promise in reducing the adverse impacts of paraquat. Similar work was carried out by Grillo and others (Grillo et al., 2014), who prepared paraquat-loaded chitosan/tripolyphosphate (CSTPP) NPs with an encapsulation efficiency of 62.66 ± 0.77%. The herbicidal activity of both free and encapsulated forms of paraquat was investigated in maize (Zea mays) and mustard (*Brassica* sp.), both of which exhibited limp leaf necrosis within 48 hours. The use of nanoherbicide in *Z. mays* resulted in more significant necrosis, possibly due to increased adhesion of the NPs to the leaf, as NPs have a higher surface area-to-volume ratio. Conversely, the impact on Brassica was less pronounced, suggesting a lower sensitivity to the herbicide. NPs enhance particle adhesion to surfaces due to their increased surface area, thereby improving the effectiveness of the associated active agents. Comparative testing revealed that nano herbicides had reduced cytotoxicity to Chinese Hamster Ovary (CHO) cells compared to the commercial herbicide MTT, while nanoparaquat induced modest chromosome damage in A. cepa (Grillo et al., 2014). In another study, paraquat was embedded in pectin, chitosan, and sodium tripolyphosphate (PEC/CS/TPP) NPs to achieve sustained pesticide release. The encapsulated herbicide was found to be less hazardous to alveolar and oral cell lines. Furthermore, the mutagenicity of the nano formulation, as assessed using a model with the Salmonella typhimurium strain, was considerably lower than that of paraquat in its commercial forms. Additionally, the soil sorption and deep soil penetration of the NP-linked herbicides decreased. After encapsulation, the herbicidal efficacy of paraquat against maize and mustard was sustained and significantly improved.

Lastly, researchers assessed the herbicidal effect of glyphosate in chitosan nano formulations on three glyphosate-resistant weed species: gallant soldier (*G. parviflora* Cav), white goosefoot (*C. album* L.), and common sorrel (*R. acetosa* L.). The chitosan dissolved in a water solution of glyphosate was found to serve as an environmentally friendly adjuvant and a biopolymeric carrier for prolonged release of glyphosate (Rychter, 2019).

Nematodes are frequently found in soil and over 4100 plant-parasitic species cause significant harm to crops in sandy soil and water-stressed conditions, resulting in approximately USD 80–118 billion in global losses (Decraemer and Hunt, 2006; Nicol et al., 2011). Nematodes primarily feed on roots and can introduce other disease-causing microbes, like bacteria and fungi, which worsen crop damage. Traditional nematicides have been identified as environmental pollutants, and new approaches involving NP applications are being investigated. Nano capsules of lansiumamide B (N-methyl-N-cis-styrylcinnamamide) with nematicidal activity against Bursaphelenehus xylophilus and Meloidogyne incognita were created using the microemulsion polymerization process (with LC50 values of 2.14 and 19.36 mg/L, respectively, after 24 hours). Comparatively, treatment with nano capsules, regular polymers, and ethoprophos (an insecticide and nematicide) resulted in reductions of 68.42, 36.84, and 26.32% in disease progression, respectively. The average number of root knots in Ipomoea aquatica decreased by 83.94, 78.03, and 63.66%, respectively, demonstrating that the nematicide nano formulation was more effective and maintained its efficacy against plant-parasitic nematodes for a longer duration (YIN et al., 2012).

The efficacy of starch-stabilized silver NPs against the root-knot nematode *M. incognita* was studied by (Cromwell et al., 2014). Under *in vitro* conditions, this nano formulation inactivated > 99% of the root nematodes within 6 hours. In the *in vivo* experiment, the population of nematodes in the soil samples treated with this nano formulation (150 g/mL) was reduced by 92 and 82% after 4 and 2 days of exposure, respectively, compared to the untreated soil samples. In a related study, researchers (Ardakani, 2013) found that administering silver NPs (AgNPs) with particle sizes of 200, 400, and 800 mg/mL to *M. incognita* resulted in 100% immobility and mortality, with a determined LC50 value of 100 mg/mL. The same nano formulation at concentrations of 0.02, 0.01, 0.005, 0.0025, 0.00125, and 0.0007% (w/w) effectively controlled *M. incognita*; however, treatment with 0.02, 0.01, and 0.005% AgNPs was lethal to tomato plants and significantly reduced tomato root and stem length and fresh weight compared to the non-treated soil samples.

Fu et al., 2018 utilized the flash nanoprecipitation technique to produce abamectin-loaded NPs with a loading capacity > 40% and encapsulation efficiency > 95% using the amphiphilic copolymers poly(lactic-co-glycolic acid)-b-poly (ethylene glycol) (PLGA-b-PEG), poly(d,l-lactide)-b-poly(ethylene glycol) (PLA-b-PEG), and poly(caprolactone)-b-poly(ethylene glycol) (PCL-b-PEG); these were tested against the southern root-knot nematode Meloidogyne incognita. There are no reports regarding the use of biopolymers as carriers of nematicides.

3.3 Biopolymer Based Nanoparticles as Nanocarrier of Nano Fertilizers and Micronutrients

The challenge in agriculture is to increase food production while using less land per person. The use of fertilizers and other agrochemicals can enhance yield but also have negative environmental consequences, such as soil nitrogen loss through volatilization and the inaccessibility of micronutrients in calcareous soil. Nanomaterials (NM) have great potential for agricultural applications due to their advantageous properties, such as their ability to be administered in small quantities and persist in the environment for extended periods (Ingle et al., 2022).

In sustainable agriculture, controlled release systems are the most suitable choice for enhancing the availability of nutrients for plants or the effectiveness of active components against diseases. These systems allow for a slower, prolonged, and more targeted release of substances. Nanomaterials engineered for controlled release are popular in sustainable agriculture due to their extended durability and the ability to transfer active ingredients to specific locations through regulated permeation (Singh et al., 2020). The improvement of crops can be accelerated through genetic engineering and genomic editing. One of the main challenges in plant engineering is getting genes through plant cell walls. Traditional gene transfer methods such as Agrobacterium-mediated gene transfer, electroporation, PEG-mediated gene transfer, and particle gun bombardment are expensive (Lusser et al., 2012) and can disrupt cell growth (Kashyap et al., 2015). However, advances in technology have made it easier to use genome editing in crop improvement, simplifying, and enhancing the precision of correction or insertion while preventing cell toxicity. Various studies have shown that nanomaterials can be used in genetic engineering and genome editing (Martin-Ortigosa et al., 2012; Torney et al., 2007), making targeted delivery into different plant species and tissues more efficient. Nanomaterials have also contributed to the controlled release of DNA and proteins, protecting them from degradation and enabling imaging of the delivery and release processes in plants (Demirer et al., 2021).

Biopolymeric nanoparticles, such as chitosan nanoparticles, have been used as carriers for delivery of DNA, RNA, and proteins in genetic engineering and the CRISPR/Cas9 system for genome editing (Qiao et al., 2019). It has also been reported that chitosan-based delivery systems can protect DNA from nuclease degradation (Borchard, 2001). Kwak et al. used chitosan-complexed single-walled nanotubes for chloroplast transformation, showcasing the potential of nanoparticle-mediated chloroplast transgene delivery tools that can be used to transform mature plants. Their chitosan nanoparticle-mediated chloroplast transgene delivery platform was recognized as simple, easy to use, cost-effective, and applicable to mature plants across different species, and did not require specialized, expensive equipment, making it suitable for widespread application in plant bioengineering and plant biology studies (Kwak et al., 2019). It was also suggested that their approach could be used in genome editing for the delivery of zinc-finger nucleases, Transcription Activator-Like Effector Nucleases (TALENs), and CRISPR-Cas9 vectors. Several studies have indicated that the Bbm and Wus2 genes can improve plant engineering efficiency and promote plant regeneration (Maher et al., 2020). Lv et al. suggested that nanoparticles can transport these genes, Wus2 and Bbm, and CRISPR/Cas9 into intact plant cells, thus generating a transgenic plant without the need for plant tissue culture processes (Lv et al., 2020).

Biopolymers are used in agriculture due to their inherent biodegradability, compatibility with living creatures, capacity to interact with biological processes, and affinity for water. Plant, animal, and microbial biopolymers are suitable alternatives to synthetic agrochemicals due to their inherent harmless nature. Polymeric nanoparticles enable the encapsulation of bioactive compounds, safeguarding them against deterioration caused by external causes, such as weather conditions (Materu et al., 2021).

Agricultural polysaccharide hydrogels serve as transporters for controlled release of nutrients, resulting in reduced fertilizer application and efficient diffusion from the root to the entire plant. The integration of polysaccharide-based hydrogels with synthetic polymers can provide polysaccharide materials that have significant use as transporters for fertilizers. The utilization of biopolymer in conjunction with SiO2, a non-reactive substance, provides numerous nanotechnological possibilities, including the delivery and encapsulation of active components at the nano scale. Biopolymers are currently considered the most efficient and minimalistic option for delivering agrochemicals due to their numerous advantages (Ghobashy, 2020; Torabi and Mohammadi, 2013).

The primary purpose of encapsulating fertilizers is to enhance the nutrient conversion rate and minimize mineral wastage. Polymer-coated controlled release fertilizers containing N, P, K, and other micronutrients, as detailed in the section "Nature-Derived Polymers for Agriculture Application," continue to be widely used in agriculture. Nanoencapsulation is more commonly used for pesticide encapsulation rather than fertilizers. In terms of nanotechnology in fertilizer applications, inorganic nanoparticles such as zinc (Zn), copper (Cu), iron (Fe), cerium (Ce), and titanium (Ti) have been extensively researched in both laboratory and field conditions compared to encapsulation of these nutrients into nanoparticulate systems (Raliya et al., 2018). A recent review on the use of engineered nanomaterials in various aspects of agriculture, including delivering nutrients to crops and functioning as insecticides and pesticides, covers the full range of currently studied organic and inorganic ENMs in this field (Adisa et al., 2019).

Biopolymers like alginate, cellulose, chitin or chitosan, hemicellulose, lignin, polypeptides, and polyesters are utilized as nanocarriers to encase nutrients and prevent dissolution and oxidation. They are considered eco-friendly due to their natural origin and biodegradability compared to traditional synthetic fertilizers (Mishra et al., 2018). Chitosan, being of natural origin, is the most widely accepted biopolymer for agricultural use. It is also a convenient matrix for controlling the adsorption and gradual release of the active ingredient (Pandey et al., 2018). Chitosan provides protection to plant cells, unlike other biopolymers that could potentially harm plants on contact (Khairy et al., 2022).

One of the key advantages of using chitosan-based nanocarriers in agriculture is their biodegradability. Chitosan breaks down naturally in the environment, leaving no harmful residues. This makes it an environmentally friendly option for sustainable agriculture. According to a study by (Ingle et al., 2022; Khati et al., 2017), the application of chitosan nanoparticles in soil not only improved plant growth but also had a positive impact on soil health by promoting beneficial microbial activity. By enhancing the efficiency of nutrient and pesticide delivery, chitosan nanocarriers can reduce the overall use of chemical inputs in agriculture. This leads to lower environmental contamination, reduced risk of developing pest resistance, and a decreased likelihood of harming non-target organisms. For instance, (Mishra et al., 2022; Sangwan et al., 2023) found that chitosan-based nanocarriers reduced the need for chemical pesticides in tomato crops by effectively controlling pest populations with lower doses.

Due to its inherent growth enhancement and antimicrobial properties, chitosan could be utilized in agriculture, either independently or as an encapsulation matrix for nutrients (Kumaraswamy et al., 2018). Because chitosan nanomaterials disperse more effectively, they are preferred to be used in the nano form rather than in the macro form. Heba et al. demonstrated the ability of chitosan nanoparticles loaded with nutrients N, P, and K to accelerate plant growth, thereby shortening the life cycle of tested wheat plants from 170 to 130 days (Abdel-Aziz et al., 2016). Field tests on barley using chitosan nanoparticles alone have shown their ability to improve grain yield through soil application to barley plants under drought stress (Behboudi et al., 2019). Hydroxyapatite (HA), the mineral component of bone and teeth, is a naturally occurring mineral form of calcium apatite and has been used for delivering macronutrients to soil. Kottegoda et al. utilized HA crystals modified with urea to increase the nitrogen utilization ratio of conventional fertilizers. The natural cellulose and lignin pores in wood stem were used as a reservoir for nano fertilizer fertilizers, and HA urea-loaded wood chips were tested for N release under different soil conditions. This nano fertilizer provided a sustained release of N for more than 60 days compared with the initial burst and fluctuating N amounts released from conventional fertilizers (Kottegoda et al., 2011). The group also conducted a follow-up field study using the HA-based nano fertilizer on rice fields and demonstrated an increase in yield compared with the conventional urea fertilizer, using only 50% of the fertilizer used in the conventional treatment (Kottegoda et al., 2017).

Chitosan nanoparticles can also be applied directly to the soil, where they slowly release their cargo to the plant roots. This method is effective for sustained nutrient delivery and improving soil health. (Choudhary et al., 2019; Kumaraswamy et al., 2021) reported that soil application of chitosan nanoparticles in maize cultivation improved nutrient uptake and increased overall plant vigor.

4. Application of Biopolymer Based Nanoparticles in Plant's Health

4.1 Biopolymer Based Nanoparticles for the Treatment of Fungal Effect

The loss of crop products due to pests and pathogens amounts to about 20–40% annually. Traditionally, toxic pesticides have been the main method of managing plant diseases, posing potential dangers to the environment and humans. Nanotechnology offers advantages such as reduced environmental and cell toxicity, extended shelf life, and enhanced water-soluble synthetic pesticide solubility, which could positively impact the environment (Worrall et al., 2018). Fungicides are primarily used to control fungal diseases in agriculture, which can diminish the quality and quantity of produce. However, the continuous use of fungicides can be detrimental to the ecosystem. Residual fungicides in soil and produce can adversely affect soil properties, microbial communities, and the health of animals and humans (Faizan et al., 2021). Therefore, alternative methods are necessary to enhance efficacy and reduce fungicide doses.

The use of nano-encapsulated pesticides decreases their phytotoxicity, improves their diffusion through the cuticle, and sustains the release of their active components at the target site. Employing biocompatible polymers for nano-entrapment of agrochemicals can further reduce toxicity by achieving targeted and controlled fungicide release (Kumar et al., 2014). Currently, there is growing interest in non-toxic and biocompatible polymers such as chitosan, carboxymethyl, sodium alginate, starch, cellulose, gelatin, and pectin (Yearla and Padmasree, 2016). Various amounts of the low-water-soluble fungicide pyraclostrobin were added to chitosan–lactide copolymer NPs. Three days after application, it was found that the nano fungicide was either as effective as commercial pyraclostrobin or less effective in preventing *C. gossypii* inhibition. However, there was an increase in inhibition observed five days after application compared to acting alone (Xu et al., 2014).

Similarly, another research group (Campos et al., 2015) utilized polymeric and Solid Lipid NPs (SLNPs) for the sustained release of the antifungal drugs tebuconazole and carbendazim, common fungicides. These polymeric NPs released approximately 47% of the fungicides over six days, while the SLNPs exhibited a higher association efficiency (> 99%), indicating a strong ionic interaction. When assessing the effect of commercial and NP formulations on plant emergence using *P. vulgaris* seeds, the germination index of the seeds (92%) was not altered by either formulation (with or without fungicides).

In another experiment, kaempferol, another low-soluble fungicide, loaded onto lecithin/chitosan, exhibited 67% inhibition efficacy after 60 days against *Fusarium oxysporum* infected petri dish (Ilk et al., 2017). When comparing the effect of carbendazim-loaded biopolymeric NPs of chitosan and pectin to carbendazim alone on *Aspergillus parasiticus* and *Fusarium oxysporum*, the authors (Sandhya et al., 2017) found an enhanced fungal inhibition rate. According to phytotoxicity studies, maize, cucumber, and tomato seeds did not experience any adverse effects from the nano formulated carbendazim when planted and allowed to grow roots. Similarly, the authors (Kaur et al., 2018) incorporated copper-oxy-chloride into chitosan–CuO/ZnO nano formulations to manage *Fusarium oxysporum* f. sp. Ciceri (FOC) in chickpeas. A similar process was used by Oh et al., 2019 to create Chitosan NPs (CNPs) from chitosan (CS). The CNPs exhibited the greatest growth-inhibitory effects on *F. oxysporum* mycelial growth, following closely behind *P. capsici*, among the phytopathogenic fungi *P. capsici*, *C. gelosporidies*, *F. oxysporum*, *S. sclerotiorum*, and *G. fujikuori*. Their antibacterial activity against Xanthomonas and Erwinia bacterial strains was also tested. The study showed that both CS and CNPs have tremendous potential as nano fungicides, as they markedly inhibited the growth of fungal and bacterial strains in tomatoes and could be used as crop protectants.

Similarly, Dong et al., 2019 found that the encapsulation of the fungicide spinosad in chitosan–isoleucine NPs protected the fungicide against photo-degradation, thereby increasing the effectiveness of this commercial fungicide. Likewise, the authors Maluin et al., 2020 demonstrated a 74.5% reduction in basal stem rot disease caused by G. boninense in palm trees by applying chitosan hexaconazole-dazomet. When encapsulated in lignin NPs, the fungicides azoxystrobin, pyraclostrobin, tebuconazole, and boscalid azoxystrobin inhibited the growth of test fungi (Phaeomoniella chlamydospora and Phaeoacremonium minimum) under *in vitro* conditions after 96 h, which cause grapevine trunk disease (Machado et al., 2020).

4.2 Biopolymer-Based Nanoparticles for the Treatment of Bacterial Effect

The antimicrobial properties of NPs have undergone comprehensive research to assess their effectiveness against fungal and bacterial diseases in plants and animals. Their ability to combat antibiotic-resistant strains of bacteria and fungi is attributed to their small size. At the nanoscale, particles function like molecules when interacting with a cell, enabling them to easily penetrate the cell membrane and positively influence the central molecular pathways. Their high surface-area-to-volume ratio enhances contact with the target, making NPs a promising option for drug delivery. These particles can be synthesized from biological entities such as bacteria, fungi, biopolymers, and lipids, rendering them valuable in various fields. The interactions of NPs with fungal pathogens are influenced by the size and shape of the particles, thereby governing disease control and toxicity.

Numerous experiments have been conducted to evaluate the antimicrobial activity of NPs. In a study by Ing et al., 2012, chitosan NPs were prepared from various concentrations of Low Molecular-Weight (LMW) and High-Molecular-Weight (HMW) chitosan. These NPs exhibited superior inhibitory activity with low Minimum Inhibitory Concentrations (MIC) for *F. solani* (MICLMW = 0.86–1.2 mg/mL and MICHMW = 0.5–1.2 mg/mL) and *C. albicans* (MICLMW = 0.25–0.86 mg/mL and MICHMW = 0.6–1.0 mg/mL) compared to the solution form (MIC = 3 mg/mL for both molecular weights and fungal strains). The inhibitory effect of the chitosan NPs was also impacted by their particle size and zeta potential. Additionally, *Aspergillus niger* was minimally affected by the chitosan NPs, except for the HMW NPs at higher concentrations. Trimethyl chitosan (TMC) NPs showed negligible antifungal activity, indicating that the parent compound, chitosan, can be utilized as a natural antifungal agent in its NP form to enhance its antifungal activity.

In another study by Chowdappa et al., 2014, low-molecular-weight chitosan was used as a reducing and stabilizing agent to produce silver NPs ranging between 10 and 15 nm in size. The production of NPs was confirmed through UV–Vis spectrum, EDS (Energy-Dispersive X-ray Spectroscopy), and FESEM (Field Emission Scanning Electron Microscopy), elucidating the structure of the chitosan and AgNPs composite. The chitosan–Ag nanocomposite significantly reduced the conidial germination of *C. gloeosporioides* (treated with 30 mM AgNO3) compared to chitosan alone. The chitosan–Ag nanocomposite inhibited conidial germination by 44, 70, and 78% at concentrations of 0.1 (0.00001), 1.0 (0.0001), and 10 g/mL (0.001%), respectively, and completely hindered spore germination at a concentration of 100 g/mL (0.01%). This effectively decreased the incidence of anthracnose in mango and suppressed the conidial germination of *C. gloeosporioides*.

The nanocomposite holds potential for preventing dormant infections of Colletotrichum in mango, thereby averting significant crop losses and facilitating the export of high-quality fruits and vegetables. Fusarium wilt, a fungal disease of chickpea caused by *Fusarium oxysporum* f. sp. ciceri (FOC), originates from seeds and soil.

In a study by Kumar et al., 2014, *in vitro* and *in vivo* investigations were conducted on chitosan (CS) and its nanocomposites as antifungal agents against FOC. Chitosan–copper oxide nanocomposites (CS-CuO) and chitosan–zinc oxide nanocomposites (CS-ZnO) demonstrated high efficacy against various FOC concentrations, specifically 50, 100, and 200 µg/mL. The CS NPs and chitosan–silver (CS-Ag) nanocomposites were more effective than standard fungicides, such as copper oxy-chloride (CuOCl). Based on the *in vitro* findings, a concentration of 100 µg/mL for all nano formulations was selected for *in vivo* plant studies under pot conditions. The greatest reduction in wilt disease (46.67%) was observed with CS-CuO nano formulations, followed by CS-ZnO (40%). Conversely, CS-Ag and CS only caused a 33.33% reduction in wilt incidence, indicating a lower impact. All synthesized nano formulations exhibited excellent antifungal efficacy, inhibiting the pathogens, and promoting the growth of chickpea plants compared to untreated plants.

Malerba and Cerana, 2019 discussed the use of chitin- and chitosan-based polymers in controlling plant diseases. Similarly, Egusa et al., 2019 observed that incorporating a chitosan nano formulation into the soil marginally enhanced the growth of tomato plants and reduced the occurrence of Fusarium wilt disease compared to untreated plants. Utilizing chitosan nano formulations complexed with proteins and CaCO3 in protein/CaCO3/chitin nanofibers proved more effective in preventing disease than using protein alone or protein/chitin nanofibers. Furthermore, Nguyen et al., 2019 employed oligochitosan-silica/carboxymethyl cellulose NPs to control Phytophthora infestans at a concentration of 800 mg/L, the lowest concentration that inhibited fungal growth.

Several experiments have been performed to determine the antimicrobial activity of NPs. Ing et al., 2012 prepared chitosan NPs from diverse concentrations of Low Molecular-Weight (LMW) and High-Molecular-Weight (HMW) chitosan. These were found to show superior inhibitory activity, i.e., a low MIC (MinimumInhibitory Concentration) for *F. solani* (MICLMW = 0.86–1.2 mg/mL and MICHMW = 0.5–1.2 mg/mL) and *C. albicans* (MICLMW = 0.25–0.86 mg/mL and MICHMW = 0.6–1.0 mg/mL) compared to the solution form (MIC = 3 mg/mL for both molecular weights and fungal strains). The particle size and zeta potential of the chitosan NPs also impacted their inhibitory effect. Additionally, *Aspergillus niger* was unaffected by the chitosan NPs, except for the HMW NPs at higher concentrations. Trimethyl chitosan (TMC) NPs showed insignificant antifungal activity. Therefore, the parent compound, i.e., chitosan, can be used as a natural antifungal agent in its NP form to enhance its antifungal activity.

Chowdappa et al., 2014 used low-molecular-weight chitosan as a reducing and stabilizing agent to create silver NPs between 10 and 15 nm in size. The production of NPs was confirmed by the UV–Vis spectrum, EDS (energy-dispersive X-ray spectroscopy), and FESEM (Field Emission Scanning Electron Microscopy), and the structure of the composite of chitosan and AgNPs was made clear. The conidial germination of *C. gloeosporioides* was significantly reduced by the chitosan–Ag nanocomposite (treated with 30 mM AgNO3) compared to chitosan alone. The chitosan–Ag nanocomposite suppressed conidial germination by 44, 70, and 78% at concentrations of 0.1 (0.00001), 1.0 (0.0001), and 10 g/mL (0.001%). The concentration of 100 g/mL (0.01%) hindered spore germination. This effectively reduced the frequency of anthracnose in mango and inhibited the conidial germination of *C. gloeosporioides*. This nanocomposite could find applications in preventing dormant infections of Colletotrichum in mango, thus avoiding huge crop losses and promoting the export of high-quality fruits and vegetables. Fusarium wilt is a fungal disease of chickpea (*Cicer arietinum*) caused by *Fusarium oxysporum* f. sp. Ciceri (FOC) and has seed- and soil-borne origins.

In vitro and *in vivo* studies of chitosan (CS) and its nanocomposites as antifungal agents against FOC were performed. Chitosan–copper oxide nanocomposites (CS-CuO) and chitosan–zinc oxide nanocomposites (CSZnO) were the most effective against various FOC concentrations, i.e., 50,

100, and 200 μg/mL. The CS NPs and chitosan–silver (CS-Ag) nanocomposites were effective and more efficient than standard fungicides, i.e., copper oxy-chloride (CuOCl). Based on the *in vitro* results, a 100 μg/mL concentration of all nano formulations was selected for *in vivo* plant studies under pot conditions. The highest wilt disease reduction (46.67%) was observed in CS-CuO nano formulations, followed by CS-ZnO (40%). In contrast, CS-Ag and CS caused only a 33.33% reduction in wilt incidence, representing the lowest impact. All synthesized nano formulations showed excellent antifungal efficacy, inhibited the pathogens, and promoted the growth of the chickpea plants compared to the untreated plants (Kaur et al., 2018).

The applications of chitin- and chitosan-based polymers in controlling plant diseases were reported (Malerba and Cerana, 2019). Similarly, Egusa et al., 2019 found that compared to untreated plants, adding a chitosan nano formulation to the soil marginally increased the growth of tomato plants and decreased the occurrence of Fusarium wilt disease. Utilizing chitosan nano formulations complexed with both proteins and CaCO3 in protein/CaCO3/chitin nanofibers proved more successful at preventing disease than protein alone or protein/chitin nanofibers. Nguyen et al., 2019 employed oligochitosansilica/carboxymethyl cellulose NPs to control Phytophthora infestans at a concentration of 800 mg/L (the lowest concentration that inhibited fungal growth).

4.3 Biopolymer-Based Nanoparticles in Plant Protection against Pathogenic disease

The utilization of nanoparticles in plant protection can involve either delivering pesticides or rapidly diagnosing plant diseases (Servin et al., 2015). Bio-based nanomaterials are more appealing for plant protection due to their ecological safety. Several research studies have demonstrated that biopolymeric nanomaterials can enhance the expression of defense enzymes/genes in plants and safeguard crops from pathogens (Chandra et al., 2015; Chun and Chandrasekaran, 2019). Notably, chitosan and other carbohydrate-based nanoparticles have proven effective in protecting crops (Divya and Jisha, 2018; Verma et al., 2020). For instance, β-d-glucan nanoparticles have displayed antifungal properties against P. aphanidermatum, a destructive fungus that affects major crop plants (Anusuya and Sathiyabama, 2015a, 2015b).

Chitosan nanoparticles have also exhibited the ability to suppress blast disease of rice, caused by the fungus *Pyricularia grisea* (Manikandan and Sathiyabama, 2016). The antifungal activity of chitosan nanoparticles was investigated against rice sheath blight disease caused by the pathogenic fungus *Rhizoctonia solani*. It was observed that the use of chitosan nanoparticles resulted in elevated levels of defense enzymes, such as peroxidase, phenylalanine ammonia-lyase, and chitinase enzymes, indicating their effectiveness as potent plant immunity boosters capable of suppressing up to 90% of disease in the detached leaf assay and 75% under greenhouse conditions (Divya et al., 2020). Choudhary et al. demonstrated that chitosan nanoparticles can inhibit the plant pathogen *Fusarium solani* under *in vitro* conditions (Choudhary et al., 2017), while Chen et al. attributed the antimicrobial activity of chitosan-based nanoparticles to their zeta potential, which plays a crucial role in binding to negatively charged microbial membranes (Chen et al., 2010).

The antifungal properties of nanocomposite thin films of pullulan and silver against *Aspergillus niger* were demonstrated by (Pinto et al., 2013). Biopolymeric nanoparticles were utilized to combat plant nematodes, including the pine wood nematode, as studied by Liang et al. They encapsulated the widely used bionematicide avermectin within nanoparticles composed of poly-γ-glutamic acid and chitosan to address its poor solubility in water and rapid photolysis. Their findings revealed that encapsulating avermectin within biopolymeric nanoparticles reduced its losses through photolysis by more than 20.0%. Moreover, the mortality rate of nematodes treated with 1 ppm of avermectin encapsulated in nanoparticles was 98.6% (Liang et al., 2018). In addition to their direct antifungal activity, polymeric nanoparticles were demonstrated by Liu et al. to control pathogens by slowly releasing fungicides tebuconazole and chlorothalonil to combat a common brown rot wood decay fungus using nanoparticle systems based on polymeric materials polyvinyl pyridine and polyvinyl pyridine-co-styrene (Liu et al., 2001).

Chitosan-based nanoparticles have been highlighted in several reports for their preparation and application in delivering and controlling the release of pesticides in agriculture. Chitosan-based nanomaterials have been utilized for delivering insecticides such as rotenone, imidacloprid, the biopesticide azadirachtin (Feng and Peng, 2012; Guan et al., 2008; Lao et al., 2010; Paula et al., 2011), and the insecticidal essential oil of Lippia sidoides. Biopolymeric nanoparticles have also been employed for the controlled release of herbicides, aiming to enhance their efficacy at reduced doses and minimize their environmental impact. Several studies have demonstrated the effectiveness of chitosan-based nanomaterials in delivering herbicides. For instance, Silva et al. used alginate/chitosan nanoparticles for paraquat herbicide delivery (dos Santos Silva et al., 2011), while Grillo et al. employed chitosan/tripolyphosphate nanoparticles as carriers of the same herbicide (Grillo et al., 2014).

5. Future Perspectives and Conclusions

5.1 Future Perspectives

The future of biopolymer-based nanoparticles in agriculture holds immense promise, particularly as the global demand for food production intensifies due to population growth, climate change, and the need for more sustainable farming practices. As researchers continue to explore and refine these technologies, several key areas are poised for development, leading to significant advancements in agriculture and beyond.

The integration of biopolymer-based nanoparticles into precision agriculture could redefine how crops are grown and managed. Precision agriculture relies on data-driven decision-making to optimize inputs such as water, fertilizers, and pesticides. Biopolymer-based nanoparticles can be engineered to deliver nutrients, agrochemicals, or antimicrobial agents in a controlled and targeted manner, enhancing the precision of agricultural interventions. In the future, these nanoparticles could be customized to respond to specific plant needs or environmental triggers, releasing their cargo only when necessary. For example, nanoparticles could be designed to release nutrients when a plant's growth stage or environmental conditions demand it, reducing overuse of inputs and minimizing waste. Coupling this with real-time sensing technology and data analytics could transform agriculture into a highly efficient, resource-conserving system.

One of the most promising aspects of biopolymer-based nanoparticles is their potential to serve as sustainable alternatives to traditional fertilizers and pesticides. Conventional agricultural chemicals often pose environmental hazards due to leaching, runoff, and accumulation in soils and water bodies. In contrast, biopolymer-based nanoparticles, due to their biodegradable nature, offer a more environmentally friendly solution. Future advancements could focus on improving the encapsulation and controlled release mechanisms of these nanoparticles to ensure that nutrients and pesticides are delivered more effectively, minimizing losses to the environment. This could drastically reduce the ecological footprint of agriculture, making farming more sustainable in the long term.

5.2 Enhanced Plant Protection and Disease Management

Plant diseases caused by pathogens such as fungi, bacteria, and viruses are major threats to global food security. Biopolymer-based nanoparticles have shown potential in combating these pathogens by either delivering antimicrobial agents or by themselves acting as antimicrobial agents. In the future, these nanoparticles could be fine-tuned for specific pathogen targets, offering tailored solutions to different crops and regions. Moreover, nanoparticles could be designed to act as immunostimulants, boosting the plant's natural defense mechanisms and providing long-lasting protection against multiple pathogens. This would not only reduce the need for chemical pesticides but also improve crop resilience, especially in regions that are prone to frequent disease outbreaks.

Another exciting future prospect is the use of biopolymer-based nanoparticles as nanocarriers for delivering genetic material or other bioactive compounds to plants. In genetic engineering and plant breeding, the ability to deliver DNA, RNA, or other molecules to specific parts of the plant is critical for developing new crop varieties with enhanced traits such as drought resistance, disease tolerance, or improved nutritional value. Biopolymer-based nanoparticles could facilitate the precise delivery of these materials, enabling the creation of genetically modified plants without the need for traditional transgenic approaches. This could revolutionize plant breeding, allowing for faster, more targeted development of crops that are better suited to changing environmental conditions.

As climate change continues to affect agricultural productivity through increased temperatures, erratic rainfall patterns, and the spread of pests and diseases, biopolymer-based nanoparticles could play a vital role in mitigating these impacts. For instance, nanoparticles could be used to deliver drought-resistant compounds or improve the efficiency of water use in plants, helping crops to survive in water-scarce regions. Nanoparticles could also be engineered to enhance the uptake of carbon dioxide, thus contributing to carbon sequestration efforts while simultaneously boosting photosynthesis and crop yields. Additionally, biopolymer-based nanoparticles could protect plants from extreme temperatures by delivering heat-shielding or cold-tolerant compounds, ensuring crop survival in adverse climatic conditions.

The role of biopolymer-based nanoparticles in enhancing soil health and interacting with the soil microbiome is another area that warrants future exploration. Healthy soils are fundamental to sustainable agriculture, as they support plant growth, nutrient cycling, and water retention. However, the overuse of chemical fertilizers and pesticides has led to soil degradation and loss of biodiversity in soil ecosystems. Biopolymer-based nanoparticles, being biodegradable, could provide a means to restore soil health by delivering nutrients in a more balanced and environmentally friendly way. Furthermore, future research could focus on the interactions between nanoparticles and beneficial soil microbes, such as nitrogen-fixing bacteria and mycorrhizal fungi, to enhance soil fertility and plant nutrient uptake. By promoting a healthy soil microbiome, biopolymer-based nanoparticles could contribute to long-term soil regeneration and sustainable farming.

The future of biopolymer-based nanoparticles in agriculture is bright, with the potential to revolutionize the industry by enhancing crop yields, improving plant health, and promoting sustainability. As research continues to evolve, these nanoparticles could become essential tools in addressing the challenges posed by climate change, population growth, and the need for more efficient resource use. By integrating advanced nanotechnology with natural biopolymers, the agricultural sector can move toward a more sustainable and resilient future, where food security is ensured without compromising the environment.

5.3 Conclusions

Biopolymer-based nanoparticles offer a groundbreaking, sustainable solution for advancing agricultural practices in the face of growing global challenges related to food security, environmental degradation, and the need for more efficient farming techniques. These nanomaterials, which merge the biodegradable and eco-friendly properties of biopolymers with the unique size-dependent characteristics of nanoparticles, have demonstrated remarkable potential to enhance plant growth, nutrient delivery, and overall plant health. Their versatility stems from the diverse types of biopolymer-based nanoparticles, including biopolymeric nanoparticles, nanocomposites, metal, and metal oxide nanoparticles, each offering tailored benefits for various agricultural applications.

One of the most significant advantages of biopolymer-based nanoparticles lies in their mechanisms of interaction with plants. Their ability to be absorbed through both root and foliar pathways and subsequently translocated and distributed within plant tissues allows for precise and efficient delivery of nutrients and agrochemicals. This targeted delivery reduces the amount of inputs required, lowering the environmental burden associated with traditional fertilizers and pesticides. Moreover, their controlled release properties further enhance the efficiency of nutrient uptake, ensuring that plants receive a steady supply of essential elements over time, promoting healthier growth and increased crop yields.

In addition to their role in enhancing plant nutrition, biopolymer-based nanoparticles have shown great promise in improving plant health through their ability to protect plants against fungal, bacterial, and viral pathogens. These nanoparticles can act as carriers for antimicrobial agents or function as standalone treatments, directly interacting with pathogens to inhibit their growth and spread. This capability not only strengthens plant defenses but also reduces the need for chemical pesticides, contributing to a more sustainable and environmentally friendly approach to pest and disease management. By incorporating biopolymer-based nanoparticles into plant protection strategies, farmers can reduce crop losses, improve food quality, and increase productivity.

The potential of these nanoparticles extends beyond immediate agricultural applications to broader implications for sustainable farming. By reducing the reliance on chemical fertilizers and pesticides, biopolymer-based nanoparticles contribute to soil health preservation and reduce the risk of water contamination and soil degradation. Their biodegradable nature ensures minimal long-term environmental impact, making them a key component in the move toward greener agricultural technologies. Additionally, the ability to engineer nanoparticles for specific functions opens up possibilities for precision agriculture, where inputs can be optimized for specific crop requirements and environmental conditions.

However, while the benefits of biopolymer-based nanoparticles are clear, several challenges remain. Further research is needed to understand the long-term effects of these nanoparticles on plant systems, soil microbiomes, and the broader ecosystem. Studies exploring the potential accumulation of nanoparticles in plants and the food chain, as well as their interactions with non-target organisms, are essential for ensuring their safe application in agriculture. Regulatory frameworks also need to be established to guide the development and use of nanotechnologies in agriculture, ensuring that they are deployed responsibly and sustainably.

In conclusion, biopolymer-based nanoparticles represent a significant leap forward in the quest for more sustainable, efficient, and productive agricultural practices. By enhancing nutrient delivery, improving plant health, and reducing environmental impact, they offer a promising solution to many of the challenges faced by modern agriculture. With continued research and development, these innovative materials have the potential to revolutionize farming, enabling the production of healthier crops while minimizing the ecological footprint of agricultural practices. Their integration into mainstream agriculture could pave the way for a future where farming is not only more productive but also more in harmony with the environment.

References

Abdel-Aziz, H. M., Hasaneen, M. N. and Omer, A. M. (2016). Nano chitosan-NPK fertilizer enhances the growth and productivity of wheat plants grown in sandy soil. Spanish Journal of Agricultural Research, 14(1), e0902-e0902.

Adisa, I. O., Pullagurala, V. L. R., Peralta-Videa, J. R., Dimkpa, C. O., Elmer, W. H., Gardea-Torresdey, J. L. et al. (2019). Recent advances in nano-enabled fertilizers and pesticides: A critical review of mechanisms of action. Environmental Science: Nano, 6(7): 2002–2030.

Ahmad, J., Qamar, S., Kausar, N. and Qureshi, M. I. (2020). Nanoparticles: the magic bullets in mitigating drought stress in plants. Nanobiotechnology in Agriculture: An Approach Towards Sustainability, 145–161.

Alghuthaymi, M. A., Abd-Elsalam, K. A., Shami, A., Said-Galive, E., Shtykova, E. V. and Naumkin, A. V. (2020). Silver/Chitosan nanocomposites: Preparation and characterization and their fungicidal activity against dairy cattle toxicosis *Penicillium expansum*. Journal of Fungi, 6(2): 51.

Ali, E., El-Shehawi, A. M., Ibrahim, O., Abdul-Hafeez, E., Moussa, M. and Hassan, F. (2021). A vital role of chitosan nanoparticles in improvisation the drought stress tolerance in *Catharanthus roseus* (L.) through biochemical and gene expression modulation. Plant Physiology and Biochemistry, 161: 166–175.

Alimohammadi, E., Bagheri, S. R., Taheri, S., Dayani, M. and Abdi, A. (2020). The impact of extended adjuvant temozolomide in newly diagnosed glioblastoma multiforme: A meta-analysis and systematic review. Oncology Reviews, 14(1).

An, C., Sun, C., Li, N., Huang, B., Jiang, J., Shen, Y. et al. (2022). Nanomaterials and nanotechnology for the delivery of agrochemicals: Strategies towards sustainable agriculture. Journal of Nanobiotechnology, 20(1): 11.

Anusuya, S. and Sathiyabama, M. (2015a). Protection of turmeric plants from rhizome rot disease under field conditions by β-D-glucan nanoparticle. International Journal of Biological Macromolecules, 77: 9–14.

Anusuya, S. and Sathiyabama, M. (2015b). β-d-Glucan nanoparticle pre-treatment induce resistance against *Pythium aphanidermatum* infection in turmeric. International Journal of Biological Macromolecules, 74: 278–282.

Ardakani, A. S. (2013). Toxicity of silver, titanium and silicon nanoparticles on the root-knot nematode, *Meloidogyne incognita*, and growth parameters of tomato. Nematology, 15(6): 671–677.

Atiwesh, G., Mikhael, A., Parrish, C. C., Banoub, J. and Le, T.-A. T. (2021). Environmental impact of bioplastic use: A review. Heliyon, 7(9).

Balusamy, S. R., Rahimi, S., Sukweenadhi, J., Sunderraj, S., Shanmugam, R., Thangavelu, L. et al. (2022). Chitosan, chitosan nanoparticles and modified chitosan biomaterials, a potential tool to combat salinity stress in plants. Carbohydrate Polymers, 284: 119189.

Bandara, S., Du, H., Carson, L., Bradford, D. and Kommalapati, R. (2020). Agricultural and biomedical applications of chitosan-based nanomaterials. Nanomaterials, 10(10): 1903.

Banerjee, A., Halder, U. and Bandopadhyay, R. (2017). Preparations and applications of polysaccharide based green synthesized metal nanoparticles: A state-of-the-art. Journal of Cluster Science, 28: 1803–1813.

Banerjee, S., Prakash, P. C. and Kadeppagari, R.-K. (2022). Zein-based nanoproducts in nutrition and food sectors. In Handbook of Consumer Nanoproducts (pp. 735–749). Springer.

Behboudi, F., Tahmasebi-Sarvestani, Z., Kassaee, M. Z., Modarres-Sanavy, S. A. M., Sorooshzadeh, A. and Mokhtassi-Bidgoli, A. (2019). Evaluation of chitosan nanoparticles effects with two application methods on wheat under drought stress. Journal of Plant Nutrition, 42(13): 1439–1451.

Bergeson, L. L. (2010). Nanosilver: US EPA's pesticide office considers how best to proceed. Environmental Quality Management, 19(3).

Borchard, G. (2001). Chitosans for gene delivery. Advanced Drug Delivery Reviews, 52(2): 145–150.

Campos, E. V., Proença, P. L., Oliveira, J. L., Melville, C. C., Della Vechia, J. F., De Andrade, D. J. et al. (2018). Chitosan nanoparticles functionalized with β-cyclodextrin: A promising carrier for botanical pesticides. Scientific Reports, 8(1): 2067.

Campos, E. V. R., Oliveira, J. L. d., da Silva, C. M. G., Pascoli, M., Pasquoto, T., Lima, R. et al. (2015). Polymeric and solid lipid nanoparticles for sustained release of carbendazim and tebuconazole in agricultural applications. Scientific Reports, 5(1): 13809.

Chakraborty, M., Hasanuzzaman, M., Rahman, M., Khan, M. A. R., Bhowmik, P., Mahmud, N. U. et al. (2020). Mechanism of plant growth promotion and disease suppression by chitosan biopolymer. Agriculture, 10(12): 624.

Chandra, S., Chakraborty, N., Dasgupta, A., Sarkar, J., Panda, K. and Acharya, K. (2015). Chitosan nanoparticles: A positive modulator of innate immune responses in plants. Scientific Reports, 5(1): 15195.

Chen, H. (2018). Metal based nanoparticles in agricultural system: Behavior, transport, and interaction with plants. Chemical Speciation & Bioavailability, 30(1): 123–134.

Chen, L.-C., Kung, S.-K., Chen, H.-H. and Lin, S.-B. (2010). Evaluation of zeta potential difference as an indicator for antibacterial strength of low molecular weight chitosan. Carbohydrate Polymers, 82(3): 913–919.

Choudhary, R. C., Kumaraswamy, R., Kumari, S., Pal, A., Raliya, R., Biswas, P. et al. (2017). Synthesis, characterization, and application of chitosan nanomaterials loaded with zinc and copper for plant growth and protection. Nanotechnology: An Agricultural Paradigm, 227–247.

Choudhary, R. C., Kumaraswamy, R., Kumari, S., Sharma, S., Pal, A., Raliya, R. et al. (2019). Zinc encapsulated chitosan nanoparticle to promote maize crop yield. International Journal of Biological Macromolecules, 127: 126–135.

Chowdappa, P., Gowda, S., Chethana, C. and Madhura, S. (2014). Antifungal activity of chitosan-silver nanoparticle composite against *Colletotrichum gloeosporioides* associated with mango anthracnose. African Journal of Microbiology Research, 8(17): 1803–1812.

Chun, S.-C. and Chandrasekaran, M. (2019). Chitosan and chitosan nanoparticles induced expression of pathogenesis-related proteins genes enhances biotic stress tolerance in tomato. International Journal of Biological Macromolecules, 125: 948–954.

Colman, S. L., Salcedo, M. F., Iglesias, M. J., Alvarez, V. A., Fiol, D. F., Casalongué, C. A. et al. (2024). Chitosan microparticles mitigate nitrogen deficiency in tomato plants. Plant Physiology and Biochemistry, 212: 108728.

Cromwell, W., Yang, J., Starr, J. and Jo, Y.-K. (2014). Nematicidal effects of silver nanoparticles on root-knot nematode in bermudagrass. Journal of Nematology, 46(3): 261.

Da Costa, M. and Sharma, P. (2016). Effect of copper oxide nanoparticles on growth, morphology, photosynthesis, and antioxidant response in *Oryza sativa*. Photosynthetica, 54: 110–119.

Decraemer, W. and Hunt, D. J. (2006). Structure and classification. In Plant nematology (pp. 3–32). CABI Wallingford UK.

Demirer, G. S., Silva, T. N., Jackson, C. T., Thomas, J. B., W. Ehrhardt, D., Rhee, S. Y. et al. (2021). Nanotechnology to advance CRISPR–Cas genetic engineering of plants. Nature Nanotechnology, 16(3): 243–250.

Dhlamini, B., Paumo, H. K., Kamdem, B. P., Katata-Seru, L. and Bahadur, I. (2022). Nano-engineering metal-based fertilizers using biopolymers: An innovative strategy for a more sustainable agriculture. Journal of Environmental Chemical Engineering, 10(3): 107729.

Divya, K. and Jisha, M. (2018). Chitosan nanoparticles preparation and applications. Environmental Chemistry Letters, 16: 101–112.

Divya, K., Thampi, M., Vijayan, S., Varghese, S. and Jisha, M. (2020). Induction of defence response in *Oryza sativa* L. against *Rhizoctonia solani* (Kuhn) by chitosan nanoparticles. Microbial Pathogenesis, 149: 104525.

Dong, E., Yang, Z., Zhou, C., Wang, C., Li, S., Ouyang, Q. et al. (2019). pH-responsive ultrasonic self-assembly spinosad-loaded nanomicelles and their antifungal activity to *Fusarium oxysporum*. Reactive and Functional Polymers, 141: 123–132.

dos Santos Silva, M., Cocenza, D. S., Grillo, R., de Melo, N. F. S., Tonello, P. S., de Oliveira, L. C. et al. (2011). Paraquat-loaded alginate/chitosan nanoparticles: preparation, characterization and soil sorption studies. Journal of Hazardous Materials, 190(1-3): 366–374.

Egusa, M., Parada, R., Aklog, Y. F., Ifuku, S. and Kaminaka, H. (2019). Nanofibrillation enhances the protective effect of crab shells against Fusarium wilt disease in tomato. International Journal of Biological Macromolecules, 128: 22–27.

Ehsan, M., Suhag, D., Rathore, R., Thakur, A. and Thakur, P. (2023). Metal nanoparticles in the field of medicine and pharmacology. In Integrated Nanomaterials and their Applications (pp. 127–152). Springer.

Ekanayake, S. A. and Godakumbura, P. I. (2021). Synthesis of a dual-functional nanofertilizer by embedding ZnO and CuO nanoparticles on an alginate-based hydrogel. ACS Omega, 6(40): 26262–26272.

El-Beltagi, H. S., Bendary, E. S., Ramadan, K. M. and Mohamed, H. I. (2022). Metallic nanoparticles and nano-based bioactive formulations as nano-fungicides for sustainable disease management in cereals. In Cereal diseases: Nanobiotechnological approaches for diagnosis and management (pp. 315–343). Springer.

El-Diehy, M. A., Farghal, I. I., Amin, M. A., Ghobashy, M. m., Nowwar, A. I. and Gayed, H. (2024). Radiation synthesis of sodium alginate/gelatin based ultra-absorbent hydrogel for efficient water and nitrogen management in wheat under drought stress. Scientific Reports, 14(1): 19463.

Elemike, E. E., Uzoh, I. M., Onwudiwe, D. C. and Babalola, O. O. (2019). The role of nanotechnology in the fortification of plant nutrients and improvement of crop production. Applied Sciences, 9(3): 499.

Elshamy, M. T., Husseiny, S. M. and Farroh, K. Y. (2019). Application of nano-chitosan NPK fertilizer on growth and productivity of potato plant. Journal of Scientific Research in Science, 36(1): 424–441.

Faizan, M., Rajput, V. D., Al-Khuraif, A. A., Arshad, M., Minkina, T., Sushkova, S. et al. (2021). Effect of foliar fertigation of chitosan nanoparticles on cadmium accumulation and toxicity in *Solanum lycopersicum*. Biology, 10(7): 666.

Faizan, M., Yu, F., Chen, C., Faraz, A. and Hayat, S. (2021). Zinc oxide nanoparticles help to enhance plant growth and alleviate abiotic stress: A review. Current Protein and Peptide Science, 22(5): 362–375.

Feba Mohan, M. and Praseetha, P. N. (2023). Prospects of biopolymers based nanocomposites for the slow and controlled release of agrochemicals formulations. Journal of Inorganic and Organometallic Polymers and Materials, 33(12): 3845–3860.

Feng, B.-H. and Peng, L.-F. (2012). Synthesis and characterization of carboxymethyl chitosan carrying ricinoleic functions as an emulsifier for azadirachtin. Carbohydrate Polymers, 88(2): 576–582.

Fernandes, M., Padrão, J., Ribeiro, A. I., Fernandes, R. D., Melro, L., Nicolau, T. et al. (2022). Polysaccharides and metal nanoparticles for functional textiles: A review. Nanomaterials, 12(6): 1006.

Fox, J.-P., Capen, J. D., Zhang, W., Ma, X. and Rossi, L. (2020). Effects of cerium oxide nanoparticles and cadmium on corn (*Zea mays* L.) seedlings physiology and root anatomy. NanoImpact, 20: 100264.

Fu, Z., Chen, K., Li, L., Zhao, F., Wang, Y., Wang, M. et al. (2018). Spherical and spindle-like abamectin-loaded nanoparticles by flash nanoprecipitation for southern root-knot nematode control: Preparation and characterization. Nanomaterials, 8(6): 449.

Gamage, A., Liyanapathiranage, A., Manamperi, A., Gunathilake, C., Mani, S., Merah, O. et al. (2022). Applications of starch biopolymers for a sustainable modern agriculture. Sustainability, 14(10): 6085.

Ghobashy, M. M. (2020). The application of natural polymer-based hydrogels for agriculture. In Hydrogels based on natural polymers (pp. 329–356). Elsevier.

Gomes, D. G., Pelegrino, M. T., Ferreira, A. S., Bazzo, J. H., Zucareli, C., Seabra, A. B. et al. (2021). Seed priming with copper-loaded chitosan nanoparticles promotes early growth and enzymatic antioxidant defense of maize (*Zea mays* L.) seedlings. Journal of Chemical Technology & Biotechnology, 96(8): 2176–2184.

Grillo, R., Pereira, A. E., Nishisaka, C. S., De Lima, R., Oehlke, K., Greiner, R. et al. (2014). Chitosan/tripolyphosphate nanoparticles loaded with paraquat herbicide: An environmentally safer alternative for weed control. Journal of Hazardous Materials, 278: 163–171.

Guan, H., Chi, D., Yu, J. and Li, X. (2008). A novel photodegradable insecticide: Preparation, characterization and properties evaluation of nano-Imidacloprid. Pesticide Biochemistry and Physiology, 92(2): 83–91.

Gupta, R. and Xie, H. (2018). Nanoparticles in daily life: Applications, toxicity and regulations. Journal of Environmental Pathology, Toxicology and Oncology, 37(3).

Haydar, M. S., Ghosh, D. and Roy, S. (2024). Slow and controlled release nanofertilizers as an efficient tool for sustainable agriculture: Recent understanding and concerns. Plant Nano Biology, 100058.

Hernández-Díaz, J. A., Garza-García, J. J., Zamudio-Ojeda, A., León-Morales, J. M., López-Velázquez, J. C. and García-Morales, S. (2021). Plant-mediated synthesis of nanoparticles and their antimicrobial activity against phytopathogens. Journal of the Science of Food and Agriculture, 101(4): 1270–1287.

Holister, P., Harper, T. E. and Vas, C. R. (2003). Nanotubes white paper. CMP Cientìfica, January.

Huang, D., Dang, F., Huang, Y., Chen, N. and Zhou, D. (2022). Uptake, translocation, and transformation of silver nanoparticles in plants. Environmental Science, Nano, 9(1): 12–39.

Hussein, H., Shaarawy, H., Hussien, N. H. and Hawash, S. (2019). Preparation of nano-fertilizer blend from banana peels. Bulletin of the National Research Centre, 43(1): 1–9.

Ilk, S., Saglam, N. and Özgen, M. (2017). Kaempferol loaded lecithin/chitosan nanoparticles: Preparation, characterization, and their potential applications as a sustainable antifungal agent. Artificial Cells, Nanomedicine, and Biotechnology, 45(5): 907–916.

Ing, L. Y., Zin, N. M., Sarwar, A. and Katas, H. (2012). Antifungal activity of chitosan nanoparticles and correlation with their physical properties. International Journal of Biomaterials, 2012(1): 632698.

Ingle, P. U., Shende, S. S., Shingote, P. R., Mishra, S. S., Sarda, V., Wasule, D. L. et al. (2022). Chitosan nanoparticles (ChNPs): A versatile growth promoter in modern agricultural production. Heliyon, 8(11).

Jiang, X., Yang, F., Jia, W., Jiang, Y., Wu, X., Song, S. et al. (2024). Nanomaterials and Nanotechnology in Agricultural Pesticide Delivery: A Review. Langmuir.

Kadam, P. M., Prajapati, D., Kumaraswamy, R., Kumari, S., Devi, K. A., Pal, A. et al. (2021). Physio-biochemical responses of wheat plant towards salicylic acid-chitosan nanoparticles. Plant Physiology and Biochemistry, 162: 699–705.

Kah, M., Beulke, S., Tiede, K. and Hofmann, T. (2013). Nanopesticides: State of knowledge, environmental fate, and exposure modeling. Critical Reviews in Environmental Science and Technology, 43(16): 1823–1867.

Kah, M. and Hofmann, T. (2014). Nanopesticide research: current trends and future priorities. Environment International, 63: 224–235.

Kamle, M., Borah, R., Bora, H., Jaiswal, A. K., Singh, R. K. and Kumar, P. (2020). Systemic acquired resistance (SAR) and induced systemic resistance (ISR): Role and mechanism of action against phytopathogens. Fungal biotechnology and Bioengineering, 457–470.

Kashyap, P. L., Xiang, X. and Heiden, P. (2015). Chitosan nanoparticle based delivery systems for sustainable agriculture. International Journal of Biological Macromolecules, 77: 36–51.

Kaur, P., Duhan, J. S. and Thakur, R. (2018). Comparative pot studies of chitosan and chitosan-metal nanocomposites as nano-agrochemicals against fusarium wilt of chickpea (*Cicer arietinum* L.). Biocatalysis and Agricultural Biotechnology, 14: 466–471.

Khairy, A. M., Tohamy, M. R., Zayed, M. A., Mahmoud, S. F., El-Tahan, A. M., El-Saadony, M. T. et al. (2022). Eco-friendly application of nano-chitosan for controlling potato and tomato bacterial wilt. Saudi Journal of Biological Sciences, 29(4): 2199–2209.

Khan, S., Zahoor, M., Khan, R. S., Ikram, M. and Islam, N. U. (2023). The impact of silver nanoparticles on the growth of plants: The agriculture applications. Heliyon, 9(6).

Khati, P., Chaudhary, P., Gangola, S., Bhatt, P. and Sharma, A. (2017). Nanochitosan supports growth of *Zea mays* and also maintains soil health following growth. 3 Biotech, 7: 1–9.

Kookana, R. S., Boxall, A. B., Reeves, P. T., Ashauer, R., Beulke, S., Chaudhry, Q. et al. (2014). Nanopesticides: guiding principles for regulatory evaluation of environmental risks. Journal of Agricultural and Food Chemistry, 62(19): 4227–4240.

Kottegoda, N., Munaweera, I., Madusanka, N. and Karunaratne, V. (2011). A green slow-release fertilizer composition based on urea-modified hydroxyapatite nanoparticles encapsulated wood. Current science, 73–78.

Kottegoda, N., Sandaruwan, C., Priyadarshana, G., Siriwardhana, A., Rathnayake, U. A., Berugoda Arachchige, D. M. et al. (2017). Urea-hydroxyapatite nanohybrids for slow release of nitrogen. ACS Nano, 11(2): 1214–1221.

Kubavat, D., Trivedi, K., Vaghela, P., Prasad, K., Vijay Anand, G. K., Trivedi, H. et al. (2020). Characterization of a chitosan-based sustained release nanofertilizer formulation used as a soil conditioner while simultaneously improving biomass production of *Zea mays* L. Land Degradation & Development, 31(17): 2734–2746.

Kumar, R., Næss, G. and Sørensen, M. (2023). Slow-release fertilizers using lignin: Challenges and future prospects. Biofuels, Bioproducts and Biorefining, 17(5): 1368–1381.

Kumar, S., Bhanjana, G., Sharma, A., Sidhu, M. and Dilbaghi, N. (2014). Synthesis, characterization and on field evaluation of pesticide loaded sodium alginate nanoparticles. Carbohydrate Polymers, 101: 1061–1067.

Kumaraswamy, R., Kumari, S., Choudhary, R. C., Pal, A., Raliya, R., Biswas, P. et al. (2018). Engineered chitosan based nanomaterials: Bioactivities, mechanisms and perspectives in plant protection and growth. International Journal of Biological Macromolecules, 113: 494–506.

Kumaraswamy, R., Saharan, V., Kumari, S., Choudhary, R. C., Pal, A., Sharma, S. S. et al. (2021). Chitosan-silicon nanofertilizer to enhance plant growth and yield in maize (*Zea mays* L.). Plant Physiology and Biochemistry, 159: 53–66.

Kwak, S.-Y., Lew, T. T. S., Sweeney, C. J., Koman, V. B., Wong, M. H., Bohmert-Tatarev, K. et al. (2019). Chloroplast-selective gene delivery and expression in planta using chitosan-complexed single-walled carbon nanotube carriers. Nature Nanotechnology, 14(5): 447–455.

Lao, S.-B., Zhang, Z.-X., Xu, H.-H. and Jiang, G.-B. (2010). Novel amphiphilic chitosan derivatives: Synthesis, characterization and micellar solubilization of rotenone. Carbohydrate Polymers, 82(4): 1136–1142.

Lateef, A., Nazir, R., Jamil, N., Alam, S., Shah, R., Khan, M. N. et al. (2019). Synthesis and characterization of environmental friendly corncob biochar based nano-composite–A potential slow release nano-fertilizer for sustainable agriculture. Environmental Nanotechnology, Monitoring & Management, 11: 100212.

Lawrencia, D., Wong, S. K., Low, D. Y. S., Goh, B. H., Goh, J. K., Ruktanonchai, U. R. et al. (2021). Controlled release fertilizers: A review on coating materials and mechanism of release. Plants, 10(2): 238.

Li, R., He, J., Xie, H., Wang, W., Bose, S. K., Sun, Y. et al. (2019). Effects of chitosan nanoparticles on seed germination and seedling growth of wheat (*Triticum aestivum* L.). International Journal of Biological Macromolecules, 126: 91–100.

Liang, W., Yu, A., Wang, G., Zheng, F., Jia, J. and Xu, H. (2018). Chitosan-based nanoparticles of avermectin to control pine wood nematodes. International Journal of Biological Macromolecules, 112: 258–263.

Liu, Y., Yan, L., Heiden, P. and Laks, P. (2001). Use of nanoparticles for controlled release of biocides in solid wood. Journal of Applied Polymer Science, 79(3): 458–465.

Lusser, M., Parisi, C., Plan, D. and Rodríguez-Cerezo, E. (2012). Deployment of new biotechnologies in plant breeding. Nature Biotechnology, 30(3): 231–239.

Lv, Z., Jiang, R., Chen, J. and Chen, W. (2020). Nanoparticle-mediated gene transformation strategies for plant genetic engineering. The Plant Journal, 104(4): 880–891.

Lynch, J. P. and Brown, K. M. (2012). New roots for agriculture: Exploiting the root phenome. Philosophical Transactions of the Royal Society B: Biological Sciences, 367(1595): 1598–1604.

Machado, T. O., Beckers, S. J., Fischer, J., Müller, B., Sayer, C., de Araújo, P. H. et al. (2020). Bio-based lignin nanocarriers loaded with fungicides as a versatile platform for drug delivery in plants. Biomacromolecules, 21(7): 2755–2763.

Machado, T. O., Beckers, S. J., Fischer, J., Sayer, C., de Araújo, P. H., Landfester, K. et al. (2021). Cellulose nanocarriers via miniemulsion allow pathogen-specific agrochemical delivery. Journal of Colloid and Interface Science, 601: 678–688.

Machado, T. O., Grabow, J., Sayer, C., de Araújo, P. H., Ehrenhard, M. L. and Wurm, F. R. (2022). Biopolymer-based nanocarriers for sustained release of agrochemicals: A review on materials and social science perspectives for a sustainable future of agri-and horticulture. Advances in Colloid and Interface Science, 303: 102645.

Maher, M. F., Nasti, R. A., Vollbrecht, M., Starker, C. G., Clark, M. D. and Voytas, D. F. (2020). Plant gene editing through *de novo* induction of meristems. Nature Biotechnology, 38(1): 84–89.

Mahmoud, N. E., Abdel-Gawad, H. and Abdelhameed, R. M. (2024). Post-synthetic modification of nano-chitosan using gibberellic acid: Foliar application on sorghum under salt stress conditions and estimation of biochemical parameters. Plant Physiology and Biochemistry, 211: 108655.

Malerba, M. and Cerana, R. (2016). Chitosan effects on plant systems. International Journal of Molecular Sciences, 17(7): 996.

Malerba, M. and Cerana, R. (2019). Recent applications of chitin-and chitosan-based polymers in plants. Polymers, 11(5): 839.

Maluin, F. N., Hussein, M. Z., Yusof, N. A., Fakurazi, S., Idris, A. S., Zainol Hilmi, N. H. et al. (2019). Preparation of chitosan–hexaconazole nanoparticles as fungicide nanodelivery system for combating Ganoderma disease in oil palm. Molecules, 24(13): 2498.

Maluin, F. N., Hussein, M. Z., Azah Yusof, N., Fakurazi, S., Idris, A. S., Zainol Hilmi, N. H. et al. (2020). Chitosan-based agronanofungicides as a sustainable alternative in the basal stem rot disease management. Journal of Agricultural and Food Chemistry, 68(15): 4305–4314.

Manikandan, A. and Sathiyabama, M. (2016). Preparation of chitosan nanoparticles and its effect on detached rice leaves infected with *Pyricularia grisea*. International Journal of Biological Macromolecules, 84: 58–61.

Martin-Ortigosa, S., Valenstein, J. S., Lin, V. S. Y., Trewyn, B. G. and Wang, K. (2012). Gold functionalized mesoporous silica nanoparticle mediated protein and DNA codelivery to plant cells via the biolistic method. Advanced Functional Materials, 22(17): 3576–3582.

Martins, N. C., Avellan, A., Rodrigues, S., Salvador, D., Rodrigues, S. M. and Trindade, T. (2020). Composites of biopolymers and ZnO NPs for controlled release of zinc in agricultural soils and timed delivery for maize. ACS Applied Nano Materials, 3(3): 2134–2148.

Materu, S. F., Heise, S. and Urban, B. (2021). Seasonal and spatial detection of pesticide residues under various weather conditions of agricultural areas of the kilombero valley ramsar site, Tanzania. Frontiers in Environmental Science, 9: 599814.

Melaj, M. A. and Daraio, M. E. (2014). HPMC layered tablets modified with chitosan and xanthan as matrices for controlled-release fertilizers. Journal of Applied Polymer Science, 131(19).

Mishra, K. K., Subbanna, A., Rajashekara, H., Paschapur, A. U., Singh, A. K., Jeevan, B. et al. (2022). Chitosan and chitosan-based nanoparticles for eco-friendly management of plant diseases and insect pests: A concentric overview. In Role of chitosan and chitosan-based nanomaterials in plant sciences (pp. 435–451). Elsevier.

Mishra, P. K., Usmani, G. A., Goswami, A. and Mondal, A. (2023). Preparation of chitosan clay sulfentrazone nanocomposite and its characterization. BioNanoScience, 13(4): 2066–2073.

Mishra, R. K., Ha, S. K., Verma, K. and Tiwari, S. K. (2018). Recent progress in selected bio-nanomaterials and their engineering applications: An overview. Journal of Science: Advanced Materials and Devices, 3(3): 263–288.

Mosa, M. A. and El-Abeid, S. E. (2023). Chitosan-loaded copper oxide nanoparticles: a promising antifungal nanocomposite against fusarium wilt disease of tomato plants. Sustainability, 15(19): 14295.

Mosavikia, A. A., Mosavi, S. G., Seghatoleslami, M. and Baradaran, R. (2020). Chitosan nanoparticle and pyridoxine seed priming improves tolerance to salinity in milk thistle seedling. Notulae Botanicae Horti Agrobotanici Cluj-Napoca, 48(1): 221–233.

Naderi, M. and Danesh-Shahraki, A. (2013). Nanofertilizers and their Roles in Sustainable Agriculture.

Naseer, I., Javad, S., Singh, A., Maqsood, S., Iqbal, S. and Jabeen, K. (2022). Alleviation mechanism of drought stress in plants using metal nanoparticles–a perspective analysis. In Sustainable Agriculture Reviews 53: Nanoparticles: A New Tool to Enhance Stress Tolerance (pp. 115–149). Springer.

Nguyen, T. N., Huynh, T. N., Hoang, D., Nguyen, D. H., Nguyen, Q. H. and Tran, T. H. (2019). Functional nanostructured oligochitosan–silica/carboxymethyl cellulose hybrid materials: Synthesis and investigation of their antifungal abilities. Polymers, 11(4): 628.

Nicol, J., Turner, S., Coyne, D. L., Nijs, L. d., Hockland, S. and Maafi, Z. T. (2011). Current nematode threats to world agriculture. Genomics and Molecular Genetics of Plant-Nematode Interactions, 21–43.

Oh, J.-W., Chun, S. C. and Chandrasekaran, M. (2019). Preparation and *in vitro* characterization of chitosan nanoparticles and their broad-spectrum antifungal action compared to antibacterial activities against phytopathogens of tomato. Agronomy, 9(1): 21.

Olad, A., Zebhi, H., Salari, D., Mirmohseni, A. and Tabar, A. R. (2018). Slow-release NPK fertilizer encapsulated by carboxymethyl cellulose-based nanocomposite with the function of water retention in soil. Materials Science and Engineering: C, 90: 333–340.

Pandey, S., Giri, K., Kumar, R., Mishra, G. and Raja Rishi, R. (2018). Nanopesticides: Opportunities in crop protection and associated environmental risks. Proceedings of the National Academy of Sciences, India Section B: Biological Sciences, 88: 1287–1308.

Paul, A. and Sharma, S. S. (2023). Biopolymer-based nanocomposites. pp. 523–550. *In*: Thomas, S., Ar, A., Jose Chirayil, C. and Thomas, B. (eds.): Handbook of Biopolymers. Springer Nature Singapore. https://doi.org/10.1007/978-981-19-0710-4_19.

Paula, H. C., Sombra, F. M., de Freitas Cavalcante, R., Abreu, F. O. and de Paula, R. C. (2011). Preparation and characterization of chitosan/cashew gum beads loaded with *Lippia sidoides* essential oil. Materials Science and Engineering: C, 31(2): 173–178.

Pereira, A. d. E. S., Oliveira, H. C. and Fraceto, L. F. (2019). Polymeric nanoparticles as an alternative for application of gibberellic acid in sustainable agriculture: A field study. Scientific Reports, 9(1): 7135.

Pimentel, C., Pina, C. M., Müller, N., Lara, L. A., Melo Rodriguez, G., Orlando, F. et al. (2023). Mineral particles in foliar fertilizer formulations can improve the rate of foliar uptake. Plants, 13(1): 71.

Pinto, R. J., Almeida, A., Fernandes, S. C., Freire, C. S., Silvestre, A. J., Neto, C. P. et al. (2013). Antifungal activity of transparent nanocomposite thin films of pullulan and silver against *Aspergillus niger*. Colloids and Surfaces B: Biointerfaces, 103: 143B.148.

Prasad, R., Bhattacharyya, A. and Nguyen, Q. D. (2017). Nanotechnology in sustainable agriculture: Recent developments, challenges, and perspectives. Frontiers in microbiology, 8: 1014.

Prashanth, K. H. and Tharanathan, R. (2007). Chitin/chitosan: modifications and their unlimited application potential—an overview. Trends in Food Science & Technology, 18(3): 117–131.

Qiao, J., Sun, W., Lin, S., Jin, R., Ma, L. and Liu, Y. (2019). Cytosolic delivery of CRISPR/Cas9 ribonucleoproteins for genome editing using chitosan-coated red fluorescent protein. Chemical communications, 55(32): 4707–4710.

Rajonee, A. A., Zaman, S. and Huq, S. M. I. (2017). Preparation, characterization and evaluation of efficacy of phosphorus and potassium incorporated nano fertilizer. Advances in Nanoparticles, 6(02): 62–74.

Rajpal, V. R., Dhingra, Y., Khungar, L., Mehta, S., Minkina, T., Rajput, V. D. et al. (2024). Current research in biotechnology. Current Research in Biotechnology, 7: 100219.

Rameshaiah, G., Pallavi, J. and Shabnam, S. (2015). Nano fertilizers and nano sensors–an attempt for developing smart agriculture. Int. J. Eng. Res. Gen. Sci., 3(1): 314–320.

Rasheed, A., Li, H., Tahir, M. M., Mahmood, A., Nawaz, M., Shah, A. N. et al. (2022). The role of nanoparticles in plant biochemical, physiological, and molecular responses under drought stress: A review. Frontiers in Plant Science, 13: 976179.

Rastogi, A., Zivcak, M., Sytar, O., Kalaji, H. M., He, X., Mbarki, S. et al. (2017). Impact of metal and metal oxide nanoparticles on plant: A critical review. Frontiers in Chemistry, 5: 78.

Rastogi, A., Zivcak, M., Tripathi, D., Yadav, S. and Kalaji, H. (2019). Phytotoxic effect of silver nanoparticles in *Triticum aestivum*: Improper regulation of photosystem I activity as the reason for oxidative damage in the chloroplast. Photosynthetica, 57(1).

Raza, A., Khandelwal, K., Pandit, S., Singh, M., Kumar, S., Rustagi, S. et al. (2024). Exploring the potential of metallic and metal oxide nanoparticles for reinforced disease management in agricultural systems: A comprehensive review. Environmental Nanotechnology, Monitoring & Management, 100998.

Rico, C. M., Peralta-Videa, J. and Gardea-Torresdey, J. (2015). Chemistry, biochemistry of nanoparticles, and their role in antioxidant defense system in plants. Nanotechnology and plant sciences: Nanoparticles and their Impact on Plants, 1–17.

Riseh, R. S., Vatankhah, M., Hassanisaadi, M. and Varma, R. S. (2024). A review of chitosan nanoparticles: Nature's gift for transforming agriculture through smart and effective delivery mechanisms. International Journal of Biological Macromolecules, 129522.

Rizwan, M., Noureen, S., Ali, S., Anwar, S., Rehman, M. Z. U., Qayyum, M. F. et al. (2019). Influence of biochar amendment and foliar application of iron oxide nanoparticles on growth, photosynthesis, and cadmium accumulation in rice biomass. Journal of Soils and Sediments, 19: 3749–3759.

Rop, K., Karuku, G. N., Mbui, D., Michira, I. and Njomo, N. (2018). Formulation of slow release NPK fertilizer (cellulose-graft-poly (acrylamide)/nano-hydroxyapatite/soluble fertilizer) composite and evaluating its N mineralization potential. Annals of Agricultural Sciences, 63(2): 163–172.

Rychter, P. (2019). Chitosan/glyphosate formulation as a potential, environmental friendly herbicide with prolonged activity. Journal of Environmental Science and Health, Part B, 54(8): 681–692.

Ryder, G. (2017). The United Nations world water development report, 2017: Wastewater: the untapped resource.

Saharan, V., Sharma, G., Yadav, M., Choudhary, M. K., Sharma, S., Pal, A. et al. (2015). Synthesis and *in vitro* antifungal efficacy of Cu–chitosan nanoparticles against pathogenic fungi of tomato. International Journal of Biological Macromolecules, 75: 346–353.

Saharan, V., Kumaraswamy, R., Choudhary, R. C., Kumari, S., Pal, A., Raliya, R. et al. (2016). Cu-chitosan nanoparticle mediated sustainable approach to enhance seedling growth in maize by mobilizing reserved food. Journal of Agricultural and Food Chemistry, 64(31): 6148–6155.

Saini, P., Gopal, M., Kumar, R., Gogoi, R. and Srivastava, C. (2015). Bioefficacy evaluation and dissipation pattern of nanoformulation versus commercial formulation of pyridalyl in tomato (*Solanum lycopersicum*). Environmental Monitoring and Assessment, 187: 1–9.

Salimi, M., Channab, B.-e., El Idrissi, A., Zahouily, M. and Motamedi, E. (2023). A comprehensive review on starch: Structure, modification, and applications in slow/controlled-release fertilizers in agriculture. Carbohydrate Polymers, 121326.

Sandhya, Kumar, S., Kumar, D. and Dilbaghi, N. (2017). Preparation, characterization, and bio-efficacy evaluation of controlled release carbendazim-loaded polymeric nanoparticles. Environmental Science and Pollution Research, 24: 926–937.

Sangwan, S., Sharma, P., Wati, L. and Mehta, S. (2023). Effect of chitosan nanoparticles on growth and physiology of crop plants. In Engineered nanomaterials for sustainable agricultural production, soil improvement and stress management (pp. 99–123). Elsevier.

Santo Pereira, A. E., Silva, P. M., Oliveira, J. L., Oliveira, H. C. and Fraceto, L. F. (2017). Chitosan nanoparticles as carrier systems for the plant growth hormone gibberellic acid. Colloids and Surfaces B: Biointerfaces, 150: 141–152.

Schwab, F., Zhai, G., Kern, M., Turner, A., Schnoor, J. L. and Wiesner, M. R. (2016). Barriers, pathways and processes for uptake, translocation and accumulation of nanomaterials in plants–Critical review. Nanotoxicology, 10(3): 257–278.

Scrinis, G. and Lyons, K. (2007). The emerging nano-corporate paradigm: nanotechnology and the transformation of nature, food and agri-food systems. The International Journal of Sociology of Agriculture and Food, 15(2): 22–44.

Sen, S. K., Chouhan, D., Das, D., Ghosh, R. and Mandal, P. (2020). Improvisation of salinity stress response in mung bean through solid matrix priming with normal and nano-sized chitosan. International Journal of Biological Macromolecules, 145: 108–123.

Servin, A., Elmer, W., Mukherjee, A., De la Torre-Roche, R., Hamdi, H., White, J. C. et al. (2015). A review of the use of engineered nanomaterials to suppress plant disease and enhance crop yield. Journal of Nanoparticle Research, 17: 1–21.

Shang, Y., Hasan, M. K., Ahammed, G. J., Li, M., Yin, H. and Zhou, J. (2019). Applications of nanotechnology in plant growth and crop protection: A review. Molecules, 24(14): 2558.

Sharma, A., Sood, K., Kaur, J. and Khatri, M. (2019). Agrochemical loaded biocompatible chitosan nanoparticles for insect pest management. Biocatalysis and Agricultural Biotechnology, 18: 101079.

Sheikhalipour, M., Esmaielpour, B., Behnamian, M., Gohari, G., Giglou, M. T., Vachova, P. et al. (2021). Chitosan–selenium nanoparticle (Cs–Se NP) foliar spray alleviates salt stress in bitter melon. Nanomaterials, 11(3): 684.

Sheteiwy, M. S., Shaghaleh, H., Hamoud, Y. A., Holford, P., Shao, H., Qi, W. et al. (2021). Zinc oxide nanoparticles: Potential effects on soil properties, crop production, food processing, and food quality. Environmental Science and Pollution Research, 28(28): 36942–36966.

Singh, M., Goswami, S. P., Sachan, P., Sahu, D. K., Beese, S. and Pandey, S. K. (2024). Nanotech for fertilizers and nutrients-improving nutrient use efficiency with nano-enabled fertilizers. Journal of Experimental Agriculture International, 46(5): 220–247.

Singh, R., Anbazhagan, P., Viswanath, H. and Tomer, A. (2020). Trichoderma species: A Blessing for Crop Production. Trichoderma: Agricultural Applications and Beyond, 127–158.

Skrzypczak, D., Izydorczyk, G., Taf, R., Moustakas, K. and Chojnacka, K. (2023). Cellulose-based fertilizers for sustainable agriculture: Effective methods for increasing crop yield and soil health. Industrial Crops and Products, 205: 117500.

Song, S., Wang, Y., Xie, J., Sun, B., Zhou, N., Shen, H. et al. (2019). Carboxymethyl chitosan modified carbon nanoparticle for controlled emamectin benzoate delivery: Improved solubility, pH-responsive release, and sustainable pest control. ACS Applied Materials & Interfaces, 11(37): 34258–34267.

Sridhar, C. (2012). Effect of nanoparticles for the maintenance of tomato seed vigour and viability. M. Sc.(Agri.) Thesis, TNAU, Coimbatore (India).

Sun, C., Zeng, Z., Cui, H. and Verheggen, F. (2020). Polymer-based nanoinsecticides: Current developments, environmental risks and future challenges-a review. Biotechnologie, Agronomie, Société et Environnement, 24.

Sun, W., Dou, F., Li, C., Ma, X. and Ma, L. Q. (2021). Impacts of metallic nanoparticles and transformed products on soil health. Critical Reviews in Environmental Science and Technology, 51(10): 973–1002.

Taher, M. A., Khojah, E., Darwish, M. S., Elsherbiny, E. A., Elawady, A. A. and Dawood, D. H. (2022). Biosynthesis of silver nanoparticles by polysaccharide of *Leucaena leucocephala* seeds and their anticancer, antifungal properties and as preservative of composite milk sample. Journal of Nanomaterials, 2022(1): 7490221.

Thakur, S., Sharma, B., Verma, A., Chaudhary, J., Tamulevicius, S. and Thakur, V. K. (2018). Recent progress in sodium alginate based sustainable hydrogels for environmental applications. Journal of Cleaner Production, 198: 143–159.

Torabi, Z. and Mohammadi, N. A. (2013). The effects of SiO_2 Nanoparticles on Mechanical and Physicochemical Properties of Potato Starch Films.

Torney, F., Trewyn, B. G., Lin, V. S.-Y. and Wang, K. (2007). Mesoporous silica nanoparticles deliver DNA and chemicals into plants. Nature Nanotechnology, 2(5): 295–300.

Torney, F. (2009). Nanoparticle mediated plant transformation. Emerging technologies in plant science research. Proceedings of the Interdepartmental Plant Physiology Major Fall Seminar Series,(IPPM'09): Iowa State University,

Van, S. N., Minh, H. D. and Anh, D. N. (2013). Study on chitosan nanoparticles on biophysical characteristics and growth of Robusta coffee in green house. Biocatalysis and Agricultural Biotechnology, 2(4): 289–294.

Vatankhah, A., Aliniaeifard, S., Moosavi-Nezhad, M., Abdi, S., Mokhtarpour, Z., Reezi, S. et al. (2023). Plants exposed to titanium dioxide nanoparticles acquired contrasting photosynthetic and morphological strategies depending on the growing light intensity: A case study in radish. Scientific Reports, 13(1): 5873.

Vejan, P., Khadiran, T., Abdullah, R. and Ahmad, N. (2021). Controlled release fertilizer: A review on developments, applications and potential in agriculture. Journal of controlled Release, 339: 321–334.

Verma, M. L., Dhanya, B. S., Rani, V., Thakur, M., Jeslin, J. and Kushwaha, R. (2020). Carbohydrate and protein based biopolymeric nanoparticles: Current status and biotechnological applications. International Journal of Biological Macromolecules, 154: 390–412.
Wang, C., Gao, X., Chen, Z., Chen, Y. and Chen, H. (2017). Preparation, characterization and application of polysaccharide-based metallic nanoparticles: A review. Polymers, 9(12): 689.
Wang, L., Yu, G., Li, J., Feng, Y., Peng, Y., Zhao, X. et al. (2019). Stretchable hydrophobic modified alginate double-network nanocomposite hydrogels for sustained release of water-insoluble pesticides. Journal of Cleaner Production, 226: 122–132.
Wang, X., Xie, H., Wang, P. and Yin, H. (2023). Nanoparticles in plants: uptake, transport and physiological activity in leaf and root. Materials, 16(8): 3097.
Wang, Z., Wang, S., Ma, T., Liang, Y., Huo, Z. and Yang, F. (2023). Synthesis of zinc oxide nanoparticles and their applications in enhancing plant stress resistance: A review. Agronomy, 13(12): 3060.
Worrall, E. A., Hamid, A., Mody, K. T., Mitter, N. and Pappu, H. R. (2018). Nanotechnology for plant disease management. Agronomy, 8(12): 285.
Wu, H., Hu, P., Xu, Y., Xiao, C., Chen, Z., Liu, X. et al. (2021). Phloem delivery of fludioxonil by plant amino acid transporter-mediated polysuccinimide nanocarriers for controlling fusarium wilt in banana. Journal of Agricultural and Food Chemistry, 69(9): 2668–2678.
Wypij, M., Trzcińska-Wencel, J., Golińska, P., Avila-Quezada, G. D., Ingle, A. P. and Rai, M. (2023). The strategic applications of natural polymer nanocomposites in food packaging and agriculture: Chances, challenges, and consumers' perception. Frontiers in Chemistry, 10: 1106230.
Xie, X., He, Z., Chen, N., Tang, Z., Wang, Q. and Cai, Y. (2019). The roles of environmental factors in regulation of oxidative stress in plant. BioMed Research International, 2019(1): 9732325.
Xu, L., Cao, L.-D., Li, F.-M., Wang, X.-J. and Huang, Q.-L. (2014). Utilization of chitosan-lactide copolymer nanoparticles as controlled release pesticide carrier for pyraclostrobin against *Colletotrichum gossypii* Southw. Journal of Dispersion Science and Technology, 35(4): 544–550.
Xu, T., Ma, C., Aytac, Z., Hu, X., Ng, K. W., White, J. C. et al. (2020). Enhancing agrichemical delivery and seedling development with biodegradable, tunable, biopolymer-based nanofiber seed coatings. ACS Sustainable Chemistry & Engineering, 8(25): 9537–9548.
Xu, T., Wang, Y., Aytac, Z., Zuverza-Mena, N., Zhao, Z., Hu, X. et al. (2022). Enhancing agrichemical delivery and plant development with biopolymer-based stimuli responsive core–shell nanostructures. ACS Nano, 16(4): 6034–6048.
Yáñez, O., Alegría-Arcos, M., Suardiaz, R., Morales-Quintana, L., Castro, R. I., Palma-Olate, J. et al. (2023). Calcium-alginate-chitosan nanoparticle as a potential solution for pesticide removal, a computational approach. Polymers, 15(14): 3020.
Yearla, S. R. and Padmasree, K. (2016). Exploitation of subabul stem lignin as a matrix in controlled release agrochemical nanoformulations: A case study with herbicide diuron. Environmental Science and Pollution Research, 23: 18085–18098.
YIN, Y.-h., GUO, Q.-m., Yun, H., WANG, L.-j. and WAN, S.-q. (2012). Preparation, characterization and nematicidal activity of lansiumamide B nano-capsules. Journal of Integrative Agriculture, 11(7): 1151–1158.
Yu, J., Wang, D., Geetha, N., Khawar, K. M., Jogaiah, S. and Mujtaba, M. (2021). Current trends and challenges in the synthesis and applications of chitosan-based nanocomposites for plants: A review. Carbohydrate Polymers, 261: 117904.
Yusefi-Tanha, E., Fallah, S., Rostamnejadi, A. and Pokhrel, L. R. (2020). Particle size and concentration dependent toxicity of copper oxide nanoparticles (CuONPs) on seed yield and antioxidant defense system in soil grown soybean (Glycine max cv. Kowsar). Science of the Total Environment, 715: 136994.
Yusefi-Tanha, E., Fallah, S., Rostamnejadi, A. and Pokhrel, L. R. (2022). Responses of soybean (*Glycine max* [L.] Merr.) to zinc oxide nanoparticles: Understanding changes in root system architecture, zinc tissue partitioning and soil characteristics. Science of the Total Environment, 835: 155348.
Zinta, R., Tiwari, J. K., Buckseth, T., Thakur, K., Goutam, U., Kumar, D. et al. (2022). Root system architecture for abiotic stress tolerance in potato: Lessons from plants. Frontiers in Plant Science, 13: 926214.

CHAPTER-5

Innovative Applications of Biopolymeric Nanoparticles in Environmental Remediation

Mani Sivakumar,[1] *Ganeshraja Ayyakannu Sundaram,*[2] *Vaishali C.V.*[3] and *Muneeswaran Thillaichidambaram*[4,*]

1. Introduction

1.1 Overview of Environmental Remediation

Environmental remediation is crucial for reversing or reducing damage caused by pollutants in soil, water, and air. The main goal is to restore these environments to a safe state for both human health and ecosystems. This need is urgent due to industrial activities, urbanization, agricultural practices, and improper waste disposal, all contributing to environmental degradation. Contamination impacts biodiversity, ecosystem functions, and public health. Polluted water can cause diseases, contaminated soil reduces agricultural productivity, and air pollution affects human respiratory and cardiovascular health (Münzel et al., 2022). As the global population grows and industrial activities increase, the strain on natural resources becomes more severe, making effective remediation essential (Boretti and Rosa, 2019).

Remediation methods include physical, chemical, and biological techniques (Wu et al., 2023). Physical methods, like excavation, remove contaminants directly. Chemical methods use reactions to detoxify pollutants. Biological methods employ microorganisms, plants, or enzymes to break down contaminants (Ojuederie and Babalola, 2017). However, traditional methods have limitations such as high costs, long treatment times, and potential secondary pollution. Thus, innovative solutions are

[1] Department of Chemical Engineering, Faculty of Engineering, Mahidol University, Nakhon Pathom 73170, Thailand. Email: sivaphy05@gmail.com
[2] Department of Research Analytics, Saveetha Dental College and Hospitals, Saveetha Institute of Medical and Technical Sciences, Poonamallee High Road, Chennai 600 077, Tamil Nadu State, India. Email: asgchem84@gmail.com
[3] JSS Academy of Higher Education and Research, JSS Medical College, Sri Shivarathreeswara Nagara, Bannimantap, Mysuru, Karnataka 570015, India. Email: vrush94mmc@hotmail.com
[4] Lab of Toxicology, Department of Health Sciences, The Graduate School of Dong-A University, Busan 49315, Republic Korea.
* Corresponding author: tmuneeswaran@hotmail.com
DOI: 10.1201/9781003558330-5

needed. Nanotechnology, especially biopolymeric nanoparticles, provides promising improvements (Sundar et al., 2010).

Biopolymeric nanoparticles are made from natural polymers. They are biodegradable, low in toxicity, and can be customized to target specific pollutants (Sreena and Nathanael, 2023). These nanoparticles offer a more efficient and eco-friendly approach to remediation. The following sections will explore how biopolymeric nanoparticles are made, how they work, and their applications in environmental remediation. This discussion will show their potential to improve the field and address pollution challenges effectively. There are several methods of environmental remediation, each designed to handle specific types of contamination. Traditional techniques include physical, chemical, and biological processes. Physical remediation involves removing contaminants through methods like excavation, dredging, and soil vapor extraction, often used for solid waste or concentrated pollutants. Chemical remediation uses chemicals to neutralize or transform pollutants, with methods like chemical oxidation, reduction, and precipitation to break down hazardous substances into less harmful compounds. Biological remediation, or bioremediation, employs microorganisms, plants, or enzymes to degrade or detoxify pollutants through techniques such as phytoremediation, bioaugmentation, and bio stimulation, favored for their eco-friendly and sustainable nature (Azubuike et al., 2016; Kafle et al., 2022).

Despite the variety of existing remediation techniques, challenges such as high costs, long treatment times, and potential secondary pollution still exist. These limitations have led to the search for new and more efficient remediation technologies. In recent years, nanotechnology has become a promising field for improving environmental remediation processes. Biopolymeric nanoparticles, made from natural polymers, have gained significant attention due to their unique properties and eco-friendly nature (Arif et al., 2022). These nanoparticles are biodegradable, low in toxicity, and can be engineered for specific functions (Kučuk et al., 2023). This chapter explores the innovative uses of biopolymeric nanoparticles in environmental remediation. It looks at how these advanced materials can overcome the limitations of traditional methods and provide new solutions for cleaning up contaminated environments. By examining how these nanoparticles are made, how they work, and their specific uses, this chapter aims to highlight their potential to transform the field of environmental remediation.

This design and mechanism of biopolymer-based nanocarriers highlight their role in enhancing the efficiency and longevity of agrochemical applications. These nanocarriers are engineered to provide a controlled and sustained release of nutrients and pesticides, which reduces the frequency of application and minimizes environmental impact (Machado et al., 2022). Additionally, the figure showcases the crucial role of these nanocarriers in environmental remediation, where they adsorb and degrade pollutants, thus improving the soil and water quality. By demonstrating the dual functionality of these innovative nanomaterials, the figure emphasizes their significant potential in advancing both agricultural productivity and environmental protection. The integration of biopolymer-based nanocarriers in agrochemical delivery systems and environmental cleanup strategies represents a sustainable approach to modern agricultural and environmental management.

1.2 Definition and Importance of Biopolymeric Nanoparticles

Biopolymeric nanoparticles are tiny particles composed of natural polymers that range in size from 1 to 100 nanometers. These natural polymers include materials like chitosan, alginate, cellulose, and proteins. Due to their nanoscale size, biopolymeric nanoparticles possess unique physical and chemical properties that distinguish them from their bulk counterparts (Reddy et al., 2021). Their small size, large surface area, and ability to be functionalized make them highly effective in a variety of applications, including environmental remediation (Saraswat et al., 2023). The importance of biopolymeric nanoparticles lies in their combination of efficacy and eco-friendliness (Jafarzadeh et al., 2024). Unlike many synthetic nanoparticles, biopolymeric nanoparticles are biodegradable and often derived from renewable resources, making them a more sustainable option.

This biodegradability ensures that they break down into harmless byproducts after their functional life, reducing the risk of long-term environmental impact (Moshood et al., 2022).

In the context of environmental remediation, biopolymeric nanoparticles offer several advantages:

i. **Biodegradability and Low Toxicity:** Biopolymeric nanoparticles are made from natural polymers, which make them non-toxic and environmentally friendly. This characteristic ensures that they break down into harmless byproducts, minimizing any long-term environmental impact (Gupta et al., 2022; Machado et al., 2022). As a result, they are particularly suitable for where human and ecological safety is paramount. Moreover, their use can reduce the risk of bioaccumulation and persistence in the environment, issues often associated with synthetic nanoparticles. The natural origin of these nanoparticles also means they are less likely to provoke adverse biological reactions, making them applications safer for widespread use (Mitchell et al., 2021). Additionally, the production of biopolymeric nanoparticles often involves fewer harmful chemicals and processes, further enhancing their eco-friendly profile (Kumar et al., 2022; Vodyashkin et al., 2022).

ii. **High Surface Area and Reactivity:** Due to their nanoscale dimensions, biopolymeric nanoparticles possess a significantly larger surface area compared to their volume. This characteristic enhances their reactivity and efficiency in capturing and degrading pollutants in environmental remediation processes (Rather et al., 2023). The increased surface area allows for more interactions with contaminants, facilitating faster and more effective remediation. This heightened reactivity is particularly advantageous in scenarios where pollutants are present at low concentrations or in complex matrices. Additionally, the small size of these nanoparticles enables them to penetrate into tight spaces and reach inaccessible areas, further improving their remediation capabilities. Furthermore, their enhanced reactivity reduces the amounts of nanoparticles required for remediation, resulting in more cost-effective and sustainable solutions (Thomas et al., 2022).

iii. **Customizability:** Biopolymeric nanoparticles offer the advantage of customization, allowing them to target specific contaminants effectively. Through adjustments to their surface properties or the inclusion of specific functional groups, these nanoparticles can be tailored to interact with particular types of pollutants (Kora and Rastogi, 2016). This tailored approach enhances the efficiency and selectivity of remediation processes, ensuring that only the target pollutants are captured or degraded. Furthermore, the ability to engineer biopolymeric nanoparticles for specific applications maximizes their utility in diverse environmental remediation scenarios (Ashraf et al., 2024). This targeted design minimizes unnecessary environmental impact and reduces the amount of remediation material required, contributing to more sustainable and efficient remediation strategies.

iv. **Versatility:** Biopolymeric nanoparticles exhibit versatility across a spectrum of remediation techniques. They excel in adsorbing heavy metals, degrading organic pollutants, and even participating in disinfection processes (Simelane et al., 2022). This adaptability extends their utility to diverse environmental contexts, including soil, water, and air remediation efforts. Their ability to address multiple types of contaminants makes them valuable tools in combating complex pollution challenges. Furthermore, their versatility allows for the development of integrated remediation strategies that can simultaneously target different pollutants present in contaminated environments (Russo et al., 2021). This multifunctionality streamlines remediation efforts, providing comprehensive solutions for environmental cleanup while minimizing the need for multiple remediation agents.

v. **Sustainability:** Utilizing renewable natural polymers in biopolymeric nanoparticles aligns with sustainable development objectives. This approach reduces reliance on finite fossil fuels and diminishes the carbon footprint linked to the manufacture and application of remediation materials. By utilizing resources that can be replenished naturally, such as chitosan and cellulose, biopolymeric nanoparticles contribute to environmental conservation efforts (Kanmani et al., 2017). Additionally, their biodegradability ensures minimal long-term impact on ecosystems, further enhancing their sustainability profile. Embracing biopolymeric nanoparticles in environmental remediation not only addresses immediate pollution concerns but also supports long-term ecological resilience and resource conservation (Noreen et al., 2024).

Using biopolymeric nanoparticles for cleaning up the environment is a big step forward. These tiny particles have special traits that can make cleanup methods work better and be safer for the environment. By using these nanoparticles, we can make cleanup methods more effective, more eco-friendly, and safer for everyone. This helps tackle pollution problems we face today and opens up new ways to protect our environment and restore ecosystems for the future.

1.3 Scope and Objectives of the Chapter

This chapter provides a thorough investigation into the diverse uses of biopolymeric nanoparticles for cleaning up the environment. It explores how these nanoparticles are made, what makes them special, and how they are applied to tackle pollution in soil, water, and air. Beginning with an introduction to biopolymeric nanoparticles, it emphasizes their importance in environmental cleanup efforts. The chapter then delves into the processes involved in synthesizing and modifying these nanoparticles to suit different cleanup tasks, focusing on factors like scalability and cost-effectiveness. Furthermore, it highlights the unique qualities of biopolymeric nanoparticles that make them effective in capturing and breaking down pollutants, demonstrating their applications in soil remediation, water purification, and air quality enhancement. Through real-life examples, it showcases the impact of these nanoparticles in combatting pollution challenges. Additionally, the chapter examines the hurdles and future directions in this field, aiming to encourage further exploration and advancement. By addressing these critical aspects, this chapter aims to deepen readers' understanding of biopolymeric nanoparticles' role in environmental remediation and spur continued research and development in this promising field.

2. Fundamentals of Biopolymeric Nanoparticles

Biopolymeric nanoparticles show unique properties that make them highly valuable in various applications, including environmental remediation. These nanoparticles offer significant advantages such as biocompatibility, low toxicity, and biodegradability, making them safe and eco-friendly options for tackling environmental challenges (Saraswat et al., 2023). Understanding these fundamental properties and processes is crucial for exploring their diverse applications in environmental cleanup and other fields of science and technology.

Biopolymeric nanoparticles are derived from natural polymers found in living organisms or renewable resources. These natural polymers include chitosan, alginate, cellulose, and proteins (Das et al., 2023). What sets biopolymeric nanoparticles apart is their biocompatibility, biodegradability, and low toxicity, making them environmentally friendly and safe for use in a wide range of applications. Additionally, their nanoscale size grants them large surface area-to-volume ratios, enhancing their reactivity and efficacy in various processes, including pollutant capture and degradation.

2.1 Types of Biopolymeric Nanoparticles

i. **Chitosan Nanoparticles:** Chitosan, derived from chitin found in the exoskeletons of crustaceans such as shrimp and crabs, is one of the most extensively studied biopolymer materials (Pellis et al., 2022). Chitosan nanoparticles, which are nanoparticles made from chitosan, have garnered significant attention due to their remarkable properties and versatile applications. These nanoparticles possess excellent biocompatibility, meaning they are well-tolerated by living organisms, including humans. This makes them ideal for various biomedical applications, such as drug delivery systems, where they can transport therapeutic agents to targeted sites in the body effectively. Additionally, chitosan nanoparticles exhibit potent antimicrobial properties, making them effective against a wide range of bacteria and fungi. This antimicrobial activity renders them valuable in wound healing applications, where they can help prevent infections and promote tissue regeneration (Ul-Islam et al., 2024; Yan et al., 2021). Furthermore, chitosan nanoparticles have shown promise in environmental remediation efforts. Their ability to adsorb heavy metals and other pollutants from water and soil makes them valuable tools for cleaning up contaminated environments (Ahmed and Mohamed, 2023). By capturing and immobilizing pollutants, chitosan nanoparticles contribute to restoring the ecosystem health and safeguarding human health (Bashir et al., 2022; Karamchandani et al., 2024). Overall, the unique combination of biocompatibility and antimicrobial properties makes chitosan nanoparticles versatile and effective materials for applications ranging from medicine to environmental protection.

Crab and shrimp shell waste serves as the primary raw material for commercial chitosan production. This seafood waste is repurposed to create an environmentally friendly and stable biomaterial. Crustacean shells predominantly consist of chitin, proteins, minerals, and lipids. During industrial processing, the shells undergo acid treatment to dissolve calcium carbonate, followed by alkali extraction to remove proteins. Subsequent deproteinization, decolorization, and deacetylation steps yield a colorless chitosan product. The detailed processing steps are illustrated in Fig. 1.

ii. **Alginate Nanoparticles:** Alginate, a biopolymer derived from seaweed, is recognized for its exceptional biocompatibility and biodegradability. This natural polymer has been extensively investigated for its diverse applications in various fields, including biomedicine and environmental science (Fernando et al., 2020). Alginate nanoparticles, which are nanoparticles made from alginate, have garnered considerable interest due to their unique properties and versatile capabilities. One of the key advantages of alginate nanoparticles lies in their ability to encapsulate and release active compounds in a controlled manner (Dodero et al., 2021). This controlled release feature makes them invaluable in drug delivery systems, where they can protect sensitive drugs from degradation and deliver them to specific targets within the body at

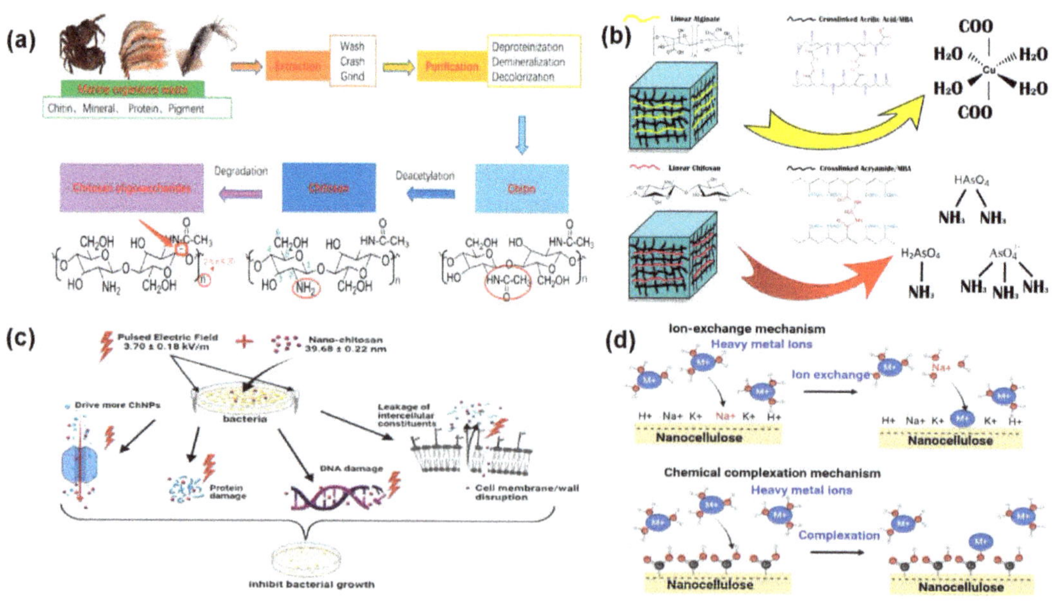

Figure 1. Overview of production and structure of chitosan (Li and Wu, 2020). (b) PAA/alginate and PAAM/chitosan of hydrogels for adsorption of arsenic and copper (ALSamman and Sánchez, 2023). (c) Nano-biopolymer antibacterial activity by pulsed electric fields (Emmanuel et al., 2024). (d) Nanocelluloses using heavy metal removal mechanism from wastewater systems (Salama et al., 2021).

the desired rate. Additionally, alginate nanoparticles have shown promise in tissue engineering applications, where they can serve as scaffolds for cell growth and tissue regeneration. Their biocompatibility and ability to mimic the extracellular matrix make them suitable substrates for supporting cell adhesion, proliferation, and differentiation. Moreover, alginate nanoparticles have been explored for environmental applications, particularly in water treatment and remediation (Venkatesan et al., 2017). Their ability to encapsulate contaminants and release them in a controlled manner offers potential solutions for removing pollutants from water sources effectively. By harnessing the unique properties of alginate nanoparticles, researchers aim to develop innovative strategies for addressing biomedical and environmental challenges, ultimately contributing to improved healthcare and environmental sustainability (Lee and Mooney, 2012).

iii. **Cellulose Nanoparticles:** Cellulose, derived from plant cell walls, stands out as a renewable and abundant biopolymer. Its widespread availability, along with its unique properties, has attracted significant interest in the field of materials science and engineering (Seddiqi et al., 2021). Cellulose nanoparticles, which are nanoparticles made from cellulose, offer a range of advantageous characteristics that make them highly promising for various applications (Abdelhamid and Mathew, 2022). One notable feature of cellulose nanoparticles is their exceptional mechanical strength, which surpasses that of many synthetic polymers. This attribute makes them ideal candidates for reinforcing composite materials, enhancing their durability and performance in diverse industries such as automotive, aerospace, and construction. Moreover, cellulose nanoparticles exhibit excellent biocompatibility, meaning they are well-tolerated by living organisms. This property makes them attractive for biomedical applications, including drug delivery systems, where they can safely transport therapeutic agents to targeted sites in the body (Patil et al., 2022). Additionally, cellulose nanoparticles possess optical transparency, enabling their use in optical devices, sensors, and biomedical imaging applications. Furthermore, cellulose nanoparticles have shown promise in environmental remediation efforts (Khorsandi et al., 2024). Their ability to adsorb pollutants from water and soil, combined with their biodegradability, makes them valuable tools for cleaning up contaminated environments (Bala et al., 2022). By harnessing the renewable and versatile nature of cellulose nanoparticles, researchers aim to develop innovative solutions for a wide range of challenges in materials science, biomedicine, and environmental engineering, ultimately contributing to sustainable development and improved quality of life.

Nanocrystalline cellulose (NCC) refers to cellulose particles that are scaled down to the nanometer range. This material represents cellulose in its highest crystalline form, comprising its nanocrystalline domains. NCC exhibits remarkable mechanical strength, biocompatibility, and tunable surface chemistry, making it a versatile material for various applications (Ahmed Mohamed et al., 2020). This structure and the properties of NCC, highlight its nanoscale dimensions and crystalline nature. Additionally, the figure demonstrates how NCC can be incorporated into composites and other materials to enhance their performance and functionality (Mikaela and Gunnar, 2015).

iv. **Protein-Based Nanoparticles:** Proteins, abundant in living organisms, provide a rich source of materials for nanoparticle synthesis. Albumin, gelatin, silk fibroin, and other proteins can be harnessed to create biopolymeric nanoparticles with tailored properties for various applications (Kučuk et al., 2023). One of the key advantages of protein-based nanoparticles lies in their high biocompatibility, ensuring minimal adverse effects when introduced into living systems. This property is particularly crucial for biomedical applications, where compatibility with biological tissues is essential for safety and efficacy. Additionally, protein-based nanoparticles exhibit excellent stability, maintaining their structural integrity and functionality under diverse environmental conditions (Saif et al., 2023). This stability enables their long-term storage and use in practical applications, enhancing their utility in drug delivery systems, tissue engineering scaffolds, and environmental remediation technologies. Moreover, protein-based nanoparticles possess the ability to undergo specific molecular interactions, allowing for precise targeting and controlled release of therapeutic agents in drug delivery applications (Mitchell et al., 2021). This targeted delivery approach minimizes off-target effects and maximizes therapeutic efficacy, making protein-based nanoparticles highly valuable in precision medicine. Furthermore, the biodegradability of protein-based nanoparticles ensures their environmentally friendly disposition after use, minimizing their impact on ecosystems. By leveraging the unique properties of protein-based nanoparticles, researchers aim to develop advanced solutions for addressing complex challenges in healthcare, biotechnology, and environmental sustainability, ultimately improving human health and environmental quality.

2.2 Synthesis Methods

Top-Down Approaches: Top-down methods involve breaking down bulk materials into nanoparticles through mechanical, chemical, or physical means. Examples include high-pressure homogenization, milling, and lithography (Iqbal et al., 2012). These methods are advantageous for their scalability and ability to produce nanoparticles with precise size control.

Bottom-Up Approaches: Bottom-up methods involve building nanoparticles from molecular or atomic components, gradually assembling them into the desired nanoparticle structure. Techniques such as emulsion polymerization, self-assembly, and electrospinning fall under this category. Bottom-up methods provide precise control over nanoparticle properties and composition, making them suitable for designing nanoparticles with tailored functionalities for specific applications (Kumar et al., 2018).

Biopolymeric nanoparticles can be synthesized using top-down or bottom-up approaches, each with distinct advantages. Top-down methods involve breaking down bulk materials into nanoparticles through mechanical, chemical, or physical means. These methods provide scalability and precise size control, making them practical for large-scale production. Abid et al. (2022). Specific examples of top-down methods include high-pressure homogenization, which forces the bulk material through a narrow space to break it into nanoparticles, and milling, where mechanical forces grind the material into nanoscale particles. Lithography, often used in microelectronics, can also be adapted to produce nanoparticles with precise shapes and sizes (Tran and Nguyen, 2017).

On the other hand, bottom-up methods involve assembling nanoparticles from molecular or atomic components. These methods offer precise control over the properties and composition of the nanoparticles, allowing for the design of materials with tailored functionalities (El-Khawaga et al., 2023). Examples of bottom-up techniques include emulsion polymerization, where monomers disperse in liquid form nanoparticles as they polymerize; self-assembly, where molecules organize themselves into structured nanoparticles under specific conditions; and electrospinning, which creates nanofibers from polymer solutions through an electrically charged process (Kumar et al., 2018).

Understanding the fundamentals of biopolymeric nanoparticles, including their types and synthesis methods, lays the groundwork for exploring their diverse applications in environmental remediation and beyond. With their unique properties and versatility, biopolymeric nanoparticles hold promise for addressing environmental challenges and advancing various fields of science and technology (Annu et al., 2024). For instance, cellulose nanoparticles synthesized via milling can be used in water filtration systems to remove contaminants, while protein-based nanoparticles produced through self-assembly can be tailored for targeted drug delivery in medical applications (Balistreri et al., 2024). The ability to control size, shape, and functional properties through different synthesis methods makes biopolymeric nanoparticles versatile tools in both environmental and biomedical fields (Vodyashkin et al., 2022). Understanding the fundamentals of biopolymeric nanoparticles, including their types and synthesis methods, lays the groundwork for exploring their diverse applications in environmental remediation and beyond (Chincholikar et al., 2023).

3. Mechanisms of Action in Environmental Remediation

Biopolymeric nanoparticles offer several mechanisms of action that make them effective in environmental remediation (Sundar et al., 2010). These mechanisms include adsorption, degradation, catalysis, and microbial interaction, each contributing uniquely to the removal or neutralization of pollutants from soil, water, and air.

3.1 Adsorption

Adsorption is a process where pollutants adhere to the surface of nanoparticles, effectively removing them from the environment. Biopolymeric nanoparticles, with their high surface area-to-volume ratio, provide extensive surface sites for adsorption. For example, chitosan nanoparticles can adsorb heavy metals such as lead, cadmium, and arsenic from water due to their amino and hydroxyl functional groups (Omer et al., 2022). Similarly, cellulose nanoparticles can adsorb dyes and organic pollutants from industrial effluents, making them useful in water purification (Voisin et al., 2017).

Biopolymers can be combined with synthetic polymers to create novel materials with enhanced properties for metal ion removal. Among the most effective biopolymers are alginate and chitosan, which have been shown to play a significant role in removing heavy metals, metalloids, and other ions from wastewater. These biopolymer-synthetic polymer composites leverage the natural adsorption capabilities of biopolymers while benefiting from the mechanical strength and durability of synthetic polymers. Figure 1b illustrates he synergistic effects of these composites in wastewater treatment, demonstrating their improved efficiency and stability in various environmental conditions. Research indicates that such combinations can lead to more sustainable and cost-effective solutions for water purification (ALSamman and Sánchez, 2023).

3.2 Degradation

Degradation involves the breakdown of pollutants into less harmful substances. Biopolymeric nanoparticles can catalyze or facilitate this process through various mechanisms (Visan et al., 2021). For instance, alginate nanoparticles can encapsulate enzymes that degrade organic pollutants like pesticides and herbicides in soil and water (Wang et al., 2019). These enzymes break down complex organic molecules into simpler, non-toxic forms, thus detoxifying contaminated sites. Additionally, protein-based nanoparticles, such as those made from silk fibroin, can encapsulate and stabilize enzymes that degrade environmental toxins, enhancing their stability and effectiveness (Pham and Tiyaboonchai, 2020).

Figure 1c outlines the process for synthesizing BMNPs, which involves the reduction of metal salts using biopolymers such as chitosan or alginate as stabilizing and capping agents. The biopolymers not only facilitate the formation of nanoparticles but also enhance their stability and dispersibility (Emmanuel et al., 2024). The figure also illustrates the application of these BMNPs in the photocatalytic degradation of dye pollutants in aqueous solutions (Sarkar et al., 2020). Under UV light irradiation, the BMNPs act as catalysts, accelerating the breakdown of dye molecules into less harmful substances. This process highlights the potential of BMNPs in environmental remediation, particularly in treating industrial wastewater contaminated with synthetic dyes. The enhanced degradation efficiency and the environmentally friendly synthesis method underscore the significance of BMNPs in sustainable environmental technologies.

3.3 Catalysis

Catalysis refers to the acceleration of chemical reactions by nanoparticles, leading to the transformation of pollutants into harmless substances. Biopolymeric nanoparticles can act as catalysts or support catalytic reactions. For example, chitosan nanoparticles can serve as supports for catalytic metals such as palladium or platinum, facilitating the reduction of nitrates and nitrites in water (Jin et al., 2020). Moreover, cellulose nanoparticles can be functionalized with catalytic groups to enhance the breakdown of organic pollutants under mild conditions, reducing the need for harsh chemicals (Çalışkan and Baran, 2022).

3.4 Microbial Interaction

Microbial interaction involves the use of biopolymeric nanoparticles to enhance the activity of microorganisms that degrade pollutants (Remya et al., 2022). These nanoparticles can serve as carriers for beneficial microbes or provide nutrients that stimulate microbial growth. For example, alginate nanoparticles can encapsulate microorganisms that biodegrade oil spills, ensuring their survival and activity in harsh environmental conditions (Pete et al., 2021). Additionally, chitosan nanoparticles can enhance the growth of pollutant-degrading bacteria by providing essential nutrients and protection, thereby accelerating bioremediation processes (Chaudhary et al., 2023). By leveraging these mechanisms—adsorption, degradation, catalysis, and microbial interaction—biopolymeric nanoparticles offer versatile and effective solutions for environmental remediation. They can address a wide range of contaminants, including heavy metals, organic pollutants, and pathogens, contributing to the restoration and preservation of natural ecosystems (Annu et al., 2024). Through ongoing research and development, the potential of biopolymeric nanoparticles in environmental cleanup continues to expand, promising more efficient and sustainable remediation technologies. (Annu et al., 2024). Chronic wounds often harbor bacteria that hinder the healing process and repair capacity. This section explores the combined use of nano-chitosan, a biodegradable and non-toxic biopolymer with broad-spectrum bacteriostatic activity, and pulsed electromagnetic fields (PEMFs), known for enhancing wound healing and antibacterial activity (El-Kaliuoby et al., 2021).

4. Applications in Water Treatment

Biopolymeric nanoparticles are gaining significant attention in water treatment due to their efficiency, versatility, and eco-friendly nature (Annu et al., 2024). Their unique properties enable them to address a wide range of contaminants in water, including heavy metals, organic pollutants, and pathogens. These nanoparticles, derived from natural polymers, provide a sustainable alternative to conventional water treatment methods (Khan et al., 2024). This section explores the applications of biopolymeric nanoparticles in water treatment, highlighting their roles in the removal of heavy metals, degradation of organic pollutants, and disinfection (Saiyad et al., 2024). It also presents case studies and examples to illustrate their practical use. The increasing demand for clean water and the limitations of traditional water treatment methods underscores the importance of innovative solutions such as biopolymeric nanoparticles (Savage and Diallo, 2005). Their ability to be engineered for specific applications further enhances their potential in creating more effective and targeted water treatment strategies. Moreover, various production strategies for polymeric biomolecule-based nanomaterials and their applications in the removal of pollutants from wastewater. The depicted processes highlight the efficiency and sustainability of these nanomaterials in treating contaminated water, showcasing their potential to significantly improve water quality and support environmental protection efforts (Muthukumaran et al., 2022).

4.1 Removal of Heavy Metals

Heavy metals such as lead, mercury, cadmium, and arsenic are toxic and pose serious health risks even at low concentrations (Ali et al., 2023). Biopolymeric nanoparticles provide an effective solution for the removal of these contaminants from water sources (Al-Gethami et al., 2024; Mamba et al., 2021). Chitosan nanoparticles, for instance, are particularly effective due to their high affinity for heavy metals. The amino and hydroxyl groups on chitosan nanoparticles can chelate metal ions, forming stable complexes that can be easily removed from water (Benettayeb et al., 2023; Hsu et al., 2024). This chelation process is highly efficient, allowing for the removal of multiple heavy metals simultaneously.

Similarly, alginate nanoparticles can encapsulate heavy metals, reducing their mobility and bioavailability (Sutirman et al., 2021). The carboxyl groups in alginate molecules interact with metal ions, creating a strong binding affinity that traps the contaminants within the nanoparticle matrix (Heuer-Jungemann et al., 2019). This encapsulation not only immobilizes the heavy metals but also prevents them from leaching back into the environment, ensuring long-term water purification (Wang and Lu, 2023). Moreover, the nanocellulose-based adsorbents primarily remove heavy metals through ion exchange and chemical complexation mechanisms Fig. 1d. In the ion exchange mechanism, hazardous metal ions (Mn^{+}) are adsorbed by displacing ions such as K^{+}, Na^{+}, or H^{+} that are initially present on the nanocellulose surface. In the chemical complexation mechanism, the carboxyl (-COO-) and hydroxyl (-OH) groups on the nanocellulose interact specifically with hazardous metal ions (Mn^{+}), forming stable complexes. The maximum adsorption capacity of nanocelluloses is determined by their surface area, functional groups, and the stoichiometric limits of their ionic sites, which typically restrict the adsorption to no more than half of the available surface ionic sites.

These nanoparticles can be applied in various water treatment systems, including filtration units and adsorption columns, providing a practical and efficient method for heavy metal removal. For example, in a filtration system, biopolymeric nanoparticles can be incorporated into filter media, enhancing the filter's capacity to capture and retain heavy metals from water passing through it (Faria et al., 2022; Pachaiappan et al., 2022). In adsorption columns, these nanoparticles can be packed into columns where contaminated water flows through, allowing the nanoparticles to adsorb and sequester heavy metals effectively.

Furthermore, biopolymeric nanoparticles are not only effective in removing heavy metals but are also cost-effective and easy to synthesize. They can be produced from abundant natural resources, making them an economically viable option for large-scale water treatment applications. Their biodegradability ensures that they do not introduce additional pollutants into the environment, aligning with the principles of green chemistry and sustainable development (Sundar et al., 2010). By incorporating biopolymeric nanoparticles into existing water treatment technologies, it is possible to enhance the efficiency and effectiveness of these systems, providing safer drinking water and protecting aquatic ecosystems from heavy metal contamination (Al-Hazmi et al., 2024; Annu et al., 2024). The development of hybrid systems combining biopolymeric nanoparticles with other treatment methods, such as activated carbon or membrane filtration, further improves the overall performance of water purification processes, making them robust against a variety of contaminants (Manikandan et al., 2022).

4.2 Degradation of Organic Pollutants

Organic pollutants such as pesticides, herbicides, pharmaceuticals, and dyes are challenging to remove from water due to their complex chemical structures. Biopolymeric nanoparticles can facilitate the degradation of these pollutants into less harmful substances, providing an innovative and effective solution to this persistent problem (Ejraei et al., 2019; Manikandan et al., 2022). Alginate nanoparticles, for example, can encapsulate enzymes that break down organic pollutants through biocatalysis (Weng et al., 2023). This encapsulation not only stabilizes the enzymes but also protects them from harsh environmental conditions such as extreme pH and temperatures, thus enhancing their longevity and activity (Weng et al., 2023). The enzymes within the alginate nanoparticles can specifically target and degrade complex organic molecules, converting them into simpler, non-toxic compounds that are easier to manage and dispose of. For instance, alginate-encapsulated laccase enzymes have been used to degrade various dyes and phenolic compounds in wastewater, showing significant reductions in pollutant concentrations (Kyomuhimbo and Brink, 2023). Furthermore, the synthesis of metal sulfide-incorporated biopolymer nanoparticles exhibits enhanced photocatalytic dye degradation efficiency (Sarkar et al., 2020). The biopolymer acts as a stabilizing and crystallizing agent, facilitating the formation of metal sulfide nanoparticles.

Specifically, the biopolymer aids in the crystallization of zinc sulfide (ZnS) nanoparticles (Tiwari and Dhoble, 2016).

ZnS-incorporated chitosan nanomaterials, with an average size of 4 nm, exhibit significant photocatalytic activity. These nanomaterials are effectively utilized in the photodegradation of organic dyes in wastewater treatment, showcasing their potential for environmental remediation. The photocatalytic properties of these nanoparticles enable the breakdown of complex dye molecules into less harmful substances under light irradiation, thus improving water quality (Das et al., 2017). Additionally, other metals such as cadmium (Cd) and lead (Pb) can be used to form similar metal sulfide nanomaterials. These metal sulfide-incorporated biopolymeric nanoparticles also demonstrate promising applications in various fields, including environmental cleanup, antimicrobial treatments, and catalytic processes. The figure highlights the synthesis steps, from the initial mixing of metal precursors with the biopolymer to the final formation of uniformly distributed nanoparticles. It also emphasizes the environmental and technological benefits of using such nanomaterials, particularly in addressing pollution and enhancing sustainable practices. Similarly, protein-based nanoparticles, such as those derived from silk fibroin, can be functionalized with catalytic groups that accelerate the breakdown of organic molecules. These functional groups can be tailored to target specific types of organic pollutants, making the degradation process highly efficient (Miao et al., 2022; Pham and Tiyaboonchai, 2020). Silk fibroin nanoparticles, for instance, have been used to degrade pesticides like atrazine and pharmaceuticals such as ibuprofen, demonstrating their versatility and effectiveness in various water treatment scenarios (Li et al., 2016; Marin et al., 2014).

Additionally, chitosan nanoparticles have shown promise in the degradation of organic pollutants. Chitosan can be modified to carry catalytic sites or to host microbial communities that degrade organic pollutants (Al-Gethami et al., 2024; Jiménez-Gómez and Cecilia, 2020). For example, chitosan nanoparticles functionalized with titanium dioxide (TiO_2) can act as photocatalysts under UV light, breaking down pollutants such as dyes and pharmaceuticals into harmless by-products (Al-Nemrawi and Nimrawi, 2021). This process, known as photocatalysis, harnesses the power of light to drive chemical reactions, making it a sustainable and energy-efficient method for water purification.

These biopolymeric nanoparticles can be easily incorporated into existing water treatment processes, such as filtration systems, adsorption columns, and membrane bioreactors (Alcalde-Garcia et al., 2023). In filtration systems, nanoparticles can be embedded within filter materials, enhancing the system's ability to degrade pollutants as water passes through (Anjum et al., 2019). In adsorption columns, they can be packed in conjunction with other adsorbents, providing a dual function of adsorption and degradation. By reducing the need for additional chemical treatments, biopolymeric nanoparticles help lower the operational costs and environmental impact of water purification processes (Al-Gethami et al., 2024; Elgarahy et al., 2023). Their use aligns with green chemistry principles, promoting the use of environmentally friendly materials and processes. Furthermore, the ability to engineer these nanoparticles for specific pollutants ensures that they can be tailored to meet the unique challenges of different contaminated water sources. Incorporating biopolymeric nanoparticles into water treatment not only enhances the degradation of organic pollutants but also represents a significant step toward more sustainable and efficient water purification technologies (Al-Gethami et al., 2024; Simelane et al., 2022). Their application can improve water quality and safety, contributing to better public health outcomes and environmental protection (Nagar and Pradeep, 2020). As research continues to advance, the development of more sophisticated and targeted biopolymeric nanoparticles will likely expand their role in environmental remediation, offering new solutions to tackle the ever-evolving challenge of water pollution.

4.3 Disinfection and Antimicrobial Properties

Waterborne pathogens, including bacteria, viruses, and fungi, pose significant health risks and necessitate effective disinfection methods. Biopolymeric nanoparticles exhibit strong antimicrobial properties, making them suitable for disinfection purposes. Chitosan nanoparticles, in particular, have demonstrated broad-spectrum antimicrobial activity against various pathogens (El-Naggar et al., 2022; Kumar et al., 2020; Yilmaz Atay, 2019). The positively charged surface of chitosan nanoparticles interacts with the negatively charged microbial cell membranes, leading to cell lysis and death (Chandrasekaran et al., 2020). These nanoparticles can be used in water disinfection systems to provide a safe and effective means of eliminating pathogens without the use of harmful chemicals. Additionally, protein-based nanoparticles, such as those made from albumin, can be functionalized with antimicrobial peptides to target specific pathogens, enhancing the specificity and efficiency of water disinfection (Habibi et al., 2022; Verma et al., 2018).

5. Applications in Soil Remediation

Soil erosion, both physical and chemical, has intensified in various regions due to climate change impacts such as heavy rain, drought, and river flooding. This erosion can lead to the geochemical dispersion of heavy metals, contaminating large areas of water and soil. Heavy metal contamination in water poses a serious threat to the preservation and safety of water resources. Therefore, effective management of water and soil is crucial for resource security. Techniques such as precipitation, ion exchange, and electrochemical reactions can be employed to remove heavy metals from solutions. Many studies focus on the sorption process, with previous research emphasizing biological adsorbents for their environmental benefits and low operational costs. The use of biopolymer nanoparticles aims to explore their ability to bind with heavy metal ions, degrade organic pollutants, and improve soil fertility.

5.1 Heavy Metal Immobilization

Heavy metal ions in wastewater enter the environment and ecological cycles in various forms, including inorganic metal ions, hydrated metal ions, hydroxyl complexes, carbonate complexes, and complexes mixed with natural matter. These ions are absorbed and concentrated by aquatic plants, animals, and microorganisms (Zhao et al., 2023). They can also interact with inorganic colloids, organic colloids, and organic-inorganic complexes in water or sediment, eventually agglomerating and settling at the bottom of water bodies. Heavy metals like cadmium, copper, and nickel accumulate in aquatic organisms and, due to their high toxicity, permeability, and persistence, cause physiological and developmental disorders, posing significant threats to human health and ecosystems.

Notably, these ions show promising potential for treating water contaminated with heavy metals. Polysaccharide-based hydrogels, a key category of biopolymer hydrogels, include cellulose, chitosan, agar, and Hyaluronic Acid (HA) hydrogels. From 2016 to 2021, polysaccharide-based hydrogels represented about 30% of all hydrogel-related publications, with chitosan hydrogels being the most frequently studied (Zhao et al., 2023). Additionally, various types of lignin have demonstrated effective absorption of heavy metal ions. Lignin-based adsorbents can adsorb both heavy metal ions and dyes, with their adsorption mechanisms and models summarized. The ion selectivity and recyclability of lignin-based adsorbents have also been reviewed (Zhang et al., 2023). A simple microfluidic method for fabricating PEI-CS biosorbents exhibit excellent selectivity for target metal ions under complex conditions and an exceptionally high adsorption capacity for Cu ions, reaching up to 146 mg g^{-1}. The mechanical strength of the PEI-CS biosorbents is over 0.34 MPa, and their re-adsorption efficiency remains above 71.2% even after five adsorption-desorption cycles (Wang et al., 2017).

A biopolymer-based heavy metal filter for phytoremediation applications was developed using a scalable immersion bath process. The functionalized cellulose filters, treated with polyamine polymers (PEI and chitosan), were evaluated for their heavy metal adsorption efficiency and tested in a phytoremediation Floating Treatment Wetland (FTW) setup. At pH 6.5, the chitosan-functionalized filters adsorbed similar or even higher amounts of 5 ppm heavy metal sulfate solutions compared to the PEI filters. Furthermore, these filters can be reused through simple acidic washing to desorb the heavy metals, as demonstrated over five consecutive cycles (Gholami et al., 2020).

Remediating heavy metal-contaminated soils is crucial for mitigating risks, making land available for agriculture, enhancing food security, and resolving land tenure issues related to changing land use patterns (Wuana and Okieimen, 2011). The impact of pristine biochar and biochar modified with manganese dioxide (BC@MnO_2) and zinc oxide (BC@ZnO) nanoparticles on the immobilization and bioavailability of Pb, Cd, Zn, and Ni in soil under ryegrass (*Lolium perenne* L.) cultivation was investigated. Treatments with BC, BC@MnO_2, and BC@ZnO significantly increased the dry weight of ryegrass shoots and roots compared to the control. Additionally, ryegrass demonstrated a strong capacity to accumulate all studied heavy metals in its roots, indicating its suitability for phytostabilization (Ghandali et al., 2024).

Biochar was used to immobilize Cd (II) and Cr (VI) while sequestering carbon in the soil. Its application improved soil fertility, increased carbon content, and boosted chili (Capsicum annuum) crop production compared to untreated soil. Sequential extractions were performed to analyze the immobilization of Cr (VI) and Cd (II) in soils amended with biochar (Mohan et al., 2024). Additionally, biochar was used to remediate several heavy metal ions in polluted soil while simultaneously promoting plant growth, thereby addressing global environmental concerns (Yang et al., 2021).

5.2 Organic Pollutant Degradation

In Fig. 2, porous biopolymers have emerged as suitable candidates among POPs (persistent organic pollutants). Biopolymers provide significant benefits, including sustainable consumption, environmental protection, and reduced reliance on toxic chemicals, making them well-suited for creating effective composites for pollutant removal. Composites incorporating porous biopolymers can address various pollutants through mechanisms such as absorption, membrane filtration, and oxidative or photocatalytic processes (Tadayoni et al., 2024). Moreover, POPs present a significant threat to human health and environmental integrity in soil and water ecosystems. This comprehensive analysis examines innovative remediation techniques aimed at addressing POP pollution. POPs are dangerous substances that resist natural degradation processes, remaining in the environment for extended periods. The review encompasses various remediation methods, including nanotechnology, phytoremediation, advanced oxidation processes, and bioremediation, evaluating their effectiveness, cost-efficiency, and environmental sustainability (Zhou et al., 2024). The global increase in plastic use results in prolonged environmental persistence due to the slow decomposition

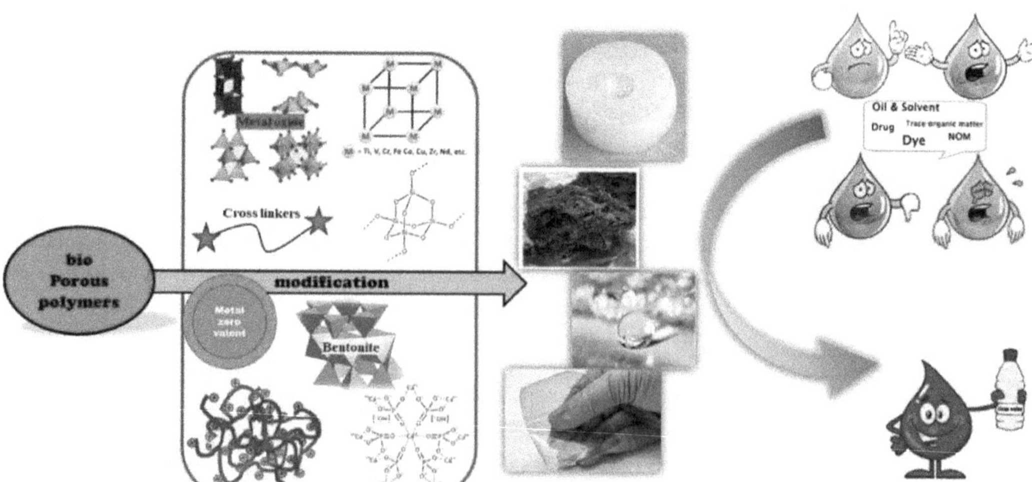

Figure 2. The porous biopolymer-based composites for the remove soil pollution (Tadayoni et al., 2024).

and degradation of plastics. Excessive polymer use endangers marine, freshwater, and terrestrial organisms, while improper waste management exacerbates environmental threats. Bioremediation processes involving microbes such as fungi and bacteria are crucial for breaking down plastics. Extremophiles, which secrete extremozymes, play a vital role in polymer disaggregation under extreme conditions, facilitating plastic degradation in challenging environments.

The environmental benefits of various biopolymers, including tannin, pectin, agar, alginate, acrylamide, carrageenan, starch, dextran, polylactic acid, and polyhydroxyalkanoates, are substantial. The key points include:

- Effective Adsorbents: Biopolymers serve as efficient adsorbents for removing both organic and inorganic pollutants such as fluorides, nitrates, phosphates, heavy metals, hydrocarbons, dyes, and pesticides. Biopolymer-based hydrogels and nanocomposite films have been employed in these applications.
- Coagulation and Flocculation: The coagulating and flocculating properties of biopolymers enhance the clarification of effluents. Pro-Environment Applications: In agriculture and construction, biopolymers are used for soil strengthening, anti-desertification efforts, and sealing concrete leaks.
- Catalysts and Membranes: They are also used in producing catalysts for hydrogen generation and proton-conducting membranes for electrochemical devices.

The classification of biopolymers such as pectin, alginate, chitosan, lignin, cellulose, chitin, carrageenan, certain proteins, and other microbial biomass compounds focuses on their sources, synthesis methods, and potential applications in treating organically polluted wastewater. The fate of biopolymers post-adsorption is crucial in treatment processes. Biopolymer degradation involves biodegradation, depolymerization, assimilation, and mineralization. Often, after sorption, biopolymers are discarded into the environment without further treatment. They are biodegradable, non-toxic, cost-effective, readily available, and produced in large quantities.

5.3 Enhancement of Soil Fertility

Enhancing soil engineering properties is crucial before construction on soft soils. The use of chemical stabilizers like cement to increase soil strength raises environmental concerns, creating a demand for more sustainable solutions. Biopolymers present a promising alternative. Research into five types of biopolymers—Xanthan Gum, Beta 1,3/1,6 Glucan, Guar Gum, Chitosan, and Alginate—has shown that soil strength improves with higher biopolymer concentrations and longer curing times. However, this improvement plateaus beyond certain concentration and curing time levels. Additionally, biopolymer-treated soils demonstrate better resistance to environmental conditions (Soldo et al., 2020). An analysis of the β-1,3/1,6-glucan biopolymer-treated Korean residual soil from an engineering perspective revealed that the inter-particle behavior of this biopolymer depends on the particle condition. For non-charged spherical particles, such as sand, biopolymer adsorption on particle surfaces and the increased contact areas between particles influence the inter-particle interactions (Chang and Cho, 2012).

Xanthan gum, a polysaccharide widely used as a food additive and rheology modifier, interacts directly with the charged surfaces of clay particles, forming matrices that resemble hard plastic between uncharged particles. This interaction enhances the soil's strength, particularly in well-graded soils with fine particles, although the effect diminishes at higher concentrations. Additionally, the strengthening effect of Xanthan gum is significantly influenced by the hydration level of the soils (Chang et al., 2015).

Biopolymers have the capability to remove heavy metals such as Cd, Cu, Pb, and Zn from solutions and can be used as additives to immobilize these metals in soil (Ko et al., 2022). A study examining the long-term effectiveness of iron oxide-loaded biochar (IOBC) in stabilizing arsenic and cationic heavy metals in a real multi-contaminated soil found that stabilizers with high alkalinity, such as RM and BC, had minimal or low effectiveness in immobilizing arsenic. However, the SPLP stabilization efficiency for arsenic was $\geq$ 74.9%, and the TCLP stabilization efficiency was $\geq$ 89.1% (Kim et al., 2022).

The interaction between biopolymer strands and fine-grained soil particles alters the soil's morphological characteristics. Stabilization using xanthan gum and guar gum notably enhances the mechanical properties of fine-grained collapsible soil, providing an eco-friendly and sustainable alternative to conventional soil additives. Xanthan gum, in particular, has demonstrated significant potential for future sustainable engineering due to its remarkable ability to strengthen various soil types (Dehghan et al., 2019; Soldo and Miletić, 2019). Testing the effects of GG and XG on the Young's modulus of silt at concentrations ranging from 0.25 to 2% revealed that the optimal concentration for GG lies between 1 and 2%, while for XG, the optimal concentration is above 2% (Ayeldeen et al., 2017).

6. Applications in Air Pollution Control

Air filtration biomaterials have emerged as a focal point of research due to the escalating concern over global air quality issues. Particulate Matter (PM) constitutes pollution comprised of tiny solid particles present in the air, including dust, dirt, and other substances. PM air pollutants originate from both natural sources and human activities. Elevated levels of PM in the surrounding atmosphere pose significant health risks to humans. PM is categorized into three groups: PM10, which consists of coarse particles and larger particles, mainly referred to inhalable particles with diameters less than 10 micrometers. Furthermore, the accumulation of discarded filters containing large amounts of trapped particulate matter (PM) creates a significant environmental threat. Beyond their toxicity, these PM filters are typically designed to capture particles within a limited concentration range, usually around 1000 mg per cubic meter. This limitation makes it challenging for such filters to effectively capture PM in heavily polluted environments. Conventional purification technologies primarily rely on size-dependent mechanisms for PM capture, which can be difficult due to the extremely small particle size and ultra-low mass of PM. Traditional filters, made from materials like fabrics, wool, and cotton, often neglect crucial factors such as pore size and removal efficiency. In recent years, air pollution control has increasingly turned to natural biopolymers like agar, agarose, alginate, carrageenan, cellulose, chitosan, dextran, gelatin, guar gum, pectin, xanthan, and others. These biopolymers are favored for their adjustable properties, flexibility, non-toxicity, biocompatibility, and biodegradability, making them well-suited for a wide range of applications.

6.1 Particulate Matter Reduction

The reuse of waste hemp noil by dissolving it in a precooled NaOH/urea system leads to the creation of porous aerogels. Once hydrophobized, these hemp aerogels display self-cleaning properties and can be utilized for air purification. They also demonstrate high efficiency in removing particulate matter (PM), such as PM2.5 and PM10, with a 94% removal rate for both. This repurposed waste hemp noil holds significant potential for environmental management and the recycling of natural resources (Lyu et al., 2023). Nanostructured biopolymer aerogels have gained attention as promising materials for particulate matter removal. Their distinctive properties, renewable nature, and high potential for customization make them appealing options for air pollution control.

Nanostructured aerogels are regarded as outstanding alternative adsorbents and filters due to their vast surface area, remarkable chemical surface properties, and wide-ranging pollutant selectivity. The pore size of these aerogels can be adjusted by altering the precursor materials, modifying the fabrication process, or incorporating reinforcement materials. Various modifications have been made to these aerogels using polymers, inorganic carbon, metal oxides, and silica. While aerogels derived from natural biomass are eco-friendly, their hydrophilic nature presents a challenge. Most biomass aerogels lack hydrophobicity, resulting in structural instability under humid conditions. Although researchers have explored chemical modifications to enhance the stability and mechanical properties of biopolymer aerogels, these approaches often increase production costs and lead to less eco friendly outcomes (Guastaferro et al., 2021; Mariana et al., 2022; Wang et al., 2020; Yahya et al., 2020). In conclusion, the development of nanostructured bioaerogels holds promise for air purification. Their modifiability and unique surface functional groups make them valuable for green air filter development, despite existing challenges.

6.2 Gas Absorption and Filtration

Biobased materials such as cellulose, chitin, silk, soy, and keratin are emerging as promising alternatives to traditional synthetic materials for filtration applications. These materials are not only cost-effective and abundant but also provide easy fabrication with customizable surface chemistry and functionality. As pollution continues to rise, effective air and water filtration has become crucial. Additionally, the need for antibacterial and antiviral fibers is vital for public health and medical settings. Biopolymers, with their inherent material properties, surface chemistry, and complex fiber structures, excel in filtering physical, chemical, and biological pollutants efficiently and in an environmentally friendly manner. Their biodegradability further enhances their appeal as sustainable and biocompatible green filters.

Air filtration typically employs two approaches: physical adsorption and chemisorption. Physical adsorption involves contaminants binding to the filter surface or embedded particles, while chemisorption entails chemical reactions rendering the pollutant inert. However, the effectiveness of both methods is constrained by the availability of active sites within the filter. Traditional air filters have depended on synthetic polymers, using mechanisms such as impaction, interception, diffusion, and electrostatic interactions to capture particles. Biopolymer-based filters operate similarly, with filtration effectiveness varying based on pollutant size and nature. Larger pollutants (> 1000 nm) are easily physically blocked, while smaller ones (100–1000 nm) are primarily filtered using electrostatics.

Effective air purification via filtration systems is crucial for maintaining optimal indoor air quality amidst worsening anthropogenic pollution. Conventional filtering materials made of glass fibers or non-biodegradable polymers struggle to completely remove fine particulate matter like PM2.5 and gaseous pollutants such as harmful chemicals and CO_2. Moreover, their non-biodegradable nature leads to landfill disposal issues. Therefore, urgent development of fully biodegradable air filtration systems is imperative. Cellulose-based filters, both pure and surface-functionalized, have emerged as high-performance solutions. This review discusses their significance, characteristics, functioning, and efficacy in removing particulate matter, volatile organic compounds, and CO_2. Additionally, it explores composite filters/membranes made using cellulose derivatives and investigates the mechanisms for pollutant absorption or removal with cellulose-based air filters (Rana et al., 2023).

6.3 Case Studies and Examples

Several case studies and examples illustrate the practical applications of biopolymeric nanoparticles in water treatment, highlighting their effectiveness and potential for widespread use (Biswas and Pal, 2021). In one study, chitosan nanoparticles were employed to remove arsenic from contaminated groundwater in a rural community (Xu et al., 2022). The application involved treating the groundwater with a suspension of chitosan nanoparticles, which effectively adsorbed the arsenic ions (Xu et al., 2022). The results showed a significant reduction in arsenic levels, making the water safe for consumption. This study demonstrates the practicality of using biopolymeric nanoparticles for addressing serious health risks associated with heavy metal contamination in drinking water (Baby et al., 2022).

Another example involves the use of alginate nanoparticles to degrade atrazine, a common herbicide found in agricultural runoff. In this case, alginate nanoparticles encapsulating specific enzymes were introduced into the contaminated water (Chaud et al., 2021; Steffens et al., 2022). The enzymes facilitated the breakdown of atrazine into non-toxic by-products. The study reported that the atrazine concentrations were reduced to below detectable levels, showcasing the potential of alginate nanoparticles for large-scale environmental remediation in agricultural settings (Chandra and Usha, 2021).

Additionally, protein-based nanoparticles have been tested in municipal water treatment plants for their antimicrobial properties (Wang et al., 2017). For instance, nanoparticles derived from silk fibroin were functionalized with antimicrobial peptides and incorporated into water filtration systems (Pham and Tiyaboonchai, 2020). These nanoparticles demonstrated high efficacy in reducing bacterial contamination, significantly lowering bacterial counts without the need for chlorine or other chemical disinfectants. This approach not only ensures the safety of drinking water but also minimizes the formation of harmful disinfection by-products often associated with chemical treatments. Furthermore, cellulose nanoparticles have been used in advanced filtration membranes to improve the removal of organic pollutants and pathogens (Carpenter et al., 2015). In one pilot study, a filtration membrane embedded with cellulose nanoparticles was installed in a wastewater treatment facility. The membrane exhibited superior performance in filtering out pharmaceuticals and pathogenic microorganisms compared to conventional membranes, leading to cleaner effluent water and reduced environmental impact (Obotey Ezugbe and Rathilal, 2020).

By leveraging the unique properties of biopolymeric nanoparticles, these water treatment applications provide efficient, sustainable, and safe solutions for addressing a wide range of contaminants (Obotey Ezugbe and Rathilal, 2020). The ongoing development and optimization of these nanoparticles continue to enhance their performance and expand their use in water treatment systems worldwide (Nishu and Kumar, 2023). These case studies and examples underscore the versatility and effectiveness of biopolymeric nanoparticles, paving the way for their broader adoption in both urban and rural water treatment infrastructures (Chen et al., 2024; Yousefi and Wnek, 2024). As research progresses, further case studies are likely to emerge, demonstrating new applications and benefits of biopolymeric nanoparticles. The integration of these advanced materials into water treatment technologies holds promise for achieving cleaner, safer water for communities around the globe, addressing both current and future water quality challenges.

7. Future Prospects and Research Directions

Biopolymeric nanoparticles (BPNs) are mainly engineered for their many advantages, such as biocompatibility, biodegradability, and low toxicity, in addition to the general benefits associated with nanoparticles. These characteristics make biopolymers, which are natural, biodegradable, and eco-friendly, particularly appealing for various biomedical applications, including drug delivery and antibacterial functions. Future research in biopolymer-based environmental remediation will likely focus on: (i) developing multifunctional biopolymers that can simultaneously adsorb, degrade, and immobilize various pollutants, (ii) enhancing the stability and reusability of biopolymer-based materials to make them more cost-effective and practical for large-scale applications, (iii) exploring new sources of biopolymers, including waste biomass and genetically engineered organisms, to ensure a sustainable supply of raw materials, and (iv) integrating biopolymer-based systems with existing remediation technologies to create hybrid approaches that maximize efficiency and effectiveness. Therefore, biopolymer-based environmental remediation represents a promising and sustainable approach to addressing pollution. Advances in nanotechnology, material science, and biotechnology are driving the development of innovative biopolymer-based materials and methods that offer effective solutions for cleaning up contaminated environments.

Challenges and opportunities: Biopolymer-based environmental remediation is a promising approach that leverages the natural properties of biopolymers to address pollution and restore ecological balance. Despite its potential, this field faces several challenges that must be addressed to fully realize its benefits. Concurrently, numerous opportunities exist for innovation and advancement. Here are the key points regarding the challenges and opportunities in biopolymer-based environmental remediation: Raising public awareness about the benefits of biopolymer-based remediation and promoting environmental education can drive demand for sustainable solutions. Community involvement and stakeholder engagement can support the implementation of biopolymer-based research. Therefore, biopolymer-based environmental remediation presents a promising avenue for addressing pollution in a sustainable and eco-friendly manner. While several challenges need to be overcome, the opportunities for innovation and advancement are significant. By addressing scalability, stability, functionalization, regulatory, and biocompatibility challenges, and leveraging advancements in green chemistry, biotechnology, and integrated approaches, the potential of biopolymer-based remediation can be fully realized. Collaborative efforts among researchers, policymakers, and industry stakeholders will be crucial in driving this field forward and achieving impactful environmental solutions.

8. Conclusion

Biopolymer-Based Environmental Remediation for Heavy Metal Immobilization: Biopolymers effectively immobilize and remove heavy metals from contaminated soils, reducing their bioavailability and toxicity. Organic Pollutant Degradation: Supporting microbial growth, biopolymers facilitate the biodegradation of complex organic pollutants in soil. Then ensuring the stability and functionality of biopolymers under varying soil conditions is crucial for their effectiveness in long-term soil remediation projects. Air pollution control is based on the two classifications, (i) Limited Direct Applications: While direct applications of biopolymers in air pollution control are less common, potential exists for use in filters and coatings that adsorb airborne pollutants. (ii) Indoor Air Quality Improvement: Further research is required to explore and optimize biopolymer applications in air pollution control, particularly for industrial emissions. Biopolymers provide significant potential for environmental remediation across various media, including water, soil, and air. Their eco-friendly nature, combined with their effectiveness in removing a wide range of pollutants, positions them as promising candidates for large-scale environmental remediation applications. Addressing challenges related to production, stability, and regulatory approval will be essential to harness their full potential and ensure their successful integration into existing environmental management systems.

The application of biopolymers in environmental remediation holds significant implications for sustainable and effective pollution management across various media. Here, we explore the specific implications for water treatment, soil remediation, and air pollution control. The use of biodegradable and renewable biopolymers reduces the environmental footprint of water treatment processes, promoting sustainability. The potential for cost-effective production of biopolymers from waste biomass can lower operational costs for water treatment facilities, making advanced treatment technologies more accessible. Continued research is needed to explore and optimize the use of biopolymers in air pollution control, including material design, efficiency, and scalability. Likewise, the implications of using biopolymers for environmental remediation are vast and transformative. Their effectiveness, sustainability, and potential for cost reduction position them as key components in future pollution management strategies. Addressing challenges related to production scalability, regulatory approvals, and application-specific research will be critical to fully realizing the benefits of biopolymers in environmental remediation. By integrating biopolymer-based technologies into existing environmental management systems, we can make significant strides toward a cleaner, healthier, and more sustainable environment. The use of biopolymers in environmental remediation is a promising and innovative approach to addressing pollution sustainably and effectively. Their inherent properties—such as biodegradability, non-toxicity, and versatility—position them as superior alternatives to synthetic materials in various remediation processes. The significant potential for biopolymer applications in water treatment, soil remediation, and air pollution control highlights their adaptability and broad environmental impact. However, to fully harness the benefits of biopolymers, several challenges need to be addressed, including scalability, economic feasibility, and regulatory compliance. Moreover, continued research is necessary to better understand the mechanisms by which biopolymers interact with pollutants and to discover new applications, particularly in less explored areas like air pollution control.

References

Abdelhamid, H. N. and Mathew, A. P. (2022). Cellulose-based nanomaterials advance biomedicine: A review. International Journal of Molecular Sciences, 23(10): 5405. https://www.mdpi.com/1422-0067/23/10/5405.

Abid, N., Khan, A. M., Shujait, S., Chaudhary, K., Ikram, M., Imran, M. et al. (2022). Synthesis of nanomaterials using various top-down and bottom-up approaches, influencing factors, advantages, and disadvantages: A review. Advances in Colloid and Interface Science, 300: 102597. https://doi.org/https://doi.org/10.1016/j.cis.2021.102597.

Ahmed, M. A. and Mohamed, A. A. (2023). The use of chitosan-based composites for environmental remediation: A review. International Journal of Biological Macromolecules, 242: 124787. https://doi.org/https://doi.org/10.1016/j.ijbiomac.2023.124787.

Ahmed Mohamed, M. R., Kao, N., Islam, M. S. and Ward, L. (2020). An overview of recent developments in hetero-catalytic conversion of cellulosic biomass. Research Communication in Engineering Science & Technology, 4: 43–54.

Al-Gethami, W., Qamar, M. A., Shariq, M., Alaghaz, A.-N. M. A., Farhan, A., Areshi, A. A. et al. (2024). Emerging environmentally friendly bio-based nanocomposites for the efficient removal of dyes and micropollutants from wastewater by adsorption: a comprehensive review [10.1039/D3RA06501D]. RSC Advances, 14(4): 2804–2834. https://doi.org/10.1039/D3RA06501D.

Al-Hazmi, H. E., Łuczak, J., Habibzadeh, S., Hasanin, M. S., Mohammadi, A., Esmaeili, A. et al. (2024). Polysaccharide nanocomposites in wastewater treatment: A review. Chemosphere, 347: 140578. https://doi.org/https://doi.org/10.1016/j.chemosphere.2023.140578.

Al-Nemrawi, N. and Nimrawi, S. (2021). A novel formulation of chitosan nanoparticles functionalized with titanium dioxide nanoparticles. Journal of Advanced Pharmaceutical Technology & Research, 12(4). https://journals.lww.com/japtr/fulltext/2021/12040/a_novel_formulation_of_chitosan_nanoparticles.15.aspx.

Alcalde-Garcia, F., Prasher, S., Kaliaguine, S., Tavares, J. R. and Dumont, M.-J. (2023). Desorption strategies and reusability of biopolymeric adsorbents and semisynthetic derivatives in hydrogel and hydrogel composites used in adsorption processes. ACS Engineering Au, 3(6): 443–460. https://doi.org/10.1021/acsengineeringau.3c00022.

Ali, Q., Zia, M. A., Kamran, M., Shabaan, M., Zulfiqar, U., Ahmad, M. et al. (2023). Nanoremediation for heavy metal contamination: A review. Hybrid Advances, 4: 100091. https://doi.org/https://doi.org/10.1016/j.hybadv.2023.100091.

ALSamman, M. T. and Sánchez, J. (2023). Adsorption of copper and arsenic from water using a semi-interpenetrating polymer network based on alginate and chitosan. Polymers, 15(9): 2192. https://www.mdpi.com/2073-4360/15/9/2192.

Anjum, M., Miandad, R., Waqas, M., Gehany, F. and Barakat, M. A. (2019). Remediation of wastewater using various nano-materials. Arabian Journal of Chemistry, 12(8): 4897–4919. https://doi.org/https://doi.org/10.1016/j.arabjc.2016.10.004.

Annu, Mittal, M., Tripathi, S. and Shin, D. K. (2024). Biopolymeric nanocomposites for wastewater remediation: An overview on recent progress and challenges. Polymers, 16(2): 294. https://www.mdpi.com/2073-4360/16/2/294.

Arif, Z. U., Khalid, M. Y., Sheikh, M. F., Zolfagharian, A. and Bodaghi, M. (2022). Biopolymeric sustainable materials and their emerging applications. Journal of Environmental Chemical Engineering, 10(4): 108159. https://doi.org/https://doi.org/10.1016/j.jece.2022.108159.

Ashraf, M. H., Hussain, N., Muneer, M. A., Arif, I. and Ali, M. R. (2024). Chapter Four - Chitosan-based nanomaterials for pharmaceutical waste remediation. pp. 83–116. *In*: Kumar, A., Bilal, M. and Ferreira, L. F. R. (Eds.). Advances in Chemical Pollution, Environmental Management and Protection (Vol. 10). Elsevier. https://doi.org/https://doi.org/10.1016/bs.apmp.2023.09.001.

Ayeldeen, M., Negm, A., El-Sawwaf, M. and Kitazume, M. (2017). Enhancing mechanical behaviors of collapsible soil using two biopolymers. Journal of Rock Mechanics and Geotechnical Engineering, 9(2): 329–339. https://doi.org/https://doi.org/10.1016/j.jrmge.2016.11.007.

Azubuike, C. C., Chikere, C. B. and Okpokwasili, G. C. (2016). Bioremediation techniques–classification based on site of application: Principles, advantages, limitations and prospects. World Journal of Microbiology and Biotechnology, 32(11): 180. https://doi.org/10.1007/s11274-016-2137-x.

Baby, R., Hussein, M. Z., Abdullah, A. H. and Zainal, Z. (2022). Nanomaterials for the treatment of heavy metal contaminated water. Polymers, 14(3): 583. https://www.mdpi.com/2073-4360/14/3/583.

Bala, S., Garg, D., Thirumalesh, B. V., Sharma, M., Sridhar, K., Inbaraj, B. S. et al. (2022). Recent strategies for bioremediation of emerging pollutants: A review for a green and sustainable environment. Toxics, 10(8): 484. https://www.mdpi.com/2305-6304/10/8/484.

Balistreri, G. N., Campbell, I. R., Li, X., Amorim, J., Zhang, S., Nance, E. et al. (2024). Bacterial cellulose nanoparticles as a sustainable drug delivery platform for protein-based therapeutics [10.1039/D3LP00184A]. RSC Applied Polymers, 2(2): 172–183. https://doi.org/10.1039/D3LP00184A.

Bashir, S. M., Ahmed Rather, G., Patrício, A., Haq, Z., Sheikh, A. A., Shah, M. Z. u. H. et al. (2022). Chitosan nanoparticles: A versatile platform for biomedical applications. Materials, 15(19): 6521. https://www.mdpi.com/1996-1944/15/19/6521.

Benettayeb, A., Seihoub, F. Z., Pal, P., Ghosh, S., Usman, M., Chia, C. H. et al. (2023). Chitosan nanoparticles as potential nano-sorbent for removal of toxic environmental pollutants. Nanomaterials, 13(3): 447. https://www.mdpi.com/2079-4991/13/3/447.

Biswas, S. and Pal, A. (2021). Application of biopolymers as a new age sustainable material for surfactant adsorption: A brief review. Carbohydrate Polymer Technologies and Applications, 2: 100145. https://doi.org/https://doi.org/10.1016/j.carpta.2021.100145.

Boretti, A. and Rosa, L. (2019). Reassessing the projections of the World Water Development Report. npj Clean Water, 2(1): 15. https://doi.org/10.1038/s41545-019-0039-9.

Çalışkan, M. and Baran, T. (2022). Palladium nanoparticles embedded over chitosan/γMnO_2 composite hybrid microspheres as heterogeneous nanocatalyst for effective reduction of nitroarenes and organic dyes in water. Journal of Organometallic Chemistry, 963: 122284. https://doi.org/https://doi.org/10.1016/j.jorganchem.2022.122284.

Carpenter, A. W., de Lannoy, C.-F. and Wiesner, M. R. (2015). Cellulose nanomaterials in water treatment technologies. Environmental Science & Technology, 49(9): 5277–5287. https://doi.org/10.1021/es506351r.

Chandra, P. N. and Usha, K. (2021). Removal of atrazine herbicide from water by polyelectrolyte multilayer membranes. Materials Today: Proceedings, 41: 622–627. https://doi.org/https://doi.org/10.1016/j.matpr.2020.05.263.

Chandrasekaran, M., Kim, K. D. and Chun, S. C. (2020). Antibacterial activity of chitosan nanoparticles: A review. Processes, 8(9): 1173. https://www.mdpi.com/2227-9717/8/9/1173.

Chang, I. and Cho, G.-C. (2012). Strengthening of Korean residual soil with β-1,3/1,6-glucan biopolymer. Construction and Building Materials, 30: 30–35. https://doi.org/https://doi.org/10.1016/j.conbuildmat.2011.11.030.

Chang, I., Im, J., Prasidhi, A. K. and Cho, G.-C. (2015). Effects of Xanthan gum biopolymer on soil strengthening. Construction and Building Materials, 74: 65–72. https://doi.org/https://doi.org/10.1016/j.conbuildmat.2014.10.026.

Chaud, M., Souto, E. B., Zielinska, A., Severino, P., Batain, F., Oliveira-Junior, J. et al. (2021). Nanopesticides in agriculture: Benefits and challenge in agricultural productivity, toxicological risks to human health and environment. Toxics, 9(6): 131. https://www.mdpi.com/2305-6304/9/6/131.

Chaudhary, P., Ahamad, L., Chaudhary, A., Kumar, G., Chen, W.-J. and Chen, S. (2023). Nanoparticle-mediated bioremediation as a powerful weapon in the removal of environmental pollutants. Journal of Environmental Chemical Engineering, 11(2): 109591. https://doi.org/https://doi.org/10.1016/j.jece.2023.109591.

Chen, L., Zhang, Y., Chen, Z., Dong, Y., Jiang, Y., Hua, J. et al. (2024). Biomaterials technology and policies in the building sector: A review. Environmental Chemistry Letters, 22(2): 715–750. https://doi.org/10.1007/s10311-023-01689-w.

Chincholikar, P., Singh, K. R. B., Natarajan, A., Kerry, R. G., Singh, J., Malviya, J. et al. (2023). Green nanobiopolymers for ecological applications: A step towards a sustainable environment [10.1039/D2RA07707H]. RSC Advances, 13(18): 12411–12429. https://doi.org/10.1039/D2RA07707H.

Das, A., Ringu, T., Ghosh, S. and Pramanik, N. (2023). A comprehensive review on recent advances in preparation, physicochemical characterization, and bioengineering applications of biopolymers. Polymer Bulletin, 80(7): 7247–7312. https://doi.org/10.1007/s00289-022-04443-4.

Das, R. K., Pachapur, V. L., Lonappan, L., Naghdi, M., Pulicharla, R., Maiti, S. et al. (2017). Biological synthesis of metallic nanoparticles: Plants, animals and microbial aspects. Nanotechnology for Environmental Engineering, 2(1): 18. https://doi.org/10.1007/s41204-017-0029-4.

Dehghan, H., Tabarsa, A., Latifi, N. and Bagheri, Y. (2019). Use of xanthan and guar gums in soil strengthening. Clean Technologies and Environmental Policy, 21(1): 155–165. https://doi.org/10.1007/s10098-018-1625-0.

Dodero, A., Alberti, S., Gaggero, G., Ferretti, M., Botter, R., Vicini, S. and Castellano, M. (2021). An up-to-date review on alginate nanoparticles and nanofibers for biomedical and pharmaceutical applications. Advanced Materials Interfaces, 8(22): 2100809. https://doi.org/https://doi.org/10.1002/admi.202100809.

Ejraei, A., Aroon, M. A. and Ziarati Saravani, A. (2019). Wastewater treatment using a hybrid system combining adsorption, photocatalytic degradation and membrane filtration processes. Journal of Water Process Engineering, 28: 45–53. https://doi.org/https://doi.org/10.1016/j.jwpe.2019.01.003.

El-Kaliuoby, M. I., Amer, M. and Shehata, N. (2021). Enhancement of nano-biopolymer antibacterial activity by pulsed electric fields. Polymers, 13(11): 1869. https://www.mdpi.com/2073-4360/13/11/1869.

El-Khawaga, A. M., Zidan, A. and El-Mageed, A. I. A. A. (2023). Preparation methods of different nanomaterials for various potential applications: A review. Journal of Molecular Structure, 1281: 135148. https://doi.org/https://doi.org/10.1016/j.molstruc.2023.135148.

El-Naggar, N. E.-A., Shiha, A. M., Mahrous, H. and Mohammed, A. B. A. (2022). Green synthesis of chitosan nanoparticles, optimization, characterization and antibacterial efficacy against multi drug resistant biofilm-forming *Acinetobacter baumannii*. Scientific Reports, 12(1): 19869. https://doi.org/10.1038/s41598-022-24303-5.

Elgarahy, A. M., Eloffy, M. G., Guibal, E., Alghamdi, H. M. and Elwakeel, K. Z. (2023). Use of biopolymers in wastewater treatment: A brief review of current trends and prospects. Chinese Journal of Chemical Engineering, 64: 292–320. https://doi.org/https://doi.org/10.1016/j.cjche.2023.05.018.

Emmanuel, S. S., Olawoyin, C. O., Adesibikan, A. A. and Opatola, E. A. (2024). A pragmatic review on bio-polymerized metallic nano-architecture for photocatalytic degradation of recalcitrant dye pollutants. Journal of Polymers and the Environment, 32(1): 1–30. https://doi.org/10.1007/s10924-023-02986-9.

Faria, M., Cunha, C., Gomes, M., Mendonça, I., Kaufmann, M., Ferreira, A. et al. (2022). Bacterial cellulose biopolymers: The sustainable solution to water-polluting microplastics. Water Research, 222: 118952. https://doi.org/https://doi.org/10.1016/j.watres.2022.118952.

Fernando, I. P. S., Lee, W., Han, E. J. and Ahn, G. (2020). Alginate-based nanomaterials: Fabrication techniques, properties, and applications. Chemical Engineering Journal, 391: 123823. https://doi.org/https://doi.org/10.1016/j.cej.2019.123823.

Ghandali, M. V., Safarzadeh, S., Ghasemi-Fasaei, R. and Zeinali, S. (2024). Heavy metals immobilization and bioavailability in multi-metal contaminated soil under ryegrass cultivation as affected by ZnO and MnO_2 nanoparticle-modified biochar. Scientific Reports, 14(1): 10684. https://doi.org/10.1038/s41598-024-61270-5.

Gholami, S., Dimde, M., Braatz, D., Müller, J., Haag, R. and Wagner, O. (2020). Reusable biopolymer based heavy metal filter as plant protection for phytoremediation. Environmental Technology & Innovation, 19: 101005. https://doi.org/https://doi.org/10.1016/j.eti.2020.101005.

Guastaferro, M., Reverchon, E. and Baldino, L. (2021). Agarose, alginate and chitosan nanostructured aerogels for pharmaceutical applications: A short review [Mini Review]. Frontiers in Bioengineering and Biotechnology, 9. https://doi.org/10.3389/fbioe.2021.688477.

Gupta, R. K., Guha, P. and Srivastav, P. P. (2022). Natural polymers in bio-degradable/edible film: A review on environmental concerns, cold plasma technology and nanotechnology application on food packaging- A recent trends. Food Chemistry Advances, 1: 100135. https://doi.org/https://doi.org/10.1016/j.focha.2022.100135.

Habibi, N., Mauser, A., Ko, Y. and Lahann, J. (2022). Protein nanoparticles: Uniting the power of proteins with engineering design approaches. Advanced Science, 9(8): 2104012. https://doi.org/https://doi.org/10.1002/advs.202104012.

Heuer-Jungemann, A., Feliu, N., Bakaimi, I., Hamaly, M., Alkilany, A., Chakraborty, I. et al. (2019). The role of ligands in the chemical synthesis and applications of inorganic nanoparticles. Chemical Reviews, 119(8): 4819–4880. https://doi.org/10.1021/acs.chemrev.8b00733.

Hsu, C.-Y., Ajaj, Y., Mahmoud, Z. H., Kamil Ghadir, G., Khalid Alani, Z., Hussein, M. M. et al. (2024). Adsorption of heavy metal ions use chitosan/graphene nanocomposites: A review study. Results in Chemistry, 7: 101332. https://doi.org/https://doi.org/10.1016/j.rechem.2024.101332.

Iqbal, P., Preece, J. A. and Mendes, P. M. (2012). Nanotechnology: The "Top-Down" and "Bottom-Up" Approaches. In Supramolecular Chemistry. https://doi.org/https://doi.org/10.1002/9780470661345.smc195.

Jafarzadeh, S., Nooshkam, M., Zargar, M., Garavand, F., Ghosh, S., Hadidi, M. et al. (2024). Green synthesis of nanomaterials for smart biopolymer packaging: challenges and outlooks. Journal of Nanostructure in Chemistry, 14(2): 113–136. https://doi.org/10.1007/s40097-023-00527-3.

Jiménez-Gómez, C. P. and Cecilia, J. A. (2020). Chitosan: A natural biopolymer with a wide and varied range of applications. Molecules, 25(17): 3981. https://www.mdpi.com/1420-3049/25/17/3981.

Jin, T., Hicks, M., Kurdyla, D., Hrapovic, S., Lam, E. and Moores, A. (2020). Palladium nanoparticles supported on chitin-based nanomaterials as heterogeneous catalysts for the Heck coupling reaction. Beilstein Journal of Organic Chemistry, 16: 2477–2483. https://doi.org/10.3762/bjoc.16.201.

Kafle, A., Timilsina, A., Gautam, A., Adhikari, K., Bhattarai, A. and Aryal, N. (2022). Phytoremediation: Mechanisms, plant selection and enhancement by natural and synthetic agents. Environmental Advances, 8: 100203. https://doi.org/https://doi.org/10.1016/j.envadv.2022.100203.

Kanmani, P., Aravind, J., Kamaraj, M., Sureshbabu, P. and Karthikeyan, S. (2017). Environmental applications of chitosan and cellulosic biopolymers: A comprehensive outlook. Bioresource Technology, 242: 295–303. https://doi.org/https://doi.org/10.1016/j.biortech.2017.03.119.

Karamchandani, B. M., Dalvi, S. G., Bagayatkar, M., Banat, I. M. and Satpute, S. K. (2024). Prospective applications of chitosan and chitosan-based nanoparticles formulations in sustainable agricultural practices. Biocatalysis and Agricultural Biotechnology, 58: 103210. https://doi.org/https://doi.org/10.1016/j.bcab.2024.103210.

Khan, R. J., Lau, C. Y., Farid, M. U., Islam, M. K., An, A. K., Yi, J. et al. (2024). Woody biomass derived nano lignin-enabled membrane with high structural stability for efficient wastewater treatment. Process Safety and Environmental Protection, 182: 595–607. https://doi.org/https://doi.org/10.1016/j.psep.2023.12.019.

Khorsandi, D., Jenson, S., Zarepour, A., Khosravi, A., Rabiee, N., Iravani, S. et al. (2024). Catalytic and biomedical applications of nanocelluloses: A review of recent developments. International Journal of Biological Macromolecules, 268: 131829. https://doi.org/https://doi.org/10.1016/j.ijbiomac.2024.131829.

Kim, D.-S., Moradi, H., Chang, Y.-Y. and Yang, J.-K. (2022). Biopolymer-Biochar matrix for long-term stabilization of arsenic in soil: Performance, mechanisms, and the effect of cationic heavy metals. Biomass Conversion and Biorefinery. https://doi.org/10.1007/s13399-022-03531-9.

Ko, M.-S., Jeon, Y.-J. and Kim, K.-W. (2022). Novel application of xanthan gum-based biopolymer for heavy metal immobilization in soil. Journal of Environmental Chemical Engineering, 10(5): 108240. https://doi.org/https://doi.org/10.1016/j.jece.2022.108240.

Kora, A. J. and Rastogi, L. (2016). Catalytic degradation of anthropogenic dye pollutants using palladium nanoparticles synthesized by gum olibanum, a glucuronoarabinogalactan biopolymer. Industrial Crops and Products, 81: 1–10. https://doi.org/https://doi.org/10.1016/j.indcrop.2015.11.055.

Kučuk, N., Primožič, M., Knez, Ž. and Leitgeb, M. (2023). Sustainable biodegradable biopolymer-based nanoparticles for healthcare applications. International Journal of Molecular Sciences, 24(4): 3188. https://www.mdpi.com/1422-0067/24/4/3188.

Kumar, R., Kumar, N., Rajput, V. D., Mandzhieva, S., Minkina, T., Saharan, B. S. et al. (2022). Advances in biopolymeric nanopesticides: A new eco-friendly/eco-protective perspective in precision agriculture. Nanomaterials, 12(22), 3964. https://www.mdpi.com/2079-4991/12/22/3964.

Kumar, S., Bhushan, P. and Bhattacharya, S. (2018). Fabrication of nanostructures with bottom-up approach and their utility in diagnostics, therapeutics, and others. pp. 167–198. *In*: Bhattacharya, S., Agarwal, A. K., Chanda, N., Pandey, A. and Sen, A. K. (eds.). Environmental, Chemical and Medical Sensors. Springer Singapore. https://doi.org/10.1007/978-981-10-7751-7_8.

Kumar, S., Mukherjee, A. and Dutta, J. (2020). Chitosan based nanocomposite films and coatings: Emerging antimicrobial food packaging alternatives. Trends in Food Science & Technology, 97: 196–209. https://doi.org/https://doi.org/10.1016/j.tifs.2020.01.002.

Kyomuhimbo, H. D. and Brink, H. G. (2023). Applications and immobilization strategies of the copper-centred laccase enzyme; a review. Heliyon, 9(2): e13156. https://doi.org/10.1016/j.heliyon.2023.e13156.

Lee, K. Y. and Mooney, D. J. (2012). Alginate: Properties and biomedical applications. Progress in Polymer Science, 37(1): 106–126. https://doi.org/https://doi.org/10.1016/j.progpolymsci.2011.06.003.

Li, B., Elango, J. and Wu, W. (2020). Recent advancement of molecular structure and biomaterial function of chitosan from marine organisms for pharmaceutical and nutraceutical application. Applied Sciences, 10(14): 4719. https://www.mdpi.com/2076-3417/10/14/4719.

Li, H., Tian, J., Wu, A., Wang, J., Ge, C. and Sun, Z. (2016). Self-assembled silk fibroin nanoparticles loaded with binary drugs in the treatment of breast carcinoma. Int. J. Nanomedicine, 11: 4373–4380. https://doi.org/10.2147/IJN.S108633.

Lyu, P., Xia, L., Liu, X., Hurren, C., Xu, W. and Wang, X. (2023). Self-cleaning superhydrophobic aerogels from waste hemp noil for ultrafast oil absorption and highly efficient PM removal. Separation and Purification Technology, 306: 122503. https://doi.org/https://doi.org/10.1016/j.seppur.2022.122503.

Machado, T. O., Grabow, J., Sayer, C., de Araújo, P. H. H., Ehrenhard, M. L. and Wurm, F. R. (2022). Biopolymer-based nanocarriers for sustained release of agrochemicals: A review on materials and social science perspectives for a sustainable future of agri- and horticulture. Advances in Colloid and Interface Science, 303: 102645. https://doi.org/https://doi.org/10.1016/j.cis.2022.102645.

Mamba, F. B., Mbuli, B. S. and Ramontja, J. (2021). Recent advances in biopolymeric membranes towards the removal of emerging organic pollutants from water. Membranes, 11(11): 798. https://www.mdpi.com/2077-0375/11/11/798.

Manikandan, S., Subbaiya, R., Saravanan, M., Ponraj, M., Selvam, M. and Pugazhendhi, A. (2022). A critical review of advanced nanotechnology and hybrid membrane based water recycling, reuse, and wastewater treatment processes. Chemosphere, 289: 132867. https://doi.org/https://doi.org/10.1016/j.chemosphere.2021.132867.

Mariana, M., H.P.S, A. K., Yahya, E. B., Olaiya, N. G., Alfatah, T., Suriani, A. B. et al. (2022). Recent trends and future prospects of nanostructured aerogels in water treatment applications. Journal of Water Process Engineering, 45: 102481. https://doi.org/https://doi.org/10.1016/j.jwpe.2021.102481.

Marin, M. A., Mallepally, R. R. and McHugh, M. A. (2014). Silk fibroin aerogels for drug delivery applications. The Journal of Supercritical Fluids, 91: 84–89. https://doi.org/https://doi.org/10.1016/j.supflu.2014.04.014.

Miao, Y., Yang, T., Yang, S., Yang, M. and Mao, C. (2022). Protein nanoparticles directed cancer imaging and therapy. Nano Convergence, 9(1): 2. https://doi.org/10.1186/s40580-021-00293-4.

Mikaela, B. and Gunnar, W. (2015). Crystalline Nanocellulose—preparation, modification, and properties. *In*: Matheus, P. and Heitor Luiz Ornaghi, J. (Eds.). Cellulose (pp. Ch. 7). IntechOpen. https://doi.org/10.5772/61899.

Mitchell, M. J., Billingsley, M. M., Haley, R. M., Wechsler, M. E., Peppas, N. A. and Langer, R. (2021). Engineering precision nanoparticles for drug delivery. Nature Reviews Drug Discovery, 20(2): 101–124. https://doi.org/10.1038/s41573-020-0090-8.

Mohan, D., Abhishek, K., Patel, M. and Pittman, C. U., Jr. (2024). Harnessing biochar in contaminated soil for heavy metal immobilization, soil health enhancement, and carbon sequestration. Industrial & Engineering Chemistry Research. https://doi.org/10.1021/acs.iecr.4c00082.

Moshood, T. D., Nawanir, G., Mahmud, F., Mohamad, F., Ahmad, M. H. and AbdulGhani, A. (2022). Biodegradable plastic applications towards sustainability: A recent innovations in the green product. Cleaner Engineering and Technology, 6: 100404. https://doi.org/https://doi.org/10.1016/j.clet.2022.100404.

Münzel, T., Hahad, O., Daiber, A. and Landrigan, P. J. (2022). Soil and water pollution and human health: What should cardiologists worry about? Cardiovascular Research, 119(2): 440–449. https://doi.org/10.1093/cvr/cvac082.

Muthukumaran, P., Suresh Babu, P., Shyamalagowri, S., Aravind, J., Kamaraj, M. and Govarthanan, M. (2022). Polymeric biomolecules based nanomaterials: Production strategies and pollutant mitigation as an emerging tool for environmental application. Chemosphere, 307: 136008. https://doi.org/https://doi.org/10.1016/j.chemosphere.2022.136008.

Nagar, A. and Pradeep, T. (2020). Clean water through nanotechnology: Needs, gaps, and fulfillment. ACS Nano, 14(6): 6420–6435. https://doi.org/10.1021/acsnano.9b01730.

Nishu and Kumar, S. (2023). Smart and innovative nanotechnology applications for water purification. Hybrid Advances, 3: 100044. https://doi.org/https://doi.org/10.1016/j.hybadv.2023.100044.

Noreen, S., Saleem, M. H., Ali, B., Khan, K. A., Hafeez, A. and Javed, M. A. (2024). Advancing plant resilience against microplastics and metals through nanotechnology. BioNanoScience. https://doi.org/10.1007/s12668-024-01301-7.

Obotey Ezugbe, E. and Rathilal, S. (2020). Membrane technologies in wastewater treatment: A review. Membranes, 10(5): 89. https://www.mdpi.com/2077-0375/10/5/89.

Ojuederie, O. B. and Babalola, O. O. (2017). Microbial and plant-assisted bioremediation of heavy metal polluted environments: A review. International Journal of Environmental Research and Public Health, 14(12): 1504. https://www.mdpi.com/1660-4601/14/12/1504.

Omer, A. M., Dey, R., Eltaweil, A. S., Abd El-Monaem, E. M. and Ziora, Z. M. (2022). Insights into recent advances of chitosan-based adsorbents for sustainable removal of heavy metals and anions. Arabian Journal of Chemistry, 15(2): 103543. https://doi.org/https://doi.org/10.1016/j.arabjc.2021.103543.

Pachaiappan, R., Cornejo-Ponce, L., Rajendran, R., Manavalan, K., Femilaa Rajan, V. and Awad, F. (2022). A review on biofiltration techniques: Recent advancements in the removal of volatile organic compounds and heavy metals in the treatment of polluted water. Bioengineered, 13(4): 8432–8477. https://doi.org/10.1080/21655979.2022.2050538.

Patil, T. V., Patel, D. K., Dutta, S. D., Ganguly, K., Santra, T. S. and Lim, K.-T. (2022). Nanocellulose, a versatile platform: From the delivery of active molecules to tissue engineering applications. Bioactive Materials, 9: 566–589. https://doi.org/https://doi.org/10.1016/j.bioactmat.2021.07.006.

Pellis, A., Guebitz, G. M. and Nyanhongo, G. S. (2022). Chitosan: Sources, processing and modification techniques. Gels, 8(7): 393. https://www.mdpi.com/2310-2861/8/7/393.

Pete, A. J., Bharti, B. and Benton, M. G. (2021). Nano-enhanced bioremediation for oil spills: A review. ACS ES&T Engineering, 1(6): 928–946. https://doi.org/10.1021/acsestengg.0c00217.

Pham, D. T. and Tiyaboonchai, W. (2020). Fibroin nanoparticles: A promising drug delivery system. Drug Delivery, 27(1): 431–448. https://doi.org/10.1080/10717544.2020.1736208.

Rana, A. K., Mostafavi, E., Alsanie, W. F., Siwal, S. S. and Thakur, V. K. (2023). Cellulose-based materials for air purification: A review. Industrial Crops and Products, 194: 116331. https://doi.org/https://doi.org/10.1016/j.indcrop.2023.116331.

Rather, M. A., Bhuyan, S., Chowdhury, R., Sarma, R., Roy, S. and Neog, P. R. (2023). Nanoremediation strategies to address environmental problems. Science of The Total Environment, 886: 163998. https://doi.org/https://doi.org/10.1016/j.scitotenv.2023.163998.

Reddy, M. S. B., Ponnamma, D., Choudhary, R. and Sadasivuni, K. K. (2021). A comparative review of natural and synthetic biopolymer composite scaffolds. Polymers, 13(7): 1105. https://www.mdpi.com/2073-4360/13/7/1105.

Remya, R. R., Julius, A., Suman, T. Y., Mohanavel, V., Karthick, A., Pazhanimuthu, C. et al. (2022). Role of nanoparticles in biodegradation and their importance in environmental and biomedical applications. Journal of Nanomaterials, 2022, 6090846. https://doi.org/10.1155/2022/6090846.
Russo, T., Fucile, P., Giacometti, R. and Sannino, F. (2021). Sustainable removal of contaminants by biopolymers: A novel approach for wastewater treatment. Current State and Future Perspectives. Processes, 9(4): 719. https://www.mdpi.com/2227-9717/9/4/719.
Saif, A., Anjum, L., Faisal, Z., Akram, N., Shah, Y. A., Irfan, R. et al. (2023). Recent advances in protein-based nanoparticles and their applications in the delivery of bioactive compounds. International Journal of Food Properties, 26(2): 2866–2880. https://doi.org/10.1080/10942912.2023.2261662.
Saiyad, M., Shah, N., Joshipura, M., Dwivedi, A. and Pillai, S. (2024). Modified biopolymers in wastewater treatment: A review. Materials Today: Proceedings. https://doi.org/https://doi.org/10.1016/j.matpr.2024.01.031.
Salama, A., Abouzeid, R., Leong, W. S., Jeevanandam, J., Samyn, P., Dufresne, A. et al. (2021). Nanocellulose-based materials for water treatment: adsorption, photocatalytic degradation, disinfection, antifouling, and nanofiltration. Nanomaterials, 11(11): 3008. https://www.mdpi.com/2079-4991/11/11/3008.
Saraswat, P., Singh, S., Prasad, M., Misra, R., Rajput, V. D. and Ranjan, R. (2023). Applications of bio-based nanomaterials in environment and agriculture: A review on recent progresses. Hybrid Advances, 4: 100097. https://doi.org/https://doi.org/10.1016/j.hybadv.2023.100097.
Sarkar, S., Ponce, N. T., Banerjee, A., Bandopadhyay, R., Rajendran, S. and Lichtfouse, E. (2020). Green polymeric nanomaterials for the photocatalytic degradation of dyes: A review. Environmental Chemistry Letters, 18(5): 1569–1580. https://doi.org/10.1007/s10311-020-01021-w.
Savage, N. and Diallo, M. S. (2005). Nanomaterials and water purification: opportunities and challenges. Journal of Nanoparticle Research, 7(4): 331–342. https://doi.org/10.1007/s11051-005-7523-5.
Seddiqi, H., Oliaei, E., Honarkar, H., Jin, J., Geonzon, L. C., Bacabac, R. G. et al. (2021). Cellulose and its derivatives: Towards biomedical applications. Cellulose, 28(4): 1893–1931. https://doi.org/10.1007/s10570-020-03674-w.
Simelane, N. P., Asante, J. K. O., Ndibewu, P. P., Mramba, A. S. and Sibali, L. L. (2022). Biopolymer composites for removal of toxic organic compounds in pharmaceutical effluents – a review. Carbohydrate Polymer Technologies and Applications, 4: 100239. https://doi.org/https://doi.org/10.1016/j.carpta.2022.100239.
Soldo, A. and Miletić, M. (2019). Study on shear strength of xanthan gum-amended soil. Sustainability, 11(21): 6142. https://www.mdpi.com/2071-1050/11/21/6142.
Soldo, A., Miletić, M. and Auad, M. L. (2020). Biopolymers as a sustainable solution for the enhancement of soil mechanical properties. Scientific Reports, 10(1): 267. https://doi.org/10.1038/s41598-019-57135-x.
Sreena, R. and Nathanael, A. J. (2023). Biodegradable biopolymeric nanoparticles for biomedical applications-challenges and future outlook. Materials, 16(6): 2364. https://www.mdpi.com/1996-1944/16/6/2364.
Steffens, C., Ballen, S. C., Scapin, E., da Silva, D. M., Steffens, J. and Jacques, R. A. (2022). Advances of nanobiosensors and its application in atrazine detection in water: A review. Sensors and Actuators Reports, 4: 100096. https://doi.org/https://doi.org/10.1016/j.snr.2022.100096.
Sundar, S., Kundu, J. and Kundu, S. C. (2010). Biopolymeric nanoparticles. Science and Technology of Advanced Materials, 11(1): 014104. https://doi.org/10.1088/1468-6996/11/1/014104.
Sutirman, Z. A., Sanagi, M. M. and Wan Aini, W. I. (2021). Alginate-based adsorbents for removal of metal ions and radionuclides from aqueous solutions: A review. International Journal of Biological Macromolecules, 174: 216–228. https://doi.org/https://doi.org/10.1016/j.ijbiomac.2021.01.150.
Tadayoni, N. S., Dinari, M., Roy, A. and Karimi Abdolmaleki, M. (2024). Recent advances in porous bio-polymer composites for the remediation of organic pollutants. Polymers, 16(11): 1543. https://www.mdpi.com/2073-4360/16/11/1543.
Thomas, M. S., Thomas, S. and Pothen, L. A. (2022). Science and technology of nanomaterials: introduction. In Nanotechnology for Environmental Remediation (pp. 1–15). https://doi.org/https://doi.org/10.1002/9783527834143.ch1.
Tiwari, A. and Dhoble, S. J. (2016). Stabilization of ZnS nanoparticles by polymeric matrices: Syntheses, optical properties and recent applications [10.1039/C6RA13108E]. RSC Advances, 6(69): 64400–64420. https://doi.org/10.1039/C6RA13108E.
Tran, K. T. M. and Nguyen, T. D. (2017). Lithography-based methods to manufacture biomaterials at small scales. Journal of Science: Advanced Materials and Devices, 2(1): 1–14. https://doi.org/https://doi.org/10.1016/j.jsamd.2016.12.001.
Ul-Islam, M., Alabbosh, K. F., Manan, S., Khan, S., Ahmad, F. and Ullah, M. W. (2024). Chitosan-based nanostructured biomaterials: Synthesis, properties, and biomedical applications. Advanced Industrial and Engineering Polymer Research, 7(1): 79–99. https://doi.org/https://doi.org/10.1016/j.aiepr.2023.07.002.
Venkatesan, J., Anil, S., Singh, S. K. and Kim, S.-K. (2017). Chapter 13 - preparations and applications of alginate nanoparticles. pp. 251–268. *In*: Venkatesan, J., Anil, S. and Kim, S.-K. (Eds.). Seaweed Polysaccharides. Elsevier. https://doi.org/https://doi.org/10.1016/B978-0-12-809816-5.00013-X.

Verma, D., Gulati, N., Kaul, S., Mukherjee, S. and Nagaich, U. (2018). Protein based nanostructures for drug delivery. Journal of Pharmaceutics, 2018, 9285854. https://doi.org/10.1155/2018/9285854.

Visan, A. I., Popescu-Pelin, G. and Socol, G. (2021). Degradation behavior of polymers used as coating materials for drug delivery—a basic review. Polymers, 13(8): 1272. https://www.mdpi.com/2073-4360/13/8/1272.

Vodyashkin, A. A., Kezimana, P., Vetcher, A. A. and Stanishevskiy, Y. M. (2022). Biopolymeric nanoparticles–multifunctional materials of the future. Polymers, 14(11): 2287. https://www.mdpi.com/2073-4360/14/11/2287.

Voisin, H., Bergström, L., Liu, P. and Mathew, A. P. (2017). Nanocellulose-based materials for water purification. Nanomaterials, 7(3): 57. https://www.mdpi.com/2079-4991/7/3/57.

Wang, B., Wan, Y., Zheng, Y., Lee, X., Liu, T., Yu, Z. et al. (2019). Alginate-based composites for environmental applications: A critical review. Critical Reviews in Environmental Science and Technology, 49(4): 318–356. https://doi.org/10.1080/10643389.2018.1547621.

Wang, B., Zhu, Y., Bai, Z., Luque, R. and Xuan, J. (2017). Functionalized chitosan biosorbents with ultra-high performance, mechanical strength and tunable selectivity for heavy metals in wastewater treatment. Chemical Engineering Journal, 325: 350–359. https://doi.org/https://doi.org/10.1016/j.cej.2017.05.065.

Wang, H.-L., Hsu, C.-Y., Wu, K. C. W., Lin, Y.-F. and Tsai, D.-H. (2020). Functional nanostructured materials: Aerosol, aerogel, and *de novo* synthesis to emerging energy and environmental applications. Advanced Powder Technology, 31(1): 104–120. https://doi.org/https://doi.org/10.1016/j.apt.2019.09.039.

Wang, L., Hu, C. and Shao, L. (2017). The antimicrobial activity of nanoparticles: Present situation and prospects for the future. Int. J. Nanomedicine, 12: 1227–1249. https://doi.org/10.2147/IJN.S121956.

Wang, Y. and Lu, Y. (2023). Sodium alginate-based functional materials toward sustainable applications: Water treatment and energy storage. Industrial & Engineering Chemistry Research, 62(29): 11279–11304. https://doi.org/10.1021/acs.iecr.3c01082.

Weng, Y., Yang, G., Li, Y., Xu, L., Chen, X., Song, H. et al. (2023). Alginate-based materials for enzyme encapsulation. Advances in Colloid and Interface Science, 318: 102957. https://doi.org/https://doi.org/10.1016/j.cis.2023.102957.

Wu, Y., Liu, Y., Kamyab, H., Rajasimman, M., Rajamohan, N., Ngo, G. H. et al. (2023). Physico-chemical and biological remediation techniques for the elimination of endocrine-disrupting hazardous chemicals. Environmental Research, 232: 116363. https://doi.org/https://doi.org/10.1016/j.envres.2023.116363.

Wuana, R. A. and Okieimen, F. E. (2011). Heavy Metals in Contaminated Soils: A Review of Sources, Chemistry, Risks and Best Available Strategies for Remediation. ISRN Ecology, 2011, 402647. https://doi.org/10.5402/2011/402647.

Xu, Z., Yu, Y., Yan, L., Yan, W. and Jing, C. (2022). Asenic removal from groundwater using granular chitosan-titanium adsorbent. Journal of Environmental Sciences, 112: 202–209. https://doi.org/https://doi.org/10.1016/j.jes.2021.05.028.

Yahya, E. B., Jummaat, F., Amirul, A. A., Adnan, A. S., Olaiya, N. G., Abdullah, C. K. et al. (2020). A review on revolutionary natural biopolymer-based aerogels for antibacterial delivery. Antibiotics, 9(10): 648. https://www.mdpi.com/2079-6382/9/10/648.

Yan, D., Li, Y., Liu, Y., Li, N., Zhang, X. and Yan, C. (2021). Antimicrobial properties of chitosan and chitosan derivatives in the treatment of enteric infections. Molecules, 26(23): 7136. https://www.mdpi.com/1420-3049/26/23/7136.

Yang, F., Wang, B., Shi, Z., Li, L., Li, Y., Mao, Z. et al. (2021). Immobilization of heavy metals (Cd, Zn, and Pb) in different contaminated soils with swine manure biochar. Environmental Pollutants and Bioavailability, 33(1): 55–65. https://doi.org/10.1080/26395940.2021.1916407.

Yilmaz Atay, H. (2019). Antibacterial activity of chitosan-based systems. pp. 457–489. *In*: Jana, S. and Jana, S. (Eds.). Functional Chitosan: Drug Delivery and Biomedical Applications. Springer Singapore. https://doi.org/10.1007/978-981-15-0263-7_15.

Yousefi, A. M. and Wnek, G. E. (2024). Poly(hydroxyalkanoates): Emerging Biopolymers in Biomedical Fields and Packaging Industries for a Circular Economy. Biomedical Materials & Devices. https://doi.org/10.1007/s44174-024-00166-4.

Zhang, Z., Chen, Y., Wang, D., Yu, D. and Wu, C. (2023). Lignin-based adsorbents for heavy metals. Industrial Crops and Products, 193: 116119. https://doi.org/https://doi.org/10.1016/j.indcrop.2022.116119.

Zhao, C., Liu, G., Tan, Q., Gao, M., Chen, G., Huang, X. et al. (2023). Polysaccharide-based biopolymer hydrogels for heavy metal detection and adsorption. Journal of Advanced Research, 44: 53–70. https://doi.org/https://doi.org/10.1016/j.jare.2022.04.005.

Zhou, M., Wang, T., Xu, K., Wang, H., Li, Z., Zhang, W.-x. et al. (2024). Broadening environmental research in the era of accurate protein structure determination and predictions. Frontiers of Environmental Science & Engineering, 18(7): 91. https://doi.org/10.1007/s11783-024-1851-0.

CHAPTER-6

Biopolymer-Based Metal Nanoparticles for Biosensing Applications

Danish Alam,[1] *Tanzeel Khan,*[2] *Ekta Kamra,*[3] *Tokeer Ahmad,*[4] *Meryam Sardar*[1,*] and *Asimul Islam*[2,*]

1. Introduction

Biosensing is considered a conventional notion which is employed for protection against adverse conditions. To add, many organic life forms possess innate biosensing, such as the presence of electro-sensitivity in sharks, toxic sensibility in several algal species and better invasive ability of canines (Ramesh et al., 2022; Naresh and Lee, 2021; Banerjee et al., 2021). Strikingly, the International Union of Pure and Applied Chemistry defines a biosensor as a device, which utilizes a specific biochemical reaction mediated by the immune system, organelles, isolated enzymes, or tissues to identify chemical compounds via monitoring thermal, optical, or electrical signals (Zhai et al., 2019). In fact, a biosensor contains a transducer and receptor (Fig. 1), wherein receptors transform physical or chemical data into an energy form, whereas a transducer converts such energy into electrical signals. Of note, the biosensor was first discovered by Clark and Lyons in 1960s (Mehrotra, 2016). To analyze various compounds and to develop a biosensor, several combinations of biopolymers and sensor elements may give promising results and thus lead to a wide range of sensing applications. Nonetheless, varieties of biopolymers may be employed for the biosensor's preparation, the complexity associated with the fabrication process of a material confines the development of biosensors (Cruz et al., 2014; Kaushik et al., 2014; Kaushik et al., 2018). Hence, it is crucial to devise a reliable and compatible biopolymer along with investigating the ease of the fabrication process.

Recently, nanotechnology has gained attention owing to its extensive applications in various areas, i.e., chemistry, biomedical sciences, and material science (Alam et al., 2023). Additionally, metal nanoparticles (NPs) are used for synthesizing various biosensors because of their spectral and optical properties (Karikalan et al., 2016). Hence, different metal NPs, e.g., Ag, Au, Pt, and Pd

[1] Department of Biosciences, Jamia Millia Islamia, Jamia Nagar, New Delhi 110025, INDIA.
[2] Centre for Interdisciplinary Research in Basic Sciences, Jamia Millia Islamia, Jamia Nagar, New Delhi 110025, INDIA.
[3] Centre for Biotechnology, Maharshi Dayanand University, Rohtak-124001, Haryana, India.
[4] Department of Chemistry, Jamia Millia Islamia, Jamia Nagar, New Delhi 110025, INDIA.
* Corresponding authors: msardar@jmi.ac.in; aislam@jmi.ac.in
DOI: 10.1201/9781003558330-6

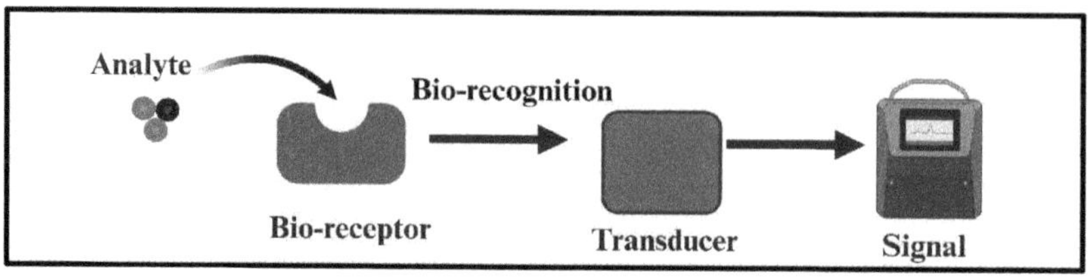

Figure 1. Schematic representation of a biosensor.

have been employed to develop different types of biosensors in various fields, including disease diagnostics as well as cancer (Lasita et al., 2022; Khan et al., 2022). Indeed, the choice of a stabilizer plays a significant role in the synthesis of metal NPs. Also, an ideal stabilizer must possess reactive functional groups to immobilize metal NPs along with biodegradability as well as biocompatibility. Amongst various stabilizers which are being used till today, biopolymers are considered as the most suitable supporting material (Rizwan et al., 2022).

A biopolymer is formed as a result of polymerization of monomers that are linked via covalent bonding. It is biodegradable and thus can be degraded by the microbial action and releases CO_2 and H_2O as by-products in the environment (Sobhan et al., 2021). On the basis of the presence of monomers, biopolymers can be categorized into 3 major groups: (1) polynucleotides (polymer of nucleotides), (2) polypeptides (polymer of amino acids) and (3) polysaccharides (polymer of monosaccharides). Amid these, polysaccharides offer polymeric matrices to immobilize metal NPs and developing biosensors, which is due to their thermal/mechanical stability, large surface area, strong adsorption ability and reactivity towards several chemical reactions. As a result, polysaccharide-based metal NPs (utilizing cellulose, gelatin, chitosan, pectin, alginate, and acacia gum as biopolymer) have been usually employed to develop various biosensors (Sawant, 2017).

2. Types of Biosensors

Biosensors are grouped into different categories based on their bio-receptor, i.e., enzyme biosensor, DNA biosensor, aptamer-based biosensor, immuno-sensors, whole-cell biosensor, microbial biosensors, etc., and transducers further classifies them as an electrochemical biosensor, calorimetric biosensor, piezoelectric biosensor and optical biosensor as shown in Fig. 2 (Madej-Kiełbik et al., 2022).

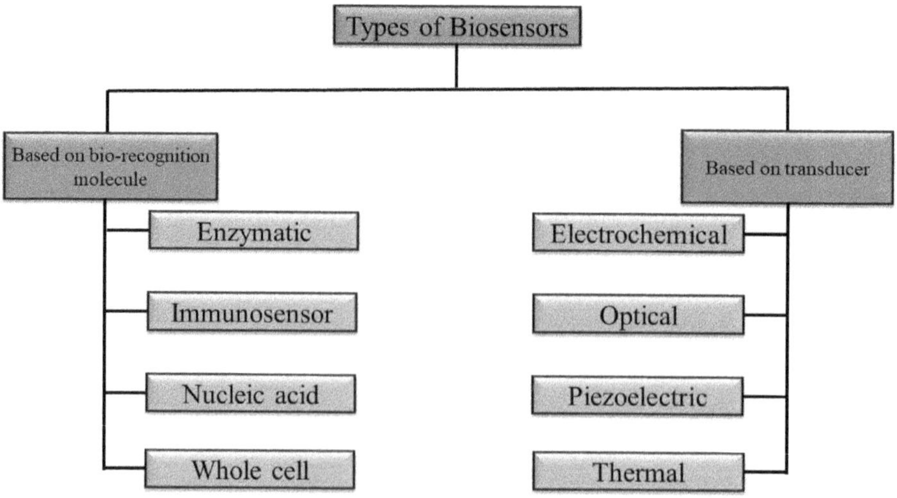

Figure 2. Classification of biosensors.

2.1 Electrochemical Biosensors

These biosensors generate a signal via an electrochemical reaction between bio-receptors and detectable analytes on the surface of transducer. Moreover, such biosensors offer label-free identification of biomolecules with high signal-to-volume ratio without causing any damage to the system (Zhou et al., 2024). Here the, bio-receptor molecules (either covalently linked or coated onto the transducer's surface) interact exclusively with analytes to be detected, and thus generating an electrical signal by means of capacitance, voltage, impedance, current, etc. (Yao et al., 2023). According to the working principle, electrochemical biosensors may act as potentiometric, capacitive, amperometric, and impedimetric transducers, transforming the biological or chemical interactions into readable signals (Kour et al., 2020).

2.2 Optical Biosensors

An optical transducer may be associated with a light source that transforms a biological incident into an electrical signal (Kumar and Singh, 2021). The major components of an optical biosensor include a light source (light-emitting diode, laser, etc.), an optical transmission media (waveguide, fiber, etc.), a bio-receptor [enzymes, antibodies (Abs), or microbes], as well as an optical detection system. In these biosensors, the biorecognition phenomenon causes physical or chemical changes, which causes alteration in the refraction, frequency, absorption, phase, transmission, amplitude, reflection, or polarization of light. Also, these biosensors can be developed by employing techniques like Surface Plasmon Resonance (SPR), fluorescence, bioluminescence, Raman scattering, and refractive index (Kaur et al., 2022).

Strikingly, optical biosensors have seen a growing demand in a variety of areas such as biomedical research, diagnostics, food quality and safety. In recent years, these biosensors were used in the detection as well as binding kinetics of interactions between the DNA and proteins. SPR was first introduced in 1980s, widely employed to study the biological and chemical interaction including nucleotide-nucleotide, drug screening, enzymatic analysis, ligand fishing, epitope mapping, signal transduction, and nucleotide-protein binding, etc. (Uniyal et al., 2022).

2.3 Piezoelectric Biosensors

In a piezoelectric biosensor (e.g., quartz crystal micro/nanobalance), the piezoelectric effect is used to quantify force, acceleration, strain, or pressure by transforming them into an electrical signal (Di Pietrantonio et al., 2019). To construct these type of sensors, the recognition molecules (enzyme, Ab, cells, etc.) are adhered on to a piezoelectric surface, wherein interaction between the target analyte and the recognition molecule occurs and induces a mechanical vibration which are converted into a readable form indicating the presence of target analyte. The change in potential may generate a standing wave in the crystal at a resonance frequency, which is dependent on the elastic nature of the crystal as well as the crystal's coating by biomarker (Pohanka, 2017). Further, change in resonance frequency (either by adsorption or desorption of the analytes on the crystal's surface) is given by

$$\Delta f = Kf^2.\Delta m/A$$

Wherein Δf, Δm, K, and A stands for change in the resonance frequency (Hz), change in the mass of absorbed material (g), a constant, and surface area (cm^2), respectively (Malhotra and Ali, 2017).

2.4 Thermal Biosensors

Thermal sensors can detect temperature change caused by the heat generated or captured during the interaction between immobilized bio-receptors and the analyte. The overall heat evolved or captured is directly proportional to the molar enthalpy and the number of moles of a product formed during a chemical reaction (Vasuki et al., 2019). The change in temperature can be defined as

$$\Delta T = -\Delta H.n_p/Cp$$

Wherein ΔT, ΔH, n_p and C_p are a change in temperature, change in enthalpy, number of moles of product, and heat capacity, respectively. These sensors are used to monitor diverse molecules, i.e., cholesterol, blood glucose, and heavy metal ions (Hg^{2+}, Cu^{2+}, Ag^{+}) (Vasuki et al., 2019).

2.5 Enzyme Biosensors

In an enzymatic biosensor, an enzyme is immobilized onto the surface of a transducer and generates a definite signal (in terms of electrochemical, or colorimetric, or optical properties) which reacts with a target analyte (Nguyen et al., 2019). The discovery of a glucose biosensor in 1962 led to development of many devices based on enzymatic biosensors. Further, enzymes were immobilized onto NPs by surface chemistry to detect desired targets, i.e., urea, blood sugar, cholesterol, etc., (Öndeş et al., 2021). To attain high sensitivity, specificity, and stability, nanomaterials act as a key factor to immobilize enzymes for fabricating various biosensors (Srivastava et al., 2013; Kumar et al., 2013).

2.6 Immunosensors

Immunosensors are affinity-based sensors that comprise of immobilizing antigen (Ag) or Ab on a transducer surface (Dupont, 2022). Owing to their high specificity and sensitivity, they can quantify various cancerous molecules, proteins, food pathogens, etc., up to 10^{-9}–10^{-15} molar concentration. Ag and Ab interactions form immuno-complex that causes alterations in resistance, current, RI, capacitance, etc., which are measured by optical or amperometric detection. One of the major issues linked with immunosensors is non-specific binding of Ab with molecules other than target Ags, which can further be improved by utilizing nanomaterials as they possess larger surface area and yield low signal-to-noise ratio (Bhardwaj et al., 2021).

2.7 Nucleic Acid Biosensors

A nucleic acid biosensor involves functionalization of nucleic acid on the surface of a transducer (Liu et al., 2023). Usually , these biosensors depend on immobilizing a single-stranded probe DNA (sspDNA) onto a transducer surface, wherein pDNA recognizes the complementary DNA sequence via hybridization, thus causes an alteration in the quantifiable signal (Rahman et al., 2015).

Recently, aptamer-based sensors have emerged, which are ssDNA/RNA molecules that may be prepared via Systematic Evolution of Ligands by Exponential enrichment method (Gan et al., 2022). In fact, aptamers have the capability to interact with an array of ligands, including proteins, viruses, small molecules, e.g., metal ions, drugs, etc. Owing to their high specificity, ease of synthesis, easy to modify, and higher stability render them appropriate for developing biosensors (Sequeira-Antunes and Ferreira, 2013). To enhance the performance of an aptamer-based biosensor, various nanomaterials like CNTs, graphene, metal NPs, quantum dots, etc., have been attached with them.

2.8 Whole-Cell Biosensors

The whole-cell biosensor is based on immobilizing alive components like cells, microbes (acting as sensing components) in spite of utilizing enzymes, Ab or DNA, under cellular conditions. While detecting pollutants, whole cell biosensors provide promising results. The interaction between the living cells and organic molecules, e.g., xenobiotics, heavy metals and change in pH within water, soil, and air generates a readable signal (Liu et al., 2022). For instance, a bacterial biosensor was previously utilized to quantify the biochemical oxygen demand (Berepiki et al., 2020). Various biopolymers and their ligands have been enlisted in Table 1.

3. Biopolymer-based metal NPs for Biosensing

Only biopolymers cannot be utilized efficiently for biosensing applications owing to their low mechanical and chemical resistance. Indeed, they are pre-made polymers that possess the tendency to absorb water. However, these issues can be addressed by creating composites of biopolymers with metal or metal oxide NPs (Aslam et al., 2023).

3.1 Chitosan-based biosensors

Previously, chitosan (CS), a natural polysaccharide, was broadly utilized to immobilize several materials, e.g., metal and metal oxide NPs owing to its high permeability, good film forming capacity, biodegradability, mechanical strength, biocompatibility, non-toxicity, easy accessibility, and economy (Mamuk et al., 2023; Radhakrishnan et al., 2015). CS possesses both NH_2 and OH groups that makes it a suitable material to be utilized for cross-linkage with various metal/metal oxide NPs, i.e., Au, Ag, Fe_2O_3, zinc oxide, and cerium oxide (Ahmad et al., 2018; Prabhakar et al., 2016). Fe_3O_4 NPs have gained significant attention as electrochemical transducers due to their ability to enhance conductivity, exhibit superparamagnetic properties, amplify signals, and provide a favorable microenvironment for protein adhesion. These characteristics contribute to raise sensitivity and selectivity. However, Fe_3O_4 NPs exhibit aggregation and fast biodegradation that can be resolved by dispersing these NPs in CS matrix, which made them suitable for biosensing applications (Prabhakar et al., 2016). For instance, Kaushik et al. (2008) designed CS based glucose biosensor, where Fe_3O_4 NPs were dispersed in a CS solution to form a nanocomposite film on indium-tin oxide glass plate and glucose oxidase (GOD) immobilized on the plate that yielded better reproducibility and stability. Notably, CS acted as a stabilizer that prevent the aggregation of Fe_3O_4 NPs without causing any modification in its optical and electrical properties, whereas Fe_3O_4NPs raise the electron transfer between glucose and CS-Fe_3O_4/indium-tin oxide electrode surface (Kaushik et al., 2008).

In another study, Au-Carbon Dots (Au-CDs)-CS-modified a glassy carbon electrode based dopamine (DA) biosensor was developed (Huang et al., 2014). The presence of carboxyl (COO^-) groups on CDs interacts with NH_3^+ group of DA, which results into electrostatic interaction thus increased recognition of DA was attained along with a higher specificity. Atif et al., (2019) synthesized a potentiometric urea biosensor based on Fe_3O_4-CS nano biocomposite along with urease enzyme (Atif et al., 2019), wherein the biosensor depicted reasonably high stability, sensitivity, selectivity, and reproducibility. Moreover, such biosensors can be utilized to monitor urea concentrations in the food industry, pharmaceutical industry, and fertilizer's production. In another work, to detect contaminants including phenolic compounds, amperometric tyrosinase biosensor using Fe_3O_4NPs-CS nanocomposite was developed (Wang et al., 2008a) with high reproducibility and stability.

Strikingly, Zeng and co-workers enhanced biocompatibility and hydrophilicity of a glucose biosensor by immobilizing graphene-CS composite on palladium NPs using *in situ* reduction (Zeng et al., 2011). To add, Yardımcı et al. (2012) devised a consistent and cost-effective approach to synthesize H_2O_2 biosensor using horseradish peroxidase (HRP) immobilized onto $CoFe_2O_4$-CS nanocomposite, where they coated HRP on to $CoFe_2O_4$-CS nanocomposite (Yardımcı et al., 2012). Moreover, the designed biosensor showed a quick amperometric signal to H_2O_2.

3.2 Cellulose-based Biosensors

Cellulose is a fibrous, renewable biopolymer composed of β-1,4-linked D-glucose units, which serves as a vital structural component of a plant cell wall. Further, it can be synthesized by fungi, marine animals, invertebrates, algae, bacteria, and amoeba, that is why, it exists as the most copious natural biopolymer (Nasrollahzadeh et al., 2020). Indeed, cellulose has different types of morphology and mechanical behavior, such as vegetable cellulose, bacterial cellulose and cellulose nanocrystals (CNC), which can be utilized to prepare nanocomposites (Hajian et al., 2017). However, they exhibit poor electrical conductivity thus demonstrating poor sensitivity to be utilized as a biosensor material. To add, being hydrophilic in nature, cellulose is not compatible with some sensing molecules thus there is an urgent need of modification so that cellulose can be utilized for biosensor applications. One such modification can be achieved by using metal/metal oxide NPs, e.g., AuNPs as they possess high catalytic activity in a variety of reactions (Biliuta and Coseri, 2019).

Of note, cellulose-based biosensors have been used in many areas, including food safety, diagnostics, and environmental control. These biosensors are advantageous as they are cost-effective, easier to dispose off, and environment-friendly in nature. To add, such materials transport liquids by capillarity without usage of any external power (Faham et al., 2019). For instance, AuNPs-Bacterial Cellulose (Au-BC) nanocomposite was synthesized by Wang et al., (2010) for amperometric recognition of glucose, where the data depicted good sensitivity and quick reaction towards glucose in the occurrence of electron mediator (Wang et al., 2010). Additionally, Zhang et al. (2010) designed a H_2O_2 biosensor in a single-step employing AuNPs-bacteria cellulose nanofiber (Au-BC) (Zhang et al., 2010).

In a study by Wang et al. (2011a) fabricated Au-BC nanocomposite for synthesizing a H_2O_2 biosensor, and applied it in the decoration of HRP, hemoglobin, and myoglobin (Mb) on a glassy carbon electrode modified with the Au-BC nanocomposite. Among these three synthesized H_2O_2 biosensors, HRP immobilized Au-BC film depicted excellent biocatalytic activity along with good sensitivity, low Limit Of Detection (LOD), and a rapid response to H_2O_2. Strikingly, Li et al. (2016) modified electrodes with AuNPs-BC nanofibers, laccase (Lac), and Nafion resulting in a biosensor that demonstrated exceptional bio eletrocatalytic activity towards a hydroquinone environmental pollutant with high sensitivity and lower LOD thus it can also be employed to detect the presence of hydroquinone within lake water.

Previously, multiwalled carbon nano tubes (MWCNT) were used to prepare biosensors. Moreover, MWCNT exhibited reasonably good physical along with chemical characteristics (Yellapu et al., 2019), yet they are associated with poor dispersibility in aqueous solutions, which could be resolved by CNC. The reason behind using CNC is its excellent biodegradability, exceptional mechanical strength, multiple reactive sites and better film-forming ability (Tang et al., 2017). In a similar fashion, Palanisamy et al. (2020) utilized the same nanocomposite for developing H_2O_2 biosensors. Moreover, the sonochemical method was used to fabricate ZnO nanorod on MWCNT-CNC nanocomposite, followed by immobilizing hemin, which demonstrated enhanced catalytic activity and a lower reduction potential for H_2O_2 as compared to hemin-modified biosensors (Palanisamy et al., 2020).

3.3 Alginate-based Biosensors

Alginate (a natural polysaccharide) is a linear unbranched biopolymer having β-D-mannuronic and α-L-guluronic acid joined together by 1,4-linkage. Alginate is considered as a suitable material due to its cost-effectiveness, hydrophilicity, biocompatibility, precise gelation abilities, non-toxic nature, etc. (Roquero and Katz, 2022). It is noteworthy to mention that alginate hydrogels act as biocompatible matrices, enabling the entrapment of biomolecules under mild conditions. In clinical diagnostics, implantable devices, e.g., biosensors, drug delivery devices, catheters, etc., are being utilized to monitor different analytes in blood and yielded promising results. Further, to increase its functionality, various additives, i.e., therapeutics and Magnetic Resonance Imaging (MRI) agents may be co-immobilized along with the implanted biosensor (Maity and Das, 2022).

For instance, Magnetic NPs (MNPs) of maghemite (Fe_2O_3) or magnetite (Fe_3O_4) were formed by coating in alginate that were utilized previously as biocompatible MRI contrast agents and for drug delivery (Boissiere et al., 2007). Likewise, Joshi et al. (2011) designed an implantable system for drug delivery and MRI, wherein a glucose biosensor, diclofenac sodium (Diclo) and MNP act as the biosensor component, anti-inflammatory agent and MRI contrast agent, respectively. Further, MNPs were prepared first using the co-precipitation method followed by its loading with the sensor as well as drug components into the alginate microspheres by using a commercial droplet generator (Joshi et al., 2011).

Of note, among NPs, copper oxide (CuO) NPs are quite attractive as they exhibit better electrical/catalytic properties, thermal stability as well as mechanical rigidity, hence, exist in different morphologies, e.g., nanoflowers, NPs, nano cubes, nanowires, etc. (SoYoon et al., 2014; Mohammadpour et al., 2014). In a similar fashion, an amperometric glucose biosensor using alginate-CuO nanobiocomposite and GOD film was synthesized by Buk et al. (2017), which demonstrated good sensitivity (7.2–30.4 μA mM^{-1} cm^{-2}), high stability as well as specificity.

Previously, Lei et al. (2005) designed an H_2O_2 amperometric biosensor by incorporating HRP-labeled AuNPs (an electrostatic composite of HRP and AuNPs) into a unique silica sol-gel/alginate hybrid film on a glassy carbon electrode. Moreover, silica sol-gel/alginate hybrid material prevents cracking of the film. In addition, HRP electrode demonstrated good catalytic activity along with high sensitivity in the reduction of H_2O_2, where hydroquinone was utilized as an electron mediator (Lei et al., 2005).

3.4 Lignin Based Biosensors

In fact, lignin exhibits an amorphous structure and is a natural aromatic polymer thus demonstrating better affinity and adsorption for carbon-based materials. Currently, various metal oxides, i.e., ZrO_2, Ga_2O_3, CeO_2, ZnO, etc., are utilized as supports in biosensors as they are stable, compatible, and safe in nature (Durmaz et al., 2023). For example, Jędrzak et al. (2019a) designed a glucose biosensor using hybrid materials composed of Ga_2O_3/lignin and ZrO_2/lignin as effective support structures. During binding with lignin, first they modified ZrO_2 NPs followed by Ga_2O_3. Subsequently, the enzyme (GOD) was immobilized on them by adsorption, wherein both Ga_2O_3/lignin-GOD and ZrO_2/lignin-GOD systems along with ferrocene and carbon paste (BASi) depicted promising results (Jędrzak et al., 2019a). Likewise, Jędrzak et al. (2019b) synthesized a glucose biosensor using multi-component material composed of Fe_3O_4/lignin/polydopamine-GOD along with ferrocene and carbon paste electrode, where this biosensor yielded good results in determining glucose concentration within different commercial products (Jędrzak et al., 2019b).

3.5 Other Biopolymer-based Biosensors

For biosensing applications, other biopolymers, e.g., starch, pectin, and collagen-based metal NPs are also being utilized. For example, Gum Arabic [(GA), a natural polysaccharide isolated from acacia trees], is being used to synthesize hydrogels but GA-bases films are breakable and hydrophilic in nature (Wolf et al., 2018; Mendes et al., 2016). Hence, blends/hybrid synthesis method has been utilized to avoid such problems. For example, Bagal-Kestwal et al. (2019) used GA and corn flour for analyzing saccharide in foods, where PtNPs were fixed with MWCNTs for immobilizing invertase and GOD modified electrodes. Further, while detecting sucrose by the synthesized invertase-GOD-based biosensor, a higher sensitivity was reported as compared to previous reports thus suggesting the role of such biosensor for multi-purpose sensing, especially in the food and beverage industry (Bagal-Kestwal et al., 2019).

In a previous report, Han et al. (2012) utilized starch (natural polysaccharide of glucose) for preparing PdNP assemblies (PdNPAs), wherein PdNPAs can reduce H_2O_2 by the electron transfer process. To add, the prepared biosensor demonstrated high performance but low limit of detection for estimating serum glucose level (Han et al., 2012). Of note, another biopolymer, i.e., collagen was utilized to develop a non-enzymatic H_2O_2 biosensor by means of Ag nanowires (AgNWs) supported on graphene nanosheets (GNs) (GN-AgNW) with collagen nanofibers serving as precursors. The data showed good electrocatalytic activity of GN-AgNW nanohybrid modified glassy carbon electrode for H_2O_2 reduction (Wang et al., 2014).

Table 1. Enlisted the various biopolymer-based biosensors and their ligands.

Biosensor	Biopolymer	Component other than biopolymer	Bio-recognition unit	Ligand	References
Enzyme	CS	CNT	GLDH	Glutamate	Hughes et al., 2015
Enzyme	Polydopamine	3D graphene foam, methylene blue-CNT	HRP	H_2O_2	Liu et al., 2015
Enzyme	Sodium alginate	Carbon nanospheres	GOD	Glucose	Han et al., 2014
Enzyme	CS	Nitrogen-doped graphene, poly (styrene sulfonate)	GOD	Glucose	Barsan et al., 2014
Enzyme	CS	Laponite clay	GOD	Glucose	Shi et al., 2008
Enzyme	Sodium alginate	PDDA, metal ions (e.g., Fe^3 or Ca^{2+}, to form AlgFe or AlgCa hydrogel)	GOD or lactate oxidase	Glucose or lactate	Qin et al., 2012
Enzyme	Chitin	Ionic liquid (1-butyl-3-methylimidazoliumh exafluorophosphate) + AuNPs	Peroxidase	Rosmarinic acid	Brondani et al., 2011
Enzyme	DNA	MWCNT	Organophosphorus hydrolase	Organophosphorus pesticide (paraoxon)	Mantha et al., 2010
Enzyme	Alginate	LDH	GOD	Glucose	Ding et al., 2009
Enzyme	CS	Ceria-based metal oxides	Tyrosinase	Dopamine	Ding et al., 2009
Enzyme	CS	Fe_3O_4 NPs	Urease and GLDH	Urea	Kaushik et al., 2009a
Enzyme	CS	Polysiloxane matrix	HRP	H_2O_2	Li et al., 2009
Enzyme	CS	MWCNT	Laccase oxidase	Catechol	Tan et al., 2009
Enzyme	CS	Fe_3O_4 NPs	GOD	Glucose	Kaushik et al., 2008
Enzyme	CS	Laponite clay	Polyphenol oxidase	Phenol	Fan et al., 2007
Enzyme	CS	CNT	GDH	Glucose	Zhang et al., 2007
Enzyme	Agarose and guargum	Dopa and dopamine	Tyrosinase	Dopa and dopamine	Tembe et al., 2006
Enzyme	CS	Cobalt hexacyanoferrate NPs, CNT	GOD	Glucose	Yang et al., 2006
Enzyme	CS	*Solanum gilo* crude extract	Peroxidase	Hydroquinone	De Oliveira et al., 2006
Aptamer based	CS	MWCNT, Ionic Liquid, PtNPs	Thiol Terminate adenosine aptamer	Adenosine	Shahdost-Fard et al., 2014
Aptamer based	CS	Graphene oxide	Amino-linked DNA aptamer	Lysozyme	Erdem et al., 2014
Aptamer based	CS	MWCNT, ionic liquid	32-Meradenosine-specific aptamer	Adenosine	Shahdost-Fard et al., 2013
Aptamer based	CS	MWCNT, ionic liquid	5′-Amino-terminated aptamer	IgE	Khezrian et al., 2013

Table 1 condt. ...

...*Table 1 condt.*

Biosensor	Biopolymer	Component other than biopolymer	Bio-recognition unit	Ligand	References
Aptamer based	CS	Graphene	Thrombin-binding aptamer	Thrombin	Wang et al., 2012
Aptamer based	CS	Co-Pt bimetal alloy NPs, reduced graphene oxide		Thrombin	Wang et al., 2011b
Aptamer based	CS	CdS QDs/CNTs	Thrombin-binding aptamer	Thrombin	Yang et al., 2011
DNA based	Trimethyl CS	Fe_3O_4, AuNPs, polythiophene	pDNA	Methylation in tumor suppres sor gene	Daneshpour et al., 2016
DNA based	CS	Fern leaf–like Fe_2O_3 microparticles	Capture pDNA	Target DNA	Xu et al., 2015
DNA based	CS	Graphene oxide-nickel ferrite	*E. coli* pDNA	*E. coli*	Tiwari et al., 2015
DNA based	CS	Iron oxide NPs	Calf thymus DNA	Pyrethroid	Kaushik et al., 2009b
Ab based	CS	Fe_3O_4 NPs	Anti-*E. coli* Ab	*E. coli*	Kozitsina et al., 2016
Ab based	Polysaccharide (CS and lectin (concanavalin A)		Anti-CA19-9 Ab	CA19-9 (biomarker for pancreatic cancer)	Soares et al., 2015
Ab based	CS	Au-MoO_3	Polyclonal Ab and HRP	Hepatitis C virus nonstructural 5A protein	Liang et al., 2013
Ab based	CS	AuNPs	Monoclonal Abs of *B. cereus*	*B. cereus*	Kang et al., 2013
Ab based	CS	TiO2, graphene, AuNPs	α-Fetoprotein Ab	α-Fetoprotein	Huang et al., 2013
Ab based	CS	AuNPs	Hepatitis B surface Ab	Hepatitis B surface Ab	Qiu et al., 2009
Ab based	Alginate	Polypyrrole	Anti-*E. coli* anti body	*E. coli*	Abu-Rabeah et al., 2009
Ab based	CS	Silica-based sol-gel, AuNPs	CA 153, CA 125, CA 199, and CEA monoclonal Ab	Tumor markers CA 153, CA 125, CA 199, and CEA	Wu et al., 2008
Ab based	Polysaccharides, humic acid, polylactic acid, etc.		Anti-Mb	Mb	Masson et al., 2004
Whole cell based	Alginate beads		Genetically modified *S. cerevisiae*	Cu^{2+}	Vopalenska et al., 2015
Whole cell based	Alginate		Genetically engineered bioluminescent bacteria	Toxic chemicals in H_2O	Jung et al., 2014
Whole cell based	Alginate hydrogel		Genetically modified *S. cerevisiae* cells	Estrogenic endocrine disrupting chemical	Fine et al., 2006
Whole cell based	Agarose		Recombinant *E. coli* cells	Organophosphate pesticides	Mulchandani et al., 1998

Strikingly, pectin (a heteropolysaccharide) can be an alternative to be utilized for developing biosensor matrix as pectin is biocompatible, thermally stabile, eco-friendly, and better absorbing capacity. Moreover, pectin exhibits good hydrogel properties, thus it might be attached with other biopolymers to form an appropriate substrate (Hasanah et al., 2019). For instance, Di Tocco et al. (2018) synthesized an electrochemical triglyceride biosensor using lipase immobilized on CS-coated MNPs on a dispersion of MWCNT/Pectin modified with CuO-NPs on a glassy carbon electrode, which could measure total triglycerides within serum samples. Additionally, such a biosensor depicted excellent performance/reproducibility, high stability, and yielded an alternate for determining triglyceride within human sera (Di Tocco et al., 2018).

4. Synthesis Strategies of Biopolymer-based NPs

Till today, numerous methods have been implemented for preparing biopolymer-based NPs comprised of various shapes as well as surface charges. Interestingly, spherical NPs possess a higher potential for cell penetration when compared to other NPs of different shapes (Vodyashkin at al., 2022). That is why, such biopolymers derived NPs can be utilized as delivery agents in our body.

4.1 Ionic Gelation Method

The transformation of a polymer solution from liquid to solid form occurs during ionic gelation, driven by ionic forces between the ionic polymers and the cross-linking reagent. Various NPs of CS, starch, alginate, cellulose, etc., were formed using this method that exhibited high encapsulation efficiency along with preservation of the bioactivity of the embedded molecules (Liu et al., 2020; Sullivan et al., 2018; Pant and Negi, 2018; Othman et al., 2018).

Previously, Fan et al., (2012) prepared 138.1 nm CS NPs employing this method, where a significant ratio of CS to sodium tripolyphosphate, pH and concentration of acetic acid solution were found to be 3.3:1, 4.7–4.8, and 0.2 mg/mL, respectively (Fan et al., 2012). However, while preparing 26 nm CS NPs, the optimal conditions were found to be the use of low molecular weight CS, 3:1 mass ratio of CS to sodium tripolyphosphate at 4.6 pH. Moreover, it is worthwhile to note that by utilizing low molecular weight CS, one could omit out the ultrasonic treatment during NPs preparation (Calvo et al., 1997). Hence, the size of the resulting NPs depends on various factors, e.g., molecular weight of CS, the pH of the CS solution, and the mass ratio of CS to sodium tripolyphosphate. In another report, the same method was also utilized by Anand et al. (2018) to prepare 8–80 nm spherical NPs, where pH of the CS solution was observed to be 6 (Anand et al., 2018).

Subsequently, Ji et al. (2018) synthesized 262 nm CS NPs by using encapsulated amylase along with ionic gelation method, where carboxymethyl short-chain amylase solution was mixed with CS dissolved in aq. acetic acid followed by stirring for 6 h at RT and centrifugation. To add, this study demonstrated that during NPs preparation via ionic gelation, loading of biopolymer NPs took place with active substances (Ji et al., 2018). This work was previously reported by Anita et al. (2011) by synthesizing starch-CuNPs complex and then loaded the complex with biomolecules by employing the similar approach (Anita et al., 2011). Further, microencapsulation was carried out to increase the functionality of the NPs, wherein NPs were coated with calcium alginate.

In another study reported by Liu et al. (2020), ionic gelation was used to create negatively charged carboxymethylene branched starch NPs. After adding starch and sodium hydroxide (NaOH) to ethanol, the mixture was agitated to 40°C for 1 h. Subsequently, the pH of the solution was regulated followed by centrifugation. Consequently, spherically shaped 50–100 nm starch NPs were obtained (Liu et al., 2020). Later, Liu et al. (2018) prepared 30–50 nm starch NPs by a different approach, where oxidized-branched chain starch was gelatinized and gradually introduced into Ca^{2+} solution while being stirred at RT followed by incubation for 2 h. The size of the formed NPs can be altered by decreasing the concentration of Ca^{2+} solution (Liu et al., 2018).

In addition, ionic gelation can be used to create nanocomposites comprised of two biopolymers. Subramanian et al. (2014) created composite NPs by dissolving CS and starch in an aq. 2% acetic acid solution, then adding Tween-80 to ensure uniform dispersion. They then added a solution of a pre-made anticancer agent and sodium tripolyphosphate to the mixture, which was then centrifuged to separate the particles, and then lyophilize them to create spherical NPs that ranged in size from 173 to 297 nm (Subramanian et al., 2014). Also, Mohamed and Laraba-Djebari (2016) prepared spherical shape alginate NPs of 150–350 nm in size by using the same method. Moreover, calcium chloride solution was gradually introduced into the sodium alginate solution at RT while continuously agitating, followed by incubation for 1 h and centrifugation (Mohamed and Laraba-Djebari, 2016). Of note, stable NPs of various biopolymers, including CS, alginate, starch, etc., can be prepared by utilizing the ionic gelation method. There are various reports that documented the capability to regulate the diameter and distribution width of the formed NPs. Further, NPs so obtained demonstrate better encapsulation efficacy, reduced diameter as well as good stability.

4.2 Nanoprecipitation Method

In contrast to emulsion-based techniques (such as emulsion-diffusion, emulsion-evaporation, and salting-out procedures), the nanoprecipitation method is a quick and easy procedure that does not require a precursor emulsion. Generally, a polar organic solvent, e.g., ethanol, acetone, etc., is used to dissolve the hydrophobic solute, such as biopolymer/lipids. The polar solvent is then combined in all ratios with a significant volume of non-solvent (often water) of the dissolved material, thus resulting into hydrophobic solute particles formation via phase separation. Further, evaporation can be used to eliminate the organic solvent. Notably, the surface of NPs is coated with protective reagents, thus it may lead to preparation of nanodrug delivery reagents (Qiu et al., 2019). Furthermore, nanoprecipitation can directly be utilized to incorporate the reagent into the NPs in a single step rather utilizing multistep procedures to generate NPs. In another report, while synthesizing starch NPs with encapsulated diclofenac sodium by utilizing the same method, El-Naggar et al. (2015) dissolved starch in NaOH at RT for 30 min, followed by adding diclofenac sodium and Tween-80 dissolved in distilled water. Subsequently, sodium tripolyphosphate was mixed, and stirred for 2 h followed by incorporating absolute ethanol. Then the resulting mixture was centrifuged and washed thus resulting into formation of spherical NPs with a size of 21.04 nm, which can be employed as a delivery model for sodium diclofenac (El-Naggar et al., 2015).

Notably, this method can be utilized to control the average diameter of the formed NPs. Indeed, the average diameter of the resulting NPs depends on the solvent system in which starch is dissolved. Previously, Chin et al. (2011) prepared starch NPs using nanoprecipitation, where starch was dissolved in NaOH and urea solutions. NaOH helps in both hydrolysis and breaking the hydrogen bonds present in starch, while urea prevents self-association of starch molecules, thus leading to an increase in its solubility. Later, 100% ethanol was introduced into the resultant followed by stirring at RT for 30 min, centrifugation and washing. As a result, the formed NPs were of sphere shape, and the range of size varied from 300–400 nm. Moreover, the size can be reduced to a range of 250–300 nm and 150–200 nm by adding Tween-80 and hexadecyl(cetyl)trimethylammonium bromide, respectively (Chin et al., 2011).

The same method was also utilized by Gutierrez et al. (2020) to prepare starch NPs, in which starch was added in aq NaOH and urea solution with varied concentrations of different substances, and agitated at 80°C for 30 min, followed by mixing it with absolute ethanol. In addition, the authors examined the factors affecting size of resulting NPs, such as the concentrations of the components in the starch-solvent system and the rate at which the starch solution was added to 100% ethanol. Further, while adding 8% (w/v) and 10% (w/v) urea within the solvent system for dissolution of starch, spherical NPs ranging in size of 13.9 ± 9.03 nm were obtained at a starch solution injection rate of 4 mL/h (Gutierrez et al., 2020).

In fact, dimethyl sulfoxide (DMSO) can also be employed as an alternative solvent for dissolving starch as it can disrupt the inter- and intramolecular hydrogen bonds of starch. In a study by Pan et al. (2022), starch NPs were produced by utilizing DMSO and ultrasound treatment to break down the starch particles. The process involved adding starch to DMSO, stirring for 10 min, and applying ultrasound for 1 h. Later, deionized water was introduced followed by ultrasound for another 1 h. Subsequently, the synthesized NPs were centrifuged, underwent dialysis and then lyophilization, thus resulting into spherical NPs with a size of 68.63 ± 9.21 nm (Pan et al., 2022).

Strikingly, nanoprecipitation is not only utilized to synthesize starch NPs but also for preparing CS NPs. Both CS as well as starch NPs can be utilized as drug carriers. Using the same method, Sotelo-Boyas et al. (2017a) obtained spherical shape CS NPs with a size of 5.8 ± 1.6 nm, where CS was mixed in acetic acid followed by adding a mixture of lime and methanol. Subsequently, the synthesized NPs were isolated by using a rotary evaporator. Further, such NPs were employed to encapsulate lime essential oil thus this composite was studied owing to being antibacterial in nature (Sotelo-Boyas et al., 2017a). Interestingly, thyme essential oils possess antibacterial properties as they comprised of volatile components. However, being volatile, thyme ester oils fail to retain their antibacterial properties with time, which can be restored by their encapsulation in nanocarriers. While utilizing CS NPs as nanocarriers for thyme essential oil, Sotelo-Boyas et al. (2017b) encapsulated them during NPs preparation. A combination of methanol and thyme essential oil was gradually mixed with CS (dissolved in acetic acid solution) while being stirred. A rotary evaporator was then used to remove the solvent and separate the resultant NPs. According to the authors, the NPs had 68% of the constituents of thyme essential oil. The resulting particles were spherical in shape and 6.4 ± 0.5 nm in size on an average (Sotelo-Boyas et al., 2017b). As a result, the nanoprecipitation technique makes it possible to create nanocarriers made of biodegradable polymers like CS, starch, etc. Drugs may be directly encapsulated in NPs during the manufacturing of NPs. Moreover, the concentration of the reaction mixture constituents, the rate at which “anti-solvent” is added to the polymer solution, and the application of ultrasound treatment are some of the variables that can regulate the size of the resultant NPs.

4.3 Microemulsion Method

Of note, microemulsions are isotropic in nature and are thermodynamically stable colloids. In fact, they comprised of three constituents: a surfactant, a non-polar and a polar phase. Moreover, microemulsions are of two types either oil-in-water (M/W) or water-in-oil (W/M) (Asgari et al., 2019). Here, W/M microemulsion is formed by the addition of aq. polymer solution to an organic solvent including a surfactant, wherein NPs get deposited (Gutiérrez et al., 2020). The implementation of this method is complicated because it involves multiple phases. However, this technique has a significant role in synthesizing biopolymers-derived nanodrugs. Further, the usage of ionic liquids helps in simplification and makes it a cost-effective process.

Strikingly, this method has been utilized to acquire NPs of biopolymers. Previously, Duyen and Hung (2021) synthesized 70–200 nm spherical shape starch NPs by using the same method, wherein alkaline starch solution was introduced into ethanol comprising Tween-80 and soybean oil. Subsequently, the solution was stirred constantly for about 1 h, which was then followed by centrifugation and lyophilization (Duyen and Hung, 2021). To add, this method can be used to synthesize nanocomposites, i.e., either cross-linking of 2 or > 2 polymers, or encapsulation of metal NPs within polymers. Further, while synthesizing magnetic CS NPs using the same, previously Wang et al., (2008b) dissolved CS in HCl, followed by addition of cyclohexane, hexanol-1, and Fe^{2+}. Subsequently, TX-100 was gradually incorporated till the mixture became transparent. Later, the mixture was agitated and heated at 60°C for 2 h followed by centrifugation, washing and lyophilization thus resulting into encapsulation of Fe_3O_4 NPs within CS NPs. Notably, the formed NPs were spherical and 10–20 nm in size and they were further employed as an adsorbent for albumin protein (Wang et al., 2008b).

Additionally, CS-alginate nanocomposites were prepared by Dubey et al. (2016) using the same technique, wherein aq. solution of alginate and acetic acid was used to pre-disperse the CS. In this method, paraffin oil was utilized as an oil phase. Further, the mixture was continually agitated for 30 min, followed by adding calcium salt solution and sodium tripolyphosphate. They were then constantly agitated for three more hours at RT. After being separated by centrifugation, the generated NPs were washed with acetone. Consequently, spherically shaped NPs with a mean diameter of 65 nm were produced. Further, the synthesized CS-alginate nanocomposites were utilized as an adsorbent for mercury ions in aq. solutions (Dubey et al., 2016). In this method, ionic liquids substitute polar, non-polar phases and surfactants.

Moreover, ionic liquids have a significant role in green chemistry as they can be reused. Also, they can be utilized as solvents because their melting point is lower than 100°C (Zhang et al., 2012). NPs of biodegradable polymer as small as 100 nm can be produced by using ionic liquids. On the other hand, to synthesize starch NPs, Zhou et al. (2014) used an ionic liquid for cross-linking in the microemulsion method. Further, L-octyl-3-methylimidazolium acetate, an ionic liquid that replaced the aq. phase, was mixed with starch powder and heated for 2.5 h. Following this, cyclohexane was added along with a mixture of the surfactant TX-100 and butanol and the solution was stirred until it became transparent. Here, epichlorohydrin was utilized as a crosslinking agent. After that, the mixture was agitated for 3 h at 50°C. Following 3 h of vigorous stirring, the starch NPs were precipitated with anhydrous ethanol and centrifuged. In order to eliminate unreacted epichlorohydrin, butanol-1, and TX-100, the precipitate was extensively washed with 100% ethanol followed by centrifugation and vacuum-drying for 24 h at 45°C. Consequently, spherical, and oval-shaped starch NPs with a mean diameter of 96.9 nm was produced (Zhou et al., 2014). Likewise, Wang et al., (2016) prepared starch NPs using the same method as reported by Zhou et al. (2014). However, the researchers employed 1-hexadecyl-3-methylimidazolium bromide as the ionic liquid rather than l-octyl-3-methylimidazolium acetate as utilized by Zhou et al., (2014). Consequently, indeterminate shape and of 80.5 nm size NPs were obtained (Wang et al., 2016).

4.4 Coacervation Method

Coacervation refers to the natural transformation of homogenous macromolecular solution (with low solvent affinity) into a superdense liquid phase. During this process, dense phase rich in polymers coexists with the supernatant as a result of liquid-liquid phase separation (occurs within a homogenous solution of charged macromolecules). These two liquid phases do not mix but interact intensely, thus facilitating the synthesis of NPs composed of various biopolymers (Vodyashkin et al., 2022). Starch NPs have been created through coacervation by mixing solutions of two types of premodified starch-positively charged (containing NH_2 group) and negatively charged (containing COOH group) at various molar ratios under agitating at RT. NPs were synthesized in 15 min , after which centrifugation and lyophilization were carried out. Such an approach produced NPs of sphere in shape, with the optimal ratio for achieving the smallest particle size of 138.9 ± 0.3 nm was estimated to be 1:3 (Barthold et al., 2016).

Various reports have demonstrated that composite NPs made from biodegradable polymers, such as CS and carboxymethyl starch may also be synthesized using the same approach (Saboktakin et al., 2011). Another coacervation approach involves saturating the polymer solution, followed by the formation of the coacervate phase. In fact, Tavares et al. (2012) successfully produced CS NPs by simply adding a sodium sulfate solution to a CS solution with stable volumes and concentrations under magnetic stirring (Tavares et al., 2012). In an alternative coacervation process, Patra et al., (2016) employed a binary mixture of acetone and ethanol to induce phase separation. This technique facilitated the creation of a more condensed polymer chain configuration, which facilitated in reducing the diameter of NPs. Of note, gelatin NPs were produced using the binary coacervation method. An aq. gelatin solution, heated to 40°C, was mixed with acetone and alcohol in a 1:1:1 volume ratio. Subsequently, an ethanol solution containing glutaraldehyde as the crosslinking agent was introduced into the reaction solution. Additionally, pH of the aq. gelatin solution was adjusted in a range of 3–11 to evaluate the impact of pH on particle size. The synthesized particles were repeatedly washed with distilled water, followed by ultrasound for 5 min and afterwards centrifugation at 2000 rpm. Consequently, the smallest particle size was attained at pH 7, resulting in particles measuring 55.67 ± 43.74 nm (Patra et al., 2016). It is worthwhile to mention that though this approach may synthesize biopolymer NPs with minimal sizes, the resulting particles exhibit a broad size distribution that could restrict some of their potential uses.

4.5 Electrospray Method

Conventionally, this approach was utilized to obtain drug nanocrystals (Pawar et al., 2018); however, this method has also been employed to prepare NPs derived from biopolymers. In this process, a solution of the substance from which the nanomaterial will be derived is passed through a nozzle under voltage and sprayed onto a collector at variable speeds, with the collector serving as a grounding agent. Indeed, specialized drums or solutions might be utilized in spite of the collector. The fundamental essentials that are used in this approach include a syringe pump, a nozzle with a cooling jacket, a high-voltage power supply, and a metal manifold. To add, this method is based on electromechanical and hydrodynamic forces. Factors which affect this process are rate of flow, electric charge applied to the polymer solution, the distance from the needle and the manifold, and the applied voltage, wherein applied electric voltage distorts the droplet interface thus leads to formation of nano-droplets (Bock et al., 2011).

Notably, this method is based on application of an electric field onto a liquid droplet, which in turn generates Coulomb force in the droplet that ultimately competes with the adhesive forces of the same. Further, nanodroplet formation takes place if the applied Coulomb forces prevail over the adhesive forces of the droplet (Kurakula et al., 2021). Interestingly, Wang et al. (2020) prepared CS NPs encapsulating tea polyphenols, where CS was prepared in acetic acid followed by addition of tea polyphenols at different mass ratios at a pH range of 4.6–4.7. Subsequently, sodium tripolyphosphate solution was prepared, the mixture of CS and tea polyphenols were introduced through a nozzle with 0.5 mm diameter. The nozzle was positioned at a distance of 5 to 15 cm, with a feed rate of 0.25–1 mL/h, and an applied voltage ranging from 10 to 20 kV. The NPs were isolated through centrifugation and subsequently lyophilization. The smallest size of CS NPs encapsulating tea polyphenols was observed to be 128 ± 13 nm, obtained under conditions including a nozzle-to-solution distance of 5 cm, a spray rate of 1 mL/h, and a voltage of 15 kV (Wang et al., 2020). Additionally, CS NPs were prepared using the electro spraying technique with a voltage generation unit, a syringe with a 0.65 mm internal needle diameter and a working distance of 10 cm. Further, CS dissolved in acetic acid was fed at a rate of 0.2–1.1 mL/h followed by vacuum drying. It is important to note that solutions with varying concentrations as well as molecular weights may also be employed to prepare CS NPs (Thien et al., 2012).

5. Conclusion and Future Perspective

Biopolymer-based metal NPs seems to be an innovative and promising tool for bio-sensing applications as they possess biocompatibility, tunable properties as well as functional flexibility. Further, the combination of biopolymers along with metal NPs led to develop such biosensors, which showed enhanced sensitivity, selectivity, and stability, thus making them suitable for various biosensing applications, from medical diagnostics to environmental monitoring. In fact, the approach to utilize naturally plentiful and biodegradable polymers aligns with our sustainability goals, which ultimately reduces relying on synthetic polymers along with potent destructive substances. To add, by subtle modifications, biopolymer-based metal NP composites have been designed that demonstrated superior performance, especially in identification of low concentrations of biological targets and in complex matrices, therefore implying their significance in next-generation biosensing.

Of note, future research should focus on novel functionalization techniques to raise the binding specificity as well as signal transduction efficiency of biopolymer-metal NPs-based biosensors. Moreover, usage of various biopolymers and metal based NPs may synthesize composites having some unique features. For example, either using biopolymers (comprised of special binding sites) or bimetallic NPs can yield improvements in sensor performance and resilience. Future developments in synthesis, functionalization, and device integration can pave the way for highly effective, environmentally friendly biosensors with widespread applications. Strikingly, incorporation of biopolymer-based NPs may lead to develop point-of-care devices, however, their stability changes with time particularly in varying environmental conditions.

References

Abu-Rabeah, K., Ashkenazi, A., Atias, D., Amir, L. and Marks, R. S. (2009). Highly sensitive amperometric immunosensor for the detection of *Escherichia coli*. Biosensors and Bioelectronics, 24(12): 3461–3466.

Ahmad, A., Siddique, J. A., Setapar, S. H., Lokhat, D., Golandaj, A. and Ramjugernath, D. (2018). Recent advances in chitosan-based films for novel biosensor. Electrically Conductive Polymer and Polymer Composites: From Synthesis to Biomedical Applications, 137–161.

Alam, D., Naaz, F., Islam, A., Sardar, M. and Ahmad, T. (2023). Role of sugar osmolytes and their nano-counterparts as inhibitors in protein fibrillation. Journal of Molecular Liquids, 122479.

Anand, M., Sathyapriya, P., Maruthupandy, M. and Beevi, A. H. (2018). Synthesis of chitosan nanoparticles by TPP and their potential mosquito larvicidal application. Frontiers in Laboratory Medicine, 2(2): 72–78.

Anita, S., Ramachandran, T., Rajendran, R., Koushik, C. V. and Mahalakshmi, M. (2011). A study of the antimicrobial property of encapsulated copper oxide nanoparticles on cotton fabric. Textile Research Journal, 81(10): 1081–1088.

Asgari, S., Saberi, A. H., McClements, D. J. and Lin, M. (2019). Microemulsions as nanoreactors for synthesis of biopolymer nanoparticles. Trends in Food Science & Technology, 86: 118–130.

Aslam, J., Zehra, S., Mobin, M., Quraishi, M. A., Verma, C. and Aslam, R. (2023). Metal/metal oxide-carbohydrate polymers framework for industrial and biological applications: Current advancements and future directions. Carbohydrate Polymers, 314: 120936.

Atif, M., Ali, A., AlSalhi, M. S. and Willander, M. (2019). Retracted Article: Effect of urea on the morphology of Fe3O4 magnetic nanoparticles and their application in potentiometric urea biosensors. Silicon, 11(3): 1371–1376.

Bagal-Kestwal, D. R. and Chiang, B. H. (2019). Platinum nanoparticle-carbon nanotubes dispersed in gum Arabic-corn flour composite-enzymes for an electrochemical sucrose sensing in commercial juice. Ionics, 25: 5551–5564.

Banerjee, A., Maity, S. and Mastrangelo, C. H. (2021). Nanostructures for biosensing, with a brief overview on cancer detection, IoT, and the role of machine learning in smart biosensors. Sensors, 21(4): 1253.

Barsan, M. M., David, M., Florescu, M., Ţugulea, L. and Brett, C. M. (2014). A new self-assembled layer-by-layer glucose biosensor based on chitosan biopolymer entrapped enzyme with nitrogen doped graphene. Bioelectrochemistry, 99: 46–52.

Barthold, S., Kletting, S., Taffner, J., de Souza Carvalho-Wodarz, C., Lepeltier, E., Loretz, B. et al. (2016). Preparation of nanosized coacervates of positive and negative starch derivatives intended for pulmonary delivery of proteins. Journal of Materials Chemistry B, 4(13): 2377–2386.

Berepiki, A., Kent, R., Machado, L. F. and Dixon, N. (2020). Development of high-performance whole cell biosensors aided by statistical modeling. ACS Synthetic Biology, 9(3): 576–589.

Bhardwaj, H., Sumana, G. and Marquette, C. A. (2021). Gold nanobipyramids integrated ultrasensitive optical and electrochemical biosensor for Aflatoxin B1 detection. Talanta, 222: 121578.

Biliuta, G. and Coseri, S. (2019). Cellulose: A ubiquitous platform for ecofriendly metal nanoparticles preparation. Coordination Chemistry Reviews, 383: 155–173.

Bock, N., Woodruff, M. A., Hutmacher, D. W. and Dargaville, T. R. (2011). Electrospraying, a reproducible method for production of polymeric microspheres for biomedical applications. Polymers, 3(1): 131–149.

Boissiere, M., Allouche, J., Chanéac, C., Brayner, R., Devoisselle, J. M., Livage, J. et al. (2007). Potentialities of silica/alginate nanoparticles as hybrid magnetic carriers. International Journal of Pharmaceutics, 344(1-2): 128–134.

Brondani, D., Zapp, E., Vieira, I. C., Dupont, J. and Scheeren, C. W. (2011). Gold nanoparticles in an ionic liquid phase supported in a biopolymeric matrix applied in the development of a rosmarinic acid biosensor. Analyst, 136(12): 2495–2505.

Buk, V., Emregul, E. and Emregul, K. C. (2017). Alginate copper oxide nano-biocomposite as a novel material for amperometric glucose biosensing. Materials Science and Engineering: C, 74: 307–314.

Calvo, P., Remuñan-López, C., Vila-Jato, J. L. and Alonso, M. J. (1997). Chitosan and chitosan/ethylene oxide-propylene oxide block copolymer nanoparticles as novel carriers for proteins and vaccines. Pharmaceutical Research, 14: 1431–1436.

Chin, S. F., Pang, S. C. and Tay, S. H. (2011). Size controlled synthesis of starch nanoparticles by a simple nanoprecipitation method. Carbohydrate Polymers, 86(4): 1817–1819.

Cruz, A. F. D., Norena, N., Kaushik, A. and Bhansali, S. (2014). A low-cost miniaturized potentiostat for point-of-care diagnosis. Biosensors and Bioelectronics, 62: 249-–254.

Daneshpour, M., Izadi, P. and Omidfar, K. (2016). Femtomolar level detection of RASSF1A tumor suppressor gene methylation by electrochemical nano-genosensor based on Fe3O4/TMC/Au nanocomposite and PT-modified electrode. Biosensors and Bioelectronics, 77: 1095–1103.

de Oliveira, I. R. W. Z. and Vieira, I. C. (2006). Immobilization procedures for the development of a biosensor for determination of hydroquinone using chitosan and gilo (*Solanum gilo*). Enzyme and microbial technology, 38(3-4): 449–456.

Di Pietrantonio, F., Cannatà, D. and Benetti, M. (2019). Biosensor technologies based on nanomaterials. In Functional nanostructured interfaces for environmental and biomedical applications (pp. 181–242). Elsevier.

Di Tocco, A., Robledo, S. N., Osuna, Y., Sandoval-Cortez, J., Granero, A. M., Vettorazzi, N. R. et al. (2018). Development of an electrochemical biosensor for the determination of triglycerides in serum samples based on a lipase/ magnetite-chitosan/copper oxide nanoparticles/multiwalled carbon nanotubes/pectin composite. Talanta, 190: 30–37.

Ding, S. N., Shan, D., Xue, H. G., Zhu, D. B. and Cosnier, S. (2009). Glucose oxidase immobilized in alginate/layered double hydroxides hybrid membrane and its biosensing application. Analytical Sciences, 25(12): 1421–1425.

Dubey, R., Bajpai, J. and Bajpai, A. K. (2016). Chitosan-alginate nanoparticles (CANPs) as potential nanosorbent for removal of Hg (II) ions. Environmental Nanotechnology, Monitoring & Management, 6: 32–44.

Dupont, D. (2022). Immunochemical methods. Encyclopedia of Dairy Sciences 3 ième édition, 1.

Durmaz, E., Sertkaya, S., Yilmaz, H., Olgun, C., Ozcelik, O., Tozluoglu, A. et al. (2023). Lignocellulosic bionanomaterials for biosensor applications. Micromachines, 14(7): 1450.

Duyen, T. T. M. and Van Hung, P. (2021). Morphology, crystalline structure and digestibility of debranched starch nanoparticles varying in average degree of polymerization and fabrication methods. Carbohydrate Polymers, 256: 117424.

El-Naggar, M. E., El-Rafie, M. H., El-Sheikh, M. A., El-Feky, G. S. and Hebeish, A. (2015). Synthesis, characterization, release kinetics and toxicity profile of drug-loaded starch nanoparticles. International journal of biological macromolecules, 81, 718-729.

Erdem, A., Eksin, E. and Muti, M. (2014). Chitosan–graphene oxide based aptasensor for the impedimetric detection of lysozyme. Colloids and Surfaces B: Biointerfaces, 115: 205–211.

Faham, S., Golmohammadi, H., Ghavami, R. and Khayatian, G. (2019). A nanocellulose-based colorimetric assay kit for smartphone sensing of iron and iron-chelating deferoxamine drug in biofluids. Analytica chimica acta, 1087: 104–112.

Fan, Q., Shan, D., Xue, H., He, Y. and Cosnier, S. (2007). Amperometric phenol biosensor based on laponite clay–chitosan nanocomposite matrix. Biosensors and Bioelectronics, 22(6): 816–821.

Fan, W., Yan, W., Xu, Z. and Ni, H. (2012). Formation mechanism of monodisperse, low molecular weight chitosan nanoparticles by ionic gelation technique. Colloids and surfaces B: Biointerfaces, 90: 21–27.

Fine, T., Leskinen, P., Isobe, T., Shiraishi, H., Morita, M., Marks, R. S. et al. (2006). Luminescent yeast cells entrapped in hydrogels for estrogenic endocrine disrupting chemical biodetection. Biosensors and Bioelectronics, 21(12): 2263–2269.

Gan, Z., Roslan, M. A. M., Abd Shukor, M. Y., Halim, M., Yasid, N. A., Abdullah, J. et al. (2022). Advances in aptamer-based biosensors and cell-internalizing SELEX technology for diagnostic and therapeutic application. Biosensors, 12(11): 922.

Gutiérrez, G., Morán, D., Marefati, A., Purhagen, J., Rayner, M. and Matos, M. (2020). Synthesis of controlled size starch nanoparticles (SNPs). Carbohydrate Polymers, 250: 116938.

Hajian, A., Lindstrom, S. B., Pettersson, T., Hamedi, M. M. and Wågberg, L. (2017). Understanding the dispersive action of nanocellulose for carbon nanomaterials. Nano Letters, 17(3): 1439–1447.

Han, E., Li, X., Cai, J. R., Cui, H. Y. and Zhang, X. A. (2014). Development of highly sensitive amperometric biosensor for glucose using carbon nanosphere/sodium alginate composite matrix for enzyme immobilization. Analytical Sciences, 30(9): 897–902.

Han, M., Liu, S., Bao, J. and Dai, Z. (2012). Pd nanoparticle assemblies—As the substitute of HRP, in their biosensing applications for H2O2 and glucose. Biosensors and Bioelectronics, 31(1): 151–156.

Hasanah, U., Setyowati, M., Efendi, R., Muslem, M., Md Sani, N. D., Safitri, E. et al. (2019). Preparation and characterization of a pectin membrane-based optical pH sensor for fish freshness monitoring. Biosensors, 9(2): 60.

Huang, K. J., Li, J., Wu, Y. Y. and Liu, Y. M. (2013). Amperometric immunobiosensor for α-fetoprotein using Au nanoparticles/chitosan/TiO2–graphene composite based platform. Bioelectrochemistry, 90: 18–23.

Huang, Q., Zhang, H., Hu, S., Li, F., Weng, W., Chen, J. et al. (2014). A sensitive and reliable dopamine biosensor was developed based on the Au@ carbon dots–chitosan composite film. Biosensors and Bioelectronics, 52: 277–280.

Hughes, G., Pemberton, R. M., Fielden, P. R. and Hart, J. P. (2015). Development of a novel reagentless, screen-printed amperometric biosensor based on glutamate dehydrogenase and NAD+, integrated with multi-walled carbon nanotubes for the determination of glutamate in food and clinical applications. Sensors and Actuators B: Chemical, 216: 614–621.

Jędrzak, A., Rębiś, T., Kuznowicz, M. and Jesionowski, T. (2019b). Bio-inspired magnetite/lignin/polydopamine-glucose oxidase biosensing nanoplatform. From synthesis, via sensing assays to comparison with others glucose testing techniques. International Journal of Biological Macromolecules, 127: 677–682.

Jędrzak, A., Rębiś, T., Kuznowicz, M., Kołodziejczak-Radzimska, A., Zdarta, J., Piasecki, A. et al. (2019a). Advanced Ga2O3/lignin and ZrO2/lignin hybrid microplatforms for glucose oxidase immobilization: Evaluation of biosensing properties by catalytic glucose oxidation. Catalysts, 9(12): 1044.

Ji, N., Hong, Y., Gu, Z., Cheng, L., Li, Z. and Li, C. (2018). Fabrication and characterization of complex nanoparticles based on carboxymethyl short chain amylose and chitosan by ionic gelation. Food & Function, 9(5): 2902–2912.

Joshi, A., Solanki, S., Chaudhari, R., Bahadur, D., Aslam, M. and Srivastava, R. (2011). Multifunctional alginate microspheres for biosensing, drug delivery and magnetic resonance imaging. Acta Biomaterialia, 7(11): 3955–3963.

Jung, I., Seo, H. B., Lee, J. E., Kim, B. C. and Gu, M. B. (2014). A dip-stick type biosensor using bioluminescent bacteria encapsulated in color-coded alginate microbeads for detection of water toxicity. Analyst, 139(18): 4696–4701.

Kang, X., Pang, G., Chen, Q. and Liang, X. (2013). Fabrication of Bacillus cereus electrochemical immunosensor based on double-layer gold nanoparticles and chitosan. Sensors and Actuators B: Chemical, 177: 1010–1016.

Karikalan, N., Velmurugan, M., Chen, S. M. and Karuppiah, C. (2016). Modern approach to the synthesis of Ni (OH) 2 decorated sulfur doped carbon nanoparticles for the nonenzymatic glucose sensor. ACS Applied Materials & Interfaces, 8(34): 22545–22553.

Kaur, B., Kumar, S. and Kaushik, B. K. (2022). Recent advancements in optical biosensors for cancer detection. Biosensors and Bioelectronics, 197: 113805.

Kaur, H. and Shorie, M. (2019). Nanomaterial based aptasensors for clinical and environmental diagnostic applications. Nanoscale Advances, 1(6): 2123–2138.

Kaushik, A., Khan, R., Solanki, P. R., Pandey, P., Alam, J., Ahmad, S. et al. (2008). Iron oxide nanoparticles–chitosan composite based glucose biosensor. Biosensors and Bioelectronics, 24(4): 676v683.

Kaushik, A., Solanki, P. R., Ansari, A. A., Sumana, G., Ahmad, S. and Malhotra, B. D. (2009a). Iron oxide-chitosan nanobiocomposite for urea sensor. Sensors and Actuators B: Chemical, 138(2): 572–580.

Kaushik, A., Solanki, P. R., Ansari, A. A., Malhotra, B. D. and Ahmad, S. (2009b). Iron oxide-chitosan hybrid nanobiocomposite based nucleic acid sensor for pyrethroid detection. Biochemical Engineering Journal, 46(2): 132–140.

Kaushik, A., Kumar, R., Huey, E., Bhansali, S., Nair, N. and Nair, M. (2014). Silica nanowires: Growth, integration, and sensing applications. Microchimica Acta, 181: 1759–1780.

Kaushik, A. and Mujawar, M. A. (2018). Point of care sensing devices: Better care for everyone. Sensors, 18(12): 4303.

Khan, T., Waseem, R., Zehra, Z., Aiman, A., Bhardwaj, P., Ansari, J. et al. (2022). Mitochondrial dysfunction: pathophysiology and mitochondria-targeted drug delivery approaches. Pharmaceutics, 14(12): 2657.

Khezrian, S., Salimi, A., Teymourian, H. and Hallaj, R. (2013). Label-free electrochemical IgE aptasensor based on covalent attachment of aptamer onto multiwalled carbon nanotubes/ionic liquid/chitosan nanocomposite modified electrode. Biosensors and Bioelectronics, 43: 218–225.

Kour, R., Arya, S., Young, S. J., Gupta, V., Bandhoria, P. and Khosla, A. (2020). Recent advances in carbon nanomaterials as electrochemical biosensors. Journal of The Electrochemical Society, 167(3): 037555.

Kozitsina, A., Svalova, T., Malysheva, N., Glazyrina, Y. and Matern, A. (2016). A new enzyme-free electrochemical immunoassay for *Escherichia coli* detection using magnetic nanoparticles. Analytical Letters, 49(2): 245–257.

Kumar, S., Kumar, S., Ali, M. A., Anand, P., Agrawal, V. V., John, R. et al. (2013). Microfluidic-integrated biosensors: Prospects for point-of-care diagnostics. Biotechnology Journal, 8(11): 1267–1279.

Kumar, S. and Singh, R. (2021). Recent optical sensing technologies for the detection of various biomolecules. Optics & Laser Technology, 134: 106620.

Kurakula, M. and Naveen, N. R. (2021). Electrospraying: A facile technology unfolding the chitosan based drug delivery and biomedical applications. European Polymer Journal, 147: 110326.
Lasita, A. L., Pabrekar, S., Wagh, N. S. and Lakkakula, J. (2022). Metallic biomaterials in biomedical applications. In Functional Biomaterials: Drug Delivery and Biomedical Applications (pp. 341–370). Singapore: Springer Singapore.
Lei, C. X., Long, L. P. and Cao, Z. L. (2005). An H2O2 Biosensor Based on Immobilization of Horseradish Peroxidase Labeled Nano-Au in Silica Sol-Gel/Alginate Composite Film. Analytical Letters, 38(11): 1721–1734.
Li, F., Chen, W., Tang, C. and Zhang, S. (2009). Development of hydrogen peroxide biosensor based on in situ covalent immobilization of horseradish peroxidase by one-pot polysaccharide-incorporated sol–gel process. Talanta, 77(4): 1304–1308.
Li, G., Sun, K., Li, D., Lv, P., Wang, Q., Huang, F. et al. (2016). Biosensor based on bacterial cellulose-Au nanoparticles electrode modified with laccase for hydroquinone detection. Colloids and Surfaces A: Physicochemical and Engineering Aspects, 509: 408–414.
Liang, M., Wang, L., Ma, C., Zhang, M. and Xie, G. (2013). Sandwich immunoassay for hepatitis C virus non-structural 5A protein using a glassy carbon electrode modified with an Au-MoO3/chitosan nanocomposite. Analytical Letters, 46(8): 1241–1254.
Liu, C., Yu, H., Zhang, B., Liu, S., Liu, C. G., Li, F. et al. (2022). Engineering whole-cell microbial biosensors: Design principles and applications in monitoring and treatment of heavy metals and organic pollutants. Biotechnology Advances, 60: 108019.
Liu, J., Wang, T., Wang, J. and Wang, E. (2015). Mussel-inspired biopolymer modified 3D graphene foam for enzyme immobilization and high performance biosensor. Electrochimica Acta, 161: 17–22.
Liu, L., Xiong, M., Rong, Q., Zhang, M. and Zhang, X. B. (2023). Nucleic acid sensors *in vivo*: Challenges and opportunities. View, 4(3): 20220064.
Liu, Q., Li, M., Xiong, L., Qiu, L., Bian, X., Sun, C. et al. (2018). Oxidation modification of debranched starch for the preparation of starch nanoparticles with calcium ions. Food Hydrocoll., 85: 86–92.
Liu, Q., Cai, W., Zhen, T., Ji, N., Dai, L., Xiong, L. et al. (2020). Preparation of debranched starch nanoparticles by ionic gelation for encapsulation of epigallocatechin gallate. International Journal of Biological Macromolecules, 161: 481–491.
Madej-Kiełbik, L., Gzyra-Jagieła, K., Jóźwik-Pruska, J., Dziuba, R. and Bednarowicz, A. (2022). Biopolymer composites with sensors for environmental and medical applications. Materials, 15(21): 7493.
Maity, C. and Das, N. (2022). Alginate-based smart materials and their application: Recent advances and perspectives. Topics in Current Chemistry, 380(1): 3.
Malhotra, B. D. and Ali, M. A. (2017). Nanomaterials for biosensors: fundamentals and applications. William Andrew.
Mamuk, A. E., Işık, C., Aslan, S. and Altuntaş, D. B. (2023). Chitosan Nanocomposites for Biosensing Applications. In Chitosan Nanocomposites: Bionanomechanical Applications (pp. 255–281). Singapore: Springer Nature Singapore.
Mantha, S., Pedrosa, V. A., Olsen, E. V., Davis, V. A. and Simonian, A. L. (2010). Renewable nanocomposite layer-by-layer assembled catalytic interfaces for biosensing applications. Langmuir, 26(24): 19114–19119.
Masson, J. F., Battaglia, T. M., Kim, Y. C., Prakash, A., Beaudoin, S. and Booksh, K. S. (2004). Preparation of analyte-sensitive polymeric supports for biochemical sensors. Talanta, 64(3): 716–725.
Mehrotra, P. (2016). Biosensors and their applications–A review. Journal of Oral Biology and Craniofacial Research, 6(2): 153–159.
Mendes, J. F., Paschoalin, R. T., Carmona, V. B., Neto, A. R. S., Marques, A. C. P., Marconcini, J. M. et al. (2016). Biodegradable polymer blends based on corn starch and thermoplastic chitosan processed by extrusion. Carbohydrate Polymers, 137: 452–458.
Mohamed, F. A. N. and Laraba-Djebari, F. (2016). Development and characterization of a new carrier for vaccine delivery based on calcium-alginate nanoparticles: Safe immunoprotective approach against scorpion envenoming. Vaccine, 34(24): 2692–2699.
Mohammadpour, R. and Ahmadvand, H. (2014). A novel field ionization gas sensor based on self-organized CuO nanowire arrays. Sensors and Actuators A: Physical, 216: 202–206.
Mulchandani, A., Kaneva, I. and Chen, W. (1998). Biosensor for direct determination of organophosphate nerve agents using recombinant *Escherichia coli* with surface-expressed organophosphorus hydrolase. 2. Fiber-optic microbial biosensor. Analytical Chemistry, 70(23): 5042–5046.
Naresh, V. and Lee, N. (2021). A review on biosensors and recent development of nanostructured materials-enabled biosensors. Sensors, 21(4): 1109.
Nasrollahzadeh, M., Shafiei, N., Nezafat, Z., Bidgoli, N. S. S. and Soleimani, F. (2020). Recent progresses in the application of cellulose, starch, alginate, gum, pectin, chitin and chitosan based (nano) catalysts in sustainable and selective oxidation reactions: A review. Carbohydrate Polymers, 241: 116353.

Nguyen, H. H., Lee, S. H., Lee, U. J., Fermin, C. D. and Kim, M. (2019). Immobilized enzymes in biosensor applications. Materials, 12(1): 121.

Öndeş, B., Akpınar, F., Uygun, M., Muti, M. and Uygun, D. A. (2021). High stability potentiometric urea biosensor based on enzyme attached nanoparticles. Microchemical Journal, 160: 105667.

Othman, N., Masarudin, M. J., Kuen, C. Y., Dasuan, N. A., Abdullah, L. C. and Md. Jamil, S. N. A. (2018). Synthesis and optimization of chitosan nanoparticles loaded with L-ascorbic acid and thymoquinone. Nanomaterials, 8(11): 920.

Palanisamy, S., Velusamy, V., Balu, S., Velmurugan, S., Yang, T. C. and Chen, S. W. (2020). Sonochemical synthesis and anchoring of zinc oxide on hemin-mediated multiwalled carbon nanotubes-cellulose nanocomposite for ultra-sensitive biosensing of H2O2. Ultrasonics Sonochemistry, 63: 104917.

Pan, X., Liu, P., Wang, Y., Yi, Y. L., Zhang, H. Q., Qian, D. W. et al. (2022). Synthesis of starch nanoparticles with controlled morphology and various adsorption rate for urea. Food Chemistry, 369: 130882.

Pant, A. and Negi, J. S. (2018). Novel controlled ionic gelation strategy for chitosan nanoparticles preparation using TPP-β-CD inclusion complex. European Journal of Pharmaceutical Sciences, 112: 180–185.

Patra, S., Basak, P. and Tibarewala, D. N. (2016). Synthesis of gelatin nano/submicron particles by binary nonsolvent aided coacervation (BNAC) method. Materials Science and Engineering: C, 59: 310–318.

Pawar, A., Thakkar, S. and Misra, M. (2018). A bird's eye view of nanoparticles prepared by electrospraying: advancements in drug delivery field. Journal of Controlled Release, 286: 179–200.

Pohanka, M. (2017). The piezoelectric biosensors: Principles and applications, a review. International Journal of Electrochemical Science, 12(1): 496–506.

Prabhakar, N., Thakur, H., Bharti, A. and Kaur, N. (2016). Chitosan-iron oxide nanocomposite based electrochemical aptasensor for determination of malathion. Analytica Chimica Acta, 939: 108–116.

Qin, C., Chen, C., Xie, Q., Wang, L., He, X., Huang, Y. et al. (2012). Amperometric enzyme electrodes of glucose and lactate based on poly (diallyldimethylammonium)-alginate-metal ion-enzyme biocomposites. Analytica chimica acta, 720: 49–56.

Qiu, C., Hu, Y., Jin, Z., McClements, D. J., Qin, Y., Xu, X. et al. (2019). A review of green techniques for the synthesis of size-controlled starch-based nanoparticles and their applications as nanodelivery systems. Trends in Food Science & Technology, 92: 138–151.

Qiu, J. D., Liang, R. P., Wang, R., Fan, L. X., Chen, Y. W. and Xia, X. H. (2009). A label-free amperometric immunosensor based on biocompatible conductive redox chitosan-ferrocene/gold nanoparticles matrix. Biosensors and Bioelectronics, 25(4): 852–857.

Radhakrishnan, K., Tripathy, J., Datey, A., Chakravortty, D. and Raichur, A. M. (2015). Mesoporous silica–chondroitin sulphate hybrid nanoparticles for targeted and bio-responsive drug delivery. New Journal of Chemistry, 39(3): 1754–1760.

Rahman, M. M., Li, X. B., Lopa, N. S., Ahn, S. J. and Lee, J. J. (2015). Electrochemical DNA hybridization sensors based on conducting polymers. Sensors, 15(2): 3801–3829.

Ramesh, M., Janani, R., Deepa, C. and Rajeshkumar, L. (2022). Nanotechnology-enabled biosensors: A review of fundamentals, design principles, materials, and applications. Biosensors, 13(1): 40.

Rizwan, K., Rahdar, A., Bilal, M. and Iqbal, H. M. (2022). MXene-based electrochemical and biosensing platforms to detect toxic elements and pesticides pollutants from environmental matrices. Chemosphere, 291: 132820.

Roquero, D. M. and Katz, E. (2022). "Smart" alginate hydrogels in biosensing, bioactuation and biocomputing: State-of-the-art and perspectives. Sensors and Actuators Reports, 4: 100095.

Saboktakin, M. R., Tabatabaie, R. M., Maharramov, A. and Ramazanov, M. A. (2011). Synthesis and in vitro evaluation of carboxymethyl starch–chitosan nanoparticles as drug delivery system to the colon. International journal of biological macromolecules, 48(3): 381–385.

Sawant, S. N. (2017). Development of biosensors from biopolymer composites. In Biopolymer composites in electronics (pp. 353–383). Elsevier.

Sequeira-Antunes, B. and Ferreira, H. A. (2023). Nucleic acid aptamer-based biosensors: a review. Biomedicines, 11(12): 3201.

Shahdost-fard, F., Salimi, A. and Khezrian, S. (2014). Highly selective and sensitive adenosine aptasensor based on platinum nanoparticles as catalytical label for amplified detection of biorecognition events through H2O2 reduction. Biosensors and Bioelectronics, 53: 355–362.

Shahdost-fard, F., Salimi, A., Sharifi, E. and Korani, A. (2013). Fabrication of a highly sensitive adenosine aptasensor based on covalent attachment of aptamer onto chitosan-carbon nanotubes-ionic liquid nanocomposite. Biosensors and Bioelectronics, 48: 100–107.

Shi, Q., Li, Q., Shan, D., Fan, Q. and Xue, H. (2008). Biopolymer-clay nanoparticles composite system (Chitosan-laponite) for electrochemical sensing based on glucose oxidase. Materials Science and Engineering: C, 28(8): 1372–1375.

Soares, A. C., Soares, J. C., Shimizu, F. M., Melendez, M. E., Carvalho, A. L. and Oliveira Jr, O. N. (2015). Controlled film architectures to detect a biomarker for pancreatic cancer using impedance spectroscopy. ACS Applied Materials & Interfaces, 7(46): 25930–25937.

Sobhan, A., Muthukumarappan, K. and Wei, L. (2021). Biosensors and biopolymer-based nanocomposites for smart food packaging: Challenges and opportunities. Food Packaging and Shelf Life, 30: 100745.

Sotelo-Boyás, M. E., Correa-Pacheco, Z. N., Bautista-Baños, S. and Corona-Rangel, M. L. (2017a). Physicochemical characterization of chitosan nanoparticles and nanocapsules incorporated with lime essential oil and their antibacterial activity against food-borne pathogens. Lwt, 77: 15–20.

Sotelo-Boyás, M., Correa-Pacheco, Z., Bautista-Baños, S. and y Gómez, Y. G. (2017b). Release study and inhibitory activity of thyme essential oil-loaded chitosan nanoparticles and nanocapsules against foodborne bacteria. International Journal of Biological Macromolecules, 103: 409–414.

SoYoon, S., Ramadoss, A., Saravanakumar, B. and Kim, S. J. (2014). Novel Cu/CuO/ZnO hybrid hierarchical nanostructures for non-enzymatic glucose sensor application. Journal of Electroanalytical Chemistry, 717: 90–95.

Srivastava, S., Ali, M. A., Solanki, P. R., Chavhan, P. M., Pandey, M. K., Mulchandani, A. et al. (2013). Mediator-free microfluidics biosensor based on titania–zirconia nanocomposite for urea detection. Rsc Advances, 3(1): 228–235.

Subramanian, S. B., Francis, A. P. and Devasena, T. (2014). Chitosan–starch nanocomposite particles as a drug carrier for the delivery of bis-desmethoxy curcumin analog. Carbohydrate Polymers, 114: 170–178.

Sullivan, D. J., Cruz-Romero, M., Collins, T., Cummins, E., Kerry, J. P. and Morris, M. A. (2018). Synthesis of monodisperse chitosan nanoparticles. Food Hydrocolloids, 83: 355–364.

Tan, Y., Deng, W., Ge, B., Xie, Q., Huang, J. and Yao, S. (2009). Biofuel cell and phenolic biosensor based on acid-resistant laccase–glutaraldehyde functionalized chitosan–multiwalled carbon nanotubes nanocomposite film. Biosensors and Bioelectronics, 24(7): 2225–2231.

Tang, J., Sisler, J., Grishkewich, N. and Tam, K. C. (2017). Functionalization of cellulose nanocrystals for advanced applications. Journal of Colloid and Interface Science, 494: 397–409.

Tavares, I. S., Caroni, A. L. P. F., Neto, A. D., Pereira, M. R. and Fonseca, J. L. C. (2012). Surface charging and dimensions of chitosan coacervated nanoparticles. Colloids and Surfaces B: Biointerfaces, 90: 254–258.

Tembe, S., Karve, M., Inamdar, S., Haram, S., Melo, J. and D'Souza, S. F. (2006). Development of electrochemical biosensor based on tyrosinase immobilized in composite biopolymeric film. Analytical Biochemistry, 349(1): 72–77.

Thien, D. V. H., Hsiao, S. W. and Ho, M. H. (2012). Synthesis of electrosprayed chitosan nanoparticles for drug sustained release. Nano Life, 2(01): 1250003.

Tiwari, I., Singh, M., Pandey, C. M. and Sumana, G. (2015). Electrochemical detection of a pathogenic *Escherichia coli* specific DNA sequence based on a graphene oxide–chitosan composite decorated with nickel ferrite nanoparticles. RSC advances, 5(82): 67115–67124.

Uniyal, S., Choudhary, K., Sachdev, S. and Kumar, S. (2022). Recent advances in K-SPR sensors for the detection of biomolecules and microorganisms: A review. IEEE Sensors Journal, 22(12): 11415–11426.

Vasuki, S., Varsha, V., Mithra, R., Dharshni, R. A., Abinaya, S., Dharshini, R. D. et al. (2019). Thermal biosensors and their applications. Am. Int. J. Res. Sci. Technol. Eng. Math, 1: 262–264.

Vodyashkin, A. A., Kezimana, P., Vetcher, A. A. and Stanishevskiy, Y. M. (2022). Biopolymeric nanoparticles–multifunctional materials of the future. Polymers, 14(11): 2287.

Vopálenská, I., Váchová, L. and Palková, Z. (2015). New biosensor for detection of copper ions in water based on immobilized genetically modified yeast cells. Biosensors and Bioelectronics, 72: 160–167.

Wang, L., Ni, P., Wei, G., Wang, J. and Li, Z. (2014). Collagen nanofiber-templated silver nanowires on graphene nanosheets for a nonenzymatic amperometric biosensor of hydrogen peroxide. Chemistry Letters, 43(4): 544–546.

Wang, S., Tan, Y., Zhao, D. and Liu, G. (2008a). Amperometric tyrosinase biosensor based on Fe3O4 nanoparticles–chitosan nanocomposite. Biosensors and Bioelectronics, 23(12): 1781–1787.

Wang, W., Li, H. Y., Zhang, D. W., Jiang, J., Cui, Y. R., Qiu, S. et al. (2010). Fabrication of bienzymatic glucose biosensor based on novel gold nanoparticles-bacteria cellulose nanofibers nanocomposite. Electroanalysis, 22(21): 2543–2550.

Wang, W., Zhang, T. J., Zhang, D. W., Li, H. Y., Ma, Y. R., Qi, L. M. et al. (2011a). Amperometric hydrogen peroxide biosensor based on the immobilization of heme proteins on gold nanoparticles–bacteria cellulose nanofibers nanocomposite. Talanta, 84(1): 71–77.

Wang, X., Cheng, J., Ji, G., Peng, X. and Luo, Z. (2016). Starch nanoparticles prepared in a two ionic liquid based microemulsion system and their drug loading and release properties. RSC Advances, 6(6): 4751–4757.

Wang, Y., Wang, X., Luo, G. and Dai, Y. (2008b). Adsorption of bovin serum albumin (BSA) onto the magnetic chitosan nanoparticles prepared by a microemulsion system. Bioresource Technology, 99(9): 3881–3884.

Wang, Y., Xiao, Y., Ma, X., Li, N. and Yang, X. (2012). Label-free and sensitive thrombin sensing on a molecularly grafted aptamer on graphene. Chemical Communications, 48(5): 738–740.

Wang, Y., Yuan, R., Chai, Y., Yuan, Y., Bai, L. and Liao, Y. (2011b). A multi-amplification aptasensor for highly sensitive detection of thrombin based on high-quality hollow CoPt nanoparticles decorated graphene. Biosensors and Bioelectronics, 30(1): 61–66.

Wang, Y., Zhang, R., Qin, W., Dai, J., Zhang, Q., Lee, K. et al. (2020). Physicochemical properties of gelatin films containing tea polyphenol-loaded chitosan nanoparticles generated by electrospray. Materials & Design, 185: 108277.

Wolf, M., Gasparin, B. C. and Paulino, A. T. (2018). Hydrolysis of lactose using β-d-galactosidase immobilized in a modified Arabic gum-based hydrogel for the production of lactose-free/low-lactose milk. International Journal of Biological Macromolecules, 115: 157–164.

Wu, J., Yan, F., Zhang, X., Yan, Y., Tang, J. and Ju, H. (2008). Disposable reagentless electrochemical immunosensor array based on a biopolymer/sol-gel membrane for simultaneous measurement of several tumor markers. Clinical chemistry, 54(9): 1481–1488.

Xu, B., Zheng, D., Qiu, W., Gao, F., Jiang, S. and Wang, Q. (2015). An ultrasensitive DNA biosensor based on covalent immobilization of probe DNA on fern leaf-like α-Fe2O3 and chitosan Hybrid film using terephthalaldehyde as arm-linker. Biosensors and Bioelectronics, 72: 175–181.

Yang, L., Zhu, J., Xu, Y., Yun, W., Zhang, R., He, P. and Fang, Y. (2011). Electrochemiluminescence aptamer biosensor for detection of thrombin based on CdS QDs/ACNTs electrode. Electroanalysis, 23(4): 1007–1012.

Yang, M., Jiang, J., Yang, Y., Chen, X., Shen, G. and Yu, R. (2006). Carbon nanotube/cobalt hexacyanoferrate nanoparticle-biopolymer system for the fabrication of biosensors. Biosensors and Bioelectronics, 21(9): 1791–1797.

Yao, Y., Huang, W., Chen, J., Liu, X., Bai, L., Chen, W. et al. (2023). Flexible and stretchable organic electrochemical transistors for physiological sensing devices. Advanced Materials, 35(35): 2209906.

Yardımcı, F. S., Şenel, M. and Baykal, A. (2012). Amperometric hydrogen peroxide biosensor based on cobalt ferrite–chitosan nanocomposite. Materials Science and Engineering: C, 32(2): 269–275.

Yellapu, G., Vishal, C. C., Kandoth, M. P., Saha, P., Bojja, R. R., Gandham, S. et al. (2019). Functionalized multi-walled carbon nanotubes based Newtonian Nano fluids for medium temperature heat transfer applications. Thermal Science and Engineering Progress, 12: 13–23.

Zeng, Q., Cheng, J. S., Liu, X. F., Bai, H. T. and Jiang, J. H. (2011). Palladium nanoparticle/chitosan-grafted graphene nanocomposites for construction of a glucose biosensor. Biosensors and Bioelectronics, 26(8): 3456–3463.

Zhai, Z., Leng, B., Yang, N., Yang, B., Liu, L., Huang, N. et al. (2019). Rational construction of 3D-networked carbon nanowalls/diamond supporting CuO architecture for high-performance electrochemical biosensors. Small, 15(48): 1901527.

Zhang, G., Zhou, H., An, C., Liu, D., Huang, Z. and Kuang, Y. (2012). Bimetallic palladium–gold nanoparticles synthesized in ionic liquid microemulsion. Colloid and Polymer Science, 290: 1435–1441.

Zhang, M., Mullens, C. and Gorski, W. (2007). Coimmobilization of dehydrogenases and their cofactors in electrochemical biosensors. Analytical Cchemistry, 79(6): 2446–2450.

Zhang, T., Wang, W., Zhang, D., Zhang, X., Ma, Y., Zhou, Y. et al. (2010). Biotemplated synthesis of gold nanoparticle–bacteria cellulose nanofiber nanocomposites and their application in biosensing. Advanced Functional Materials, 20(7): 1152–1160.

Zhou, G., Luo, Z. and Fu, X. (2014). Preparation and characterization of starch nanoparticles in ionic liquid-in-oil microemulsions system. Industrial Crops and Products, 52: 105–110.

Zhou, J., Zhou, S., Fan, P., Li, X., Ying, Y., Ping, J. and Pan, Y. (2024). Implantable electrochemical microsensors for *in vivo* monitoring of animal physiological information. Nano-Micro Letters, 16(1): 49.

CHAPTER-7

Removal of Organic Pollutants and Catalytic Applications for Environmental Sustainability

Roshini Nandagopal[1,2] and *Sujata Mandal*[1,2,*]

1. Introduction

The advent/production of newer and more complex chemicals, specifically advanced healthcare and lifestyle products, and their enhanced consumption/disposal has led to an increase in complex pollutants in all three strata of the environment (air, water and soil). Environmental pollution is a global concern for the sustainability of all living organisms. Therefore, reducing the pollution load in the air, water, and soil is essential for environmental sustainability (Owhonka et al., 2021; Protection Agency, 2023; Al-Hazmi et al., 2023; Mandal et al., 2024; Deng et al., 2024; Xuan et al., 2024; Balsalobre-Lorente et al., 2024; World Water Assessment Programme (United Nations)).

Water pollution has become a critical issue of global consequence in recent years. The contamination of water ecosystems results from the infiltration of organic, inorganic, and organometallic pollutants, as well as from pathogenic agents. Among all, organic pollutants form a group of toxic, carcinogenic and complex chemicals originating from diverse sources, including pharmaceuticals, personal care products, fire/flame retardants, solvents, plasticizers, endocrine disrupting chemicals, surfactants, agrochemicals such as pesticides and growth regulators. All these organic pollutants are generated and introduced into the water resources due to anthropogenic activities. Such pollutants pose a serious threat to public health and safety and also have negative impacts on the environment and aquatic life. Therefore, it is important to address the issue of water pollution with a sense of urgency and purpose, as it is critical to maintaining a sustainable and healthy planet. The primary and secondary water treatment plants are primarily built to handle domestic effluent and, therefore, do not efficiently remove the vast amounts of organic loads and complex molecules in industrial wastewater. So, the water treatment system requires facilities to be fitted with additional processes (tertiary-level) for handling challenging organic pollutants and toxic residues. The tertiary treatment process includes advanced water treatment techniques over the primary and secondary treatments. The tertiary treatment process can be biological,

[1] Advanced Materials Laboratory (AML), CSIR – Central Leather Research Institute (CSIR-CLRI), Sardar Patel Road, Adyar, Chennai- 600020, India.

[2] Academy of Scientific and Innovative Research (AcSIR), Ghaziabad – 201002, India.

* Corresponding author: sujata@clri.res.in; sujatamandal@rediffmail.com, ORCID ID: 0000-0002-0549-8851

DOI: 10.1201/9781003558330-7

chemical, and physical techniques (Boikanyo et al., 2018) such as activated sludge, trickling filters, rotating biological contactors, coagulation/flocculation, reverse osmosis, electrochemical technique, adsorption, ion exchange, advanced oxidation process (AOP), membrane filtration, and sonochemical methods (Faisal Ibney Hai and Fukushi, 2007; Kyzas et al., 2013; Boikanyo et al., 2018). Nevertheless, a low-cost and sustainable method with high efficiency is essential for water/wastewater purification (Schellenberg et al., 2020; Owhonka et al., 2021; Suresh et al., 2021; Alvi et al., 2023; Babu Ponnusami et al., 2023; Awewomom et al., 2024; Annu et al., 2024; World Water Assessment Programme (United Nations)). The classification of the chemical treatment and water recycling technologies is presented in Fig. 1 (Gupta et al., 2012).

Among the various tertiary-level water/wastewater treatment techniques, adsorption is the most popular because of its effectivity in the low-concentration range (in ppm/ppb level), simple operation and low-energy consuming process (Mandal and Kulkarni, 2011). A large variety of adsorbents have been developed and employed to remove various organic and inorganic contaminants from water/wastewater with high recyclability. In addition to adsorption, the advanced oxidation process (AOP) is the technique which destroys organic pollutants instead of converting/transferring them to another phase/form, as in the case with other physical treatment processes (Wen et al., 2018; Boikanyo et al., 2018). AOPs typically involve the chemical/photochemical activation of oxidizing substances, such as ozone, hydrogen peroxide, Fenton's reagent, to produce highly reactive free radicals ($HO^\bullet$, $O_2^\bullet$, $HO_2^\bullet$, etc.) that have at least one unpaired electron involved in oxidation reactions (Molnar Jazić et al. 2024). With the evolution of AOP, sulphate, persulphate, and periodate radical-based processes have also been used for water treatment. In AOPs, *in situ* generated highly reactive free radicals

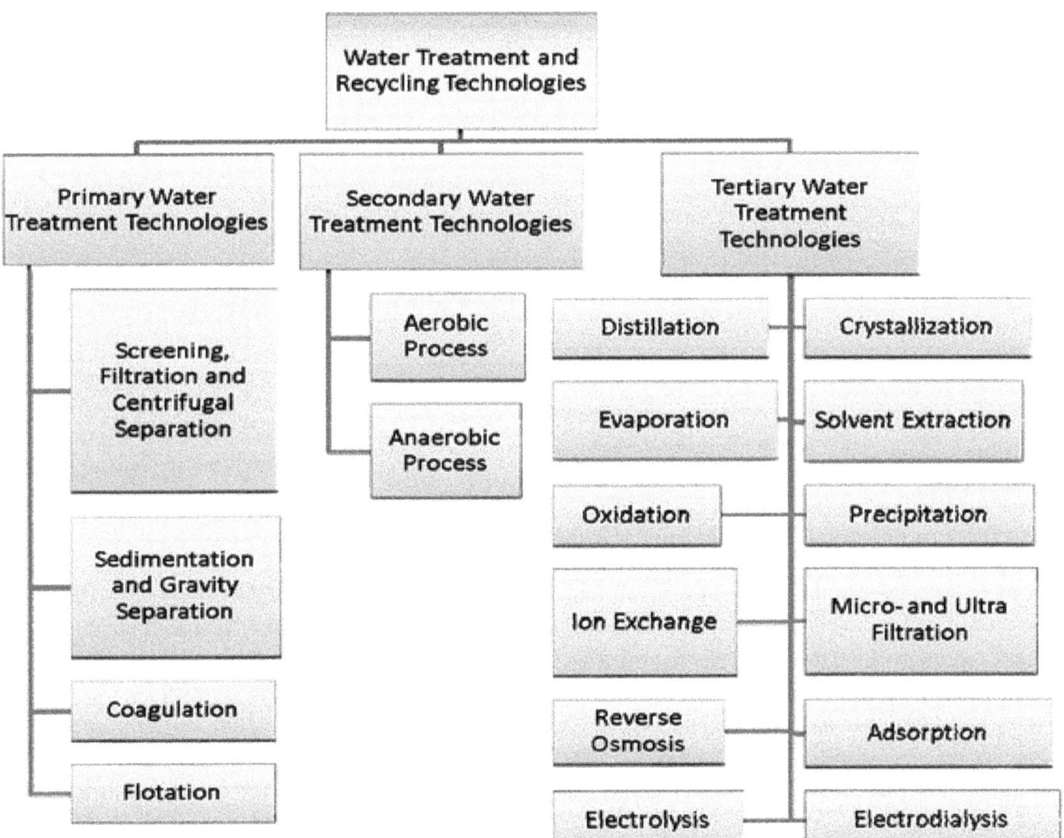

Figure 1. Classification of the chemical treatment and water recycling technologies. Reproduced with permission from Gupta et al. (2012).

could effectively mineralize most of the organic pollutants due to their high oxidizing potential (Gupta et al., 2012). A catalytic AOP involves catalyst-based activation of the oxidizing species and is in use alongside other systems such as light, ozone, hydrogen peroxide, Fenton's reagent, etc. (Molnar Jazić et al., 2024). Among different AOPs, the utilization of solar energy, an abundant renewable energy resource, is effectively done by the photocatalysis process. So, it is considered one of the promising and sustainable methods for wastewater treatment (Yan et al., 2015; Zhou et al., 2019; Huang et al., 2019; Suresh et al., 2021).

This chapter discusses the biopolymeric nanoparticles and nanocomposites made from biopolymers and various nanocatalysts and their application toward environmental remediation through adsorption, catalytic decomposition and advanced oxidation processes. Biopolymers can be classified on various scales based on their origin, degradability, polymer backbone, and type of monomers. The classification of biopolymers based on their origin is presented in Fig. 2.

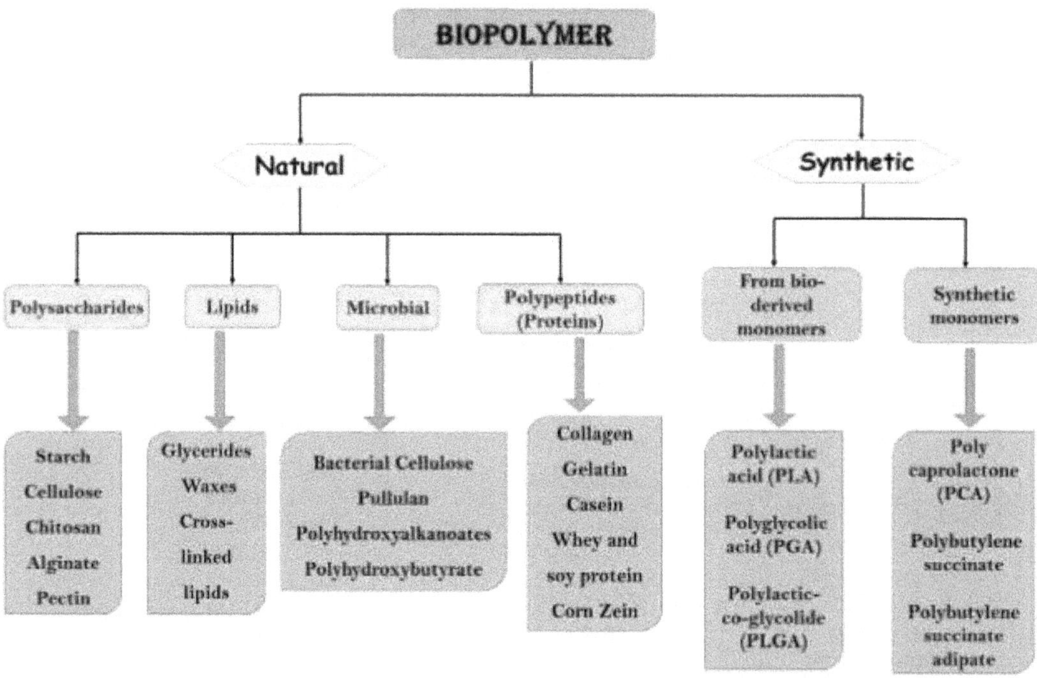

Figure 2. Classification of biopolymers based on their origin.

2. Biopolymer Nanocomposites

Biopolymeric nanocomposites are the composites prepared by the combination of biopolymer and other nanomaterials, viz. metals/metal oxides/hydroxides, CNTs, graphene/graphene oxides, ceramics, etc., where the dimension of at least one phase is in the nanometre range (1–100 nm). Nanocomposite materials show exceptional multifunctional properties due to the introduction of nanocomponents into their structure; however, the desirable properties can be achieved if only the nanomaterials from the first phase are homogenously dispersed within the matrix of the second phase. Biopolymeric nanocomposites are a special class of materials and find versatile industrial applications, including the paper industry (Vasilev et al., 2020), water purification, catalysis (Li et al., 2021; Marwani et al., 2022; Keshipour et al., 2023); biomedical (Sundar et al., 2010; Mitchell et al., 2021) antibacterial products, and the food and food packaging industries (Luo et al., 2020; Basavegowda and Baek, 2021). The biopolymer-based nanocomposites are becoming increasingly popular because they are abundant, non-toxic, cost-effective, biodegradable, biocompatible, and offer multiple functionalities. More importantly, they are a sustainable and eco-friendly alternative to synthetic polymers (Gamage et al., 2021; Nasrollahzadeh et al., 2021; Annu et al., 2024).

The synthesis of biopolymeric nanocomposites typically involves multiple steps. At first, the nanomaterials are synthesized separately by conventional techniques, such as co-precipitation, hydrothermal, solvothermal and cold vapour deposition. The synthesized nanoparticles are then introduced into the polymer matrix using the sol-gel, intercalation, template-based synthesis, graft copolymerization, and electrospinning techniques (Annu et al., 2024).

3. Applications of Biopolymeric Nanocomposites in Wastewater Remediation

Nanostructured catalysts/photocatalysts have attracted considerable research attention during past decades due to their enhanced physicochemical properties. The catalysts (photocatalysts) used in AOPs are metals, metal oxides, transition metal chalcogenides, carbon-based materials and their composites. While the nano size of the catalysts ensures enhanced performance, separation/ recovery of the nano-sized catalysts from the reaction media becomes a difficult and at times- an energy-consuming task. Moreover, the agglomeration tendency of these nanomaterials resulted in a reduced number of active sites, leading to reduced activity. In order to overcome these limitations of nanocatalysts, they are often supported and/or stabilized on inorganic or polymeric support materials. Supporting the catalysts on inorganic or polymeric materials solves the problem of its separation and recovery and leads to improved functional properties due to specific photocatalyst-support interactions (Boikanyo et al., 2018). Therefore, developing low-cost and more effective supports is necessary for the sustainable industrialization of heterogeneous photocatalysis.

Synthetic polymers, because of their non-biodegradable nature, continue to exist on the earth's surface for many years and form a massive volume of solid waste, creating serious environmental issues (Sarma and Das, 2021). Consequently, biopolymers usually have a much less shelf life than synthetic polymers; they are readily biodegradable and, therefore, have gained considerable popularity among researchers as an alternative to synthetic polymers in diverse applications (Sarma and Das, 2021). Among them, polysaccharide derivatives, such as chitosan, cellulose, starch, and alginate, have received attention in multiple applications (Lee et al., 2015; Musa and Bwatanglang, 2020). In photocatalytic decontamination systems, their ability to associate physically and chemically is beneficial considering suitable support, which needs to have an enormous affinity for the adsorption of contaminants in order to attract them to their contact with the catalyst (Singh et al., 2013; Boikanyo et al., 2018). Therefore, biopolymer-based catalyst's supports and catalysts-biopolymer nanocomposites including cellulose, starch, chitosan, alginate, guar gum, pectin, lignin, collagen, and gelatin, have gained remarkable attention owing to their abundance, unique structural and physicochemical properties, viz. excellent chemical/mechanical stability, reinforcement capability, structural variations, ease of modification, biocompatibility, biodegradability, sustainability, low cost, and non-toxic nature, for environmental sustainability (Guibal et al., 2007; Mandal and Mayadevi, 2008; Lee et al., 2015; Boikanyo et al., 2018; Musa and Bwatanglang, 2020; Jaleh et al., 2021).

3.1 Nanocellulose and Cellulose-based Nanocomposites

Cellulose is the most abundant carbohydrate polymer on earth. Cellulose possesses many advantageous properties, such as biodegradability, biocompatibility, and a high concentration of tunable hydroxyl groups, which make it the sustainable material of choice for several environmental applications. Cellulose-derived nanomaterials, including nanocellulose, cellulose nanofibres, cellulose nanofibrils (CNFs), cellulose nanocrystals (CNCs), cellulose nanowhiskers, bacterial cellulose and cellulose nanocomposites, have been explored for water/wastewater treatment through various techniques (Mandal and Mayadevi, 2008; Wei et al., 2014; Carpenter et al., 2015a, b; Moodley Kandasamy and Arumugam, 2022; Das et al., 2022; Zhang et al., 2023; Etale et al., 2023).

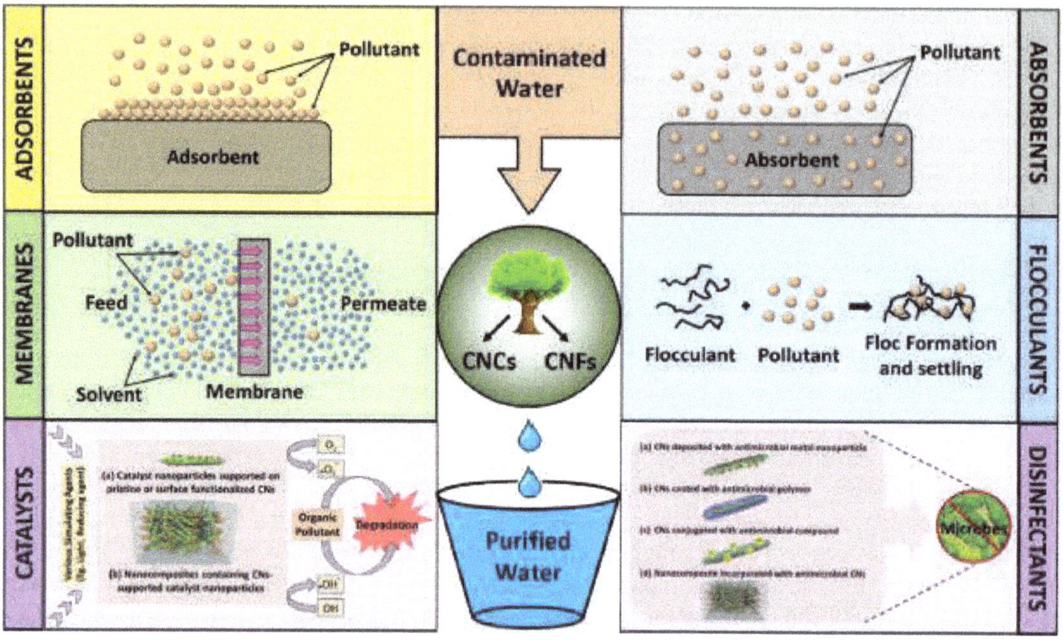

Figure 3. Schematic representation of cellulose nanomaterial-based various water/wastewater treatment processes. Reproduced with permission from Mohammed et al. (2018).

A schematic representation of cellulose nanomaterial-based various water/wastewater treatment processes is presented in Fig. 3 (Mohammed et al., 2018).

In a study by Batmaz et al. (2014), CNCs and carboxylated CNCs (CNC-OX) were synthesized and applied for the adsorptive removal of MB dye from water. The MB adsorption capacity of CNCs was 118 mg/g, which was increased to 769 mg/g on carboxylation of the CNCs (CNC-OX). The significant increase in MB adsorption capacity of CNC-OX was due to carboxylation, which contributed to the adsorption process by introducing extra negative charge through the counter-ion exchange process between the active anionic sites and the positively charged dye molecules (Batmaz et al., 2014). In another study by Qiao et al. (2015), pristine cellulose, cellulose nanocrystals (CNCs), and carboxylate-functionalized cellulose nanocrystal (CNM) adsorbents were prepared and investigated for the removal of four different cationic dyes from water (Fig. 4A). CNM exhibited superior adsorption efficiency for all four dyes compared to CNCs and pristine cellulose (Fig. 4B). The pristine cellulose showed the lowest adsorption capacity for all four dyes. The high adsorption capacity of CNCs was due to the high specific surface area of CNCs compared to the raw cellulose. In contrast, the superior adsorption capacity of CNM was a combined effect due to enhanced surface area and carboxylate functionalization that promoted electrostatic attraction for the cationic dyes (Qiao et al., 2015).

A heterogeneous Fenton catalyst, cellulose-supported Fe_3O_4 nanoparticles, was made by *in situ* synthesis in a NaOH/urea a solvent system without any crosslinking agents. Two different catalysts, Fe_3O_4@MAC and Fe_3O_4@MAFCC, were synthesized by varying the procedure of mechanical activation (MA) and applied for catalytic degradation of methylene blue (MB) dye. Both the catalysts showed very high efficiency compared to Fe_3O_4 alone for the decomposition of MB in an aqueous medium and in the presence of H_2O_2. The MB degradation efficiency of Fe_3O_4, Fe_3O_4@ MAC and Fe_3O_4@MAFCC were 42.7, 90.8 and 98.2%, respectively. The MB removal mechanism involved the adsorption of MB onto the catalyst's surface, followed by catalytic degradation of the adsorbed MB (Lu et al., 2019).

Qaisar et al. fabricated titanium-doped activated carbon–cellulose nanocomposites using industrial metal (titanium scrap) waste and organic waste (dead plant leaves) to produce metal-

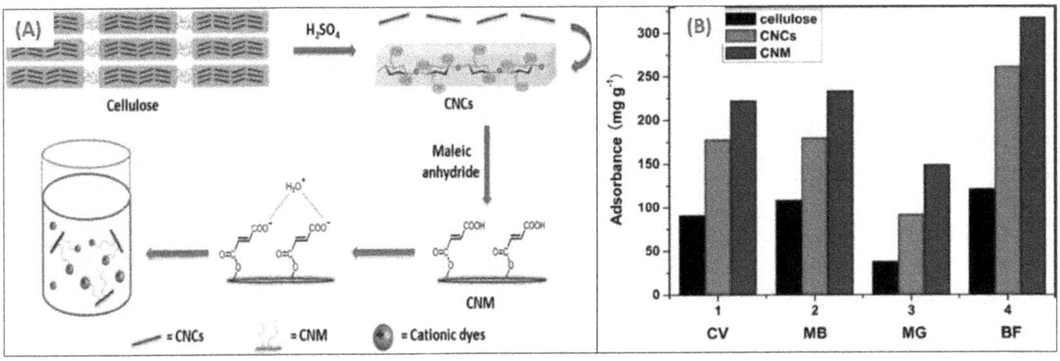

Figure 4. (A) Synthesis of cellulose nanocrystal (CNC) and carboxylate-functionalized cellulose nanocrystal (CNM) from raw cellulose for removal of four cationic dyes, Crystal Violet (CV), Methylene Blue (MB), Malachite Green (MG) and Basic Fuchsin (BF) from water by adsorption; (B) Adsorption of CV, MB, MG and BF dyes from water using cellulose, CNCs and CNM. Reproduced with permission from Qiao et al., (2015).

doped carbon nanocomposites. The authors prepared three composites, Ti-ACNs, 5% Ti-AC-Cel-NC, and 10% TiAC-Cel-NC, for the removal of organic dyes crystal violet and methyl violet in which the composites removed 96, 26.5, 31.2% of crystal violet and 92, 24, 32.4% of methyl violet dye respectively. The nanocellulose provided support as a uniform matrix. Dispersing the Ti-ACNs on the support matrix increased its available surface area by reducing the agglomeration, which increased its efficiency towards removing the organic pollutants (Maqbool et al., 2022).

Table 1. Nanocellulose and cellulose-based biopolymeric nanoparticles used for the removal of various organic pollutants in water/wastewater by adsorption and catalytic treatment technologies and their removal efficiencies.

Sl. No	Nanocellulose/cellulose nanocomposite	Organic pollutants removed and technique used	Removal efficiency	References
1.	Cellulose nanocrystal (CNC) Carboxylated cellulose nanocrystal (CNC-OX)	Methylene blue (MB) Adsorption	CNC: 118 mg/g CNC-OX: 769 mg/g	(Batmaz et al., 2014)
2.	Cellulose nanocrystal (CNC) Carboxylate-functionalized cellulose nanocrystal (CNM)	Crystal violet (CV) Methylene blue (MB) Malachite green (MG) Basic fuchsin (BF) Adsorption	CNC-CV: 185.2 mg/g CNM-CV: 243.9 mg/g	(Qiao et al., 2015)
3.	Cellulose + Titanium with Carbon nanostructures	Crystal Violet (CV) Methyl Violet (MV) Catalysis	31.2% 32.4%	(Maqbool et al., 2022)
4.	Cellulose + chitosan	Congo red (CR) Catalysis	75.8%	(Soon et al., 2019)
5.	Cellulose nanocrystal with Keratin	Reactive Black 5 Direct Red 80 Adsorption	1201 mg/g 1070 mg/g	(Song et al., 2017)
6.	Carboxymethyl cellulose + Metallic Cobalt	2,6-dinitrophenol Methylene blue Catalysis	2,6-DNP = 80% MB = 95%	(Kamal et al., 2019)
7.	Nano Fe(0)@FeS doped cellulose nanofibres	Methylene blue (MB) Congo red (CR) Adsorption	200 mg/g 111.1 mg/g	(Sankararamakrishnan et al., 2020)
8.	Co–Pt/cellulose naocomposite	p-nitrophenol Catalysis	100% reduction in 98 seconds	(Landge et al., 2022)
9.	Fe_3O_4@Cellulose Nanocomposite	Methylene blue (MB) Catalysis	91–98%	(Lu et al., 2019)

A nano adsorbent was designed by Chu et al. by incorporating nanocellulose (NCC) and chitosan (Cts) within a poly(hydroxyalkanoate) matrix via an electrospinning method. This nano adsorbent was investigated for removing Congo red dye and showed a removal efficiency of 75.8%, which is 4-fold higher than the adsorption efficiency of nanocellulose (30.9%) and chitosan (10.5%) individually. The trapping of nanocellulose and chitosan into the PHA matrix resulted in a higher porosity of the composite and, therefore, an increased adsorption of Congo red dye in water (Soon et al., 2019).

Kaili et al. prepared a cellulose nanocrystal incorporated into a keratin matrix to remove the organic pollutants Reactive Black 5 and Direct Red 80. This biosorbent with the cellulose nanofiller exhibited improved properties, such as mechanical strength, specific surface area and porosity, by suppressing the keratin structural shrinkage. The highly porous nature and interconnected structure of the CNC-reinforced keratin adsorbent showed high removal efficiency for the dyes RB5 (96%) and DR80 (98%) from water. The adsorption was favoured at low pH conditions. At a higher pH, the electrostatic interaction between the adsorbent and the dye molecules was disturbed, leading to a decrease in adsorption capacity (Song et al., 2017).

The nanocellulose and cellulose-based biopolymeric nanoparticles used for the removal of various organic pollutants in water/wastewater by adsorption and catalytic treatment technologies and their removal efficiencies are summarised in Table 1.

3.2 Starch-based nanocomposites

Starch-based adsorbents have drawn much attraction in the field of wastewater treatment. Starch-based adsorbents are fabricated via surface modification through functionalization to improve their adsorption efficiency (Abd El-Ghany et al., 2023). A large variety of nanofillers, such as graphene and its derivatives, CNTs, metal/metal oxides, and silicate nanoclays, have been added as fillers in starch to make nanocomposites. The nanofillers enhanced the mechanical and thermal stability of the composites due to their superior dispersion inside the polymer matrix (de Sousa Martinez de Freitas Ana Paula Bernardo da Silva and Lemes, 2022).

Nanoscale Zero-Valent Iron (NZVI)- modified by starch, carboxymethyl cellulose (CMC) and Sodium Dodecyl Benzene Sulphonate (SDBS) were synthesized and investigated for activation of persulphate to degrade the antibiotic sulphamethazine (SMT) in an aqueous solution. The study revealed that the biopolymer: NZVI ratio influenced the degradation rate. The starch-stabilized NZVI exhibited a removal efficiency of 71.6%, and the efficiency decreased slowly with the ageing of the nanoparticles (Dong et al., 2020). Starch was used as a support to disperse nanoscale zero-valent iron (Fe^0) to remove acid blue-25 dye from water. The study also investigated the performance of starch as a support material compared to silica gel for the nanoscale zero-valent iron in the degradation of acid blue-25 dye through adsorption–enhanced reduction, oxidative degradation, corrosion, and precipitation mechanisms. The dye removal rate was very high and followed a pseudo-first-order kinetics, and 100% dye removal was achieved within 6 min of treatment. The dye removal rate was higher for silica-Fe^0 than for starch-Fe^0 and pristine Fe^0 (Elkady et al. 2019). Starch-stabilized Nanoscale Zero-Valent Iron (NZVI) was added to the nitrate removal process. Increasing the starch-stabilized NZVI dose increased the nitrate removal efficiency and the ammonium yield (Zhou et al., 2017). In bio-based denitrification, starch-stabilized nanoscale zero-valent iron was used as an additional electron donor to remove nitrate ions from water. The nanoscale zero-valent iron showed 78.84% removal efficiency for nitrate ions in water (Rajab Beigy et al., 2018) The adsorption rate of naphthalene onto starch-modified NZVI nanoparticles was pH-dependent, with maximum adsorption of 98% achieved at a pH between 4 and 5. The adsorption kinetics followed a combination of pseudo-second-order and intra-particle diffusion models, while the adsorption equilibrium followed the Langmuir isotherm model (Malekzadeh et al., 2018).

A new magnetic nano-adsorbent, Fe_3O_4@Starch-g-poly (vinyl sulphate), was prepared by Pourjavadi et al. (2016) using a graft copolymerization technique to remove cationic dyes from water. The study revealed that the starch-based magnetic nano-adsorbent showed remarkable adsorption capacities for removing the cationic dyes, Methylene blue (621 mg/g) and Malachite green (567 mg/g). The adsorption kinetics followed the pseudo-second-order kinetic model, while the adsorption equilibrium was best fitted in the Langmuir isotherm model (Pourjavadi et al., 2016). Stan et al. (2019) synthesized starch-coated Fe_3O_4 nanoparticles, which were stable in both acidic and alkaline pH and were investigated for the removal of textile dye Optilan Blue. The adsorption capacity for Optilan Blue dye obtained from the Langmuir isotherm model was 86–125 mg/g (Stan et al., 2019).

Magnetic $Fe_3O_4@SiO_2$ starch-graft-poly(acrylic acid) nanocomposites were prepared and used as absorbents to remove Crystal violet dye from aqueous solutions. The magnetic composite showed an adsorption percentage of > 85%, even after five adsorption-desorption cycles (Pourjavadi et al., 2013). The nanoscale zero-valent iron modified by starch (starch-NZVI) was investigated for removing tetracycline in batch experiments as a function of reactant concentration, pH, time, competitive anions and ionic strength. Tetracycline was removed in two stages: the first is the degradation and adsorption process, and the second stage is the flocculation process. The removal efficiency was 99%, of which 30% was through adsorption and degradation, and 69% was due

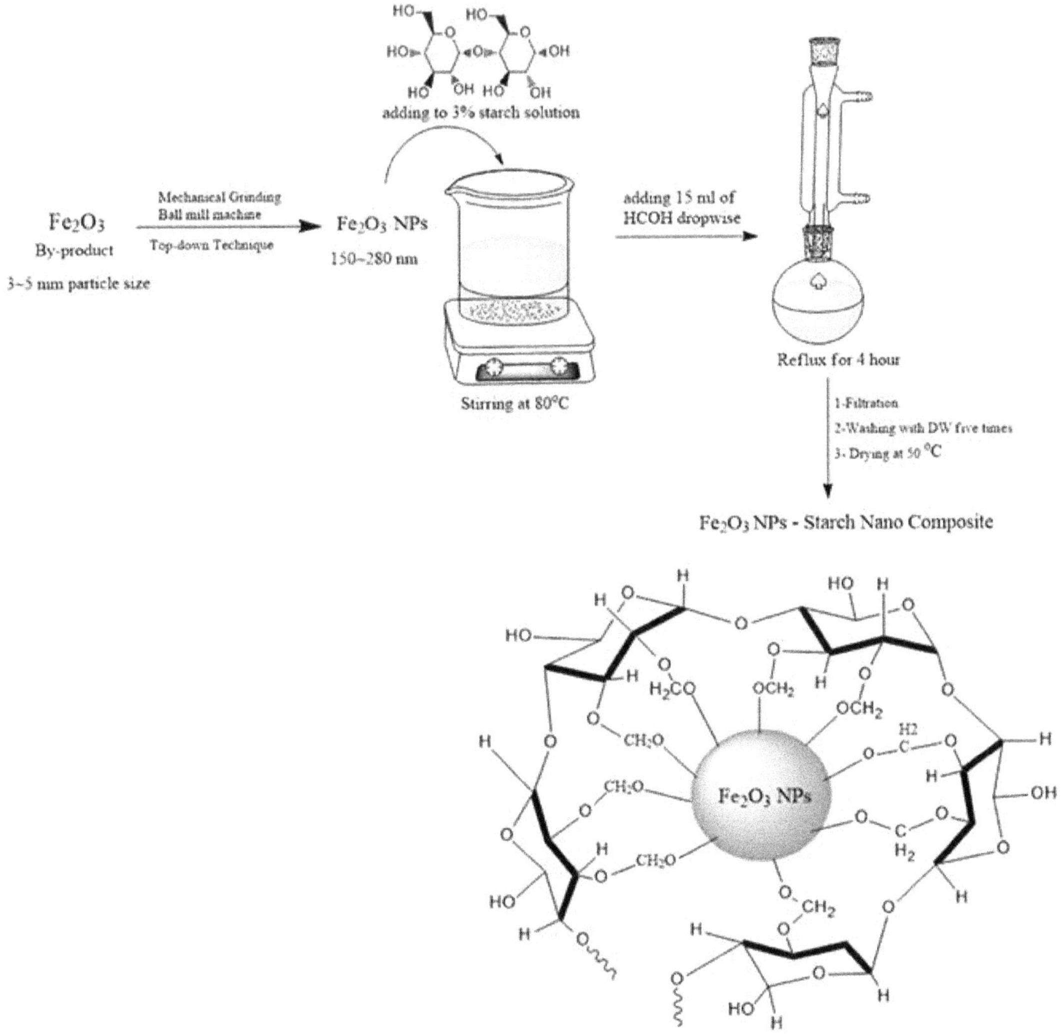

Figure 5. Preparation of Fe_2O_3 NPs-Starch nanocomposite from starch and steel industry waste-derived Fe_2O_3 NPs. Reproduced with permission from Mahmoud et al. (2019).

to flocculation. The removal efficiency was affected by the solution pH, and a higher removal of tetracycline was achieved in a weakly acidic medium (Fu et al., 2015), A nano sorbent was fabricated from nanoscale spherical biochar derived from Pisum sativum pods (N-PSPB) by encapsulation with starch hydrogel (SHGL). The starch-biochar nanosorbent was investigated for removing Naproxen drug (NAP) and chromium (VI) through spontaneous adsorption reaction. The adsorption processes were best fitted with the pseudo-second model with excellent adsorption capacity values of 420.13 and 309.82 mg/g for Cr (VI) and NAP drugs, respectively. The removal of the NAP drug by the starch-biochar nanosorbent was predicted as physical adsorption (pore filling mechanism), π-π interaction and hydrogen-bond formation between carboxylic, hydroxyl groups of drugs and carboxylic, hydroxyl groups of nanosorbent (Mohamed and Mahmoud 2020a). A schematic diagram of the preparation of Fe_2O_3 NPs-Starch nanocomposite from starch and steel industry waste-derived Fe_2O_3 NPs is presented in Fig. 5 (Mahmoud et al., 2019).

Magnetic adsorbent based on starch was prepared by encapsulating starch hydrogel matrix onto Metal-Organic Frameworks (MOFs) and doped with nano magnetite (Fe_3O_4) through microwave irradiation. The adsorbent was applied to remove the Fluvastatin statin drug from water, which showed a maximum adsorption capacity of 782.05 mg/g at equilibrium (Mohamed and Mahmoud, 2020b). Jia et al. prepared a porous adsorbent based on starch (maize)/chitosan/UiO-66-COOH-type carboxylic zirconium MOF with a facile CH-adhesive approach for the removal of sulphanilamide drug from water. The drug adsorption capacity of the porous adsorbent obtained from the Langmuir

Table 2. Starch-based biopolymeric nanocomposites used for the removal of various organic pollutants in water/wastewater by adsorption and catalytic treatment technologies and their removal efficiencies.

Sl. No	Nanocomposite composition	Organic Pollutant Removed & Technique Used	Removal efficiency	References
1.	Starch + Zero-valent iron	Sulphamethazine Catalysis	71.6%	(Dong et al., 2020)
2.	Starch + Zero-valent iron	Acid Blue-25 Catalysis	100%	(Elkady et al., 2019)
3.	Starch + Zero-valent iron	Nitrate Catalysis	78.84%	(Rajab Beigy et al., 2018)
4.	Starch +Zero-valent iron	Naphthalene Catalysis	99%	(Malekzadeh et al., 2018)
5.	Starch + Fe_3O_4	Methylene blue Malachite green Adsorption	621 mg/g 567 mg/g	(Pourjavadi et al. 2016)
6.	Starch + Fe_3O_4	Optilan Blue Adsorption	86-125 mg/g	(Stan et al., 2019)
7.	Starch +Fe_3O_4, SiO_2, poly(acrylic acid)	Crystal violet Catalysis	>85%	(Pourjavadi et al., 2013)
8.	Starch +Zero-valent iron	Tetracyline	99%	(Fu et al., 2015)
9.	Starch + Biochar	Naproxen drug Chromium (VI) Catalysis	90.07% 97.55%	(Mohamed and Mahmoud, 2020a)
10.	Starch + Fe_3O_4	Fluvastatin statin drug Catalysis	97%	(Mohamed and Mahmoud, 2020b)
11.	Starch + chitosan + UiO-66-COOH	Sulphanilamide Adsorption	46.6 mg/g	(Jia et al., 2021)
12.	Calcium modified starch + Fe–Mn binary oxide	As(III) Cd(II) Adsorption	77.1% 68.7%	(Chen et al., 2020)

isotherm model was 46.64 mg/g. The adsorption kinetics followed the pseudo-second-order model. The adsorption occurred via interaction between the Zr-O moiety in carboxylic Zr-MOF, which has a high affinity for the sulphanilamide drug (Jia et al., 2021).

Carbon-based nanomaterials are considered the most attractive nanofillers, and they are widely used to incorporate inside polymer chains to enhance the polymer's adsorption efficiency. A nanocomposite was made using a combination of Graphene Oxide (GO) nanoparticles and hydroxyethyl starch (HES), which was applied to remove Malachite green dye in an aqueous medium. The maximum adsorption capacity obtained from the Langmuir isotherm fitting was 89.3 mg/g for malachite green dye (Onder et al., 2022). Dialdehyde starch nanocrystals were combined with graphene oxide nanosheets, forming hybrid aerogel (SRGO), to prevent the self-stacking of the graphene nanosheets effectively. The SRGO aerogel was an excellent adsorbent for Rhodamine B and Crystal violet dyes and exhibited an adsorption capacity of 539 mg/g and 318 mg/g, respectively (Chen et al., 2019).

The starch-based biopolymeric nanocomposites used for the removal of various organic pollutants in water/wastewater by adsorption and catalytic treatment technologies and their removal efficiencies are consolidated in Table 2.

3.3 Chitosan-based nanocomposites

Chitin and chitosan are the second most abundant polysaccharides with distinctive chemical, physical, and biological properties, and have been studied to remove pollutants, especially water-soluble pollutants, metal ions, and organic pollutants from water. Chitosan is one of the most important derivatives of chitin and is commercially produced by deacetylation of chitin (Suyambulingam et al., 2023). Chitosan is found in natural sources such as shrimp, crab and lobster residues, fungal mycelia and green algae (Weska et al., 2007; Muslim et al., 2018; Ramos Berger et al., 2018). It is a semicrystalline white powder with four crystalline polymorphs; it is insoluble in water and most organic solvents but soluble in aqueous acid solutions such as citric, formic, acetic, and lactic acids and also in solvents like p-toluene sulfonic acid, and dimethyl sulphoxide (Suyambulingam et al., 2023). Incorporating nanoparticles in the chitosan polymeric matrix increases its underlying properties, such as mechanical, chemical, and physical stability, with high surface area. Chitosan nanocomposites are made by chemical or physical interactions of chitosan with materials like carbon, clay, polymer, and metal/metal oxide (Kumar et al., 2019). These limitations of chitosan, such as poor mechanical properties, severe shrinkage, and deformation after drying, could be overcome by making composites with inorganic materials (Cestari et al., 2005). Chitosan-based nanocomposites containing carbon nanotubes, graphene oxides, and inorganic nanoparticles such as metals and metal oxides (ZnO, Ag, Fe_3O_4, TiO_2, etc.) have been developed by introducing these nanoparticles into the chitosan matrix for their application as adsorbent/catalyst nanocomposites for the removal of various contaminants from water/wastewater (Guibal et al., 2007; Mounir et al., 2021; Leudjo Taka et al., 2021; Suyambulingam et al., 2023).

Chitosan-La^{3+}-Graphite nanocomposite (CS-La-GR) has been developed by doping lanthanum into the chitosan matrix, leading to an increased adsorption capacity for organic contaminants and also decreased the electron-hole recombination rates during photocatalytic reaction providing additional photocatalytic stability. The CS-LaGR nanocomposite was tested for the photocatalytic degradation of methylene blue (MB) dye, which exhibited 93.5% degradation of MB. The adsorption equilibrium followed the Langmuir isotherm model, and the adsorption kinetics followed a pseudo-first-order kinetic model (Sirajudheen and Meenakshi, 2019). A complex hybrid biomaterial chitosan/AgCl/ZnO hydrogel beads was synthesized for wastewater treatment application. The chitosan/AgCl/ZnO hydrogel beads were applied for the degradation of MB dye, and the MB degradation efficiency was 99% (Taghizadeh et al., 2020). Chitosan Crosslinked Graphene Oxide (CGO) nanosheets in the form of hydrogels were used for the effective removal of both cationic and anionic dyes. The CGO nanosheets with higher levels of graphene oxide favoured cationic dye adsorption, while those with higher chitosan favoured anionic dye adsorption. The adsorption of MB, Rhodamine B, Methylene orange and Congo red dyes on the CGO nanosheets was 95, 93, 75, and 73%, respectively (Vo et al., 2020). Graphene oxide crosslinked chitosan composite was employed to effectively remove a model dye, methylene blue. The GO-chitosan composite showed superior adsorption capacity for MB (402.6 mg/g) compared to GO alone (286.9 mg/g). Along with increased adsorption capacity, the GO-chitosan composite also displayed improved stability for reuse in multiple adsorption-desorption processes (Sabzevari et al., 2018).

Magnetic chitosan nanocomposite has been made with Poly (acryloxyethyltrimethyl ammonium chloride) (PDAC) and iron oxide (Fe_3O_4-CS/PDAC) and was used for the removal of chromium (VI) and Sunset Yellow (SY) dye. The nanocomposite Fe_3O_4-CS/PDAC exhibited adsorption capacities of 163.93 and 769.23 mg/g for Cr(VI) and SY, respectively (Zhang et al., 2020). A reactive brilliant red dye was removed using chitosan@SnO_2@Fe_3O_4 magnetic adsorbent, which showed a dye adsorption capacity of 981.23 mg/g. The adsorption kinetic data and equilibrium adsorption behaviours of the chitosan@SnO_2@Fe_3O_4 adsorbent for the brilliant red dye were well depicted by the pseudo-second-order kinetic model and Langmuir isotherm model (Yu et al., 2020). Magnetic Fe_3O_4-chitosan nano adsorbents were synthesized by several researchers at different periods and applied to remove different contaminants from water. The Fe_3O_4-chitosan nano adsorbent prepared by Asgari et al. (2020) showed the metronidazole (an antibiotic) adsorption capacity of 97.06 mg/g and the adsorption followed the Freundlich isotherm model while the adsorption kinetics was explained using pseudo-second-order kinetic model (Asgari et al., 2020). On the other hand, the Fe_3O_4-CS magnetic nano adsorbent prepared by Du et al. (2014) for removing Orange I dye from aqueous solutions exhibited a maximum adsorption capacity of 183.2 mg/g (Du et al., 2014). In another study, chitosan-coated magnetic nanoparticles (Fe_3O_4-CS) were synthesized by Kalkan et al. (2012) for the adsorption of a reactive textile dye (Reactive Yellow 145, RY145), which showed the maximum adsorption capacity of 47.62 mg/g. The adsorption of RY145 on Fe_3O_4-CS nanoparticles followed the Langmuir isotherm model, and the thermodynamic parameters indicated that the adsorption was endothermic and spontaneous (Kalkan et al., 2012).

A hybrid nanocomposite (TCS@MOF) was synthesized by immobilizing MIL-88(Fe) photocatalyst and TiO_2 into the chitosan matrix. The TCS@MOF nanocomposite was used to degrade monocrotophos, an insecticide and showed 98.79% degradation efficiency within 30 min of irradiation (Vigneshwaran et al., 2021). Effective adsorption of methyl parathion pesticide was carried out using the two-hybrid nanocomposite polymers made by bentonite clay intercalated with chitosan, and the adsorption capacity was compared with bentonite alone. The maximum adsorption capacities for bentonite (BE), bentonite intercalated chitosan (BE/P.CH), and bentonite/poly 2- hydroxyethyl methacrylate- poly-methyl methacrylate (BE/P.HEMA-MMA) copolymer were 287.3 mg/g, 634.5 mg/g, and 868.5 mg/g, respectively (Abukhadra et al., 2021). This study clearly depicts that composite formation with chitosan enhanced the adsorption capacity of the bentonite (Fig. 6).

The chitosan-based biopolymeric nanoparticles used for the removal of various organic pollutants in water/wastewater by adsorption and catalytic treatment technologies and their removal efficiencies are listed in Table 3.

3.4 Alginate-based Nanocomposites

Alginates are natural polysaccharide biopolymers and are derived from the cell walls of brown algae and some bacteria (Roberts and Martens, 2016). Organic and/or inorganic (nano)composite materials containing various salts of alginic acid (sodium, potassium, calcium, etc.) and other nanomaterials are useful for the adsorptive removal of heavy metal ions from water (Mandal et al., 2012; Qamar et al., 2022). Nanomaterials, such as silica, metal oxides, GO, CNT, hybrid inorganic-organic, non-magnetic-magnetic, polymeric nanocarriers, and several other materials have been used in combination with alginate biopolymers have been explored for environmental remediation (Qamar

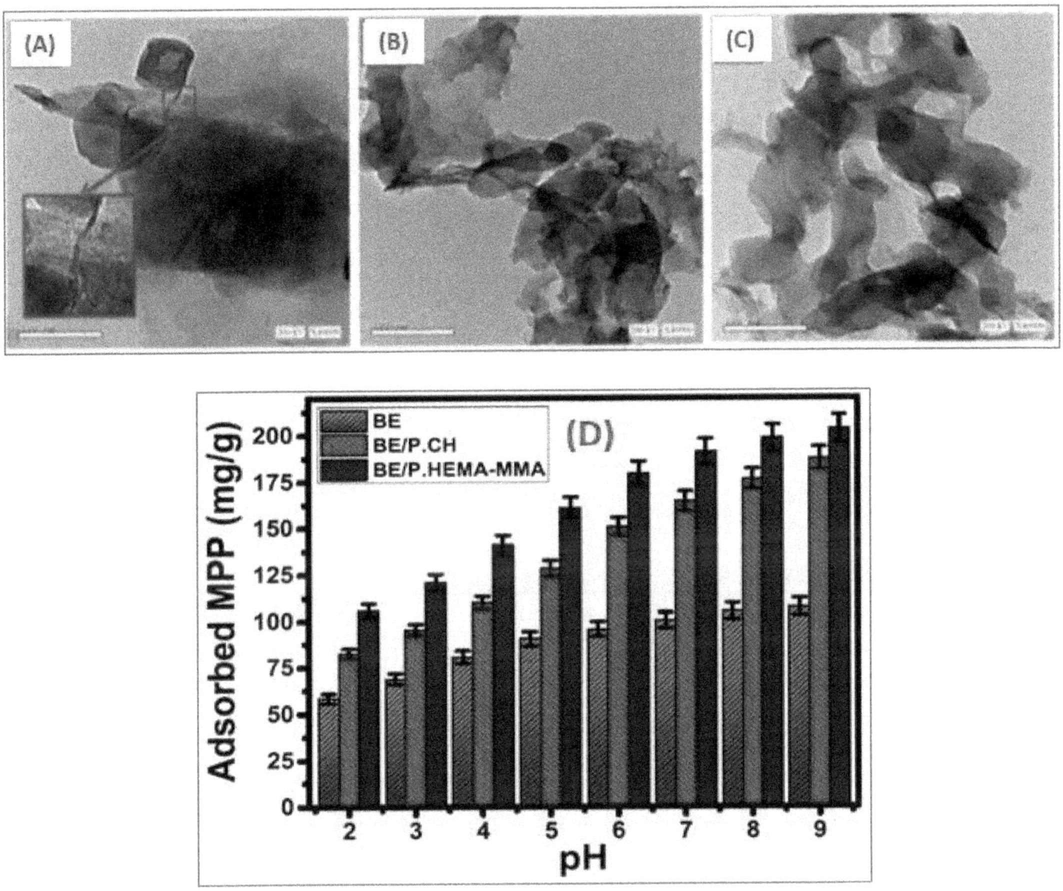

Figure 6. TEM images of (A) bentonite (BE), (B) bentonite intercalated with chitosan polymer (BE/P.CH), (C) bentonite/ poly 2- hydroxyethyl methacrylate- poly-methyl methacrylate (BE/P.HEMA-MMA), copolymer, and (D) their adsorption capacities for methyl parathion pesticide in aqueous medium and at various pH. Reproduced with permission from Abukhadra et al. (2021).

et al., 2022). A schematic diagram of various alginate-based nanocomposites and their applications in water pollution remediation is presented in Fig. 7 (Qamar et al., 2022).

Alginate-layered double hydroxide composites were synthesized from sodium alginate and LDH of two different compositions, Mg/Al and Zn/Al LDH. Both the composites, MALDH-Alg and ZALDH-Alg, exhibited enhanced adsorption capacities for orange II dye and fluoride ions in an aqueous medium compared to the respective LDH or alginate when used individually (Mandal et al., 2012). Though the specific surface area of the LDH did not improve by composite formation with alginate, the increased adsorption capacity of the LDH-Alg composites was attributed to the increased inter-layer space of the LDH, specifically in the case of ZALDH-Alg composite, which exhibited a reasonable increase in adsorption capacity for both orange II dye and fluoride ions compared to pristine ZnAl LDH (Mandal et al., 2012). Ma et al. have fabricated a 3D alginate hydrogel carbon nanotubes/L-cysteine@graphene oxide/sodium alginate (CNTs/L-cys@GO/SA) for the effective removal of antibiotic ciprofloxacin. This triple-network composite hydrogel has independent three-dimensional networks that enhance the porosity, which leads to increased active adsorption sites. This 3D alginate hydrogel showed high efficiency in removing ciprofloxacin with an adsorption capacity of 200 mg/g in an aqueous medium (Ma et al., 2020).

Copper ferrite/calcium alginate nanocomposite (CG) was synthesized to remove malachite green from wastewater using two water treatment techniques: adsorption and photocatalytic degradation. The nanocomposite CG exhibited increased surface area and total pore volume compared to the

Table 3. Chitosan-based biopolymeric nanoparticles used for the removal of various organic pollutants in water/wastewater by adsorption and by catalytic treatment technologies and their removal efficiencies.

Sl. No	Nanocomposite	Organic Pollutant Removed & Technique Used	Removal efficiency	References
1.	Chitosan + PdNPs@ chitosan-MWCNT	Congo red, Methyl orange, Methylene blue, Methyl red Catalysis	99%	(Sargin et al., 2020)
2.	Chitosan + Lanthanum + graphite	Methylene blue Photocatalysis	93.5%	(Sirajudheen and Meenakshi, 2019)
3.	Chitosan + ZnO + AgCl	Methylene Blue Photocatalysis	99%	(Taghizadeh et al., 2020)
4.	Chitosan + Fe_3O_4	Sunset yellow Adsorption	769.23 mg/g	(Zhang et al., 2020)
5.	Chitosan + Fe_3O_4	Reactive brilliant red Adsorption	981.23 mg/g	(Yu et al., 2020)
6.	Chitosan + Fe_3O_4	Metronidazole Adsorption	99.05 %	(Asgari et al., 2020)
7.	Chitosan + MIL-88(Fe) anchored TiO_2	Monocrotophos Catalysis	98.79%	(Vigneshwaran et al., 2021)
8.	Chitosan + Bentonite	Parathion Adsorption	634.5 mg/g	(Abukhadra et al., 2021)
9.	Chitosan + Graphene oxide	Methylene blue Adsorption	402.6 mg/g	(Sabzevari et al., 2018)
10.	Chitosan + Fe_3O_4	Orange I Adsorption	183.2 mg/g	(Du et al., 2014)
11.	Chitosan + Fe_3O_4	Reactive yellow 145 Adsorption	47.62 mg/g	(Kalkan et al., 2012)
12.	Chitosan + Ag NP	2-nitrophenol Acridine orange Catalysis	90%	(Khan et al., 2020)
13.	Chitosan + GO	Methylene blue Rhodamine B Methyl orange Congo red Adsorption	94.79% 92.05% 74.45% 70.72%	(Vo et al., 2020)

pristine Copper Ferrite (CF). The maximum adsorption capacity of malachite green dye at 40 °C by the CG was 297.62 mg/g compared to 84.24 mg/g by the CF. The photocatalytic degradation efficiency for the dye was 90% and 83.2% for the CG and CF, respectively, clearly showing the superiority of the alginate-nanoparticle composite over the nanoparticle alone (Hassan et al., 2024).

3.5 Other Biopolymer Nanocomposites

Humic Acid (HA) coated magnetic Fe_3O_4 was synthesized and applied as a heterogeneous photo-Fenton catalyst for the degradation of Carbamazepine, Ibuprofen, Bisphenol A and 5-Tolylbenzotriazole in real wastewater. Magnetic Particles (MPs) coated with different amounts of HA were prepared by co-precipitation under anoxic and oxygenated conditions. The study revealed that HA coating increased the catalyst efficiency of Fe_3O_4, both in the dark and under irradiation. The highest efficiency was observed at a $pH < 4$ under irradiation (Gonçalves et al., 2020). In another study, HA-functionalized Fe_3O_4 nanosorbents were synthesized through an eco-friendly route and applied for adsorptive removal of carcinogenic malachite green dye from different water sources. The sorption process was significantly influenced by the water pH, and the highest adsorption of 99.2% occurring at a pH of 6 (Gautam and Tiwari, 2020).

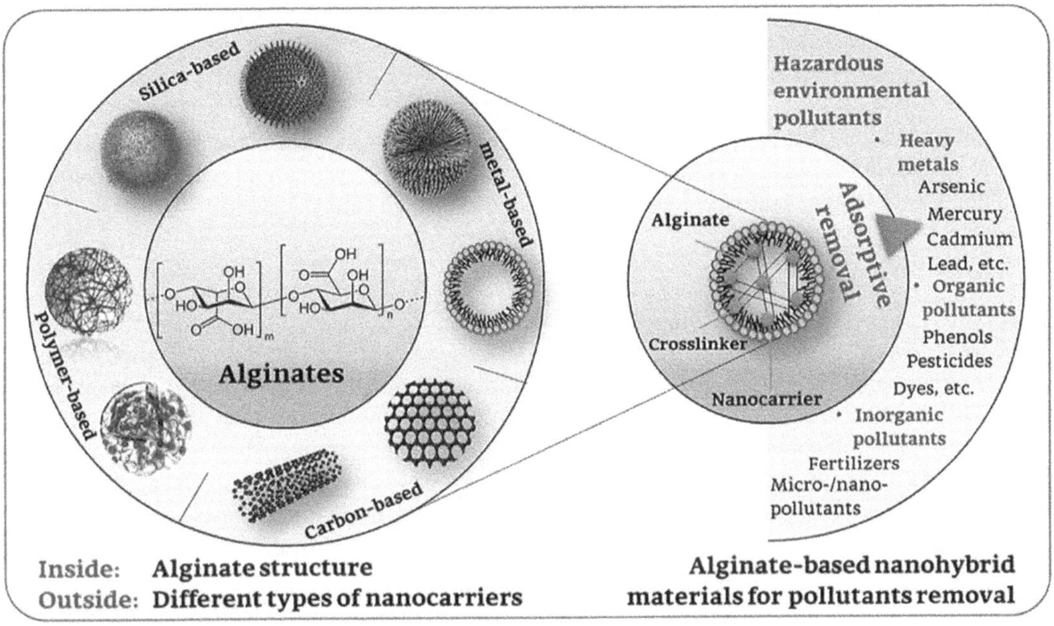

Figure 7. Schematic presentation of various alginate-based nanocomposites and their applications in water pollution remediation. Reproduced with permission from Qamar et al. (2022).

Magnetic lignosulphonate adsorbent was synthesized by coating lignosulphonate on Fe_3O_4 and applied to remove organic dyes from water. The maximum adsorption capacities of the magnetic lignosulphonate were 198.24 mg/g and 192.51 mg/g for Congo Red and Titan Yellow, respectively. The highest adsorption capacity was obtained at pH 7. The desorption performance of the material indicated its good recyclability (Hu et al., 2020).

Lignin and lignin amine-coated Fe_3O_4 nanoparticles were synthesized and studied for the adsorptive removal of organic dyes, Methylene Blue (MB) and Acid Scarlet GR (AS-GR), from water. These nanoparticles [lignin magnetic nanoparticles (LMNPs) and lignin amine magnetic nanoparticles (LAMNPs)] exhibited pH-dependent adsorption behaviour. The study showed that combining lignin with nanoparticles increased the adsorption capacities 2–5 times higher than other

Table 4. Humic acid and lignocellulosic biopolymer nanocomposites used for the removal of various organic pollutants in water/wastewater by catalytic treatment technologies and their removal efficiencies.

Sl. No.	Nanocomposite	Organic Pollutant Removed & Technique Used	Removal efficiency	References
1.	Humic acid + Fe_3O_4	Carbamazepine (CBZ) Ibuprofen (IBP) 5-tolylbenzotriazole (5-TBA) Bisphenol A (BPA) Catalysis	> 80% > 80% > 80% IBP = 35%	(Gonçalves et al., 2020)
2.	Humic acid + Fe_3O_4	Malachite green Adsorption	85%.	(Gautam and Tiwari, 2020)
3.	Lignosulphonate + Fe_3O_4	Congo red (CR) Titan yellow (TY) Adsorption	83% 80%	(Hu et al., 2020)
4.	Lignin + Fe_3O_4	Methylene blue (MB) by LMNPs Acid scarlet GR (AS-GR) by LAMNPs Adsorption	211.42 mg/g 176.49 mg/g	(Li et al., 2018)

lignin-based adsorbents. The MB adsorption capacity of LMNP was 211.42 mg/g, and the AS-GR adsorption capacity of LAMNP was 176.49 mg/g. About 90% regeneration of the adsorbent was possible by adjusting the medium pH at 12 and 2 for the dyes AS-GR and MB, respectively (Li et al., 2018).

Humic acid and lignocellulosic biopolymer nanocomposites used for the removal of various organic pollutants in water/wastewater by catalytic treatment technologies and their removal efficiencies are listed in Table 4.

4. Summary

Biopolymeric nanoparticles and biopolymeric nanocomposites made from biopolymers and their application toward environmental remediation through adsorption, catalytic decomposition and advanced oxidation processes are discussed. Among all natural biopolymers, polysaccharide derivatives, such as cellulose, starch, chitosan, and alginate, have received the most attention, specifically in photocatalytic decontamination systems. The ability of these polysaccharide derivatives to associate with various nanocatalysts physically and/or chemically made them the material of choice as support for the nanocatalysts. Moreover, these biopolymers, cellulose, starch, chitosan and alginate, have good adsorption properties. This enormous adsorption affinity of these biopolymers attracts the contaminants to their surface, and bringing the pollutants near the catalyst surface facilitates the catalytic reaction. Moreover, dispersion of the adsorbent/catalysts on the biopolymer surface increases the available surface area of the active material through reduced agglomeration leading to increased adsorption/catalytic efficiency. Furthermore, supported nanocatalysts and nanocomposites exhibited improved recycling ability than the unsupported/individual nanocatalysts. Subsequently, for real-life application of the developed biopolymeric nanocomposite adsorbent and/or catalysts, the research focus should further emphasize the cost-effectiveness of the materials and the process and regeneration and reuse of the adsorbent/catalyst.

In biopolymeric nanocomposites, though the support material is biodegradable, environmentally safe disposal of non-biodegradable nanocatalysts should be ensured for environmental sustainability.

Acknowledgement

RN acknowledged the financial assistance received from the NWP0100 project. CSIR-CLRI communication number 2065.

References

Abd El-Ghany, N. A., Elella, M. H. A., Abdallah, H. M., Mostafa, M. S. and Samy, M. (2023). Recent Advances in Various Starch Formulation for Wastewater Purification via Adsorption Technique: A Review. J. Polym. Environ. 31: 2792–2825. https://doi.org/10.1007/s10924-023-02798-x.

Abukhadra, M. R., El-Sherbeeny, A. M., El-Meligy, M. A. and Luqman, M. (2021). Insight into carbohydrate polymers (chitosan and 2- hydroxyethyl methacrylate/methyl methacrylate) intercalated bentonite-based nanocomposites as multifunctional and environmental adsorbents for methyl parathion pesticide. Int. J. Biol. Macromol., 167: 335–344. https://doi.org/10.1016/J.IJBIOMAC.2020.11.209.

Al-Hazmi, H. E., Mohammadi, A., Hejna, A., Majtacz, J., Esmaeili, A., Habibzadeh, S. et al. (2023) Wastewater reuse in agriculture: Prospects and challenges. Environ. Res., 236: 116711. https://doi.org/10.1016/J.ENVRES.2023.116711.

Alvi, M., Batstone, D., Mbamba, C. K., Keymer, P., French, T., Ward, A. et al. (2023). Deep learning in wastewater treatment: a critical review. Water Res., 245: 120518. https://doi.org/10.1016/J.WATRES.2023.120518.

Annu, Mittal, M., Tripathi, S. and Shin, D. K. (2024). Biopolymeric Nanocomposites for Wastewater Remediation: An Overview on Recent Progress and Challenges. Polymers (Basel). 16. https://doi.org/10.3390/polym16020294.

Asgari, E., Sheikhmohammadi, A. and Yeganeh, J (2020). Application of the Fe3O4-chitosan nano-adsorbent for the adsorption of metronidazole from wastewater: Optimization, kinetic, thermodynamic and equilibrium studies. Int. J. Biol. Macromol., 164: 694–706. https://doi.org/10.1016/J.IJBIOMAC.2020.07.188.

Awewomom, J., Dzeble, F., Takyi, Y. D., Ashie, W. B., Ettey, E. N. Y. O., Afua, P. E. et al. (2024). Addressing global environmental pollution using environmental control techniques: a focus on environmental policy and preventive environmental management. Discover Environment, 2: 8. https://doi.org/10.1007/s44274-024-00033-5.

Babu Ponnusami, A., Sinha, S., Ashokan, H., V Paul, M., Hariharan, S. P., Arun, J., Gopinath, K. P. et al. (2023). Advanced oxidation process (AOP) combined biological process for wastewater treatment: A review on advancements, feasibility and practicability of combined techniques. Environ. Res. 237: 116944. https://doi.org/10.1016/J.ENVRES.2023.116944.

Balsalobre-Lorente, D., Nur, T., Topaloglu, E. E. and Evcimen, C. (2024) Assessing the impact of the economic complexity on the ecological footprint in G7 countries: Fresh evidence under human development and energy innovation processes. Gondwana Research, 127: 226–245. https://doi.org/10.1016/J.GR.2023.03.017.

Basavegowda, N. and Baek, K. -H. (2021). Advances in functional biopolymer-based nanocomposites for active food packaging applications. Polymers (Basel), 13. https://doi.org/10.3390/polym13234198.

Batmaz, R., Mohammed, N., Zaman, M., Minhas, G., Berry, R. M. and Tam, K. C. (2014). Cellulose nanocrystals as promising adsorbents for the removal of cationic dyes. Cellulose, 21: 1655–1665. https://doi.org/10.1007/s10570-014-0168-8.

Boikanyo, D., Mishra, A. K., Mishra, S. B. and Mhlanga, S. D. (2018). Biopolymers: A Natural Support for Photocatalysts Applied to Pollution Remediation. In: Nanotechnology in Environmental Science. pp. 649–684.

Carpenter, A. W., de Lannoy, C.-F. and Wiesner, M. R. (2015a). Cellulose nanomaterials in water treatment technologies. Environ. Sci. Technol., 49: 5277–5287. https://doi.org/10.1021/es506351r.

Cestari, A. R., Vieira, E. F. S., Pinto, A. A. and Lopes, E. C. N. (2005). Multistep adsorption of anionic dyes on silica/chitosan hybrid: 1. Comparative kinetic data from liquid- and solid-phase models. J. Colloid Interface Sci., 292: 363–372. https://doi.org/10.1016/J.JCIS.2005.05.096.

Chen, H., Xu, F., Chen, Z., Jiang, O., Gustave, W. and Tang, X. (2020). Arsenic and cadmium removal from water by a calcium-modified and starch-stabilized ferromanganese binary oxide. Journal of Environmental Sciences, 96:186–193. https://doi.org/10.1016/J.JES.2020.03.060.

Chen, Y., Dai, G. and Gao, Q. (2019). Starch nanoparticles–graphene aerogels with high supercapacitor performance and efficient adsorption. ACS Sustain Chem. Eng., 7: 14064–14073. https://doi.org/10.1021/acssuschemeng.9b02594.

Das, R., Lindström, T., Sharma, P. R., Chi, K. and Hsiao, B. S. (2022). Nanocellulose for sustainable water purification. Chem. Rev., 122:8936–9031. https://doi.org/10.1021/acs.chemrev.1c00683.

Deng, Q. S., Cuesta, L., Alvarado, R., Murshed, M., Tillaguango, B., Işık, C. and Rehman, A. (2024). Nexus between government stability and environmental pollution. J. Clean Prod., 434: 140061. https://doi.org/10.1016/J.JCLEPRO.2023.140061.

de Sousa Martinez de Freitas Ana Paula Bernardo da Silva LSMIANNKCVS de FNB dos S., Lemes, A. P. (2022). Thermoplastic starch nanocomposites: sources, production and applications – A review. J Biomater Sci Polym Ed. 33: 900–945. https://doi.org/10.1080/09205063.2021.2021351.

Dong, H., Ning, Q., Li, L., Wang, Y., Wang, B., Zhang, L. et al. (2020). A comparative study on the activation of persulfate by bare and surface-stabilized nanoscale zero-valent iron for the removal of sulfamethazine. Sep. Purif. Technol., 230: 115869. https://doi.org/10.1016/J.SEPPUR.2019.115869.

Du, Y., Pei, M., He, Y., Yu, F., Guo, W. and Wang, L. (2014). Preparation, Characterization and Application of Magnetic Fe3O4-CS for the Adsorption of Orange I from Aqueous Solutions. PLoS One, 9:e108647. https://doi.org/10.1371/journal.pone.0108647.

Elkady, M., Shokry, H., El-Sharkawy, A., El-Subruiti, G. and Hamad, H. (2019). New insights into the activity of green supported nanoscale zero-valent iron composites for enhanced acid blue-25 dye synergistic decolorization from aqueous medium. J. Mol. Liq., 294: 111628. https://doi.org/10.1016/J.MOLLIQ.2019.111628.

Etale, A., Onyianta, A. J., Turner, S. R. and Eichhorn, S. J. (2023). Cellulose: A review of water interactions, applications in composites, and water treatment. Chem. Rev., 123: 2016–2048. https://doi.org/10.1021/acs.chemrev.2c00477.

Faisal, Ibney Hai, K. Y. and Fukushi, K. (2007). Hybrid treatment systems for dye wastewater. Crit. Rev. Environ. Sci. Technol., 37: 315–377. https://doi.org/10.1080/10643380601174723.

Fu, Y., Peng, L., Zeng, Q., Yang, Y., Song, H., Shao, J. et al. (2015). High efficient removal of tetracycline from solution by degradation and flocculation with nanoscale zerovalent iron. Chemical Engineering Journal, 270: 631–640. https://doi.org/10.1016/J.CEJ.2015.02.070.

Gamage, A., Punniamoorthy, T. and Madhujith, T. (2021). Starch-based hybrid nanomaterials for environmental remediation. *In*: Emeje, M. O. (Ed.). Starch. IntechOpen, Rijeka.

Gautam, R. K. and Tiwari, I. (2020). Humic acid functionalized magnetic nanomaterials for remediation of dye wastewater under ultrasonication: Application in real water samples, recycling and reuse of nanosorbents. Chemosphere, 245: 125553. https://doi.org/10.1016/J.CHEMOSPHERE.2019.125553.

Gonçalves, N. P. F., Minella, M., Fabbri, D., Calza, P., Malitesta, C., Mazzotta, E. et al. (2020). Humic acid coated magnetic particles as highly efficient heterogeneous photo-Fenton materials for wastewater treatments. Chemical Engineering Journal, 390: 124619. https://doi.org/10.1016/J.CEJ.2020.124619.

Guibal, E., Vincent, T. and Blondet, F. (2007). Biopolymers as Supports for Heterogeneous Catalysis: Focus on Chitosan, a Promising Aminopolysaccharide. *In*: Ion Exchange and Solvent Extraction. pp. 151–292.

Gupta, V. K., Ali, I., Saleh, T. A., Nayak, A. and Agarwal, S. (2012). Chemical treatment technologies for waste-water recycling—an overview. RSC Adv., 2: 6380–6388. https://doi.org/10.1039/C2RA20340E.

Hassan, A. F., El-Naggar, G. A., Braish, A. G., El-Latif, M. M. A., Shaltout, W. A. and Elsayed, M. S. (2024). Utilization of synthesized copper ferrite/calcium alginate nanocomposite for adsorption and photocatalytic degradation of malachite green. J. Inorg. Organomet. Polym. Mater, 34: 190–206. https://doi.org/10.1007/s10904-023-02806-6.

Huang, J., Yi, S., Zheng, C. and Lo, I. M. C. (2019). Persulfate activation by natural zeolite supported nanoscale zero-valent iron for trichloroethylene degradation in groundwater. Science of The Total Environment, 684: 351–359. https://doi.org/10.1016/J.SCITOTENV.2019.05.331.

Hu, L., Guang, C., Liu, Y., Su, Z., Gong, S., Yao, Y. and Wang, Y. (2020). Adsorption behavior of dyes from an aqueous solution onto composite magnetic lignin adsorbent. Chemosphere, 246: 125757. https://doi.org/10.1016/J.CHEMOSPHERE.2019.125757.

Jaleh, B., Nasrollahzadeh, M., Nasri, A., Eslamipanah, M., Moradi, A. and Nezafat, Z. (2021). Biopolymer-derived (nano)catalysts for hydrogen evolution via hydrolysis of hydrides and electrochemical and photocatalytic techniques: A review. Int. J. Biol. Macromol., 182: 1056–1090. https://doi.org/10.1016/J.IJBIOMAC.2021.04.087.

Jia, X., Zhang, B., Chen, C., Fu, X. and Huang, Q. (2021). Immobilization of chitosan grafted carboxylic Zr-MOF to porous starch for sulfanilamide adsorption. Carbohydr. Polym., 253: 117305. https://doi.org/10.1016/J.CARBPOL.2020.117305.

Kalkan, N. A., Aksoy, S., Aksoy, E. A. and Hasirci, N. (2012). Adsorption of reactive yellow 145 onto chitosan coated magnetite nanoparticles. J. Appl. Polym. Sci., 124: 576–584. https://doi.org/https://doi.org/10.1002/app.34986.

Kamal, T., Ahmad, I., Khan, S. B. and Asiri, A. M. (2019). Bacterial cellulose as support for biopolymer stabilized catalytic cobalt nanoparticles. Int. J. Biol. Macromol., 135: 1162–1170. https://doi.org/10.1016/J.IJBIOMAC.2019.05.057.

Keshipour, S., Hadidi, M. and Gholipour, O. (2023). A review on hydrogen generation by photo-, electro-, and photoelectro-catalysts based on chitosan, chitin, cellulose, and carbon materials obtained from these biopolymers. Advances in Polymer Technology, 2023:8835940. https://doi.org/https://doi.org/10.1155/2023/8835940.

Khan, M. S. J., Khan, S. B., Kamal, T. and Asiri, A. M. (2020). Catalytic application of silver nanoparticles in chitosan hydrogel prepared by a facile method. J. Polym. Environ., 28: 962–972. https://doi.org/10.1007/s10924-020-01657-3.

Kumar, S., Ye, F., Dobretsov, S. and Dutta, J. (2019). Chitosan nanocomposite coatings for food, paints, and water treatment applications. Applied Sciences, 9: 2409. https://doi.org/10.3390/app9122409.

Kyzas, G. Z., Kostoglou, M., Lazaridis, N. K. and Bikiaris, D. N. (2013). Decolorization of dyeing wastewater using polymeric absorbents - An overview. *In*: Günay, M. (Ed.). Eco-Friendly Textile Dyeing and Finishing. IntechOpen, Rijeka.

Landge, V. K., Hakke, V. S., Kakunuri, M., Babu, G. U. B., Boczkaj, G. and Sonawane, S. H. (2022). Synthesis of bimetallic Co–Pt/cellulose nanocomposites for catalytic reduction of p-nitrophenol. React. Chem. Eng. 7: 641–652. https://doi.org/10.1039/D1RE00422K.

Lee, M., Chen, B.-Y. and Den, W. (2015). Chitosan as a natural polymer for heterogeneous catalysts support: a short review on its applications. Applied Sciences, 5: 1272–1283. https://doi.org/10.3390/app5041272.

Leudjo Taka, A., Klink, M. J., Yangkou Mbianda, X. and Naidoo, E. B. (2021). Chitosan nanocomposites for water treatment by fixed-bed continuous flow column adsorption: A review. Carbohydr. Polym., 255: 117398. https://doi.org/10.1016/J.CARBPOL.2020.117398.

Li, X., He, Y., Sui, H. and He, L. (2018). One-Step Fabrication of Dual Responsive Lignin Coated Fe3O4 Nanoparticles for Efficient Removal of Cationic and Anionic Dyes. Nanomaterials 8. https://doi.org/10.3390/nano8030162.

Li, Y. H., Li, J. Y. and Xu, Y. J. (2021). Bimetallic nanoparticles as cocatalysts for versatile photoredox catalysis. Energy Chem., 3:100047. https://doi.org/10.1016/J.ENCHEM.2020.100047.

Luo, Y., Wang, Q. and Zhang, Y. (2020). Biopolymer-based nanotechnology approaches to deliver bioactive compounds for food applications: A perspective on the past, present, and future. J. Agric. Food Chem., 68: 12993–13000. https://doi.org/10.1021/acs.jafc.0c00277.

Lu, Q., Zhang, Y., Hu, H., Wang, W., Huang, Z., Chen, D. et al. (2019). *In situ* synthesis of a stable Fe3O4@cellulose nanocomposite for efficient catalytic degradation of methylene blue. Nanomaterials, 9. https://doi.org/10.3390/nano9020275.

Mahmoud, M. E., Nabil, G. M., Zaki, M. M. and Saleh, M. M. (2019). Starch functionalization of iron oxide by-product from steel industry as a sustainable low cost nanocomposite for removal of divalent toxic metal ions from water. Int. J. Biol. Macromol., 137: 455–468. https://doi.org/10.1016/J.IJBIOMAC.2019.06.170.

Ma, J., Jiang, Z., Cao, J. and Yu, F. (2020). Enhanced adsorption for the removal of antibiotics by carbon nanotubes/graphene oxide/sodium alginate triple-network nanocomposite hydrogels in aqueous solutions. Chemosphere, 242: 125188. https://doi.org/10.1016/J.CHEMOSPHERE.2019.125188.

Malekzadeh, M., Nejaei, A., Baneshi, M. M., Kokhdan, E. P. and Bardania, H. (2018). The use of starch-modified magnetic Fe0 nanoparticles for naphthalene adsorption from water samples: Adsorption isotherm, kinetic and thermodynamic studies. Appl. Organomet. Chem., 32:e4434. https://doi.org/https://doi.org/10.1002/aoc.4434.

Mandal, S. and Kulkarni, B. D. (2011) Separation strategies for processing of dilute liquid streams. International Journal of Chemical Engineering.

Mandal, S. and Mayadevi, S. (2008). Cellulose supported layered double hydroxides for the adsorption of fluoride from aqueous solution. Chemosphere, 72: 995–998. https://doi.org/10.1016/J.CHEMOSPHERE.2008.03.053.

Mandal, S., Patil, V. S. and Mayadevi, S. (2012). Alginate and hydrotalcite-like anionic clay composite systems: Synthesis, characterization and application studies. Microporous and Mesoporous Materials, 158: 241–246. https://doi.org/10.1016/J.MICROMESO.2012.03.046.

Mandal, S., Stephen, D. and Janardhanan, S. K. (2024). Activated carbon with composite pore structures made from peanut shell and areca nut fibers as sustainable adsorbent material for the efficient removal of active pharmaceuticals from aqueous media. RSC Sustain, 2: 3022–3035. https://doi.org/10.1039/D4SU00262H.

Maqbool, Q., Barucca, G., Sabbatini, S., Parlapiano, M., Ruello, M. L. and Tittarelli, F. (2022). Transformation of industrial and organic waste into titanium doped activated carbon – cellulose nanocomposite for rapid removal of organic pollutants. J. Hazard Mater, 423: 126958. https://doi.org/10.1016/J.JHAZMAT.2021.126958.

Marwani, H. M., Ahmad, S. and Rahman, M. M. (2022). Catalytic reduction of environmental pollutants with biopolymer hydrogel cross-linked gelatin conjugated tin-doped gadolinium oxide nanocomposites. Gels, 8. https://doi.org/10.3390/gels8020086.

Mitchell, M. J., Billingsley, M. M., Haley, R. M., Wechsler, M. E., Peppas, N. A. and Langer, R. (2021). Engineering precision nanoparticles for drug delivery. Nat. Rev. Drug Discov., 20: 101–124. https://doi.org/10.1038/s41573-020-0090-8.

Mohamed, A. K. and Mahmoud, M. E. (2020a). Nanoscale *Pisum sativum* pods biochar encapsulated starch hydrogel: A novel nanosorbent for efficient chromium (VI) ions and naproxen drug removal. Bioresour. Technol., 308: 123263. https://doi.org/10.1016/J.BIORTECH.2020.123263.

Mohamed, A. K. and Mahmoud, M. E. (2020b). Encapsulation of starch hydrogel and doping nanomagnetite onto metal-organic frameworks for efficient removal of fluvastatin antibiotic from water. Carbohydr. Polym., 245: 116438. https://doi.org/10.1016/J.CARBPOL.2020.116438.

Mohammed, N., Grishkewich, N. and Tam, K. C. (2018). Cellulose nanomaterials: promising sustainable nanomaterials for application in water/wastewater treatment processes. Environ. Sci. Nano., 5: 623–658. https://doi.org/10.1039/C7EN01029J.

Molnar Jazić, J., Gross, A., Glaser, B., Agbaba, J., Simetić, T., Nikić, J. et al. (2024). Boosting advanced oxidation processes by biochar-based catalysts to mitigate pesticides and their metabolites in water treatment: A meta-analysis. J. Environ. Chem. Eng., 12: 114260. https://doi.org/10.1016/J.JECE.2024.114260.

Moodley Kandasamy, G. and Arumugam, V and B. A. (2022). Nanocellulose-based materials for wastewater treatment. pp. 809–841. *In*: Barhoum, A. (Ed.). Handbook of Nanocelluloses: Classification, Properties, Fabrication, and Emerging Applications. Springer International Publishing, Cham.

Mounir, C., Ahlafi, H., Aazza, M., Moussout, H. and Mounir, S. (2021). Kinetics and Langmuir–Hinshelwood mechanism for the catalytic reduction of para-nitrophenol over Cu catalysts supported on chitin and chitosan biopolymers. Reaction Kinetics, Mechanisms and Catalysis, 134: 285–302. https://doi.org/10.1007/s11144-021-02066-w.

Musa, Y. and Bwatanglang, I. B. (2020). Current role and future developments of biopolymers in green and sustainable chemistry and catalysis. Sustainable Nanocellulose and Nanohydrogels from Natural Sources, 131–154. https://doi.org/10.1016/B978-0-12-816789-2.00006-7.

Muslim, S. N., Kadmy, I. M. S. A., Ali, A. N. M., Salman, B. K., Ahmad, M., Khazaal, S. S. et al. (2018). Chitosan extracted from *Aspergillus flavus* shows synergistic effect, eases quorum sensing mediated virulence factors and biofilm against nosocomial pathogen Pseudomonas aeruginosa. Int. J. Biol. Macromol., 107: 52–58. https://doi.org/10.1016/J.IJBIOMAC.2017.08.146.

Nasrollahzadeh, M., Bidgoli, N. S. S., Nezafat, Z. and Shafiei, N. (2021). Catalytic applications of biopolymer-based metal nanoparticles. Biopolymer-Based Metal Nanoparticle Chemistry for Sustainable Applications: Volume 2: Applications, 423–516. https://doi.org/10.1016/B978-0-323-89970-3.00007-X.

Onder, A., Kıvanç, M. R., Durmuş, S., Ilgin, P., Ozay, H. and Ozay, O. (2022). Adsorption of malachite green from aqueous solution using hydroxyethyl starch hydrogel improved by graphene oxide. J. Polym. Environ., 30: 2928–2942. https://doi.org/10.1007/s10924-022-02410-8.

Owhonka, A., Fubara, E. F. and Justice, O. B. (2021). Wastewater quality- it's impact on the environment and human physiology: A review. International Journal of Advance Research and Innovation, 9: 43–58. https://doi.org/10.51976/ijari.942107.

Pourjavadi, A., Abedin-Moghanaki, A. and Tavakoli, A. (2016). Efficient removal of cationic dyes using a new magnetic nanocomposite based on starch-g-poly(vinylalcohol) and functionalized with sulfate groups. RSC Adv., 6: 38042–38051. https://doi.org/10.1039/C6RA02517J.

Pourjavadi, A., Hosseini, S. H., Seidi, F. and Soleyman, R. (2013). Magnetic removal of crystal violet from aqueous solutions using polysaccharide-based magnetic nanocomposite hydrogels. Polym. Int., 62: 1038–1044. https://doi.org/https://doi.org/10.1002/pi.4389.

Protection Agency E. (2023). The work of the EPA can be divided into three main areas.

Qamar, S. A., Qamar, M., Basharat, A., Bilal, M., Cheng, H. and Iqbal, H. M. N. (2022) .Alginate-based nano-adsorbent materials – Bioinspired solution to mitigate hazardous environmental pollutants. Chemosphere, 288: 132618. https://doi.org/10.1016/J.CHEMOSPHERE.2021.132618.

Qiao, H., Zhou, Y., Yu, F., Wang, E., Min, Y., Huang, Q. e al. (2015). Effective removal of cationic dyes using carboxylate-functionalized cellulose nanocrystals. Chemosphere, 141: 297–303. https://doi.org/10.1016/J.CHEMOSPHERE.2015.07.078.

Rajab Beigy, M., Rasekh, B., Yazdian, F., Aminzadeh, B. and Shekarriz, M. (2018). High nitrate removal by starch-stabilized Fe0 nanoparticles in aqueous solution in a controlled system. Eng Life Sci., 18: 187–195. https://doi.org/https://doi.org/10.1002/elsc.201700127.

Ramos Berger, L. R., Montenegro Stamford, T. C., de Oliveira, K. Á. R., de Miranda Pereira Pessoa, A., de Lima, M. A. B., Estevez Pintado, M. M. et al. (2018). Chitosan produced from Mucorales fungi using agroindustrial by-products and its efficacy to inhibit Colletotrichum species. Int. J. Biol. Macromol., 108: 635–641. https://doi.org/10.1016/J.IJBIOMAC.2017.11.178.

Roberts, J. J. and Martens, P. J. (2016). Engineering biosynthetic cell encapsulation systems. Biosynthetic Polymers for Medical Applications, 205–239. https://doi.org/10.1016/B978-1-78242-105-4.00009-2.

Sabzevari, M., Cree, D. E. and Wilson, L. D. (2018). Graphene oxide–chitosan composite material for treatment of a model dye effluent. ACS Omega, 3: 13045–13054. https://doi.org/10.1021/acsomega.8b01871.

Sankararamakrishnan, N., Singh, N. and Srivastava, I. (2020). Hierarchical nano Fe(0)@FeS doped cellulose nanofibres derived from agrowaste – Potential bionanocomposite for treatment of organic dyes. Int. J. Biol. Macromol., 151: 713–722. https://doi.org/10.1016/J.IJBIOMAC.2020.02.155.

Sargin, I., Baran, T. and Arslan, G. (2020). Environmental remediation by chitosan-carbon nanotube supported palladium nanoparticles: Conversion of toxic nitroarenes into aromatic amines, degradation of dye pollutants and green synthesis of biaryls. Sep. Purif. Technol., 247: 116987. https://doi.org/10.1016/J.SEPPUR.2020.116987.

Sarma, A. and Das, M. K. (2021). Improving the sustainable performance of Biopolymers using nanotechnology. Polymer-Plastics Technology and Materials, 60: 1935–1965. https://doi.org/10.1080/25740881.2021.1937645.

Schellenberg, T., Subramanian, V., Ganeshan, G., Tompkins, D. and Pradeep, R. (2020). Wastewater Discharge Standards in the Evolving Context of Urban Sustainability–The Case of India. Front Environ. Sci., 8.

Singh, S., Mahalingam, H. and Singh, P. K. (2013). Polymer-supported titanium dioxide photocatalysts for environmental remediation: A review. Appl. Catal. A Gen., 462–463: 178–195. https://doi.org/10.1016/J.APCATA.2013.04.039.

Sirajudheen, P. and Meenakshi, S. (2019). Facile synthesis of chitosan-La3+-graphite composite and its influence in photocatalytic degradation of methylene blue. Int. J. Biol. Macromol., 133: 253–261. https://doi.org/10.1016/J.IJBIOMAC.2019.04.073.

Song, K., Xu, H., Xu, L., Xie, K. and Yang, Y. (2017). Cellulose nanocrystal-reinforced keratin bioadsorbent for effective removal of dyes from aqueous solution. Bioresour. Technol., 232: 254–262. https://doi.org/10.1016/J.BIORTECH.2017.01.070.

Soon, C. Y., Rahman, N. A., Tee, Y. B., Talib, R. A., Tan, C. H., Abdan, K. and Chan, E. W. C. (2019). Electrospun biocomposite: nanocellulose and chitosan entrapped within a poly(hydroxyalkanoate) matrix for Congo red removal. Journal of Materials Research and Technology, 8: 5091–5102. https://doi.org/10.1016/J.JMRT.2019.08.030.

Stan, M., Lung, I., Soran, M. L., Opris, O., Leostean, C., Popa, A., Copaciu, F. et al. (2019). Data on the removal of Optilan Blue dye from aqueous media using starch-coated green synthesized magnetite nanoparticles. Data Brief 25: 104165. https://doi.org/10.1016/J.DIB.2019.104165.

Sundar, S., Kundu, J. and Kundu. S. C. (2010.) Biopolymeric nanoparticles. Sci. Technol. Adv. Mater., 11: 014104. https://doi.org/10.1088/1468-6996/11/1/014104.

Suresh, R., Rajendran, S., Hoang, T. K. A., Vo, D. V. N., Siddiqui, M. N. and Cornejo-Ponce, L. (2021). Recent progress in green and biopolymer based photocatalysts for the abatement of aquatic pollutants. Environ. Res., 199: 111324. https://doi.org/10.1016/J.ENVRES.2021.111324.

Suyambulingam, I., Gangadhar, L., Sana, S. S., Divakaran, D., Siengchin, S., Kurup, L. A., Iyyadurai, J. et al. (2023). Chitosan biopolymer and its nanocomposites: emerging material as adsorbent in wastewater treatment. Advances in Materials Science and Engineering, 2023:9387016. https://doi.org/10.1155/2023/9387016.

Taghizadeh, M. T., Siyahi, V., Ashassi-Sorkhabi, H. and Zarrini, G. (2020). ZnO, AgCl and AgCl/ZnO nanocomposites incorporated chitosan in the form of hydrogel beads for photocatalytic degradation of MB, *E. coli* and *S. aureus*. Int. J. Biol. Macromol., 147: 1018–1028. https://doi.org/10.1016/J.IJBIOMAC.2019.10.070.

Vasilev, S., Vodyashkin, A., Vasileva, D., Zelenovskiy, P., Chezganov, D., Yuzhakov, V. et al. (2020). An Investigative Study on the Effect of Pre-Coating Polymer Solutions on the Fabrication of Low Cost Anti-Adhesive Release Paper. Nanomaterials 10. https://doi.org/10.3390/nano10081436.

Vigneshwaran, S., Sirajudheen, P., Karthikeyan, P., Nikitha, M., Ramkumar, K. and Meenakshi, S. (2021). Immobilization of MIL-88(Fe) anchored TiO2-chitosan(2D/2D) hybrid nanocomposite for the degradation of organophosphate pesticide: Characterization, mechanism and degradation intermediates. J. Hazard Mater, 406: 124728. https://doi.org/10.1016/J.JHAZMAT.2020.124728.

Vo, T. S., Vo, T. T. B. C., Suk, J. W. and Kim, K. (2020). Recycling performance of graphene oxide-chitosan hybrid hydrogels for removal of cationic and anionic dyes. Nano Converg, 7: 4. https://doi.org/10.1186/s40580-019-0215-0.

Wei, H., Rodriguez, K., Renneckar, S. and Vikesland, P. J. (2014). Environmental science and engineering applications of nanocellulose-based nanocomposites. Environ. Sci: Nano, 1: 302–316. https://doi.org/10.1039/C4EN00059E.

Wen, D., Wu, Z., Tang, Y., Li, M. and Qiang, Z. (2018). Accelerated degradation of sulfamethazine in water by VUV/UV photo-Fenton process: Impact of sulfamethazine concentration on reaction mechanism. J. Hazard Mater, 344: 1181–1187. https://doi.org/10.1016/J.JHAZMAT.2017.10.032.

Weska, R. F., Moura, J. M., Batista, L. M., Rizzi, J. and Pinto, L. A. A. (2007). Optimization of deacetylation in the production of chitosan from shrimp wastes: Use of response surface methodology. J. Food Eng., 80: 749–753. https://doi.org/10.1016/J.JFOODENG.2006.02.006.

World Water Assessment Programme (United Nations) Wastewater : the untapped resource : the United Nations world water development report 2017.

Xuan, V. N., Hoa, P. X., Thu, N. T. P. and Huong, L. M. (2024). Factors affecting environmental pollution for green economy: The case of ASEAN countries. Environmental Challenges, 14: 100827. https://doi.org/10.1016/J.ENVC.2023.100827.

Yan, J., Han, L., Gao, W., Xue, S. and Chen, M. (2015). Biochar supported nanoscale zerovalent iron composite used as persulfate activator for removing trichloroethylene. Bioresour. Technol. 175: 269–274. https://doi.org/10.1016/J.BIORTECH.2014.10.103.

Yu, S., Wang, J. and Cui, J. (2020). Preparation of a novel chitosan-based magnetic adsorbent CTS@SnO2@Fe3O4 for effective treatment of dye wastewater. Int. J. Biol. Macromol., 156: 1474–1482. https://doi.org/10.1016/J.IJBIOMAC.2019.11.194.

Zhang, M., Zhang, Z., Peng, Y., Feng, L., Li, X., Zhao, C. et al. (2020). Novel cationic polymer modified magnetic chitosan beads for efficient adsorption of heavy metals and dyes over a wide pH range. Int. J. Biol. Macromol., 156: 289–301. https://doi.org/10.1016/J.IJBIOMAC.2020.04.020.

Zhang, Z., Ahmed, A. I. S., Malik, M. Z., Ali, N., khan, A., Ali, F. et al. (2023). Cellulose/inorganic nanoparticles-based nano-biocomposite for abatement of water and wastewater pollutants. Chemosphere, 313: 137483. https://doi.org/10.1016/J.CHEMOSPHERE.2022.137483.

Zhou, J., Sun, Q., Chen, D., Wang, H. and Yang, K. (2017). Ochrobactrum anthropi used to control ammonium for nitrate removal by starch-stabilized nanoscale zero valent iron. Water Science and Technology, 76: 1827–1832. https://doi.org/10.2166/wst.2017.286.

Zhou, Z., Ma, J., Liu, X., Lin, C., Sun, K., Zhang, H. et al. (2019). Activation of peroxydisulfate by nanoscale zero-valent iron for sulfamethoxazole removal in agricultural soil: Effect, mechanism and ecotoxicity. Chemosphere, 223: 196–203. https://doi.org/10.1016/J.CHEMOSPHERE.2019.02.074.

CHAPTER-8

Biopolymeric Nanocomposites for Water Purification

Simi Asma Salim, Mahima S. Mohan, Paramita Pakhira, Nayana Mulayath and *Siddhardha Busi**

1. Introduction

Water is one of the most prominent resources in the universe (Rathi and Kumar, 2021). Fresh water plays a critical role in various aspects of human life and global economy. It is mandatory for multiple purposes, including agriculture, human consumption, industrial activities, and recreation. Availability of potable water is inevitable for maintaining human health. Although it is a valuable resource, a freshwater resource can be vulnerable to contamination because of anthropogenic actions besides being a natural process. Due to the technological advancement and industrial expansion, freshwater resources are at global threat (Sharma and Bhattacharya, 2017). Over 700 organic compounds have been found in drinking water, originated from different sources such as municipal, industrial, and agricultural operations besides from naturally decomposed animal and vegetable matter (Pollard et al., 1992). Regardless of its source, water contains natural contaminants (Fawell and Nieuwenhuijsen, 2003). The natural contaminants in ground water depends on the geological substances through which it passes and the quality of the recharge water. As the water flows through rocks and soils, it can take up various metals including magnesium, calcium, chloride, arsenate, fluoride etc. (Hlongwane et al., 2019). Moreover, water can be polluted by microorganisms and chemicals due to human activities (anthropogenic pollution) (Fawell and Nieuwenhuijsen, 2003).

Broadly, three types of contaminants in water are: (1) Organic contaminants; (2) Inorganic contaminants; (3) Pathogenic microorganisms (Sharma and Bhattacharya, 2017).

1.1 Organic Contaminants

Anthropogenic activities lead to organic contamination in the environment. The sources include the use of pesticides, domestic and industrial waste etc. These organic materials can cause serious health consequences like hormonal disruptions, cancers, and nervous system disorder. Pesticides contaminate the environment through both agriculture and public hygienic sources (Sharma and Bhattacharya, 2017). Dyes account as one among the major categories of organic compounds that

Department of Microbiology, School of Life Sciences, Pondicherry University, Puducherry-605014, India.
* Corresponding author: siddhardha.busi@gmail.com
DOI: 10.1201/9781003558330-8

are a significant contributor to environmental problems (Singha et al., 2021). Pharmaceutical, tanning, printing, textile industries and farming activities are the significant contributors of organic compounds that pose toxicity risk. Approximately 70% of the pollutants in sewage represent organic compounds (Hlongwane et al., 2019). Bioaccumulation of organic pollutants is due to their slow degradation rate, regardless many organic pollutants are biodegradable (Gu et al., 2016). All these organic contaminants present in the water possess the potential to induce adverse effects on human health or ecological systems (Sharma and Bhattacharya, 2017).

1.2 Inorganic/Heavy Metal Contaminants

Usual inorganic compounds found in water resources are heavy metals such as lead, arsenic, mercury, cadmium, chromium, nickel, and polyatomic compounds. They exhibit low degradation rate and remain longer in aqueous system. These toxic heavy metals can contaminate water through both natural processes like volcanic activity, soil erosion, corrosion of metal, evaporation of metal from soil and water and sediment re-suspension, natural weathering, as well as anthropogenic activities (Briffa et al., 2020). Wastewater originating from mines, smelters, battery dyes, and electronic factories represent major anthropogenic sources of heavy metals. Also, about 30% of pollutants in sewage water comprises of inorganic contaminants (Hlongwane et al., 2019).

1.3 Pathogens

Water contamination by pathogens is the most hazardous amongst various water contaminants (Sharma et al., 2021). These include organisms such as bacteria, virus, protozoa, algae, and certain parasites (Ashbolt, 2015). Pathogens that pose significant risks to human health and frequently observed in drinking water sources are protozoa such as *Cryptosporidium* and *Giardia*, bacteria such as *Campylobacter* and *E. coli*, viruses such as enterovirus and norovirus, etc. (Dechesne and Soyeux, 2007). Pathogens are commonly introduced into water sources through contaminated soil and faecal contamination (Hlongwane et al., 2019). Microorganisms will degrade water quality and cause significant risks to public health through diseases like diarrhoea, haemorrhagic colitis, and haemolytic uremic syndrome, etc (Hlongwane et al., 2019).

2. Water Purification

Different physical, chemical and biological water treatment/purification strategies have been introduced by various researchers. All these strategies have both merits and limitations. For instance, the ion exchange method is efficient but economic constraints limits its use, microbial treatment is eco-friendly but it is slow and difficult to standardize, chemical treatment methods are fast but not recommended due to toxic properties of chemicals. Moreover, adequate water purification using single processes are challenging. Therefore, a combination of techniques is recommended to ensure satisfactory water quality. Recently researchers are making serious efforts to integrate different techniques to cover up the disadvantages and to exploit the advantages of different purification techniques. One such strategy is the polymeric nanocomposite. It exhibits superior physico-chemical properties compared to individual components. Because of its processibility, polymers can act as a good host material. As said earlier, incorporation or reinforcement of nanomaterials (nanofillers) in polymer matrix heightens the polymer characteristics and yield polymeric nanocomposites with preferred qualities and thereby helps in efficient purification of water (Pandey et al., 2017). Importantly, more than normal polymeric nanocomposites, biopolymeric nanocomposites have additional advantages of being biodegradable and eco-friendly in nature. Different types of biopolymers, their nature of origin, and properties are explained below.

3. Biopolymers and Types

Biopolymers are large molecules made up of monomer units. The most prevalent macromolecules encompass biopolymers like proteins and carbohydrates (Baranwal et al., 2022). They can be characterized as materials that undergo microbial breakdown, leading to processes like photodegradation, oxidation, and hydrolysis, thus altering the polymer during degradation. These polymers break down into CO_2, CH_4, and inorganic compounds due to microbial enzyme activity within a short timeframe (Verma and Fortunati, 2019).

Biopolymers exhibit compatibility with living organisms and their ability to degrade naturally is highly beneficial in various applications. These applications include the manufacture of emulsions, edible films, and food packaging materials. Additionally, they are employed as drug transport mediums, implants used for medicine, agents for healing wounds, and dressing materials in medical fields (Othman, 2014).

Biopolymers can be grouped into natural and synthetic types depending on their source. Polysaccharides and proteins give rise to natural biopolymers. At the same time, those synthesized through microbial fermentation or chemical methods are classified as synthetic biopolymers. For instance polyhydroxyalkanoate (PHA), polyhydroxybutyrate (PHB), etc., which are synthesized by microbial fermentation and chemically synthesized polycaprolactone from petroleum products (Fig. 1).

Starch, cellulose, chitosan, agar, gluten, whey protein, gelatin, alginate, and collagen are among the natural biopolymers widely utilized, derived from polysaccharides and proteins. Synthetic biopolymer offers several advantages, including promoting sustainability within industries and improving qualities such as durability, flexibility, glossy appearance, transparency, and tensile strength (Othman, 2014).

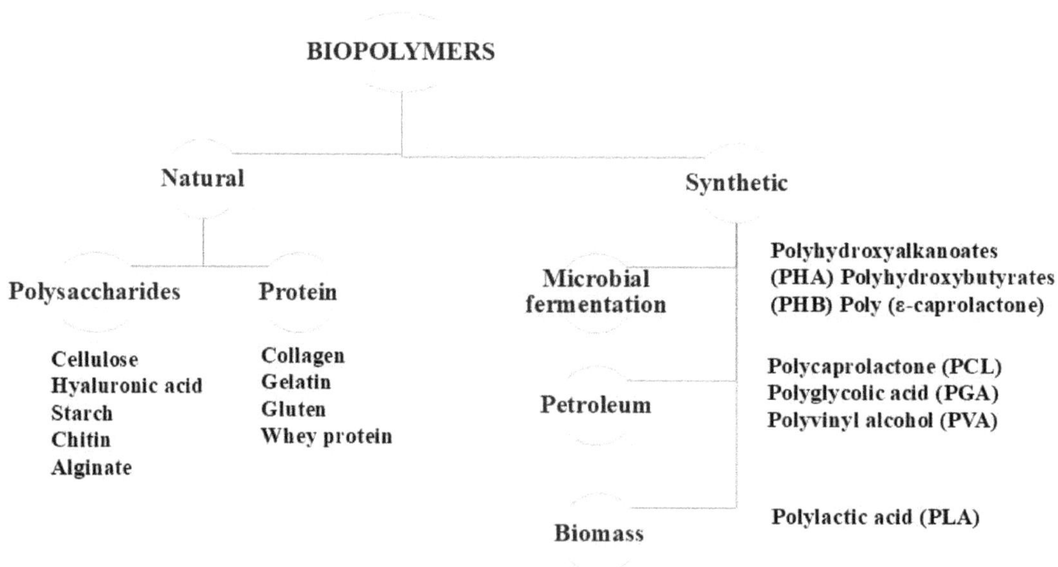

Figure 1. Classification of biopolymers. Natural biopolymers are polymers which are present as such in nature. Synthetic biopolymers are biopolymers synthesized by microbial fermentation, derived from biomass and petroleum.

3.1 Polysaccharides

3.1.1 Starch

Starch serves as the primary carbohydrate storage in higher plants. It has significant uses beyond its role as a dietary component. It has been historically utilized as an adhesive in industries such as paper and wood, as well as gum in the textile industry (Carvalho, 2008). Starch typically consists of α-amylose and amylopectin. Amylose consists of glucose molecules arranged in a linear chain, whereas amylopectin is largely branched. The hydroxy moieties present on the starch backbone can react with alcohols, leading to the formation of hydrogen bonds, ethers, and esters. This hydrophilic nature is useful for enhancing the degradation of certain hydrophobic polymers. Microorganisms or enzymes can hydrolyse the starch into glucose which can then be metabolized (Park et al., 2017). Starch can be converted into chemicals for synthetic polymer production, used in fermentative processes for biopolymers, or grafted to produce new polymeric materials or fillers (Carvalho, 2008). Starch-based materials have extensive uses in biomaterials like bone replacement, tissue regeneration, and therapeutic delivery systems. However, challenges like high water sensitivity, poor barrier properties, and weak mechanical strength restrict their usage in medical applications (Park et al., 2017).

3.1.2 Cellulose

Cellulose is a commonly available biopolymer in nature. It primarily consists of linear sequence of glucose molecules linked by β-1, 4 glycosidic bonds, commonly found in plants and bacteria. Cellulose molecules can form microfibrils through the arrangement of glucose monomer units and the establishment of inter- and intra-molecular hydrogen bonds (Verma and Fortunati, 2019b). It can be obtained from different sources such as macroalgae, rice hull, sugarcane pulp from agricultural residues, sugar beet, wood, banana rachis, bamboo, potato tubers, cotton, hemp, etc. (Baranwal et al., 2022). Numerous commercially available cellulose derivatives exist, with Cellulose Acetate (CA) being the predominant choice for food packaging applications, including fresh produce and baked goods (Peelman et al., 2013). Cellulose acetate (CA) is a thermoplastic material made by esterifying cellulose. Renowned for its compostability and impressive optical and mechanical traits, CA finds applications across various fields including fibres, films, laminates, adhesives, coatings, and plastic products (Li et al., 2010). As a vital structural component in plant walls, cellulose boasts remarkable chemical stability (Flaris and Singh, 2009).

3.1.3 Chitin and Chitosan

After cellulose, chitin is most abundant biopolymer in nature. It is a highly basic polysaccharide with a nitrogen concentration of 5–8%. It forms organized crystalline microfibrils within the framework of arthropods and insects, as well as in fungal and microbial cells. Chitin is primarily an acetylated polysaccharide consisting of the monomer units- N-acetyl-D-glucosamine units linked by β-1, 4 linkages. By deacetylating chitin, chitosan is derived, characterized by poly-β-(1, 4)-2-amino-2-deoxy-D-glucose units. Chemically active groups in chitin include free amine and hydroxyl groups (Verma and Fortunati, 2019). Chemical cross-linking, often using reactive dialdehydes or carbodiimide compounds, modifies chitosan's properties, such as molecular weight and degree of deacetylation (Park et al., 2017). Different chitosan forms include gels, films, membranes, particles, and scaffolds. Chitosan is derived from various organisms, including crustaceans, molluscs, insects, and fungi, with crustacean shells being the main source. It has distinct characteristics, including biodegradability, nontoxicity, chelating capabilities, anti-oxidative properties, anticoagulant effects, antimicrobial action, and biocompatibility, make it valuable for producing bioactive materials with diverse applications (Baranwal et al., 2022).

3.1.4 Alginate

Alginate, a naturally sourced polymer, is derived from brown algae by processing it with aqueous alkaline solutions. It is an anionic and hydrophilic polymer having carboxyl end groups. Alginate consists of linear polysaccharides comprising of copolymers β-D-mannuronic acid and α-L-guluronic acid units. The arrangement and length of units can differ based on the alginate's source. Alginate hydrogels are formed by the gelation of alginate molecules, which is facilitated by multivalent cations, particularly divalent calcium. The physical properties and formation of alginate hydrogels are affected by factors like the sequence of composition, length of G-blocks, and molecular weight. It is seizing attention in biomedical applications because of its biocompatibility, minimal toxicity, and economic efficiency. Alginate can be chemically modified to produce derivatives with diverse structures and functionalities (Park et al., 2017). Alginate primarily consists of alginic acid and its derivatives (ALG). These compounds contain carboxylic groups, which contribute to their hydrophilic nature and ability to bind water (Álvarez-Castillo et al., 2021).

3.2 Proteins

3.2.1 Collagen

Collagen, primarily type I, is the highest prevalent type of collagen (Zhang et al., 2013), which serves as a primary constituent of animal connective tissues and the major protein in mammals (about 30% of the entire content of protein in the body) (Park et al., 2017). It is structured as triple helices. Within the triple helix, stability is maintained through intramolecular and intermolecular hydrogen bonds (Zhang et al., 2013). Collagen has various uses in different sectors, including leather goods, cosmetics, and biomedical fields such as wound dressings, implants, and systems designed for drug delivery. The most favourable attributes of collagen are biocompatibility, minimal antigenicity, biodegradability, ability to form cross-links, and exceptional mechanical strength.

3.2.2 Gelatin

Gelatin is derived from the partial breakdown of collagen, forming a soluble protein. Its properties can vary based on the collagen source, animal age, and the collagen type used. Gelatin is a biodegradable, biocompatible, and is a thermo reversible gelling agent, rendering it appropriate for medical and pharmaceutical uses. The extent of collagen transformation into gelatin is reliant on pre-treatment intensity and aqueous extraction conditions. Thermal processing is applied to disrupt the hydrogen and covalent bonds within the triple helical organisation of collagen, leading to a conversion from helical to unfolded conformations and the formation of soluble gelatin (Park et al., 2017). Gelatin is produced through the thermal denaturation of collagen, which is extracted from animal skin and bones using a weak acid solution. Type A gelatin, resulting from acid pretreatment, is commonly sourced from pigskin, whereas type B gelatin, resulting from alkaline treatment, is sourced from cattle skin and bones. The collagen conversion into gelatin occurs in two stages: first, heat-induced structural alteration at around 40°C, disrupting hydrogen and ionic bonds, followed by the enzymatic cleavage of covalent bonds. In pharmaceutical and food industries, gelatin films are utilized for microencapsulation and the production of coatings for tablets and capsules (Zhao et al., 2008).

3.3 Biomass and Microbial fermentation derived biopolymers

3.3.1 PLA

Lactic acid, the monomer of biopolymer polylactic acid, can be sourced from various renewable sources like corn, wheat, sugarcane, whey, molasses, etc. The characteristics of PLA are determined by the L and D lactic acid monomers present. L-PLA tends to have a high melting point and crystallinity. Both the monomers are blended to form an amorphous structure with a lower glass transition temperature (Sorrentino et al., 2007). PLA possesses several advantages, including biodegradability, and its lifespan can be adjusted accordingly (Othman, 2014). PLA is derived from biomass and is commonly referred to as bioplastic. Biodegradable property distinguishes it from traditional plastics like PS, PE, or PP. Consequently, PLA has gained significant attention as a biodegradable alternative, gradually replacing polymers derived from petrochemicals in multiple applications, including medical and other plastic industries. Due to its lack of toxicity and carcinogenic properties, polylactic acid is widely utilized as a biocompatible polymer. PLA requires a significant amount of time for degradation. It can decompose into CO_2 and water in controlled composting conditions, typically achieved at temperatures around 140°F. However, degradation processes are considerably prolonged in tightly packed landfills or composting containers lacking sufficient light and oxygen (Ilyas et al., 2020).

3.3.2 PHB/PHA

Polyhydroxybutyrate (PHB) are accumulated by many bacteria as reserves of energy and carbon. This bio-polyester possesses biodegradable and biocompatible features, making it appropriate for various commercial applications (Sorrentino et al., 2007). PHA is a thermoplastic material with physical properties and chemical structure closely resembling that of polypropylene and polyethylene and is utilized in various processes such as drawing, pressing film, and injection moulding. It can degrade quickly and reliably, which reduces the amount of space needed in landfills for waste disposal. PHA possesses the capability to biodegrade in various settings, including both terrestrial and aquatic environments (Ilyas et al., 2020). PHA can be produced through either chemical synthesis or biological methods. Various microorganisms are involved in PHA synthesis, such as *Pseudomonas putida* CA-3 and *Cupriavidus necator*. The growth and synthesis of PHAs rely on specific nutrients like starch, glucose, fructose, and fatty acids. Based on the quantity of carbon sources present, synthesized PHAs may vary in chain length, ranging from medium to short chain lengths (Bhushan and Kumar, 2019). PHB undergoes natural and efficient degradation into CO_2 and water by numerous soil bacteria. Under favourable environmental conditions, bacteria such as *Alcaligenes eutrophus* can accumulate up to 90% of their cellular mass as PHB (Flaris and Singh, 2009).

3.4 Petroleum

3.4.1 PCL

Polycaprolactone (PCL) is a polymer material formed by the polymerization of 3-caprolactone via ring opening, resulting in a linear structure. With a considerable crystallinity level of approximately 50%, the polymer possesses a semi-crystalline structure. PCL's remarkable stretchability and low stiffness make it highly appealing for various commodity applications. Additionally, its suitability extends to medical and agricultural sectors, presenting potential outcomes for diverse applications. Because of its low melting point, PCL is combined with other polymers for various applications (Ishiaku et al., 2002; Sorrentino et al., 2007).

4. Biopolymeric Nanocomposites

Nanocomposites are multiphasic materials, of which any one part can have nanoscale dimensions. It can be nanoparticles, nano whiskers, nanotubes, nanofibres, etc. (called nanofillers). Because of their dissimilarities in structure and chemistry, they possess unique and different properties including high mechanical strength, thermal stability, electrical conductivity, higher resistance to chemicals, etc. Depending on the presence or absence of polymeric material there are two kinds of nanocomposites-polymer and non-polymer nanocomposites. In the polymer nanocomposites—the nanoparticles or nanofillers are reinforced/incorporated/ embedded in a polymer matrix (Sen, 2020).

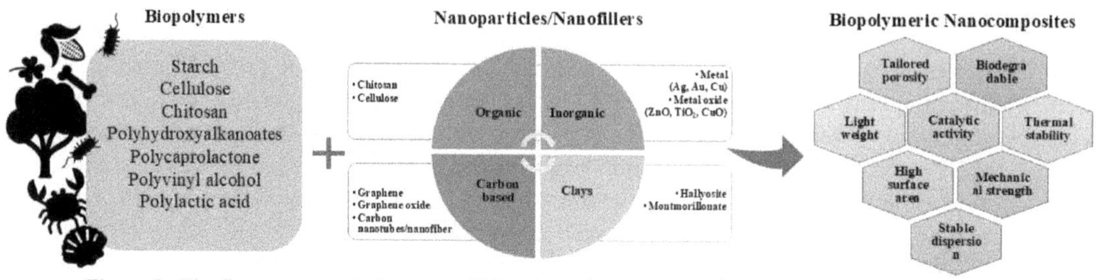

Figure 2. The figure represents the types of biopolymeric nanocomposites and their unique properties.

From the above section, we have come across various biopolymers and their properties. In essence, in a defined time period, biodegradable polymers break down easily to simple molecules like CO_2 and H_2O by enzyme action. Hence biopolymers can be used as bulk matrix for nanocomposites particularly to heighten the mechanical as well as barrier properties while conserving their biodegradability (Tang et al., 2012). Reinforced nanofillers can be of different types. It can be nanoparticles, nano clays, nanotubes, or other nanoscale materials, and contribute to the advantageous properties of nanocomposites. The nanofillers can be organic such as nanostructures of chitosan and cellulose, inorganic such as nanostructures of metals or metaloxides, clay nanostructures such as halloysite, montmorillonite, carbon nanostructures including graphene, fullerene, or carbon nanotubes and carbon nanofibers (Wawrzyńczak et al., 2024) (Fig. 2). Recently, efforts were made to invent nanocomposites having dual or mixed nanofillers. Here more than one nanostructure is reinforced into biopolymers for efficient functionality. Properties of different nanostructures together make the nanocomposite more efficient.

Organic nanofillers: Cellulose nanofibers (CNF) were reinforced in polycaprolactone based biopolymer matrix to develop organic membranes having the desired mechanical properties for removing pollutants from water (Palacios Hinestroza et al., 2020). Similarly, two nanoforms of cellulose- nanocrystalline cellulose and cellulose nano whisker incorporated poly (D-lactic acid) membranes were developed by Kian et al. (2021).

Carbon based nanofillers: Carbon based nanofillers like graphene, graphene oxide and derivatives, carbon nanotubes, has proved to be an outstanding candidate owing to their large surface area, noncorrosive and light weight nature and has excellent mechanical strength and can be used as a good absorbent (Sharma et al., 2018).

Biocompatible Graphene Oxide (GO), the oxidized form of graphene has unique and adjustable physical-chemical properties such as optical, electrical, mechanical, antimicrobial properties, and catalytic activity. It displays affinity to a wide variety of contaminants as well. Because of these properties graphene oxide is used as nanofiller of many biopolymeric nanocomposites for water treatment/purification applications. For instance, polycaprolactone/graphene oxide nanocomposite was developed for water treatment by Maio et al. (2020). Gelatin-chitosan biopolymer-based nanocomposite fabricated for removal of micropollutants from drinking water by Kovtun et al. using graphene oxide as coating as well as embedding nanomaterial (Kovtun et al., 2020).

Carbon nanotubes (CNTs) due to their excellent physical, mechanical, electrical, and optical properties act as an adsorbent and co-catalyst enhancing water treatment membranes (Esfahani et al., 2015). CNTs are widely used as nanofillers of polymeric/biopolymeric nanocomposites. Carboxylic acid functionalized Multi-walled carbon nano tubes (MWCNTs) and functionalized graphene oxide were reinforced to polylactic acid polymer membrane for efficient treatment of waste water (Khalil et al., 2023).

Clay based nanofillers: Clay minerals are widely available and have unique layer structure with high cation and anion absorption capacity via ion exchange, good adsorption power, chemical and physical stability, altogether makes it appropriate for removal of inorganic impurities from water (Irandoost et al., 2019). Clay (montmorillonite) supported silver nanoparticles (AgNPs) was used as a nanofiller of poly (ε-caprolactone) biopolymer matrix nanocomposite for the ultrafiltration of wastewater. Reduction of various organic and inorganic contaminants and antimicrobial properties of the developed nanocomposite was confirmed by authors (Benhacine et al., 2018). Irandoost et al. (2019) developed polycaprolactone based adsorbant modified by nanoclay and nanozeolite for the reduction of lead from aqueous solution.

Metal and Metaloxide based nanofillers: Metal nanoparticles have proved to possess antimicrobial properties. Copper and silver nanoparticles (CuNPs and AgNPs) were integrated into chitosan polymer matrix and effective antimicrobial activity was achieved (Morsi et al., 2017). Oxide nanoparticles, like other nanoparticles, display unique chemical and physical properties and are vastly recommended in removal of contaminants owing to their small size and high density on their edge surface sites. Titanium dioxide (TiO_2) nanoparticles were reinforced in polycaprolactone based ultrafiltration membrane for the wastewater treatment (Peter et al., 2020) and effluents from dairy industry (Nivedita and Joseph, 2021). An aerogel based on gelatin with oil/water separation and photocatalytic efficacy was fabricated by incorporating TiO_2 nanoparticles (Jiang et al., 2019). Mixed aluminium and cerium oxide nanoparticles were entrapped in calcium alginate beads fabricated for adsorption of fluoride (Swain et al., 2013). Likewise, zinc oxide nanoparticles were encapsulated in calcium alginate matrix for the removal of organic pollutants (Zhao et al., 2020).

5. Biopolymeric Nanocomposites as Adsorbents in Water Purification Systems

In the water purification process, one of the crucial techniques applied for the elimination of contaminants is adsorption. Impurities attaches the periphery of the adsorbent material and therefore its removal from water. The adsorbents formulated as pellets, granules or powder provides greater surface area for maximum adsorption. Activated carbon, zeolites, clay minerals, silica, etc., are generally used as commercial adsorbents for the water purification process. However, due to their higher cost, they are generally used in a large-scale process. Recently, bio nanocomposites became a promising approach developed in order to overcome the limitations of commercial adsorbents. Bio nanocomposites are a fusion of biopolymers and nanoparticles, which enhances adsorption process. Bio nanocomposites possess characteristics of both nanoparticles and biopolymers which includes higher surface area, efficacy of nanoparticles, biocompatibility, and renewable properties (Kolya and Kang, 2023). They are reported for their ability to remove soluble particles, metal ions, microbes, dyes, etc., from the water. These composites have bonus features such as it can be made into films with activated functionalities and variable dimensions. The major characteristics of nanocomposites are catalytic properties, great adsorption capacity and mechanical properties (Udayakumar et al., 2021).

The effluents from industries contain higher quantities of heavy metals, and are discharged into the milieu (Udayakumar et al., 2021) including water resources. In a study, a novel magnetic nanocomposite was developed from Walnut Shell Ash (WSA) along with starch and Fe_3O_4 to eliminate Cu (II) from water. The nanocomposite adsorbed 45.4 mg/g copper to it. Moreover, it removed Cu from three water samples indicating its reusability for at least 3 times with appropriate efficacy (Foroutan et al., 2022).

Ground water widely used for drinking purposes were contaminated with excess nitrate becomes a big concern due to carcinogenic nitrosamine production and blue-baby syndrome. A group of researchers developed chitosan-polystyrene-Zn nanocomposites for the nitrate removal.

The study showed that 90% of the nitrate ions were removed by 0.5 g of nanocomposite from 25 ml of water after 30 mins of treatment (pH = 3) using the batch adsorption technique, whereas in fixed bed column removed nitrate by 82.5% under similar conditions. They found that chitosan-polystyrene-Zn nanocomposites can be reusable for 3 times (Keshvardoostchokami et al., 2017).

Heavy metal cations are the most harmful pollutants because of their higher toxicity, water solubility, poor detection without advanced systems and stability. Mercury is one among the heavy metal cations, become detrimental after the conversion into Hg^{2+} by the microorganisms in the water. The exclusion of mercury from water has fascinated globally due to the tragedy happened in Minamata, Japan. Recently, a group of researchers synthesized monodisperse functional magnetic dialdehyde starch nanocomposite without any toxic cross-linking agent. The nanocomposite showed good adsorption towards mercuric ions at higher and lower concentrations. Due to the economic and green nature, the nanocomposite likely to treat mercury contamination in wastewater (Wang et al., 2015). Hassan and Hrdina, 2017 prepared chitosan/nanohydroxyapatite nanocomposites to investigate the mercuric adsorption property using static and dynamic adsorption studies. In this study, they used different ratios of chitosan and nanohydroxyapatite. In static adsorption method, Hg^{2+} was effectively adsorbed after 60 mins of treatment with 1.2 g/L of nanocomposite at pH 7 and the one with 1:2 ratio exhibited more cationic adsorption owing to its increased surface area and surface function groups. While, in the dynamic method, the same showed most efficient adsorption at 2 ml/min (starting concentration of 300 mg/L).

Recently, chitosan combined with magnetite nanocomposite (CMNC) was exploited for the removal of heavy metals from contaminated water. CMNC showed higher adsorption rate of 53.19 and 30.3 mg/g for Co^{2+} and Ni^{2+} correspondingly under ideal conditions (pH 6, 2 h time and dosage = 2 g/L). In addition, interaction between CMNC was stronger with Co^{2+} compared to Ni^{2+}, indicating its effective removal from the water (Sharifi et al., 2023).

Organic pollutants are highly toxic and non-biodegradable in nature, widely used in agriculture, textile and dyeing, chemical industry, etc. These pollutants are carcinogenic and mutagenic and may have impacts on human health, hence, can be categorized as dangerous pollutants. Due to strong bonds in their structure, these contaminants are difficult to undergo decomposition and affects the food chain. The major organic substances that contaminate water sources are dyes, aromatic substances, oil spills, pesticides, organic acids, pharmaceutical residues, and herbicides (Tadayoni et al., 2024). Like biopolymers, biopolymeric nanocomposites are acknowledged for their capacity to adsorb organic pollutants from the medium.

Waterbodies polluted by dyes from industries such as paper, textiles, printing, cosmetics, pharmacy, food, etc., has become a major environmental concern in developing nations. Other than environmental issues, the release of waste water contaminated by dyes without prior treatment causes health issues such as skin irritation, sore eyes, central nervous system dysfunction and so on to the humans.

Starch based composites are reported to be well-known adsorbents for the clearance of impurities and aromatic compounds from wastewater . Amiri and coworkers have prepared biopolymer-based adsorbent for the effective elimination of methylene blue dyes from wastewater. Researchers investigated different materials such as native Corn Starch (CS), starch nanocrystals (SNCs) and phosphate corn starch nanocrystals (PSNCs), where PSNCs showed effective adsorption potential for the clearance of dye from waste water (Amiri et al., 2023).

Chitosan co-polymerized with Poly Vinyl Alcohol (PVA) are generally used in the water treatment process to eradicate dyes due to their thermal stability, good adsorption, economic, non-toxic, and sustainable nature. Chitosan-PVA polymeric composite in combination with CuO nanowires (CS-PVA@CuO) effectively removed Acid blue 25, an anionic dye from aqueous environment (Lakkaboyana et al., 2021). Chitosan composite tailored with zinc-doped hydroxyapatite was used for the exclusion of dye, Methylene Orange (MO). The maximum adsorption efficiency of Zn^{2+}@Hap@Cs was 453.15 mg/g at 60°C (Billah et al., 2024). Hassan and coworkers prepared chitosan/

silica/ZnO_2 nanocomposite for the adsorption of methylene blue from water. The adsorption ability of nanocomposite was almost 300 mg/g in a slight acidic condition. This composite is economic and effective for reducing the concentration of organic dyes from aqueous solution (Hassan et al., 2019).

Mahmoodi and Javanbhakt, 2021 synthesized magnetic nanocomposite, zeolitic imidazolate framework-8 (ZIF-8) using mucilaginous polysaccharide of basil seeds for the exclusion of azo dyes. ZIF-8 adsorbed the cationic dye methylene blue and anionic Eriochrome black T. The former was adsorbed to 9.09 mg/g and the latter was adsorbed to 13.21 mg/g on to the nanocomposite. Magnetic separation property and its reusability makes the ZIF-8 as a better adsorbent.

Pesticides are one of the key elements of organic pollutants and stimulates the growth of toxic algae. In a study, bi-functional ternary nanocomposites were prepared using chitosan enveloped carbon nanofibres with silver doped Co_3O_4 for the detection and degradation of water contaminated with nitroaromatic compounds (4-nitrophenol) at nanomolar concentrations (Balram et al., 2022). Similarly, highly toxic pesticides such as chlorpyrifos and diazinon were detected in aquatic environment. Chlorpyrifos (CPF) is extensively used in agriculture for decades can disrupts the nervous system and ultimately kills the insects. Diazinon is an Organic Phosphorous Pesticide (OPPs) widely used in animal husbandry and agricultural practices. The nanocomposite hydrogel, prepared by Salehi and group has shown efficient adsorption capacity of diazinon and chlorpyrifos (Salehi et al., 2024). Acetamiprid, a neonicotinoid pesticide is generally used in crops, fruits, tea, and vegetables to prevent aphids and other pests. This pesticide has strong insecticidal activity, highly soluble in water and has low impact on humans. Due to its high solubility, there are higher possibility of leaching pesticides into the water system through rain water. Mosavi et al. developed a bio nanocomposite, carboxymethyl tragacanth-grafted-poly(3-aminophenol)/zinc oxide@iron oxide (CMT-g-P3AP/ZnO@Fe3O4) for the ejection of acetamiprid from contaminated water under optimum conditions (Mosavi et al., 2024).

The pharmaceutical sector releases high concentration of antibiotic wastes in to the environment. Effluents from hospitals and domestic sewages often contain pharmaceutical residues which contaminates the water bodies. Moreover, pharmaceutical products affect environment by (i) development of antibiotic resistance; (ii) disrupting ecological balance by killing beneficial organisms in the environment; (iii) health menaces to humans and animals through contaminated water or crops. The use of conventional wastewater treatment practices may not effectively remove such residues from the wastewater. Therefore, bio adsorbents could be a potential strategy to mitigate pollution due to pharmaceutical residues in the environment. A group of researchers developed iron nanoparticles supported in clay biopolymer prepared as carboxymethyl cellulose and acrylamide-based beads for the exclusion of antibiotics, ciprofloxacin, and levofloxacin from industrial effluents. Under optimum conditions, the nanocomposite removed ciprofloxacin and levofloxacin at the rate of $>$ 92 and 93% respectively. The adsorbent possesses higher adsorption rate, good stability, eco-friendly and good regeneration ability (Gopal et al., 2022; Al-Hazmi et al., 2024). Various biopolymeric nanocomposites fabricated and studied for their adsorption ability and thereby to eliminate contaminants from water are tabulated (Table 1).

6. Biopolymeric Nanocomposites as Filters/Filter Supports for Water Purification Systems

Filtration techniques are widely used for the purging of contaminants and particulate matters from water. The membranes used in the filtration is of two types: symmetric/homogenous and asymmetric/heterogenous. They vary in the pore structure and uniformity. Symmetric membranes are one layered polymer with homogeneity in their pore sizes, whereas, asymmetric membranes are made up of two different layers; outer layer (thin) controls selectivity and the inner layer (porous & thick) provides strength to the membrane. Asymmetric membranes are applicable for the large-scale water treatment process due to the thinner structure, which offers higher water flux compared

Table 1. Different types of adsorption properties of biopolymer nanocomposites and the respective contaminants in water were mentioned.

Biopolymer nanocomposites	Pollutants	References
	Inorganic Pollutants	
Chitosan and $CaCO_3$ nanoparticles	Cd, Cu, Pb, Zn, Cr and Ni	Hassan et al., 2024
Walnut shell ash/starch/Fe_3O_4	Cu (II)	Foroutan et al., 2022
Chitosan-polystyrene-Zn nanocomposite	Nitrate	Keshvardoostchokami et al., 2017
Fe_3O_4-chitosan	Nickel and Cobalt	Sharifi et al., 2023
Grafted TEMPO-oxidised cellulose nanofiber (TEMPO-CNF)	Pb	Abou-Zeid et al., 2021
Chitosan-silver nanoparticles-carbon nanotubes (Chit-SNPs-CNTs)	Ni^{2+}	Azzam et al., 2023
Chitosan-g-PMMA/silica	Cr (VI)	Sethy and Sahoo, 2019
Chitosan/Ag-hydroxyapatite beads (CS/Ag-HA)	Cu^{2+}	Li et al., 2018
Ni-Al-CO_3-layered double hydroxide (LDH)	Cd^{2+} and Cu^{2+}	El Rouby et al., 2020
Chitosan/polyvinyl alcohol, poly(ethyleneimine), Fe_3O_4 impregnated beads (PEI@AC@Fe_3O_4)	Cr (VI)	Singh et al., 2022
Pec-g-PHEAA/Fe_3O_4	Cu(II) and Hg(II) ions	Kulal et al., 2020
	Organic pollutants	
Chitosan-Alginate nanocomposite	Malathion	Sabbagh et al., 2023
Copper-coated chitosan nanocomposite (CuCH)	Organophosphorus pesticides-malathion, methyl parathion and parathion	Jaiswal et al., 2012
Chitosan/Zero-valent iron nanoparticle (CS/nZVI) composite	Diazinon	Farhadi et al., 2021
Arabic gum-grafted-polyamidoxime and $CuFe_2O_4$ magnetic nanoparticles (AG-PAO/$CuFe_2O_4$)	Chlorpyrifos	Hassanzadeh-Afruzi et al., 2022
Magnetic-carboxymethyl cellulose nanofiber (Fe_3O_4-cmCNF)	Organophosphorus pesticides	Mohamed et al., 2022
Chitosan/Zinc phthalocyanine composite (ZnPc-CS)	4-Nitrophenol Azo dyes (Methyl orange and Congo red)	Ali et al., 2017
Magnetic graphene oxide and carboxymethyl cellulose (GO-CMC-Fe)	Herbicide, Atrazine	Khawaja et al., 2021
Fe clay cellulose-acrylamide beads	Ciprofloxacin and levofloxacin	Gopal et al., 2022
Chitosan capped ZnS QDs+ Nickel ferrite NPs	Tetracycline & Minocycline	Kaur et al., 2024
Bismuth oxyiodide-chitosan (BiOl-Ch)	Amoxicillin	Bisaria et al., 2022
Ag (NP) decorated clay polymer composite	Ciprofloxacin and 17α-ethinylestradiol	Goswami et al., 2024
Chitosan+ lignin + photosynthesized Fe MNPs (Lg/Cht MNPs)	Diesel oil	El Rabey et al., 2023

Table 1 contd. ...

...Table 1 contd.

Chitosan/Ag-hydroxyapatite beads (Chitosan/ Ag-HA)	Rhodamine B and Microbial load	Li et al., 2018
Biopolymer nanocomposites	**Pollutants**	**References**
Chitosan/polyvinyl alcohol, poly(ethyleneimine), Fe3O4 impregnated beads PEI@AC@Fe_3O_4	Azo dyes (tartrazine, sunset yellow, erythrosine)	Singh et al., 2022
Pec-g-PHEAA/Fe_3O_4	Rhodamine 6G dye	Kulal et al., 2020
Clinoptilolite (CLT)/Starch/$CoFe_2O_4$	Methylene blue, methyl violet and crystal violet dyes	Foroutan et al., 2021
Nickel oxide/chitosan/polyvinyl alcohol nanocomposite films	Methyl orange	Ali et al., 2021
Microorganisms		
Ag (NP) decorated clay polymer composite	*Bacillus subtilis, Pseudomonas aeruginosa*	Goswami et al., 2024
Nickel oxide/chitosan/polyvinyl alcohol nanocomposite films	*Staphylococcus aureus, Bacillus cereus, Escherichia coli, Salmonella typhimurium*	Ali et al., 2021
Chitosan/Zinc phthalocyanine composite (ZnPc-CS)	*E. coli*	Ali et al., 2017
Xanthan gum + poly (vinyl imidazole) + Montmorillonite (MMT) nano clays (XG-g-PVI/MMT)	*E. coli* *S. aureus*	Elella et al., 2021

to symmetric membranes (Kolya and Kang, 2023). Based on the size of the pollutants, membranes used for the filtration are categorized as: (i) microfiltration removes suspended matters and bacteria with a pore size of 10 μm–100 nm; (ii) ultrafiltration membranes are effective for the removal of nanoparticles and viruses, size range-100 to 2 nm; (iii) nanofiltration has a pore size of 2 nm to 1 nm, used to remove organic pollutants; (iv) reverse osmosis used to remove heavy metal ions, having nanopore size of 0.1–1 nm (Salama et al., 2021) (Fig. 3).

The main disadvantage of membrane filtration is fouling, denotes to the buildup of undesirable substances on the pores. This hinders the filtration capacity, increases energy consumption, and lifespan of the membranes. Biopolymeric nanocomposites were reported to alleviate fouling in the filtration process (Kolya and Kang, 2023). In a study, Poly (D-Lactic Acid) (PDLA) membranes decorated with nanocrystalline cellulose (NSS) and cellulose nano whisker (CNW) were used at different loading concentrations such as 1 and 2% wt. They reported that CNWs-filled PDLA membranes have potential to be used as a water filtration system (Kian et al., 2021). Table 2 represents biopolymeric nanocomposites in filter/filtration system for the deduction of various inorganic, organic and microbes from the water.

Recently, polysulphone membranes incorporated with carbon nanotubes and chitosan for the elimination of heavy metals from the aqueous solution. Nanocomposites such as COOH-functionalized carbon nanotubes/chitosan and normal carbon nanotubes/chitosan were used as a layer over the polysulphone (PS) to yield CNTs-COOH/CHIT/PS and CNTs/CHIT/PS membranes. Heavy metal rejection capacities of both nanocomposites were measured at different pH and

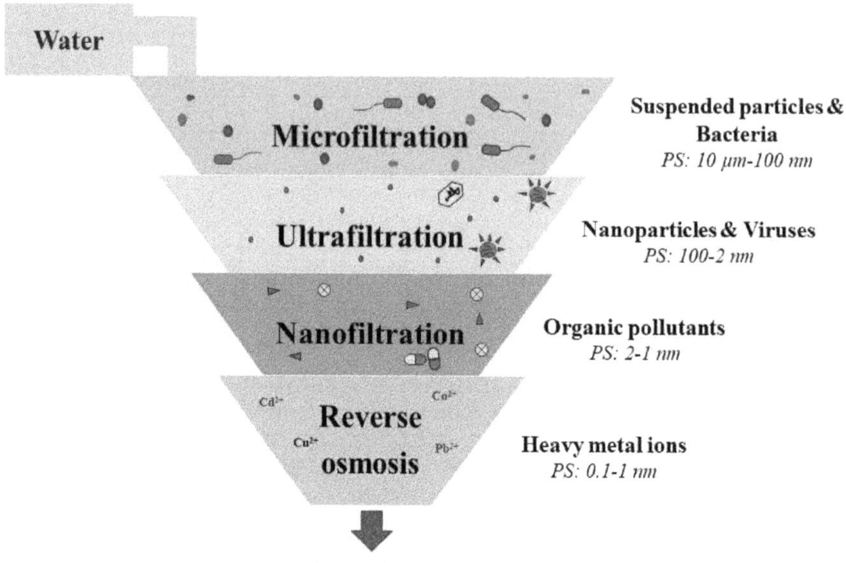

Figure 3. Types of filtrations or filters used in water treatment process.

pressures. Both membranes rejected Cu^{+2}, Ni^{+2}, Pb^{+2}, Cd^{+2} and Co^{+2} ions from aqueous phase in the range of 71–92.2% at pH 3. The COOH functionalized formulation showed more rejection than the other (Alawady et al., 2020).

Elessawy and co-workers fabricated a nanocomposite membrane supported on polyvinyl alcohol and carrageenan polymers reinforced with graphene oxide nanoparticles for the removal of cadmium and lead from water (Elessawy et al., 2022). Palacios Hinestroza et al., 2020 prepared membrane filters comprised of polycaprolactone (PCL) and cellulose nanofibres (CNF) for the elimination of iron and chromium. PCL/CNF membrane removed 100% turbidness, 100% conductivity and heavy metals by 75 to 99%.

In another study, researchers fabricated a nanocomposite membrane combined with cellulose nanocrystals for the exclusion of copper and lead. The membrane rejected toxic heavy metals such as $CuSO_4$, $CuCl_2$ and $PbCl_2$ by more than 90% (Hoang et al., 2020).

Cellulose based filtration methods are commonly utilized for the decontamination of organic impurities such as dyes. Dogra et al. 2022 fabricated an economical graphene oxide-cellulose (GO-C-NC) filter for the effective removal of a cationic dye in water. The membrane reduced the dye concentration with an efficiency of 99.41% in the laboratory condition even after 12 cycles. Combined effect of the GO-C-NC with peepal leaf extract enhances the process.

Gopi and co-workers synthesized PVDF membrane functionalized with chitin nano whisker for the rejection of indigo carmine dye. The dye removal efficacy by the nanocomposite was 88.9% in comparison to the non-functionalized PVDF (Gopi et al., 2017). Mousavi et al., 2023 reported chitosan linked graphene oxide incorporated into poly(ether-b-amide) thin film nanocomposite membrane coated on polyether sulphone for water purification. They reported the rejection capacities of CSGO TFC for malachite green dye and the antifouling properties of the membrane.

Decontamination of oil from wastewater from sources such as food, pharmaceutical and petrochemical industries is necessary to safeguard water resources. In a study, researchers used Bacterial Cellulose (BC) and cellulose nanofibres (CNF) were applied for the elimination of oil from oil-in-water emulsions. Compared to the CNF membranes, BC showed higher efficacy and water flux for the removal of oil emulsion (Hassan et al., 2017). Similarly, Gopi et al. 2018 prepared

Table 2. Pollutants removed by biopolymeric nanocomposites through filtration mechanism are enlisted.

Bionanocomposites	**Pollutants**	**References**
	Inorganic pollutants	
CNTs-COOH/CHIT/PS and CNTs/CHIT/PS membranes	Cu^{+2}, Ni^{+2}, Pb^{+2}, Cd^{+2} and Co^{+2} ions	Alawady et al., 2020
Fe_3O_4@SiO_2-Chitosan	Pb^{2+}, Cu^{2+} and Cd^{2+}	Kamari and Shahbazi, 2021
Dual layer membrane, poly (lactic acid) (PLA)+poly (butylene succinate) (PBS) with cellulose nano whisker (CNWs)	Co^{2+} and Ni^{2+}	Kian et al., 2021
Activated charcoal-banana rachis cellulose nanocrystal composites	Ni (II), Pb (II), Cr (III), Cd (II) and Co (II)	Rahman et al., 2024
Macroporous alginate/akageneite composite (MAAC)	Arsenic	Sinha et al., 2018
Cellulose nanocrystals/polyether sulphone (CNC/PES)	Cu^{2+} and Pb^{2+}	Hoang et al., 2020
Chitosan-EDTA-mGO	Cd^{2+}, Pb^{2+}	Salahshoor et al., 2021
	Organic Pollutants	
Fe_3O_4@SiO_2-Chitosan	Methylene blue, Congo red and reactive black 5 dyes	Kamari and Shahbazi, 2021
Fe-MOF crosslinked with chitosan based nanofiltration membrane	Cationic and anionic dyes	Maraddi et al., 2023
Activated charcoal-banana rachis cellulose nanocrystal composites	Methyl orange dye	Rahman et al., 2024
Chitosan-EDTA-mGO	Methylene blue, RR195 azoic dye	Salahshoor et al., 2021
Chitosan-graphene oxide (CSGO)-poly(ether-b-amide) (PEBA1657/PES TFC)	Malachite green	Mousavi et al., 2023
Bacterial cellulose (BC) + cellulose nanofibres (CNF)	Oil	Hassan et al., 2017
Chitin nano whisker + polyvinylidene fluoride	Oil	Gopi et al., 2018
	Microorganisms	
Lignin capped AgNPs + Cork pieces	*E. coli, S. aureus* and *B. cereus* spores	Bermeo et al., 2022

chitin nano whiskers coated on PVDF membrane showed improved separation of oil and water with enhanced flux rate, indicating its ability as a filtration system.

Another major contaminant present in water sources are microorganisms. In a study, silver nanoparticles capped with lignin (AgLNP) was incorporated into cork pieces as a filter for the ejection of pathogenic microorganisms from the water. The combination effectively diminished the load of *E. coli* and *S. aureus*. Remarkably, the constructed filtering system eliminated *E. coli* cells within a treatment period of 6 h onwards and *Bacillus cereus* spores from 16–48 h of water treatment. Hence, the results proved the efficacy of nanocomposite as a filtering system for water disinfection (Bermeo et al., 2022). Wang et al., 2021 reported that cellulose/dopamine-TiO_2 composite based filtration membrane showed antifouling activity in the water treatment process. The composite was formulated by linking chitosan (CS) and dopamine modified TiO_2 (DA- TiO_2) NPs, where CS and DA- TiO_2 have more than 90% antibacterial activity against *E. coli* and *S. aureus.* Ahmed and co-workers fabricated nanomembranes made up of chitosan (CS)/polyvinyl alcohol (PVA) incorporated with graphene oxide (GO) with gum rosin and AgNPs used for the decontamination of microbes from water (Ahmed et al., 2023).

7. Antimicrobial Property of Biopolymeric Nanocomposites

Bio nanocomposites have several applications due to their antimicrobial properties they can be used in water purification and in the food packaging industry. They are useful in decreasing the growth of various contaminant microorganisms. Nanocomposites are used for photocatalysis, drug delivery, tissue engineering and surface coated nanocomposites have been used to inhibit the microbial growth. Biopolymer-based nanocomposites are preferable due of their excess availability, cost effectiveness and renewability. There are several antimicrobial bio-nanocomposite used in the water purification process. It was found that biopolymeric nanocomposites used in water purification have antimicrobial properties (Sharma et al., 2020).

Alginate is used for the preparation of different nanocomposites due to its non-toxic, cost effective and good biodegradability as well as biocompatibility and stable nature, gel producing and thickening nature. Whey protein isolates film showed maximum mechanical strength. Past studies showed that antimicrobial potential of PBS (polyesters of polybutylene succinate) with Zinc Oxide (ZnO) against *E. coli* and *S. aureus* and results showed clear zone of 1.31 and 1.25 cm respectively (Sivakanthan et al., 2020; Xie, 2024; Petchwattana et al., 2016).

SiO_2 nanoparticles filled with PBAT composite tested against the same organisms showed zone of inhibition of 17.2 mm and 16.7 mm when these nanocomposites were prepared by the method of solvent casting (Venkatesan and Rajeswari, 2019). Nanocomposite prepared by PLA/PBAT/nanocrystal cellulose with silver nano hybrids showed antibacterial activity against various bacteria (Ma et al., 201). PBAT when combined with silver nanoparticle films showed good activity against *E. coli* as well as against *Listeria monocytogenes* (Shankar and Rhim, 2016). PBAT with ZnO nanofilms significantly increased both mechanical and thermal stability of the nanocomposite and exhibited tremendous antibacterial activity against several bacteria (Venkatesan and Rajeswari, 2017).

Polydimethyloxane/montmorillonite- chlorhexidine acetate bionanocomposite showed efficient antimicrobial activity against various bacteria (Meng et al., 2009).

Cellulose acetate with CuNPs (2%) was tested for antimicrobial activity using zone of inhibition assay against various microorganisms viz *S. aureus* (21 mm), *P. aeruginosa* (18mm), *Candida albicans* (19mm) and *Aspergillus niger* (15mm) (Abou-Yousef et al., 2018). Essential oils extracted from different plants when used in a bio nanocomposite showed antimicrobial activity against different microorganisms (Sivakanthan et al., 2020).

From several studies it was also observed that chitin based AgNC, chitosan based AgNC, chitosan-starch based AgNC, alginate based AgNC, rubber-based nanocomposite have an activity against *E. coli* and *S. aureus* (Zahran et al., 2019).

Methylcellulose-Ag nanocomposites having antimicrobial activity against various Gram-positive and negative bacteria (Maity et al., 2012). Fe_3O_4 levan nanocomposite having antibacterial capability against *E. coli* and *S. aureus* (Taran et al., 2017).

Chitosan nanocomposites have several antimicrobial properties due to the interaction of negative charged microbial cell membrane with positive charged amine group in chitosan or due to its chelating properties, it can selectively bind with different metals, inhibit various metabolic enzymes of microbial cells by entering their active centres that leads to reduced bacterial growth (Kumar et al., 2020).

Nanocomposite films made up of pullulan and AgNPs showed good antibacterial activity against food borne pathogens. The maximum zone of inhibition was observed against *S. typhimurium* and *E. coli* (Wypij et al., 2023). Furcellaran films (FUR)+AgNP nanocomposite showed antimicrobial potential against *P. aeruginosa*, *E. facecalis* and *S. aureus* with a good zone of inhibition. Polyvinyl alcohol/cellulose nanocrystal/graphene oxide (PVA/CNC1/GO) nanocomposite loaded with AgNPs exhibited significant antibacterial activity (zone of inhibition 13.5 mm) against *E. coli* and *S. aureus*.

(Bai et al., 2020). Graphene incorporating CFx exhibited bacterial inhibition against Gram-positive *Micrococcus luteus*. BIO/CFx showed inhibition zone of about 42mm and BIO/gnP-5/CFx have 35mm zone against *Micrococcus luteus* (Scaffaro et al., 2016). C-AuNPs-Amp (chitosan capped gold nanoparticles coupled with ampicillin) nanocomposite have antibacterial activity against *E. coli* (Chamundeeswari et al., 2010).

Organic rectorite and quaternized chitosan (OREC/HTCC) nanocomposite showed good inhibition potential against different bacteria and fungi while the optimum condition is determined in the presence of a weak base and weak acid. Minimum inhibitory concentration of the nanocomposites was less than 0.00313% (w/v) against *S. aureus* and *B. subtilis*. It was also observed that inhibition rate on *S. aureus* reached above 90% within 30 mins (Wang et al., 2009). PVA/TrGO (polyvinyl alcohol/thermally reduced graphene oxide) nanocomposite possessing good antibacterial activity against *E. coli* (Cho et al., 2021).

Nanocomposite made of sodium alginate with polyvinyl alcohol, and g-acrylamide which is known as NaAlg-PVA-g-Am exhibited antibacterial effect against *E. coli* and *S. aureus* (Novientri et al., 2019). Rozilah and coworkers fabricated sugar palm starch nanocrystalline cellulose biocomposites. The nanocomposites exhibited antibacterial activity against *E. coli*, Salmonella cholerasuis, and *S. aureus*.

Chitosan biopolymer incorporated with different types of nanofillers such as montmorillonite (Na-MMT, Cloisite30B), Nano-silver and Ag-Ions were fabricated as films and tested against four food pathogens including *S. aureus*, *L. monocytogenes*, *S. typhimurium*, and *E. coli*. Except Na-MMT incorporated bionanocomposite film, others showed antibacterial activity (Rhim et al., 2006). Ag glycogen nanocomposite showed higher reduction against *S. aureus* whereas *C. albicans* was more resistant (Bozanic et al., 2011). Methyl Cellulose (MC) nanocomposite films with pediocin and ZnO oxide nanoparticles was effective against *S. aureus* and *L. monocytogenes* (Espitia et al., 2013).

ZnO biopolymer nanocomposite coated with cotton, the foam used against *A. niger* NCIM563. Nanocomposite showed good antifungal activity against the test fungus (Sanoop et al., 2012). Another nanocomposite made up of silver nanoparticles which is stabilized in Guar Gum Alkyl amine that is known as GGAA combine with Ciprofloxacin exhibited effective antibacterial activity against different microorganisms such as *P. aeruginosa*, *E. coli*, *B. cereus*, and *S. aureus* (Islan et al., 2015). Chitosan-based nanocomposite film combined with copper and zinc oxide nanoparticles extracted from *Urtica dioica* showed bactericidal activity against *Enterobacter cloacae*, *S. typhi*, *S. aureus*, *Campylobacter jejuni* (Kalia et al., 2021). Nanocomposite films composed of sodium alginate and chitosan and copper or Zinc oxide nanoparticles (SA-CS@ CuO/ZnO) exhibited antibacterial effect against various bacteria. From results it was established that treatment with SA-CS@CuO/ZnO nanocomposite inhibited *E. coli* and *S. aureus* growth above 60% in the dark condition and above 90% under light condition (Guan et al., 2021).

Pectin Zirconium (Iv) selenotungstophosphate (PC/ZSWP) exhibited antibacterial activity against *S. aureus* at different concentrations tested. 240 μg/ml is more effective in inhibition of the bacteria (Sharma et al. 2014). Gelatin or tragacanth combined with ZnO-NPs nanocomposite film (GEL-TGC-ZnO) showed effective antibacterial activity against various organisms (Shahvalizadeh et al., 2021).

Alginate- cupric oxide bio nanocomposite showed activity against *E. coli* and *S. aureus* (Safaei et al., 2018). Chitosan-polyaniline nanocomposite has antimicrobial properties against *A. niger* and *E. coli* (Mohammadi et al., 2019). Nanocomposite hydrogel made up of fully exfoliated titanate layered supported AgNPs active against *E. coli* and *B. subtilis* (Sarkar et al., 2017). Gelatine- Zr phosphate nanocomposite exhibited antimicrobial activity against *E. coli* at a concentration of 7mg/ml (Thakur et al., 2017).

AgCNCS exhibited antimicrobial activity against *E. faecalis* (MIC = 1.56 µg/ml), *S. typhimurium* (MIC = 3.125 µg/ml), *L. monocytogenes* (MIC = 7.8 µg/ml) and *P. aeruginosa* (MIC = 12.5 µg/ml) (Raza et al., 2021). Chitosan with the combination of zinc oxide-polyvinyl pyrrolidone nanocomposite inhibited several Gram-positive and Gram-negative bacteria. Inhibition against Gram-positive bacteria was higher as compared to Gram negative bacteria due to cell wall differences (Karpuraranjith et al., 2017). Gelatin/ AgNps/Clay nanocomposite showed inhibitory activity. Reduction of the bacterial pathogen was observed within 12 hr of incubation (Kanmani et al., 2014).

Guargum/gelatin/ silver nanocomposite showed the zone of inhibition against *E. coli* (14.5 mm), *S. aureus* (14.5 mm) and *P. aeruginosa* (15 mm) (Khan et al., 2020). The strongest antibacterial effect against *S. aureus* and *E. coli* was promoted by St-SiO2-4% AgNP bio composite (Hajizadeh et al., 2020). Among various nanocomposites CS (chitosan) AgNPs and HPMC (hydroxypropyl methylcellulose) exhibited good antibacterial activity (Rojas et al., 2023). Chitosan coated zinc oxide nanocomposite showed the highest inhibition against *E. coli* followed by *K. pneumoniae*, *S. aureus* and least against *B. subtilis* (Bharathi et al., 2019). Pc/ZSPNC (pectin zirconium silicon phosphate) nanocomposite showed good antibacterial potential against *E. coli* and *S. aureus* (Pathania et al., 2015).

Ag-NPs embedded TG hydrogel solution was tested against *B. subtilis* and *E. coli* (Rao et al., 2018). Compared to normal nanoparticles biopolymeric nanocomposites like Furcellaran and gelatin composite film reinforced with selenium nanoparticles have more antibacterial activity against *E. coli* and *S. aureus* (Jamroz et al., 2018). Cellulose nanofibrils and silver nanoparticles with polyvinyl alcohol nanocomposite film showed some degree of inhibition across different microorganisms (Madivoli et al., 2022). Sodium alginate montmorillonite nanocomposite films incorporated with 1.5% (v/v) *Origanum majorana* essential oil inhibited the growth of *E. coli*, *L. monocytogens*, *B. cereus*, *S. aureus* (Alboofetileh et al., 2018).

Chitosan silver nanocomposite exhibited a better antibacterial activity against *S. aureus* than silver nanoparticles (Regiel et al., 2012). Crosslinking of lignin with polyvinyl alcohol combine nanocomposite showed 99.9% reduction rate against *S. aureus* but no reduction against *E. coli* (Lee et al., 2019). Starch Reduced Graphene Oxide (SRGO)-polyiodide nanocomposite exhibited potential antibacterial activities against *E. coli* and *S. aureus* with IC50 value 0.45 and 0.41 mg/ml (Narayanan et al. 2021). Poly-L-lactic acid composite with Ti/Au nanostructure was effective against *E. coli*, where the appearance of small colonies suggested its bacteriostatic effect (Slepicka et al., 2021).

8. Conclusion

Biopolymeric nanocomposites have the potential in water purification/treatment strategies. Because of its processibility as well as biodegradable property, biopolymers are an excellent support system for the nanoparticles/nanofillers. Unique adsorbent, antibacterial, catalytic, and pollutant removal properties of the reinforced, encapsulated, or entrapped nanoparticles/nanofillers can be improved and effectively utilized when in combination (as a bio nanocomposite) than its performance alone. However, research importance should be given to rectify the limitations such as scalability and stability. Furthermore, the biopolymeric nanocomposite production should be economical. Solving these problems will certainly pave the way in obtaining sustainable and efficient water purification systems.

References

Abou-Yousef, H., Saber, E., Abdel-Aziz, M. S. and Kamel, S. (2018). Efficient alternative of antimicrobial nanocomposites based on cellulose acetate/Cu-NPs. Soft Materials, 16(3): 141–150.

Abou-Zeid, R. E., Kamal, K. H., Abd El-Aziz, M. E., Morsi, S. M. and Kamel, S. (2021). Grafted TEMPO-oxidized cellulose nanofiber embedded with modified magnetite for effective adsorption of lead ions. Int. J. Biol. Macromol. 2021 Jan 15; 167: 1091–1101. doi: 10.1016/j.ijbiomac.2020.11.063. Epub 2020 Nov 11. PMID: 33186652.

Ahmed, E.M., Isawi, H., Morsy, M., Hemida, M.H. and Moustafa, H. (2023). Effective nanomembranes from chitosan/PVA blend decorated graphene oxide with gum rosin and silver nanoparticles for removal of heavy metals and microbes from water resources. Surfaces and Interfaces, 39: 102980.

Alawady, A.R., Alshahrani, A.A., Aouak, T.A. and Alandis, N.M. (2020). Polysulfone membranes with CNTs/Chitosan biopolymer nanocomposite as selective layer for remarkable heavy metal ions rejection capacity. Chemical Engineering Journal, 388: 124267.

Alboofetileh, M., Rezaei, M., Hosseini, H. and Abdollahi, M. (2014). Antimicrobial activity of alginate/clay nanocomposite films enriched with essential oils against three common foodborne pathogens. Food Control, 36(1): 1–7.

Alboofetileh, M., Rezaei, M., Hosseini, H. and Abdollahi, M. (2018). Morphological, physico-mechanical, and antimicrobial properties of sodium alginate-montmorillonite nanocomposite films incorporated with marjoram essential oil. Journal of Food Processing and Preservation, 42(5): e13596.

Al-Hazmi, H. E., Łuczak, J., Habibzadeh, S., Hasanin, M. S., Mohammadi, A., Esmaeili, A. et al. (2024). Polysaccharide nanocomposites in wastewater treatment: A review. Chemosphere. 2024 Jan; 347: 140578. doi: 10.1016/j.chemosphere.2023.140578. Epub 2023 Nov 6. PMID: 37939921.

Ali, F., Khan, S. B., Kamal, T., Anwar, Y., Alamry, K. A. and Asiri, A. M. (2017). Anti-bacterial chitosan/zinc phthalocyanine fibers supported metallic and bimetallic nanoparticles for the removal of organic pollutants. Carbohydr. Polym. 2017 Oct 1; 173: 676–689. doi: 10.1016/j.carbpol.2017.05.074. Epub 2017 Jun 4. PMID: 28732913.

Ali, H., Tiama, T. M. and Ismail, A. M. (2021). New and efficient NiO/chitosan/polyvinyl alcohol nanocomposites as antibacterial and dye adsorptive films. Int. J. Biol. Macromol. 2021 Sep 1;186: 278–288. doi: 10.1016/j.ijbiomac.2021.07.055. Epub. 2021 Jul 8. PMID: 34246674.

Álvarez-Castillo, E., Aguilar, J. M., Bengoechea, C., López-Castejón, M. L. and Guerrero, A. (2021). Rheology and Water Absorption Properties of Alginate–Soy Protein Composites. Polymers, 13(11): 1807. https://doi.org/10.3390/polym13111807.

Amiri, M.J., Raayatpisheh, M., Radi, M. and Amiri, S. (2023). Preparation and characterization of biopolymer-based adsorbents and their application for methylene blue removal from wastewater. Scientific Reports, 13(1): 17263.

Ashbolt, N. J. (2015). Microbial contamination of drinking water and human health from community water systems. Current Environmental Health Reports, 2(1): 95–106. https://doi.org/10.1007/s40572-014-0037-5.

Azzam, E. M. S., Elsofany, W. I., Abdulaziz, F., AlGhamdi, H. A., Al Alhareth, A. Y. (2023). Ecofriendly Elimination of Ni (II) using fabricated nanocomposite based on chitosan/silver nanoparticles/carbon nanotubes. Polymers (Basel). 2023 Jun 21;15(13): 2759. doi: 10.3390/polym15132759. PMID: 37447405; PMCID: PMC10347260.

Bai, H., Liang, Z., Wang, D., Guo, J., Zhang, S., Ma, P. and Dong, W. (2020). Biopolymer nanocomposites with customized mechanical property and exceptionally antibacterial performance. Composites Science and Technology, 199: 108338.

Balram, D., Lian, K.Y., Sebastian, N., Al-Mubaddel, F.S. and Noman, M.T. (2022). Bi-functional renewable biopolymer wrapped CNFs/Ag doped spinel cobalt oxide as a sensitive platform for highly toxic nitroaromatic compound detection and degradation. Chemosphere, 291: 132998.

Baranwal, J., Barse, B., Fais, A., Delogu, G. L. and Kumar, A. (2022). Biopolymer: A sustainable material for food and medical applications. Polymers, 14(5): 983. https://doi.org/10.3390/polym14050983.

Benhacine, F., Abdellaoui, N., Arous, O. and Hadj-Hamou, A. S. (2018). Behaviours of poly (ε-caprolactone)/silver-montmorillonite nanocomposite in membrane ultrafiltration for wastewater treatment. Environmental technology.

Bermeo, L., Ivanova, K., Pérez, L.M., Forés, E., Pérez-Rafael, S., Casas-Zapata, J.C. et al. (2022). Sono-enzymatically embedded antibacterial silver-lignin nanoparticles on cork filter material for water disinfection. International journal of Molecular Sciences, 23(19): 11679.

Bharathi, D., Ranjithkumar, R., Chandarshekar, B. and Bhuvaneshwari, V. (2019). Preparation of chitosan coated zinc oxide nanocomposite for enhanced antibacterial and photocatalytic activity: As a bionanocomposite. International Journal of Biological Macromolecules, 129: 989–996.

Bhushan, B. and Kumar, R. (2019). Plasma treated and untreated thermoplastic biopolymers/biocomposites in tissue engineering and biodegradable implants. In Materials for Biomedical Engineering (pp. 339–369). Elsevier. https://doi.org/10.1016/B978-0-12-816901-8.00011-0.

Billah, R. E. K., Zaghloul, A., Oladoja, N. A., Algethami, J. S., Soufiane, A., López-Maldonado, E. A. et al. (2024). Multifunctional biocomposites based on cross-linked Shrimp waste-derived chitosan modified Zn2+@ Calcium apatite for the removal of methyl orange and antibacterial activity. Materials Today Sustainability, 25: 100660.

Bisaria, K., Wadhwa, S., Mathur, A., Roy, S., Dixit, A. and Singh, R. (2022). New bismuth oxyiodide/chitosan nanocomposite for ultrasonic waves expedited adsorptive removal of amoxicillin from aqueous medium: Kinetic, isotherm and thermodynamic investigations. Environ. Sci. Pollut. Res. Int. 2022 Dec; 29(57): 86260–86276. doi: 10.1007/s11356-021-17546-8. Epub 2022 Jan 7. PMID: 34993771.

Božanić, D. K., Dimitrijević-Branković, S., Bibić, N., Luyt, A. S. and Djoković, V. (2011). Silver nanoparticles encapsulated in glycogen biopolymer: Morphology, optical and antimicrobial properties. Carbohydrate polymers, 83(2): 883–890.

Briffa, J., Sinagra, E. and Blundell, R. (2020). Heavy metal pollution in the environment and their toxicological effects on humans. Heliyon, 6(9): e04691. https://doi.org/10.1016/j.heliyon.2020.e04691.

Carvalho, A.J., 2008. Starch: Major sources, properties and applications as thermoplastic materials. Monomers, Polymers and Composites from Renewable Resources, pp. 321–342.

Chamundeeswari, M., Sobhana, S. L., Jacob, J. P., Kumar, M. G., Devi, M. P., Sastry, T. P. et al. (2010). Preparation, characterization and evaluation of a biopolymeric gold nanocomposite with antimicrobial activity. Biotechnology and Applied Biochemistry, 55(1): 29–35.

Cho, B. G., Joshi, S. R., Lee, S., Kim, S. K., Park, Y. B. and Kim, G. H. (2021). Enhanced mechanical and antibacterial properties of nanocomposites based on poly (vinyl alcohol) and biopolymer-derived reduced graphene oxide. Polymers, 13(4): 615.

Dechesne, M. and Soyeux, E. (2007). Assessment of source water pathogen contamination. Journal of Water and Health, 5(S1): 39–50. https://doi.org/10.2166/wh.2007.133.

Dogra, I., Kumar, B. R., Etika, K. C., Chavali, M., Khalifa, A. S., Gharib, A. F. et al. 2022. Environmentally friendly low-cost graphene oxide-cellulose nanocomposite filter for dye removal from water. Journal of King Saud University-Science, 34(5): 102122.

El Rabey, H. A., Almutairi, F. M., Tayel, A. A., Alalawy, A. I., Mohammed, G. M., Aljohani, M. M. et al. 2023. Magnetic biopolymers' nanocomposites from chitosan, lignin and phycosynthesized iron nanoparticles to remediate water from polluting oil. International Journal of Biological Macromolecules, 251: 126318.

El Rouby, W. M. A., El-Dek, S. I., Goher, M. E. and Noaemy, S. G. (2020). Efficient water decontamination using layered double hydroxide beads nanocomposites. Environ. Sci. Pollut. Res. Int. 2020 Jun; 27(16): 18985–19003. doi: 10.1007/s11356-018-3257-7. Epub 2018 Oct 2. PMID: 30280341.

Elella, M. H. A., Goda, E. S., Abdallah, H. M., Shalan, A. E., Gamal, H. and Yoon, K. R. (2021). Innovative bactericidal adsorbents containing modified xanthan gum/montmorillonite nanocomposites for wastewater treatment. International Journal of Biological Macromolecules, 167: 1113–1125.

Elessawy, N. A., Gouda, M. H., Elnouby, M., Ali, S. M., Salerno, M. and Youssef, M. E. (2022). Sustainable microbial and heavy metal reduction in water purification systems based on PVA/IC nanofiber membrane doped with PANI/GO. Polymers, 14(8): 1558.

Esfahani, M. R., Tyler, J. L., Stretz, H. A. and Wells, M. J. (2015). Effects of a dual nanofiller, nano-TiO2 and MWCNT, for polysulfone-based nanocomposite membranes for water purification. Desalination, 372: 47–56.

Espitia, P. J. P., Soares, N. D. F. F., Teófilo, R. F., dos Reis Coimbra, J. S., Vitor, D. M., Batista, R. A. et al. (2013). Physical–mechanical and antimicrobial properties of nanocomposite films with pediocin and ZnO nanoparticles. Carbohydrate Polymers, 94(1): 199–208.

Farhadi, S., Sohrabi, M. R., Motiee, F. and Davallo, M. (2021). Organophosphorus diazinon pesticide removing from aqueous solution by zero-valent iron supported on biopolymer chitosan: RSM optimization methodology. Journal of Polymers and the Environment, 29: 103–120.

Fawell, J. and Nieuwenhuijsen, M. J. (2003). Contaminants in drinking water. British Medical Bulletin, 68(1): 199–208. https://doi.org/10.1093/bmb/ldg027.

Flaris, V. and Singh, G. (2009). Recent developments in biopolymers. Journal of Vinyl and Additive Technology, 15(1): 1–11. https://doi.org/10.1002/vnl.20171.

Foroughnia, A., Khalaji, A. D., Kolvari, E. and Koukabi, N. (2021). Synthesis of new chitosan Schiff base and its Fe2O3 nanocomposite: Evaluation of methyl orange removal and antibacterial activity. International Journal of Biological Macromolecules, 177: 83–91.

Foroutan, R., Peighambardoust, S. J., Hemmati, S., Khatooni, H. and Ramavandi, B. (2021). Preparation of clinoptilolite/starch/CoFe2O4 magnetic nanocomposite powder and its elimination properties for cationic dyes from water and wastewater. Int J Biol Macromol. 2021 Oct 31; 189:432–442. doi: 10.1016/j.ijbiomac.2021.08.144. Epub 2021 Aug 24. PMID: 34450143.

Foroutan, R., Peighambardoust, S. J., Mohammadi, R., Peighambardoust, S. H. and Ramavandi, B. (2022). Development of new magnetic adsorbent of walnut shell ash/starch/Fe3O4 for effective copper ions removal: Treatment of groundwater samples. Chemosphere. 2022 Jun;296:133978. doi: 10.1016/j.chemosphere.2022.133978. Epub 2022 Feb 14. PMID: 35176297.

Gopal, G., Natarajan, C. and Mukherjee, A. (2022). Adsorptive removal of fluoroquinolone antibiotics using green synthesized and highly efficient Fe clay cellulose-acrylamide beads. Environmental Technology & Innovation, 28: 102783.

Gopi, S., Balakrishnan, P., Pius, A. and Thomas, S. (2017). Chitin nanowhisker (ChNW)-functionalized electrospun PVDF membrane for enhanced removal of Indigo carmine. Carbohydrate Polymers, 165: 115–122.

Gopi, S., Kargl, R., Kleinschek, K. S., Pius, A. and Thomas, S. (2018). Chitin nanowhisker–Inspired electrospun PVDF membrane for enhanced oil-water separation. Journal of Environmental Management, 228: 249–259.

Goswami, S., Dutta, D., Pandey, S., Chattopadhyay, P., Dubey, R. and Tiwari, D. (2024). Novel fibrous Ag (NP) decorated clay-polymer composite: Implications in water purification contaminated with predominant micro-pollutants and bacteria. Journal of Environmental Management, 359: 121063.

Gu, J., Zhou, W., Jiang, B., Wang, L., Ma, Y., Guo, H. et al. (2016). Effects of biochar on the transformation and earthworm bioaccumulation of organic pollutants in soil. Chemosphere, 145: 431–437. https://doi.org/10.1016/j.chemosphere.2015.11.106.

Guan, G., Zhang, L., Zhu, J., Wu, H., Li, W. and Sun, Q. (2021). Antibacterial properties and mechanism of biopolymer-based films functionalized by CuO/ZnO nanoparticles against *Escherichia coli* and Staphylococcus aureus. Journal of Hazardous Materials, 402: 123542.

Hajizadeh, H., Peighambardoust, S. J., Peighambardoust, S. H. and Peressini, D. (2020). Physical, mechanical, and antibacterial characteristics of bio-nanocomposite films loaded with Ag-modified SiO2 and TiO2 nanoparticles. Journal of Food Science, 85(4): 1193–1202.

Hassan, A. F. and Hrdina, R. (2018). Chitosan/nanohydroxyapatite composite based scallop shells as an efficient adsorbent for mercuric ions: Static and dynamic adsorption studies. Int. J. Biol. Macromol. 2018 Apr 1; 109:507–516. doi: 10.1016/j.ijbiomac.2017.12.094. Epub 2017 Dec 19. PMID: 29269012.

Hassan, H., Salama, A., El-Ziaty, A. K. and El-Sakhawy, M. (2019). New chitosan/silica/zinc oxide nanocomposite as adsorbent for dye removal. Int J Biol Macromol. 2019 Jun 15;131:520–526. doi: 10.1016/j.ijbiomac.2019.03.087. Epub 2019 Mar 14. PMID: 30880055.

Hassan, E., Hassan, M., Abou-Zeid, R., Berglund, L. and Oksman, K. (2017). Use of bacterial cellulose and crosslinked cellulose nanofibers membranes for removal of oil from oil-in-water emulsions. Polymers, 9(9): 388.

Hassan, S.S., Abd El-Aziz, M.E., Fayez, A.E.S., Kamel, A.H. and Youssef, A.M. (2024). Synthesis and characterization of bio-nanocomposite based on chitosan and CaCO3 nanoparticles for heavy metals removal. International Journal of Biological Macromolecules, 255: 128007.

Hassanzadeh-Afruzi, F., Maleki, A. and Zare, E. N. (2022. Efficient remediation of chlorpyrifos pesticide from contaminated water by superparamagnetic adsorbent based on Arabic gum-grafted-polyamidoxime. International Journal of Biological Macromolecules, 203: 445–456.

Hlongwane, G. N., Sekoai, P. T., Meyyappan, M. and Moothi, K. (2019). Simultaneous removal of pollutants from water using nanoparticles: A shift from single pollutant control to multiple pollutant control. Science of The Total Environment, 656: 808–833. https://doi.org/10.1016/j.scitotenv.2018.11.257.

Hoang, M. T., Pham, T. D., Verheyen, D., Nguyen, M. K., Pham, T. T., Zhu, J. et al. 2020. Fabrication of thin film nanocomposite nanofiltration membrane incorporated with cellulose nanocrystals for removal of Cu (II) and Pb (II). Chemical Engineering Science, 228: 115998.

Ilyas, R. A., Sapuan, S. M., Kadier, A., Kalil, M. S., Ibrahim, R., Atikah, M. S. N. et al. (2020). Properties and characterization of PLA, PHA, and other types of biopolymer composites. In Advanced Processing, Properties, and Applications of Starch and Other Bio-Based Polymers (pp. 111–138). Elsevier. https://doi.org/10.1016/B978-0-12-819661-8.00008-1.

Irandoost, M., Pezeshki-Modaress, M. and Javanbakht, V. (2019). Removal of lead from aqueous solution with nanofibrous nanocomposite of polycaprolactone adsorbent modified by nanoclay and nanozeolite. Journal of Water Process Engineering, 32: 100981.

Ishiaku, U. S., Pang, K. W., Lee, W. S. and Ishak, Z. M. (2002). Mechanical properties and enzymic degradation of thermoplastic and granular sago starch filled poly (ε-caprolactone). European Polymer Journal, 38(2): 393–401.

Islan, G. A., Mukherjee, A. and Castro, G. R. (2015). Development of biopolymer nanocomposite for silver nanoparticles and ciprofloxacin controlled release. International Journal of Biological Macromolecules, 72: 740–750.

Jaiswal, M., Chauhan, D. and Sankararamakrishnan, N. (2012). Copper chitosan nanocomposite: synthesis, characterization, and application in removal of organophosphorous pesticide from agricultural runoff. Environmental Science and Pollution Research, 19: 2055–2062.

Jamróz, E., Khachatryan, G., Kopel, P., Juszczak, L., Kawecka, A., Krzyściak, P. et al. (2020). Furcellaran nanocomposite films: The effect of nanofillers on the structural, thermal, mechanical and antimicrobial properties of biopolymer films. Carbohydrate Polymers, 240: 116244.

Jamróz, E., Kopel, P., Juszczak, L., Kawecka, A., Bytesnikova, Z., Milosavljević, V. et al. (2018). Development and characterisation of furcellaran-gelatin films containing SeNPs and AgNPs that have antimicrobial activity. Food Hydrocolloids, 83: 9–16.

Jiang, J., Zhang, Q., Zhan, X. and Chen, F. (2019). A multifunctional gelatin-based aerogel with superior pollutants adsorption, oil/water separation and photocatalytic properties. Chemical Engineering Journal, 358: 1539–1551.

Kalia, A., Kaur, M., Shami, A., Jawandha, S. K., Alghuthaymi, M. A., Thakur, A. et al. (2021). Nettle-leaf extract derived ZnO/CuO nanoparticle-biopolymer-based antioxidant and antimicrobial nanocomposite packaging films and their impact on extending the post-harvest shelf life of guava fruit. Biomolecules, 11(2): 224.

Kamari, S. and Shahbazi, A. (2021). High–performance nanofiltration membrane blended by Fe3O4@ SiO2–CS bionanocomposite for efficient simultaneous rejection of salts/heavy metals ions/dyes with high permeability, retention increase and fouling decline. Chemical Engineering Journal, 417: 127930.

Kanmani, P. and Rhim, J. W. (2014). Physical, mechanical and antimicrobial properties of gelatin based active nanocomposite films containing AgNPs and nanoclay. Food Hydrocolloids, 35: 644–652.

Karpuraranjith, M. and Thambidurai, S. (2017). Chitosan/zinc oxide-polyvinylpyrrolidone (CS/ZnO-PVP) nanocomposite for better thermal and antibacterial activity. International journal of biological macromolecules, 104: 1753–1761.

Kaur, J., Kaur, M., Aggarwal, D., Kumar, V., Tikoo, K., Kaushik, A. et al. 2024. Unveiling the multifaceted applications of magnetically responsive chitosan capped ZnS QDs for sensing and annihilation of pharmaceutical drugs. Talanta, 266: 125084.

Keshvardoostchokami, M., Babaei, S., Piri, F. and Zamani, A. Nitrate removal from aqueous solutions by ZnO nanoparticles and chitosan-polystyrene-Zn nanocomposite: Kinetic, isotherm, batch and fixed-bed studies. Int. J. Biol. Macromol. 2017 Aug;101:922–930. doi: 10.1016/j.ijbiomac.2017.03.162. Epub 2017 Mar 30. PMID: 28365288.

Khalil, H., Wadi, V. S., Hegab, H. M., Nassar, L., Naddeo, V., Yousef, A. F. et al. (2023). High-performance f-GO/MWCNTs-COOH nanohybrid-based polylactic acid mixed matrix membrane for wastewater treatment. Journal of Water Process Engineering, 53: 103784.

Khan, N., Kumar, D. and Kumar, P. (2020). Silver nanoparticles embedded guar gum/gelatin nanocomposite: Green synthesis, characterization and antibacterial activity. Colloid and Interface Science Communications, 35: 100242.

Khawaja, H., Zahir, E., Asghar, M. A., Rafique, K. and Asghar, M. A. (2021). Synthesis and application of covalently grafted magnetic graphene oxide carboxymethyl cellulose nanocomposite for the removal of atrazine from an aqueous phase. Journal of Macromolecular Science, Part B, 60(12): 1025–1044.

Kian, L. K., Jawaid, M., Alamery, S. and Vaseashta, A. (2021). Fabrication and characterization of novel poly (D-lactic acid) nanocomposite membrane for water filtration purpose. Nanomaterials, 11(2): 255.

Kian, L.K., Jawaid, M., Nasef, M.M., Fouad, H. and Karim, Z. (2021). Poly (lactic acid)/poly (butylene succinate) dual-layer membranes with cellulose nanowhisker for heavy metal ion separation. International Journal of Biological Macromolecules, 192: 654–664.

Klangmuang, P. and Sothornvit, R. (2016). Barrier properties, mechanical properties and antimicrobial activity of hydroxypropyl methylcellulose-based nanocomposite films incorporated with Thai essential oils. Food Hydrocolloids, 61: 609–616.

Kolya, H. and Kang, C. W. (2023). Next-generation water treatment: Exploring the potential of biopolymer-based nanocomposites in adsorption and membrane filtration. Polymers, 15(16): 3421.

Kovtun, A., Campodoni, E., Favaretto, L., Zambianchi, M., Salatino, A., Amalfitano, S. et al. (2020). Multifunctional graphene oxide/biopolymer composite aerogels for microcontaminants removal from drinking water. Chemosphere, 259: 127501.

Kulal, P. and Badalamoole, V. (2020). Magnetite nanoparticle embedded Pectin-graft-poly(N-hydroxyethylacrylamide) hydrogel: Evaluation as adsorbent for dyes and heavy metal ions from waste water. Int. J. Biol. Macromol. 2020 Aug 1;156:1408–1417. doi: 10.1016/j.ijbiomac.2019.11.181. Epub 2019 Nov 22. PMID: 31760033.

Kumar, S., Mukherjee, A. and Dutta, J. (2020). Chitosan based nanocomposite films and coatings: Emerging antimicrobial food packaging alternatives. Trends in Food Science & Technology, 97: 196–209.

Lakkaboyana, S. K., Soontarapa, K., Marella, R. K. and Kannan, K. (2021). Preparation of novel chitosan polymeric nanocomposite as an efficient material for the removal of Acid Blue 25 from aqueous environment. International Journal of Biological Macromolecules, 168: 760–768.

Lee, E., Song, Y. and Lee, S. (2019). Crosslinking of lignin/poly (vinyl alcohol) nanocomposite fiber webs and their antimicrobial and ultraviolet-protective properties. Textile Research Journal, 89(1): 3–12.

Li, L., Iqbal, J., Zhu, Y., Zhang, P., Chen, W., Bhatnagar, A. et al. (2018). Chitosan/Ag-hydroxyapatite nanocomposite beads as a potential adsorbent for the efficient removal of toxic aquatic pollutants. Int. J. Biol. Macromol. 2018 Dec;120(Pt B):1752–1759. doi: 10.1016/j.ijbiomac.2018.09.190. Epub 2018 Sep 30. PMID: 30282013.

Li, M., Kim, I. and Jeong, Y. G. (2010). Cellulose acetate/multiwalled carbon nanotube nanocomposites with improved mechanical, thermal, and electrical properties. Journal of Applied Polymer Science, 118(4): 2475–2481. https://doi.org/10.1002/app.32591.

Ma, P, Jiang, L., Yu, M., Dong, W. and Chen, M. (2016). Green antibacterial nanocomposites from poly (lactide)/poly (butylene adipate-co-terephthalate)/nanocrystal cellulose-silver nanohybrids. ACS Sustainable Chemistry & Engineering, 4(12): 6417–6426.

Madivoli, E. S., Kareru, P. G., Gichuki, J. and Elbagoury, M. M. (2022). Cellulose nanofibrils and silver nanoparticles enhances the mechanical and antimicrobial properties of polyvinyl alcohol nanocomposite film. Scientific Reports, 12(1): 19005.

Mahmoodi, M. and Javanbakht, V. (2021). Fabrication of Zn-based magnetic zeolitic imidazolate framework bionanocomposite using basil seed mucilage for removal of azo cationic and anionic dyes from aqueous solution. Int. J. Biol. Macromol. 2021 Jan 15;167:1076–1090. doi: 10.1016/j.ijbiomac.2020.11.062. Epub 2020 Nov 11. PMID: 33186651.

Maio, A., Gammino, M., Gulino, E. F., Megna, B., Fara, P. and Scaffaro, R. (2020). Rapid one-step fabrication of graphene oxide-decorated polycaprolactone three-dimensional templates for water treatment. ACS Applied Polymer Materials, 2(11): 4993–5005.

Maity, D., Molick, M. M. R., Mondal, D., Bhowmick, B., Bain, M. K., Bankura, K. et al. (2012). Synthesis of methylocellouse-silver nanocomposite and investigation of mechanical and antimicrobial properties. Carbohydrate Polymers, 90(4): 1818–1825.

Malagurski, I., Levic, S., Nesic, A., Mitric, M., Pavlovic, V. and Dimitrijevic-Brankovic, S. (2017). Mineralized agar-based nanocomposite films: Potential food packaging materials with antimicrobial properties. Carbohydrate Polymers, 175: 55–62.

Maraddi, A., Halakarni, M., Halanur, M. M. and Nataraj, S. K. (2023). Fe-MOF induced biopolymer-based sustainable self-cleaning membranes for effective selective separation and wastewater treatment. Sustainable Materials and Technologies, 35: e00537.

Meng, N., Zhou, N. L., Zhang, S. Q. and Shen, J. 2009. Synthesis and antimicrobial activities of polymer/montmotrillonite–chlorhexidine acetate nanocomposites films. Applied Clay Science, 42(3-4): 667–670.

Mohamed, A. H., Yahaya, N., Mohamad, S., Kamaruzaman, S., Osman, H., Nishiyama, N. et al. (2022). Synthesis of oil palm empty fruit bunch-based magnetic-carboxymethyl cellulose nanofiber composite for magnetic solid-phase extraction of organophosphorus pesticides in environmental water samples. Microchemical Journal, 183: 108045.

Mohammadi, B., Pirsa, S. and Alizadeh, M. (2019). Preparing chitosan–polyaniline nanocomposite film and examining its mechanical, electrical, and antimicrobial properties. Polymers and Polymer Composites, 27(8): 507–517.

Moradi, G., Zinadini, S. and Rahimi, M. (2023). Designing of the green γ-AlOOH@ Naringin thin film composite PVDF based nanofiltration membrane and application for pharmaceutical wastewater treatment. Journal of Environmental Chemical Engineering, 11(3): 109952.

Morsi, R. E., Alsabagh, A. M., Nasr, S. A. and Zaki, M. M. (2017). Multifunctional nanocomposites of chitosan, silver nanoparticles, copper nanoparticles and carbon nanotubes for water treatment: antimicrobial characteristics. International Journal of Biological Macromolecules, 97: 264–269.

Mosavi, S. S., Zare, E. N., Behniafar, H., Nezhad, S. M. and Neisiany, R. E. (2024). Remediation of acetamiprid pesticide from contaminated water by antibacterial biosorbent based on carboxymethyl tragacanth-grafted-poly (3-aminophenol) decorated with ZnO@ Fe3O4. Environmental Research, 252: 118893.

Mousavi, S. R., Asghari, M., Mahmoodi, N. M. and Salahshoori, I. (2023). Water decolorization and antifouling melioration of a novel PEBA1657/PES TFC membrane using chitosan-decorated graphene oxide fillers. Journal of Environmental Chemical Engineering, 11(3): 109955.

Narayanan, K. B., Park, G. T. and Han, S. S. (2021). Antibacterial properties of starch-reduced graphene oxide–polyiodide nanocomposite. Food Chemistry, 342: 128385.

Nivedita, S. and Joseph, S. (2021). Performance of polycaprolactone/TiO2 composite membrane for the effective treatment of dairy effluents. Water Science and Technology, 83(10): 2477–2485.

Novientri, G., Abbas, G. H. and Budianto, E. (2019). Nanocomposite hydrogel-based biopolymer modified with silver nanoparticles as an antibacterial material for wound treatment. Journal of Applied Pharmaceutical Science, 9(11): 001–009.

Othman, S. H. (2014). Bio-nanocomposite materials for food packaging applications: Types of biopolymer and nano-sized filler. Agriculture and Agricultural Science Procedia, 2: 296–303. https://doi.org/10.1016/j.aaspro.2014.11.042.

Palacios Hinestroza, H., Urena-Saborio, H., Zurita, F., Guerrero de Leon, A.A., Sundaram, G. and Sulbarán-Rangel, B. (2020). Nanocellulose and polycaprolactone nanospun composite membranes and their potential for the removal of pollutants from water. Molecules, 25(3): 683.

Pandey, N., Shukla, S. K. and Singh, N. B. (2017). Water purification by polymer nanocomposites: An overview. Nanocomposites, 3(2): 47–66.

Park, S.-B., Lih, E., Park, K.-S., Joung, Y. K. and Han, D. K. (2017). Biopolymer-based functional composites for medical applications. Progress in Polymer Science, 68: 77–105. https://doi.org/10.1016/j.progpolymsci.2016.12.003.

Pathania, D., Sharma, G. and Thakur, R. (2015). Pectin@ zirconium (IV) silicophosphate nanocomposite ion exchanger: photo catalysis, heavy metal separation and antibacterial activity. Chemical Engineering Journal, 267: 235–244.

Peelman, N., Ragaert, P., De Meulenaer, B., Adons, D., Peeters, R., Cardon, L. et al. (2013). Application of bioplastics for food packaging. Trends in Food Science & Technology, 32(2): 128–141. https://doi.org/10.1016/j.tifs.2013.06.003.

Petchwattana, N., Covavisaruch, S., Wibooranawong, S. and Naknaen, P. (2016). Antimicrobial food packaging prepared from poly (butylene succinate) and zinc oxide. Measurement 93: 442–448.

Peter, J., Jose, J., Thomas, A., Joy, J. C., Nivedita, S. and Joseph, S. (2020). Synthesis of TiO2 incorporated polycaprolactone based ultrafiltration membranes for water treatment. In Global Challenges in Energy and Environment: Select Proceedings of ICEE 2018 (pp. 119–130). Springer Singapore.

Pollard, S. J. T., Fowler, G. D., Sollars, C. J. and Perry, R. (1992). Low-cost adsorbents for waste and wastewater treatment: A review.

Preethi, S., Abarna, K., Nithyasri, M., Kishore, P., Deepika, K., Ranjithkumar, R. et al. (2020). Synthesis and characterization of chitosan/zinc oxide nanocomposite for antibacterial activity onto cotton fabrics and dye degradation applications. International Journal of Biological Macromolecules, 164: 2779–2787.

Rahman, O., Rahman, M. M. and Maniruzzaman, M. (2024). Removal of dye and heavy metals from industrial wastewater by activated charcoal-banana rachis cellulose nanocrystal composites filter. International journal of Environmental Analytical Chemistry, 104(7): 1478–1496.

Rao, K. M., Kumar, A., Krishna Rao, K. S. V., Haider, A. and Han, S. S. (2018). Biodegradable tragacanth gum based silver nanocomposite hydrogels and their antibacterial evaluation. Journal of Polymers and the Environment, 26: 778–788.

Rathi, B. S. and Kumar, P. S. (2021). Application of adsorption process for effective removal of emerging contaminants from water and wastewater. Environmental Pollution, 280: 116995. https://doi.org/10.1016/j.envpol.2021.116995.

Raza, S., Ansari, A., Siddiqui, N. N., Ibrahim, F., Abro, M. I. and Aman, A. (2021). Biosynthesis of silver nanoparticles for the fabrication of non cytotoxic and antibacterial metallic polymer based nanocomposite system. Scientific Reports, 11(1): 10500.

Regiel, A., Irusta, S., Kyzioł, A., Arruebo, M. and Santamaria, J. (2012). Preparation and characterization of chitosan–silver nanocomposite films and their antibacterial activity against *Staphylococcus aureus*. Nanotechnology, 24(1): 015101.

Rhim, J. W., Hong, S. I., Park, H. M. and Ng, P. K. (2006). Preparation and characterization of chitosan-based nanocomposite films with antimicrobial activity. Journal of agricultural and Food Chemistry, 54(16): 5814–5822.

Rhim, J.-W., Park, H.-M. and Ha, C.-S. (2013). Bio-nanocomposites for food packaging applications. Progress in Polymer Science, 38(10–11): 1629–1652. https://doi.org/10.1016/j.progpolymsci.2013.05.008.

Rojas, M. A., Amalraj, J. and Santos, L. S. (2023). Biopolymer-based composite hydrogels embedding small silver nanoparticles for advanced antimicrobial applications: experimental and theoretical insights. Polymers, 15(16): 3370.

Rozilah, A., Jaafar, C. A., Sapuan, S. M., Zainol, I. and Ilyas, R. A. (2020). The effects of silver nanoparticles compositions on the mechanical, physiochemical, antibacterial, and morphology properties of sugar palm starch biocomposites for antibacterial coating. Polymers, 12(11): 2605.

Sabbagh, N., Tahvildari, K. and Sharif, A. A. M. (2023). Highly efficient and rapid removal of malathion using crosslinked chitosan-alginate nanocomposites and optimization of parameters by box–behnken design: Isotherms and kinetic studies. Journal of Polymers and the Environment, 31(6): 2595–2611.

Safaei, M. and Taran, M. (2018). Optimized synthesis, characterization, and antibacterial activity of an alginate–cupric oxide bionanocomposite. Journal of Applied Polymer Science, 135(2): 45682.

Salahshoor, Z., Shahbazi, A. and Maddah, S. (2021). Magnetic field–influenced nanofiltration membrane blended by CS–EDTA–mGO as multi–functionality green modifier to enhance nanofiltration performance, efficient removal of Na2SO4/Pb2+/RR195 and cyclic wastewater treatment. Chemosphere, 278: 130379.

Salama, A., Abouzeid, R., Leong, W. S., Jeevanandam, J., Samyn, P., Dufresne, A. et al. (2021). Nanocellulose-based materials for water treatment: Adsorption, photocatalytic degradation, disinfection, antifouling, and nanofiltration. Nanomaterials, 11(11): 3008.

Salehi, M. M., Hassanzadeh-Afruzi, F., Esmailzadeh, F., Choopani, L., Rajabi, K., Kuzekanan, H. N. et al. (2024). Chlorpyrifos and diazinon elimination through pAAm-g-XG/HKUST-1@ Fe3O4 biopolymer nanoadsorbent hydrogel from wastewater: Preparation, characterization, kinetics and isotherm. Separation and Purification Technology, 334: 126097.

Sanoop, P. K., Mahesh, K. V., Nampoothiri, K. M., Mangalaraja, R. V. and Ananthakumar, S. (2012). Multifunctional ZnO-biopolymer nanocomposite coatings for health-care polymer foams and fabrics. Journal of Applied Polymer Science, 126(S1): E233–E244.

Sarkar, A. K., Saha, A., Midya, L., Banerjee, C., Mandre, N., Panda, A. B. et al. (2017). Cross-linked biopolymer stabilized exfoliated titanate nanosheet-supported AgNPs: a green sustainable ternary nanocomposite hydrogel for catalytic and antimicrobial activity. ACS Sustainable Chemistry & Engineering, 5(2): 1881–1891.

Scaffaro, R., Botta, L., Maio, A., Mistretta, M. C. and La Mantia, F. P. (2016). Effect of graphene nanoplatelets on the physical and antimicrobial properties of biopolymer-based nanocomposites. Materials, 9(5): 351.

Sen, M. (2020). Nanocomposite materials. Nanotechnology and the Environment, 1–12.

Sethy, T. R. and Sahoo, P. K. (2019). Highly toxic Cr (VI) adsorption by (chitosan-g-PMMA)/silica bionanocomposite prepared via emulsifier-free emulsion polymerisation. Int. J. Biol. Macromol. 2019 Feb 1;122:1184–1190. doi: 10.1016/j.ijbiomac.2018.09.069. Epub 2018 Sep 12. PMID: 30218734.

Shahvalizadeh, R., Ahmadi, R., Davandeh, I., Pezeshki, A., Moslemi, S. A. S., Karimi, S. et al. (2021). Antimicrobial bio-nanocomposite films based on gelatin, tragacanth, and zinc oxide nanoparticles–Microstructural, mechanical, thermo-physical, and barrier properties. Food Chemistry, 354: 129492.

Shankar, S. and Rhim, J. W. (2016). Tocopherol-mediated synthesis of Silver nanoparticles and preparation of anitimicrobial PBAT/silver nanoparticles composite films. LWT-Food Science and Technology 72: 149–156.

Sharifi, M. J., Nouralishahi, A. and Hallajisani, A. (2023). Fe3O4-chitosan nanocomposite as a magnetic biosorbent for removal of nickel and cobalt heavy metals from polluted water. Int. J. Biol. Macromol. 2023 Sep 1;248:125984. doi: 10.1016/j.ijbiomac.2023.125984. Epub 2023 Jul 27. PMID: 37506786.

Sharma, B., Malik, P. and Jain, P. (2018). Biopolymer reinforced nanocomposites: A comprehensive review. Materials Today Communications, 16: 353–363.

Sharma, G., Pathania, D. and Naushad, M. (2014). Preparation, characterization and antimicrobial activity of biopolymer based nanocomposite ion exchanger pectin zirconium (IV) selenotungstophosphate: Application for removal of toxic metals. Journal of Industrial and Engineering Chemistry, 20(6): 4482–4490.

Sharma, R., Agrawal, P. R., Kumar, R., Gupta, G. and Ittishree. (2021). Current scenario of heavy metal contamination in water. In Contamination of Water (pp. 49–64). Elsevier. https://doi.org/10.1016/B978-0-12-824058-8.00010-4.

Sharma, R., Jafari, S. M. and Sharma, S. (2020). Antimicrobial bio-nanocomposites and their potential applications in food packaging. Food Control, 112: 107086.

Sharma, S. and Bhattacharya, A. (2017). Drinking water contamination and treatment techniques. Applied Water Science, 7(3): 1043–1067. https://doi.org/10.1007/s13201-016-0455-7.

Singh, S., Arputharaj, E., Dahms, H. U., Patel, A. K. and Huang, Y. L. (2022). Chitosan-based nanocomposites for removal of Cr(VI) and synthetic food colorants from wastewater. Bioresour Technol. 2022 May;351:127018. doi: 10.1016/j.biortech.2022.127018. Epub 2022 Mar 17. PMID: 35307519.

Singha, K., Pandit, P., Maity, S. and Sharma, S. R. (2021). Harmful environmental effects for textile chemical dyeing practice. In Green Chemistry for Sustainable Textiles (pp. 153–164). Woodhead Publishing.

Sinha, A., Cha, B. G. and Kim, J. (2018). Three-dimensional macroporous alginate scaffolds embedded with akaganeite nanorods for the filter-based high-speed preparation of arsenic-free drinking water. ACS Applied Nano Materials, 1(4): 1940–1948.

Sivakanthan, S., Rajendran, S., Gamage, A., Madhujith, T. and Mani, S. (2020). Antioxidant and antimicrobial applications of biopolymers: A review. Food Research International, 136: 109327.

Slepička, P., Fajstavr, D., Krejčová, M., Rimpelová, S., Slepičková Kasálková, N., Kolská, Z. et al. (2021). Biopolymer Composites with Ti/Au Nanostructures and Their Antibacterial Properties. Pharmaceutics, 13(6): 826.

Sorrentino, A., Gorrasi, G. and Vittoria, V. (2007). Potential perspectives of bio-nanocomposites for food packaging applications. Trends in Food Science & Technology, 18(2): 84–95. https://doi.org/10.1016/j.tifs.2006.09.004.

Srivastav, A. L. and Ranjan, M. (2020). Inorganic water pollutants. In Inorganic Pollutants in Water (pp. 1–15). Elsevier. https://doi.org/10.1016/B978-0-12-818965-8.00001-9.

Susilowati, E. (2020, June). Fabrication of silver-chitosan nanocomposite films and their antibacterial activity. In IOP Conference Series: Materials Science and Engineering (Vol. 858, No. 1, p. 012042). IOP Publishing.

Swain, S. K., Patnaik, T. and Dey, R. K. (2013). Efficient removal of fluoride using new composite material of biopolymer alginate entrapped mixed metal oxide nanomaterials. Desalination and Water Treatment, 51(22–24): 4368–4378.

Tadayoni, N. S., Dinari, M., Roy, A. and Karimi Abdolmaleki, M. (2024). Recent advances in porous bio-polymer composites for the remediation of organic pollutants. Polymers, 16(11): 1543.

Tang, X. Z., Kumar, P., Alavi, S. and Sandeep, K. P. (2012). Recent advances in biopolymers and biopolymer-based nanocomposites for food packaging materials. Critical Reviews in Food Science and Nutrition, 52(5): 426–442.

Taran, M., Etemadi, S. and Safaei, M. (2017). Microbial levan biopolymer production and its use for the synthesis of an antibacterial iron (II, III) oxide–levan nanocomposite. Journal of Applied Polymer Science, 134(12).

Thakur, M., Sharma, G., Ahamad, T., Ghfar, A. A., Pathania, D. and Naushad, M. (2017). Efficient photocatalytic degradation of toxic dyes from aqueous environment using gelatin-Zr (IV) phosphate nanocomposite and its antimicrobial activity. Colloids and Surfaces B: Biointerfaces, 157: 456–463.

Udayakumar, G. P., Muthusamy, S., Selvaganesh, B., Sivarajasekar, N., Rambabu, K., Banat, F. et al. (2021). Biopolymers and composites: Properties, characterization and their applications in food, medical and pharmaceutical industries. Journal of Environmental Chemical Engineering, 9(4): 105322.

Udayakumar, G. P., Muthusamy, S., Selvaganesh, B., Sivarajasekar, N., Rambabu, K., Sivamani, S. et al. (2021). Ecofriendly biopolymers and composites: Preparation and their applications in water-treatment. Biotechnology Advances, 52:107815.

Vatanpour, V., Madaeni, S. S., Rajabi, L., Zinadini, S. and Derakhshan, A. A. (2012). Boehmite nanoparticles as a new nanofiller for preparation of antifouling mixed matrix membranes. Journal of Membrane Science, 401: 132–143.

Venkatesan, R. and Rajeswari, N. (2017). ZnO/PBAT nanocomposites films: Investigation on the mechanical and Biology activity for food packaging. Polymers for Advanced Technologies, 28(1): 20–27.

Venkatesan, R. and Rajeswari, N. (2019). Preparation, mechanical and antimicrobial properties of SoiO 2/poly (butylene adipate-co-terephthalate) films for active food packaging. Silicon, 115(5): 2233–2239.

Verma, D. and Fortunati, E. (2019). Biopolymer processing and its composites. In Biomass, Biopolymer-Based Materials, and Bioenergy (pp. 3–23). Elsevier. https://doi.org/10.1016/B978-0-08-102426-3.00001-1.

Wang, Y., Zhang, Y., Hou, C., Qi, Z., He, X. and Li, Y. (2015). Facile synthesis of monodisperse functional magnetic dialdehyde starch nano-composite and used for highly effective recovery of Hg(II). Chemosphere. 2015 Dec;141:26–33. doi: 10.1016/j.chemosphere.2015.06.019. Epub 2015 Jun 15. PMID: 26086563.

Wang, D., Lin, J., Huang, J., Zhang, H., Lin, S., Chen, L. et al. (2021). A chitosan/dopamine-TiO 2 composite nanofiltration membrane for antifouling in water purification. Cellulose, 28: 4959–4973.

Wang, X., Du, Y., Luo, J., Yang, J., Wang, W. and Kennedy, J. F. (2009). A novel biopolymer/rectorite nanocomposite with antimicrobial activity. Carbohydrate Polymers, 77(3): 449–456.

Wawrzyńczak, A., Chudzińska, J. and Feliczak-Guzik, A. (2024). Metal and metal oxides nanoparticles as nanofillers for biodegradable polymers. ChemPhysChem, e202300823.

Wypij, M., Rai, M., Zemljic, L. F., Bracic, M., Hribernik, S. and Golinska, P. (2023). Pullulan-based films impregnated with silver nanoparticles from the Fusarium culmorum strain JTW1 for potential application in the food industry and medicine. Frontiers in Bioengineering and Biotechnolgy, 11: 1241739.

Xie, F. (2024). Alginate-based nanocomposites for food preservation: Recent progress showcasing heightened material properties and functionalities. Advanced Nanocompositics, 1(1): 248–274.

Zahran, M. and Marei, A. H. (2019). Innovative natural polymer metal nanocomposites and their antimicrobial activity. International journal of Biological Macromolecules, 136: 586–596.

Zhang, Y., Olsen, K., Grossi, A. and Otte, J. (2013). Effect of pretreatment on enzymatic hydrolysis of bovine collagen and formation of ACE-inhibitory peptides. Food Chemistry, 141(3): 2343–2354. https://doi.org/10.1016/j.foodchem.2013.05.058.

Zhao, R., Torley, P. and Halley, P. J. (2008). Emerging biodegradable materials: starch- and protein-based bio-nanocomposites. Journal of Materials Science, 43(9): 3058–3071. https://doi.org/10.1007/s10853-007-2434-8.

Zhao, W. B., Du, M. R., Liu, K. K., Zhou, R., Ma, R. N., Jiao, Z. et al. (2020). Hydrophilic ZnO nanoparticles@calcium alginate composite for water purification. ACS Applied Materials & Interfaces, 12(11): 13305–13315.

CHAPTER-9

Biopolymer Nanoparticle-Mediated for Soil Remediation

Sumira Malik,[1,*] *Ajay Mahalka,*[1] *Abhinav Kumar,*[1] *Sagar Mondal,*[1] *Smita lata,*[1] *Jutishna Bora*[1] and *Seema Ramniwas*[2]

1. Introduction

Soil contamination is a growing concern globally due to the adverse environmental and human health effects it poses. Heavy metals, pesticides, petroleum hydrocarbons, and other pollutants find their way into the soil from various sources, including industrial activities, agricultural practices, and urban development (Khan et al., 2021; Monib et al., 2024). The widespread presence of these contaminants necessitates effective soil remediation strategies to mitigate their impact on ecosystems and human populations.

Traditional soil remediation methods, such as excavation and landfill disposal, incineration, and soil washing, have been employed for decades. While these methods have been somewhat effective, they often entail high costs, and significant energy consumption, and may not be suitable for large-scale or environmentally sensitive sites and however recently nanotechnology and nanomaterials have significant contribution in environmental cleanup (Guerra et al., 2018; Galdames; 2020; Del Prado-Audelo, 2021). Furthermore, they may not address the root causes of contamination, leading to recurrent pollution. In response to these challenges, there is a growing interest in innovative and sustainable soil remediation approaches. One such approach involves the use of biopolymer nanoparticles (BNPs) as a remediation tool. BNPs are a class of nanomaterials derived from natural polymers like chitosan, cellulose, and starch (Muthukumaran et al., 2022). These nanoparticles have gained attention due to their unique properties, including high surface area, biocompatibility, and the ability to be functionalized for specific applications (Ahmad et al., 2022)

The use of BNPs in soil remediation provides several advantages. First, BNPs can adsorb and sequester contaminants through physical and chemical interactions. Their high surface area and surface charge allow them to bind to a wide range of pollutants, including heavy metals, pesticides, and organic compounds (Li et al., 2019). Second, BNPs can promote chemical reactions and transformation processes in soil, facilitating the breakdown of contaminants into less harmful forms. Third, BNPs can enhance microbial activity in the soil, promoting biodegradation of organic pollutants. Finally, BNPs can serve as carriers for nutrients and beneficial microbes, thereby improving soil quality and fertility (Zhang et al., 2017; Song et al., 2019).

[1] Amity Institute of Biotechnology, Amity University Jharkhand, Ranchi, India.
[2] Department of Biotechnology, University Center for Research & Development (UCRD) Chandigarh University, NH-05 Chandigarh- Ludhiana Highway, Mohali, Punjab, India.
* Corresponding author: smalik@rnc.amity.edu
DOI: 10.1201/9781003558330-9

The purpose of this review is to provide an overview of the role of BNPs in soil remediation. It will explore the mechanisms by which BNPs facilitate contaminant removal and soil restoration, discuss their applications in various soil types and contamination scenarios, and examine the challenges and opportunities associated with their use.

1.1 Soil Contamination and Remediation Fundamentals

Soil contamination, resulting from various anthropogenic and natural sources, presents a significant threat to environmental quality and human well-being (Smith et al., 2019). Understanding the types, sources, and consequences of soil pollution is essential for appreciating the importance of innovative remediation methods.

Types and Sources of Soil Contamination

Soil contamination is a multifaceted issue, with different types of contaminants stemming from distinct sources:

- **Heavy Metal Contamination:** Heavy metals such as lead, cadmium, and mercury are often introduced into soil through industrial processes, mining activities, and the use of agrochemicals (Khan et al., 2021). These toxic elements accumulate in soils, posing long-term risks to ecosystems and human health.
- **Organic Pollutants:** Organic contaminants, including pesticides, herbicides, and petroleum hydrocarbons, primarily arise from agricultural practices, improper waste disposal, and industrial discharges. These compounds persist in the environment, causing harm to both terrestrial and aquatic ecosystems.
- **Nutrient Imbalance:** Soil can also suffer from nutrient imbalances, such as excessive nitrogen or phosphorus, due to agricultural fertilization practices (Saha et al., 2024). These imbalances can lead to water pollution, algal blooms, and ecosystem disruption.
- **Salinity and pH:** Soil salinity and pH imbalances may occur naturally due to geological factors or result from irrigation practices (Gupta et al., 2019). Elevated salinity levels can render soil unsuitable for agriculture, while extreme pH levels can affect nutrient availability.

Environmental Consequences of Soil Pollution

Soil pollution engenders a host of adverse environmental consequences, extending to terrestrial and aquatic ecosystems as well as human populations:

- **Ecological Impact:** Soil contaminants can disrupt soil ecosystems by harming beneficial microorganisms, plants, and animals, ultimately leading to reduced biodiversity (Smith et al., 2019). This disruption can have cascading effects on entire ecosystems.
- **Water Contamination:** Soil contaminants can leach into groundwater, resulting in the contamination of drinking water sources and aquatic habitats (EPA, 2020). Contaminated water poses health risks to humans and aquatic life, with long-lasting consequences.
- **Airborne Contaminant Release:** Certain soil contaminants, particularly volatile organic compounds, can volatilize into the air, contributing to air pollution (Khan et al., 2021). This airborne contamination can have respiratory and environmental impacts.
- **Food Chain Contamination:** Soil pollutants can accumulate in plants, entering the food chain and posing health risks to humans and wildlife (Gupta et al., 2019). This bioaccumulation can result in long-term exposure to toxic substances.

Traditional Soil Remediation Methods

Historically, conventional soil remediation methods have been employed to address contamination issues. However, these methods have limitations in terms of sustainability and effectiveness:

- **Excavation and Landfill Disposal:** This method involves removing contaminated soil and transporting it to landfills or disposal sites (EPA, 2020). While effective in isolating contaminants, it is often costly and can create new pollution problems.
- **Incineration:** Some contaminants can be destroyed through incineration, but this approach consumes significant energy and may release harmful emissions (Tang et al., 2016).
- **Soil Washing:** Soil washing employs water or solvents to remove pollutants, but it generates wastewater that must be treated (Kumar et al., 2017).
- **Bioremediation:** Microbial processes are used to degrade contaminants in bioremediation, but they may be slow and ineffective for certain pollutants (Gupta et al., 2019).

In view of these challenges, there is a growing need for innovative and sustainable soil remediation approaches that address the root causes of soil contamination and provide long-term solutions.

1.2 Biopolymer Nanoparticles

Biopolymer nanoparticles (BNPs), derived from natural polymers such as chitosan, cellulose, and starch, depicted in Fig. 1, have gained increasing attention for their unique properties and versatility in various fields, including environmental remediation (Salem et al., 2020).

Definition and Characteristics of Biopolymer

Biopolymers are naturally occurring polymers produced by living organisms. They are characterized by their biodegradability, biocompatibility, and low environmental impact (Sharma et al., 2018). Examples of biopolymers include proteins, nucleic acids, and polysaccharides. These materials offer several advantages, including their renewable nature and reduced toxicity compared to synthetic polymers.

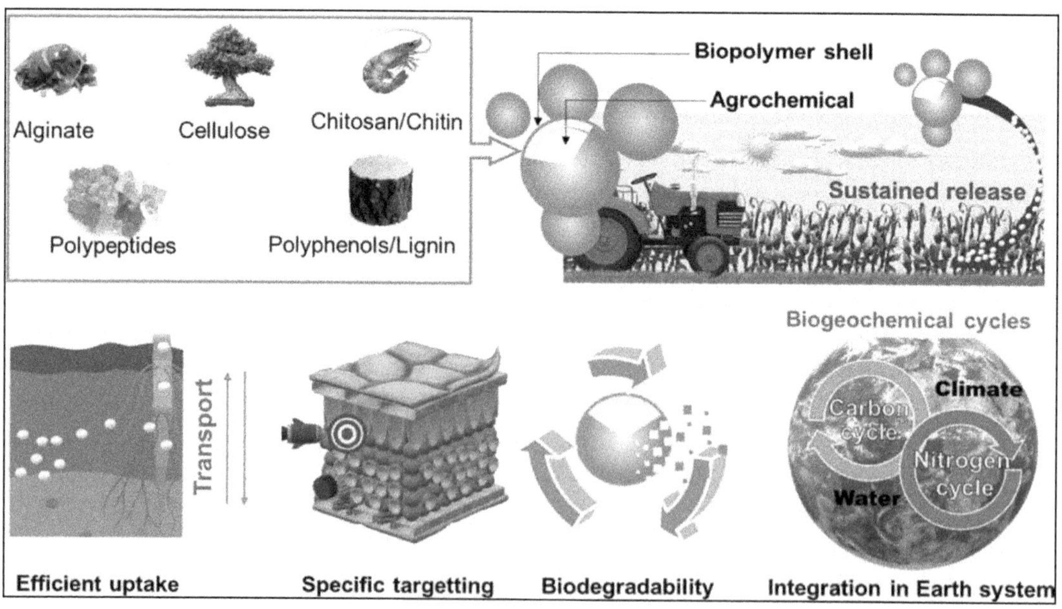

Figure 1. BNPs properties (Machado et al., 2022).

Nanoparticles

Nanoparticles are particles with dimensions in the nanometre range. They can be engineered from various materials, including metals, metal oxides, and polymers. Nanoparticles exhibit unique properties due to their small size, such as high surface area-to-volume ratio and enhanced reactivity (El-Eskandarany et al., 2020). These properties make them suitable for a wide range of applications, including drug delivery, imaging, and environmental remediation.

Advantages of Using Biopolymer Nanoparticles in Soil Remediation

The application of BNPs in soil remediation provides several advantages:

- **Biocompatibility:** BNPs are inherently biocompatible, making them safe for use in soil ecosystems and minimizing the risk of adverse ecological effects (El-Eskandarany et al., 2020).
- **High Surface Area:** BNPs possess a high surface area, enabling efficient adsorption and sequestration of contaminants from soil (Li et al., 2019). Their large surface area facilitates the binding of various pollutants, including heavy metals and organic compounds.
- **Functionalization:** BNPs can be easily functionalized or modified to enhance their affinity for specific contaminants (Nasrollahzadeh 2021). This adaptability allows for tailored solutions to different soil contamination scenarios.

Common Biopolymers Used in Soil Remediation

Several biopolymers have shown promise in soil remediation applications:

- **Chitosan:** Chitosan, derived from chitin, exhibits strong metal-binding capabilities and has been employed to remediate heavy metal-contaminated soils (El-Eskandarany et al., 2020).
- **Cellulose:** Cellulose-based BNPs are biodegradable and have been used for the removal of organic pollutants from soil (Zhang et al., 2017).
- **Starch:** Starch-based BNPs have been investigated for their potential in nutrient delivery and soil enhancement (Zhang et al., 2017).

Synthesis and Modification of Biopolymer Nanoparticles

The synthesis and modification of BNPs are crucial aspects of their application in soil remediation. Various methods, such as chemical cross-linking and nanoprecipitation, are employed to produce BNPs with controlled size and properties (Salem et al., 2020). Additionally, the functionalization of BNPs can be achieved by attaching specific molecules or ligands to enhance their contaminant-binding capabilities (Sidhu et al., 2025).

The multifaceted nature of BNPs, stemming from their biocompatibility, high surface area, and tunable properties, positions them as promising candidates for addressing soil contamination challenges.

1.3 Mechanisms of Biopolymer Nanoparticle-Mediated Soil Remediation

1.3.1 Adsorption and sequestration of contaminants diet and nutrition

One of the fundamental mechanisms underlying BNP-mediated soil remediation is the adsorption and sequestration of contaminants as mentioned in Figure 2 (Zhang et al., 2017; Qian et al., 2020). BNPs, with their high surface area and functional groups, are well-suited for capturing a wide range of pollutants.

- **Physical Adsorption:** BNPs can physically adsorb contaminants onto their surfaces, essentially trapping them (Li et al., 2019). This adsorption occurs through electrostatic interactions, van der Waals forces, and hydrogen bonding.

- **Chemical Adsorption:** In some cases, BNPs chemically interact with contaminants, forming stable complexes or chelates. This chemical adsorption is particularly effective for heavy metals like lead and cadmium.
- **Contaminant Sequestration:** Once adsorbed, contaminants are sequestered within the BNP matrix. This sequestration prevents the contaminants from leaching into groundwater or affecting surrounding ecosystems.

1.3.2 Chemical reactions and transformation processes

BNPs also contribute to soil remediation through chemical reactions and transformation processes (Kumar et al., 2017). These processes involve the modification or degradation of contaminants.

- **Redox Reactions:** Certain BNPs, when appropriately functionalized, can engage in redox reactions with contaminants (El-Eskandarany et al., 2020). This can lead to the reduction or oxidation of pollutants, converting them into less harmful forms.
- **Complexation:** BNPs can form stable complexes with contaminants, altering their chemical properties and reducing their toxicity (Li et al., 2019). Complexation is particularly effective for metal contaminants.
- **Catalytic Activity:** Some BNPs possess catalytic properties that can accelerate the degradation of organic pollutants (Zhang et al., 2017). This catalytic activity is valuable for breaking down complex organic compounds.

1.3.3 Microbial interactions and biodegradation

BNPs play a role in enhancing microbial interactions and biodegradation in soil (Salem et al., 2020). Microbes, such as bacteria and fungi, are natural agents of soil remediation, and BNPs can promote their activity.

- **Microbial Adhesion:** BNPs can act as carriers, facilitating the adhesion and colonization of beneficial microbes on their surfaces (Li et al., 2019). This enhances microbial degradation of organic contaminants.
- **Nutrient Delivery:** BNPs can serve as nutrient carriers, providing essential nutrients to soil microorganisms (Zhang et al., 2017). This nutrient delivery enhances microbial growth and metabolic activity.
- **Biofilm Formation:** BNPs can aid in the formation of microbial biofilms, which are critical for the degradation of complex organic compounds (Sidhu et al., 2025). Biofilms create microenvironments conducive to biodegradation.

1.3.4 Nutrient delivery and soil enhancement

In addition to contaminant remediation, BNPs contribute to soil improvement and nutrient management (Zhang et al., 2017). This dual functionality enhances soil quality and fertility.

- **Nutrient Carrier:** BNPs can encapsulate and release nutrients slowly, ensuring a sustained nutrient supply to plants (El-Eskandarany et al., 2020). This is particularly beneficial for nutrient-poor soils.
- **Soil Structure Enhancement:** BNPs can improve soil structure, increasing water retention and aeration. This leads to enhanced plant growth and root development.
- **pH Modification:** Some BNPs can influence soil pH, making it more suitable for specific crops (Li et al., 2019). pH adjustment can optimize nutrient availability.

1.4 Applications of Biopolymers Nanoparticles-Mediated Soil Remediation

BNPs, with their unique properties and versatile mechanisms, find utility in addressing a wide range of soil contamination scenarios, contributing to sustainable environmental management.

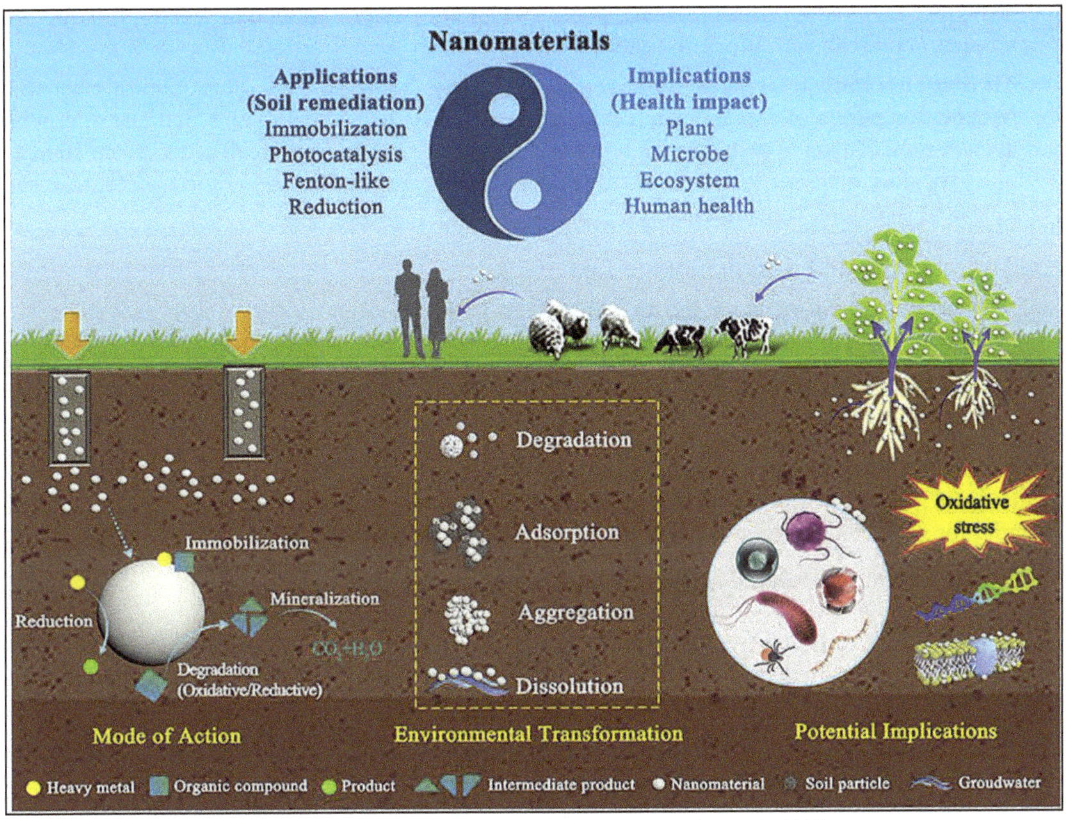

Figure 2. Mechanism of NPs in soil Remediation (Qian et al., 2020).

1.4.1 *Contaminant-specific applications*

BNPs are tailored to address specific contaminants found in soil, offering targeted remediation solutions:

- **Heavy Metal Remediation:** BNPs, particularly chitosan-based nanoparticles, excel in sequestering heavy metals like lead, cadmium, and copper (Li et al., 2019). These nanoparticles effectively adsorb metal ions, reducing their bioavailability and ecological impact.
- **Organic Pollutant Removal:** Cellulose-based BNPs have demonstrated efficacy in removing organic pollutants, such as pesticides and hydrocarbons, from soil (Zhang et al., 2017). Their high surface area and affinity for organic compounds make them valuable tools in treating contaminated sites.
- **Nutrient Delivery for Agriculture:** Starch-based BNPs can be used to deliver essential nutrients to agricultural soils, enhancing crop growth and yield (Sidhu et al., 2025). Nutrient-loaded BNPs release nutrients gradually, ensuring sustained availability.

1.4.2 *Land and site-specific applications*

BNPs are adaptable to various soil types and contamination scenarios, making them suitable for a range of environments:

- **Agricultural Soil Remediation:** BNPs can be integrated into agricultural practices to mitigate nutrient imbalances and enhance soil structure (El-Eskandarany et al., 2020). These applications improve crop health and productivity while reducing nutrient runoff.
- **Urban Soil Remediation:** Urban environments often face soil contamination from industrial and residential activities. BNPs offer a sustainable approach to addressing contaminants like

heavy metals in urban soils (A G, M., et al., 2017). They can be applied to green infrastructure projects and urban agriculture initiatives.

- **Mining and Industrial Sites:** Abandoned mines and industrial sites often require extensive remediation efforts. BNPs, with their capacity to sequester heavy metals and facilitate microbial degradation, can aid in restoring such sites (Gupta et al., 2019). **Fat:** Healthy fats from sources like avocados, nuts, seeds, and olive oil are essential for heart health. Monitoring fat intake can help control calorie consumption.

1.4.3 Agriculture and environment sustainability

BNPs contribute to long-term soil health and sustainability:

Enhanced Soil Fertility: BNPs improve soil fertility by promoting nutrient retention and availability (Li et al., 2019). This benefit is particularly valuable in agricultural contexts, reducing the need for excessive fertilizer use.

Reduced Runoff and Water Contamination: BNPs help prevent nutrient runoff into water bodies, reducing the risk of eutrophication and water pollution (El-Eskandarany et al., 2020). This is essential for safeguarding aquatic ecosystems and drinking water sources.

Mitigation of Soil Erosion: Soil erosion can result in the loss of topsoil and degradation of agricultural land. BNPs, when incorporated into the soil, enhance its stability, and reduce erosion, preserving valuable soil resources (Sidhu et al., 2025).

Case studies showcasing successful applications

Numerous case studies demonstrate the effectiveness of BNPs in real-world soil remediation scenarios:

Chitosan-Based BNPs in Heavy Metal-Contaminated Soils: Chitosan-based BNPs, effectively reduced the bioavailability of cadmium and lead in contaminated agricultural soils, leading to reduced metal uptake by crops (Karamchandani et al. 2024).

Cellulose-Based BNPs for Pesticide Removal: A study highlighted the use of cellulose-based BNPs for the removal of organochlorine pesticides from soil. The BNPs demonstrated high adsorption capacity and selectivity for the target pollutants (Hassanisaadi et al., 2023).

Starch-Based BNPs for Sustainable Agriculture: A study applied starch-based BNPs loaded with micronutrients in agricultural soils. The nanoparticles improved nutrient availability and crop yields while reducing nutrient leaching (Abd-Elsalam et al.,2024).

These case studies underscore the practicality and versatility of BNPs in addressing a variety of soil contamination challenges and advancing environmental sustainability.

1.5 Challenges and Considerations in Biopolymer Nanoparticle-Mediated Soil Remediation

1.5.1 Environmental and ecological impact assessment

Before implementing BNPs for soil remediation, it is crucial to conduct a thorough environmental and ecological impact assessment (Li et al., 2019). BNPs are biocompatible and generally considered safe, but potential unintended consequences must be evaluated. Assessments should consider factors such as nanoparticle persistence in the environment and potential effects on non-target organisms.

Risk to Beneficial Microorganisms: While BNPs can enhance microbial activity, their presence may also impact non-target microorganisms in the soil (El-Eskandarany et al., 2020). Understanding the potential disruptions to soil microbial communities is essential.

Bioavailability of Nanoparticles: Assessments should investigate the bioavailability of BNPs to plants and the potential for BNPs to be taken up by crops (Zhang et al., 2017). This ensures that the BNP application does not inadvertently introduce nanoparticles into the food chain.

1.5.2 Regulatory and safety considerations

BNP use in soil remediation may be subject to regulatory oversight, and adherence to safety guidelines is paramount. Understanding and complying with relevant regulations is essential to ensure a responsible BNP application.

- **Regulatory Approvals:** Depending on the jurisdiction, the use of BNPs in soil remediation may require regulatory approvals and permits (Sidhu et al., 2025). Researchers and practitioners should be aware of the legal framework governing nanomaterials.
- **Safety Protocols:** Safety protocols must be established to minimize potential human exposure to BNPs during application (Gupta et al., 2019). This includes appropriate personnal protective equipment and handling procedures.

1.5.3 Long-term environmental impacts

The long-term environmental impacts of BNPs in soil must be considered. Research into the fate and transport of BNPs, as well as their potential accumulation in soil, is necessary to assess any long-term consequences.

Nanoparticle Persistence: Understanding how long BNPs remain active in soil and whether they eventually degrade or accumulate is crucial (Sidhu et al., 2025). This information helps gauge the duration of BNP effectiveness.

Secondary Contamination: Assessments should also explore the potential for BNPs which lead to secondary contamination, such as the release of other pollutants or alterations in soil chemistry (Gupta et al., 2019). Evaluating these impacts is vital for sustainable remediation.

1.5.4 Public perception and acceptance

Public perception and acceptance of BNP-mediated soil remediation may influence project success (Li et al., 2019). Engaging with the community and stakeholders to address concerns and provide information is essential.

- **Transparency:** Transparency in project planning, risk assessments, and safety protocols can foster public trust (Salem et al., 2020). Open communication can address misconceptions about nanomaterials.
- **Public Participation:** Involving the community and stakeholders in decision-making processes can lead to more informed and acceptable remediation strategies (Zhang et al., 2017). Public input can contribute to responsible and sustainable practices.

As BNPs continue to evolve as a tool for soil remediation, addressing these challenges and considerations is vital to ensure effective, safe, and sustainable applications. Responsible research and practices will contribute to the advancement of BNP-mediated soil remediation while minimizing potential risks and environmental impacts.

Staying Consistent with Exercise

Staying consistent with exercise can be challenging, but it is crucial for long-term benefits. Finding enjoyable activities, setting achievable goals, and enlisting the support of friends or family can help maintain motivation.

2. Future Trends and Innovations in BNP-Mediated Soil Remediation

2.1 Nanoparticle Engineering for Enhanced Remediation

Future developments in BNP engineering will focus on enhancing their remediation capabilities:

Smart Nanoparticles: Researchers are working on smart BNPs that respond to environmental cues, such as pH or contaminant levels (Nasrollahzadeh, 2021). These nanoparticles can release their payload precisely when and where needed, increasing remediation efficiency.

Multifunctional BNPs: BNPs with multiple functionalities, such as simultaneous adsorption of heavy metals and degradation of organic pollutants, are under investigation (El-Eskandarany et al., 2020). This integrated approach can address complex contamination scenarios.

2.2 Nanoparticles for Emerging Contaminants

The remediation of emerging contaminants, such as pharmaceuticals and microplastics, presents a growing challenge. BNPs are poised to play a crucial role:

Pharmaceutical Residue Remediation: BNPs can be designed to selectively adsorb pharmaceutical residues from soil (Li et al., 2019). As pharmaceutical use continues to rise, this application will become increasingly relevant.

Microplastic Removal: BNPs with the ability to capture microplastics in soil are a developing area of research (Gupta et al., 2019). These nanoparticles can contribute to mitigating the pervasive issue of microplastic pollution.

2.3 Precision Soil Remediation

Future trends in soil remediation will include a move towards precision and site-specific approaches:

Sensors and Monitoring: Advanced sensors and monitoring technologies will enable real-time assessment of soil conditions. This data can inform precise BNP application strategies.

Machine Learning and Modelling: Machine learning algorithms and modelling will aid in predicting the behaviour of BNPs in different soil types and under various conditions (Zhang et al., 2017). This knowledge will optimize remediation strategies.

2.4 Sustainable BNP Production

The sustainability of BNP production will be a key focus:

Green Synthesis: Researchers are developing green synthesis methods that use environmentally friendly processes and reagents to produce BNPs (Sidhu et al., 2025). This reduces the environmental footprint of BNP production.

Biodegradable BNPs: BNPs designed for biodegradability will minimize long-term environmental impacts (El-Eskandarany et al., 2020). These nanoparticles will break down naturally over time, leaving no trace in the environment.

2.5 Integration with Soil Health Management

BNP-mediated soil remediation will increasingly be integrated with soil health management:

Soil Microbiome Engineering: BNPs may be used to engineer the soil microbiome to enhance natural remediation processes (Li et al., 2019). This approach can restore soil health and ecosystem function.

Soil Carbon Sequestration: BNPs can play a role in sequestering carbon in soil, contributing to carbon capture and climate change mitigation efforts (Gupta et al., 2019). The integration of soil carbon sequestration and remediation is a promising avenue for sustainable agriculture.

2.6 International Collaboration and Knowledge Sharing

As the field of BNP-mediated soil remediation progresses, international collaboration and knowledge sharing will become increasingly important:

Global Data Repositories: Establishing global data repositories on BNP applications and their effectiveness in different regions will facilitate best practices and informed decision-making (Kumar et al., 2017).

Interdisciplinary Research: Collaboration between scientists, engineers, environmentalists, and policymakers will foster holistic approaches to soil remediation (Zhang et al., 2017). Multidisciplinary teams can tackle complex contamination issues more effectively.

2.7 Public Awareness and Engagement

Engaging the public in understanding the benefits and safety of BNP-mediated soil remediation will be essential for its acceptance:

Education and Outreach: Public awareness campaigns and educational programs can inform communities about the advantages and safety measures associated with BNP use (Sidhu et al., 2018). An informed public is more likely to support BNP-based remediation projects. AI based new techniques has also been deployed to remediate mines-based pollutants (Ji et al., 2022).

Ethical Considerations: Ethical discussions surrounding the use of nanomaterials in the environment should continue, ensuring responsible and ethical practices (El-Eskandarany et al., 2020).

The future of BNPs in soil remediation holds tremendous potential for addressing soil contamination challenges sustainably and effectively. By embracing innovation, precision, sustainability, and interdisciplinary collaboration, BNPs will continue to emerge as a powerful tool in restoring and preserving soil health.

3. Conclusion

Throughout this review, we have explored the diverse applications and unique mechanisms of BNPs in soil remediation (Li et al., 2019). We have highlighted BNPs' role in enhancing soil health and environmental sustainability and future techniques in Figure 3 (Souza et al., 2020). Additionally, the challenges and considerations involved in their application, from environmental impact assessment to regulatory compliance have been discussed. BNPs hold significant promise for addressing soil contamination challenges while aligning with sustainability goals. They can restore contaminated sites, boost agricultural productivity, and protect water bodies. Combining BNPs with emerging nanotechnological advances, such as nanosensors, and nanorobots, can revolutionize soil monitoring and remediation. Collaborative efforts across borders can lead to standardized protocols, data sharing, and a global knowledge network, advancing the field. Establishing training programs and educational initiatives will equip professionals with skills and knowledge required for responsible BNP-mediated soil remediation (Zhang et al., 2017). Ongoing public engagement is crucial for building trust and garnering support for BNP applications. Prioritize safety throughout BNP production, application, and disposal (Li et al., 2019). Maintain transparency in research, development, and project implementation to foster trust with stakeholders. As we embrace the

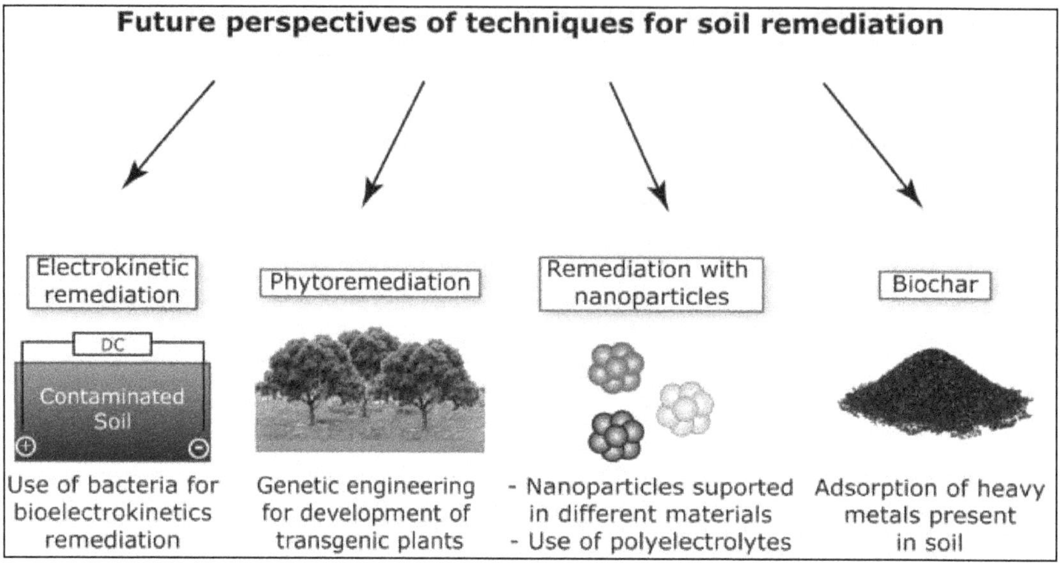

Figure 3. Future perspective of techniques for soil remediation (Souza et al., 2020).

potential of BNPs in soil remediation, it is paramount to do so with a commitment to ethical considerations and responsible practices. The future holds great promise for BNPs as a powerful tool in restoring and preserving soil health (Souza et al., 2020).

References

Abd-Elsalam, K. A., Hashim, A. F., Abdelkhalek, T. E. and Hassan, R. K. (2024). Polysaccharide Biopolymer Nanoparticles for Soil Nutrient Delivery. In Biopolymeric Nanoparticles for Agricultural Applications (pp. 79–105). Cham: Springer Nature Switzerland.

Ahmad, F., Salem-Bekhit, M. M., Khan, F., Alshehri, S., Khan, A., Ghoneim, M. M. et al. (2022). Unique properties of surface-functionalized nanoparticles for bio-application: functionalization mechanisms and importance in application. Nanomaterials, 12(8): 1333.

A G, M., Kaleeswari, R. K., S, M., M, B., T, U., M, S. et al. (2025). Innovative nanoremediation techniques for soil contamination: exploring metal based nanomaterial approaches – a review. Soil and Sediment Contamination: An International Journal, 1–69. https://doi.org/10.1080/15320383.2025.2477818.

Del Prado-Audelo, M. L., García Kerdan, I., Escutia-Guadarrama, L., Reyna-González, J. M., Magaña, J. J. and Leyva-Gómez, G. (2021). Nano remediation: nanomaterials and nanotechnologies for environmental cleanup. Frontiers in Environmental Science, 9: 793765.

El-Eskandarany, M. S., Ali, N. and Al-Salem, S. M. (2020). Solid-state conversion of magnesium waste to advanced hydrogen-storage nanopowder particles. Nanomaterials, 10(6): 1037.

Galdames, A., Ruiz-Rubio, L., Orueta, M., Sánchez-Arzalluz, M. and Vilas-Vilela, J. L. (2020). Zero-Valent Iron Nanoparticles for Soil and Groundwater Remediation. International Journal of Environmental Research and Public Health, 17(16): 5817.

Guerra, F. D., Attia, M. F., Whitehead, D. C. and Alexis, F. (2018). Nanotechnology for environmental remediation: materials and applications. Molecules, 23(7): 1760.

Gupta, N., Yadav, K. K., Kumar, V., Krishnan, S., Kumar, S., Nejad, Z. D. et al. (2021). Evaluating heavy metals contamination in soil and vegetables in the region of North India: Levels, transfer and potential human health risk analysis. Environmental Toxicology and Pharmacology, 82: 103563.

Hassanisaadi, M., Riseh, R. S., Rabiei, A., Varma, R. S. and Kennedy, J. F. (2023). Nano/micro-cellulose-based materials as remarkable sorbents for the remediation of agricultural resources from chemical pollutants. International Journal of Biological Macromolecules, 246: 125763.

Ji, M., Al-Huqail, A. A., Majdi, A., Liu, H., Ali, I., & Marzouki, R. (2022). Remediation of mine polluted soil with nano-enhanced materials: Development of extreme learning machine approaches. Chemosphere, 307(Pt 3), 135772. https://doi.org/10.1016/j.chemosphere.2022.135772.

Karamchandani, B. M., Dalvi, S. G., Bagayatkar, M., Banat, I. M. and Satpute, S. K. (2024). Prospective applications of chitosan and chitosan-based nanoparticles formulations in sustainable agricultural practices. Biocatalysis and Agricultural Biotechnology, 103210.

Khan, S., Naushad, M., Lima, E. C., Zhang, S., Shaheen, S. M. and Rinklebe, J. (2021). Global soil pollution by toxic elements: Current status and future perspectives on the risk assessment and remediation strategies–A review. Journal of Hazardous Materials, 417: 126039.

Li, C., Zhou, K., Qin, W., Tian, C., Qi, M., Yan, X. et al. (2019). A review on heavy metals contamination in soil: effects, sources, and remediation techniques. Soil and Sediment Contamination: An International Journal, 28(4): 380–394.

Machado, T. O., Grabow, J., Sayer, C., de Araújo, P. H., Ehrenhard, M. L. and Wurm, F. R. (2022). Biopolymer-based nanocarriers for sustained release of agrochemicals: A review on materials and social science perspectives for a sustainable future of agri-and horticulture. Advances in Colloid And Interface Science, 303: 102645.

Monib, A. W., Niazi, P., Azizi, A., Sediqi, S. and Baseer, A. Q. (2024). Heavy metal contamination in urban soils: Health impacts on humans and plants: A review. European Journal of Theoretical and Applied Sciences, 2(1): 546–565.

Muthukumaran, P., Babu, P. S., Shyamalagowri, S., Aravind, J., Kamaraj, M. and Govarthanan, M. (2022). Polymeric biomolecules-based nanomaterials: production strategies and pollutant mitigation as an emerging tool for environmental application. Chemosphere, 307: 136008.

Nasrollahzadeh, M. (2021). Biopolymer-based metal nanoparticle chemistry for sustainable applications: volume 2: applications. Elsevier.

Qian, Y., Qin, C., Chen, M. and Lin, S. (2020). Nanotechnology in soil remediation− applications vs. implications. Ecotoxicology and Environmental Safety, 201: 110815.

Saha, B., Fatima, A., Saha, S., Sahoo, S. K. and Poddar, P. (2024). Environmental Pollution Due to Improper Use of Chemical Fertilizers and Their Remediation. In Environmental Contaminants (pp. 203–219). Apple Academic Press.

Sidhu, A. K., Verma, N., Kaushal, P., Agrawal, S. B. and Bhatia, S. (2025). Green synthesis of polymeric nanoparticles: agricultural applications and toxicological implications. Discover Applied Sciences, 7(6): 1–20.

Song, C., Ye, F., Zhang, H., Hong, J., Hua, C., Wang, B. et al. (2019). Metal (loid) oxides and metal sulfides nanomaterials reduced heavy metals uptake in soil cultivated cucumber plants. Environmental Pollution, 255: 113354.

Souza, L. R. R., Pomarolli, L. C. and Da Veiga, M. A. M. S. (2020). From classic methodologies to application of nanomaterials for soil remediation: an integrated view of methods for decontamination of toxic metal (oid) s. Environmental Science and Pollution Research, 27(10): 10205–10227.

CHAPTER-10

Biopolymer-based Metal Nanoparticles in Environmental Monitoring

Taranga Jyoti Baruah,[1,#] *Jiwasmika Baishya,*[1] *Khrote-Ü-Lasushe,*[2] *Partha Pratim Das,*[3] *Seema Ramniwas,*[4] *Ajay Mahalka,*[5] *Smita Lata,*[5] *Jutishna Bora*[5,#] and *Sumira Malik*[5,*]

1. Environmental Pollution

1.1 Overview

With a rapid advent in newer technologies and discovery of new innovations, the world today is heading towards a briskly progressive one. While a progressive world promises an enhancement in the quality of life, it also alters the natural state of the surrounding environment vastly. Hence the brunt of any advancement is often borne by the environment around us—air, water, and soil. Each of these environmental elements have undergone huge deterioration in its quality in the recent past (Roy et al., 2021; Tofighty and Mohammadi, 2011; Masojídek et al., 2011, Artiola et al., 2019). For example, the average air quality of the world is under 'Moderate' category on the World's Air Pollution Real-time Air Quality Index Scale with a major number of Asian cities heading towards the 'Unhealthy' category. Such shifts in the state of our environment are alarming and should be treated as a major concern for the planet.

Like air, water pollution is also an area of major concern (Vasilachi et al., 2021). Water is an abundantly available and an unexceptionally required resource with only < 1% of its global supply available and safe for human consumption (Tofighty and Mohammadi, 2011). This is due to the presence of harmful pollutants, such as heavy metals, pesticides, and other pharmaceutical and personal care products (Chen et al., 2003; Qu et al., 2013) in most water bodies. According to the

[1] Department of Biochemistry, Assam Royal Global University, Guwahati, Assam.
[2] Department of Botany, St. Joseph's College (Autonomous), Kohima, Nagaland.
[3] Multidisciplinary Research Unit (MRU), Fakhruddin Ali Ahmed Medical College, Barpeta, Assam.
[4] Department of Biotechnology, University Center for Research & Development (UCRD) Chandigarh University, NH-05 Chandigarh- Ludhiana Highway, Mohali, Punjab, India.
[5] Amity institute of Biotechnology, Amity University Jharkhand, Ranchi.
* Corresponding author: smalik@rnc.amity.edu
# Equal contribution
DOI: 10.1201/9781003558330-10

World Health Organization, over 760 million people have been deprived of adequate drinking water supply in 2011 (Li et al., 2008). This scarcity of drinking water as a result of pollution is compounded by growing populations, climate changes and increasing energy costs leading to a major upheaval in the cost of potable water (Tofighty and Mohammadi, 2011; Crane and Scott, 2012).

The deterioration of the environment is primarily a result of industrial, agricultural, and manmade activities that either lead to release of toxic compounds in the environment or increase the level of existing toxic compounds in the environment (Adeleye et al., 2016). The pollutants such as air pollutants (CO, SO_2 and dioxins) and water pollutants (heavy metals and perfluoro alkylated acids) added to the environment by these activities can cause short-term illnesses to severe health hazards such as chronic poisoning caused by CO and lead contamination (Mancebo and Wang, 2015; Roy and Bharadvaja, 2021). The types of environmental pollution are explained in Fig. 1.

1.2 Types of Environmental Pollutants

Environmental pollutants can be an outcome of both natural and manmade processes and can be broadly classified as dyes, pesticides, heavy metals, and Polycyclic Aromatic Hydrocarbons (PAHs) (Seewagen, 2020). Dyes are Persistent Organic Pollutants (POPs) that are primarily used in the textile industries. They are xenobiotic and pollution associated with them are a result of manmade processes such as their manufacturing and fabric dyeing (Kant, 2012). These processes generate large volumes of industrial environmental contaminants into the environment. For example, synthetic dyes detected in textile wastewater have been shown to decrease light penetration into rivers and hence, alter photosynthetic efficiency of aquatic vegetation. Dyes like Azure-B, Disperse Red 1and Disperse Orange 1 can be mutagenic in nature and can potentially alter DNA and RNA structures in living organisms (Lellis et al., 2019).

Pesticides are chemicals which are largely used in various parts of the world, especially in regions where agriculture is a primary occupational and economical sector of the society. Despite the protection provided by these chemicals to crops and animals against insect infestations, their contamination into the environmental systems can pose a threat to life around it (Hassan and El Nemr, 2020). For example, ingestion of dichlorodiphenyltrichloroethane, commonly known as DDT, by humans can lead to accumulation of the molecule within body lipids and complications in adipose fat tissues (Jia et al., 2020). In addition to that, concentration of pesticides via biomagnification and bioconcentration can cause several other adversities in living organisms such as, hormonal imbalances and immune suppression in humans, tumour formation in animals and fish, intergenerational impacts of teratogens, muck on fish gills and crusts, etc. (Zhen et al., 2019; Hwang et al., 2018). Pesticides include herbicides and photosynthesis inhibitor herbicides such as urea, triazines, and phenolics are a huge threat to our environment. They often influence non-target habitats, such as the neighbouring vegetation and freshwater ecosystems (Masojídek et al., 2011; Tortolini et al., 2016) and can act as endocrine disruptors, especially in cases of long-term exposure (Combarnous, 2017).

Heavy metals pollutants include titanium, iron, vanadium, chromium, manganese, cobalt, copper, zinc, arsenic, molybdenum, nickel, cadmium, silver, tin, platinum, gold, mercury, and lead (Tchounwou et al., 2012). These metals infiltrate into our environment via the use of fertilizers and pesticides, inappropriate waste management and various anthropogenic activities. Metal-dependent industrial processes such as metal mining, steelmaking, extraction from dumped landfills, road repair, etc., allow add to the presence of these pollutants via metal leaching (Gautam et al., 2016). PAHs are emitted by use of petroleum products and combustion of materials, such as coal and firewood and include molecules such as naphthalene, phenanthrene, anthracene, fluoranthene, pyrene, chrysene, benzo anthracene and benzo fluoranthene (Pandey et al., 2017; Masih et al., 2010). These are often stable molecules and exhibit resistance to decomposition, making them harmful pollutants of the environment. These above points have been summarised in Fig. 1.

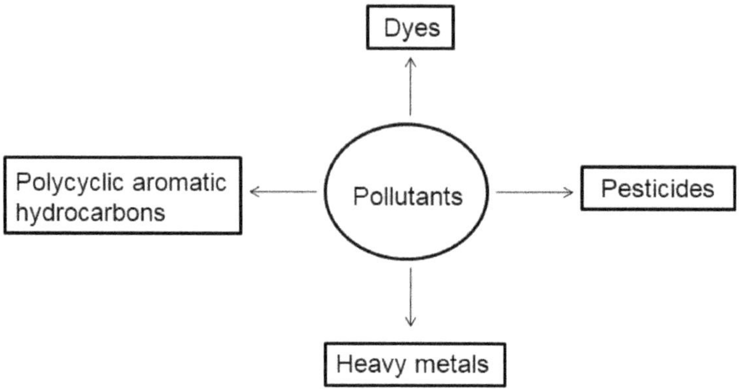

Figure 1. Types of environmental pollution.

2. Monitoring of Environmental Pollution

2.1 Consequences of Environmental Pollution

Environmental pollution can have widespread detrimental effects on life on earth. These effects worsen with progressive accumulation of contaminants and can have long-term and irreversible effects if left unchecked. In aquatic habitats, the major toxic elements have been shown to be heavy metals and Persistent Organic Pollutants (POPs), including perfluoro alkylated acids (PFAs) being a concern worldwide (Vasilachi et al., 2021; Altenburger et al., 1996; Groffen et al., 2021). In addition to these, contaminated water bodies may also contain pesticides, chemically synthesized fertilizers, dyes, detergents, and heavy metals that can adversely affect human health (Jayaswal et al., 2018). A high nitrogen level in contaminated water can be a risk to infants who are < 6 months old, causing Blue Baby syndrome (characterized by a decrease in the oxygen-carrying capacity of haemoglobin) in them (Ghangrekar and Chatterjee, 2018). Heavy metals can disrupt endocrine function by mimicking, blocking and/or by interfering with hormones present in the human body and causing neurological and reproductive developmental delays. Similarly, PFAs have been linked with infertility, thyroid dysfunction, cancer, and obesity (Quinete and Hauser-Davis, 2021).

In addition to that, human-induced phenomenon; such as; cultural eutrophication also contaminates aquatic environments. Cultural eutrophication refers to the expansion of harmful algal blooms, as a result of large population size and increased rate of food production that discharge large amounts of nitrogen and phosphorus-rich compounds in water bodies (Tilley et al., 2005). Algal blooms are toxic to the native aquatic flora and fauna and their emergence dysregulates the biogeochemical cycle of organic pollutants that are persistently present in the natural environment (Dachs et al., 2000). These blooms lead to a decrease in oxygen percentages in conjunction to an increase in poisonous compounds preventing the use of consumable potable water and limiting water use by fisheries (Khan and Mohammad, 2014). Eutrophic waters can also cause skin-eye irritation, gastric inflammation and vomiting in humans. Like eutrophication, ocean acidification caused by anthropogenic pollutants also threaten aquatic life by decreasing their chances of survival in the habitat (Ivanina and Sokolova, 2015). One of ways it does so is by elevating carbon dioxide and trace metal levels and causing imbalances in the acid-base homeostasis within aquatic organisms (Billé et al., 2013).

The major contributors to environmental pollution in terrestrial ecosystems are industries. Industrial pollutants such as aromatic hydrocarbons, plasticizers and various xenobiotic compounds are regularly dumped without any regulations (Phale et al., 2019). These pollutants have mutagenic and/or carcinogenic properties and can cause detrimental effects on soil and life on land. For example, accumulation of industrial pollutants has been shown to decrease agricultural land production and induce seed dormancy. Accumulation of industrial pollutants that are radioactive in nature have been shown to increase the risks of forest fires in zones contaminated with them (Álvarez et al., 2017).

Increase in pollutant levels by municipal sewage and farm fertilization have led to leaching of heavy metals, chemical pesticides, fertilizers and other micropollutants into ground water and thereby compromising its quality (Phale et al., 2019). And finally, increase in industrial pollutant levels in land have accelerated the rate of deforestation leading to significantly high soil erosion rates and high probability of flash flood events in many areas of the world (Artiola et al., 2019).

With regards to the aerial environment, major contaminants include greenhouse gases, fuel soot particles, methane, halocarbons, ozone, and multiple nitrous oxides. Air pollutants can destabilize an ecosystem, contaminate soil, and water and threaten food reserves (Agrawal et al., 2003). This is specifically true when pollutants such as the nitrogenous and sulphur oxides interact with the surrounding atmosphere and cause acid deposition on earth in the form of rain, fog, and/or snow (Singh and Agrawal, 2007). In fact, nitrogenous oxides, particulate fine and ultrafine matter have been known to be able to cross the human blood-brain barrier, modulate vasoregulation and even triggering neural inflammation (Roberts et al., 2019). Other oxides such as carbon monoxide can decrease oxygen delivery to the body's organs, impair vision and can be fatal at extremely high concentrations (Kelishadi and Poursafa, 2010).

Air pollutants can lead to an increase in multiple health problems with affecting populations such as a decrease in lung function, asthma, inflammation of bronchi and bronchioles, infant mortality, perinatal effects, allergy, cardiovascular and mental disorders, endothelial dysfunction, etc. (Siddique et al., 2011). Additionally, high particulate matter concentration of air pollutants can cause emission changes as well as physical and chemical changes in the atmosphere (Orru et al., 2017). These contaminants not only pollute the air we breathe, but also contribute to global warming and is the sixth leading cause of death in the world. Global warming leads to ocean acidity, increase in the sea level (due to rapid melting of glaciers and snow), frequent tropical calamities such as storms and loss of viable agricultural grounds and animal habitats (Orru et al., 2017; Michaelis, 1992; Jacobson et al., 2009). Some air pollutants can aid in the generation of reactive oxygen species that cause oxidative stress and predispose exposed individuals to cancer development (Ismail et al., 2019). Increased oxidative stress and poor cardiovascular circulation cause inflammation in organs, which can also cause mental health problems.

2.2 Significance of Monitoring Environmental Pollutants

Environmental pollution is a looming calamity that threatens every aspect of life on earth. With each passing day, our day-to-day activities move us closer to an environmental catastrophe and we ought to take preventive measures against it sooner than later. There has been a rapid decline in the quality of our air, water and soil and hence environmental pollutants and their sources must be strictly regulated and continuously monitored (Adeleye et al., 2016). The main sources of both organic and inorganic pollutants present in the environment are credited from the industries, agricultural farms, and human activities (Khanam et al., 2020). Hence, the environment needs to be monitored for the presence of pesticides, toxins, pathogens, heavy metals, volatile toxins, and endocrine-disrupting chemical compounds (Justino et al., 2017).

3. Classical Methods of Removing Environmental Pollutants

Traditionally different environmental pollutants are removed with processes that aim to eliminate contaminants via physical alterations methods compounded by specific electrochemical and biological procedures. For example, conventional methods like coagulation, flocculation, AC adsorption, froth floatation, sand filtration are commonly used to reduce air pollutants. Coagulation and flocculation are traditional pre-treatment methods used for decontaminating polluted water systems (Amuda and Amoo, 2007). These methods remove nutrient pollutants like phosphorus and nitrogen from water as well as prepare water for biological treatments that improve the biodegradability of waste products and remove macro pollutants (Aguilar et al., 2002). This treatment is a physiochemical approach aimed at removing heavy metal contaminants and soluble compounds from industrial wastewater by forming large flocs (Iwuozor, 2019). This is achieved by the use of ferric chloride and polymers that destabilize colloidal materials and aids in the agglomeration of small particles to form big flocs that can settle. It also aids in the elimination of suspended solids and some organic material. Coagulation in conjunction with ozonation has been used to treat sanitary landfill leachates (Poznyak et al., 2008).

Adsorption is a surface phenomenon for removal of organic and inorganic pollutants. Adsorption treatments can be done in different ways depending on the chemical properties of the pollutants. For example, in the membrane adsorption method contaminants are adsorbed to the membranes while the pollutant carrying the source flows through it. Chromium (VI), zinc (II) and lead (II) are examples of contaminants that can be removed by membrane-based adsorption methods (as well as polymer membrane-based adsorption methods) especially polypyrrole (Khulbe and Maatsuura, 2018). Organic dyes and discharges from pharmaceutical companies are examples of contaminants that can be removed by activated carbon adsorption method (Zambianchi et al., 2017; El-Geundi, 1997). In the zeolite adsorption method heavy metals are largely selected for adsorption by natural zeolites that have a high surface, large pores, are thermally stable, cheap, chemically inert, and easily modified surface functional group that enables it to have excellent selectivity against metal pollutants (El-Geundi, 1997). Biosorption is yet another traditional method wherein adsorbents are either prepared naturally or engineered artificially rendering them cost efficient and highly selective with regenerative abilities. The monitoring performance of this method is comparable to other conventional techniques and is mainly used for the treatment of secondary and tertiary effluent pollutants (Chaukura et al., 2016).

Along with these physical methods of pollutant removal, a few biological and chemical methods are also used to monitor surrounding environmental pollution. Of these, biological methods are more cost effective and generate environment-friendly by products from pollutant degradation. These methods are often combined with chemical methods like wet air oxidation, ozonation, etc. (Jeworski and Heinzle, 2000). Some of the reaction-based chemical methods are also well established for removing volatile and reactive pollutants from the environment. This includes processes like oxidation, ozonation, Fenton reaction, UV irradiation, peroxidation, and a combination of therein. Complete degradation of organic pollutants with mineralization in the environment can be achieved by advanced oxidation processes. Reduction of contaminated aromatic compounds and chemical oxygen by oxidizing and/or degrading pesticides (belonging to the classes organochlorine, organophosphorus, and triazine) can be achieved by ozonation treatment (Chedeville et al., 2009). Polyphenols can be treated by AOP (almost up to 90%), Fenton and solar photo-fentons (Lindholm-Lehto et al., 2015). In addition to that, photo-fentons can also treat reoccurring organic compounds. Pollutants such as cyanide can be removed using a combination of NaClO and hydrogen peroxide that form singlet oxygen and hydroxide ions, and pesticides can be removed via photolysis (the process of degradation by light or, photons) combined with hydrogen peroxide treatment and with the assistance of solvents like trichloromethane and carbon tetrachloride (Suty et al., 2004; James et al., 2014; Oliveira et al., 2014). Heterogenous photocatalysis treatment is yet another chemical method of pollutant removal that is cost effective and coverts pollutants to molecules that have a less toxic footprint in the environment. This treatment is done by semiconductors such as titanium dioxide, zinc oxide, ferric oxide, copper sulphide and cadmium sulphide that remove and convert pollutant molecules into easily crystallizable forms that are phytochemically and have stable and non-toxic behaviour.

A range of electrochemical processes have also been used as traditional methods for removing environmental pollutants. For example, the process of electrocoagulation is used to dose water using a metal anode and introduce cations without increasing the salinity of the solution (Elazzouzi et al., 2017). This process is easily operatable, has rapid sedimentation rates, reduced sludge production and has a smaller number of coagulant ions required for treatment. The process of electrochemical oxidation is used to remove organic pollutants via direct or indirect oxidation, with the latter one shown to be more effective (Martínez-Huitle and Panizza, 2018; Panizza and Cerisola, 2005).

Furthermore, electrochemical advanced oxidation processes have been found to not only remove pollutants and completely mineralize them but also to do so in an environment-friendly method (Dirany et al., 2012). Electrokinetic remediation is yet another potential electrochemical technique that remediates areas contaminated with organic and inorganic pollutants via electromigration, electroosmosis and electrophoresis combined with or without electrolysis (Hassan et al., 2015). Finally, the Electro-Fenton method breaks organic compounds by generating hydroxyl radicals via electrochemical oxidation indirectly, alongside electro generation of both hydrogen peroxide and ferrous ions under *in situ* conditions (Martínez-Huitle and Panizza, 2018; Fernandes, 2015).

4. Nanoparticles in Environmental Monitoring

As industrialization and urbanization continue to escalate, the resulting rise in pollution levels and the pressure on the environment pose unprecedented challenges. These processes, driven by population growth, economic development, and technological advancement, have profound implications for ecosystems, biodiversity, and human health (Humbal et al., 2023; Power et al., 2018; Hussain and Zhou, 2022). Therefore, it is imperative to address these challenges and implement pollution control measures which are crucial for building a sustainable and resilient future. Among the innovative solutions, nanoparticles have emerged as powerful tools that are capable of accurate detection, analysis, and mitigation of pollutants. With their unique chemical and physical properties, nanoparticles offer immense potential in combating pollution and environmental monitoring (Asghar et al., 2024). Pollution sensors that utilize nanoparticles can detect pollutants with remarkable sensitivity and precise detection even at trace levels. This capability is pivotal in monitoring air quality, water purity, or soil contamination, where contaminants like heavy metals, organic pollutants, and pathogens pose significant threats to ecosystems and human health (Gobinath et al., 2021).

In water monitoring, nanoparticles find diverse applications ranging from detecting industrial pollutants in wastewater to monitoring natural water bodies for microbial contaminants. Quantum dots, for instance, offer exceptional fluorescence properties, enabling the detection of heavy metals like mercury and lead with unparalleled sensitivity (Xue et al., 2017). Similarly, magnetic nanoparticles facilitate the rapid separation and detection of oil spills or hydrocarbon pollutants, aiding in timely response and mitigation efforts (Mehmood et al., 2021).

In air quality monitoring, nanoparticles play a pivotal role in detecting particulate matter, Volatile Organic Compounds (VOCs), and other airborne pollutants. Carbon nanotubes, owing to their high aspect ratio and conductivity, serve as sensitive sensors for detecting gases like nitrogen dioxide and ozone, which are key indicators of air pollution (Naishadham et al., 2023). Additionally, nanomaterial-based filters exhibit enhanced efficiency in capturing fine particulate matter, mitigating respiratory health risks associated with air pollution (Chitranshi et al., 2022).

Beyond pollutant detection, nanoparticles contribute to the monitoring of environmental parameters critical for ecosystem health. Nanosensors integrated into autonomous platforms enable real-time monitoring of parameters like pH, temperature, and dissolved oxygen levels in aquatic environments, providing invaluable insights into ecosystem dynamics and facilitating timely interventions to prevent ecological crises (Sugiharto et al., 2023). These have been summarized in Fig. 2.

Various applications of nanomaterials include:

- Iron Oxide: Used for water purification and soil remediation (Zhang et al., 2020).
- Manganese Dioxide: Effective in removing heavy metals from contaminated water (Zhang et al., 2020).
- Indium Phosphide Nanowires: Detect specific gases and pollutants (Zafar and Iqbal, 2016).
- Platinum and Rhodium: Employed in catalytic converters to reduce vehicle emissions (Balaram, 2020).

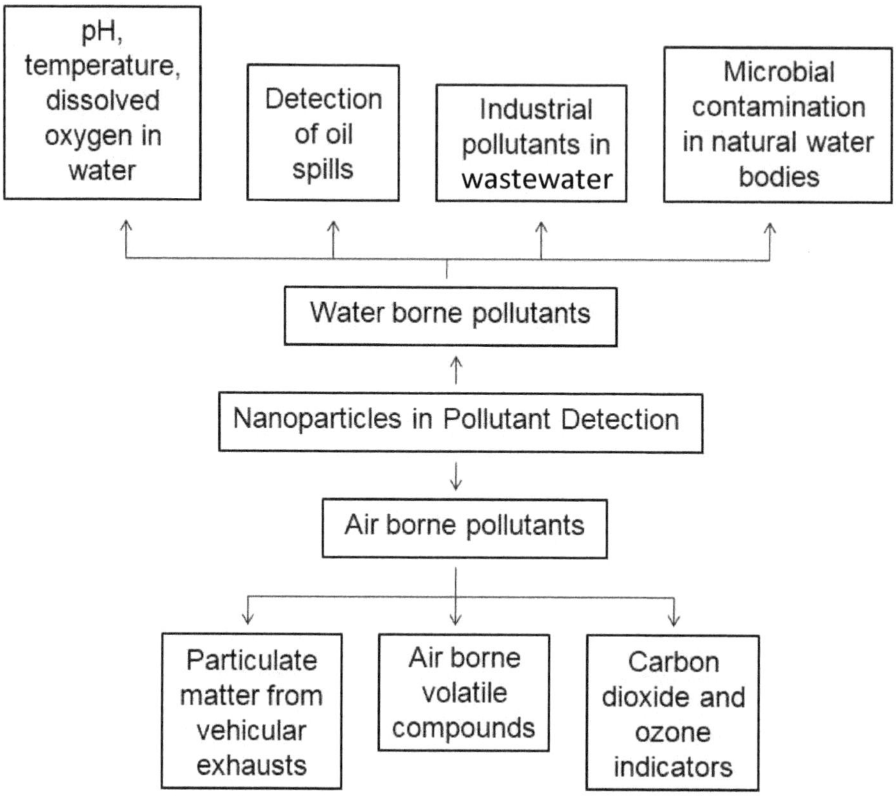

Figure 2. Applications of nano particles in pollutant detection.

- Zinc Oxide and Titanium Dioxide: Used for UV protection and air purification (Yang et al., 2023).
- Silica and Graphene: Enhance pollutant adsorption and removal (Zhang et al., 2020).

At the heart of nanoparticle-based monitoring lies nano sensors. These devices detect physical, chemical, or biological phenomena at the nanoscale. Their large surface area-to-volume ratio enables exceptional responsiveness to environmental stimuli (Khanna, 2012). Some specific applications include:

- Capped Nanoparticles: Tailored for targeted pollutant detection.
- Iron-Based Nanoparticles: Efficiently remove contaminants from water.
- Gold Nanoparticles: Used in biosensors for detecting pathogens.
- Metal Oxide Nanoparticles: Employed in gas sensors.
- Metal Nanoparticles: Enhance catalytic processes for pollution abatement.

Moreover, the advent of nanotechnology has revolutionized environmental monitoring through the miniaturization and integration of sensing systems. Nanoparticle-based sensors, compact and portable, facilitate on-site monitoring in remote or inaccessible regions, overcoming logistical challenges associated with traditional analytical methods. This capability is particularly beneficial in monitoring biodiversity hotspots, fragile ecosystems, and areas prone to natural disasters, where timely detection and response are imperative for conservation efforts (Srikanth et al., 2021; Su et al., 2012).

In conclusion, nanoparticles represent a paradigm shift in environmental monitoring, offering unparalleled sensitivity, selectivity, and portability in detecting and analyzing pollutants.

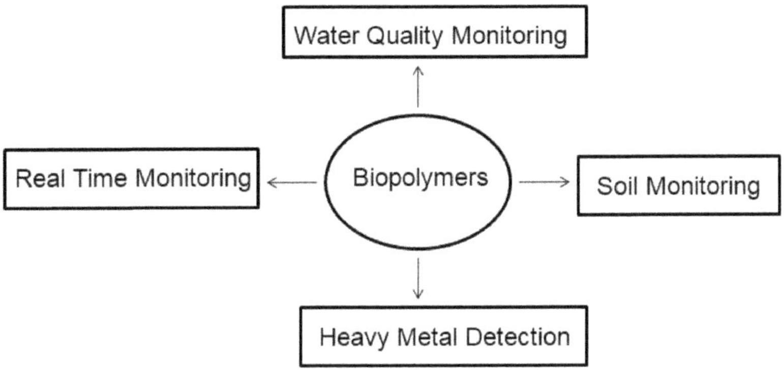

Figure 3. Scope of biopolymers in environmental monitoring.

From water quality assessment to air pollution monitoring and beyond, nanoparticles empower scientists, policymakers, and environmental stewards with the tools needed to safeguard our planet's precious ecosystems for generations to come. As we continue to confront the complex challenges of environmental sustainability, the integration of nanotechnology holds immense promise in shaping a healthier and more resilient future for our planet.

5. Scope of Biopolymer-based Environmental Monitoring

5.1 Biopolymers in Environmental Monitoring

Biopolymers are organic substances derived from biological sources or biosynthesis within living organisms primarily of natural origin. Biopolymers are found in various compounds produced by plants biomass, animals and microorganisms that are essential in the functioning and maintaining structure of living organisms (Baranwal et al., 2022). The major biopolymers comprise polysaccharides, polypeptides, fats, and polynucleotides bound together by covalent bonds (Baranwal et al., 2022; Stryer, 2002). Due to their biodegradable and biocompatible properties, biopolymers find various applications across different fields like the food industry, biomedical industry, medicine and drug delivery, tissue engineering, cosmetics, bio sensors and environmental applications (Baranwal et al., 2022; Gheorghita et al., 2021; Pires et al., 2023). Indeed, biopolymers can be utilized in environmental monitoring through applications such as biopolymer-based sensors, water quality control and monitoring, bioindicators, biodegradable sampling, remediation techniques and assessing environmental changes (Yadav et al., 2015). Biopolymers can be incorporated into sensor platforms to detect specific environmental parameters such as pH, heavy metals, organic pollutants, or biological contaminants (Madej-Kiełbik et al., 2022). These sensors can be utilized for the selective interactions of biopolymers with target analytes to get accurate and sensitive detection. Biopolymers can also be used as substrates for immobilizing biological indicators such as enzymes, antibodies, or microorganisms (Mohamad et al., 2015; Ghosh et al., 2015). These bioindicators can detect the presence of specific pollutants or toxins in environmental samples. Bioindicators can reflect and respond to the changes in the condition of the environment and thus provide valuable insights into the health of ecosystems. For example, mussels have been proposed as a global bioindicator of microplastic pollution while insects can be valuable bioindicators for environmental monitoring purposes that act as an indicator of changes in soil quality, air composition and water conditions (Li et al., 2019; Chowdhury et al., 2023). Biopolymers can also be used to manufacture biodegradable sampling materials such as filters or membranes for collecting environmental samples (Sorokulova et al., 2015). These bio-sustainable materials minimize environmental impact during sample collection and ensure environmental protection in different stages. The biopolymer-based nanomaterials can be added to soil or water to enhance bioremediation or phytoremediation monitoring changes in pollutant concentrations over time and assess the remediation progress

(Dhillon et al., 2012). Besides that, biopolymer-based sensors cells, enzymes, antibodies, aptamer, DNA-based bio sensors and biomimetic sensors can be used to continuously monitor environmental parameters (Gavrilaş et al., 2022). Overall, the diverse properties and applications of biopolymers make them valuable tools for environmental monitoring, offering sustainable solutions by assessing and managing environmental quality at different levels. The range of biopolymers in environmental monitoring have been summarized in Fig. 3.

The escalating presence of different environmental pollutants has become a significant cause for concern. Biopolymers offer a wide range of beneficial properties, such as adsorption, flocculation, and coagulation, which make them invaluable for treating environmental pollutants. Biopolymers can adsorb pollutants by effectively binding contaminants to their surfaces and removing them from water or soil. This process occurs through various mechanisms, such as electrostatic interactions, hydrogen bonding, and hydrophobic interactions, depending on the nature of the biopolymer and the pollutants involved. The large surface area and functional groups present in biopolymers help in adsorption capacity, allowing them to effectively capture and immobilize a wide range of pollutants, including heavy metals, organic compounds, and dyes. Biopolymer-based composites have become an extensive choice for applications as an adsorbent due to sustainability, cost-effectiveness, and environmental advantages (TG et al., 2023). Materials commonly used in composites for adsorption purposes include activated carbon, zeolites, polymers and various nanomaterials like graphene and carbon nanotubes (Sadegh et al., 2017; Gupta et al., 2016). The biopolymer-based flocculant offers several advantages including a broad range of pH tolerance, easy modification, adaptability, and high efficiency (Jiang, 2015; Jiang et al., 2021). The biopolymer-based flocculant such as chitosan, starch, cellulose, and lignin can be used in water treatment processes to remove suspended solids, organic matter, and contaminants from water sources by facilitating aggregation. The flocculation property enables the trapping and immobilization of pollutants like heavy metals in wastewater streams. Chitosan with improved phytochemical properties can be used to immobilize biological indicators, such as enzymes or microorganisms for detecting pollutants and monitoring environmental changes specifically heavy metals (Aranaz et al., 2021; Goci et al., 2023). Additionally, biopolymer-based flocculants can aid water treatment by aggregating suspended solids and contaminants, facilitating their removal from water sources. Flocculation enhances the sedimentation process by destabilized particles, thereby increasing their weight, which facilitates their removal through filtration. In water purification, particularly in surface water treatment, this stage is crucial for eliminating organic impurities such as viruses and bacteria. Monitoring the effectiveness of these flocculants can provide insights into water quality and treatment efficiency. The utilization of biopolymers in environmental monitoring is continuously advancing. Indeed, biopolymers provide sustainable solutions for environmental monitoring and ecosystem assessment, contributing to the preservation and restoration of natural habitats in various forms.

Biopolymers, derived from renewable sources, offer sustainable alternatives to conventional materials in environmental monitoring applications. Biopolymer-based environmental monitoring holds immense promise for sustainable solutions to environmental challenges (Palanisamy et al., 2019; Muralikrishna et al., 2022). From water and soil quality assessment to air pollution monitoring, biopolymers offer eco-friendly alternatives with diverse applications (Moola et al., 2024). The integration of biopolymers with nanotechnology and advanced sensing techniques further enhances their capabilities.

Biopolymers in Water Quality Monitoring: Biopolymer-based sensors have shown promise in detecting various contaminants in water sources. For example, chitosan, a biopolymer derived from crustacean shells, has been utilized for heavy metal detection due to its excellent adsorption properties (Ravikumar et al., 2019). Cellulose-based membranes have been effective in removing particulate matter and microbial contaminants from water (Azimi et al., 2024). These biopolymer-based solutions contribute to improving water quality and ensuring safe drinking water for communities.

Soil Monitoring with Biopolymers: Biopolymers play a significant role in soil monitoring, aiding in soil health assessment and contamination detection. Alginate-based sensors have been used for detecting soil pollutants and degradation processes (Rizwan et al., 2022). Additionally, biopolymer-based hydrogels have been employed for controlled release of fertilizers, enhancing soil fertility while minimizing environmental impacts (Tariq et al., 2023). These applications demonstrate the potential of biopolymers in sustainable agriculture and land management practices.

The integration of biopolymers with nanomaterials has led to the development of advanced sensing platforms with improved performance. Biopolymer-based nanocomposites offer enhanced sensitivity and selectivity for detecting environmental pollutants. For instance, nanocomposites of chitosan and nanoparticles have been utilized for the detection of heavy metals and organic compounds in water (Morsi et al., 2017). These biopolymer-based nanocomposites represent a promising avenue for sensitive and reliable environmental monitoring.

Biopolymer Hydrogels for Real-time Monitoring: Biopolymer hydrogels have emerged as versatile matrices for the immobilization of sensing elements, enabling real-time monitoring of environmental parameters. Hydrogel-based sensors exhibit rapid response kinetics and can be tailored for specific analytes. Research has shown the utility of hydrogel-based sensors in monitoring pH, temperature, and dissolved oxygen levels in aquatic environments (Tavakoli and Tang, 2017). These biopolymer hydrogel sensors provide insights into ecosystem dynamics and aid in conservation efforts. The different futuristic scopes of biopolymers in various aspects of environmental monitoring are discussed in Table 1 and Fig. 2.

Biopolymer-based environmental monitoring bridges the gap between scientific principles, sustainable resource management, and pollution risk assessment. As we harness the potential of biopolymers, we move towards a greener, healthier planet—one where monitoring systems contribute to informed decision-making and ecological well-being.

Table 1. Table showing scope of biopolymers in various aspects of environmental monitoring.

Scope	Role
Pollutant Detection and Assessment	Biopolymer-based sensors play a pivotal role in detecting pollutants. These sensors utilize biopolymer matrices to interact with specific analytes, providing real-time data on environmental contamination.
	Monitoring systems designed using biopolymers allow us to estimate exposure levels at both individual and population scales. By synthesizing monitoring data with toxicological and epidemiological information, we gain insights into pollution risks to humans and ecosystems
Renewable Resource Management	Biopolymers find applications in managing renewable natural resources such as agriculture, fisheries, and forests. Monitoring systems help assess the impact of human activities on these resources, guiding sustainable practices
Integration with Health Data	Biopolymer-based monitoring extends beyond environmental parameters. It integrates with health data, enabling us to evaluate the health risks associated with pollution exposure
	By combining monitoring results with pre-market screening data, we enhance our understanding of potential hazards
Land Use Changes and Biodiversity	Geographic Information System (GIS) analyses and remote sensing studies fall within the purview of biopolymer-based monitoring. These techniques assess anthropogenic impacts on land cover changes, biodiversity, and global climate change
Scientific Basis and Methodology	Research focuses on scaling methods, biological indicators, dynamic models, and pollution indices
	Optimization of monitoring networks, data handling, and quality assurance procedures are critical aspects

5.2 Biopolymers and Metal-based Nanoparticles

Although metal-based nanoparticles have reported harmful effects on the human body but metal based nanoparticles are still highly relevant in various applications. Metal-based Nanoparticles (MN) have the ability to cross cellular membranes leading to damages of cellular components like proteins, DNA etc. However, keeping in mind, the huge range of applications of metal-based nanoparticles , various methods based on green chemistry have been applied to obtain metal nanoparticles which would be more environmentally safe. These methods improve the biodegradability of the nanoparticles and also increase their adaptability for many reactions. A major variety of biodegradable nanoparticles are the Biopolymer-based Nanoparticles (BN). Biopolymer based nanoparticles have the advantage of being environmentally safe but they suffer from having lower adaptability across various functional parameters. As such the integration of MN with the BN have been carried out for increasing the range of applications of the combined BN-MN nanoparticles (Nasrollahzadeh et al., 2021).

Biopolymers mostly used for the synthesis of nanoparticles include polysaccharides like starch, cellulose, dextran and proteins like collagen, gelatin, and albumin. However, reports on applications of the different types of BN, the BN derived from polysaccharides have observed to have clear functional advantages in comparsion to protein-based BN. Polysaccharides can act as stabilizers in the synthesis of MN. They are important in the green synthesis of NM as they can carry out reduction reaction of the MN during the synthesis process, and also give several desirable properties to the MN (Vodyashkin et al., 2022). Some of the most widely available polysaccharides are starch and cellulose and both are utilized in the green synthesis of MN.

Starch is present in many cereal crops and vegetables like potatoes. Starch has been widely used in the process of synthesis of silver MN. Under alkaline conditions, starch is made to react with $AgNO_3$ to obtain silver MN. The size and shape of the MN produced are influenced by the extent of the basicity of the reaction media. Starch has also been used and a reducing agent and as a stabilizer in the synthesis of Gold and Platinum MN. In all these processes, starch is initially dissolved in water or in alkaline solution like NaOH, and then heated to 80°C for the dissolving to be clear. The salts of the metals like Ag, Au, Pt are subsequently reduced by the numerous hydroxyl and carbonyl groups present in starch molecules leading to the formation of the corresponding metals. These metals form aggregates of various sizes and shapes depending on the reaction conditions and the stabilizing effects of starch, leading to the formation of their corresponding MN. The synthesized MN can then be characterized by techniques like electron microscopy, X-ray diffraction etc. The process of synthesis of MN using cellulose follows a similar protocol. Cellulose acts as a reducing agent via its carbonyl and hydroxyl groups leading to the production of the metals. These metallic aggregates are further stabilized by cellulose preventing their precipitation which leads to the production of the corresponding MN along with regulation of their size. Cellulose for the process of MN synthesis is obtained from diverse sources like water hyacinth or bacterial cellulose (Fakayode et al., 2016).

Biopolymer based MN are being used for various functions. They have reported to have antimicrobial activity, drug delivery and catalytic activities. Another function of the biopolymer-based MN is in the field of the environmental monitoring. Biopolymer based MN is widely used in the detection of both organic and inorganic environmental contaminants. These can include pesticides, vehicular exhaust, etc., which can be present in the air, water, or soil. Besides chemicals, these MN can also be used to detect toxins, pathogens, and other harmful bioactive pollutants. Pesticides, herbicides and other agriculture-based contaminants are major sources of environmental pollution as they can prevent photosynthesis in a wide variety of plants and also mimic endocrine components in animals and humans leading to endocrine disbalance. Alginate based MN are used for the detection of several pesticides (Madej-Kiełbik et al., 2022).

The detection of heavy metals in the environment is another situation that requires monitoring and biopolymer-based MN are used for the purpose. Chitosan based MN are used to detect heavy metals like K^+ in a concentration range of 9 nM, Cd^+ in a range of 0.063 μM and Hg^{2+} in a range of 0.5 nM. Cellulose based MN have been used for the detection of heavy metals like Pb^{2+} and

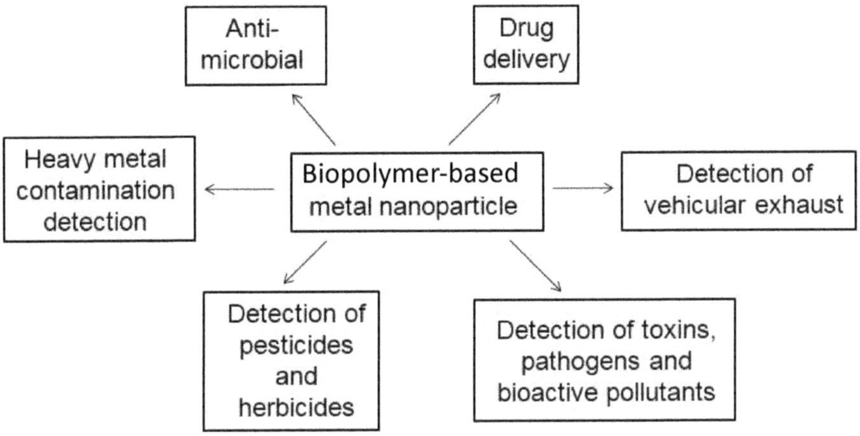

Figure 4. Functions of biopolymer-based metal nanoparticles.

Fe^{3+} (Ramdzan et al., 2020). The various functions of biopolymer-based metal nanoparticles have been summarized in Fig. 4.

6. Conclusion

Environmental pollution is a big challenge to the health and progress of nations and it affects the systems on which humans are dependent on for survival. Pollution can be manifested in various forms and it is highly imperative that the levels of the different pollutants be regularly monitored also strategies devised for the removal of pollutants from the environment. Metal-based nanoparticles are important in the process of monitoring various pollutants, but they themselves are not environmentally friendly. As such natural biopolymer-based nanoparticles are used for monitoring and removing the pollutants from the environment. Furthermore, the fusion of biopolymers like polysaccharides and metals into nanoparticles are also a useful option in the monitoring of several pollutants even when they are present at very low concentrations. This particular field of fusion nanoparticles involving biopolymers and metals have potentially large scope and research has shown promising results.

References

Adeleye, A. S., Conway, J. R., Garner, K., Huang, Y., Su, Y. and Keller, A. A. (2016). Engineered nanomaterials for water treatment and remediation: Costs, benefits, and applicability. Chemical Engineering Journal, 286: 640–662.

Agrawal, M., Singh, B., Rajput, M., Marshall, F. and Bell, J. N. B. (2003). Effect of air pollution on peri-urban agriculture: a case study. Environmental Pollution, 126(3): 323–329.

Aguilar, M. I., Saez, J., Llorens, M., Soler, A. and Ortuno, J. F. (2002). Nutrient removal and sludge production in the coagulation–flocculation process. Water Research, 36(11): 2910–2919.

Altenburger, R., Boedeker, W., Faust, M. and Grimme, L. H. (1996). Regulations for combined effects of pollutants: Consequences from risk assessment in aquatic toxicology. Food and Chemical Toxicology, 34(11-12): 1155–1157.

Álvarez, X., Valero, E., Santos, R. M., Varandas, S. G. P., Fernandes, L. S. and Pacheco, F. A. L. (2017). Anthropogenic nutrients and eutrophication in multiple land use watersheds: Best management practices and policies for the protection of water resources. Land Use Policy, 69: 1–11.

Amuda, O. S. and Amoo, I. A. (2007). Coagulation/flocculation process and sludge conditioning in beverage industrial wastewater treatment. Journal of Hazardous Materials, 141(3): 778–783.

Aranaz, I., Alcántara, A. R., Civera, M. C., Arias, C., Elorza, B., Heras Caballero, A. et al. (2021). Chitosan: An overview of its properties and applications. Polymers, 13(19): 3256.

Artiola, J. F., Walworth, J. L., Musil, S. A. and Crimmins, M. A. (2019). Soil and land pollution. In Environmental and pollution science (pp. 219–235). Academic Press.

Asghar, N., Hussain, A., Nguyen, D. A., Ali, S., Hussain, I., Junejo, A. et al. (2024). Advancement in nanomaterials for environmental pollutants remediation: a systematic review on bibliometrics analysis, material types, synthesis pathways, and related mechanisms. Journal of Nanobiotechnology, 22(1): 1–10.

Azimi, B., Sepahvand, S., Ismaeilimoghadam, S., Kargarzadeh, H., Ashori, A., Jonoobi, M. et al. (2024). Application of cellulose-based materials as water purification filters; a state-of-the-art review. Journal of Polymers and the Environment, 32(1): 345–366.

Balaram, V. (2020). Environmental impact of platinum, palladium, and rhodium emissions from autocatalytic converters–a brief review of the latest developments. Handbook of Environmental Materials Management, 1–37.

Baranwal, J., Barse, B., Fais, A., Delogu, G. L. and Kumar, A. (2022). Biopolymer: A sustainable material for food and medical applications. Polymers, 14(5): 1–8.

Billé, R., Kelly, R., Biastoch, A., Harrould-Kolieb, E., Herr, D., Joos, F. et al. (2013). Taking action against ocean acidification: a review of management and policy options. Environmental Management, 52(4): 761–779.

Chaukura, N., Gwenzi, W., Tavengwa, N. and Manyuchi, M. M. (2016). Biosorbents for the removal of synthetic organics and emerging pollutants: Opportunities and challenges for developing countries. Environmental Development, 19: 84–89.

Chedeville, O., Debacq, M. and Porte, C. (2009). Removal of phenolic compounds present in olive mill wastewaters by ozonation. Desalination, 249(2): 865–869.

Chen, J., Liu, M., Zhang, L., Zhang, J. and Jin, L. (2003). Application of nano TiO2 towards polluted water treatment combined with electro-photochemical method. Water Research, 37(16): 3815–3820.

Chitranshi, M., Chen, D. R., Kosel, P., Cahay, M. and Schulz, M. (2022). Flexible and lightweight carbon nanotube composite filter for particulate matter air filtration. Nanomaterials, 12(22): 1–12.

Chowdhury, S., Dubey, V. K., Choudhury, S., Das, A., Jeengar, D., Sujatha, B. et al. (2023). Insects as bioindicator: A hidden gem for environmental monitoring. Frontiers in Environmental Science, 11: 1–14.

Combarnous, Y. (2017). Endocrine Disruptor Compounds (EDCs) and agriculture: The case of pesticides. Comptes Rendus. Biologies, 340(9-10): 406–409.

Crane, R. A. and Scott, T. B. (2012). Nanoscale zero-valent iron: Future prospects for an emerging water treatment technology. Journal of Hazardous Materials, 211: 112–125.

Dachs, J., Eisenreich, S. J. and Hoff, R. M. (2000). Influence of eutrophication on air—water exchange, vertical fluxes, and phytoplankton concentrations of persistent organic pollutants. Environmental Science & Technology, 34(6): 1095–1102.

Dhillon, G. S., Kaur, S., Verma, M. and Brar, S. K. (2012). Biopolymer-based nanomaterials: potential applications in bioremediation of contaminated wastewaters and soils. In Comprehensive Analytical Chemistry (Vol. 59, pp. 91–129). Elsevier.

Dirany, A., Sirés, I., Oturan, N., Özcan, A. and Oturan, M. A. (2012). Electrochemical treatment of the antibiotic sulfachloropyridazine: kinetics, reaction pathways, and toxicity evolution. Environmental Science & Technology, 46(7): 4074–4082.

Elazzouzi, M., Haboubi, K. and Elyoubi, M. S. (2017). Electrocoagulation flocculation as a low-cost process for pollutants removal from urban wastewater. Chemical Engineering Research and Design, 117: 614–626.

El-Geundi, M. S. (1997). Adsorbents for industrial pollution control. Adsorption Science & Technology, 15(10): 777–787.

Fakayode, O. J., Oladipo, A. O., Oluwafemi, O. S. and Songca, S. P. (2016). Biopolymer-mediated green synthesis of noble metal nanostructures. Recent Advances in Biopolymers, 19–44.

Fernandes, A., Pacheco, M. J., Ciríaco, L. and Lopes, A. J. A. C. B. E. (2015). Review on the electrochemical processes for the treatment of sanitary landfill leachates: Present and future. Applied Catalysis B: Environmental, 176: 183–200.

G Gupta, V. K., Tyagi, I. and Agarwal, S. (2015). Study on the removal of heavy metal ions from industry waste by carbon nanotubes: Effect of the surface modification-A review. Critical Reviews in Environmental Science and Technology, 46(2): 93–118.

Gautam, P. K., Gautam, R. K., Banerjee, S., Chattopadhyaya, M. C. and Pandey, J. D. (2016). Heavy metals in the environment: fate, transport, toxicity and remediation technologies. Nova Sci Publishers, 60: 101–130.

Gavrilaş, S., Ursachi, C. Ş., Perţa-Crişan, S. and Munteanu, F. D. (2022). Recent trends in biosensors for environmental quality monitoring. Sensors, 22(4): 1–13.

Ghangrekar, M. M. and Chatterjee, P. (2018). Water pollutants classification and its effects on environment. Carbon Nanotubes for Clean water, 1: 11–26.

Gheorghita, R., Anchidin-Norocel, L., Filip, R., Dimian, M. and Covasa, M. (2021). Applications of biopolymers for drugs and probiotics delivery. Polymers, 13(16): 1–12.

Ghosh, S., Lahiri, D., Nag, M., Dey, A., Sarkar, T., Pathak, S. K. et al. (2021). Bacterial biopolymer: Its role in pathogenesis to effective biomaterials. Polymers, 13(8): 1–12.

Gobinath, R., Bandeppa, Manasa, V., Rajendiran, S., Kumar, K., Paul, R. et al. (2021). Nanoparticle-mediated adsorption of pollutants: A way forward to mitigation of environmental pollution. Microbial Rejuvenation of Polluted Environment, 2: 317–348.

Goci, M. C., Leudjo Taka, A., Martin, L. and Klink, M. J. (2023). Chitosan-based polymer nanocomposites for environmental remediation of mercury pollution. Polymers, 15(3): 1–14.

Groffen, T., Rijnders, J., van Doorn, L., Jorissen, C., De Borger, S. M., Luttikhuis, D. O. et al. (2021). Preliminary study on the distribution of metals and persistent organic pollutants (POPs), including perfluoroalkylated acids (PFAS), in the aquatic environment near Morogoro, Tanzania, and the potential health risks for humans. Environmental Research, 192: 1–9.

Hassaan, M. A. and El Nemr, A. (2020). Pesticides pollution: Classifications, human health impact, extraction and treatment techniques. The Egyptian Journal of Aquatic Research, 46(3): 207–220.

Hassan, I., Mohamedelhassan, E. and Yanful, E. K. (2015). Solar powered electrokinetic remediation of Cu polluted soil using a novel anode configuration. Electrochimica Acta, 181: 58–67.

Humbal, A., Chaudhary, N. and Pathak, B. (2023). Urbanization trends, climate change, and environmental sustainability. In Climate Change and Urban Environment Sustainability (pp. 151–166). Singapore: Springer Nature Singapore.

Hussain, J. and Zhou, K. (2022). Globalization, industrialization, and urbanization in Belt and Road Initiative countries: implications for environmental sustainability and energy demand. Environmental Science and Pollution Research, 29(53): 80549–80567.

Hwang, J. I., Zimmerman, A. R. and Kim, J. E. (2018). Bioconcentration factor-based management of soil pesticide residues: Endosulfan uptake by carrot and potato plants. Science of the Total Environment, 627: 514–522.

Ismail, I. N., Jalaludin, J., Bakar, S. A., Hisamuddin, N. H. and Suhaimi, N. F. (2019). Association of traffic-related air pollution (TRAP) with DNA damage and respiratory health symptoms among primary school children in Selangor. Asian Journal of Atmospheric Environment, 13(2): 106–116.

Ivanina, A. V. and Sokolova, I. M. (2015). Interactive effects of metal pollution and ocean acidification on physiology of marine organisms. Current Zoology, 61(4): 653–668.

Iwuozor, K. O. (2019). Prospects and challenges of using coagulation-flocculation method in the treatment of effluents. Advanced Journal of Chemistry-Section A, 2(2): 105–127.

Jacobson, M. Z. (2009). Review of solutions to global warming, air pollution, and energy security. Energy & Environmental Science, 2(2): 148–173.

James, C. P., Germain, E. and Judd, S. (2014). Micropollutant removal by advanced oxidation of microfiltered secondary effluent for water reuse. Separation and Purification Technology, 127: 77–83.

Jayaswal, K., Sahu, V. and Gurjar, B. R. (2018). Water pollution, human health and remediation. Water Remediation, 1: 11–27.

Jeworski, M. and Heinzle, E. (2000). Combined chemical-biological treatment of wastewater containing refractory pollutants. Biotechnology Annual Review, 6: 163–196.

Jia, Z. Q., Zhang, Y. C., Huang, Q. T., Jones, A. K., Han, Z. J. and Zhao, C. Q. (2020). Acute toxicity, bioconcentration, elimination, action mode and detoxification metabolism of broflanilide in zebrafish, Danio rerio. Journal of Hazardous Materials, 394: 1–12.

Jiang, J. Q. (2015). The role of coagulation in water treatment. Current Opinion in Chemical Engineering, 8: 36-44.

Jiang, X., Li, Y., Tang, X., Jiang, J., He, Q., Xiong, Z. and Zheng, H. (2021). Biopolymer-based flocculants: A review of recent technologies. Environmental Science and Pollution Research, 28: 46934–46963.

Justino, C. I., Duarte, A. C. and Rocha-Santos, T. A. (2017). Recent progress in biosensors for environmental monitoring: A review. Sensors, 17(12): 1–12.

Kant, R. (2011). Textile dyeing industry an environmental hazard. Natural Science, 4(1): 1–5.

Kelishadi, R. and Poursafa, P. (2010). Air pollution and non-respiratory health hazards for children. Archives of Medical Science, 6(4): 483–495.

Khan, M. N. and Mohammad, F. (2014). Eutrophication: challenges and solutions. Eutrophication: Causes, Consequences and Control, 2: 1–15.

Khanam, Z., Gupta, S. and Verma, A. (2020). Endophytic fungi-based biosensors for environmental contaminants-A perspective. South African Journal of Botany, 134: 401–406.

Khanna, V.K. (2012). Nanosensors: Physical, Chemical, and Biological (1st ed.). CRC Press.

Khulbe, K. C. and Matsuura, T. (2018). Removal of heavy metals and pollutants by membrane adsorption techniques. Applied water science, 8: 1–30.

Lellis, B., Fávaro-Polonio, C. Z., Pamphile, J. A. and Polonio, J. C. (2019). Effects of textile dyes on health and the environment and bioremediation potential of living organisms. Biotechnology Research and Innovation, 3(2): 275–290.

Li, J., Lusher, A. L., Rotchell, J. M., Deudero, S., Turra, A., Bråte, I. L. N. et al. (2019). Using mussel as a global bioindicator of coastal microplastic pollution. Environmental Pollution, 244: 522–533.
Li, Q., Mahendra, S., Lyon, D. Y., Brunet, L., Liga, M. V., Li, D. et al. (2008). Antimicrobial nanomaterials for water disinfection and microbial control: potential applications and implications. Water Research, 42(18): 4591–4602.
Lindholm-Lehto, P. C., Knuutinen, J. S., Ahkola, H. S. and Herve, S. H. (2015). Refractory organic pollutants and toxicity in pulp and paper mill wastewaters. Environmental Science and Pollution Research, 22: 6473–6499.
Madej-Kiełbik, L., Gzyra-Jagieła, K., Jóźwik-Pruska, J., Dziuba, R. and Bednarowicz, A. (2022). Biopolymer composites with sensors for environmental and medical applications. Materials, 15(21): 1–7.
Malik, A. and Grohmann, E. (Eds.). (2012). Environmental protection strategies for sustainable development (Vol. 520). Dordrecht, The Netherlands: Springer.
Mancebo, S. E. and Wang, S. Q. (2015). Recognizing the impact of ambient air pollution on skin health. Journal of the European Academy of Dermatology and Venereology, 29(12): 2326–2332.
Martínez-Huitle, C. A. and Panizza, M. (2018). Electrochemical oxidation of organic pollutants for wastewater treatment. Current Opinion in Electrochemistry, 11: 62–71.
Masih, J., Masih, A., Kulshrestha, A., Singhvi, R. and Taneja, A. (2010). Characteristics of polycyclic aromatic hydrocarbons in indoor and outdoor atmosphere in the North central part of India. Journal of hazardous Materials, 177(1-3): 190–198.
Masojídek, J., Souček, P., Máchová, J., Frolík, J., Klem, K. and Malý, J. (2011). Detection of photosynthetic herbicides: Algal growth inhibition test vs. electrochemical photosystem II biosensor. Ecotoxicology and Environmental Safety, 74(1): 117–122.
Mehmood, A., Khan, F. S. A., Mubarak, N. M., Mazari, S. A., Jatoi, A. S., Khalid, M. et al. (2021). Carbon and polymer-based magnetic nanocomposites for oil-spill remediation—a comprehensive review. Environmental Science and Pollution Research, 28(39): 54477–54496.
Michaelis, P. (1992). Global warming: Efficient policies in the case of multiple pollutants. Environmental and Resource Economics, 2: 61–77.
Mohamad, N. R., Marzuki, N. H. C., Buang, N. A., Huyop, F. and Wahab, R. A. (2015). An overview of technologies for immobilization of enzymes and surface analysis techniques for immobilized enzymes. Biotechnology & Biotechnological Equipment, 29(2): 205–220.
Moola, A. K., Prabhakar, M. R., Dey, B., Paramasivan, B., Vangala, S. M., Jakkampudi, R. et al. (2024). Biopolymeric composite materials for environmental applications. Physical Sciences Reviews, 9(6): 2153–2174.
Morsi, R. E., Alsabagh, A. M., Nasr, S. A. and Zaki, M. M. (2017). Multifunctional nanocomposites of chitosan, silver nanoparticles, copper nanoparticles and carbon nanotubes for water treatment: antimicrobial characteristics. International journal of Biological Macromolecules, 97: 264–269.
Muralikrishna, S., Kempahanumakkagari, S., Thippeswamy, R. and Surareungchai, W. (2022). Functional polymer materials for environmental monitoring and safety applications. Polymer-Based Advanced Functional Materials for Energy and Environmental Applications, 177–204.
Naishadham, K., Naishadham, G., Cabrera, N. and Bekyarova, E. (2023). Response surface modeling of the steady-state impedance responses of gas sensor arrays comprising functionalized carbon nanotubes to detect ozone and nitrogen dioxide. Sensors, 23(20): 1–9.
Nasrollahzadeh, M., Bidgoli, N. S. S., Nezafat, Z. and Shafiei, N. (2021). Catalytic applications of biopolymer-based metal nanoparticles. Biopolymer-based metal nanoparticle chemistry for sustainable applications. Elsevier, 423–516.
Oliveira, A. G. D., Ribeiro, J. P., Oliveira, J. T. D., Keukeleire, D. D., Duarte, M. S. and Nascimento, R. F. D. (2014). Degradation of the pesticide chlorpyrifos in aqueous solutions with UV/H2O2: optimization and effect of interfering anions. Journal of Advanced Oxidation Technologies, 17(1): 133–138.
Orru, H., Ebi, K. L. and Forsberg, B. (2017). The interplay of climate change and air pollution on health. Current Environmental Health Reports, 4: 504–513.
Palanisamy, K., Jeyaseelan, A., Murugesan, K. and Palanisamy, S. B. (2019). Biopolymer technologies for environmental applications. Nanoscience and Biotechnology for Environmental Applications, 55–83.
Pandey, R., Masood, F., Singh, H. P. and Batish, D. R. (2017). Polycyclic aromatic hydrocarbons as environmental pollutants: a review. International Journal of Advanced Research in Science, Engineering and Technology, 6: 1361–1369.
Panizza, M. and Cerisola, G. J. E. A. (2005). Application of diamond electrodes to electrochemical processes. Electrochimica Acta, 51(2): 191–199.
Phale, P. S., Sharma, A. and Gautam, K. (2019). Microbial degradation of xenobiotics like aromatic pollutants from the terrestrial environments. In Pharmaceuticals and personal care products: waste management and treatment technology (pp. 259–278). Butterworth-Heinemann.

Pires, P. C., Mascarenhas-Melo, F., Pedrosa, K., Lopes, D., Lopes, J., Macário-Soares, A. et al. (2023). Polymer-based biomaterials for pharmaceutical and biomedical applications: A focus on topical drug administration. European Polymer Journal, 187: 1–11.

Power, A. L., Tennant, R. K., Jones, R. T., Tang, Y., Du, J., Worsley, A. T. et al. (2018). Monitoring impacts of urbanisation and industrialisation on air quality in the Anthropocene using urban pond sediments. Frontiers in Earth Science, 6: 1–11.

Poznyak, T., Bautista, G. L., Chaírez, I., Córdova, R. I. and Ríos, L. E. (2008). Decomposition of toxic pollutants in landfill leachate by ozone after coagulation treatment. Journal of Hazardous Materials, 152(3): 1108–1114.

Qu, X., Alvarez, P. J. and Li, Q. (2013). Applications of nanotechnology in water and wastewater treatment. Water Research, 47(12): 3931–3946.

Quinete, N. and Hauser-Davis, R. A. (2021). Drinking water pollutants may affect the immune system: Concerns regarding COVID-19 health effects. Environmental Science and Pollution Research, 28: 1235–1246.

Ramdzan, N. S. M., Fen, Y. W., Anas, N. A. A., Omar, N. A. S. and Saleviter, S. (2020). Development of biopolymer and conducting polymer-based optical sensors for heavy metal ion detection. Molecules, 25(11): 1–11.

Ravikumar, R., Chen, L. H., Xin Hui, M. M. and Chan, C. C. (2019). Ion-imprinted chitosan-based interferometric sensor for selective detection of heavy metal ions. Journal of Lightwave Technology, 37(11): 2778–2783.

Rizwan, K., Rasheed, T. and Bilal, M. (2022). Alginate-based nanobiosorbents for bioremediation of environmental pollutants. In Nano-Biosorbents for decontamination of water, air, and soil pollution (pp. 479–502). Elsevier.

Roberts, S., Arseneault, L., Barratt, B., Beevers, S., Danese, A., Odgers, C. L. et al. (2019). Exploration of NO2 and PM2. 5 air pollution and mental health problems using high-resolution data in London-based children from a UK longitudinal cohort study. Psychiatry Research, 272: 8–17.

Roy, A. and Bharadvaja, N. (2021). Efficient removal of heavy metals from artificial wastewater using biochar. Environmental Nanotechnology, Monitoring & Management, 16: 1–10.

Roy, A., Sharma, A., Yadav, S., Jule, L. T. and Krishnaraj, R. (2021). Nanomaterials for remediation of environmental pollutants. Bioinorganic Chemistry and Applications, 2021(1): 1–10.

Sadegh, H., Ali, G. A., Gupta, V. K., Makhlouf, A. S. H., Shahryari-Ghoshekandi, R., Nadagouda, M. N. et al. (2017). The role of nanomaterials as effective adsorbents and their applications in wastewater treatment. Journal of Nanostructure in Chemistry, 7: 1–14.

Seewagen, C. L. (2020). The threat of global mercury pollution to bird migration: potential mechanisms and current evidence. Ecotoxicology, 29(8): 1254–1267.

Siddique, S., Ray, M. R. and Lahiri, T. (2011). Effects of air pollution on the respiratory health of children: a study in the capital city of India. Air Quality, Atmosphere & Health, 4: 95–102.

Singh, A. and Agrawal, M. (2007). Acid rain and its ecological consequences. Journal of Environmental Biology, 29(1): 1–5.

Sorokulova, I., Olsen, E. and Vodyanoy, V. (2015). Biopolymers for sample collection, protection, and preservation. Applied Microbiology and Biotechnology, 99(13): 5397–5406.

Srikanth, V., Shastri, M., Sindhu Sree, M., Navya Rani, M., Shivaramu, P. D. and Rangappa, D. (2021). Nanostructure material-based sensors for environmental applications. Nanostructured Materials for Environmental Applications, 565–589.

Stryer, L. (2002). The Citric Acid Cycle: The Citric Acid Cycle Must Be Capable of Being Rapidly Replenished. U: Stryer L., Berg JM, Tymozcko JL (ur). Biochemistry.

Su, S., Wu, W., Gao, J., Lu, J. and Fan, C. (2012). Nanomaterials-based sensors for applications in environmental monitoring. Journal of Materials Chemistry, 22(35): 18101–18110.

Sugiharto, W. H., Susanto, H. and Prasetijo, A. B. (2023). Real-Time Water Quality Assessment via IoT: Monitoring pH, TDS, Temperature, and Turbidity. Journal homepage: http://iieta. org/journals/isi, 28(4), 823–831.

Suty, H., De Traversay, C. and Cost, M. (2004). Applications of advanced oxidation processes: Present and future. Water Science and Technology, 49(4): 227–233.

Tariq, Z., Iqbal, D. N., Rizwan, M., Ahmad, M., Faheem, M. and Ahmed, M. (2023). Significance of biopolymer-based hydrogels and their applications in agriculture: a review in perspective of synthesis and their degree of swelling for water holding. RSC Advances, 13(35): 24731–24754.

Tavakoli, J. and Tang, Y. (2017). Hydrogel based sensors for biomedical applications: An updated review. Polymers, 9(8): 1–13.

Tchounwou, P. B., Yedjou, C. G., Patlolla, A. K. and Sutton, D. J. (2012). Heavy metal toxicity and the environment. Molecular, clinical and environmental toxicology: volume 3: Environmental Toxicology, 133–164.

TG, Y. G., Ballupete Nagaraju, S., Puttegowda, M., Verma, A., Rangappa, S. M. and Siengchin, S. (2023). Biopolymer-based composites: an eco-friendly alternative from agricultural waste biomass. Journal of Composites Science, 7(6): 1–12.

Tilley, S. J., Orlova, E. V., Gilbert, R. J., Andrew, P. W. and Saibil, H. R. (2005). Structural basis of pore formation by the bacterial toxin pneumolysin. Cell, 121(2): 247–256.

Tofighy, M. A. and Mohammadi, T. (2011). Adsorption of divalent heavy metal ions from water using carbon nanotube sheets. Journal of Hazardous Materials, 185(1): 140–147.

Tortolini, C., Bollella, P., Antiochia, R., Favero, G. and Mazzei, F. (2016). Inhibition-based biosensor for atrazine detection. Sensors and Actuators B: Chemical, 224: 552–558.

Tran, M. H., Choi, T. R., Yang, Y. H. and Lee, E. Y. (2024). Synthesis and characterization of self-healing bio-based polyurethane from microbial poly (3-hydroxybutyrate) produced in methanotrophs. International Journal of Biological Macromolecules, 281: 1–12.

Vasilachi, I. C., Asiminicesei, D. M., Fertu, D. I. and Gavrilescu, M. (2021). Occurrence and fate of emerging pollutants in water environment and options for their removal. Water. 2021; 13: 1–15.

Vodyashkin, A. A., Kezimana, P., Vetcher, A. A. and Stanishevskiy, Y. M. (2022). Biopolymeric nanoparticles–multifunctional materials of the future. Polymers, 14(11): 1–12.

Xue, X. Y., Cheng, R., Shi, L., Ma, Z. and Zheng, X. (2017). Nanomaterials for water pollution monitoring and remediation. Environmental Chemistry Letters, 15(1): 23–27.

Yadav, P., Yadav, H., Shah, V. G., Shah, G. and Dhaka, G. (2015). Biomedical Biopolymers, their Origin and Evolution in Biomedical Sciences: A Systematic Review. Journal of clinical and diagnostic research : JCDR, 9(9): ZE21–ZE25.

Yang, P., Wang, H., Chen, Y., Li, Y., Zhang, J., Zhang, C. et al. (2023). Ball-milling of titanium dioxide and zinc oxide for enhanced UV protection. Frontiers in Materials, 10: 1–12.

Zafar, F. and Iqbal, A. (2016). Indium phosphide nanowires and their applications in optoelectronic devices. Proceedings of the Royal Society A: Mathematical, Physical and Engineering Sciences, 472(2187): 1–11.

Zambianchi, M., Durso, M., Liscio, A., Treossi, E., Bettini, C., Capobianco, M. L. et al. (2017). Graphene oxide doped polysulfone membrane adsorbers for the removal of organic contaminants from water. Chemical Engineering Journal, 326: 130–140.

Zhang, H., Xu, F., Xue, J., Chen, S., Wang, J. and Yang, Y. (2020). Enhanced removal of heavy metal ions from aqueous solution using manganese dioxide-loaded biochar: Behavior and mechanism. Scientific Reports, 10(1): 1–14.

Zhen, X., Liu, L., Wang, X., Zhong, G. and Tang, J. (2019). Fates and ecological effects of current-use pesticides (CUPs) in a typical river-estuarine system of Laizhou Bay, North China. Environmental Pollution, 252: 573–579.

CHAPTER-11

Biopolymer-Based Metal Nanoparticles for Photocatalytic Degradation of Dyes

Vicente de Oliveira Sousa Neto,[1,*] *Antonio Joel Ramiro de Castro,*[3] *Gilberto Dantas Saraiva,*[2] *Paulo de Tarso Cavalcante Freire*[5] and *Ronaldo Ferreira do Nascimento*[4]

1. Introduction

1.1 Short-Term Expectations: Industrial Dyes Market Forecast to 2028

According to The Business Research Company, the market for synthetic dyes and pigments has expanded significantly in the last few years. At a Cumulative Annual Growth Rate (CAGR) of 10.1%, it will increase from US $67.11 billion in 2023 to US $73.91 billion in 2024 (The Business Research Company, 2024).

In the upcoming years, a significant expansion in the size of the synthetic dye and pigment market is anticipated. By 2028, it will have increased to US $ 105.18 billion, with a 9.2% compound annual growth rate, CAGR (The Business Research Company, 2024).

A rising need for high-performance pigments, strong demand from end-user industries, and growing demand from the packaging industry are all factors contributing to the growth in the projection period. Investing in a wide range of high wet-fast disperse dyes, adopting IoT[a] and big data analytics to improve the manufacturing process and lower production costs, moving their manufacturing capacities from the West to low-cost countries to save manufacturing costs, and investing in R&D and technological innovation to develop products with improved physical and

[1] Laboratory of Study and Research in Pollutants Removal by Adsorption - LERPAD, Department of Chemistry - State University of Ceará (UECE-FECLESC), Rua José de Queiroz Pessoa, N° 2554 - University Plateau - CEP: 63.900-000 - Quixadá/CE.

[2] Laboratory of Synthesis and Characterization of Materials - LASCAM, Department of Physics, State University of Ceará (UECE-FECLESC), Rua José de Queiroz Pessoa, N° 2554 - University Plateau - CEP: 63.900-000 - Quixadá/CE.

[3] Federal University of Ceará, Campus Quixadá, Av. José de Freitas Queiroz, 5003, Quixadá - CE, 63902-580

[4] Trace Analysis Laboratory (LAT) - Department of Analytical Chemistry and Physical Chemistry, Federal University of Ceará – UFC, Campus do Pici, CP 12100, CEP 60451-970 Fortaleza – CE, Brazil.

[5] Department of Physics, Federal University of Ceará, CEP 60.455–970, Fortaleza, CE, Brazil.

* Corresponding author: vicente.neto@uece.br

[a] The term "Internet of Things" (IoT) describes a network of real-world things, such as cars, appliances, and other machinery, that are integrated with software, sensors, and network connectivity to enable data collection and sharing.

DOI: 10.1201/9781003558330-11

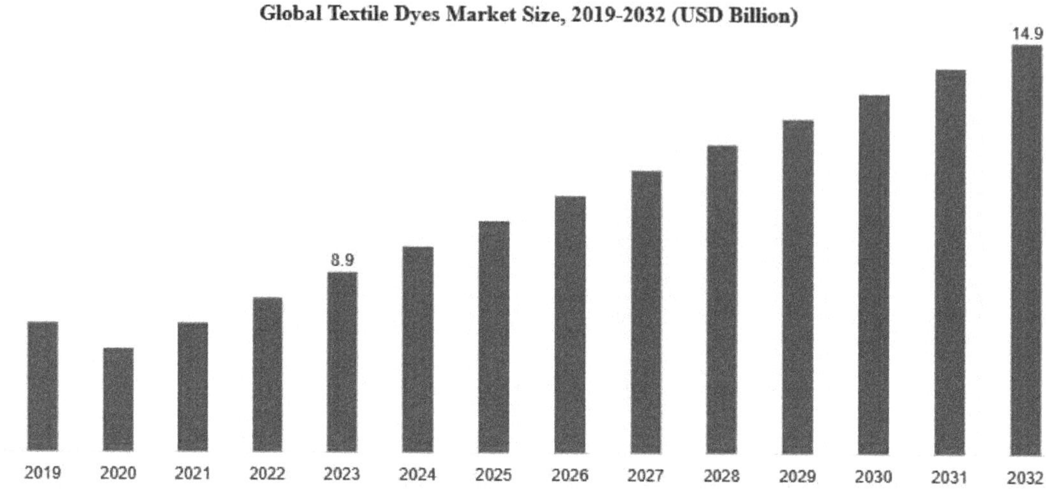

Figure 1. Global Textile Dyes Market valued at USD 8.9 billion in 2023 and USD 14.9 billion by 2032, at a CAGR of 5.89% during the forecast period 2024–2032. Source https://www.astuteanalytica.com/industry-report/tyextile-dyes-market.

chemical properties are some of the major trends that will occur during the forecast period (The Business Research Company, 2024). According to Astute Analytica, with a Compound Annual GrowthRate (CAGR) of 5.89% from 2024 to 2032, the global textile dyes market is expected to reach a worth of US$ 14.9 billion by 2032, up from US$ 8.9 billion in 2023 (Astute Analytica, 2024).

The global market for textile dyes is an intricate industry with broad ramifications for many economies, particularly those that largely depend on the textile sector (Fig. 1). The growing need for colored textiles for a variety of uses, such as apparel, home furnishings, and industrial textiles, is the reason for this expansion (Astute Analytica, 2024; Markets and Markets, 2024). The strong expansion of the textile sector in Asian nations, especially China and India, is one of the main factors propelling this market. The market for textile dyes is significantly influenced by China, the largest textile exporter in the world (Markets and Markets, 2024). The rise in new markets, government efforts, lower interest rates, population increase worldwide, and the accessibility of a large variety of synthetic dyes are all responsible for the historical period's growth.

1.3 Dyer and Aquatic Environments

Water contamination has become a significant worldwide problem in recent decades. Water ecosystems are contaminated by diseases, as well as organic, inorganic, and organometallic contaminants (Suresh et al., 2021; Hader et al., 2020; Kumar et al., 2019; Schmeller et al., 2018). Numerous tons of organic dyes are released into surrounding aquatic environments annually by the textile industry (Berradi et al., 2019). The physicochemical properties of natural water are negatively impacted by organic dyes. For example, the presence of organic dye in water reduces the amount of sunlight that reaches the surface, hindering aquatic plants' ability to photosynthesize (Moorthy et al., 2021). Aquatic animals are ultimately impacted (Tkaczyk et al., 2020). Additionally, dyes raise the biological oxygen demand, which eventually disrupts aquatic ecosystems. It has been observed that Scenedesmus quadricauda growth may be inhibited by the high concentration of indigo dye (Chia and Musa, 2014). According to Chia and Musa (2014), methylene blue, a widely used textile dye, is hazardous to microalgae. By inhibiting the synthesis of chlorophyll, it also inhibits the growth of Chlorella vulgaris and Spirulina platensis (Ma et al., 2020). Furthermore, organic dyes are carcinogenic and mutagenic to living organisms. N-methylated diaminotriphenylmethane, or malachite green, is a prime example of a substance hazardous to mammalian cells. Additionally, it irritates the human gastrointestinal tract, interferes with rabbit reproduction, and induces tumors in mice (Ma et al., 2020). Many organic pollutants, such as cosmetics, pharmaceuticals, microplastics, sweeteners, sunscreens, insecticides, fertilizers, and pesticides, are released into natural water ecosystems from the agriculture and healthcare sectors (Li et al., 2020; Rosal et al., 2010; Wang et al., 2020). Current water treatment facilities are unable to remove a large number of these pollutants (Petrie et al., 2015). Living creatures are harmed by the contaminants' persistent traces (1–100 ng L^{-1}) (Bueno et al., 2012). For instance, 3-benzophenone exhibits toxicity to people and causes endocrine disruption in water (Ghazipura et al., 2017). The anti-diabetic drug metformin may have negative effects on people's health (Velazquez and Oliván, 2020). Additionally, the presence of heavy metals in water supplies can have a significant negative impact on health (Egorova and Ananikov, 2017). Heavy metals from anthropogenic sources are entering water ecosystems and do not naturally decompose (Ledezma et al., 2021). Instead, they cause bioaccumulation or biomagnification, negatively impacting ecosystems on the whole (Mathew et al., 2019).

Furthermore, according to the 2019 The Joint Monitoring Program Report, 4.2 billion people lacked access to sanitation services and 2.2 billion people could not obtain safe drinking water (UNICEF & WHO, 2019). A shortage of clean water and inadequate hygiene contribute to the global spread of diseases like cholera, typhoid, diarrhea, and others, resulting in many deaths. Therefore, before being released into aquatic environments, contaminants must be completely eliminated. One of the main objectives for researchers working on wastewater purification is contaminant elimination. Furthermore, there is a pressing need to investigate sustainable and affordable materials for wastewater treatment due to increasing pollution and the energy demand.

Various treatment techniques, including coagulation/flocculation, reverse osmosis, electrochemical methods, adsorption, ion exchange, membrane filtration, biological degradation, and sonochemical approaches, have been employed to eliminate contaminants from wastewater (Jiang, 2015; Unal et al., 2020; Sousa Neto and Nascimento, 2019; dos Santos et al., 2021; Taka et al., 2021; Hansima et al., 2021). Nevertheless, these techniques have their own disadvantages, including low removal efficiency, high power requirements, costly chemicals and instruments, high maintenance costs, convoluted processes, and secondary pollution (Hitam and Jalil, 2020). For wastewater purification, a low-cost, environmentally friendly technique with high removal effectiveness is needed. This chapter provides a detailed analysis of current developments in photocatalysis for the removal of dye pollutants from water bodies. Specifically, emphasis is placed on biopolymer-based metal nanoparticles for the photocatalytic degradation of dyes.

2. Photocatalysis in Water Treatment

In order to address the current and upcoming energy problems, technologies for harvesting solar energy using nano semiconducting materials have recently been developed and implemented (Lakhera and Neppolian, 2021). Significant effort has gone into creating metal oxide nanoparticles with improved solar energy gathering capabilities. Regarded as an effective and environmentally friendly oxidation process, photocatalysis is very promising for use in wastewater treatment (Fig. 2). Photocatalysis is becoming the most advanced sustainable wastewater treatment method due to the rapid advancement of new material development (Lakhera and Neppolian, 2021).

For the treatment of wastewater, photocatalytic methods, either homogeneous or heterogeneous, have been widely used (Maniakova et al., 2020). Furthermore, it has been demonstrated that hybrid processes such as Fenton, sonolysis, membrane filtering, and ozonation are more effective when combined with photocatalytic processes. These reactions best explain the photocatalytic pollutant degradation process.

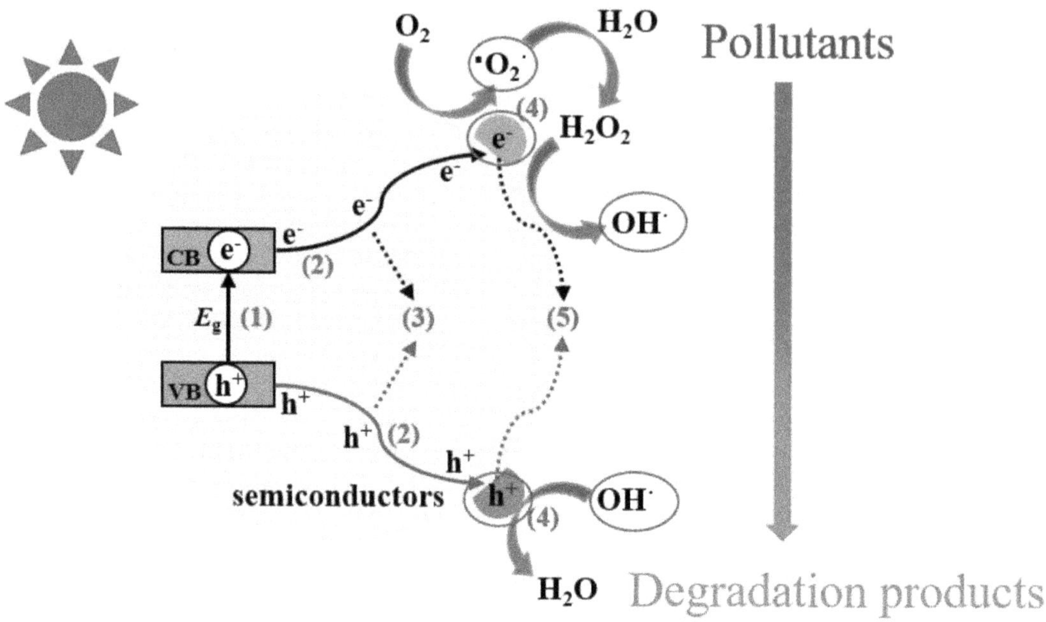

Figure 2. Photocatalytic processes over a heterogeneous photocatalyst. This figure by Ren et al., 2021 is licensed under the Creative Commons Attribution 4.0 International license. Source: https://www.mdpi.com/nanomaterials/nanomaterials-11-01804/article_deploy/html/images/nanomaterials-11-01804-g001.png.

$$\text{Semiconductor} + h\nu \rightarrow h^+ + e^- \tag{1}$$

$$e^- + O_2 \rightarrow {}^{\bullet}O_2^- \tag{2}$$

$${}^{\bullet}O_2^- + H^+ \rightarrow {}^{\bullet}OOH \tag{3}$$

$$2{}^{\bullet}OOH \rightarrow O_2 + H_2O_2 \tag{4}$$

$$H_2O_2 + {}^{\bullet}O_2^- \rightarrow {}^{\bullet}OH + OH^- + O_2 \tag{5}$$

$$H_2O_2 + h\nu \rightarrow 2{}^{\bullet}OH \tag{6}$$

$$h^+ + H_2O \rightarrow {}^{\bullet}OH + H^+ \tag{7}$$

$$h^+ + OH^- \rightarrow {}^{\bullet}OH \tag{8}$$

$$\text{Pollutants} + ({}^{\bullet}OH, h^+, e^-, {}^{\bullet}OOH \text{ or } {}^{\bullet}O_2^-) \rightarrow \text{degradation products} \tag{9}$$

$$Fe(OH)^{2+} + h\nu \rightarrow Fe^{2+} + {}^{\bullet}OH;\ \lambda < 580 \text{ nm} \tag{10}$$

$$H_2O_2 + h\nu \rightarrow 2{}^{\bullet}OH;\ \lambda < 310 \text{ nm} \tag{11}$$

A brief discussion of photocatalytic processes, including hybrid, homogeneous, and heterogeneous types, will be covered in the following topic.

2.1 Homogenous Photocatalysis

The catalysts and reactants are equally distributed and in the same phase in homogeneous photocatalysis. In the chemical industries of polymers, pharmaceuticals, and wastewater treatment, homogeneous catalysts account for around 20% of the total (van Leeuwen et al., 2004). Solar photo-Fenton with ethylenediamine-N,N'-disuccinic acid (EDDS) complex (Fe^{+2}/H_2O_2/EDDS/sunlight), solar photo-Fenton with O_3/H_2O_2, ultrasonic, light/O_3 or light/O3/H_2O_2 are examples of homogeneous solar-driven AOPs (Lakhera and Neppolian, 2021; Kim et al., 2004; Stan et al., 2012).

As demonstrated further (Lemeune et al., 2004), hydroxyl radicals are generated in the case of the O_3/H_2O_2 process.

$$H_2O_2 + 2O_3 \rightarrow 2{*}OH + 3O_2 \tag{12}$$

Hydrogen peroxide photolyzes in the presence of sunlight or H_2O_2 to create hydroxyl radicals (Jones, 1999). Homogeneous catalysts have high catalytic activity due to their uniform distribution and are very selective for the breakdown of contaminants. Nevertheless, they have temperature restrictions and are challenging to separate when the reaction is finished. By combining heterogeneous and homogeneous photocatalysts, hybrid methods have been created to overcome the drawbacks of homogeneous catalysts.

2.2 Heterogeneous Photocatalysis

Heterogeneous photocatalyst wastewater treatment is a promising technique due to its cost-effectiveness, simplicity, and access to abundant sunlight. A semiconducting material's valence and conduction band edges need to be suitably positive and negative on the Normalized Hydrogen Electrode (NHE), respectively, for it to be photo catalytically active. Reactive Oxygen Species (ROS) are necessary for a photocatalyst to break down pollutants, and they cannot be produced without the right band structure. The redox potential needed to produce ROS such as superoxide radical ($O_2^{\bullet-}$), hydroxyl ions (OH^-), and hydroxyl radical (•OH) is shown in Eqs. (13) and (14) (Nonell and Flores, 2016). The lifespan of this ROS is extremely brief-micro to nanoseconds (Table 1).

$$O_2 \xrightarrow{\text{energy transfer}} {}^1O_2 \tag{13}$$

$$O_2 \xrightarrow{e^-(-0.16\ V)} O_2^{\bullet-} \tag{14}$$

$$O_2^{\bullet-} \xrightarrow{H^+} HO_2^{\bullet} \tag{15}$$

$$O_2^{\bullet-} \xrightarrow{e^- + 2H^+(+0.94\ V)} H_2O_2 \tag{16}$$

$$H_2O_2 \xrightarrow{e^- + H^+(+0.38\ V)} {\bullet}OH \tag{17}$$

$$^{\bullet}OH \xrightarrow{e^- + H^+(+2.33\ V)} H_2O \tag{18}$$

Over time, several visible light-active photocatalysts, including $BiVO_4$, $ZnFe_2O_4$, Bi_2O_3, Fe_3O_4, Bi_2WO_6, WO_3, Cu_2O, and $ZnCo_2O_4$, have been identified for their ability to degrade pollutants by photocatalysis (Tien and Chen, 2021; Lakhera et al., 2018). Unfortunately, the rapid recombination rate of the charge carriers or the inappropriate redox potential of these photocatalysts restricts their photocatalytic activity. For the treatment of wastewater, modifications of these photocatalysts with appropriate cocatalysts have demonstrated better performance (Table 2). For extended use and economic viability, the photocatalysts must be inexpensive, non-toxic, and aqueously stable. To increase the efficiency of the photocatalytic process, rapid separation, and transfer of produced

Table 1. The lifetime of photocatalytically generated reactive oxygen species (Horikoshi and Serpone, 2020).

Name	Formula	Lifetime
Superoxide anion radical	$O_2^{\bullet-}$	10^{-6} s
Hydroxyl radical	$^{\bullet}OH$	10^{-9} s
Singlet oxygen	$^1O_2$	$2 * 10^{-6}$ s
Hydroperoxy radical	$HO_2^{\bullet}$	10^{-9} s
Hydroxyl ions	OH^-	$70 * 10^{-6}$ s
Ozone	O_3	

Table 2. Some of important biopolymers which have been used as part of photocatalysts for various environmental applications.

Biopolymer	Structure	Functional groups	Properties	Obtaining/Extracting	References
Chitin	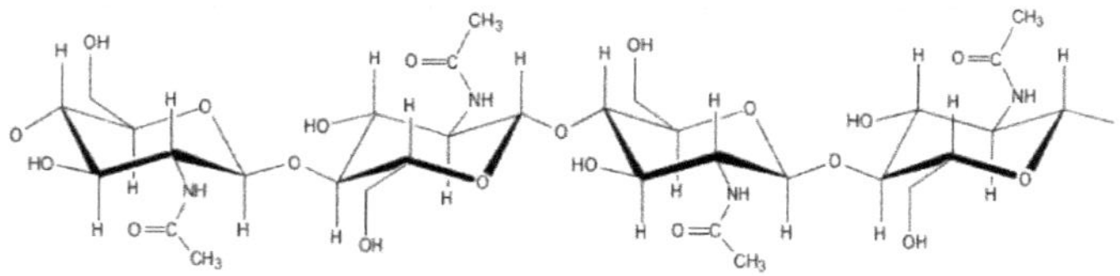	N-Acetyl group; Hydroxyl groups	It is a water-insoluble polymer with a higher number of D-glucosamine monomers than N-acetyl-D-glucosamine monomers.	It is obtained from crab and shrimp shells, originating from waste produced by the industry that processes these crustaceans as food.	Amiri et al, 2022
Chitosan	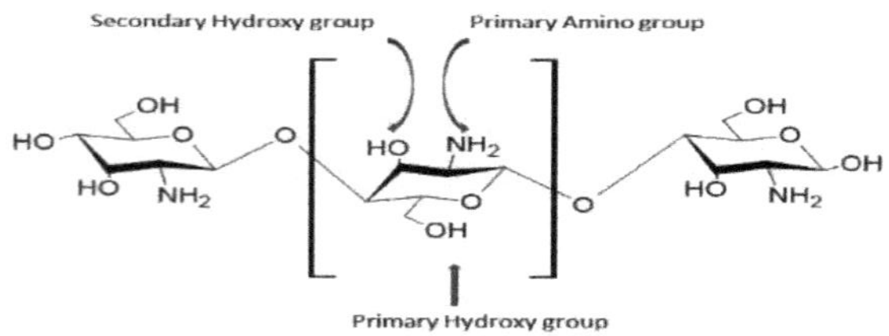	Amino groups Hydroxyl groups	It is a biodegradable, natural bioactive biopolymer that has significant antioxidant and antimicrobial properties and is the ideal foundation for enhancing the stability.	Obtained from the deacetylation of chitin.	Taghipour et al., 2024
Cellulose	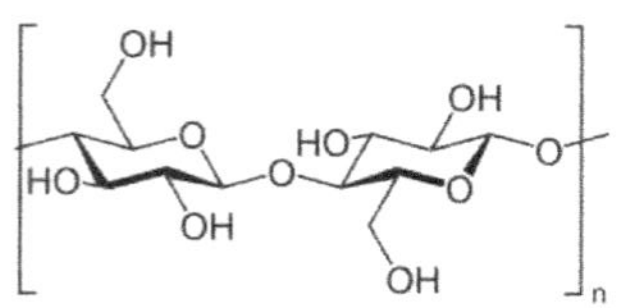	Hydroxyl groups	It is a chemical molecule composed of a complex chain of glucose. Soluble with chemical degradation in sulfuric acid, and in concentrated solutions of zinc chloride. Soluble in aqueous solutions of cupric am monium hydroxide ($Cu(NH_3)_4(OH)_2$).	Cellulose forms the structure of plant cell walls.	Alves, 2015
Alginate	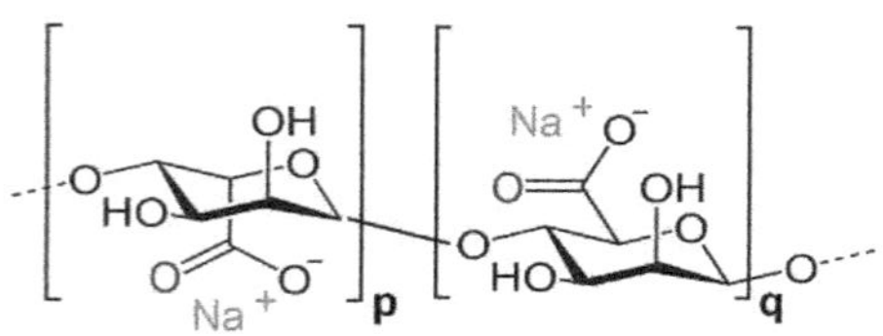	Hydroxyl groups; Acetyl group	It is a linear polymer compound composed of β-(1,4) d-mannuronic acid (M) and α-(1,4) l-guluronic acid (G). Natural alginate is a derivative of alginic acid.	Alginate, mainly derived from brown algae.	Kumar et al., Kumar, 2024

electrons from the semiconductor's conduction band are essential, in addition to visible light absorption.

2.3 Green Photocatalysts

To expand relevance in the near future, the synthesis of materials from natural, renewable, and biological resources should continue. These natural resources are used to create materials such as semiconductors, noble metals, and ceramics, which require little energy during the creation process and produce fewer byproducts (Prusty et al., 2016). Al-Othman et al. (2012) stated that renewable resources such as sugarcane bagasse, coconut coir, peanuts, bamboo wood, groundnut shell, and coal are used to create materials like carbon-based nanostructures. Carbon-based nanostructures can be produced with less energy by calcining and filtering naturally occurring materials. According to the concept of green technology, more goods can be produced in an environmentally beneficial manner without endangering the environment or society. The green chemical technique uses comparatively less energy and emits fewer harmful emissions. The three current technologies for distributing energy throughout the world are hydroelectricity, nuclear energy, and wind energy, all of which are more environmentally friendly than fossil fuels (Bilgili et al., 2015; Abas et al., 2015). Greener methods of implementing alternative energy production techniques are necessary to balance the world population's overall energy needs. Generally speaking, green photocatalysis is a method of synthesis that aids in the manufacture of various photocatalysts through the utilization of biomasses, natural resources, and biological extracts (Fox and Dulay, 1993). For example, the removal of pollutants from water bodies, the reduction of toxic molecules, and the production of hydrogen fuels are examples of how these precursors or resource materials are a more environmentally friendly, sustainable, and commercially viable method of converting light energy to chemical energy than semiconducting materials (Fox and Dulay, 1993; Herrmann, 1999).

2.4 Hybrid Processes

Both homogeneous and heterogeneous procedures offer benefits in eliminating harmful wastewater. Nevertheless, there are certain drawbacks to photocatalysis, Fenton, membrane filtration, and ozonation procedures that reduce their effectiveness in fully mineralizing hazardous contaminants. For instance, the ozonation process suffers from inadequate mineralization, and the photocatalytic process is sluggish due to a poor oxidation rate. These technologies have limited economic and practical feasibility due to their flaws. The hybrid process involves combining techniques like photocatalysis and ozonation to overcome the limitations of each technique alone and increase the effectiveness of Advanced Oxidation Processes (AOPs) methods.

2.5 Photodegradation of Dyes

Coloring agents are compounds like dye and pigment that give color. Auxochromes (-OH, -COOH, -SO3H, and -NH3) and chromophores (-C=N-, -C=O-, -C=C-, -N=N-, and quinone) are found in dyes. These compounds alter the total energy of the electron system to either induce or amplify the color of the chromophore. The fact that dyes are obvious pollutants gives them an aesthetic appeal. Despite its modest quantity (few ppm), it is detectable. Synthetic organic dyes are discarded into watery environments by a number of industries, including textile, apparel, leather, plastic, paper, food processing, and cosmetics.

Water is used extensively in the textile industry during the dyeing process. Wastewater from the textile industry is among the most polluting in the overall industrial sector when volumes released and effluent composition are taken into account. According to Houas et al. (2001), around 15% of the world's total dye production is wasted during the dyeing process and is subsequently emitted as effluents.

2.6 Biopolymers Based Photocatalysts

Despite the widespread use of synthetic polymers in photocatalysts, the need for sustainable development and environmental protection drives us to "return to nature." Due to their availability and versatility, biopolymers such as cellulose, chitosan, alginate, pectin, dextran, cyclodextrins, natural gums, agar, gelatin, chitin, and collagen have garnered interest over the years for various biological and environmental uses. High functionality, surface area, non-toxicity, antimicrobial activity, biocompatibility, flexibility, and ease of processing are just a few of the adaptable qualities that have found uses in food packaging (Noorbakhsh-Soltani et al., 2018), pharmaceuticals (Mu et al., 2020), biofuels (Hassan Hassan Abdellatif et al., 2018), water treatment (Tabriz et al., 2019), energy storage (Zhang et al., 2019), and biomedical applications (Du et al., 2019). Table 2 lists significant biopolymers along with their composition, features, and functions. The number of articles pertaining to materials based on biopolymers has surged recently. The greatest contribution to materials science comes from the diffusion of articles about biopolymers across several scientific fields. This indicates that biomaterials have been receiving increasing attention.

Compared to other biomaterials, chitosan, alginate, and cellulose are more selective and reactive as biopolymers (Sharma et al., 2017, Rathore et al., 2014). Furthermore, by appropriately functionalizing them, their adsorption capabilities can be altered to varying degrees. Additionally, several researchers report using plant and fruit pieces as inexpensive adsorbents to remove pollutants (Jawad et al., 2018a, b). Biopolymers have many applications such as immobilization (Fakhri et al., 2018), support matrix (Preethi et al., 2017), surface capping agents (Jiang et al., 2009), composites (Virkutyte et al., 2012; Habiba et al., 2016), and core-shell catalysts (Agorku et al., 2014).

Diverse methodologies and approaches have been utilized to fabricate composites incorporating biopolymers for photocatalytic purposes. These include composite films and membranes, photocatalytic hydrogels and aerogels, various hybrid materials, and nanocomposites with metal oxides, chlorides, nitrides, and other elements as support for catalysts and heterojunctions. Table 3 lists several synthesis pathways, benefits, and examples of different biopolymer-based photocatalysts. In the next section, we will discuss these techniques in more detail.

3. Techniques for Synthesizing Biopolymer Based Photocatalysts

Considering the creation of photocatalyst nanoparticles within a biopolymer matrix, there are various proven methodologies. Nanoscale pore spaces can be organized and arranged in an array to provide substantial benefits in several applications, especially photocatalysis. Numerous synthesis techniques, including sol-gel, hydrothermal, electrospinning, plasma vapor deposition, atomic layer deposition, and hot press techniques, have enhanced our understanding of how to improve the properties of nanomaterials for their intended uses. Promising techniques for creating cellulose-based supporting materials and cellulose-photocatalyst composites are reviewed in summary form.

3.1 Atomic Layer Deposition (ALD): The Atomic Layer Deposition (ALD) approach, which combines the advantages of Chemical Vapor Deposition (CVD) and sol-gel methods in the creation of fibrous structures, is recognized as a self-limiting gas-phase technique for thin-film deposition. In ALD, surface reactions cause the film to grow while the precursor is in the vapor state. Using this method, inert gas purges are used to separate the precursor pulses after the precursor vapors are introduced into the reactor. A monolayer of that precursor covers the substrate surface following each precursor feed and the ensuing purge. Only the adsorbed surface layer reacts with the subsequent precursor since the gas phase is no longer attached to the precursor. As a result, the film grows layer by layer, and the number of deposition cycles allows for precise control over the film thickness. This self-limiting growth mechanism produces thin films from ALD that are uniformly thick and well-organized over wide regions. Because of these characteristics, ALD is an effective technique for evenly covering intricately structured surfaces. Furthermore, ALD has been successfully used to

Table 3. Various techniques for synthesizing biopolymer based photocatalysts. In addition to the below listed methods, combinations of these and other novel methods have also been reported.

Techniques	Summary	Advantages s	Limitations	References
Atomic layer deposition (ALD)	The advantages of chemical vapor deposition and the sol-gel approach are combined in the atomic layer deposition process. Precursors are delivered into the reactor in gaseous or vapor phases, where they remain until the substrate surface is saturated with monolayers.	Films are structured and their thickness is controllable. This method has been applied to a variety of materials, including nitrides and oxides.	Slow Deposition Rate, High Precursor Cost, Film Quality, Temperature Restrictions	Kemell et al. (2005) Pore et al. (2008)
Hot press	To turn fine powder into a compact form, the hot press process uses both heat and pressure simultaneously. Both hard and soft materials, such as ceramics, polymers, rubbers, and carbon-metal composites, have been created using this process.	This enhances the composite moieties' interfacial interaction. It is also employed in the production of homogeneous films that retain their shape and structure.	High Cost, Extended Cycle Time, Material Restrictions, Precision Control, Requirements for Post-Processing	Mohamed et al., 2017 Li et al., 2013
Electrospinning (ES)	This is a simple, versatile technique that uses polymers to create fibers (ranging in size from nanometers to micrometers) or one-dimensional material from different precursors. This causes the liquid to be injected from the spinneret and charges the polymer surface using an electric field..	The surface to volume ratio, porosity, thickness, and shape of the electro spun nanofibers can all be tuned.	Poor Cell Penetration, Low Mechanical Strength, Needle Blockages, Material Limitations, Uniformity Issues, Environmental Sensitivity	Bonakdar and Rodrigue (2024) Habiba et al., 2019
Hydrothermal method/ Solvothermal	Using this process, synthesis is done at a high temperature in a steel autoclave. As the vapor pressure reaches saturation, nanomaterials begin to aggregate. Stabilizing agents can be used to dissolve the aggregation.	This approach has the benefit of directly synthesizing additional metal compounds as well as micro and nano metal oxides. Reagents like ammonium fluoride can also be used to obtain high degree shape control.	High Pressure Requirements, Temperature Control, Long Reaction Times, Limited Solvent Choices, Scalability Issues	Demazeau (2011) Li et al., (2016).
Sol-gel	Regulated hydrolysis and condensation of precursors (metal chlorides, nitrates, and alkoxides). The hydroxyl and amino groups found in biopolymers, including cellulose, chitosan, and pectin, provide an excellent template for the synthesis of metal oxides. The polymer adsorbs the metal salts, which are then hydrolyzed.	Tunable particle size, specific surface area, low-temperature synthesis, and content and shape. The polymer matrix preserves the characteristics of the nanoparticles.	Control over Nanoparticle Size and Shape Toxic Byproducts, Cost of Raw Materials, Hazard of Contamination, Limitation on Film Thickness, Stability and Reproducibility	Youssef et al. (2017) Paul (2023)

structure a range of materials, such as noble metals, nitrides, and metal oxides (Kemell et al., 2005; Pore et al., 2008).

Water is used as the oxygen precursor during the reactions in this procedure, and metal alkoxides or halides are the most commonly used precursors to generate the required metal oxides. The deposition temperature employed during preparation determines the degree of crystallinity of the resulting metal oxide nanoparticles. For example, Kemell et al., 2005 prepared TiO_2/cellulose composite films by depositing TiO_2 thin films (30–55 nm) at 150°C and 250°C on ashless filter paper. According to Cremers et al., (2019), in this technique, the substrate is subjected to a series of vapor phase reactants in alternating order. The film thickness can be regulated at the atomic level due to the self-saturating nature of the surface processes. A typical ALD process consists of multiple ALD cycles, each comprising four distinct phases that make up the prototypical trimethylaluminum (TMA)/H2O.process depicted in Fig. 3. The schematic illustration in Fig. 3 refers to one reaction cycle of the atomic layer deposition (ALD) process, using the trimethylaluminium (TMA)-water ALD process to make thin aluminium oxide films as an example. Here, the starting surface contains hydroxyls (OH groups) as reactive sites. Step 1 is the reaction with TMA, Step 2 is a purge or evacuation step, Step 3 is the reaction with water, and Step 4 is a purge or evacuation step.

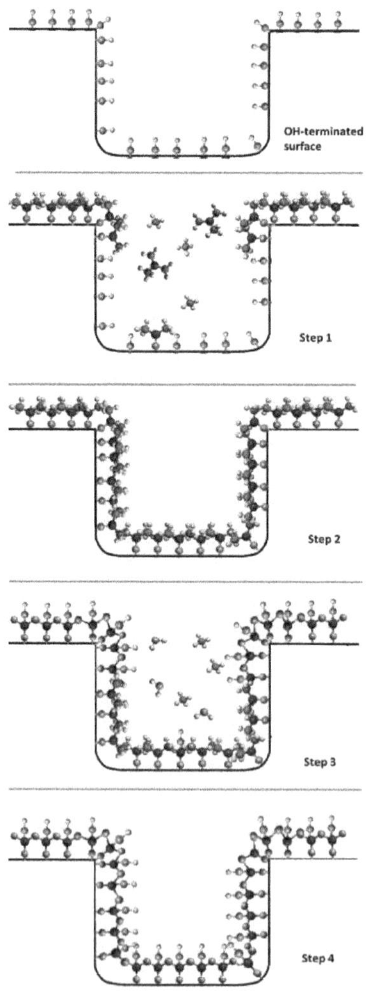

Figure 3. Schematic illustration of one reaction cycle of the atomic layer deposition (ALD) process, using the trimethylaluminium-water ALD process to make thin aluminium oxide films as example. This scheme has been first published by Cremers et al., in Applied Physics Reviews, vol. 6, art. 021302, year 2019; https://doi.org/10.1063/1.5060967,

3.2 Sol-Gel method: The sol-gel method, which targets a specific surface area, nanoparticle size, and degree of aggregation, is attractive because it provides excellent compositional and morphological control over the product properties. It is based on the controlled hydrolysis and condensation of appropriate precursors, usually nitrates, metal alkoxides, or chlorides. The precipitates or gels produced with the sol-gel process are often amorphous, have a large specific surface area, and occasionally display mesoporous behavior. The salient characteristics of the metal oxide nanoparticles made by the sol-gel process are important factors to consider when creating an inorganic nanoparticle/biopolymer dope. Because it can control the growth of the particles, maintain their unique morphology, and improve stability, cellulose is a good option for serving as their host material. In the polymer matrix, inorganic nanoparticles' superior electrical, magnetic, and optical qualities can be maintained. In this case, hydroxyl groups and a porous structure in wet cellulose make it an excellent template or scaffold for metal oxide nanoparticles. In theory, the metal alkoxides in solution will adsorb onto hydroxylated substrate surfaces and then hydrolyze to produce nanoparticles.

Weon et al., 2023, used the sol–gel transition of biopolymer solutions co-dissolved in [Emim][Ac] to create blended biopolymer-based photocatalytic hydrogel beads. Using a range of model dyes with varying charges and hydrophobicities, they assessed the blended biopolymer/TiO_2 hydrogel beads' adsorption capability and photodegradation activity. Carbon nanotubes (CNT), chitosan, and carrageenan were added to the cellulose/TiO_2 hydrogel beads to modify their surface properties. Moreover, to improve the effectiveness of dye removal from aqueous solutions, the authors created biopolymer/TiO_2 magnetic hydrogel microbeads using a [Emim][Ac]-in-oil emulsion (Fig. 4).

The study examined the impact of various component ratios, biopolymer charge, TiO_2 content, and magnetic particle content on the composite hydrogel microbeads' adsorption ability and photodegradation activity.

3.3 Hot press: The process of hot pressing involves applying pressure (compressive stress) and heat (elevated temperature) simultaneously to compact small powder and produce partially or fully sintered components. This method was developed for the low-strain-rate powder metallurgy process used in industrial settings. It has been effectively used to manufacture both hard and soft materials, including ceramic components and cutting tools made of diamond-metal composites (Sarin et al., 2003). In theory, pressure enhances the driving force for densification by causing a rearrangement of particles, which lowers the processing temperature needed for the subsequent sintering process.

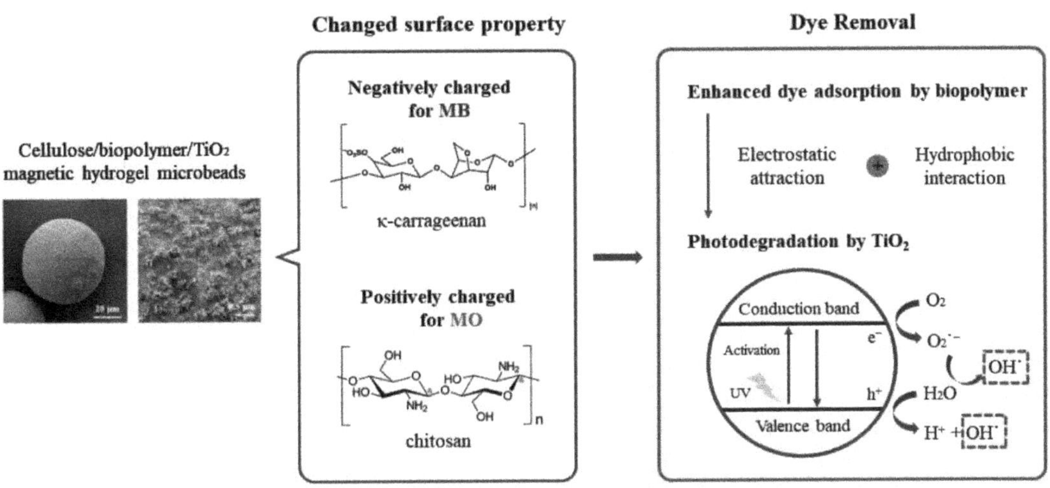

Figure 4. Schematic representation of the preparation of cellulose/biopolymer/TiO_2 composite hydrogel beads and their dye removal mechanism. This figure from Weon et al., 2023 is an open access article distributed under the terms and conditions of the Creative Commons Attribution (CC BY) license (https://creativecommons.org/licenses/by/4.0/). Link: https://pub.mdpi-res.com/gels/gels-09-00630/article_deploy/html/images/gels-09-00630-g001.png?1691240481.

Additionally, the compact powder produced has reduced particle sizes and is flexible enough to functionally grade the ceramic layers while maintaining fine control over the microstructure (Winandy and Krzysik, 2007). It is best to optimize a material's densification settings for each layer in a single phase to avoid several batch processing steps and to ensure an efficient and economical production process. Furthermore, this method's flexibility enables compositionally grading the surfaces to decrease heat stresses and polarization losses while also improving interfacial contact (Sarin et al,. 2003; Winandy and Krzysik, 2007).

Nor et al., (2016) used a hot-press process to successfully synthesize a PVDF-based nanocomposite membrane with TiO_2 nanofibers for the breakdown of organic pollutants in wastewater exposed to UV radiation. This method's benefit is that it can maintain the morphological structure of the produced nanofiber photocatalyst while ensuring a homogeneous functional composite layer. To create a stable, unbroken layer, powder coating traditionally involves the deposition of powdered coating ingredients onto the surface of the target substrate followed by a high-temperature curing phase. To encourage film production, low-temperature melting additives are occasionally added (Shen et al., 2016; Gutierrez et al., 20013).

3.3 Hydrothermal method: An intriguing method for directly creating micro- and nano-sized crystalline metal oxides with excellent control over structure is hydrothermal synthesis. The synthesis is usually carried out at high temperatures in a steel autoclave vessel until the vapor pressure reaches saturation, causing the nanoparticles to aggregate. Additives, such as stabilizing agents, are sometimes employed to prevent agglomeration. When working with heterogeneous processes in the presence of aqueous or non-aqueous solvents or mineralizers, hydrothermal techniques typically use high temperatures (> 100°C) or pressures (> 1 atm) to dissolve and recrystallize materials that are insoluble under typical conditions in a closed system.

The fabrication of hierarchical nanofibrous structured metal oxide–biopolymer composite sheets is advantageous due to its straightforward and easy method. According to earlier research focused on cellulose as a matrix, it is possible to create ultrathin amorphous metal oxide (such as TiO_2) gel films by first depositing nanoparticles on each nanofiber of the bulk cellulose precursor and then allowing them to change into fine nanoparticles. The diameter of each precursor nanofiber varies from 20 to 50 nm, and the tiny nanoparticles are uniformly attached to each one. The metal oxide–cellulose composite materials produced during the process are bulk substances and do not contain any functional groups or binders, indicating that the hierarchical network topologies are genuinely maintained (Luo et al., 2014).

The sol-gel and hydrothermal methods were combined to create the nanotubular Na_2MoO_4/TiO_2 composites from natural cellulose (such as commercially available filter paper) are shown in Fig. 5. This material consists of cellulose microfiber assemblies made up of several randomly intertwined cellulose nanofibers. Utilizing the surface sol-gel method, ultrathin titania gel films were deposited onto the cellulose nanofiber surfaces of the filter paper by applying Ti $(O^nBu)_4$ precursor, as per an earlier study by Li and Huang (2016). After being placed in suction filtration, a piece of regular quantitative filter paper was first rinsed with ethanol and then dried for fifteen minutes. To ensure that the precursor was sufficiently adsorbed on the cellulose nanofibers, the Ti $(O^nBu)_4$ solution (10.0 mL, 100 mM in toluene/ethanol solvent (v:v = 1:1)) was poured into the funnel and allowed to sit for three minutes. After the physically adsorbed Ti $(O^nBu)4$ was removed, 10.0 mL of ethanol was added and filtered. Next, 20.0 mL of water was filtered to encourage hydrolysis. The titania-cellulose-nanofiber composite with a negatively charged surface was effectively created when the deposition cycle was repeated five times (Fig. 5a). After that, 20 mL of PDDA aqueous solution (1.0 g L^{-1} in water) was added to the filter funnel. Half of the mixture was gradually suction-filtered, and the remaining portion was left for eight minutes. For thorough and complete absorption on the surfaces of the titania gel layers, which were deposited by electrostatic contact, a small amount of PDDA

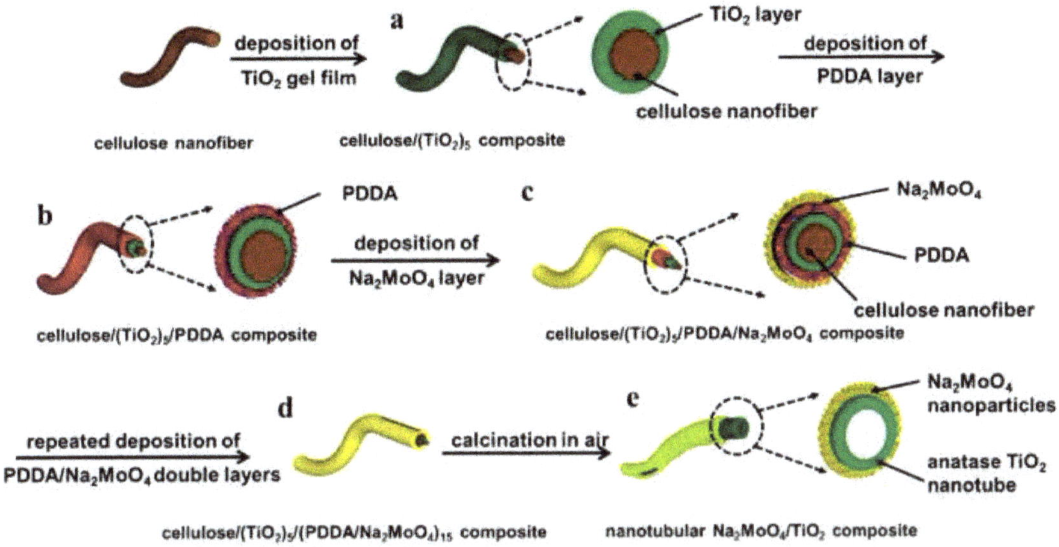

Figure 5. Schematic illustration of the synthetic processes for the nanotubular Na_2MoO_4/TiO_2 composite derived from natural cellulose substance. This figure from Yu et al. (2021) is an open access article distributed under the terms and conditions of the Creative Commons Attribution (CC BY) license (https://creativecommons.org/licenses/by/4.0/). Link: https://pub.mdpi-res.com/materials/materials-14-00357/article_deploy/html/images/materials-14-00357-sch001.png?1610588222.

solution was added and left for an additional eight minutes. To remove the unassembled reagent, 20.0 mL of water was filtered afterwards. At the end of drying in airflow for eighteen minutes, the composite sheet became PDDA/titania/cellulose nanofiber (Fig. 5b). The Na_2MoO_4/PDDA/titania/cellulose nanofiber composites were created by depositing the Na_2MoO_4 layers closely behind the PDDA layers in the same manner at different concentrations of the Na_2MoO_4 solution (0.02 M, 0.05 M, and 0.1 M) (Fig. 5c). Subsequently, the PDDA/Na_2MoO_4 bi-layer sedimentation process was carried out fifteen times. The resulting as-deposited (PDDA/Na_2MoO_4)15/titania/cellulose-nanofiber composite sheets (Fig. 5d) were burned at 450°C for six hours (heating speed: 2°C min^{-1}) to separate the PDDA ingredient and the primitive cellulose substance. This resulted in nanotubular Na_2MoO_4/TiO_2 composites (Fig. 5e), which had varying Na_2MoO_4 contents.

3.4 Electrospinning: With diameters ranging from 2 nm to several micrometers, electrospinning (ES) is a straightforward, economical, and versatile method for synthesizing 1D nanofibers from various materials, including composites, polymers, and ceramics. Polymer solutions are used in this process. ES is also regarded as a non-mechanical and electrostatic approach that charges the surface of a polymer droplet with an electric field or high voltage before causing a liquid jet to be ejected through a spinneret. In summary, an electrical potential is placed between a grounded target and a polymer droplet. An expelled charged jet of polymer solution occurs when the surface tension of the droplet is overcome by the applied electric field. High tunable porosity, a high surface-to-volume ratio, malleability to conform to a wide range of sizes and forms, and composition control are just a few benefits of spun nanofibers. Specifically, 1D nanofibers produced by electrospinning metal oxides and similar materials have demonstrated significant promise in enhancing photocatalytic performance by encouraging rapid electron transfer and delaying their recombination.

4. Biopolymer-Based Metal Nanoparticle: Dye Photodegradation

According to Sarkar et al, 2020, studies indicate that biopolymer-containing nanomaterials show promise for the photocatalytic degradation of organic contaminants. This topic, will discuss the creation and uses of biopolymeric nanomaterials for the photocatalytic azo dye degradation process.

4.1 Chitosan Bionanocomposite: Removal of Industrial Dyes through Photocatalytic Degradation

Because of its high adsorption capacity, chitosan is a popular biosorbent for wastewater treatment. It is a non-toxic, reasonably priced, and biodegradable biopolymer. Adsorbed dyes are hard to remove; hence utilizing chitosan for dye removal can be challenging because long contact durations are needed (Grzybek et al., 2022). Researchers have investigated creating compounds based on chitosan that enable phase separation after adsorption as a solution to this problem. To improve chitosan's adsorption properties, one method is to mix it with inorganic substances like TiO_2 (Pal and Pal, 2019; Rasoulifard et al., 2017) TiO_2 is a photocatalyst that can break down organic pollutants by generating reactive oxygen species (ROS). By adding TiO_2 nano powder to chitosan, you can efficiently replenish the adsorbent and shorten the time it takes for dyes adsorbed on chitosan to degrade. Colors can be removed from chitosan thanks to its functional groups (Bahrudin and Nawi, 2019).

Hoang and Nguyen (2023) used chitosan/TiO_2/glycerol films made by casting processes outside in the sun to produce a very effective and sustainable approach for treating synthetic effluent from the textile sector (Fig. 6). The degradation of Near Wash Yellow (NWY) was investigated by the authors in relation to several experimental parameters, including pH, TiO_2/chitosan ratio, glycerol/chitosan ratio, and water height. They carried out batch tests with and without light irradiation to assess the film's performance and ascertain the pseudo-first-order and Langmuir-Hinshelwood kinetics of NWY degradation. Researchers also used a specially made setup to evaluate the film's effectiveness for continuous-flow NWY deterioration. The use of chitosan-based materials for wastewater treatment and their potential for the economical and environmentally acceptable removal of contaminants from industrial effluents are both well-explained in this study.

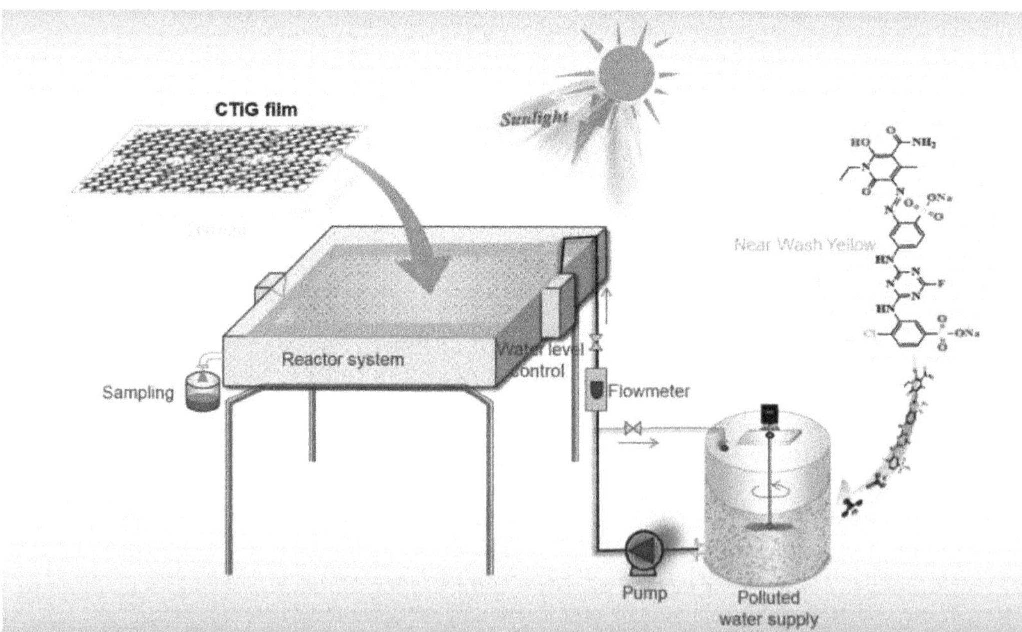

Figure 6. The continuous-flow design for NWY degradation using CTiG under sunlight irradiation. This figure from Hoang and Nguyen (2023) is an open access article distributed under the terms and conditions of the Creative Commons Attribution (CC BY) license (https://creativecommons.org/licenses/by/4.0/). Link: https://pub.mdpi-res.com/sustainability/sustainability-15-06979/article_deploy/html/images/sustainability-15-06979-g003.png?1682061760.

4.2 Cellulose as a Nanostructure with Dual Applications: Industrial Dye Degradation and Bactericidal Properties

Ikram et al., 2022 developed a cellulose nanostructure with dual applications. This study used a straightforward coprecipitation approach to create cellulose nanocrystals (CNC) and CuO nanostructures doped with Ag. The authors enhanced the photocatalytic, catalytic, and antibacterial activity by doping Ag into fixed quantities of CNC and CuO. The production of nanorods and a pH effect were shown to be responsible for the observed rise in catalytic activity upon doping, but the opposite trend was reported in photocatalytic activity. The bactericidal activity against *S. aureus* and *E. coli* was enhanced by the addition of Ag and CNC dopants to CuO. According to Ikram et al., 2022, to create CNCs, 10 g of Avicel was first added to 200 mL of DI water and a 50% solution of concentrated H_2SO_4, which was then agitated at 45°C for 30 minutes (Fig. 7a).

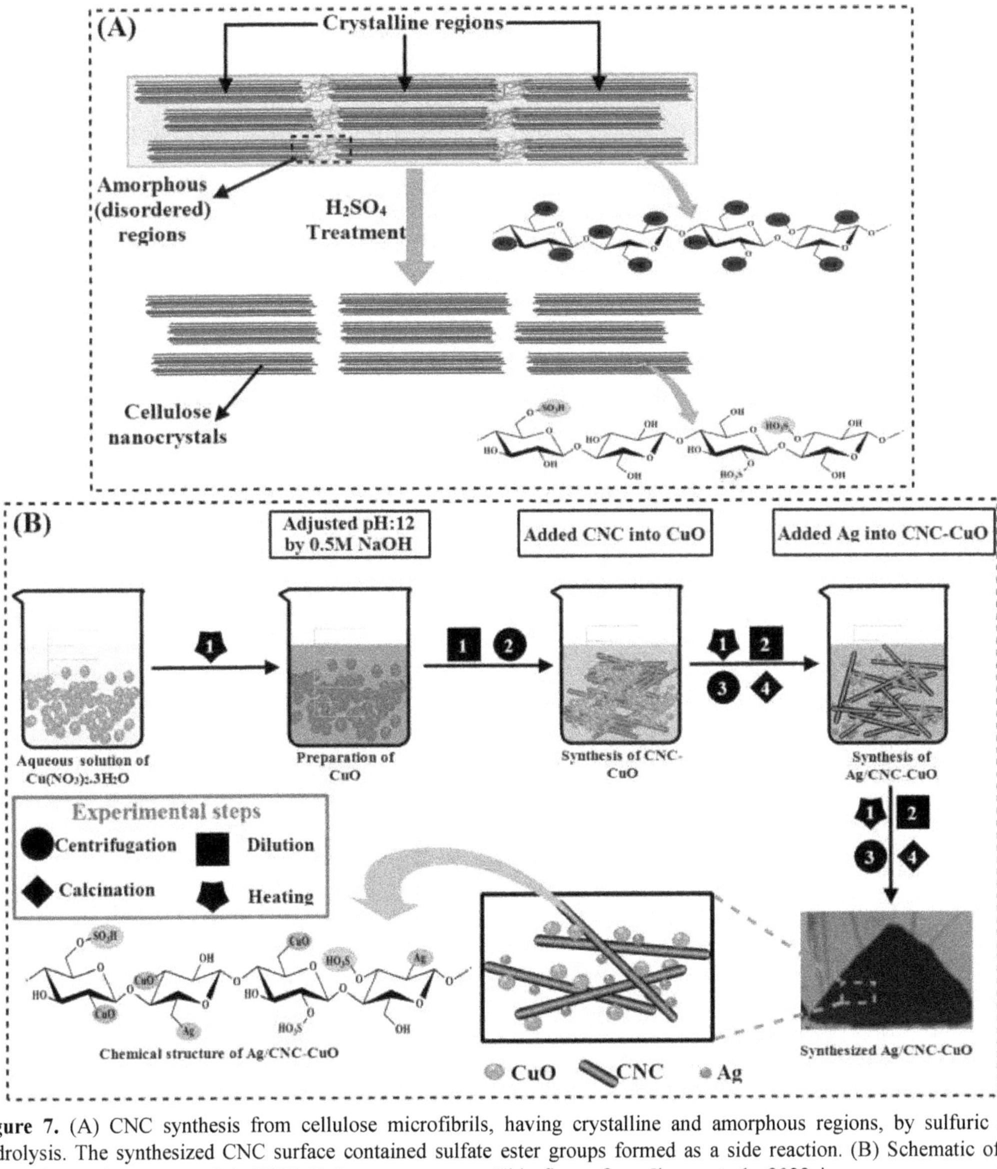

Figure 7. (A) CNC synthesis from cellulose microfibrils, having crystalline and amorphous regions, by sulfuric acid hydrolysis. The synthesized CNC surface contained sulfate ester groups formed as a side reaction. (B) Schematic of the preparation and structure of Ag/CNC-CuO nanostructures. This figure from Ikram et al., 2022 is an open access article distributed under the terms and conditions of the Creative Commons Attribution (CC BY) license (https://creativecommons.org/licenses/by/4.0/). Link: https://pubs.acs.org/cms/10.1021/acsomega.2c00240/asset/images/medium/ao2c00240_0001.gif.

Then, the H_2SO_4 was diluted with 5 L of DI water, and the supernatant was gathered. An NaOH (0.5 M) solution was used to change the pH of the collected solution from acidic to neutral. This neutral solution underwent three centrifugations for nine minutes at 7100 rpm. For additional research, the centrifuged CNC solution with a neutral pH was dried for seven hours at 100 °C. The concentration of the obtained CNC in powder form was 117 mg/mL.

A coprecipitation technique was used to create the CuO nanomaterial (Fig. 7b). Cu $(NO_3)_2 \cdot 3H_2O$, 0.1 M, was first prepared as the Cu source while being constantly stirred. The pH was then fixed at 12 by gradually adding a 0.5 M NaOH solution drop by drop to the colloidal solution while vigorously stirring at 90°C for 30 minutes. To remove the nitrate from the mixture, 36 milliliters of DI water were added. The mixture was then centrifuged three times. The produced CuO solution was mixed with 2.5 weight percent CNC, and the combination was heated for 30 minutes at 95°C. The doped solution was then mixed with 150 mL of water, centrifuged three times, and heated to 120°C for calcination. Ultimately, $AgNO_3$ was added as a precursor to dope 2% Ag into the CNC-CuO solution. The combination was then agitated at about 90°C and centrifuged twice. After seven hours of drying at 120°C, the precipitates were thoroughly crushed using a mortar and pestle.

4.3 Oxide Nanotubes/Alginate Microfiber Composites for Photodegradation of Methylene Blue

Bsoul et al., 2024, developed ZnO nanotubes /Alginate. This study uses a substrate-free hydrothermal synthesis method to create Zinc Oxide (ZnO) nanotubes to examine their effectiveness in cleaning contaminated water. To prevent secondary contamination, the synthesized tubes are impregnated into calcium alginate microfibers, which are selected for their high surface area to improve photocatalytic degradation performance and for easy handling during decontamination and subsequent collection. According to this study, Methylene Blue (MB), a typical organic contaminant, is photocatalytically degraded by this composite to demonstrate its effectiveness. After five hours of exposure to a solar simulator, MB had degraded by 88%. Cycle tests show constant levels of degradation in the first four cycles (81–89%), then a decline to 72% in the fifth cycle, which corresponds with the microfibers breaking down into smaller pieces. This work creatively presents a method for displaying the findings of photocatalytic degradation by using normalized pollutant concentration plotted against irradiated energy rather than time, as energy encompasses surface area, time, and irradiation power (Fig. 8). This shows that irradiating the sample with an energy of about 18 kJ is sufficient to cause the 88% destruction of MB.

4.4 Final Considerations

Biopolymers have been employed in the advancement of green nanotechnology for a long time. Through industrially feasible techniques, they offer promising improvements to nonbiodegradable materials in the environment. Biopolymers have completely changed how biocatalysts are made (Tahyade et al., 2018). Additionally, by serving as both reducing and stabilizing agents in aqueous conditions, the majority of these biopolymers are employed in the environmentally friendly production of various noble metal nanoparticles. The biodegradability and biocompatibility of biopolymers increase the range of applications for which they can be used. To create efficient catalysts, biopolymers can be mixed with various components, such as other biopolymers or biodegradable synthetic polymers. Metal or metal oxide nanoparticles (NPs) (such Cu, ZnO, TiO_2, CuO, and Fe_3O_4) are added to the surface of biopolymers to increase their catalytic activity, which results in efficient catalysts for a range of chemical processes (Nasrollahzadeh et al., 2021).

Because of their good dispersion of NPs and particles, high porosity, small size, and large surface area, biopolymer-based metal NPs exhibit significant catalytic activity in various processes. These catalysts are often most useful in the reduction and degradation processes of environmental

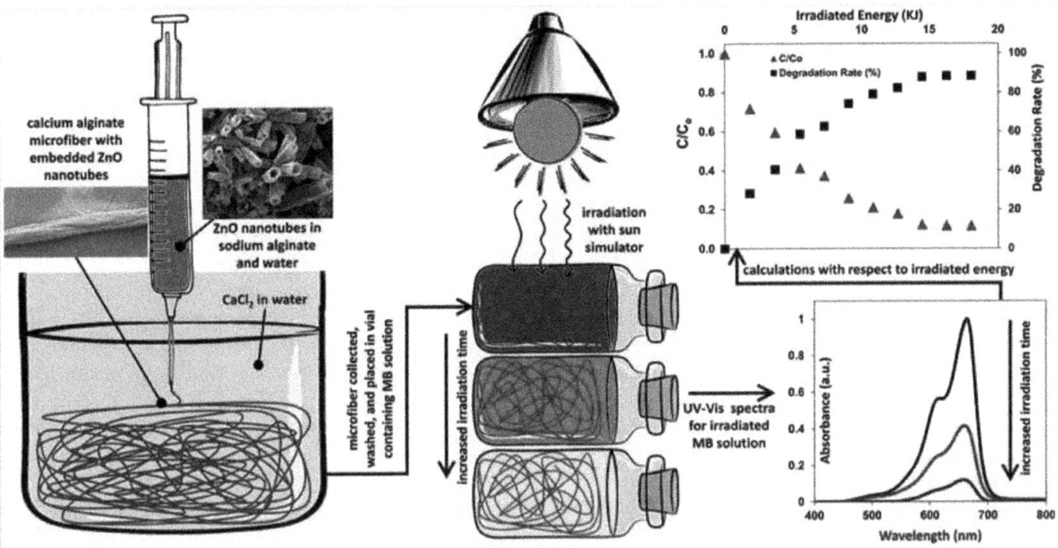

Figure 8. The synthesis of ZnO nanotubes-alginate microfibers for MB photocatalytic degradation with innovative results presentation. This figure from Bsoul et al., 2024 is an open access article distributed under the terms and conditions of the Creative Commons Attribution (CC BY) license (https://creativecommons.org/licenses/by/4.0/). Link: https://pubs.rsc.org/en/Image/Get?imageInfo.ImageType=GA&imageInfo.ImageIdentifier.ManuscriptID=D4RA01229A&imageInfo.ImageIdentifier.Year=2024.

contaminants. Nevertheless, despite these recent advancements in biopolymer-based metal NPs, several important issues need to be considered in further research (Nasrollahzadeh et al., 2021):

- ✓ The creation of efficient techniques to make the raw materials for these biopolymers easier to prepare.
- ✓ Creation of efficient processes for producing magnetic biopolymer-based catalysts that are easily separated from reaction fluids.
- ✓ Utilizing various biowaste materials to create biopolymers.
- ✓ The majority of catalyst preparation techniques are small-scale. Consequently, actions need to be done to industrially produce these catalysts.
- ✓ Techniques for decorating metallic nanoparticles on biopolymers using plant extracts have been proposed, but not much has been accomplished and not much progress has been made.

References

Abas, N., Kalair, A. and Khan, N. (2015). Review of fossil fuels and future energy technologies. Futures, 69: 31–49.

Agorku, E. S., Mittal, H., Mamba, B. B., Pandey, A. C. and Mishra, A. K. (2014). Fabrication of photocatalyst based on Eu3+-doped ZnS-SiO2 and sodium alginate core shell nanocomposite. International Journal of Biological Macromolecules, 70: 143–149. https://doi.org/10.1016/j.ijbiomac.2014.06.037.

Al-Othman, Z. A., Ali, R. and Naushad, M. (2012). Hexavalent chromium removal from aqueous medium by activated carbon prepared from peanut shell: adsorption kinetics, equilibrium and thermodynamic studies. Chem. Eng. J. 184: 238–247. https://doi.org/10.1016/j.cej.2012.01.048.

Alves, L. C. H. (2015). Cellulose solutions: Dissolution, regeneration, solution structure and molecular interactions. 144. https://estudogeral.uc.pt/handle/10316/87580.

Amiri, H., Aghbashlo, M., Sharma, M., Gaffey, J., Manning, L., Moosavi Basri, S. M. et al. (2022). Chitin and chitosan derived from crustacean waste valorization streams can support food systems and the UN Sustainable Development Goals. Nature Food, 3(10): 822–828. https://doi.org/10.1007/s10311-020-01021-w.

Astute Analytica, (2024): Textile Dyes Market - Industry Dynamics, Market Size, And Opportunity Forecast To 2032. Link: https://www.astuteanalytica.com/industry-report/tyextile-dyes-market. Accessed on 06/18/2024.

Bahrudin, N. N. and Nawi, M. A. (2019). Mechanistic of photocatalytic decolorization and mineralization of methyl orange dye by immobilized TiO2/chitosan-montmorillonite. J. Water Process Eng., 31: 100843.

Berradi, M., Hsissou, R., Khudhair, M., Assouag, M., Cherkaoui, O., Bachiri, A. E. et al. (2019). Textile finishing dyes and their impact on aquatic environs. Heliyon 5: e02711. https://doi.org/10.1016/j.heliyon.2019.e02711.

Bilgili, M, Ozbek, A, Sahin, B. and Kahraman, A. (2015). An overview of renewable electric power capacity and progress in new technologies in the world. Renew Sust. Energ. Rev., 49: 323–334.

Bonakdar, M. A. and Rodrigue, D. (2024). Electrospinning: Processes, Structures, and Materials. Macromol. 4: 58–103. https://doi.org/10.3390/macromol4010004.

Bsoul, A., Alkhaldi, I., Albiss, B., Ocak, Y. S. and Mohamed Ali, M. S. (2024). Synthesis and performance evaluation of zinc oxide tubes/alginate microfibre composites for photodegradation of methylene blue: a novel reporting approach. RSC Advances, 14(28): 20182–20190. https://doi.org/10.1039/d4ra01229a.

Bueno, M. M., Gomez, M. J., Herrera, S., Hernando, M. D., Agüera, A. and Fern'andez-Alba, A. R. (2012). Occurrence and persistence of organic emerging contaminants and priority pollutants in five sewage treatment plants of Spain: Two years pilot survey monitoring. Environ. Pollut. 164: 267–273. https://doi.org/10.1016/j.envpol.2012.01.038.

Chia, M. A. and Musa, R. I. (2014). Effect of indigo dye effluent on the growth, biomass production and phenotypic plasticity of *Scenedesmus quadricauda* (Chlorococcales). An Acad. Bras Ciˆencias, 86(1): 419–428. https://doi.org/10.1590/0001-3765201420130225.

Cremers, V., Puurunen, R. L. and Dendooven, J. (2019). Conformality in atomic layer deposition: Current status overview of analysis and modelling. Applied Physics Reviews, 6(2). https://doi.org/10.1063/1.5060967.

Demazeau, G. (2011). Solvothermal and hydrothermal processes: The main physico-chemical factors involved and new trends. Res. Chem. Intermed., 37: 107–123. https://doi.org/10.1007/s11164-011-0240-z.

dos Santos, A. J., Kronka, M. S., Fortunato, G. V. and Lanza, M. R. V. (2021). Recent advances in electrochemical water technologies for the treatment of antibiotics: a short review. Curr. Opin. Electrochem. 100674. https://doi.org/10.1016/j.coelec.2020.100674.

Du, H., Liu, W., Zhang, M., Si, C., Zhang, X. and Li, B. (2019). Cellulose nanocrystals and cellulose nanofibrils based hydrogels for biomedical applications. Carbohydrate Polymers, 209: 130–144. https://doi.org/10.1016/j.carbpol.2019.01.020.

Egorova, K. S. and Ananikov, V. P. (2017). Toxicity of metal compounds: knowledge and myths. Organometallics, 36: 4071–4090. https://doi.org/10.1021/acs.organomet.7b00605.

Fakhri, A., Gupta, V. K., Rabizadeh, H., Agarwal, S., Sadeghi, N. and Tahami, S. (2018). Preparation and characterization of WS2 decorated and immobilized on chitosan and polycaprolactone as biodegradable polymers nanofibers: Photocatalysis study and antibiotic-conjugated for antibacterial evaluation. International journal of biological macromolecules, 120(Pt B): 1789–1793. https://doi.org/10.1016/j.ijbiomac.2018.09.207.

Fox, M. A. and Dulay, M. T. (1993). Heterogeneous photocatalysis. Chem. Rev., 93(1): 341–357.

Ghazipura, M., McGowan, R., Arslan, A. and Hossaind, T. (2017). Exposure to benzophenone-3 and reproductive toxicity: a systematic review of human and animal studies. Reprod. Toxicol., 73: 175–183. https://doi.org/10.1016/j.reprotox.2017.08.015.

Grzybek, P., Jakubski, Ł. and Dudek, G. (2022). Neat chitosan porous materials: a review of preparation, structure characterization and application. International Journal of Molecular Sciences, 23(17): 9932. https://doi.org/10.3390/ijms23179932.

Gutierrez, J., Fernandes, S. C. M., Mondragon, I. and Tercjak, A. (2013). Multifunctional hybrid nanopapers based on bacterial cellulose and sol–gel synthesized titanium/vanadium oxide nanoparticles. Cellulose, 20(3): 1301–1311. https://doi.org/10.1007/s10570-013-9898-2.

Habiba, U., Islam, M. S., Siddique, T. A., Afifi, A. M. and Ang, B. C. (2016). Adsorption and photocatalytic degradation of anionic dyes on Chitosan/PVA/Na-Titanate/TiO2 composites synthesized by solution casting method. Carbohydrate Polymers, 149: 317–331. https://doi.org/10.1016/j.carbpol.2016.04.127.

Habiba, U., Lee, J. J. L., Joo, T. C., Ang, B. C. and Afifi, A. M. (2019). Degradation of methyl orange and congo red by using chitosan/polyvinyl alcohol/TiO2 electrospun nanofibrous membrane. International Journal of Biological Macromolecules, 131: 821–827. https://doi.org/10.1016/j.ijbiomac.2019.03.132.

Hader, D. P., Banaszak, A. T., Villafañe, V. E., Narvarte, M. A., González, R. A. and Helbling, E. W. (2020). Anthropogenic pollution of aquatic ecosystems: Emerging problems with global implications. Sci. Total Environ. 713: 136586. https://doi.org/10.1016/j. scitotenv.2020.136586.

Hansima, M. A. C. K., Makehelwala, M., Jinadasa, K. B. S. N., Wei, Y., Nanayakkara, K. G. N., Herath, A. C. etal. (2021). Fouling of ion exchange membranes used in the electrodialysis reversal advanced water treatment: Areview. Chemosphere, 263: 127951. https://doi.org/10.1016/j.chemosphere.2020.127951.

Hassan Hassan Abdellatif, F., Babin, J., Arnal-Herault, C., David, L. and Jonquieres, A. (2018). Grafting cellulose acetate with ionic liquids for biofuel purification membranes : Influence of the anion. Carbohydrate polymers, 196: 176–186. https://doi.org/10.1016/j.carbpol.2018.05.008.

Herrmann, J.-M. (1999). Heterogeneous photocatalysis: fundamentals and applications to the removal of various types of aqueous pollutants. Catal Today, 53(1): 115–129.

Hitam, C. N. C. and Jalil, A. A. (2020). A review on exploration of Fe2O3 photocatalyst towards degradation of dyes and organic contaminants. J. Environ. Manag. 258: 110050. https://doi.org/10.1016/j.jenvman.2019.110050.

Hoang, N. T. T. and Nguyen, D. D. (2023). Improving the degradation kinetics of industrial dyes with chitosan/TiO_2/ Glycerol films for the sustainable recovery of chitosan from waste streams. Sustainability, 15(8), 6979. https://doi.org/10.3390/su15086979.

Horikoshi, S. and Serpone, N. (2020). Can the photocatalyst TiO2 be incorporated into a wastewater treatment method? Background and prospects. Catalysis Today, 340: 334–346. https://doi.org/10.1016/j.cattod.2018.10.020.

Houas, A., Lachheb, H., Ksibi, M., Elaloui, E., Guillard, C. and Herrmann, J. M. (2001). Photocatalytic degradation pathway of methylene blue in water. Applied Catalysis B: Environmental, 31(2): 145–157. https://doi.org/10.1016/S0926-3373(00)00276-9.

https://www.ecodebate.com.br/2023/10/30/corantes-industriais-ameacam-saude-humana-e-ambiental/ cosultar.

Ikram, M., Hafeez, I., Naz, M., Haider, A., Naz, S., Ul-Hamid, A. et al. (2022). Highly efficient industrial dye degradation, bactericidal properties, and *in silico* molecular docking analysis of Ag/cellulose-doped CuO nanostructures. ACS omega, 7(20): 17043–17054. https://doi.org/10.1021/acsomega.2c00240.

Jawad, A. H., Kadhum, A. M. and Ngoh, Y. S. (2018a). Applicability of dragon fruit (*Hylocereus polyrhizus*) peels as low-cost biosorbent for adsorption of methylene blue from aqueous solution: Kinetics, equilibrium and thermodynamics studies. Desalination and Water Treatment, 109: 231–240. https://doi.org/10.5004/dwt.2018.21976.

Jawad, A. H., Waheeb, A. S., Rashid, R. A., Nawawi, W. I. and Yousif, E. (2018b). Equilibrium isotherms, kinetics, and thermodynamics studies of methylene blue adsorption on pomegranate (*Punica granatum*) peels as a natural low-cost biosorbent. Desalination and Water Treatment, 105: 322–331. https://doi.org/10.5004/dwt.2018.22021.

Jiang, J. Q. (2015). The role of coagulation in water treatment. Curr. Opin. Chemi. Eng. 8: 36–44. https://doi.org/10.1016/j.coche.2015.01.008.

Jiang, R., Zhu, H., Li, X. and Xiao, L. (2009). Visible light photocatalytic decolourization of C. I. Acid Red 66 by chitosan capped CdS composite nanoparticles. Chemical Engineering Journal, 152(2-3): 537–542. https://doi.org/10.1016/j.cej.2009.05.037.

Jones, C. W. (1999). Applications of hydrogen peroxide and derivatives (Vol. 2). Royal Society of Chemistry. https://doi.org/10.1039/9781847550132.

Kemell, M., Pore, V., Ritala, M., Leskelä, M. and Lindén, M. (2005). Atomic layer deposition in nanometer-level replication of cellulosic substances and preparation of photocatalytic TiO2/cellulose composites. Journal of the American Chemical Society, 127(41): 14178–14179. https://doi.org/10.1021/ja0532887.

Kim, S., Park, H. and Choi, W. (2004). Comparative study of homogeneous and heterogeneous photocatalytic redox reactions: PW12O403- vs TiO2. Journal of Physical Chemistry B, 108(20): 6402–6411. https://doi.org/10.1021/jp049789g.

Kumar, B., Singh, N. and Kumar, P. (2024). A review on sources, modification techniques, properties and potential applications of alginate-based modified polymers. European Polymer Journal, 213: 113078. https://doi.org/10.1016/j.eurpolymj.2024.113078.

Kumar, V., Parihar, R. D., Sharma, A., Bakshi, P., Sidhu, G. P. S., Bali, A. S. et al. (2019). Global evaluation of heavy metal content in surface water bodies: a meta-analysis using heavy metal pollution indices and multivariate statistical analyses. Chemosphere, 236: 124364. https://doi.org/10.1016/j.chemosphere.2019.124364.

Lakhera, S. K. and Neppolian, B. (2021). Visible light photocatalysis: Case study (process). In Handbook of Nanomaterials for Wastewater Treatment: Fundamentals and Scale up Issues. Elsevier Inc. https://doi.org/10.1016/B978-0-12-821496-1.00001-5.

Lakhera, S. K., Watts, A., Hafeez, H. Y. and Neppolian, B. (2018). Interparticle double charge transfer mechanism of heterojunction α-Fe2O3/Cu2O mixed oxide catalysts and its visible light photocatalytic activity. Catalysis Today, 300: 58–70. https://doi.org/10.1016/j.cattod.2017.03.020.

Ledezma, C. Z., Bolagay, D. N., Figueroa, F., Ledezma, E. Z., Ni, M., Alexis, F. et al. (2021). Heavy metal water pollution: A fresh look about hazards, novel andconventional remediation methods. Environ. Technol. Innovat. 22: 101504. https://doi.org/10.1016/j.eti.2021.101504.

Lemeune, S., Jameel, H., Chang, H. M. and Kadla, J. F. (2004). Effects of ozone and chlorine dioxide on the chemical properties of cellulose fibers. Journal of Applied Polymer Science, 93(3): 1219–1223.

Li, G., Sun, D., Guo, L., Yuan, J., Gao, K., Jie, Z. et al. (2023). Microstructure evolution and tensile behavior of hot pressed powder metallurgy superalloys. Journal of Materials Research, 38(17): 4118–4129. https://doi.org/10.1557/s43578-023-01127-1.

Li, J., Wu, Q., Wu, J. (2016). Synthesis of nanoparticles via solvothermal and hydrothermal methods. *In*: Aliofkhazraei, M. (Eds.). Handbook of Nanoparticles. Springer, Cham. https://doi.org/10.1007/978-3-319-15338-4_17.

Li, S. and Huang, J. (2016). Cellulose-Rich Nanofiber-Based Functional Nanoarchitectures. Advanced materials (Deerfield Beach, Fla.), 28(6): 1143–1158. https://doi.org/10.1002/adma.201501878.

Li, Y., Cao, T., Mei, Z., Li, X. and Sun, D. (2020). Separating type I heterojunction of NaBi (MoO4)2/Bi2MoO6 by TiO2 nanofibers for enhanced visible-photocatalysis. Chem. Phys. 533 https://doi.org/10.1016/j.chemphys.2020.110696.110696.

Luo, Y., Xu, J. and Huang, J. (2014). Hierarchical nanofibrous anatase-titania-cellulose composite and its photocatalytic property. CrystEngComm, 16(3): 464–471. https://doi.org/10.1039/c3ce41906a.

Ma, R., Fang, W., Zhang, H., Sun, J., Su, H., Chen, T. et al. (2020). Transcriptome analysis of zebra fish (Danio rerio) eggs following treatment with malachite green. Aquaculture, 514: 734500. https://doi.org/10.1016/j.aquaculture.2019.734500.

Maniakova, G., Kowalska, K., Murgolo, S., Mascolo, G., Libralato, G., Lofrano, G. et al. (2020). Comparison between heterogeneous and homogeneous solar driven advanced oxidation processes for urban wastewater treatment: Pharmaceuticals removal and toxicity. Separation and Purification Technology, 236: 116249. https://doi.org/10.1016/j.seppur.2019.116249.

Markets and Markets, 2024: Textile Dyes Market. Link: https://www.marketsandmarkets.com/Market-Reports/textile-dye-market-226167405.html. Accessed on 06/18/2024.

Mathew, B. B., Biju, V. G., Nideghatta Beeregowda, K. (2019). Accumulation of lead (Pb II) metal ions by Bacillus toyonensis SCE1 species, innate to industrial-area ground water and nanoparticle synthesis. Appl. Nanosci., 9: 49–66. https://doi.org/10.1007/s13204-018-0892-8.

Mohamed, M. A., Abd Mutalib, M., Mohd Hir, Z. A., M. Zain, M. F., Mohamad, A. B., Jeffery Minggu, L. et al. (2017). An overview on cellulose-based material in tailoring bio-hybrid nanostructured photocatalysts for water treatment and renewable energy applications. International Journal of Biological Macromolecules, 103: 1232–1256. https://doi.org/10.1016/j.ijbiomac.2017.05.181.

Moorthy, A. K., Rathi, B. G., Shukla, S. P., Kumar, K. and Bharti, V. S. (2021). Acute toxicity of textile dye Methylene blue on growth and metabolism of selected freshwater microalgae. Environ. Toxicol. Pharmacol., 82: 103552. https://doi.org/10.1016/j.etap.2020.103552.

Mu, Y., Fu, Y., Li, J., Yu, X., Li, Y., Wang, Y. et al. (2019). Multifunctional quercetin conjugated chitosan nano-micelles with P-gp inhibition and permeation enhancement of anticancer drug. Carbohydrate Polymers, 203: 10–18. https://doi.org/10.1016/j.carbpol.2018.09.020.

Nasrollahzadeh, M., Nezafat, Z., Shafiei, N. and Soleimani, F. (2021). Polysaccharides in the food industry. In Biopolymer-Based Metal Nanoparticle Chemistry for Sustainable Applications: Volume 2: Applications. https://doi.org/10.1016/B978-0-323-89970-3.00002-0.

Nonell, S. and Flors, C. (Eds.). (2016). Singlet Oxygen: Applications in Biosciences and Nanosciences, Volume 1. Royal Society of Chemistry.

Noorbakhsh-Soltani, S. M., Zerafat, M. M. and Sabbaghi, S. (2018). A comparative study of gelatin and starch-based nano-composite films modified by nano-cellulose and chitosan for food packaging applications. Carbohydrate polymers, 189: 48–55. https://doi.org/10.1016/j.carbpol.2018.02.012.

Nor, N. A. M., Jaafar, J., Ismail, A. F., Mohamed, M. A., Rahman, M. A., Othman, M. H. D. et al. (2016). Preparation and performance of PVDF-based nanocomposite membrane consisting of TiO2 nanofibers for organic pollutant decomposition in wastewater under UV irradiation. Desalination, 391: 89–97. https://doi.org/10.1016/j.desal.2016.01.015.

Pal, P. and Pal, A. (2019). Dye removal using waste beads: Efficient utilization of surface-modified chitosan beads generated after lead adsorption process. J. Water Process Eng., 31: 100882. https://doi.org/10.1016/j.jwpe.2019.100882.

Paul, R. (2023). Sol-Gel Method : Overcoming the Limitations in Nanoparticle Synthesis. Research & Reviews: Journal of Pharmaceutics and Nanotechnology e-RRJPN|, 11(1):2023. https://doi.org/10.4172/23477857.10.1.008.

Petrie, B., Barden, R. and Kasprzyk-Hordern, B. (2015). A review on emerging contaminants in wastewaters and the environment: current knowledge, understudied areas and recommendations for future monitoring. Water Res. 72: 3–27. https://doi.org/ 10.1016/j.watres.2014.08.053.

Pore, V., Kivelä, T., Ritala, M. and Leskelä, M. (2008). Atomic layer deposition of photocatalytic TiO2 thin films from TiF4 and H2O. Dalton Transactions, 45: 6467–6474. https://doi.org/10.1039/b809953g.

Preethi, J., Farzana, M. H. and Meenakshi, S. (2017). Photo-reduction of Cr(VI) using chitosan supported zinc oxide materials. International Journal of Biological Macromolecules, 104(Pt B): 1783–1793. https://doi.org/10.1016/j.ijbiomac.2017.02.082.

Prusty, J. K., Patro, S. K. and Basarkar, S. (2016). Concrete using agro-waste as fine aggregate for sustainable built environment – a review. Int. J. Sustain. Built. Environ., 5(2): 312–333.

Rasoulifard, M. H., Seyed Dorraji, M. S. and Mozafari, V. (2017). Visible light photocatalytic activity of chitosan/poly(vinyl alcohol)/TiO2nanocomposite for dye removal: Taguchi-based optimization. Environ. Prog. Sustain. Energy, 36: 66–72. https://doi.org/10.1002/ep.12438.

Rathore, B. S., Sharma, G., Pathania, D. and Gupta, V. K. (2014). Synthesis, characterization and antibacterial activity of cellulose acetate-tin (IV) phosphate nanocomposite. Carbohydrate Polymers, 103: 221–227. https://doi.org/10.1016/j.carbpol.2013.12.011.

Ren, G., Han, H., Wang, Y., Liu, S., Zhao, J., Meng, X. et al. (2021). Recent advances in photocatalysis applications in water treatment: A review. Nanomaterials, 11(7)1804. https://doi.org/10.3390/nano11071804.

Ronaldo Fereira do Nascimento, Odair Pastor Ferreira, Amauri Jardim De Paula and Vicente de Oliveira Sousa Neto (Eds.). (2019). In Advanced Nanomaterials,Nanomaterials Applications for Environmental Matrices, Elsevier, pp. 489–508, ISBN 9780128148297,https://doi.org/10.1016/B978-0-12-814829-7.00027-6.

Rosal, R., Palomares, I. R., Boltes, K., Piñas, F. F., Legan´es, F. and Petre, A. (2010). Ecotoxicological assessment of surfactants in the aquatic environment: combined toxicity of docusate sodium with chlorinated pollutants. Chemosphere, 81: 288–293. https://doi.org/10.1016/j.chemosphere.2010.05.050.

Sarin, V., Pal, U. and Gopalan, S. (2003). Development of Hot Pressing as a Low Cost Processing Technique for Fuel Cell Fabrication. 1–41.

Sarkar, S., Ponce, N. T., Banerjee, A., Bandopadhyay, R., Rajendran, S. and Lichtfouse, E. (2020). Green polymeric nanomaterials for the photocatalytic degradation of dyes: a review. Environmental Chemistry Letters, 18(5): 1569–1580. https://doi.org/10.1007/s10311-020-01021-w.

Schmeller, D. S., Loyau, A., Bao, K., Brack, W., Chatzinotas, A., Vleeschouwer, F. D. et al. (2018). People, pollution and pathogens – global change impacts in mountain freshwater ecosystems. Sci. Total Environ., 622-623:756–763. https://doi.org/10.1016/j.scitotenv.2017.12.006.

Sharma, G., Naushad, M., Al-Muhtaseb, A. H., Kumar, A., Khan, M. R., Kalia, S. et al. (2017). Fabrication and characterization of chitosan-crosslinked-poly(alginic acid) nanohydrogel for adsorptive removal of Cr(VI) metal ion from aqueous medium. International Journal of Biological Macromolecules, 95: 484–493. https://doi.org/10.1016/j.ijbiomac.2016.11.072.

Shen, L., Yu, X., Cheng, C., Song, C., Wang, X., Zhu, M. et al. (2016). High filtration performance thin film nanofibrous composite membrane prepared by electrospraying technique and hot-pressing treatment. Journal of Membrane Science, 499: 470–479. https://doi.org/10.1016/j.memsci.2015.11.004.

Sousa Neto, V. and Nascimento, R. F. (2019). Introduction. In Nanomaterials Applications for Environmental Matrices: Water, Soil and Air. Elsevier Inc. https://doi.org/10.1016/B978-0-12-814829-7.00001-X.

Stan, C. D., Cretescu, I., Pastravanu, C., Poulios, I. and Dragan, M. (2012). Treatment of pesticides in wastewater by heterogeneous and homogeneous photocatalysis. International Journal of Photoenergy, 2012, 6.

Suresh, R., Rajendran, S., Hoang, T. K. A., Vo, D. V. N., Siddiqui, M. N. and Cornejo-Ponce, L. (2021). Recent progress in green and biopolymer based photocatalysts for the abatement of aquatic pollutants. Environmental Research, 199(February), 111324. https://doi.org/10.1016/j.envres.2021.111324.

Tabriz, A., Ur Rehman Alvi, M. A., Khan Niazi, M. B., Batool, M., Bhatti, M. F., Khan, A. L. et al. (2019). Quaternized trimethyl functionalized chitosan based antifungal membranes for drinking water treatment. Carbohydrate Polymers, 207: 17–25. https://doi.org/10.1016/j.carbpol.2018.11.066.

Taghipour, S., Ehtesham Nia, A., Hokmabadi, H. and Yahia, E. M. (2024). Quality evaluation of fresh pistachios (*Pistacia vera* L.) cultivars coated with chitosan/TiO2 nanocomposite. International Journal of Biological Macromolecules, 258(P2): 129055. https://doi.org/10.1016/j.ijbiomac.2023.129055.

Taka, A. L., Klink, M. J., Mbianda, X. Y. and Naidoo, E. B. (2021). Chitosan nanocomposites for water treatment by fixed-bed continuous flow column adsorption: A review. Carbohydr. Polym., 255: 117398. https://doi.org/10.1016/j.carbpol.2020.117398.

Tayade, R. J., Gandhi, V., Conversion, S. E., Reduction, C. D., Generation, H., Phosphate, S. et al. (2018). Nano-Photocatalytic Materials for Environmental Applications. In Aquananotechnology. https://doi.org/10.1201/b17455-16.

The Business Research Company, 2024: Synthetic Dye And Pigment Market Definition And Segments. Link: https://www.thebusinessresearchcompany.com/report/synthetic-dye-and-pigment-global-market-report. Accessed on 06/18/2024.

Tien, T. M. and Chen, E. L. (2023). Interfacial Charge Transfer Effects of MoS2/α-Fe2O3 Nano-Heterojunction and Efficient Photocatalytic Hydrogen Evolution under Visible-Light Irradiation. Nanomaterials (Basel). 2023 Oct 15; 13(20): 2763. https://doi.org/10.3390/nano13202763.

Tkaczyk, A., Mitrowska, K. and Posyniak, A. (2020). Synthetic organic dyes as contaminants of the aquatic environment and their implications for ecosystems: A review. Sci. Total Environ. 717: 137222. https://doi.org/10.1016/j.scitotenv.2020.137222.

Unal, B. O., Omwene, P. I., Yagcioglu, M., Canbolat, Ç. B., Karagunduz, A., Keskinler, B. et al. (2020a). Treatment of organized industrial zone wastewater by microfiltration/reverse osmosis membrane process for water recovery: from lab to pilot scale. J. Water Process. Eng. 38: 101646. https://doi.org/10.1016/j.jwpe.2020.101646.

Unal, B. O., Omwene, P. I., Yagcioglu, M., Canbolat, Ç. B., Karagunduz, A., Keskinler, B. et al. (2020). Treatment of organized industrial zone wastewater by microfiltration/reverse osmosis membrane process for water recovery: from lab to pilot scale. J. Water Process Eng., 38: 101646. https://doi.org/10.1016/j.

UNICEF and WHO. (2019). Progress on household drinking water, sanitation and hygiene 2000–2017 Special focus on inequalities. In Launch version July 12 Main report Progress on Drinking Water, Sanitation and Hygiene. https://www.who.int/publications/i/item/9789241516235.

van Leeuwen, P. W. and Leeuwen, P. W. N. M. (2004). Homogeneous Catalysis: Understanding the Art. Alemanha: Kluwer Academic Publishers.

Vel´azquez, G. A.E., Oliv´an, L.M.G., 2020. Occurrence, toxic effects and removal of metformin in the aquatic environments in the world: recent trends and perspectives. Sci. Total Environ., 702: 134924. https://doi.org/10.1016/j.scitotenv.2019.134924.

Virkutyte, J., Jegatheesan, V. and Varma, R. S. (2012). Visible light activated TiO2/microcrystalline cellulose nanocatalyst to destroy organic contaminants in water. Bioresource Technology, 113: 288–293. https://doi.org/10.1016/j.biortech.2011.12.090.

Wang, T., Wang, L., Chen, Q., Kalogerakis, N., Ji, R. and Ma, Y. (2020). Interactions between microplastics and organic pollutants: effects on toxicity, bioaccumulation, degradation, and transport. Sci. Total Environ., 748: 142427. https://doi.org/10.1016/j.scitotenv.2020.142427.

Weon, S. H., Han, J., Choi, Y. K., Park, S. and Lee, S. H. (2023). Development of Blended Biopolymer-Based Photocatalytic Hydrogel Beads for Adsorption and Photodegradation of Dyes. Gels, 9(8). https://doi.org/10.3390/gels9080630.

Winandy, J. E. and Krzysik, A. M. (2007). Thermal degradation of wood fibers during hot-pressing of MDF composites: Part I. Relative effects and benefits of thermal exposure. Wood and Fiber Science, 39(3): 450–461.

Youssef, A. M., El-Nahrawy, A. M. and Hammad, A. B. A. (2017). Sol-gel synthesis and char- acterizations of hybrid chitosan-PEG/calcium silicate nanocomposite modified with ZnO-NPs and (E102) for optical and antibacterial applications. International Journal of Biological Macromolecules. 97: 561–567. https://doi.org/10.1016/j.ijbiomac.2017.01.059.

Yu, B., Lin, Z. and Huang, J. (2021) A Bio-Inspired Nanotubular Na2MoO4/TiO2 composite as a high-performance anodic material for lithium-ion batteries. Materials, 14: 357. https://doi.org/10.3390/ma14020357.

Zhang, K., Tao, P., Zhang, Y., Liao, X. and Nie, S. (2019). Highly thermal conductivity of CNF/AlN hybrid films for thermal management of flexible energy storage devices. Carbohydrate Polymers, 213: 228–235. https://doi.org/10.1016/j.carbpol.2019.02.087.

Zhou, Z., Peng, X., Zhong, L., Wu, L., Cao, X. and Sun, R. C. (2016). Electrospun cellulose acetate supported Ag@AgCl composites with facet-dependent photocatalytic properties on degradation of organic dyes under visible-light irradiation. Carbohydrate Polymers, 136: 322–328. https://doi.org/10.1016/j.carbpol.2015.09.009.

CHAPTER-12

Waste-derived Biopolymers (WD-BPs)

An Emerging Material for the Degradation of Environmental Pollutants

Rishabh Anand Omar,[1] *Neetu Talreja,*[2,*] *Divya Chauhan*[3] and *Mohammad Ashfaq*[5,*]

1. Introduction

As the increment in the world population, the demand for primary crops also increases. In the last two decades, crop production increased by 52%, which may increase by 9.3 billion tons in 2030. However, this increment has also made agriculture a significant contributor to waste generation, producing large amounts of waste during creation and processing. This excessive resource use and waste production can lead to deteriorating environmental conditions, potentially disrupting global environmental stability. The adoption of modern technology in agriculture has enhanced crop yields, which also carries some negative impacts, leaving enduring marks on the environment, society, and economy (Singh et al., 2020; Omar et al., 2024). Agricultural waste poses a significant challenge, for humans generating 15 trillion metric tons of such waste each year via cultivation, farming, harvesting, and industrial activities. Regrettably, only a small part of this waste is repurposed, with the majority decided on through landfilling or burning. This environmental contamination and resource misuse has emerged as a major issue (Kumar et al., 2021).

Current circumstances highlight concerns such as the eternal impact of plastics, the limited convenience of landfill space, the depletion of petroleum resources, incineration emissions, and living organism's risks, all of which underscore the importance of environmental protection (Kumar et al., 2021). Additionally, population growth and need for lightweight, affordable, and easily available materials have led to an increase in plastic production. The production and use of plastics have significant environmental implications, particularly regarding the change of climate and the

[1] Centre for Environmental Science and Engineering, Indian Institute of Technology Kanpur, Kanpur, 208016, India.
[2] Faculty of Science and Technology, Department of Science, Alliance University, Bengaluru, Karnataka 562106, India.
[3] Department of Drinking Water and Sanitation, Ministry of Jal Shakti, 1208-A, Pandit Deendayal Antyodaya Bhawan, CGO Complex, Lodhi Road, New Delhi 110003.
[4] School of Sciences, Woxsen University, Telangana, Hyderabad 502345, India.
* Corresponding authors: neetutalreja99@gmail.com; mohdashfaqbiotech@gmail.com
DOI: 10.1201/9781003558330-12

emission of greenhouse gases. Pollution by plastic waste also affects socio-economic and ecological aspects (Ford et al., 2022). Given the threat of climate change, it is essential to transition to a more environmentally conscious economy, not as a choice, but as a responsibility. To achieve sustainable development, it is crucial to employ knowledge-based production and utilize biological resources, processes, and techniques across all economic sectors to provide goods and services sustainably. From a materials perspective, the solution lies in embracing more eco-friendly trends, leading to the emergence of "biocomposites ." Biocomposites are materials made by combining natural fibers with a matrix material (Edebali, 2020). These natural fibers, sourced from flax, hemp, jute, sisal, bamboo, and others, can be combined with various types of matrix materials, including biodegradable, non-biodegradable, or synthetic ones. Biodegradable matrix materials, such as starch, cellulose, and chitosan, degrade naturally. Non-biodegradable matrix materials resist degradation and are suitable for long-lasting applications. Synthetic matrix materials, like epoxy resins, are often blended with natural fibers to create hybrid biocomposities. Using bio composites could improve resource efficiency and waste stream management by replacing fossil fuel-based resources (Zwawi, 2020).

One strategy for achieving sustainability goals involves replacing traditional polymers with biodegradable polymers in bio composites to create more environmentally friendly materials known as green biocomposites or Waste Derived BPs (WD-BPs). WD-BPs are a type of bio composite where both the matrix and the reinforced fibers come from renewable sources. WD-BPs enable the production of new products that benefit nature, either because they are biodegradable or, if not, because they can sequester carbon from CO_2 in the form of biological macrostructures, while there has been extensive research on bio-composites based on agricultural materials, the focus has largely been on natural fibers. There is a research gap in the development of sustainable WD-BPs from waste of agriculture/food industries (Thomas and Balakrishnan, 2021). This chapter seeks to enhance awareness about the economy by advocating for the adoption of WD-BPs synthesized by green synthesis as a practical substitute for conventional materials in lightweight applications. It will evaluate the potential, economic viability, and environmental benefits of WD-BPs in sustainable development and recommend their integration into the food/agricultural sector. Figure 1 Shows the schematic illustration of the different WD-BPs.

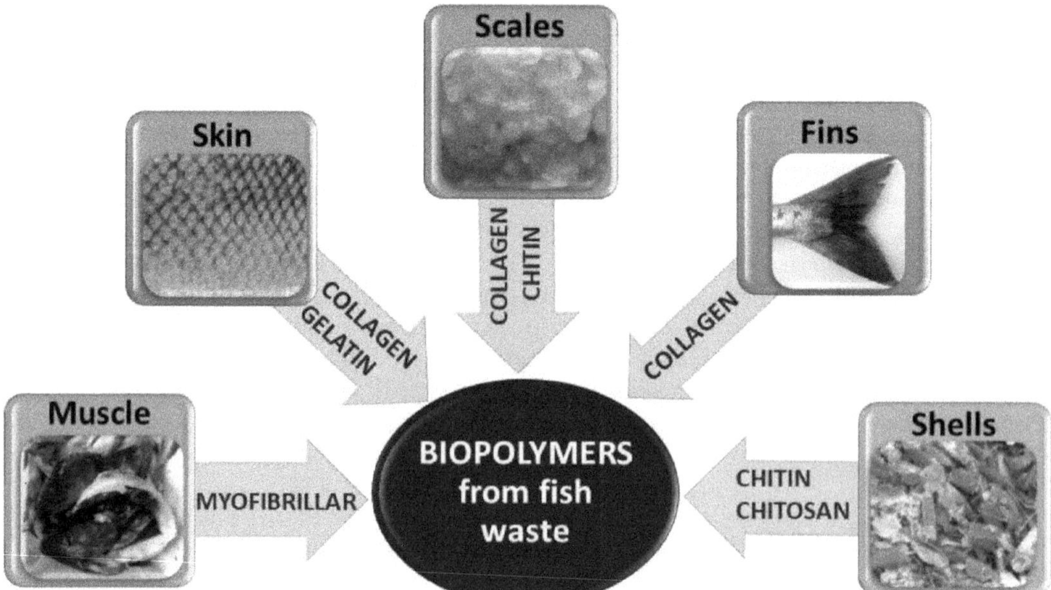

Figure 1. A schematic illustration of the different WD-BPs. The image was taken with permission (Lionetto et al., 2021).

2. Waste derived BPs (WD-BPs)

WD-BPs are complex structures, containing monomeric chains which are bonded covalently and derived from biomolecules. Unlike synthetic polymers, bio derived polymers also have an allocated molecular structure which leads to a well-precise 3D arrangement. The "bio" term confirms that these polymers can naturally degrade through biological activity, making them biodegradable compounds. This characteristic makes BPs an attractive alternative to fossil-based polymers and a potential solution to urgent environmental issues. On the other hand, human-made are called as from the petrochemical products are called synthetic polymers. These polymers are non-biodegradable. Petroleum resources are limited and add radically to universal pollution. Unlike synthetic polymers, BPs follow a circular approach in designing of the material, its manufacturing, their application, and proper disposal (Visco et al., 2022). Figure 2 Shows the schematic illustration of the different source of WD-BPs and their applications.

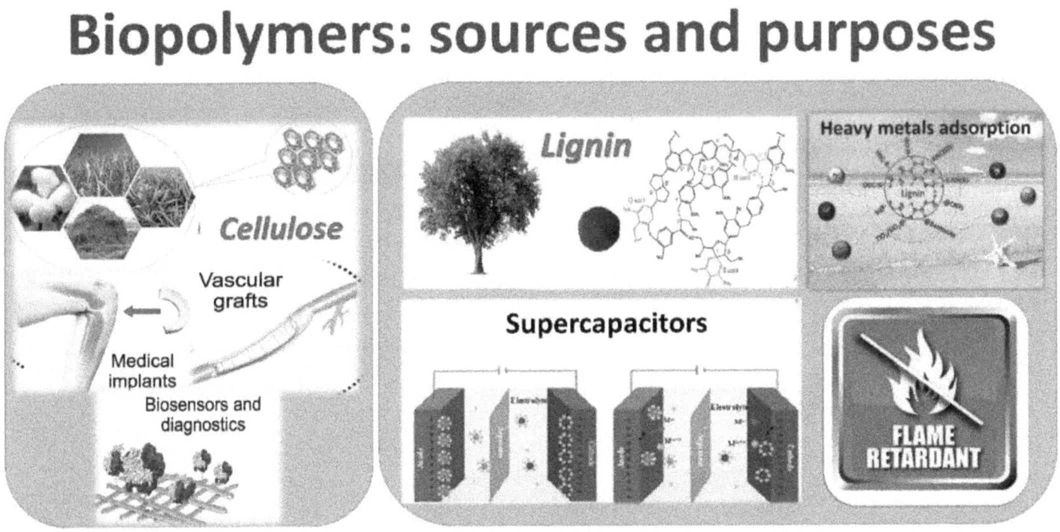

Figure 2. A schematic illustration of the different source of WD-BPs and their applications. The image was reproduced with the permission (Ferrari et al., 2022).

3. Types of BPs

BPs can be classified in various ways, with one common method being based on their source. They can originate from plants, animals, or microbes, each offering unique properties. Another classification approach is based on their monomeric units; for instance, chitin and cellulose consist of amino sugars and glucose, respectively, while nucleic acid and proteins are made up of nucleotides and amino acids. Degradability is a significant classification criterion, with some WD-BPs, such as cellulose and starch, being biologically degradable and thus environmentally friendly. In contrast, synthetic polymers are non-decomposable and can endure in the environment for extended periods. Heat responsiveness is also a crucial factor in biopolymer classification. For example, polylactic acid (PLA) and thermoplastic starch can be cast into different structures when heated, whereas cellulose acetate is heat-resistant and suitable for high-temperature applications (George et al., 2020). There are multiple categories being used to categorize BPs, and every category is further split based on special properties. Defining these groups decides the application and development of these bioderived polymers, ultimately enhancing product sustainability (Baranwal et al., 2022).

Globally, a signification amount of food being consumed in a year which is directly or indirectly sourced from plants. After harvesting fruits and vegetables from the plants huge amounts of organic waste is left behind, which is used as a burning fuel or animal feed. On the other hand, the waste derived after harvesting of crops including husk, bagasse dust, etc., are the nonedible parts of the fruits and enhance plastic and environmental pollution (Lin and Begho, 2022). These materials can serve as base materials for producing WD-BPs, with starch and cellulose being examples of WD-BPs that can be derived from various biological or agricultural sources. These WD-BPs provide excellent abilities that can be compared to those of conventional synthetic polymers, which contribute to ecosystem destruction and environmental crises (George et al., 2020).

3.1 Polylactic acid polymer

Polylactic acid (PLA), a WD-BPs, is manufactured through fermentation, involving the polymerization of lactic acid monomers in a controlled manner sourced from renewable materials like maize, sugarcane, corn, wheat, and by-products from the agricultural or food industries. In plants and animals, a common metabolic byproduct product is lactic acid (Sabee et al., 2021). Similarly, lactic acid and PLA's hydrolytic byproducts are non-toxic. Compared to several synthetic polymers, PLA production consumes lesser energy and emits fewer greenhouse gases. Therefore, in comparison to synthetic polymers, PLA can be synthesized on a large scale and can replace synthetic polymers in multiple applications. PLA is classified as a thermoplastic because of its reaction to heat, unlike thermosets. Thermoplastics can be heated to their melting point, cooled, and reheated without significant degradation, making them highly valuable. When heated, thermoplastics, including PLA, liquefy instead of burning, enabling them to be easily injection molded and recycled. During crystallization, PLA forms three structural forms: a, b, and g. The cold crystallization forms a-PLA, however, b form is synthesized by deformation and mechanical effect. PLA exhibits an excellent crystallization rate due to rapidly growing semi-crystalline regions. The degree of crystallinity influences various polymer properties, including strength, melting temperature, stiffness, hardness, etc., as noted by Lim et al. (Lim et al., 2008).

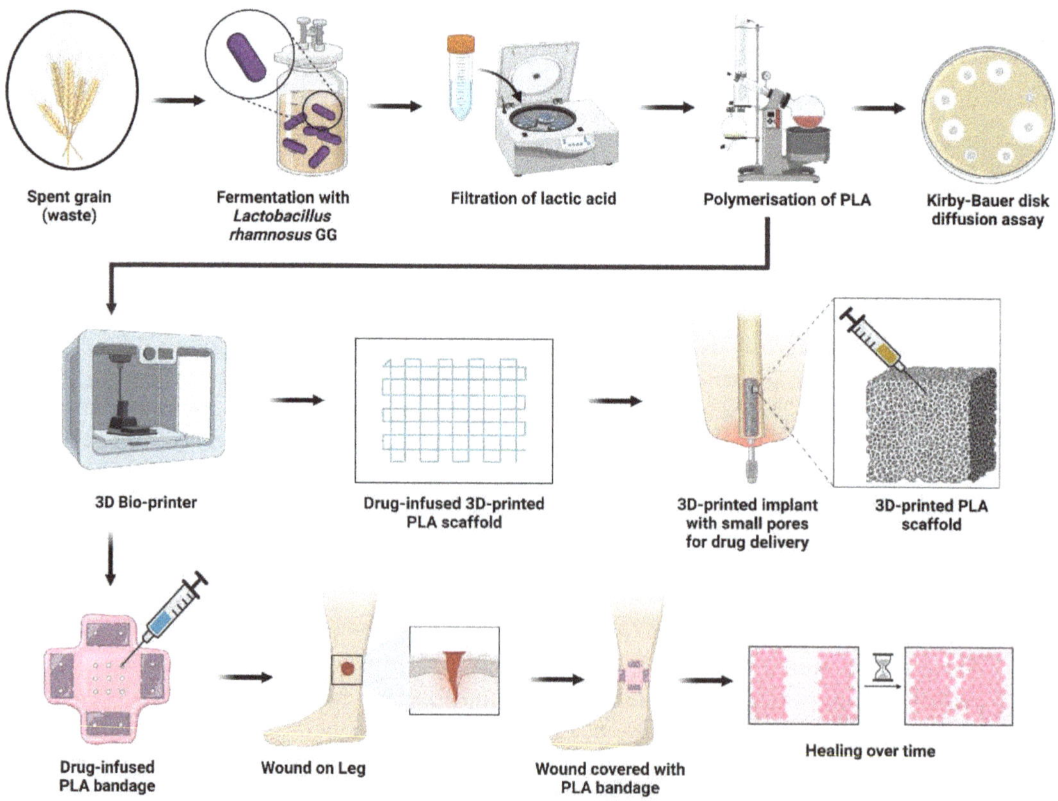

Figure 3. A graphical representation of the synthesis of WD-BPs, PLA and their biomedical applications. The image was reproduced with the permission (Vakati et al., 2024).

PLA is renowned for its various advantages, such as biocompatibility, good mechanical strength, and cost-efficiency which also exhibit exceptional ultraviolet light barrier properties. Although, these films are brittle, fragile, and have low toughness (Tang et al., 2020). These characteristics make PLA unsuitable for high-stress applications, and its inertness poses challenges for surface modifications for different purposes. PLA also faces thermal stability issues, as it can undergo molecular weight loss during high-thermal treatments due to ester linkage degradation. This degradation can occur even at temperatures lower than PLA's melting point, and it accelerates significantly after melting. To mitigate this issue, different techniques are employed, including hydrolysis, esterification, hydrolysis, oxygen-based degradation, and depolymerization. The extent of thermal degradation depends on factors such as moisture levels, lactic acid concentration, and particle size (Li et al., 2020). Previously, the primary application of PLA was in medical fields including tissue scaffolds, implant devices, and internal sutures because cost inefficiency, less molecular weight, and limited availability. However, in recent years, new methods have been developed to produce PLA with high molecular weight, cost-effectively, expanding its potential applications (Sreekumar et al., 2021). Additionally, composite of PLA is made by blending with other materials for improvement in properties. Some of the key materials blended with PLA include chitosan, cellulose, starch, and hydroxyapatite.

3.2 Polyhydroxyalkanoates

Polyhydroxyalkanoates (PHAs) are biological degradable polyesters that are derived naturally and are synthesized through the fermentation process using bacterial cells (Vigneswari et al., 2021). Bacteria generate PHAs for their survival through the accumulation of PHAs granules via imbalanced growth during bacterial fermentation. There are PHAs composed of three to five monomers, known as short-chain PHAs, which include polyhydroxyvalerate (PHV), polyhydroxybutyrate (PHB), and their copolymers. Among these, PHB has been largely studied and used polymer because of their extensive mechanical strength. PHAs with medium length of chains are produced using 6 to 14 carbon atoms monomers containing. Examples of these include polyhydroxyoctanoate (PHO), polyhydroxydecanoate (PHD), and polyhydroxylhexanoate (PHHx) (McAdam et al., 2020). Pryadko et al. and Guzik et al. have reported that PHAs can incorporate over 150 monomers to create materials with a wide range of characteristics and applications. The properties of PHAs, both mechanically and biologically, can be altered by modification on the surface through blending, or combination with other polymers, different enzymes, or inorganic materials, thus expanding its applications (Pryadko et al., 2021; Guzik et al., 2020). Furthermore, the characteristics of PHAs can be improved by creating nanocomposite of nanofibers of nanofiber with PHAs, leading to the formation of multiphase materials, via modifications with chemicals. Jo et al. studied the PHA surface modification for sticky applications, focusing on binary salinization of cellulose nanofibers, which improved the strongness flowing, and dispersion of the bio composite (Jo et al., 2022). In another study, Cataldi et al. made thermally stable PHA biopolymer doped with gold nanoparticles having an electromagnetic shielding effect and thermal conductivity which was utilized for electric applications (Cataldi et al., 2020). Figure 4 A schematic illustration of the WD-BPs, PHA. PHAs are primarily used in parceling, including boxes and layers, as well as in finishes, medications, and pharma applications.

3.3 Starch

Starch, a polysaccharide which is abundantly present in biomass of plants. It comprises two forms: linear polymer which is called amylose, and a branched polymer known as mylopectin. Starch has been used for various purposes for centuries, with paper manufacturing being a prominent application. Moreover, the textile industry employs substantial quantities of starch to enhance thread strength during weaving (Gamage et al., 2022). Starch, sourced mainly from cereals, tubers, and legumes, is primarily derived from corn, potatoes, rice, and wheat. Having granular morphology.

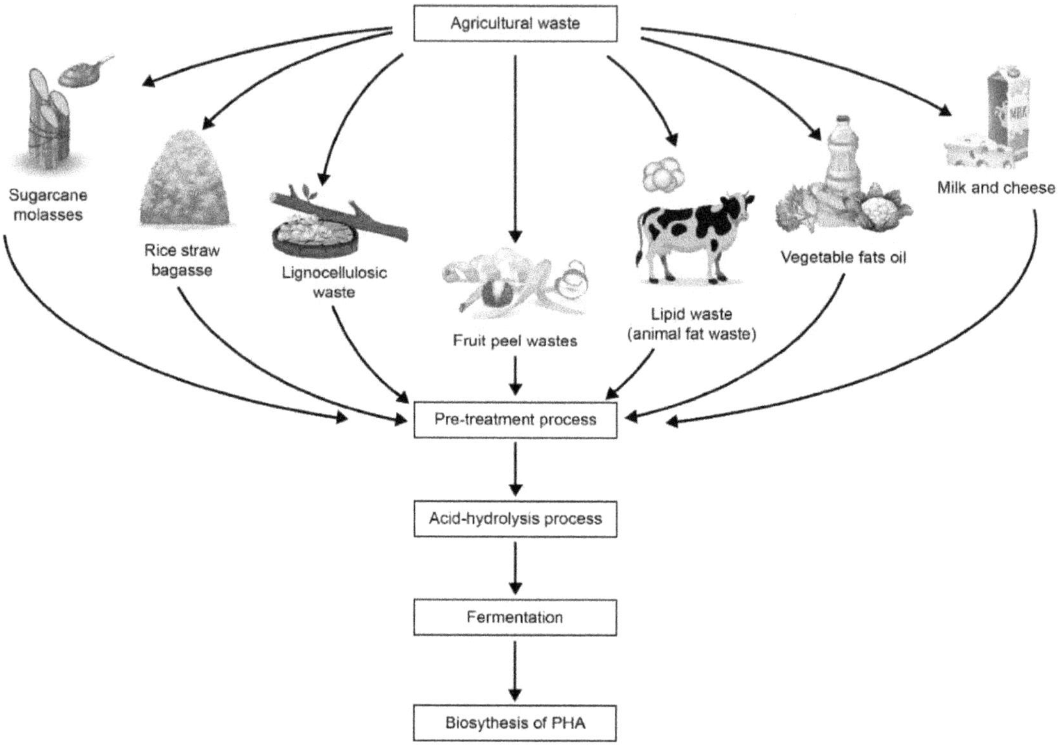

Figure 4. A schematic illustration of the WD-BPs, PHA. The image was reproduced with the permission (Vigneswari et al., 2021).

Due to presence of hydrogen bonding, molecules of starch show cyclic structure leading to provide ordered crystalline regions with a rigidity in structure (Gadhave et al., 2018). This characteristic contributes to thermal stability and enhanced glass transition temperature for starch, also enhanced solubility in alcohol and cold water. However, its solubility is limited to hot water only. In the breakdown mechanism, starch starts to lose its semi crystalline structure on contact with water, leading to release the two components amylopectin and amyloses which further forms a network with high water retaining capacity. This complete process, known as starch gelatinization, in which thermoplastic starch (TPS) is a thermoplastic substance is formed by transformation of starch. Many biocomposites can be formed by blending the TPS with other molecules such as fillers or reinforcements (Balart et al., 2021). Grylewicz et al. studied the mechanical characteristics of composites of TPS though wood fiber and potato starch, and determined the effect of eutectic solvents on the composites mechanical strength (Grylewicz et al., 2019). A correlation was determined in the characteristics of the composites and deep eutectic solvents. Osman et al., loaded dolomite into starch using solvent casting method at different concentrations. They concluded that the tensile strength of biocompositiess enhanced due starch matrix loaded with filler loading leading to reduce the size of particles and enhanced distribution (Osman et al., 2021).

Starch granules can be used like a filler to improve the biodegradation of common polymers like polyethylene, polypropylene, and polystyrene. To enhance alignment with polyolefins, granular starch is still treated with chemicals for its surface modification and enhancing hydrophobicity. Moreover, starch serves as an appropriate filler material for synthesizing ecofriendly biocomposites due to its non-impactable melt flow properties and thermal stability in comparison with synthetic plastics. The properties of starch biocomposites are significantly affected by the shape, quantity, and dimension of the filler, as well as its alignment with the matrix of polymers (Siraj et al., 2022). In a study by Piyada et al., the strength of films of rice starch was enhanced by starch nanocrystals which were produced through the acid hydrolysis process. The addition of starch in nano form enhanced the film's crystallinity, although excessive amounts decreased it. Increasing the starch nanocrystal content also improved the films' water obstacle properties and temperature stability (Piyada et al., 2013). Similarly, Dai et al. used nanoparticles of taro starch and gelatinized starch for the corn starch film reinforcements, leading to enhanced film thermal stability (Dai et al., 2015).

3.4 Cellulose

Cellulose, the largely abundant and virtually unlimited naturally originated polymer globally, is the structural component of cell walls of plants at the primary level, in photosynthetic organisms constituting approximately partial biomass. Green plants transform CO_2 and H_2O into glucose via photosynthesis. Cellulose's distinct structure, which significantly influences its chemical reactions, makes it a versatile raw material for various chemical conversions. A range of functional and sustainable BPs, both, can be derived from cellulose monomers (Moohan et al., 2019). Consequently, cellulosic polymers serve as substitutes to petroleum polymers in numerous treatments. Cellulosic structures including nanocellulose, natural fiber, and cellulose derivatives provide a wide range of favorable results for biocomposite development (Omar and Jain, 2022). Utilizing matrices of cellulose biopolymeric offers materials with multifunctionality with various applications on advanced biotechnology. The properties of composites can be enhanced by using these materials as fillers. Natural cellulose fibers are increasingly used as alternatives to talc and glass fibers for reinforcing biodegradable polymer composites (Shaghaleh et al., 2018). Xie et al. studied the effects of mixing cellulosic fibers of bamboo to composites of cement, finding that the optimal fiber content significantly enhances fracture strength and influences energy absorption in the cement (Xie et al., 2019). Sumesh et al. used peanut oil cake for extracting cellulose and used as a filler, increasing the mechanical and thermal properties in natural fiber composites and getting positive results in mechanical property assessments (Sumesh et al., 2020). However, SScellulosic fiber shows less strength in comparison to carbon and glass fibers to glass fiber and carbon fiber composites due to nonbonding between polymer matrix which is hydrophobic in nature and natural polymers which are hydrophilic leading to reduced thermal stability. Moreover, the mechanical characteristics of cellulose fibers tend to weaken when they absorb moisture. Therefore, modification on the surface of cellulose fibers is necessary either by chemical (chemical grafting) or by physical (physicochemical treatment methods. Liu et al., suggested that chemical treatment can eliminate non-cellulose parts from fibers of organic material, enhancing their mechanical and thermal properties (Liu et al., 2019).

4. Synthesis of BPs from Various Waste Sources

BPs which have been artificially altered through biotic chemical processes are known as synthetic BPs. Various methods and techniques can be used to synthesize BPs, with some typical techniques aimed at achieving well-defined structural and mechanical properties are as follows: Esterification, dehydration, polycondensation, hydrolysis, and granulation of BPs.

4.1 Esterification of BPs

Esterification of BPs involves a chemical process which occurs during the synthesis of esters. This reaction combines alcohol (ROH) with an organic acid (RCOOH) to produce an ester (RCOOR) and water. Specifically, when a carboxylic acid reacts with an alcohol, one ester forms. The esterification of BPs such as hemicelluloses, starch, and cellulose are as follows:

4.1.1 Cellulose esterification

Cellulose esterification can occur either throughout the polymeric chain or on the outer threads of cellulose to create cellulose esters. When large quantities of cellulose are produced using a heterogeneous and homogeneous process, modification of the surface is done. For the inorganic easter of cellulose, the most commonly used compound is Cellulose Nitrated (CN)s, therefore it is produced in a large quantity. The frequent applications of cellulose nitrate are for coating purposes propellants, plastic industries, and explosives. In the heterogeneous type equilibrium, cellulose nitrate is esterified by cellulose along with a mixture of sulfuric acid and nitric acid, resulting 1.8–2.8 degree of substitution. CS esterification is a process in which sulfuric acid and cellulose react through direct heterogeneous reaction, where synthetic CS is produced by displacing an ether or ester group in cellulose (Zwawi, 2021).

4.1.2 Starch esterification

Acetic acid and acetic anhydride are utilized as esterification agents to produce starch acetate. The addition of acetyl groups to starch molecules is influenced by factors such as pH, time duration, and the presence of a catalyst. Starch acetate can have low (0.01–0.2), medium (0.2–1.5), or high (1.5–3) Degrees of Substitution (DS), indicating the number of acetyl groups incorporated. Alkenyl succinic anhydride DDSA is a distinct type of alkenyl succinic anhydride. Esterifying DDSA starch in an aqueous solution result in a water-resistant product widely used in the film and paper industries (Thomas et al., 2021).

4.1.3 Hemicellulose esterification

Hemicellulose is a plant cell and cell wall polymer. It is utilized to produce films and hydrogels and finds applications in fields of biomedical sciences such as the delivery of drugs and the engineering of tissues. Synthesis of hemicellulose by the chemical method results in xylitol furfural, lactic acid, and ethanol. Various biomaterials, including lauroylated hemicelluloses, cationic hemicelluloses, cylated hemicelluloses, and carboxymethyl hemicelluloses, provide improved cost effectiveness, less hydrophobicity, higher stability on electrochemical and thermal properties, and quicker reaction rates when hemicellulose is homogeneously upgraded. Esterification of maleic anhydride with hemicelluloses rich in xylan along with LiOH catalyst in an ionic aqueous medium under conditions of homogeneity produces a product with a degree of substitution (DS) ranging from 0.095 to 0.75. This product has a wide range of uses in the treatment of wastewater, crop production drugs and food industries (Visco et al., 2022).

4.2 Dehydration of BPs

Trehalose is known for its outstanding ability to protect fluctuating systems from dehydration. Cyclodextrins (CDs) was utilized for dehydration to enhance the mechanical strength of proteins through dehydration. Chemical bonding between enzymes and derivatives of CD in a liquid environment result in improved thermal stability of enzymes due to their superior supramolecular interaction on the enzyme surface. Dehydration of BPs effectively protects biomolecules. Polysaccharide BPs exhibit glass transition characteristics, which significantly affect the thermal properties of naturally dehydrated BPs. The speed of dehydration is reduced by free water evaporation. Diffusion or water takes place in either the aqueous or vapor phase, or both phases at the time of dehydration of the biopolymer. The film's contraction during dehydration alters the polymer film's structure. The kinetics of dehydration of various BPs have been utilized in applications such as adhesives, coatings, the food industry, and pharmaceutical drug delivery (George et al., 2020).

4.3 BPs Polycondensation

Polycondensation is a process of polymerization where multiple monomers are combined to form polymers, often escorted by the proclamation of various byproducts having low molecular weight like salt, alcohol, and water. This synthesis method provides a versatile approach for modifying BPs. Phosphorus-containing polymers can be produced through phosphoric dihalides, phosphoric acid, or H-phosphonic acid diesters or diols polycondensation. For example, melamine/formaldehyde polycondensation in acidic conditions, creates a gel with a high spherical cavity. For the synthesis of fiber resin and flexible foam this process is efficient. Additionally, through the polycondensation process the microspheres of formaldehyde or monodisperse melamine can also be synthesized. In Knoevenagel polycondensation synthesis, backbone double bond substitution occurs, leading to modified PPV containing cyano groups that exhibit higher electropolymerizing activity (Baranwal et al., 2022).

4.4 BPs Hydrolysis

The toxic less nature of synthetic BPs is attributed to their ability to break down without forming hydrocarbons or carbon dioxide, instead producing water. However, the final breakdown product, carbon dioxide, can be harmful to both human health and the environment. Azotobacter plays a crucial character in the formation of Poly-3-hydroxybutyrate (PHB), a synthetic biopolymer via hydrolytic process modification. The hydrolysis of PHB synthetic biopolymer, results in film formation at 70°C in the presence of a phosphate buffer, is influenced by factors such as temperature of the reaction, medium of the incubation, biopolymer's chemical composition, and molecular weight. At the time of the hydrolytic synthesis the hydrolysis of PHB occurs mainly at the surface area only. Cellulose, a biopolymer, having robust intermolecular arrangement in its structural backbone, leads to high strength and resistance for the process of hydrolysis. In contrast, hemicellulose is amorphous and random, with small cohesion (DeStefano et al., 2020). Thermal cleavage of the cellulose biopolymeric chain is initiated through dehydration followed by a hydrolysis reaction under critical water conditions. Cellulose hydrolysis occurs at temperatures ranges in between 200 – 350°C, with or without a catalyst, in a nitrogen atmosphere, leading to bond breakage. Lignin and hemicellulose can also be produced through hydrolysis, involving the breakdown of bonds in the presence of critical water. Enzymatic hydrolysis of cellulose biopolymer in an aqueous solution produces xylitol and furfural. Additionally, hydrolysis plays a crucial role in lignin degradation, as the high ion product of water promotes the reaction medium.

4.5 Granulation of BPs

The method for producing granules of BPs has become hypothetical and unreliable. Biological granulation of BPs is seen as a promising approach, offering advantages such as greater tolerance to toxicity, treatment of pollutants with high carbon/nutrient loads, improved settleability, and increased biomass retention, among others. Granulation results in the formation of Extracellular Polymeric Substances (EPS), which consist of nucleic acids, humic acids, polysaccharide, lipids, and proteins that help in binding of microorganisms collected. For completion of biopolymer granulation, it requires a lengthy period, approximately 308 days, that progresses through two different phases which are phase 1 and phase 2. Phase 1 process occurs from day 0 to approximately 182 – 220 days and represents the initial stages of low granular formation, while phase 2 occurs after 182 – 220 days and represents the development of a stable granular structure. Granulation under specific microbial compositions enhances the stability of Aerobic Granular Sludge (AGS), facilitating further qualitative research for industrial applications (Udoetok et al., 2024; Šuput et al., 2023). Table 1 describes different methods for the synthesis of BPs.

5. Application of WD-BPs

As various industries such as tanneries, textiles and power plants produce pollutants which are harmful to the environmental, for examples multiple dyes which are hazardous to the environment, effluents to toxic chemicals, heavy metals, pharmacological products, and some others, the amount of these pollutants has increased in the past few years. These pollutants might have serious adverse effects on the health of humans as well as the environment. Dyes in water are affected by the photosynthetic activity of plants as well is it affects human health in the context of neurological, reproductive, and other organ system (Velusamy et al., 2021). It can also cause rhinitis, cancer, occupational asthma, dermatitis, and allergic conjunctivitis. This has led to a growing requirement

Table 1. Various methods of synthesis of BPs.

S. No.	Raw material	Synthesis method	Properties	Applications	References
1	Starch, eggshell, chitosan	Thermal treatment and dry	Tensile strength increased and water absorption decreased due to eggshell	Synthesis of Bioplastic	Kasmuri and Zait, 2018
2	PLA, starch, and cellulose	Casting then injection	Enhanced tensile strength	Used in coating, medical devices, and electronic devices	Sidek et al., 2019
3	Flex fibers and cellulose acetate	Acetylation then cooling	Enhanced tensile strength	Synthesis of Bioplastic	Chalermthai et al., 2020
4	PLA, starch, and cellulose	Extraction then heat-cool	Enhanced tensile strength	Synthesis of Bioplastic	Yaradoddi et al., 2016
5	Polymers and wastewater sludge	Microbial degradation	Early and easily degradable	Applicable in textile and packaging industry	Liu et al., 2019
6	Waste paper, lactic acid, and cellulose	Thermal alkaline leaching	Improved tensile strength	Synthesis of Bioplastic with reduced carbon emission	Joshi et al., 2016
7	Chitin, food waste and PHA	Valorization	Increases flexibility and tensile strength	Used in chemical media, spray industry and bioplastic synthesis	Tsang et al., 2019
8	Castor oil, pumpkin starch, and chitosan	Thermal treatment then dried	Increases young modules, tensile strength, and biodegradability	Synthesis of bioplastic	Hasan et al., 2018
9	Acetic acid and peels of banana	Thermal treatment then dried	Improved flexibility	Synthesis of bioplastic	Ghamande et al., 2018
10	Chitosan and starch	Thermal treatment then dried	Improved tensile strength	Synthesis of household and bioplastic	Cacique et al., 2017

for efficient removal of such contaminants, prompting researchers to explore the use of biomaterials as potential adsorbents to reduce pollutant levels in the environment (Zhou et al., 2019).

5.1 Heavy Metal Removal

The previous studies suggested that food waste derived biomaterials are the efficient alternatives for the other traditional adsorbents because of their efficient adsorption capacity, affinity towards the contaminants and functionalized surface enriched with -COOH, $-NH_2$ and -OH functional groups. These functional groups have selective affinity toward metal ions. Several studies have indicated that biomaterials extracted from food wastes are effective alternatives to traditional adsorbents due to their higher affinity, capacity, and presence of functional groups like -OH, -COOH, and -NH2, which are selective towards metal ion contaminants. These biomaterials can reduce or detoxify the concentration of ionic contamination of heavy metal in contaminated water. For example, bar-like hydroxyapatite (HAP) with a good surface area which is 159.1 m^2/g and excellent pore volume, which is 1.0 cm^3/g was synthesized from shells of eggs using a pump of syringe via a neutralization reaction for adsorption of liquid Pb (II) ions, following the Langmuir reaction model of a pseudo-second order. Similarly, shells of eggs were also used for the synthesis of HAP for the removal of lead and nickel ions from aqueous solutions (Pang et al., 2021). Chitin extracted from crustacean shells was used by as an absorbent for removing Cu^{2+}, Pb^{2+}, Zn^{2+}, and Cd^{2+} ions from the effluent of an industry. Zubair et al. extracted keratin biomaterial from chicken feathers and synthesized keratin-graphene oxide by crosslinking keratin with graphene oxide to make a biosorbent for removing metal oxyanions such as Se, As, Cr and other cationic metals such as Pb, Ni, Zn and Co from the stock solution in water up to a level of 99%. In another study by Biswas et al., used chitosan derived from the shells of prawns and clay enriched with kaolinite for synthesis of bio-composite beads. These beads were used for removing cationic hexavalent chromium, methylene blue, and lead, heavy metals from the effluents of textile and tanneries (Biswas et al., 2020). Orange waste-derived pectin and cellulose microfibers were utilized for the formation of a composite adsorbent for multiple heavy metal removal application including cadmium, copper, and ion in wastewater (Lessa et al., 2020). Similarly, Corncob cellulose peel extracted pectin showed higher affinity towards lead ions and removed by physiochemical adsorption in wastewater. Cellulose extracted from corn cob waste was utilized as an adsorbents for the removal of Cu and Cd (Isa et al., 2020). Nanocrystals of cellulose were synthesized using waste cellulosic biomaterial showed efficient binding affinity to cadmium and led ions in liquid solutions at the time of incorporation of moieties of succinic acid onto nanocrystals. Nanocomposites of pectin or graphene oxide were utilized as an efficient material for trivalent chromium ions and later as an efficient photocatalyst for organic dyes degrading (such as methyl orange and methylene blue) in wastewater (Kaushal et al., 2020). HAP Nanocrystals were synthesized from a scale of a fish known as *Tilapia nilotica* and used for adrotpive removal of selenium in a liquid medium, showing higher adsorption capacity in comparison to the commercially available chitosan, HAP, and fish scale (Kongsri et al., 2013). Mercury adsorption in Gorontalo Province's Bone River was determined by chitosan synthesized from shrimp shells (Lukum et al., 2020) However, chitosan synthesized from shell waste of *F. indicus* efficiently adsorbed approximately 50% of the cadmium Cd from Vembanad Lake. Moreover, chitosan-supported biomaterial synthesized from chitin that was produced from the shells of Moroccan shrimp, can effectively adsorb arsenic (V) from the wastewater sample prepared artificially with a maximum adsorption capacity of 12.32 mg/g of the adsorbent (Billah et al., 2022).

5.2 Organic Dye Removal

Adeogun et al., produced hydroxyapatite from waste eggshells of poultry using a wet reaction of chemical precipitation. They employed this material as an adsorbent to remove Reactive Yellow 4 (RY4) dye from water. The adsorbent, which contains anionic functional groups like carbonate, phosphate, and hydroxyl, enhances the process of adsorption (Adeogun et al., 2018). In a study, Swamiappan et al., synthesized HAP in nano form from biowastes of fish scales of *Labeo rohita* using a method of heat treatment in alkaline environment. They used the produced HAP powder in nano form to remove cationic methylene blue dye from water, achieving an adsorption efficiency of 66.67% (Swamiappan et al., 2021). Chicken eggshells derived HAP was mixed with three different clays – namely, kaolin, bentonite, and halloysite – and the produced composite of HAP-clay showed higher methylene blue adsorption in wastewater (Annan et al., 2021). Chitin, a biopolymer derived naturally from the scales of *Labeo rohita* fish, was utilized in methylene blue removal because of the amino groups present in it (Das et al., 2018). A nanocomposite of magnetic collagen, synthesized using collagen powder of fish scale and magnetic iron oxide nanoparticles, effectively removed 1000 ppm crystal violet. In a study, chitosan was extracted from prawn shell waste for bioadsorption of acid black azo dye via decolorization (Ravi et al., 2015). Chitosan from Fenneropenaeus indicus (shrimp) shell waste removed approximately 93.23% of methylene blue dye through bioadsorption (Raiyaan et al., 2021). Keratin synthesized from chicken feather-derived, incorporated with cellulose nanocrystals, exhibited excellent biosorption performance in removing Reactive Black 5 and Direct Red 80 dyes (Song et al., 2017). Keratin molecules, with functional groups of amino and carboxyl can be used as adsorption of organic dyes (Xu and Yang, 2014).

Waste of kitchen like residue of tea, peels of onion was utilized to produce lignin and cellulose which were employed as adsorber for crystal violet dye adsorption (Ganguly et al., 2021). Another research used cellulose extracted from discarded orange, sweet lime, rinds of banana, and pomegranate processed by hydrolysis in an alkaline environment, bleaching, and treatment in acid, which was then mixed with bismuth vanadate to create a composite in nano form for methyl orange degrading in visible light (Tavker et al., 2020). Pomelo pith extracted pectin was used for adsorption of a toxic dye named Acid Blue 25 (AB25) in wastewater. The physisorption occurs due to presence of carboxylate, ester and hydroxylate functional groups in pectin's GalA unit, which directly interacts with AB255 (Shahrin et al., 2021). Hastuti et al. developed a microporous membrane of pectin for adsorptive removal of methylene blue dyes, showed 19.39 mg/g adsorption capacity of (Hastuti et al., 2020). Trivalent chromium was effectively adsorbed on a nanocomposite of graphene oxide and pectin, later a similar material was used for degradation of organic dyes such as methyl orange and methylene blue through photocatalytic process (Kaushal et al., 2020). Joshi et al., synthesized a composite of graphene oxide and cellulose with a high porosity. Cellulose was extracted from the peel of pomelo fruit. The material was used for adsorption of multiple dyes including rhodamine 6G, methylene blue, malachite green methyl orange, and rose Bengal (Joshi et al., 2022). Lignin precured from rice straw using a method of organosolv with N, N-dimethylformamide, water, and aniline was determined as a promising biowaste derived material for adsorption of methylene blue in wastewater showed 20.6 mg/g adsorption capacity (/Zhang et al., 2016).

5.3 Other Type of Pollutants

Waste of eggshells was utilized to produce nanoparticles of HAP through the wet method of chemical precipitation using a phosphate solution. These nanoparticles were then used as an adsorbent to remove cephalexin antibiotic from aqueous solutions, achieving an adsorption efficiency of 70.70% (Alhasan et al., 2022). In another study, Inbaraj et al. synthesized a carbon nanocomposite from the waste of magnetite green tea for adsorbing hydrocarbons including Chrysene, Benzo[α] anthracene, and Benzo[α] pyrene in wastewater. In a study, Hamzah et al. extracted HAP from fish scales by multiple methods including irradiation by microwave, decomposition by heat, ant treatment in an alkaline environment. They then mixed rice husk with the synthesized HAP for developing an adsorber to remove ammonia contamination wastewater of aquaculture. The authors showed 79% ammonia removal using HAP produced by thermal degradation process (Hamzah et al., 2019). Table 2 describes the degradation of different pollutants using BPs.

Table 2. Degradation of various pollutants using different BPs.

S. No.	Name of Biopolymer	Name of pollutant	Degradation (%)/adsorption efficiency (mg g^{-1})	References
1	Chitosan film	Black dye (T)	413 mg g^{-1}	Boudouaia et al., 2019
2	Hydroxyapatite	Fluoride removal	81%	Workeneh et al., 2019
3	Hydrogels	Methylene blue	92.7%	Liu et al., 2017
4	Powder of HAP	Cadmium	22.94 mg g^{-1}	Foroutan et al., 2021
5	HAP	Congo red	98%	Das et al., 2019
6	MOF integrated pectin hydrogel	Hexavalent chromium	99.87%	Mahmoud and Mohamed, 2020
7	Cellulose hydrogel composite	Methylene blue	328.36 mg g^{-1}	Lei et al., 2022
8	HAP	Hexavalent chromium	61.4%	Selimin et al., 2022
9	Fibers of cellulose with modification	Lead	41.2 mg g^{-1}	Daochalermwong et al., 2020

6. Challenges in WD-BPs

While biomaterials derived from food wastes provide numerous applications, including bioremediation and solving the issue of bio-waste disposal, challenges exist in their conversion processes and application for pollutant remediation. Ensuring a continuous and sufficient source of biowaste for biomaterial withdrawal at a commercial scale is a concern, as particular food waste sources may not always be readily available in large quantities. Therefore, a combination of various food wastes could be necessary. Additionally, most biowaste is lignocellulosic, which poses challenges in effective prior treatment or pre-processing. The chemical composition of biological wastes, such as waste from the kitchen, vegetable waste from multiple vegetables or fruits, and the agricultural industry, is complex and miscellaneous, influenced by factors like climate, demographics, and species diversity. This complexity makes designing effective extraction methods difficult and often requires waste segregation, leading to increased processing costs. Furthermore, the assortment, shipping, arrangement, wiping, and packing of food waste can be costly for production at a large scale. High water content in some waste food, around 90%, requires sophisticated handling and preservation techniques to prevent biological degradation. The complexity of waste food and its fast biological degradation further complicates large-scale handling (Peinemann and Pleissner, 2020; Haldar et al., 2022).

Various methods of extraction, such as extraction through soxhlet, hydro distillation, hydrolysis using acid, and extraction of solid and liquid, were employed to get biomaterials from the wastage of food. However, these techniques often involve excessive solvent use, leading to degradation through heat and loss of components that are sensitive to heat, requirement of energy at higher levels, and high time for extraction (Dhua et al., 2022). While waste of food in a dried form is suitable for conversion through the thermochemical method, dehydrating of high-water-containing waste biomass before valorization needs energy in a significant manner (Haldar et al., 2022). Additionally, these methods of extraction typically use solvents that are toxic, making it a hazardous process. Hence, there is a need for a more environmentally friendly solvent system for extraction. Green extraction methods, such as enzymatic extraction and the use of deep eutectic solvents, either individually or combined with microwave or ultrasonication extraction, could provide a solution (Azmir et al., 2013). However, these methods are still in the early stages of study. Understanding the mechanisms of these methods is crucial to facilitate the transition of waste from food, valorization techniques from the laboratory to the large industrial scale. Yet, due to the complex characteristics of food waste, accurately determining the paths and processes of these conversion methods remains challenging (Cheng et al., 2020). Creating customized bio-based materials using food waste by activation using physical or a chemical method creates a challenge of enhanced cost leading to an increase in materials cost. Moreover, the uncertain or difficult chemistry of these bio-based materials impedes the regeneration of spent sorbents after the removal of pollutants (Chaukura et al., 2016). Most studies on biomaterials-based adsorption have been based on experiments at the laboratory scale in batch mode, which are not readily applicable to real environmental parameters. Therefore, it is essential to conduct studies using actual wastewater samples to assess the effectiveness of pollutant remediation using biomaterials derived from food waste.

7. Future Scope

This chapter explores the potential applications of biopolymeric materials in degradation of environmental pollutants and bioengineering, focusing on their physical, mechanical, degradability, and biocompatibility properties. It discusses promising technologies for synthesizing smart BPs and their various eco-friendly uses in remediation. Future challenges for humanity include energy and resource management, food security, healthcare, transportation, infrastructure, and communication. BPs are expected to play a crucial role in addressing these challenges. They are poised to become the dominant materials of the next century, providing global access to innovative solutions. Synthetic BPs have already made significant contributions to adsorbents as catalytic material development. Future applications may also benefit from BPs, which can help conserve energy and promote the use of renewable resources. As the cost of manufacturing bioplastics decreases and new solid adaptations are developed, the value of BPs is expected to rise. Bioplastics can replace traditional plastics in multiple applications, such as packaging of food, utensils, and storage containers, thereby reducing an environmental impact. In the electrical industry, plastics are increasingly being used for insulation, plugs, sockets, and housing electrical equipment. Research on the development and modification in BPs and filler nanoparticles could replace the use of synthetic material for the removal of contaminants at the industrial level or large scale. Apart from pollutant degradation, advances in bioengineering, biochemistry, and molecular biology have led to the development of innovative cellulose-based products for biomedical applications. Cellulose-based materials exhibit good physical and biological properties, making them suitable for engineering tissues, wound dressings, drug delivery, and medical implants.

8. Conclusion

In recent years, there has been a growing interest in the industrial application of BPs and eco-friendly products derived from them. BPs offer several advantages such as renewability, non-toxicity, environmental friendliness, biodegradability, lightweight, ease of manufacturing, and high structural performance, enhancing mechanical strength and stiffness. The properties of hybrid biopolymer-based materials are influenced by various factors including their composition of chemicals, environmental conditions, temperature, size of particles, and materials surface area. Extensive research has shown that BPs are suitable for use in pollutant degradation, dye removal, heavy metal removal, other remediation processes, drug delivery, agriculture, pharmaceuticals, biomedical applications, bioengineering, and the food industry, indicating their versatility. Further advancements in biopolymer-based products could lead to the replacement of traditional non-biodegradable materials. As these bioproducts become more durable and efficient, new markets may emerge. Ongoing research focusing on recital and life-cycle assessment is crucial to determine the full potential and utility of BPs.

References

Adeogun, A. I., Ofudje, A. E., Idowu, M. A. and Kareem, S. O. (2018). Facile development of nano-size calcium hydroxyapatite based ceramic from eggshells: synthesis and characterization. Waste and Biomass Valorization, 9: 1469–1473.

Alhasan, H. S., Alahmadi, N., Yasin, S. A., Khalaf, M. Y. and Ali, G. A. (2022). Low-cost and eco-friendly hydroxyapatite nanoparticles derived from eggshell waste for cephalexin removal. Separations, 9(1): 10.

Annan, E., Arkorful, G. K., Konadu, D. S., Asimeng, B., Dodoo-Arhin, D. and Egblewogbe, M. (2021). Synthesis and characterization of hydroxyapatite-(HAP-) clay composites and adsorption studies on methylene blue for water treatment. Journal of Chemistry, pp. 1–15.

Azmir, J., Zaidul, I. S. M., Rahman, M. M., Sharif, K. M., Mohamed, A., Sahena, F. et al. (2013). Techniques for extraction of bioactive compounds from plant materials: A review. Journal of Food Engineering, 117(4): 426–436.

Balart, R., Garcia-Garcia, D., Fombuena, V., Quiles-Carrillo, L. and Arrieta, M. P. (2021). Biopolymers from natural resources. Polymers, 13(15): 2532.

Baranwal, J., Barse, B., Fais, A., Delogu, G. L. and Kumar, A. (2022). Biopolymer: A sustainable material for food and medical applications. Polymers, 14(5): 983.

Billah, R. E. K., Islam, M. A., Lgaz, H., Lima, E. C., Abdellaoui, Y. et al. (2022). Shellfish waste-derived mesoporous chitosan for impressive removal of arsenic (V) from aqueous solutions: A combined experimental and computational approach. Arabian Journal of Chemistry, 15(10): 104123.

Biswas, S., Rashid, T. U., Debnath, T., Haque, P. and Rahman, M. M. (2020). Application of chitosan-clay biocomposite beads for removal of heavy metal and dye from industrial effluent. Journal of Composites Science, 4(1): 16.

Boudouaia, N., Bengharez, Z. and Jellali, S. (2019). Preparation and characterization of chitosan extracted from shrimp shells waste and chitosan film: Application for Eriochrome black T removal from aqueous solutions. Applied Water Science, 9(4): 91.

Cacique, P. P., Rios, M. N., Barbosa, I. O. and Wentz, A. P. (2017). Bioplastics production from starch and chitosan blends. Revista Eletrônica Perspectivas da Ciência e Tecnologia-ISSN: 1984-5693, 9: 46–53.

Cataldi, P., Steiner, P., Raine, T., Lin, K., Kocabas, C., Young, R. J. et al. (2020). Multifunctional biocomposites based on polyhydroxyalkanoate and graphene/carbon nanofiber hybrids for electrical and thermal applications. ACS Applied Polymer Materials, 2(8): 3525–3534.

Chalermthai, B., Ashraf, M.T., Bastidas-Oyanedel, J.R., Olsen, B.D., Schmidt, J.E. and Taher, H., 2020. Techno-economic assessment of whey protein-based plastic production from a co-polymerization process. Polymers, 12(4): 847.

Chaukura, N., Gwenzi, W., Tavengwa, N. and Manyuchi, M. M. (2016). Biosorbents for the removal of synthetic organics and emerging pollutants: opportunities and challenges for developing countries. Environmental Development, 19: 84–89.

Cheng, S. Y., Tan, X., Show, P. L., Rambabu, K., Banat, F., Veeramuthu, A. et al. (2020). Incorporating biowaste into circular bioeconomy: A critical review of current trend and scaling up feasibility. Environmental Technology & Innovation, 19: 101034.

Dai, L., Qiu, C., Xiong, L. and Sun, Q. (2015). Characterisation of corn starch-based films reinforced with taro starch nanoparticles. Food Chemistry, 174: 82–88.

Daochalermwong, A., Chanka, N., Songsrirote, K., Dittanet, P., Niamnuy, C. and Seubsai, A. (2020). Removal of heavy metal ions using modified celluloses prepared from pineapple leaf fiber. ACS omega, 5(10): 5285–5296.

Das, M. P., Renuka, M., Vijaylakshmi, J. V. and Suguna, P. R. (2018). Removal of methylene blue by adsorption using fish scale chitin. Nature Environment and Pollution Technology, 17(3): 993–998.

Das, M. P., Vijaylakshmi, J. V., Sugunaw, J. V., Renuka, M. and Prasad, K. (2019). Efficient dye decolorization of an azo dye on fish scale hydroxyapatite. Research Journal of Pharmacy and Technology, 12(6): 2917–2921.

DeStefano, V., Khan, S. and Tabada, A. (2020). Applications of PLA in modern medicine. Engineered Regeneration, 1: 76–87.

Dhua, S., Kumar, K., Sharanagat, V. S. and Nema, P. K. (2022). Bioactive compounds and its optimization from food waste: Review on novel extraction techniques. Nutrition & Food Science, 52(8): 1270–1288.

Edebali, S. (2021). Methods of engineering of biopolymers and biocomposites. In Advanced green materials (pp. 351–357). Woodhead Publishing.

Ferrari, F., Striani, R., Fico, D., Alam, M. M. and Greco, A. (2022). Esposito Corcione C. An Overview on Wood Waste Valorization as Biopolymers and Biocomposites: Definition, Classification, Production, Properties and Applications. Polymers, 14(24): 5519. https://doi.org/10.3390/polym14245519.

Ford, H. V., Jones, N. H., Davies, A. J., Godley, B. J., Jambeck, J. R., Napper, I. E. et al. (2022). The fundamental links between climate change and marine plastic pollution. Science of the Total Environment, 806: 150392.

Foroutan, R., Peighambardoust, S. J., Hosseini, S. S., Akbari, A. and Ramavandi, B. (2021). Hydroxyapatite biomaterial production from chicken (femur and beak) and fishbone waste through a chemical less method for Cd2+ removal from shipbuilding wastewater. Journal of Hazardous Materials, 413: 125428.

Gadhave, R. V., Das, A., Mahanwar, P. A. and Gadekar, P. T. (2018). Starch based bio-plastics: The future of sustainable packaging.

Gamage, A., Liyanapathiranage, A., Manamperi, A., Gunathilake, C., Mani, S., Merah, O. et al. (2022). Applications of starch biopolymers for a sustainable modern agriculture. Sustainability, 14(10): 6085.

Ganguly, P., Khan, A., Das, P. and Bhowal, A. (2021). Cellulose from lignocellulose kitchen waste and its application for energy and environment: bioethanol production and dye removal. Indian Chemical Engineer, 63(2): 161–171.

George, A., Sanjay, M. R., Srisuk, R., Parameswaranpillai, J. and Siengchin, S. (2020). A comprehensive review on chemical properties and applications of biopolymers and their composites. International Journal of Biological Macromolecules, 154: 329–338.

Ghamande, M., Kulkarni, A., Shah, N., Kothari, S. and Bhosale, S. (2018). Bio-plastic (generating plastic from banana peels). In International Conference on New Frontiers of Engineering, Management, Social Science and Humanities (pp. 39–42).

Grylewicz, A., Spychaj, T. and Zdanowicz, M. (2019). Thermoplastic starch/wood biocomposites processed with deep eutectic solvents. Composites Part A: Applied Science and Manufacturing, 121: 517–524.

Guzik, M., Witko, T., Steinbüchel, A., Wojnarowska, M., Sołtysik, M. and Wawak, S. (2020). What has been trending in the research of polyhydroxyalkanoates? A systematic review. Frontiers in Bioengineering and Biotechnology, 8: 959.

Haldar, D., Shabbirahmed, A. M., Singhania, R. R., Chen, C. W., Dong, C. D., Ponnusamy, V. K. et al. (2022). Understanding the management of household food waste and its engineering for sustainable valorization-A state-of-the-art review. Bioresource Technology, 358: 127390.

Hamzah, S., Yatim, N. I., Alias, M., Ali, A., Rasit, N. and Abuhabib, A. (2019). Extraction of hydroxyapatite from fish scales and its integration with rice husk for ammonia removal in aquaculture wastewater. Indonesian Journal of Chemistry, 19(4): 1019–1030.

Hasan, M., Rahmayani, R. F. I. and Munandar. (2018, March). Bioplastic from chitosan and yellow pumpkin starch with castor oil as plasticizer. In IOP Conference Series: Materials Science and Engineering (Vol. 333: 012087). IOP Publishing.

Hastuti, B., Afifah, S. N., Mulyani, B. and Susilowati, E. (2020), July. Adsorption of Methylene Blue dyes using pectin membrane. In Journal of Physics: Conference Series (Vol. 1503, No. 1: 012031). IOP Publishing.

Inbaraj, B. S., Sridhar, K. and Chen, B. H. (2021). Removal of polycyclic aromatic hydrocarbons from water by magnetic activated carbon nanocomposite from green tea waste. Journal of Hazardous Materials, 415: 125701.

Isa, I., Setiawati, E., Mohammad, E. and Kunusa, W. (2020, November). Utilization of corncob cellulose isolate (Zea mays) as adsorbent of heavy metal copper and cadmium. In IOP Conference Series: Earth and Environmental Science (Vol. 589, No. 1: 012035). IOP Publishing.

Jo, J., Jeong, S. Y., Lee, J., Park, C. and Koo, B. (2022). Green and sustainable hot melt adhesive (HMA) based on polyhydroxyalkanoate (PHA) and silanized cellulose nanofibers (SCNFs). Polymers, 14(23): 5284.

Joshi, P., Sharma, O. P., Ganguly, S. K., Srivastava, M. and Khatri, O. P. (2022). Fruit waste-derived cellulose and graphene-based aerogels: Plausible adsorption pathways for fast and efficient removal of organic dyes. Journal of Colloid and Interface Science, 608: 2870–2883.

Joshi, S. and Ujjawal Sharm, G. G. (2016). Bio-plastic from waste newspaper. Inter. J. Eng. Res. Technol., 2(3).

Kasmuri, N. and Zait, M. S. A. (2018). Enhancement of bio-plastic using eggshells and chitosan on potato starch based. Int. J. Eng. Technol, 7: 110–115.

Kaur, P., Sharma, N., Munagala, M., Rajkhowa, R., Aallardyce, B., Shastri, Y. et al. (2021). Nanocellulose: Resources, physio-chemical properties, current uses and future applications. Frontiers in Nanotechnology, 3: 747329.

Kaushal, S., Kaur, N., Kaur, M. and Singh, P. P. (2020). Dual-Responsive Pectin/Graphene Oxide (Pc/GO) nano-composite as an efficient adsorbent for Cr (III) ions and photocatalyst for degradation of organic dyes in waste water. Journal of Photochemistry and Photobiology A: Chemistry, 403: 112841.

Kaushal, S., Kaur, N., Kaur, M. and Singh, P. P. (2020). Dual-Responsive Pectin/Graphene Oxide (Pc/GO) nano-composite as an efficient adsorbent for Cr (III) ions and photocatalyst for degradation of organic dyes in waste water. Journal of Photochemistry and Photobiology A: Chemistry, 403: 112841.

Kongsri, S., Janpradit, K., Buapa, K., Techawongstien, S. and Chanthai, S. (2013). Nanocrystalline hydroxyapatite from fish scale waste: Preparation, characterization and application for selenium adsorption in aqueous solution. Chemical Engineering Journal, 215: 522–532.

Kumar, R., Verma, A., Shome, A., Sinha, R., Sinha, S., Jha, P. K. et al. (2021). Impacts of plastic pollution on ecosystem services, sustainable development goals, and need to focus on circular economy and policy interventions. Sustainability, 13(17): 9963.

Lei, C., Bian, Y., Zhi, F., Hou, X., Lv, C. and Hu, Q. (2022). Enhanced adsorption capacity of cellulose hydrogel based on corn stalk for pollutants removal and mechanism exploration. Journal of Cleaner Production, 375: 134130.

Lessa, E. F., Medina, A. L., Ribeiro, A. S. and Fajardo, A. R. (2020). Removal of multi-metals from water using reusable pectin/cellulose microfibers composite beads. Arabian Journal of Chemistry, 13(1): 709–720.

Li, G., Zhao, M., Xu, F., Yang, B., Li, X., Meng, X. e al. (2020). Synthesis and biological application of polylactic acid. Molecules, 25(21): 5023.

Lionetto, F. and Esposito Corcione, C. (2021). Recent applications of biopolymers derived from fish industry waste in food packaging. Polymers. 13(14): 2337. https://doi.org/10.3390/polym13142337.

Lim, L. T., Auras, R. and Rubino, M. (2008). Processing technologies for poly (lactic acid). Progress in Polymer Science, 33(8): 820–852.

Lin, M. and Begho, T. (2022). Crop residue burning in South Asia: A review of the scale, effect, and solutions with a focus on reducing reactive nitrogen losses. Journal of Environmental Management, 314: 115104.

Liu, F., Li, J. and Zhang, X. L. (2019, October). Bioplastic production from wastewater sludge and application. In IOP conference series: earth and environmental science (Vol. 344, No. 1: 012071). IOP Publishing.

Liu, Y., Xie, J., Wu, N., Ma, Y., Menon, C. and Tong, J. (2019). Characterization of natural cellulose fiber from corn stalk waste subjected to different surface treatments. Cellulose, 26: 4707–4719.

Liu, Z., Li, D., Dai, H. and Huang, H. (2017). Enhanced properties of tea residue cellulose hydrogels by addition of graphene oxide. Journal of Molecular Liquids, 244: 110–116.

Lukum, A., Paramata, Y., Botutihe, D. N., Akume, J., Sukamto, K. and Paramata, A. R. (2020, November). Development of Bioadsorbent Chitosan from Shrimp Shell Waste to Mercury Absorption Efficiency. In IOP Conference Series: Earth and Environmental Science (Vol. 589, No. 1: 012018). IOP Publishing.

Mahmoud, M. E. and Mohamed, A. K. (2020). Novel derived pectin hydrogel from mandarin peel based metal-organic frameworks composite for enhanced Cr (VI) and Pb (II) ions removal. International Journal of Biological Macromolecules, 164: 920–931.

McAdam, B., Brennan Fournet, M., McDonald, P. and Mojicevic, M. (2020). Production of polyhydroxybutyrate (PHB) and factors impacting its chemical and mechanical characteristics. Polymers, 12(12): 2908.

Moohan, J., Stewart, S. A., Espinosa, E., Rosal, A., Rodríguez, A., Larrañeta, E. et al. (2019). Cellulose nanofibers and other biopolymers for biomedical applications. A review. Applied Sciences, 10(1): 65.

Mostafa, N. A., Farag, A. A., Abo-dief, H. M. and Tayeb, A. M. (2018). Production of biodegradable plastic from agricultural wastes. Arabian Journal of Chemistry, 11(4): 546–553.

Nurazzi, N., Asyraf, M. R. M., Rayung, M., Norrrahim, M. N. F., Shazleen, S. S., Rani, M. S. A. et al. (2021). Thermogravimetric analysis properties of cellulosic natural fiber polymer composites: A review on influence of chemical treatments. Polymers, 13(16): 2710.

Omar, R. A. and Jain, M. (2022). Preparation and applications of chitosan–gold bionanocomposites. In Biodegradable and Environmental Applications of Bionanocomposites (pp. 67–97). Cham: Springer International Publishing.

Omar, R. A., Talreja, N., Chuhan, D. and Ashfaq, M. (2024). Waste-derived carbon nanostructures (WD-CNs): An innovative step toward waste to treasury, Environmental Research, 246: 118096, https://doi.org/10.1016/j.envres.2023.118096.

Osman, A. F., Siah, L., Alrashdi, A. A., Ul-Hamid, A. and Ibrahim, I. (2021). Improving the tensile and tear properties of thermoplastic starch/dolomite biocomposite film through sonication process. Polymers, 13(2): 274.

Pang, Y. L., Tan, J. H., Lim, S. and Chong, W. C. (2021). A State-of-the-Art Review on Biowaste Derived Chitosan Biomaterials for Biosorption of Organic Dyes: Parameter Studies, Kinetics, Isotherms and Thermodynamics. Polymers, 13(17): 3009.

Patel, N., Feofilovs, M. and Blumberga, D. (2022). Agro biopolymer: A sustainable future of agriculture–state of art review. Environmental and Climate Technologies, 26(1): 499–511.

Peinemann, J. C. and Pleissner, D. (2020). Continuous pretreatment, hydrolysis, and fermentation of organic residues for the production of biochemicals. Bioresource Technology, 295: 122256.

Piyada, K., Waranyou, S. and Thawien, W. (2013). Mechanical, thermal and structural properties of rice starch films reinforced with rice starch nanocrystals.

Pryadko, A., Surmeneva, M. A. and Surmenev, R. A. (2021). Review of hybrid materials based on polyhydroxyalkanoates for tissue engineering applications. Polymers, 13(11): 1738.

Raiyaan, G. I. D., Khalith, S. B. M., Sheriff, M. A. and Arunachalam, K. D. (2021). Bio-adsorption of methylene blue dye using chitosan-extracted from *Fenneropenaeus indicus* shrimp shell waste. Journal of Aquaculture & Marine Biology, 10(4): 146–150.

Ravi, T., Stanley, A. L., Narendrakumar, G. and Rao, A. K. S. (2015). Extraction of chitosan from prawn shell waste and its application in dye decolorization. J. Chem. Pharm. Res., 7: 695–703.

Sabee, M. M. S. M., Uyen, N. T. T., Ahmad, N. and Hamid, Z. A. A. (2021). Plastics packaging for pharmaceutical products. Reference Module in Materials Science and Materials Engineering; Elsevier: Amsterdam, The Netherlands.

Selimin, M. A., Latif, A. F. A., Lee, C. W., Muhamad, M. S., Basri, H. and Lee, T. C. (2022). Adsorption efficiency of hydroxyapatite synthesised from black tilapia fish scales for chromium (VI) removal. Materials Today: Proceedings, 57: 1142–1146.

Shaghaleh, H., Xu, X. and Wang, S. (2018). Current progress in production of biopolymeric materials based on cellulose, cellulose nanofibers, and cellulose derivatives. RSC Advances, 8(2): 825–842.

Shahrin, E. W. E., Narudin, N. A. H., Padmosoedarso, K. M., Kusrini, E., Mahadi, A. H., Shahri, N. N. M. et al. (2021). Pectin derived from pomelo pith as a superior adsorbent to remove toxic Acid Blue 25 from aqueous solution. Carbohydrate Polymer Technologies and Applications, 2: 100116.

Shalaby, M., Farag, S., Haddad, A., Hamouda, D., Hassan, A. and Saeed, H. (2021). Collagen polymer and magnetic collagen nanocomposite recycled from waste to reduce polluted water toxicity. Polymers and Polymer Composites, 29(9): 1515–1527.

Sidek, I. S., Draman, S. F. S., Abdullah, S. R. S. and Anuar, N. (2019). Current development on bioplastics and its future prospects: an introductory review. INWASCON Technol. Mag, 1: 03–08.

Singh, R. B., Srinagesh, B. and Anand, S. (eds.). (2020). Urban health risk and resilience in Asian cities. Singapore: Springer.

Siraj, S., Al-Marzouqi, A. H., Iqbal, M. Z. and Ahmed, W. (2022). Impact of micro silica filler particle size on mechanical properties of polymeric based composite material. Polymers, 14(22): 4830.

Song, K., Xu, H., Xu, L., Xie, K. and Yang, Y. (2017). Cellulose nanocrystal-reinforced keratin bioadsorbent for effective removal of dyes from aqueous solution. Bioresource Technology, 232: 254–262.

Sreekumar, K., Bindhu, B. and Veluraja, K. (2021). Perspectives of polylactic acid from structure to applications. Polymers from Renewable Resources, 12(1-2): 60–74.

Sumesh, K. R., Kanthavel, K. and Kavimani, V. (2020). Peanut oil cake-derived cellulose fiber: Extraction, application of mechanical and thermal properties in pineapple/flax natural fiber composites. International Journal of Biological Macromolecules, 150: 775–785.

Šuput, D., Popović, S., Hromiš, N., Rakita, S., Spasevski, N., Lončar, B. et al. (2023). The influence of oil cake granulation and ultrasonic pretreatment on the properties of biopolymer films based on Camelina sativa oilseed cake. Food and Feed Research, 50(2): 61–75.

Swamiappan, S., Ganesan, S., Sekar, V., Devaraj, S., Subramanian, A., Ponnusamy, V. K. et al. (2021). Effective removal of cationic methylene blue dye using nano-hydroxyapatite synthesized from fish scale bio-waste. International Journal of Applied Ceramic Technology, 18(3): 902–912.

Tang, T. O. and Simon, G. P. (2020). Biodegradation of 3D-printed polylactic acid milliprojections under physiological conditions. Journal of Applied Polymer Science, 137(38): 49129.

Tavker, N., Gaur, U. and Sharma, M. (2020). Cellulose supported bismuth vanadate nanocomposite for effective removal of organic pollutant. Journal of Environmental Chemical Engineering, 8(4): 104027.

Thomas, S. and Balakrishnan, P. (eds.). (2021). Green composites. Springer.

Tsang, Y. F., Kumar, V., Samadar, P., Yang, Y., Lee, J., Ok, Y. S. et al. (2019). Production of bioplastic through food waste valorization. Environment International, 127: 625–644.

Udoetok, I. A., Karoyo, A. H., Ubuo, E. E. and Asuquo, E. D. (2024). Granulation of lithium-ion sieves using biopolymers: A review. Polymers. 16(11): 1520. https://doi.org/10.3390/polym16111520.

Vakati, S. R., Vanderlaan, G., Gacura, M. D., Ji, X., Chen, L. and Piovesan, D. (2024). Synthesis of poly-lactic acid by ring open polymerization from beer spent grain for drug delivery. Polymers, 16(4): 483. https://doi.org/10.3390/polym16040483.

Velusamy, S., Roy, A., Sundaram, S. and Kumar Mallick, T. (2021). A review on heavy metal ions and containing dyes removal through graphene oxide-based adsorption strategies for textile wastewater treatment. The Chemical Record, 21(7): 1570–1610.

Vigneswari, S., Noor, M. S. M., Amelia, T. S. M., Balakrishnan, K., Adnan, A., Bhubalan, K. et al. (2021). Recent advances in the biosynthesis of polyhydroxyalkanoates from lignocellulosic feedstocks. Life, 11(8): 807. https://doi.org/10.3390/life11080807.

Visco, A., Scolaro, C., Facchin, M., Brahimi, S., Belhamdi, H., Gatto, V. et al. (2022). Agri-food wastes for bioplastics: European prospective on possible applications in their second life for a circular economy. Polymers, 14(13): 2752.

Workeneh, K., Zereffa, E. A., Segne, T. A. and Eswaramoorthy, R. (2019). Eggshell-derived nanohydroxyapatite adsorbent for defluoridation of drinking water from bofo of Ethiopia. Journal of Nanomaterials.

Xie, X., Zhou, Z. and Yan, Y. (2019). Flexural properties and impact behaviour analysis of bamboo cellulosic fibers filled cement based composites. Construction and Building Materials, 220: 403–414.

Xu, H. and Yang, Y. (2014). Controlled de-cross-linking and disentanglement of feather keratin for fiber preparation via a novel process. ACS Sustainable Chemistry & Engineering, 2(6): 1404–1410.

Yaradoddi, J., Patil, V., Ganachari, S., Banapurmath, N., Hunashyal, A., Shettar, A. et al. (2016). Biodegradable plastic production from fruit waste material and its sustainable use for green applications. Int. J. Pharm. Res. Allied Sci., 5(4): 72–81.

Zhang, S., Wang, Z., Zhang, Y., Pan, H. and Tao, L. (2016). Adsorption of methylene blue on organosolv lignin from rice straw. Procedia Environmental Sciences, 31: 3–11.

Zhou, Y., Lu, J., Zhou, Y. and Liu, Y. (2019). Recent advances for dyes removal using novel adsorbents: a review. Environmental Pollution, 252: 352–365.

Zubair, M., Roopesh, M. S. and Ullah, A. (2022). Nano-modified feather keratin derived green and sustainable biosorbents for the remediation of heavy metals from synthetic wastewater. Chemosphere, 308: 136339.

Zwawi, M. (2021). A review on natural fiber bio-composites, surface modifications and applications. Molecules, 26(2): 404.

Zynudheen, A. A., Ninan, G., Mathew, P. T. and Joseph, J. (2009). Removal of heavy metals from water by the direct addition of chitosan prepared from prawn and squilla shells.

CHAPTER-13

Cyclodextrin Based Nanoparticles for Environmental Pollution

Benay Tuncsoy[1,*] and *Mustafa Tuncsoy*[2]

1. Introduction

Environmental pollution occurs because of the accumulation and spread of various pollutants (waste materials, chemicals, radioactive substances, etc.) in water, air, and soil. It can be caused by industrial activities, agriculture, energy production, transport, domestic wastes, and other human activities. These pollutants can be released directly into water resources, atmosphere, or soil. For example, toxic gases emitted from factories can cause air pollution, discharge of domestic wastes into natural water resources without treatment can cause water pollution, excessive use of chemical fertilisers can cause soil pollution. Environmental pollution has been increasing rapidly in recent years and has become uncontrollable. By 2030, the Global Sustainable Development Goals (SDGs) of the United Nations stipulate that everyone should have affordable, equitable, and universal access to secure, clean water (Pandey et al., 2022). It is our duty to protect the environment in line with the principles of reduce-reuse-recycle.

Types of environmental pollution are generally classified according to the types of pollution that affect different surroundings in the environment. The main types of environmental pollution are as follows:

Air Pollution: Harmful gases, particles and chemicals released into the atmosphere from sources such as industrial plants, vehicle exhausts, agricultural activities, and combustion of domestic fuels cause air pollution. This can lead to respiratory diseases, lung problems and serious effects on human health in general.

Water Pollution: Water pollution caused by mixing of domestic wastes, industrial wastes, agricultural pesticides, and fertilisers into water resources adversely affects drinking water resources and aquatic ecosystems. Water pollution may cause the death of aquatic life forms, render water unusable and cause the formation of water sources that threaten human health.

Soil Pollution: Soil pollution, which occurs as a result of mixing chemicals from various sources such as pesticides, industrial wastes, oil spills and mining activities into the soil, reduces soil fertility, affects vegetation and disrupts the ecosystem balance.

[1] Adana Alparslan Turkes Science and Technology University, Faculty of Engineering, Department of Bioengineering.
[2] University of Cukurova, Faculty of Science and Letter, Department of Biology.
* Corresponding author: btuncsoy@atu.edu.tr
DOI: 10.1201/9781003558330-13

Water pollution constitutes an important problem among environmental pollution. As a result of human activities, organic pollutants such as dyes, pesticides, heavy metals cause pollution in wetlands. Especially algae blooms that occur as a result of the increase in air temperature negatively affect the species living in these areas to a significant extent. In recent years, solutions can be obtained by utilising adsorption techniques in water pollution studies (Rathi and Kumar, 2021; Rangappa et al., 2024; Ahammad et al., 2023). Adsorption is the process of adhesion of atoms, ions, or molecules in a substance (gas or liquid) to the surface of the absorbent substance. This technology offers superior performance due to its cost-effectiveness, stability, recyclability, and reliability in maintaining environmental and health standards for toxic pollutants (Rangappa et al., 2024). Adsorption takes place at the interface between the solid adsorbent and the contaminated water during water treatment. The most important feature of this process is that when the solution and adsorbent come into contact with each other, the solute molecules are attracted to the adsorbent surface through functional groups on the surface. Adsorbate is defined as the solute that accumulates in the adsorption processes taking place at the solid interface, while the solid on which it is stored is called adsorbent (Gehrke et al., 2015). To remove water pollutants, the most used adsorbents are Activated Carbon (AC), nanomaterials, natural adsorbents, and bio-sorbents. Bio-sorbents are one of the most widely used adsorbents for the removal of environmental contaminants from aquatic environments. The most preferred bio-sorbents in recent years are materials such as cellulose, chitosan, cyclodextrin, tree bark, sawdust, resin, hard fruit shells and seed pulp, grain, tea and coffee pulp, agricultural peelings.

Cyclodextrins (CDs) are cyclic oligosaccharides consisting of α-1,4-linked glycopyranose units. CDs are natural polymers and are produced by degradation of starch (mostly corn) with the enzyme glucosyltransferase. As they are biocompatible, natural polymers, they have a wide range of applications. Apart from pharmaceutical use, they are also used in agriculture, food, textile, and cosmetics industries. It was first discovered by Antoine Villiers in 1891 and in 1953, Freudenberg et al. obtained the first patent for the use of CDs in drug formulations. The nomenclature of CDs depends on the amount of glycopyranose subunit. The most commonly used are α, β, γCDs consisting of 6, 7 and 8 glucopyranose subunits, respectively (Messner et al., 2010). In recent studies, cyclodextrin polymers and cyclodextrin-based composites are frequently used as biosorbents for the removal of environmental pollutants from aquatic environments (Tian et al., 2021; Okasha et al., 2023; Saravanan et al., 2023; Safapour et al., 2022; Ozelcaglayan and Parker, 2023; Liu et al., 2024).

The main aim of this chapter is to provide a summary to examine the cyclodextrin based materials as an adsorbent for removal of environmental pollutant from wastewater. Also, this chapter focuses on how and in which conditions cyclodextrin based materials adsorb the pollutants in the water.

2. Polymer Nanoparticles

Polymers, which are obtained naturally or synthetically, consisting of very long chain molecules by repeating the addition of small-sized molecules called monomers, which is a chemical that has found many uses for itself; It has gained a place in many important fields, especially in medicine, pharmacy, engineering, and industry (McCrum et al., 1997).

Nanoparticles composed of polymers are the most widely investigated systems for drug carriers due to the fact that the polymer surface and dimensions can be modified in order to deliver the active substance to the desired region, in other words, controlled release (active or passive targeting), biocompatible, biodegradable, reduce toxicity, increase the solubility of drugs with low water solubility and increase the drug effect in the tissue (Marcucci and Lefoulon, 2004; Cohen-Sela et al., 2009). Based on these advantages, it has been proven in studies that amphiphilic cyclodextrin nanoparticles (ACD), which constitute an area of use for artificially synthesised polymerisation, has the ability to encapsulate highly lipophilic drugs, provide a controlled release profile, maintain the stability of the drug under physiological conditions and increase mucosal penetration and cellular uptake of drugs by using them as drug carrier systems (Wang et al., 2012).

In general terms, nanoparticles are particles with at least one dimension in the range of 1–100 nm. Those with one dimension below 100 nm are called nanosheets, two dimensions are called nanotubes, and three dimensions below 100 nm are called nanoparticles (International Organization of Standardization; ASTM International).

Particles can be basically classified as follows:

- Semiconductor-based nanoparticles (gold, quantum dots, iron oxide).
- Metal-based nanoparticles (nanoshells, nanosticks, superparamagnetic iron oxide nanoparticles).
- Carbon-based nanoparticles (fullerenes, multilayer carbon nanotubes, etc.) (Lewinski et al., 2008).
- Polymeric nanoparticles: synthetic polymers (polylactic acid, polycaprolactone, polymaltic acid, polyglutamic acid and their copolymers) (Gilding and Reed, 1979); natural polymers (proteins, alginate, gelatin, albumin, nucleic acids, polysaccharides (chitosan), oligosaccharides (cyclodextrin).

3. Cyclodextrin Nanoparticles

Polymer-based nanoparticles are prominent due to their biodegradable and biocompatible properties. They also provide controlled drug release and pharmacokinetic control with chemical surface modifications (Baratt, 2003; Senel and McClure, 2004). They can be in natural and synthetic structure. As a result of the breakdown of starch catalysed by the enzyme CGT-ase, a ring molecular structure is formed as shown in Fig. 1. They have a conical structure. Their inner cavities have a hydrophobic structure. The opposite structure prevails on the outer surface. Due to the steric

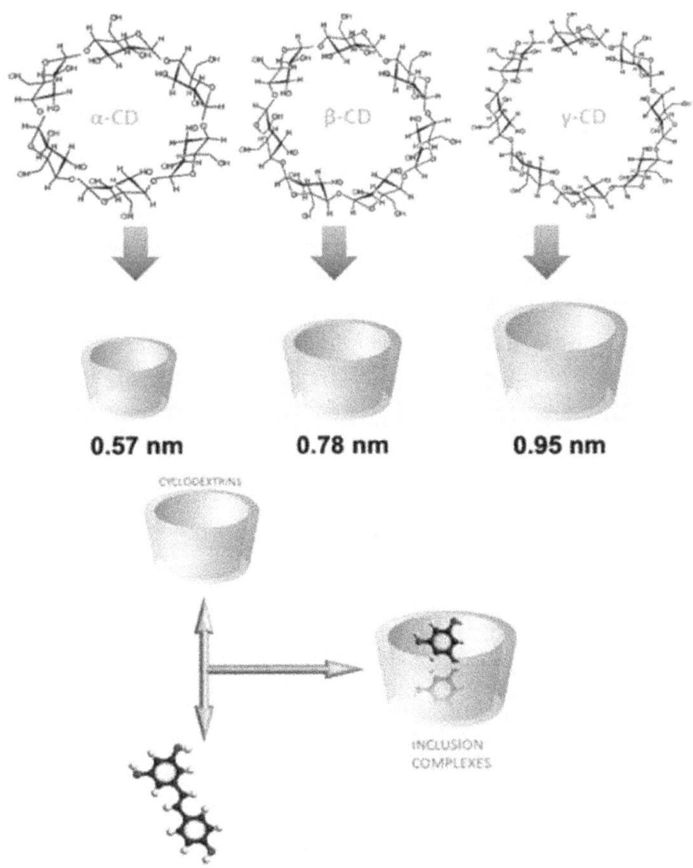

Figure 1. Cyclodextrins chemical structure (Almagro and Pedreño, 2020).

hindrance, cyclodextrins require at least six glucopyranose units to be linked by α-1,4 bonds (Urooj et al., 2024). Cyclodextrins with seven glucopyranoses are called β- cyclodextrins and those with eight glucopyranoses are called γ cyclodextrins; cyclodextrins containing a large number of glucopyranoses are not preferred due to their low complex formation and increased solubility in water. It has been reported that cyclodextrins function as a transporter that provides membrane passage by keeping lipophilic active substances in aqueous solution and due to the low affinity of the lipophilic cell membrane to hydrophilic cyclodextrins, cyclodextrins remain in the aqueous outer membrane and increase the uptake of sufficiently lipophilic active substances into the cell (Loftsson et al., 2005).

3.1 Cyclodextrin Based Nano Sponges

Due to the high hydroxyl groups in the structure of cyclodextrins, they can easily form reticulated structures by copolymerisation with suitable crosslinkers (Fig. 2). In recent years, cross-linked material type nano sponges have become widespread. Nano sponges are defined as hydrophilic water-insoluble molecules with high stability in a wide temperature and pH range. In addition, having a supramolecular 3D hyperreticulated nanoporous structure is their most important feature. (Cecone et al., 2020; Rizzi et al., 2021; Sadjadi and Koohestani, 2021; Suvarna et al., 2021; Utrezi et al., 2022). Nano sponges are obtained by coupling cyclodextrins with diisocyanate crosslinkers. If the mole ratio of cyclodextrin and crosslinker is greater than 7, these polymers can be produced. In general, cyclodextrin polymers consist of chains containing aliphatic or aromatic sites depending on the type of crosslinker used. Cyclodextrin nano sponges are used in environmental remediation applications. Active cross-linkers convert the nano space at the molecular level into three-dimensional nanoporous polymers. The environment of the cyclodextrin cage is lipophilic and traps small organic molecules. It also optimises van der Waals interactions. However, water tends to form hydrogen bonds and is hydrophilic. Thus, an affinity environment is provided for organic molecules. When solid powder forms of nanoporous polymeric materials are deposited in water containing organic pollutants, foreign organic molecules are retained. Thus, two separate phases are formed as an organic solid phase and aqueous phase with completely different properties. Organic foreign molecules are present in solid lipophilic polymers and the aqueous phase is organically free (Salgın et al., 2017; Kumari et al., 2020; Utrezi et al., 2022). By optimising host-host interactions, organic molecules are adsorbed via physical inclusion. These interactions are highly effective non-covalent bonding and are effective in the removal of organic molecules even at very low concentrations. The adsorption of organic molecules is due to the inclusion effect between cyclodextrins and organic molecules. Hence, cyclodextrin application in environmental remediation processes lead to remarkable results (Utrezi et al., 2022).

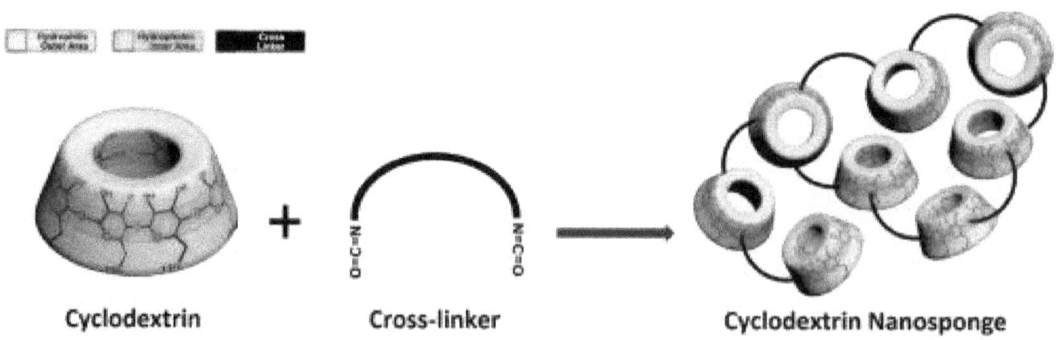

Figure 2. Cyclodextrins are linked with cross-liking polymer to form cyclodextrin nano sponge (Sherje et al., 2017).

3.2 Application of Cyclodextrin Based Materials as an Adsorbent

Cyclodextrin-based materials have been widely preferred in recent years for the removal of various organic pollutants from aquatic environments due to their unique physical and chemical properties such as size matching effect and hydrophobic interaction. As a result of the unconscious release of industrial wastes in aquatic environments, heavy metals such as Pb (Lead), Cd (Cadmium), Cr (Chromium), Ar (Arsenic) have been reaching rivers, and underground water systems and oceans in recent years. Heavy metals remain either soluble or in a suspension form in aquatic environments and eventually settle to the bottom and are taken up by organisms and transferred to other living species through the food chain. This leads to bioaccumulation in each organism (Chuin et al., 2024). Therefore, although many methods such as chemical precipitation, reverse osmosis, electro dialysis, ion exchange and photocatalysis are used to remove heavy metals from water, satisfactory results cannot be achieved due to the disadvantages of these methods, such as high cost and insufficient results. In recent years, there has been an increase in research on the use of cyclodextrin-based adsorbents. Jemli et al. (2024) reported that the use of C-β cyclodextrin polymers resulted in an increase in the number of Hg^{+2} per site at 55°C. Moreover, in another study by Wang et al. (2024), a cyclodextrin-based adsorbent PANI@PCDP β-cyclodextrin was produced because of the integration of cyclodextrin polymers with polyaniline and as a result, it was determined that it provided the removal of both Cr (IV) and bisphenol A from wastewater with its multiple binding sites.

In addition to heavy metals, pesticides constitute the second most common pollutant group in aquatic environments. In a study by Yao et al. (2024), β-cyclodextrin was obtained with reduced graphene oxide aerogel and it was determined that this aerogel had a high adsorption capacity. It has also been reported that it provides the elimination of pesticide pollution in water systems. There are also many studies in which cyclodextrin-based nanosponges have been used to remove pesticides from the environment; atrazine (Romita et al., 2019; Wang et al., 2022), imidocloprid (Utzeri et al., 2022), carbendozim (Rizzi et al., 2021). Apart from these, there are studies in which cyclodextrin-based adsorbents are used to remove dyes, pharmaceutical drugs and other aromatic compounds from aquatic environments and positive results are reported (Murcia-Salvador et al., 2019; Lellis et al., 2019; Varan et al., 2020; Huang et al., 2020).

4. Conclusions and Future Outlooks

Water is vital for the proper functioning of the ecosystem. Usable water resources are becoming polluted and unusable due to increasing urbanisation, industrialisation and environmental pollution, and the discharge of wastewater in industrial environments to the receiving environment without being treated or reduced to the desired standards. Failure to be aware that water is a natural resource and pollution of water during the use/consumption stages have disrupted the ecological balance. Pollutants in water consist of organic pollutants, inorganic pollutants, microorganisms, radioactive substances, and turbid substances. Some organic and inorganic pollutants are highly toxic and carcinogenic. Some of the pollutants in water can be biodegraded or removed by converting them into simpler compounds. However, pollutants that are not biodegradable or non-biodegradable remain in the water environment for a long time. The water must be purified to remove them. The most important of these methods used for the removal of environmental pollutants from the water environment is the use of adsorbents and cyclodextrin is the most widely investigated of these adsorbents in recent years. Cyclodextrins have many different application areas such as food industry and cancer treatment. There are still wide ranges of application areas to be investigated.

References

Ahammad, N. A., Ahmad, M. A., Hameed, B. H. and Mohd, Din A. T. (2023). A mini review of recent progress in the removal of emerging contaminants from pharmaceutical waste using various adsorbents. Environ. Sci. Pollut. Res. Int., 30(60): 124459–124473.

Almagro, L., Pedreño, M.Á. (2020). Use of cyclodextrins to improve the production of plant bioactive compounds. Phytochem. Rev., 19: 1061–1080.

ASTM International: E 2456-06 Terminology for Nanotechnology. ASTM International, West Conshohocken, USA, 2008.

Baratt, G. (2003). Colloidal Drug Carriers: Achievements and Perspectives, Cellular and Molecular Life Sciences, 60: 21–37.

Cecone, C., Hoti, G., Krabicová, I., Appleton, S. L., Caldera, F., Bracco, P. et al. (2020). Sustainable Synthesis of Cyclodextrin-Based Polymers by Exploiting Natural Deep Eutectic Solvents. Green. Chem. 22: 5806–5814.

Chuin, L. M., Kamaruzaman, S., Praveena, S. M. and Yahaya, N. (2024). Recent applications of β-cyclodextrin in selective adsorption of pesticides, heavy metals, and organic pollutants from water samples: Mini review, Microchemical Journal, 206: 111583.

Cohen-Sela, E., Chorny, M., Koroukhov, N., Danenberg, H. D. and Golomb, G. (2009). A new double emulsion solvent diffusion technique for encapsulating hydrophilic molecules in PLGA nanoparticles. Journal of Control Release, 133: 90–95.

Gehrke, I., Geiser, A. and Somborn-Schulz, A. (2015). Innovations in nanotechnology for water treatment. Nanotechnol. Sci. Appl., 8: 1–17.

Gianluca, U., Matias Pedro M. C., Dina, M. and Valente Artur J. M. (2022). Cyclodextrin-Based Nanosponges: Overview and Opportunities. Frontiers in Chemistry. 10.

Gilding, D. and Reed, A. (1979). Biodegradable Polymers for Use in Surgery, Polyglycolic/poly (Lactic acid) Homo- and Copolymers: 1, Polymer, 20: 1459–1464.

Huang, Q., Chai, K., Zhou, L. and Ji, H. (2020). A Phenyl-Rich β-cyclodextrin porous crosslinked polymer for efficient removal of aromatic pollutants: Insight into adsorption performance and mechanism. Chem. Eng. J., 387: 124020.

International Organization of Standardization: Nanotechnologies – Terminology and Definitions for Nano-Objects, ISO/TS 27687:2008 (E). ISO, Geneva, Switzerland, 2008.

Jemli, Lütke, S. F., Chamtouri, F., Amara, F. B., S., Bejar, Oliveira, M. L. S., Knani, S. et al. (2024). A novel cartoon crosslinked β-cyclodextrin (C-β-CD) polymer for effective uptake of Hg from aqueous solutions: Kinetics, equilibrium, thermodynamics, and statistical physics approach, Separation and Purification Technology, 330, Part C,125578,

Kumari, P., Singh, P. and Singhal, A. (2020). Cyclodextrin-based nanostructured materials for sustainable water remediation applications. Environ. Sci. Pollut. Res., 27: 32432–32448.

Lellis, B., Fávaro-Polonio, C. Z., Pamphile, J. A. and Polonio, J. C. (2019). Effects of textile dyes on health and the environment and bioremediation potential of living organisms. Biotechnol. Res. Innovation, 3: 275–290.

Lewinski, N., V. Colvin and R. Drezek. (2008). Cytotoxicity of Nanoparticles. Nano, Small, Micro, 4: 26–49.

Liu, C., Crini, G., Wilson, L. D., Balasubramanian, P. and Li, F. (2024). Removal of contaminants present in water and wastewater by cyclodextrin-based adsorbents: A bibliometric review from 1993 to 2022. Environ Pollut., 1; 348: 123815.

Loftsson, T., Jarho, P., Masson, M. and Jarvinen, T. (2005). Cyclodextrins in Drug Delivery, Expert Opinion in Drug Delivery, 2(2): 1–17.

Marcucci, F. and Lefoulon, F. (2004). Active Targeting with Particulate Drug Carriers in Tumor Therapy: Fundamentals and Recent Progress, Drug Discovery Today, 9(5): 219–228.

McCrum, N. G., Buckley, C. P. and Bucknall, C. B. (1997). Principles of Polymer Engineering, Oxford University Press, New York, 1-2.

Messner, M., Kurkov, S. V., Jansook, P. and Loftsson, T. (2010). Self-assembled cyclodextrin aggregates and nanoparticles. Int. J. Pharm., 387: 199–208.

Murcia-Salvador, A., Pellicer, J. A., Fortea, M. I., Gómez-López, V. M., Rodríguez- López, M. I., Núñez-Delicado, E., et al. (2019). Adsorption of Direct Blue 78 Using Chitosan and Cyclodextrins as Adsorbents. Polymers, 11: 1003.

Okasha, A. T., Abdel-Khalek, A. A., Alenazi, N. A., AlHammadi, A. A., Al Zoubi, W., Alhammadi, S. et al. (2023). Progress of synthetic cyclodextrins-based materials as effective adsorbents of the common water pollutants: comprehensive review. J. Environ. Chem. Eng. 11: 109824.

Ozelcaglayan, E. D. and Parker, W. J. (2023). β-Cyclodextrin functionalized adsorbents for removal of organic micropollutants from water. Chemosphere, 320: 137964.

Pandey, S., Twala, B., Singh, R., Gehlot, A., Singh, A., Montero, E. C. et al. (2022). Wastewater treatment with technical intervention inclination towards smart cities. Sustainability, 14(18): 11563.

Pin Wang, Guangyu An, Peter Jarvis, Wen Liu, Shunke Ding, Ruixin Qu et al. (2024). Preparation of β-cyclodextrin-reduced graphene oxide aerogel and its application for adsorption of herbicides. Journal of Cleaner Production, 468: 143109.

Rangappa, H. S., Herath, I., Lin, C. and Ch S. (2024). Industrial waste-based adsorbents as a new trend for removal of water-borne emerging contaminants. Environ, Pollut. 15; 343: 123140.

Rathi, B. S. and Kumar, P. S. (2021). Application of adsorption process for effective removal of emerging contaminants from water and wastewater. Environ. Pollut. 1; 280: 116995.

Rizzi, V., Gubitosa, J., Signorile, R., Fini, P., Cecone, C., Matencio, A. et al. (2021). Cyclodextrin Nanosponges as Adsorbent Material to Remove Hazardous Pollutants from Water: The Case of Ciprofloxacin. Chem. Eng. J. 411: 128514.

Romita, R., Rizzi, V., Semeraro, P. et al. (2019). Operational parameters affecting the atrazine removal from water by using cyclodextrin based polymers as efficient adsorbents for cleaner technologies. Environmental Technology & Innovation, 16: 100454.

Sadjadi, S. and Koohestani, F. (2021). Composite of cross-linked chitosan beads and a cyclodextrin nanosponge: a metal-free catalyst for promoting ultrasonic-assisted chemical transformations in aqueous media. J. Phys. Chem. Sol., 156: 110157.

Safapour, S., Mazhar, M., Nikanfard, M. and Liaghat, F. (2022). Recent advancements on the functionalized cyclodextrin-based adsorbents for dye removal from aqueous solutions. Int. J. Environ. Sci. Technol. 19: 5753–5790.

Salgın, S., Salgın, U. and Vatansever, Ö. (2017). Synthesis and characterization of β-Cyclodextrin Nanosponge and Its Application for the Removal of *p*-Nitrophenol from Water. Clean – Soil, Air, Water, 45: 1500837.

Saravanan, A., Karishma, S., Kumar, P. S., Thamarai, P. and Yaashikaa, P. R. (2023). Recent insights into mechanism of modified bio-adsorbents for the remediation of environmental pollutants. Environ. Pollut. 339: 122720.

Senel, S. and McClure, S. J. (2004). Potential Applications of Chitosan in Veterinary Medicine, Advance Drug Delivery Reviews, 56: 1467–1480.

Sherje, A. P., Dravyakar, B. R., Kadam, D. and Jadhav, M. (2017). Cyclodextrin-based nanosponges: A critical review, Carbohydrate Polymers, 173: 37–49.

Suvarna, V., Singh, V., Sharma, D. and Murahari, M. (2021). Experimental and computational insight of the supramolecular complexes of Irbesartan with βcyclodextrin Based Nanosponges. J. Drug Deliv. Sci. Tech., 63: 102494.

Szejtli, J. (2004). Cyclodextrins and Encapsulations, Encyclopedia of Nanoscience and Nanotechnology H.S. Nalwa (Eds.), 2 283–304.

Tian, B., Hua, S., Tian, Y. and Liu, J. (2021). Cyclodextrin-based adsorbents for the removal of pollutants from wastewater: a review. Environ. Sci. Pollut. Res. 28: 1317–1340.

Urooj, T., Mishra, M. and Pandey, S. (2024). Unlocking environmental solutions: a review of cyclodextrins in pollutant removal. Discov. Environ 2: 65.

Utzeri, G., Matias, P. M. C., Murtinho, D. and Valente, A. J. M. (2022). Cyclodextrin-based nanosponges: overview and opportunities. Frontiers in Chemistry, 10: 859406.

Varan, C., Anceschi, A., Sevli, S., Bruni, N., Giraudo, L., Bilgiç, E. et al. (2020). Preparation and characterization of cyclodextrin nanosponges for organic toxic molecule removal. Int. J. Pharmaceutics, 585: 119485.

Wang, A. Z. and R. Langer, O. C. (2012). Farokhzad, Nanoparticle Delivery of Cancer Drugs, Annual Review of Medicine, 63: 185–198.

Wang, Q., Shao, J., Xu, J., Dong, F., Xiong, Y. and Chen, Q. (2022). *In-situ* formed Cyclodextrin-functionalized graphene oxide / poly(N-isopropylacrylamide) nanocomposite hydrogel as a recovery adsorbent for phenol and microfluidic valve. Journal of Colloid and Interface Science, 607: 253–268. doi:10.1016/j.jcis.2021.08.130.

Wang, P., An, G., Jarvis, P., Liu, W., Ding, S., Qu, R. et al. (2024). Simultaneous removal of organic micropollutants and metals from water by a multifunctional β-cyclodextrin polymer-supported-polyaniline composite, Chemical Engineering Journal, 482: 148826,

Yao, T., Chen, H., Luo, Y., Li, H., Shao, X., Zheng, Q. et al. (2024). Unlocking environmental solutions: A review of cyclodextrins in pollutant removal. Discover Environment, 2: 65.

CHAPTER-14

A Comparative Study on the Effect of Micro and Nano-Biopolymers for Geotechnical Applications

Evangelin Ramani Sujatha[1,*] and *Govindarajan Kannan*[2]

1. Introduction

Biopolymers are natural macromolecules consisting of several repeating units and are produced from natural sources, which can either be living organisms like plants, animals and microbes or synthesized from materials of biological origin like sugar, natural oils/fats, amino acids, etc. The monomeric units are connected by covalent bonds forming large structures (Singh et al., 2021). Biopolymers have a complex structure and well-defined 3D shapes in general. The primary advantage of the choice of the biopolymer is their appreciable biodegradability and biocompatibility which makes them suitable for various applications in food, pharmaceutical, cosmetic, agriculture and construction industries. The interest in biopolymers has increased in the last decade as they provide sustainable solutions with neutral or negative carbon footprints. The use of biopolymers as building/construction material is among the recent additions to the application of biopolymers. This is particularly of great significance in the construction industry. The use of conventional materials, particularly cement contributes to 7% anthropogenic production of greenhouse gases (GHG) as a result of both chemical reactions and energy consumption (Miller et al., 2021) with a major share of these gases from fossil fuels. Cement alone among construction materials contributes to approximately 60% of GHG emissions during the manufacturing process. Urbanization, infrastructure development and the need to replace existing infrastructure owing to ageing in the near future will increase the demand for construction materials like cement which is not a sustainable option. This mandates the need for new, sustainable and environmentally friendly materials like biopolymers for use in the construction industry. Similarly, biopolymers like chitosan, gelatin, cellulose, soluble starch, etc., are used as capping and reducing agents in the synthesis of nanoparticles (Mohan et al., 2016) and their use in the process replaces toxic reagents that are used for various nanoparticle synthesis.

Biopolymers like guar gum, gellan gum, xanthan gum, chitosan, starch, cellulose, alginates, lignosulphates, agar, gelatin, casein, etc., are increasingly used for different applications in civil engineering. They are widely adopted as water-retaining agents, binders, antisag agents, dispersants,

[1] Centre for Advanced Research in Environment, School of Civil Engineering, SASTRA Deemed University, Thanjavur 613401, Tamil Nadu, India.
[2] Department of Civil Engineering, Mepco Schlenk Engineering College, Sivakasi 626005, Tamil Nadu, India.
* Corresponding author: r.evangelin@gmail.com; sujatha@civil.sastra.edu
DOI: 10.1201/9781003558330-14

superplasticizers, vicosifiers, coatings, set retarders, air-entraining agents, pore-forming material and finishing material. For example, carboxy methyl cellulose can act as a dispersing agent and/or retarder while lignosulphonates are used as dispersants in cement slurries (Niaounakis, 2015). The choice of lignosulphonates though can retard the setting of cement. Biopolymers like casein, gelatin and gum Arabic are reported to be excellent natural retarders with binder gypsum and find application in plastering work (Gypsum Plaster for Concrete or Cement Surfaces, 1973). Modified guar and clay of rheological grade can be used to improve the water retention capacity of gypsum and mortars (Berenguer et al., 1999). Similarly, rosin can be used as an air-entraining agent in concrete (Inoue et al., 1982), and agar can be used for the production of porous concrete (Horinoue et al., 2011).

Biopolymers also find application in subterranean construction. Biodegradable polymers have been used in cement compositions that are used for sealing subterranean formations. Biopolymers have been successfully used in water purification, particularly chitosan which shows superior performance by removing the metals present in water by chelation. The long polymer chain structure of chitosan allows several binding sites to aid the chelation process (Mohan et al., 2016). Shimizu Construction Co. Ltd. patented a biodegradable mesh synthesized from aliphatic polyester produced by bacteria and recommends it to prevent soil erosion (Horiuchi et al., 1994). The mesh decomposes due to bacterial action over time and does not impair the natural landscape. Despite the several applications of biopolymers in civil engineering, geotechnical and geo-environmental applications of biopolymers are limited but there is a huge potential for biopolymers in this area. Biopolymers like xanthan gum, guar gum, gellan gum, cellulose, chitosan, and gelatin, etc., have shown a significant effect in improving the strength of soil and have modified soil properties to suit the geo-environmental application (Kannan and Sujatha, 2023b). For example, Kwon et al. (2023) used xanthan gum-treated soil to control internal erosion on an earthen embankment at a field scale. Kumar and Sujatha (2021) emphasized the choice of biopolymers xanthan gum, guar gum and β glucan for use as additives to modify clayey sand into landfill liners as biopolymers fill the voids by the formation of hydrogel and reduce the permeability of the soil . Similarly, of the limited literature available on the geotechnical applications of biopolymers, there is a minimal collection of nano-biopolymers. This study attempts to summarize the role of nano-sized biopolymers and nano-particles in geo-environmental applications. Also, the study provides a comparison of the efficacy of micro and nano-sized biopolymers on the geotechnical properties of the soil.

2. Biopolymers—Types and Classification

Biopolymers are broadly classified based on their origin, biodegradability, structure and polymer backbone as outlined in Fig. 1. Biopolymers can be classified as natural and synthetic biopolymers based on their origin. Biopolymers that are produced by living organisms are natural biopolymers while those synthesized from biological sources are synthetic biopolymers. Synthetic biopolymers are produced from resins, proteins, fats, sugars, amino acids, etc. Examples of natural biopolymers are starch, cellulose and gelatin while examples of synthetic biopolymers are chitosan, chitin, polycaprolactone, and polyvinyl alcohol, etc. Biodegradable biopolymers are obtained from renewable sources allowing for the replacement or substitution of polymers based on fossil fuel and are synthesized usually from natural fibres, starch or sugar (Mohan et al., 2016), for example, acacia gum, an edible polymer is composed of proteins and carbohydrates. All biological matter is assumed to be biodegradable in general but there are also examples of biological compounds that are not readily degraded. For example, lignin, an abundant biopolymer in nature can be degraded by specific microbes and the process time is extremely slow (Steinbüchel, 2005). Such biopolymers are termed non-biodegradable biopolymers.

Based on the repeating units, biopolymers can be grouped into protein, keratin (a principal example of protein) and nucleic acid (a biomolecule). Keratin can further be classified into primary and secondary keratin. As shown in Fig. 1 based on the polymer backbone, the biopolymers can

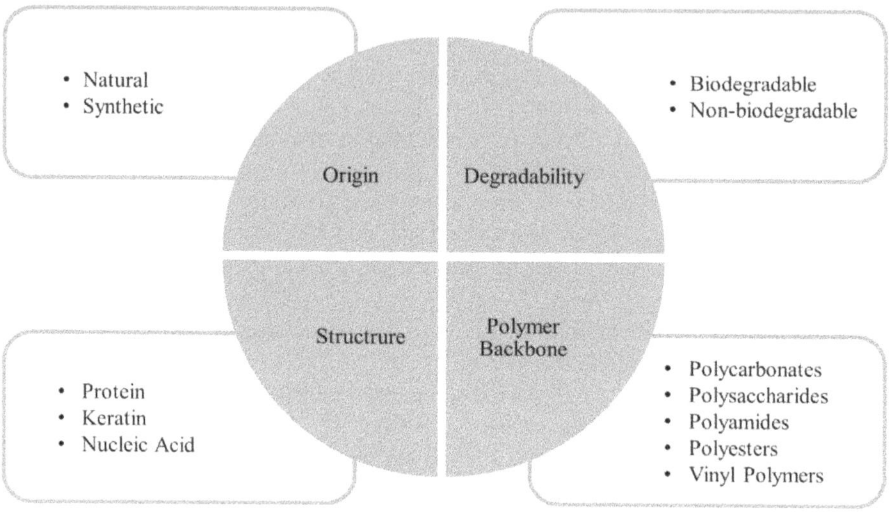

Figure 1. Classification of Biopolymers (Mohammed et al., 2021; Steinbüchel, 2005).

be classified into five groups, namely, polycarbonates, polysaccharides, polyamides, polyesters and vinyl polymers. Polysaccharides like xanthan gum, guar gum, etc., are widely adopted for geotechnical and geo-environmental applications in research and also for field-scale studies, though field-scale studies are limited. Natural polysaccharides can be obtained from plants, animals, microbes and algae. Polysaccharides can further be classified into homopolysaccharides and heteropolysaccharides based on the type of monosaccharide building blocks and also physiological properties (Mohammed et al., 2021).

3. Nanoparticles and their Role in Geo-environmental Applications

Treating soil with nano additive is another contemporary approach to ground improvement which has been under study for over a decade. As discussed earlier, cement and lime as such are neither harmful nor toxic to the soil, the real concern is with the enormous carbon emissions with their production. As discussed earlier, cement and lime are both harmful to the soil environment and also there is a significant concern with carbon emissions during their production process. On average, to treat plastic clay with cement or lime, 10% of total soil mass of additive is required. This accounts for almost 150 kg of cement or lime for 1 m^3 of soil. If one option to overcome this issue is the use of organic biopolymers, the other option is the use of additives which are required in fewer quantities. This is the prime advantage of nano additives, which are required in lesser quantities than conventional cement or lime.

3.1 Strength Improvement

Similar to biopolymers, literature has proven that nano additives will be effective in various applications of soil stabilization, and strength improvement is primary among them. Early reported research on nano additives started with the application of colloidal silica for enhancing the cohesive strength of coarse-grained sand (Arora et al., 2019; Miwa and Yonekura, 1994). Since then, several materials have been investigated for stabilizing soil with various nano additives. These smaller particles have unique advantages over large-sized particles. To prove this effect, Bahmani et al., (2014) stabilized cemented clay with nano silica of different sizes (15 nm and 80 nm). Results indicated that soil stabilized with 0.4% nano-silica (15 nm) provided better strength and reduced HC when compared to 0.4% nano-silica (80 nm) treatment. This enhanced performance accounts for the better interaction of nanoparticles with the soil due to their high specific surface area. The choice of these nano additives for soil stabilization depends on the soil type and their particle size.

3.1.1 Coarse-grained soil

Coarse-grained soil lacks natural cohesion and hence needs a cohesive nano additive for strength improvement. Unlike fine-grained soil, which can establish better interaction with most nano additives, literature reported so far claims only colloidal nano silica and nano clay are potentially suitable for coarse-grained soil like sand.

Colloidal nano silica is a hydrophilic low viscous suspension of amorphous silica nanoparticles. Ghadr et al. (2020) investigated the effect of colloidal nano-silica on sand stabilization using 15% colloidal nano-silica and showed that the effective cohesion increased by 750%, whereas the effective friction angle increased only by 5%. The colloidal nano-silica hydrosol fills the voids in sand and hardens with time, resulting in enhanced shear strength parameters.

Nano clays are stacked silicates prepared from conventional clay with minerals like kaolinite, halloysite, montmorillonite, etc. (Tabarsa et al., 2018). Among several nano-clay types, the montmorillonite variety is the most commonly available nano-clay in the market (Ali et al., 2020). Very little literature has reported on the applicability of nano-clay to stabilize coarse-grained soil as the soil lacks natural cohesion and chemical interaction with the additive. Baziar et al. (2018) ran extensive laboratory tests to assess the suitability of nano-clay for stabilizing dam core soil. The addition of 3% nano clay to clayey sand increased the shear strength by 30%.

The shear strength increased only in the drained condition, whereas excess moisture diminished the strength in the undrained condition. Kholghifard and Behbahani (2022) treated clayey sand with nano clay for strength improvement. Results showed that the addition of 1% nano clay increased the cohesion by 50.5% and friction angle by 22.26%. Upon adding water, the nanoparticles become viscous and bind to the sand grains to increase their strength. Similar to the observations of Baziar et al. (2018), the addition of nano clay beyond the optimum dosage reduced the shear strength parameters.

The discussion shows that both the colloidal nano-silica and nano clay can enhance the strength in coarse-grained soils. However, their higher dosage requirements imply that they cannot be an effective soil stabilizer when compared to biopolymers.

3.1.2 Fine-grained soil

Most of the reported research on nanomaterial-based soil stabilization pertains to fine-grained soils owing to their poor shear strength. Some of the common nano additives used for soil stabilization are nano carbons, nano clay, nano silica, nano calcium carbonate, and nano alumina.

Carbon nanomaterials

Carbon nanomaterials include nanotubes, nano fibres, and nanosheets made of carbon atoms connected by covalent bonds. Alsharef et al., (2019) investigated the potential of low-plastic and high-plastic clays treated with carbon nanotubes (CNT) and carbon nanofibres (CNF). For low plastic clay, a dosage of 0.2% CNT and CNF increased the UCS by 2.01 times and 2.38 times. Similarly, in high plastic clay, 0.2% CNT and CNF increased the strength by 0.39 times and 0.49 times. Results show that carbon nanomaterials are more effective in low-plastic clays than high-plastic clay. Microstructural analysis showed that the nanofibrous nature of carbon nanomaterials filled the voids and bridged the cracks to increase the strength. Despite these advantageous effects, nanocarbons are toxic and are not recommended for sustainable and environmentally friendly ground improvement (Maiti et al., 2019).

Nano clay

Nano clay is the only natural nanomaterial that is suitable for both coarse-grained and fine-grained soils. Emmanuel et al. (2019) compared the effect of halloysite nanotubes sourced from different suppliers to assess their potential to improve the geotechnical properties of highly plastic clay. Treating the soil with a 5% halloysite nanotube increased the Unconfined Compression Strength (UCS) by 67%. The authors analyzed the reason behind the enhanced behaviour and stated that the interlocking of halloysite nanotubes in between the soil grains resisted the sliding among the particles and caused the strength improvement. Similarly, Khalid et al. (2015) studied the performance of intermediate plastic clay treated with 3% nano clay. The nano clay was prepared by ball-milling the study soil and improved the strength by 15%. Despite these reported proofs of the efficiency of nano clay, the material is prepared by either physically or chemically processing the existing clay mineral. In general, excess clay hampers the bearing capacity of the soil. In the case of fine-grained soil with existing clay content, the addition of nano-clay may lead to undesirable results. A similar justification was given byTaha and Taha (2012) when nano clay was used to stabilize clay, among other alternatives.

Nano alumina

Most reported literature on soil stabilization with nano alumina used it as a secondary stabilizer to complement the hydration reaction with cement (Ahmadi and Shafiee, 2019; Luo et al., 2012). However, Taha and Taha (2012) used nano alumina as one of the three nano additives to compare their performance on clay stabilization. They reported that nano alumina halved the hydraulic conductivity HC but did not involve any chemical interaction with the highly plastic clay.

Nano silica

Nano silica marks the most-researched nanomaterial for ground improvement with proven results. Although colloidal silica has viscous properties that could contribute to the stabilization of fine-grained soil, viscous material could pose a difficulty in achieving homogeneous blending with the soil (Chang et al., 2015). Nano silica has been used for the strength improvement of a variety of soils, (reported in detail by the authors elsewhere in another publication (Kannan and Sujatha, 2022a)). Changizi and Haddad (2017) reported that a minimal quantity of 0.7% nano-silica increases the UCS by 56% in high-plastic clays. Similarly, Kannan and Sujatha (2022b) reported that 0.6% nano-silica increased the strength of low plastic organic silt by 2.21 times. Literature also holds proof that these nano-silica blends well with other materials like cement (Bahmani et al., 2016) and fibres (Changizi and Haddad, 2015; Kannan and Sujatha, 2023c) contributing to strength improvement. However, a contradictory opinion on the reaction mechanism of nano silica-treated soil prevails with a different opinion by various researchers. The common mechanism of strength improvement on soil treated with nano silica is either by void filling and interlocking between soil particles or by viscous gel formation. Changizi and Haddad (2017) suggested that viscous gel formation complemented the interlocking facilitated by nano-silica in between the soil. However, the viscous gel did not form any new material. A similar viscous gel formation was reported by Joju and Chandrakaran (2012) also. Kannan and Sujatha (2022b) on the other hand showed that this viscous gel on specific conditions resembles the formation of Calcium Silicate Hydrate (CSH) forming cementitious bonds in the soil. In contrast, Ahmadi and Shafiee (2019) suggested that no viscous gel or new chemical action was observed in their investigation. The strength improvement was due to particle dispersion and rearrangement. Hence, it is clear that the clay minerals have a crucial role in improving the strength, while Ahmadi and Shafiee (2019) did not observe any CSH formation, the presence of vaterite (an intermediate form of calcium carbonate) in the work of Kannan and Sujatha (2022b) reacted with the nano-silica forming CSH. Despite the mechanism, in all the cases, nano-silica ensured strength improvement at an optimal dosage of less than 1% in fine-grained soils.

Nano calcium carbonate

Nano calcium carbonate started emerging as a potential soil stabilizer in recent times. Choobbasti et al. (2019) proved that adding 1.2% nano calcium carbonate to low plasticity clay can increase the friction angle and cohesion by 21.7 and 59.7%, respectively, after 28 days curing. The authors stated that cementitious binding of nano calcium carbonate improved with ageing and enhanced the shear strength parameters. Mohammadi et al. (2021, 2022) studied the performance of various sand-clay blends using nano calcium carbonate. Results indicate that 0.7% nano calcium carbonate for lower bentonite content soil and 1.1% nano calcium carbonate for higher bentonite content soil yield increased strength and suppressed the swelling in the soil. Kannan et al. (2023) showed that nano calcium carbonate works well with the strength improvement of organic silts as well. On average, the strength increased by 1.94 times with just 0.4% of nano calcium carbonate in 90 days. In all three works, the enhanced performance is attributed to the viscous CSH formed from soil-additive interaction (Kannan et al., 2023; Mohammadi et al., 2021, 2022).

3.2 Erosion Control

The advantage of nano additives is their small particle size. They blend in with the soil to bring enhanced properties. However, the smallest particle size would be the reason for their ineffectiveness as a dust suppressant, as the additive itself will be carried away if not blended properly. However, researchers tried blending nano additives with soil to prevent soil erosion. Unlike the studies on strength improvement, not many nano additives were tested for their erosion resistance potential and the studies are still in the developmental phase. The reason is the bulk requirement of the nano additive for practical applications and complexities in the application techniques.

3.2.1 Water erosion

The runoff from rainfall tends to erode the soil particles in slope which would affect the productivity of the soil and also cause landslides. Attempts were made to subside the erodibility of soil by blending nano additives. Shourijeh et al. (2022) tested seven different soil types treated with a wide range of additives like sodium bentonite, calcium bentonite, lime, silica fume and nano clay. Silica fume treatment did not alter the erosion resistance behaviour of the soil. On the other hand, nano clay treatment showed negligible variations up to a dosage of 4% and mild variations beyond it. The authors inferred that the presence of montmorillonite-based nano clay did not bring any visible variations to the erosion resistance of the soil unlike other additives studied. On the other hand, Zomorodian et al. (2019) showed that the addition of 1.5% nano silica solution in sand provided a 92% reduction in water erosion within a period of just 1 day, which is much faster than the conventional treatment methods. The authors stated that the viscous nano silica gel formed bounded the soil grains and enhanced their shear strength, resulting in the soil's erosion resistance. Similarly, Leone et al. (2021) showed that nano silica solution could offer better resistance against storm erosion on dunes near the seaside. So it is clear that nano silica in its viscous solution state would be effective for erosion resistance whereas in its powdered form is of no use.

Among the conventional nano additives, very few researchers experimented with nature-friendly solutions for erosion resistance. For instance, Zhou et al. (2018) simulated a rainfall model to study the erosion resistance potential of nanocarbon (from coconut shells). The study results revealed that 0.7% nanocarbon would effectively reduce the water percolation rate, improve the runoff, and reduce the sediment loss. Chen et al. (2020) experimented with an innovative solution of treating loess with nano biochar which reduced both sediment loss and nitrate loss during runoff. With just 1.0% nano biochar, the treatment promoted nitrate migration to a deeper depth.

3.2.2 Wind erosion

Wind erosion is another major issue in arid and semi-arid regions, where the soil gets carried away by wind destroying the soil nutrients and fertility in these regions. Izadi et al., (2024) tested sandy soil treated with a blend of nano clay and blast furnace slag subjected to a wind tunnel erosion test. Results showed that at an optimal combination of 3% nano clay and 1% blast furnace slag, the soil showed a 50% increase in moisture resistance capacity and resisted wind erosion even at 100 kmph.

Studies on erosion control show that similar to shear strength enhancement characteristics, nano additives can be used to subside erosion as well. However, the restriction is on their lack of commercial and bulk-scale availability. Also, to prove that the benefits would be more on the nanoscale, a detailed comparative study is required with micro-sized additives.

3.3 Landfill Liner

Similar to the studies on soil erosion, most investigations on the applications of nano additives as landfill liners are limited to nano clay. The reason is that they can be an eco-friendly substitution for conventional liner materials. Mukri et al. (2018) pointed out the scarcity of pure kaolin for field use and suggested that nano kaolin replacement can effectively reduce the quantity of kaolin clay as a landfill liner, pointing out its ability to reduce permeability. Sarand et al. (2022) pulverized the locally available soil in Iran to its nano form and mixed it with the native soil along with bentonite. At an optimum dosage of 2% of nano soil, the permeability is reduced by 13.75 times which is a desired quality in designing landfill liners. Tao et al. (2023) proposed the idea of using the combination of nano silica and sisal fibres as a composite liner material where the nano silica would reduce the hydraulic conductivity and the sisal fibres would reduce the cracking dimensions of the liner bottom. At an optimum dosage of 0.75% nano silica and 0.3% sisal fibre the fissure length was reduced by 19.94 times and the cracking factor was reduced by 27.25 times, respectively. Similarly, Nikbakht et al. (2023) combined montmorillonite nano clay and cellulose nanofibres with low-plastic clay to study their suitability for a landfill liner. Both nano clay and nano fibres showed a reduction in permeability at an optimal dosage of 9%, individually. However, the composite blend of nano clay and nano fibre reduced the permeability to a maximum extent of around 669 times. Similar to the previous case, the nano clay filled the pores whereas the nano fibres controlled the cracking. However, to date, most studies have been done only on nano clay and extension of this concept for other nano additives, particularly bio nano additives are needed to incorporate environmentally-friendly options in landfill lining.

3.4 Chemical Adsorption

Although one reason for using nano additives as landfill liners is their ability to reduce permeability, the added advantage of using nano additives is their efficiency in adsorbing metals that might be harmful to the environment. For instance, Mahmoud et al. (2020) reported that 0.2% nano biochar treated saline soil has 15.22% higher phosphorous adsorption capacity due to the surface precipitation phenomenon. Though sewage sludge is known for its nutrient value, it may sometimes impart toxic characteristics to the soil. Feizi et al. (2019) experimented with the ability of nanometal oxides to adsorb cadmium, copper, nickel and zinc in sewage sludge-treated alkaline soil. Although the results varied depending on the chemical characteristics of the sludge, it was proved that nanoparticles of zinc oxide and magnesium oxide showed better sorption than conventional geomaterials like zeolite and bentonite. In a similar experiment, Özçoban and Acarer (2022) tested the heavy metal attenuation characteristics of 5% nano iron oxide, nano aluminium oxide and nano clay-treated kaolin. Results showed that the treated soil had a minimum of 90% iron, 47% manganese, 39% zinc, 33% copper and 36% lead removal rates. The study on the application of nanomaterial for soil metal adsorption is vast, but it is still explored a little because of the constraints with the large requirements and consequently the cost for such applications.

4. Nano size Effect and its Advantages for Environmental Applications

Although several literature proofs were discussed to show the ability of nanomaterial in geotechnical applications, the advantages of such smaller particle sizes still need to be clearly established to promote this future-ready additive for various field applications and research projects. The growing concern about environmental carbon pollution with conventional additives like cement and lime demanded the need for techniques that would emit less in the environment (Kannan and Sujatha, 2023a). Lesser emissions can either be achieved by entirely changing the source of such emissions or reducing the use of such emitting sources. In this present era, completely avoiding lime and cement or any other additive in that case is not possible. Hence, nano additive came as a handy solution to

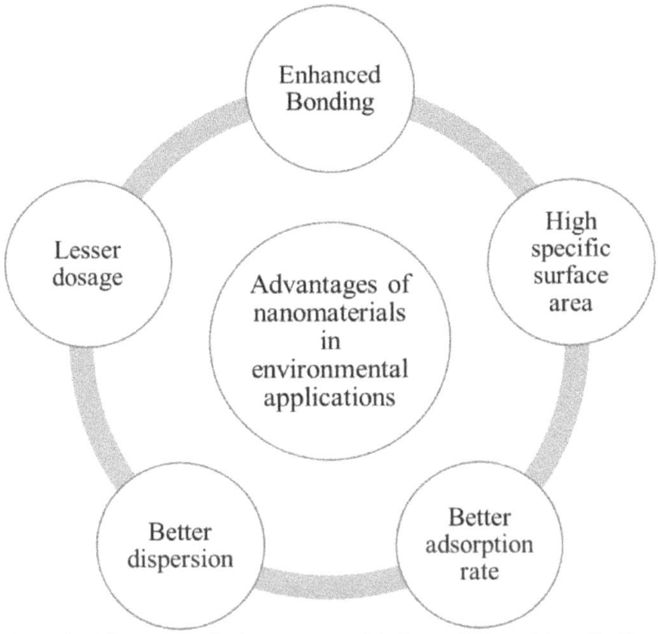

Figure 2. Advantages of using nanomaterials for environmental applications.

achieve a lesser emission with limited usage of material that would yield effective results. Figure 2 shows the most important advantages of using nanomaterials in environmental applications.

Nanomaterials, by its definition, are any particles with a size less than 100 nm in at least any one of its dimensions (Kannan and Sujatha, 2022a). With a decrease in the particle size, the specific surface area of the material starts to increase as there would be more particles per unit gram and hence more surface area per unit gram. With more exposed surfaces, the sites for interaction will also be increased. As a result, the bonding between soil and the nanoparticles will be stronger. Also, rather than conventional micro-sized additives which will be smaller in quantity per unit mass, more nanoparticles would ensure better dispersion and hence the bonding will be more uniform throughout. Because of this uniform dispersion, the desired results can be achieved in a relatively lesser dosage, thereby achieving a better adsorption rate than a similar quantity of micro-sized additives.

5. Comparative Study on the Effect of Size of Biopolymers

Biopolymer stabilization is not a new method in geotechnical ground improvement. However, the concept of nanobiopolymers is a relatively newer topic in the field. This concept is the key to emphasize the size effect in biopolymer stabilization. So far, not many biopolymers have been explored in their nano form. However, there are a few proven results that would confirm the advantages of using nanobiopolymers as a futuristic soil stabilizer. Cellulose is a major polysaccharide biopolymer known for its abundance. Kannan and Sujatha (2023a) and Sujatha and Kannan (2022) studied the geotechnical improvement in low-plastic organic silt treated with sodium carboxymethyl cellulose (CMC) in its micro form and nano form respectively. Both these investigations were carried out on the same soil to ensure only the nano size is the varying factor and the optimum dosage of additive in both micro and nano-sized forms was found to be 0.5%.

5.1 Strength

Based on the experimental outcomes, it was observed that soil treated with nano CMC showed enhanced strength in soil than its micro-sized counterpart (as observed in Table 1). It should be

Table 1. Strength enhancement in nano CMC treated soil (Kannan and Sujatha, 2023a; Sujatha and Kannan, 2022).

Curing period	Unconfined Compression Strength (kPa)		Strength improvement (%)
	Soil + 0.5% CMC	Soil + 0.5% nano CMC	
0 days	219.67	264.64	20.47
7 days	288.59	343.76	19.12
14 days	296.25	346.67	17.02
28 days	304.7	349.07	14.56

noted that the strength enhancement was consistent even at higher curing periods of up to 28 days. Also, despite the reduction in the size of the additive, the mechanism of strength improvement was still the same by bridging the soil with viscous CMC gel, where only the size effect brought the changes.

5.2 Permeability

Unlike strength, permeability showed a negligible variation with nanoscale implementation. For instance, the permeability of CMC-treated silt at the optimum dosage of 0.5% was 0.12×10^{-8} cm/s on the initial day which was slightly increased to 0.72×10^{-8} cm/s. Similarly, the permeability of 0.5% nano CMC treated silt was 0.58×10^{-8} cm/s and 1.39×10^{-8} cm/s, during the initial day and 28 days respectively. Here, the increase in permeability was rather considered as an advantageous effect than a drawback. Both CMC and nano CMC in general tend to absorb and retain more water which reflects the increased permeability at higher curing periods (Kannan and Sujatha, 2023a; Sujatha and Kannan, 2022). Having said so, the increased permeability of nano CMC-treated soil indicates that nano CMC absorbed more water than conventional CMC.

5.3 Consolidation

Both CMC and nano CMC treated soil maintained the same coefficient of consolidation of 0.06 m^2/year at the highest dosage of 0.75%. However, the compression index and swell index reduced from 0.84 to 0.68 and 0.08 to 0.06 with 0.75% CMC and 0.75% nano CMC respectively. The moisture-retaining capacity of nano CMC prevents the expulsion of water and deters the consolidation process, resulting in the reduction of compression magnitude whereas the viscous nature of the gel prevents further swelling.

Although the numerical data are typical outcomes based on the material characteristics, these comparison data clearly establish that nano CMC have beneficial effects that conventional micro-sized CMC. This analogy will fit almost all nano additives if taken under consideration.

6. Conclusion

Biopolymers find extensive applications in numerous fields like pharmaceutics, medicine, food, cosmetics and construction industries. The construction industry is one of the major contributors to the emission of GHG. The environmental guidelines recommend all industries to shift their production and supply chain activities to a more carbon-neutral side. In such a scenario, biopolymers are a boon to the construction industry. Though biopolymers find many applications in the building and construction material field, their use in the geotechnical and geo-environmental fields is mostly at the research level. Also, in most research findings the effect of size of the biopolymer is not discussed. This study highlights the effectiveness of nano-biopolymers in geotechnical applications, particularly in the improvement of strength. The study points the direction of the choice of nanobipoylmers in geotechnical applications.

References

Ahmadi, H. and Shafiee, O. (2019). Experimental comparative study on the performance of nano-SiO2 and microsilica in stabilization of clay. European Physical Journal Plus, 134(9): 1–14. https://doi.org/10.1140/epjp/i2019-12918-1.

Ali, S. W., Basak, S. and Shukla, A. (2020). Nanoparticles: a potential alternative to classical fire retardants for textile substrates. In Handbook of Nanomaterials for Manufacturing Applications (pp. 265–278). Elsevier. https://doi.org/10.1016/B978-0-12-821381-0.00011-9.

Alsharef, J. M. A., Taha, M. R., Govindasamy, P., Firoozi, A. A. and Al-Mansob, R. A. (2019). Effect of nanocarbons on physical and mechanical properties of soils. In Carbon Nanomaterials for Agri-food and Environmental Applications (Issue 1997). Elsevier Inc. https://doi.org/10.1016/B978-0-12-819786-8.00020-7.

Arora, A., Singh, B. and Kaur, P. (2019). Performance of Nano-particles in stabilization of soil: A comprehensive review. Materials Today: Proceedings, 17: 124–130. https://doi.org/10.1016/j.matpr.2019.06.409.

Bahmani, S. H., Huat, B. B. K., Asadi, A. and Farzadnia, N. (2014). Stabilization of residual soil using SiO2 nanoparticles and cement. Construction and Building Materials, 64: 350–359. https://doi.org/10.1016/j.conbuildmat.2014.04.086.

Bahmani, S. H., Farzadnia, N., Asadi, A. and Huat, B. B. K. (2016). The effect of size and replacement content of nanosilica on strength development of cement treated residual soil. Construction and Building Materials, 118: 294–306. https://doi.org/10.1016/j.conbuildmat.2016.05.075.

Baziar, M. H., Saeidaskari, J. and Alibolandi, M. (2018). Effects of Nanoclay on the Treatment of Core Material in Earth Dams. Journal of Materials in Civil Engineering, 30(10): 1–11. https://doi.org/10.1061/(asce)mt.1943-5533.0002415.

Berenguer, A. A., Limpo Orozco, F. J., Del Valle Alvarez, B. E. and Hidalgo Martin, M. (1999). Additive for special plasters and mortars, compositions containing the additive and utilization of the additive in special plasters and mortars (WO9959933 A1).

Chang, I., Im, J., Prasidhi, A. K. and Cho, G.-C. C. (2015). Effects of Xanthan gum biopolymer on soil strengthening. Construction and Building Materials, 74(x): 65–72. https://doi.org/10.1016/j.conbuildmat.2014.10.026.

Changizi, F. and Haddad, A. (2015). Strength properties of soft clay treated with mixture of nano-SiO2 and recycled polyester fiber. Journal of Rock Mechanics and Geotechnical Engineering, 7(4): 367–378. https://doi.org/10.1016/j.jrmge.2015.03.013.

Changizi, F. and Haddad, A. (2017). Improving the geotechnical properties of soft clay with nano-silica particles. Proceedings of the Institution of Civil Engineers: Ground Improvement, 170(2): 62–71. https://doi.org/10.1680/jgrim.15.00026.

Chen, X., Zhou, B., Wang, Q., Tao, W. and Lin, H. (2020). Nano-biochar reduced soil erosion and nitrate loss in sloping fields on the Loess Plateau of China. CATENA, 187: 104346. https://doi.org/10.1016/j.catena.2019.104346.

Choobbasti, A. J., Samakoosh, M. A. and Kutanaei, S. S. (2019). Mechanical properties soil stabilized with nano calcium carbonate and reinforced with carpet waste fibers. Construction and Building Materials, 211: 1094–1104. https://doi.org/10.1016/j.conbuildmat.2019.03.306.

Emmanuel, E., Lau, C. C., Anggraini, V. and Pasbakhsh, P. (2019). Stabilization of a soft marine clay using halloysite nanotubes: A multi-scale approach. Applied Clay Science, 173(October 2018): 65–78. https://doi.org/10.1016/j.clay.2019.03.014.

Feizi, M., Jalali, M., Antoniadis, V., Shaheen, S. M., Ok, Y. S. and Rinklebe, J. (2019). Geo- and nano-materials affect the mono-metal and competitive sorption of Cd, Cu, Ni, and Zn in a sewage sludge-treated alkaline soil. Journal of Hazardous Materials, 379: 120567. https://doi.org/10.1016/j.jhazmat.2019.04.050.

Ghadr, S., Assadi-Langroudi, A., Hung, C., O'Kelly, B. C., Bahadori, H. and Ghodsi, T. (2020). Stabilization of sand with colloidal nano-silica hydrosols. Applied Sciences (Switzerland), 10(15). https://doi.org/10.3390/app10155192.

Gypsum plaster for concrete or cement surfaces (JPS4860123 A). (1973).

Horinoue, M., Okada, N. and Nomura, T. (2011). Method for producing porous concrete (JP2011105579 A).

Horiuchi, S., Kosaka, N., Wakabayashi, S., Yonemura, S., Iizuka, Y., Yamaguchi, H. et al. (1994). Mesh form for molding biodegradable plastic concrete and production of concrete molded object using the same (JPH06297420 A).

Inoue, Y., Irie, H., Oouchi, T., Kojima, T. and Ishigami, M. (1982). Air entraining agent for concrete (JPS57166352 A).

Izadi, R., Mahinroosta, M., Allahverdi, A. and Ghadir, P. (2024). Stabilizers based on nanoclay and blast furnace slag to reduce wind erosion of sandy soil green stabilization of sandy soil. International Journal of Environmental Science and Technology. https://doi.org/10.1007/s13762-024-05630-8.

Joju, M. R. and Chandrakaran, S. (2012). Comparative Study on Stabilisation of Marine Clay Using Nano - Silica and. Indian Geotechnical Conference, Bengakuru, India 13 - 15 December 2018, 1–6.

Kannan, G. and Sujatha, E. R. (2022a). A review on the Choice of Nano-Silica as Soil Stabilizer. Silicon, 14(12): 6477–6492. https://doi.org/10.1007/s12633-021-01455-z.

Kannan, G. and Sujatha, E. R. (2022b). Geotechnical behaviour of nano-silica stabilized organic soil. Geomechanics and Engineering, 28(3): 239–253. https://doi.org/10.12989/gae.2022.28.3.239.

Kannan, G. and Sujatha, E. R. (2023a). A study on the efficacy of low viscous nanosized biopolymer on the mechanical and hydraulic properties of organic silt. Geomechanics and Engineering, 34(3): 221–231. https://doi.org/10.12989/gae.2023.34.3.221.

Kannan, G. and Sujatha, E. R. (2023b). Crustacean polysaccharides for the geotechnical enhancement of organic silt: A clean and green alternative. Carbohydrate Polymers, 299: 120227. https://doi.org/10.1016/j.carbpol.2022.120227.

Kannan, G. and Sujatha, E. R. (2023c). Effect of nano additive on mechanical properties of natural fiber reinforced soil. Journal of Natural Fibers, 20(1): 1–15. https://doi.org/10.1080/15440478.2022.2143980.

Kannan, G., O'Kelly, B. C. and Sujatha, E. R. (2023). Geotechnical investigation of low-plasticity organic soil treated with nano-calcium carbonate. Journal of Rock Mechanics and Geotechnical Engineering, 15(2): 500–509. https://doi.org/10.1016/j.jrmge.2022.05.004.

Khalid, N., Mukri, M., Kamarudin, F., Abdul Ghani, A. H., Arshad, M. F., Sidek, N. et al. (2015). Effect of nanoclay in soft soil stabilization. In InCIEC 2014 (pp. 905–914). Springer Singapore. https://doi.org/10.1007/978-981-287-290-6_79.

Kholghifard, M. and Amini Behbahani, B. (2022). Shear strength of clayey sand treated by nanoclay mixed with recycled polyester fiber. Journal of Central South University, 29(1): 259–269. https://doi.org/10.1007/s11771-022-4895-y.

Kumar, S. A. and Sujatha, E. R. (2021). An appraisal of the hydro-mechanical behaviour of polysaccharides, xanthan gum, guar gum and β-glucan amended soil. Carbohydrate Polymers, 265(April): 118083. https://doi.org/10.1016/j.carbpol.2021.118083.

Kwon, Y.-M., Moon, J.-H., Cho, G.-C., Kim, Y.-U. and Chang, I. (2023). Xanthan gum biopolymer-based soil treatment as a construction material to mitigate internal erosion of earthen embankment: A field-scale. Construction and Building Materials, 389(January): 131716. https://doi.org/10.1016/j.conbuildmat.2023.131716.

Leone, E., Kobayashi, N., Francone, A., Bartolo, S. De, Strafella, D., D'Alessandro, F. et al. (2021). Use of nanosilica for increasing dune erosion resistance during a sea storm. Journal of Marine Science and Engineering, 9(6): 620. https://doi.org/10.3390/jmse9060620.

Luo, H. L., Hsiao, D. H., Lin, D. F. and Lin, C. K. (2012). Cohesive soil stabilized using sewage sludge ash/cement and nano aluminum oxide. International Journal of Transportation Science and Technology, 1(1): 83–99. https://doi.org/10.1260/2046-0430.1.1.83.

Mahmoud, E., El Baroudy, A., Ali, N. and Sleem, M. (2020). Spectroscopic studies on the phosphorus adsorption in salt-affected soils with or without nano-biochar additions. Environmental Research, 184: 109277. https://doi.org/10.1016/j.envres.2020.109277.

Maiti, D., Tong, X., Mou, X. and Yang, K. (2019). Carbon-Based Nanomaterials for Biomedical Applications: A Recent Study. Frontiers in Pharmacology, 9(March), 1–16. https://doi.org/10.3389/fphar.2018.01401.

Miller, S. A., Habert, G., Myers, R. J. and Harvey, J. T. (2021). Achieving net zero greenhouse gas emissions in the cement industry via value chain mitigation strategies. One Earth, 4(10): 1398–1411. https://doi.org/10.1016/j.oneear.2021.09.011.

Miwa, M. and Yonekura, R. (1994). Fundamental properties of sodium silicate based grouts. Soils and Foundations, 34(3): 113–121. https://doi.org/10.3208/sandf1972.34.3_113.

Mohammadi, M., Khodaparast, M. and Rajabi, A. M. (2021). Effect of nano calcium carbonate (nano CaCO3) on the strength and consolidation properties of clayey sand soil. Road Materials and Pavement Design. https://doi.org/10.1080/14680629.2021.1976255.

Mohammadi, M., Rajabi, A. M. and Khodaparast, M. (2022). Experimental and numerical evaluation of the effect of nano calcium carbonate on geotechnical properties of clayey sand soil. KSCE Journal of Civil Engineering, 26(1), 35–46. https://doi.org/10.1007/s12205-021-1914-8.

Mohammed, A. S. A., Naveed, M. and Jost, N. (2021). Polysaccharides; Classification, Chemical Properties, and Future Perspective Applications in Fields of Pharmacology and Biological Medicine (A Review of Current Applications and Upcoming Potentialities). Journal of Polymers and the Environment, 29(8): 2359–2371. https://doi.org/10.1007/s10924-021-02052-2.

Mohan, S., Oluwafemi, O. S., Kalarikkal, N., Thomas, S. and Songca, S. P. (2016). Biopolymers – Application in Nanoscience and Nanotechnology. In Recent Advances in Biopolymers (Vol. 11, Issue tourism, p. 13). InTech. https://doi.org/10.5772/62225.

Mukri, M., Zainuddni, A. N., Abdullah, N. A. and Ibrahim, N. (2018). Performance of different percentage on nano-kaolin as additives in soil liner application. Materials Today: Proceedings, 5(10): 21604–21611. https://doi.org/10.1016/j.matpr.2018.07.009.

Niaounakis, M. (2015). Introduction. In Biopolymers: Processing and Products. https://doi.org/10.1016/b978-0-323-26698-7.00001-5.

Nikbakht, M., Sarand, F. B., Dabiri, R. and Hajialilue Bonab, M. (2023). Investigation of the Leachate Effect on Permeability and Geotechnical Characteristics of Fine-Grained Soil Modified Using Nanoclay–Nanofiber Composites. Water, 15(2): 294. https://doi.org/10.3390/w15020294.

Özçoban, M. Ş. and Acarer, S. (2022). Investigation of the Effect of Leachate on Permeability and Heavy Metal Removal in Soils Improved with Nano Additives. Applied Sciences, 12(12): 6104. https://doi.org/10.3390/app12126104.

Sarand, F. B., Azarnia, A., Jigheh, H. S., Ebrahimiasl, S. and Dabiri, R. (2022). Physio-mechanical evaluation of Nano-soil as an additive to the sand-bentonite mixture for Tabriz city landfill liner Effects of Bentonite on RCCP View project Comparisons of Liquefaction Potential Evaluation Methods View project. AUT Journal of Civil Engineering, 6(3): 359–368. https://doi.org/10.22060/ajce.2023.21129.5793.

Shourijeh, P. T., Soroush, A. and Daneshi-Sadr, A.-H. H. (2022). The effects of lime, bentonite and nano-clay on erosion characteristics of clay soils. European Journal of Environmental and Civil Engineering, 26(9): 3762–3787. https://doi.org/10.1080/19648189.2020.1818629.

Singh, R., Gautam, S., Sharma, B., Jain, P. and Chauhan, K. D. (2021). Biopolymers and their classifications. In Biopolymers and their Industrial Applications (pp. 21–44). Elsevier. https://doi.org/10.1016/B978-0-12-819240-5.00002-X.

Steinbüchel, A. (2005). Non-biodegradable biopolymers from renewable resources: Perspectives and impacts. Current Opinion in Biotechnology, 16(6): 607–613. https://doi.org/10.1016/j.copbio.2005.10.011.

Sujatha, E. R. and Kannan, G. (2022). An Investigation on the Potential of Cellulose for Soil Stabilization. Sustainability (Switzerland), 14(23): 16277. https://doi.org/10.3390/su142316277.

Tabarsa, A., Latifi, N., Meehan, C. L. and Manahiloh, K. N. (2018). Laboratory investigation and field evaluation of loess improvement using nanoclay – A sustainable material for construction. Construction and Building Materials, 158: 454–463. https://doi.org/10.1016/j.conbuildmat.2017.09.096.

Taha, M. R. and Taha, O. M. E. (2012). Influence of nano-material on the expansive and shrinkage soil behavior. Journal of Nanoparticle Research, 14(10). https://doi.org/10.1007/s11051-012-1190-0.

Tao, G., Guo, E., Yuan, J., Chen, Q. and Nimbalkar, S. (2023). Permeability and Cracking of Compacted Clay Liner Improved by Nano-SiO2 and Sisal Fiber. KSCE Journal of Civil Engineering, 27(12): 5109–5122. https://doi.org/10.1007/s12205-023-1435-8.

Zhou, B., Chen, X., Wang, Q., Wei, W. and Zhang, T. (2018). Effects of nano carbon on soil erosion and nutrient loss in a semi-arid loess region of Northwestern China. International Journal of Agricultural and Biological Engineering, 11(1): 138–145. https://doi.org/10.25165/j.ijabe.20181101.2775.

Zomorodian, S. M. A., Soleymani, A. and O'Kelly, B. C. (2019). Briefing: Improving erosion resistance of sand using nano-silica additive. Proceedings of the Institution of Civil Engineers: Ground Improvement, 172(1): 3–11. https://doi.org/10.1680/jgrim.17.00086.

CHAPTER-15

Sustainable use of Biopolymer-metal Nanoparticles and Commercialization

Sneha Kumari Binha and *Dinesh Prasad**

1. Introduction

In recent years, there has been growing interest in the sustainable use of biopolymer-metal nanoparticle composites and their potential for commercialization. Biopolymers, derived from natural sources such as plants, animals, and microorganisms, offer several advantages over synthetic polymers, including biodegradability, renewability, and reduced environmental impact. Metal nanoparticles, on the other hand, exhibit unique physical, chemical, and optical properties that make them valuable for various applications, ranging from healthcare to environmental remediation.

The use of biodegradable polymers, or biopolymers, presents a notable solution to address the widespread reliance on synthetic plastics (Lim et al., 2022). Derived primarily from renewable sources such as biomass, corn, sugarcane, and molasses, these bio-based polymers have garnered considerable attention. As these polymers undergo degradation, they undergo changes such as a decrease in pH and molecular weight, along with an increase in crystallinity (Chiulan et al., 2021). These polymers are renewable resources that undergo swift degradation by natural microorganisms under suitable conditions, including adequate moisture, oxygen, and temperature (Li et al., 2022). They decompose without leaving behind any harmful or discernible residues. Incorporating various natural fillers or additives notably boosts the physical and mechanical characteristics of these biomaterials (Khalid et al., 2021). Over the past decade, the significance of these polymers has grown exponentially, leading to expanded applications across diverse engineering sectors. They are extensively utilized in a wide array of industries, including pharmaceuticals, food packaging, electronics, textiles, agriculture, construction, biomedical, aerospace, and automotive (Li et al., 2022; Khalid et al., 2021; Arif et al., 2022).

The combination of biopolymers and metal nanoparticles presents exciting opportunities for creating advanced materials with enhanced properties and functionalities. These composite materials have shown promise in diverse fields such as biomedicine, catalysis, sensing, packaging, and wastewater treatment. Moreover, their sustainable nature aligns with the increasing global focus on environmentally friendly solutions and the transition towards a circular economy. Table 1 shows recently adopted different biopolymers-based materials and their applications.

Department of Bioengineering and Biotechnology, Birla Institute of Technology, Mesra, Ranchi, India-835215.
* Corresponding author: dinesh@bitmesra.ac.in
DOI: 10.1201/9781003558330-15

Table 1. Recently adopted different biopolymers-based materials and their applications.

Biopolymers	Added materials	Applications	References
polylactic acid/ polyethylene oxide (PLA/PEO)	CNT	Degradation-based nanobiosensor	Zare et al., 2022
Poly L Lactic Acid (PLLA)	Xeolitic imidazolate framework-8 (ZIF-8)	Bone scaffold	Yang et al., 2018
Polycaprolactone (PCL)	Organo modified montmorillonite clay	Fixation devices	Malik et al., 2018
Polycaprolactone (PCL)	nHAP/MWCNT	Bone regeneration	Díaz et al., 2021
polylactic acid (PLA)	MWCNTs	Chemo, photothermal therapy and tissue engineering	Das et al., 2019
Chitosan	Nano-sized BG	Bone scaffold and bone tissue regeneration	Singh et al., 2019
polylactic acid (PLA)	HAP	Drug delivery and tissue engineering	Mondal et al., 2020
Polyvinyl alcohol/Carboxymethyl cellulose (PVA/CMC)	Na_2CO_3	Paper processing, adhesives and coatings, water-soluble films, emulsifiers, textiles	Zhu et al., 2020
Polyhydroxybutyrate (PHB)	ZrO_2	Surgical implant	Ferlic et al., 2020
Polyhydroxybutyrate (PHB)	MWCNT	Tissue engineering	Parvizifard and Karbasi, 2019

This introduction will explore the sustainable aspects of utilizing biopolymer-metal nanoparticle composites, including their synthesis methods, applications, environmental benefits, challenges, and the pathway towards commercialization. By harnessing the unique properties of biopolymers and metal nanoparticles in a sustainable manner, these innovative materials hold the potential to drive positive societal and economic impacts while minimizing environmental harm.

2. Biopolymers: Properties and Sustainable Sourcing

Biopolymers are a class of polymers derived from natural sources such as plants, animals, or microorganisms. They are characterized by their biodegradability, renewable nature, and often exhibit properties that make them attractive for various applications in materials science, medicine, and biotechnology.

2.1 Natural Origin: Biopolymers are synthesized by living organisms through biochemical pathways. They can be found in a wide range of natural sources, including cellulose from plants, chitin from crustaceans and insects, proteins from animals, and nucleic acids from microorganisms. In contemporary biopolymeric composites, a variety of natural fibres are extensively utilized as reinforcement materials (Rangappa and Sienchin, 2022). These fibres come from diverse sources such as plants (like jute, cotton, bamboo, and flax), animals (including silk and wood), minerals (such as asbestos), and microbes (like polycyclic aromatic hydrocarbons and bacterial cellulose). The incorporation of these fibres not only enhances mechanical properties but also contributes to a sustainable environment due to their renewable nature and availability (Grozdanov et al., 2010). As both the fibres (reinforcement) and biopolymers (matrix) are derived from renewable sources and are compostable, biocomposities serve as excellent alternatives to carbon- and glass-reinforced polymer composites (Khalid et al., 2022).

2.2 Renewable: In recent decades, growing environmental concerns have spurred increased attention towards bio-based materials, including biofibres, biopolymers, and biocomposities. This trend has led researchers to focus on these materials as viable replacements for synthetic alternatives

(Liu et al., 2021; Mohan et al., 2023). Natural fibres are categorized into plant, animal, and mineral fibres based on their extraction source. These fibres are utilized as reinforcement within polymer matrices to create bio-based composites and polymer composites (Mazzanti et al., 2019; Sari et al., 2019). Researchers are attracted to natural fibres due to their advantageous properties, such as low density, cost-effectiveness, widespread availability, biodegradability, and ease of processing. Additionally, natural fibres exhibit notable mechanical, thermal, and acoustic properties, along with high fracture resistance. Numerous studies have highlighted the integration of natural fibres into polymer composites and bio-based composites as substitutes for synthetic materials. Consequently, various industries are embracing the adoption of bio-based materials (Abbasi et al., 2019; Battegazzore et al., 2019; Dominici et al., 2019). Unlike synthetic polymers derived from fossil fuels, biopolymers are derived from renewable resources. This makes them more sustainable and environmentally friendly alternatives, as they reduce dependency on finite resources and contribute to the development of a bio-based economy.

2.3 Biodegradability: One of the key advantages of biopolymers is their ability to degrade naturally into simpler compounds through enzymatic processes when exposed to environmental conditions such as heat, moisture, and microbial activity. Biodegradable polymers, sourced from both renewable and petroleum origins, have garnered significant interest for their contribution to environmental sustainability. They undergo microbial degradation, breaking down into minerals, biomass, CO_2, H2O, and CH4, thus supporting ecological balance (Stephen et al., 2023). Biopolymers, large molecular structures derived from animals, plants, or microorganisms, typically encompass proteins, lipids, and polysaccharides (Gao et al., 2022). Protein-based biopolymers encompass substances like gelatin, gluten, keratin, and casein. Lipid-based biopolymers include fatty acids or esters, as well as oils derived from plants (such as soybean and sunflower oil) or animals (Morin et al., 2019). Polysaccharides, another category, are found in various forms such as chitosan, starch, chitin/chitosan, and alginate (Mochane et al., 2020). These biopolymers have gained considerable interest because of their environmentally friendly nature, lack of toxicity, compatibility with living organisms, versatile chemical properties, and inherent functionality (Horue et al., 2021; Daria et al., 2020; Vinod et al., 2020).

2.4 Diverse Properties: Biopolymers exhibit a wide range of physical and chemical properties, which can be tailored and modified through genetic engineering, chemical modifications, or processing techniques. These properties include mechanical strength, flexibility, biocompatibility, and thermal stability, making them suitable for a variety of applications in different industries. Biopolymers typically exhibit properties that can be categorized into three main groups (George et al., 2020): (i) relative properties, self-reflected by the polymer; (ii) processing properties, which determine behaviour during formation; and (iii) functional properties, which refer to its practical capabilities. These properties from each category are interconnected, contributing to the overall desirability of the biopolymer.

2.5 Versatility: Biopolymers have diverse applications across various fields, including food packaging, biomedical devices, drug delivery systems, tissue engineering, and environmental remediation. Their versatility stems from their inherent properties as well as their ability to be processed into different forms such as films, fibres, hydrogels, and nanoparticles (George et al., 2020; Rangappa and Siengchin, 2022; Stephen et al., 2023).

2.6 Challenges: Despite their numerous advantages, biopolymers also pose challenges such as limited mechanical properties, higher production costs compared to conventional polymers, and potential competition with food resources. Addressing these challenges requires ongoing research and innovation in areas such as material engineering, processing technologies, and sustainable production methods (Joye et al., 2016).

3. Metal Nanoparticles: Synthesis and Applications

Metal nanoparticles are tiny particles of metal, typically ranging in size from 1 to 100 nanometres (You et al., 2013). They possess unique optical, electronic, and catalytic properties due to their small size and high surface area-to-volume ratio (Heiligtag et al., 2013). Synthesis methods for metal nanoparticles have evolved significantly, allowing precise control over their size, shape, and surface properties (Auffan et al., 2009). Some common synthesis techniques include:

3.1 Methods for Synthesizing Metal Nanoparticles

3.1.1 Chemical Reduction: In the chemical reduction method, an ionic salt undergoes reduction in a suitable medium in the presence of a surfactant, employing various reducing agents (Guzmán et al., 2009). For instance, sodium borohydride is employed in an aqueous solution to fabricate metal nanoparticles. The resulting metal nanoparticles are coated with trisodium citrate (TSC) or Sodium Lauryl Sulphate (SLS). Occasionally, a stabilizing agent is combined with the reducing agent. The stability of the metal nanoparticles in the dispersion is assessed through absorbance analysis (Chattopadhyay et al., 2014). Different reducing agents, such as sodium borohydride (NaBH4), glucose, ethylene glycol, ethanol, sodium citrate, and hydrazine hydrate, are utilized in the synthesis of silver nanoparticles (Landage et al., 2014).

3.1.2 Chemical Vapour Deposition (CVD): Chemical Vapour Deposition (CVD) was initially documented and patented during the late 19th century. It was primarily utilized for creating carbon fibres as filaments and carbon powder for colouring electric lamp filaments. This deposition process involves the application of a thin film of the desired material onto a surface through chemical reactions involving gaseous molecules carrying atoms essential for film formation (Pedersen et al., 2014). The material to be deposited is released in a volatile form, acting as a precursor. Subsequently, a sequence of chemical reactions occurs among the precursor fragments, the precursor itself, and the surface of the substrate, resulting in the formation of a thin film. Typically, Atomic Layer Deposition (ALD) thin films are generated through surface chemical reactions in this method (Miikkulainen et al., 2013; George et al., 2010; Pedersen and Elliott, 2014).

3.1.3 Electrochemical Deposition: Metal nanoparticles can be synthesized by applying an electric current to a solution containing metal ions. The ions are reduced at the electrode surface, leading to the formation of nanoparticles. Various techniques such as electrodeposition synthesis or template synthesis are employed in the fabrication of nanomaterials (Dikusar et al., 2009). Porous alumina membranes are manufactured using ion etching techniques (Furneaux et al., 1989). Moreover, electrochemical deposition methods like electrochemical anode dissolving processes are favoured for producing porous semiconductor structures (Dikusar et al., 2009). Metal nanoparticles are synthesized through Electrochemical Deposition (ED) utilizing baths comprising metal salts. These baths are either acidic or basic and utilize a three-terminal potentiostat. This method necessitates a cathode electrode where the metal nanoparticles are deposited, along with an anode electrode and a reference electrode such as Ag\AgCl or calomel. A slight voltage is applied for a suitable duration in an electrolytic bath containing metal salt. This technique finds extensive application in various processes like cyclic voltammetry, double pulse, and potential step deposition (Islam and Islam, 2013).

3.1.4 Microwave-Assisted Synthesis: Microwave irradiation can accelerate the reduction of metal ions, leading to faster synthesis of nanoparticles. Currently, microwave methods are favoured over traditional thermal heating for nanoparticle synthesis. By applying microwave frequencies ranging from 300 MHz to 300 GHz, polar molecules like H2O align with the electric field. The reorientation of these dipolar molecules in response to the alternating electric field generates molecular friction, leading to the dissipation of energy as heat. This technique has been recently utilized in the production of silver nanoparticles, where a solution of silver nitrate is irradiated with carboxymethyl chitosan, serving both as a reducing agent and a stabilizer (Wang et al., 2010).

3.1.5 Green Synthesis: The emergence of green synthesis methods for nanoparticle preparation represents a growing trend in nanotechnology. These approaches aim to address challenges encountered with conventional methods, such as complex reactions, high costs, and safety concerns. Green chemistry strategies integrate innovative techniques into synthesis processes and utilize various chemical substances to minimize risks to both human health and the environment (Bhosale et al., 2014). Biological approaches to nanoparticle preparation utilize various microorganisms, enzymes, and plant-derived materials such as isolates and extracts. These methods offer several advantages compared to physical and chemical techniques, including cost-effectiveness, environmental friendliness, and scalability for large-scale production. Moreover, green synthesis avoids the need for high pressure, energy, elevated temperatures, and toxic chemicals.

3.1.5.1 Nanoparticle synthesis using bacteria: Interest in utilizing natural resources, like microorganisms, for synthesizing nanoparticles is on the rise globally. Prokaryotes in particular, have garnered attention for their capacity to produce metallic nanoparticles due to their abundance in various environments and their adaptability to extreme conditions. Bacteria, with their rapid reproduction rates and ease of cultivation and manipulation, offer distinct advantages (Pantidos and Horsfall, 2014). Their growth can be regulated by adjusting factors such as oxygen levels, temperature, and incubation duration (He et al., 2007) suggested that altering the pH of the growth medium during incubation leads to the formation of nanoparticles with varied sizes and shapes (He et al., 2007).

3.1.5.2 Nanoparticle synthesis using fungi: Fungi secrete enzymes and proteins that serve as reducing agents, enabling the synthesis of metal nanoparticles from metal salts. Metals like silver (Ag) can bind to the cytoplasmic membrane via electrostatic interactions, undergoing reduction to form silver nuclei, which then accumulate and eventually lead to the formation of silver nanoparticles (Mukherjee et al., 2001; Ahmad et al., 2003). Various fungi, including *Fusarium oxysporum, Aspergillus fumigatus, Trichoderma reesei*, and others, secrete specific reducing agents that facilitate the reduction of these metal ions. However, practical implementation of microorganism-based nanoparticle synthesis faces challenges due to the necessity of maintaining stringent aseptic conditions (Pantidos and Horsfall, 2014).

3.1.5.3 Nanoparticle synthesis using plants and plant products: Lately, there has been a surge of interest worldwide in the process of synthesizing nanoparticles through plant mediation (Prathna et al., 2011). Various plant extracts, abundant in secondary metabolites like phenolic acids, flavonoids, terpenoids, and alkaloids, are being utilized to create a range of metallic nanoparticles including gold, silver, copper, and zinc. These extracts selectively reduce metallic ions, facilitating the formation of metallic nanoparticles. The inherent redox properties of primary and secondary plant metabolites play a crucial role in this process, acting as both reducing and capping agents, thus promoting the eco-friendly synthesis of nanoparticles (Kuppusamy et al., 2016; Iravani, 2011). Certain plant species such as *Brassica juncea, Ilex crenata, Sesbania drummondii, Clethra barbinervis,* and *Acanthopanax scidophylloides* demonstrate the capacity for photoremediation of heavy metals (Iravani, 2011). The synthesis of nanoparticles involves purifying the bio-reducing agent from the biological extract and carefully mixing it with an aqueous solution containing the relevant metal precursor. This controlled process initiates spontaneous reactions at room temperature, yielding nanoparticles. Occasionally, the addition of Cd, along with stirring and heating, may be necessary to expedite the synthesis process. Plant extracts are preferred for their easy availability, suitability for bulk production, and the environmentally friendly nature of the byproducts or waste produced (Pantidos et al., 2014).

3.1.6 Supercritical fluid technology: Supercritical fluids (SCFs) find extensive applications across various domains of material and chemical sciences. Their distinctive property lies in the ability to adjust their physicochemical attributes between gas and liquid states by simple adjustments in pressure or temperature. This flexibility enables the manipulation of the reaction environment, affecting parameters such as density, viscosity, diffusivity, or surface tension (Aymonier et al., 2006). SCFs are particularly valued for their rapid expansion capabilities, making them selectively advantageous for synthesizing fine powders like nanoparticles (Su et al., 2021). The physical characteristics of SCFs, notably their density, influence many aspects of their behaviour. Commonly employed SCFs include carbon dioxide ($scCO_2$), nitrous oxide, water, methanol, ethanol, ethane, propane, n-hexane, and ammonia (Sekhon, 2010). In this process, the formation of non-aggregated nanoparticles is facilitated by enveloping them in a protective layer that adheres to their surfaces. This coating serves to inhibit particle growth and provides a steric hindrance against aggregation. The synthesis of nanostructures takes place within a reactor equipped with high-pressure pumps and pressurized by a backpressure regulator valve. The precursor, dissolved in various solvents, is introduced into the reactor. The resulting product (nanomaterial) can be retrieved either as a powdered substance using a particle filter located prior to the backpressure regulator valve or as a suspension at the process outlet (Adschiri and Yoko, 2018; Tsang et al., 2018).

3.2 Properties and unique characteristics of metal nanoparticles

Metal nanoparticles exhibit a range of unique characteristics owing to their small size and high surface area-to-volume ratio. Some key characteristics include:

1. **Size-dependent properties:** As the size of metal nanoparticles decreases, their optical, electronic, and catalytic properties can change dramatically due to quantum effects (Xiong et al., 2011; Deepa et al., 2012).
2. **Surface Plasmon Resonance (SPR):** Metal nanoparticles, particularly noble metals like gold and silver, can exhibit SPR, where free electrons collectively oscillate in resonance with incident electromagnetic radiation, leading to strong absorption and scattering of light. This property finds applications in sensing, imaging, and catalysis (Fathi et al., 2019; Mahmudin et al., 2015).
3. **High surface area:** Metal nanoparticles have a significantly high surface area-to-volume ratio compared to bulk materials, making them excellent candidates for catalysis, sensing, and surface-enhanced spectroscopy (Abdul et al., 2019).
4. **Chemical reactivity:** The high surface energy of metal nanoparticles makes them highly reactive, enabling efficient catalytic reactions and surface modification (Larmier et al., 2016).
5. **Optical properties:** Metal nanoparticles can exhibit vibrant colours due to the interaction of light with their surface electrons, leading to applications in colourimetric sensing, imaging, and display technologies (Zhang and Wang, 2017).
6. **Catalytic activity:** Metal nanoparticles serve as efficient catalysts due to their high surface area and unique electronic structure. They find applications in various catalytic reactions, including hydrogenation, oxidation, and carbon-carbon bond formation (Eisa et al., 2018).
7. **Biocompatibility:** Certain metal nanoparticles, such as gold and silver, are biocompatible and can be used in biomedical applications such as drug delivery, imaging, and therapy (Augustine and Hasan, 2020).
8. **Surface functionalization:** Metal nanoparticles can be functionalized with various ligands or biomolecules to control their stability, solubility, and interaction with other molecules, enabling tailored applications in nanomedicine, sensing, and environmental remediation (Hu et al., 2016).

9. **Magnetic properties:** Some metal nanoparticles, such as gold and iron nanoparticles, exhibit magnetic behaviour, making them useful in Magnetic Resonance Imaging (MRI), data storage, and magnetic separation techniques (Gréget et al., 2012).
10. **Thermal conductivity:** Metal nanoparticles can enhance the thermal conductivity of materials when dispersed within them, leading to applications in thermal interface materials, nanocomposites, and thermal management systems (Liu et al., 2015).

3.3 Diverse Applications of Metal Nanoparticles in Various Fields

1. **Catalysis:** Metal nanoparticles serve as highly efficient catalysts in various chemical reactions due to their large surface area and unique electronic properties. They are used in hydrogenation, oxidation, and carbon-carbon bond formation reactions (Astruc, 2020).
2. **Biomedical Applications:** Gold nanoparticles, for example, are used in biomedical imaging, drug delivery, and photothermal therapy. Silver nanoparticles exhibit antimicrobial properties and find use in wound dressings, coatings for medical devices, and water purification systems (McNamara and Tofail, 2017).
3. **Electronics:** Metal nanoparticles are used in the fabrication of electronic devices such as sensors, conductive inks, and printed electronics due to their excellent electrical properties (Gautam et al., 2021).
4. **Optics:** Metal nanoparticles exhibit Localized Surface Plasmon Resonance (LSPR), which can be tuned by controlling their size, shape, and composition. This property is exploited in applications such as Surface-Enhanced Raman Spectroscopy (SERS), photonic devices, and optical sensing (Wang et al., 2020).
5. **Environmental Remediation:** Metal nanoparticles can be used for the degradation of organic pollutants in water and soil, as well as for the removal of heavy metals through adsorption and catalytic reduction processes (Saravanan et al., 2021).

4. Biopolymer-Metal Nanoparticle Composites

Biopolymer-metal nanoparticle composites represent a fascinating intersection of nanotechnology and biomaterials science. These materials combine the unique properties of biopolymers, such as biocompatibility, biodegradability, and structural versatility, with the remarkable properties of metal nanoparticles, such as high conductivity, catalytic activity, and optical properties. The synthesis of these composites typically involves incorporating metal nanoparticles, such as gold, silver, or platinum, into a biopolymer matrix through various methods, including chemical reduction, physical deposition, or electrochemical techniques. Biopolymers commonly used in these composites include chitosan, cellulose, alginate, gelatin, and silk fibroin, Table 2 shows an overview of various biopolymers.

Research in the field of biopolymer-metal nanoparticle composites is ongoing, with efforts focused on optimizing synthesis techniques, understanding structure-property relationships, and exploring new applications across diverse domains. However, challenges such as scalability, stability, and biocompatibility need to be addressed to realize the full potential of these innovative materials.

4.1 Introduction to Biopolymer-Metal Nanoparticle Composites

Biopolymer-metal nanoparticle composites represent a burgeoning field at the intersection of nanotechnology and biomaterials science. These composite materials harness the unique properties of both biopolymers and metal nanoparticles, offering a diverse array of applications across various domains.

Table 2. An overview encompassing various biopolymers, including their origins, distinct attributes, and rates of degradation.

Biopolymer	Origins	Properties	Degradation Time	Applications	References
Polycaprolactone **(PCL)**	Petroleum-based products	(i) Excellent chemical and solvent resistance (ii) Highly employed in composite material (iii) Highly elastic in nature	6–28 months	(i) Surgical implants (ii) Drug delivery devices (iii) Regenerative medicine	Pfau et al., 2020
Polyvinyl acetate **(PVAc)**	Polyvinyl acetate	(i) Nontoxic and thermoplastic polymer (ii) Partially crystalline (iii) High tensile strength and flexibility (iv) Soluble in wate	Several months to years	(i) Textile and paper industry (ii) Food packaging (iii) Biomedical applications	Jain et al., 2017
Polybutylene succinate **(PBS)**	Sugarcane, sugar beet, corn, potato, wheat	(i) High heat resistance among the general biodegradability resin (ii) PBS has excellent compatibility with a fibre.	2–3 months	(i) Bags, food, and cosmetic packaging. (ii) Disposable products (iii) Agriculture industries	Delamarche et al., 2020
Poly-gamma-glutamic acid **(PGA)**	Aliphatic polyester	(i) High degree of crystallinity (45–55%), (ii) High thermal stability (220–230°C) (iii) High gas barrier (three times higher than ethylene vinyl alcohol)	5 months	(i) Drug delivery, (ii) Dental, and orthopedic applications (iii) Tissue engineering	Samantaray et al., 2020
Polylactic acid **(PLA)**	Corn starch, wheat, sugarcane	(i) Very low in price (ii) Good resistance to UV radiation (iii) Low thermal stability (iv) Excellent use as matrix in composite material (v) Non-toxic to the environment	20 months	(i) Food packaging (ii) Drug delivery (iii) Highly used in composite materials	Aliotta et al., 2017
Cellulose	Plant fibers	(i) Abundantly available in nature (ii) Highly crystalline in nature (iii) Good insulation and anti-static properties (iv) Hydrophilic in nature	Weeks to months	(i) Biomedical (ii) Drug delivery (iii) Cosmetics (iv) Clothes (v) Food packaging	Chmolowska et al., 2017
Hyaluronic acid **(HA)**	Animal	(i) Abundantly available polysaccharides (ii) Biocompatibility (iii) High viscosity	4 months	(i) Tissue engineering (ii) Cosmetic industry (iii) Wound dressing (iv) Food packaging	Taniguchi et al., 2019
Silk	Animal	(i) Naturally available protein (ii) Excellent mechanical properties (iii) High luster	6 weeks	(i) Tissue engineering (ii) Food applications (iii) Textile industry	Rastogi et al., 2019
Alginate	Plant	(i) Naturally available polysaccharides (ii) Excellent antimicrobial properties	80 days	(i) Textile industry (ii) Drug delivery system (iii) Tissue engineering (iv) Wound dressing	Karimi et al., 2020

Table 2 contd. ...

...Table 2 contd.

Biopolymer	Origins	Properties	Degradation Time	Applications	References
Collagen	Animal	(i) Naturally available protein (ii) Biocompatibility (iii) Excellent physiochemical properties	12 h	(i) Tissue engineering (ii) Cosmetic industry (iii) Food industry	Polman et al., 2021
Starch	Plant	(i) Naturally available polysaccharides (ii) Excellent water vapour properties (iii) High viscosity (iv) Excellent storage modulus	Several weeks	(i) Drug delivery systems (ii) Textile industry (iii) Paper industry (iv) Food packaging	Tarique et al., 2021

Biopolymers are naturally derived polymers, often sourced from renewable biological resources such as plants, animals, or microorganisms. Examples include chitosan, cellulose, alginate, gelatin, and silk fibroin. Biopolymers are favoured for their biocompatibility, biodegradability, and structural versatility, making them ideal candidates for a range of biomedical and environmental applications.

Metal nanoparticles possess remarkable physical and chemical properties due to their nanoscale dimensions. Common examples include gold, silver, platinum, and palladium nanoparticles. These nanoparticles exhibit unique characteristics such as high surface area-to-volume ratio, catalytic activity, plasmonic behaviour, and tunable optical properties.

The synthesis of biopolymer-metal nanoparticle composites involves the incorporation of metal nanoparticles into a biopolymer matrix through various techniques, including chemical reduction, physical deposition, or electrochemical methods. The choice of the synthesis method depends on factors such as desired nanoparticle size, distribution, and compatibility with the biopolymer matrix.

4.2 Methods for Incorporating Metal Nanoparticles into Biopolymers

4.2.1 In situ Synthesis: Metal nanoparticles can be synthesized directly within the biopolymer matrix. This can be achieved through methods such as chemical reduction, electrochemical deposition, or green synthesis using plant extracts. In this method, metal precursors are added to the biopolymer solution, and reduction agents are introduced to initiate the formation of nanoparticles (Ho et al., 2021; Pan et al., 2020; Asgari et al., 2019).

4.2.2 Physical Mixing: Metal nanoparticles can be physically mixed with biopolymer solutions or matrices. This method involves dispersing pre-synthesized nanoparticles into the biopolymer solution through techniques like sonication, stirring, or homogenization. The challenge here is achieving uniform dispersion and preventing aggregation of nanoparticles (Rahmani et al., 2024; Hassan et al., 2014; Bang et al., 2010).

4.2.3 Electrospinning: Electrospinning is a technique used to produce nanofibres from polymer solutions. Metal nanoparticles can be incorporated into biopolymer nanofibres during the electrospinning process by adding the nanoparticles to the polymer solution before electrospinning. This method enables the fabrication of composite nanofibres with enhanced properties (Segala et al., 2015; Ahmed et al., 2020).

4.2.4 Layer-by-Layer Assembly: This method involves the sequential adsorption of polyelectrolytes and metal nanoparticles onto a substrate or template. Biopolymers with charged functional groups, such as chitosan or alginate, can be alternately deposited with metal nanoparticles through electrostatic interactions. This technique allows precise control over the thickness and composition of the multilayer films (Taketa et al., 2021; Barsan et al., 2016).

4.2.5 Microfluidics: Microfluidic techniques enable the precise control of fluids in microscale channels. Metal nanoparticles can be synthesized and incorporated into biopolymer matrices within microfluidic devices. This approach provides advantages such as rapid mixing, precise control over reaction conditions, and the ability to produce nanoparticles with uniform size and shape (Yao et al., 2023; Niculescu et al., 2021).

4.2.6 Sol-Gel Method: In this method, metal nanoparticles are synthesized in a sol-gel matrix composed of biopolymers. The sol-gel process involves the transition of a solution (sol) into a gel phase, where metal precursors are dispersed and subsequently transformed into nanoparticles. Biopolymers like gelatin or cellulose can serve as the matrix for the sol-gel process (Hamrayev and Shameli, 2021).

4.2.7 Chemical Modification: Biopolymers can be chemically modified to introduce functional groups that facilitate the binding or nucleation of metal nanoparticles. For example, carboxyl, amino, or thiol groups can be introduced onto the biopolymer backbone, enabling coordination with metal ions or direct binding to preformed nanoparticles (Muir and Burdick, 2020; Hoque et al., 2023).

4.2.8 Spray Pyrolysis: This approach involves introducing the initial substance into a solvent for suspension, followed by ultrasonic treatment. A carrier gas is then introduced into a heated container, causing vapourization at high temperatures, leading to the breakdown of oxide materials and subsequent reduction (Choa et al., 2003). This deposition method is highly economical and suitable for applications such as thin films and coatings (Narayan et al., 2019).

4.2.9 Sol Gel Approach and Colloidal Processing: This chemical process can be executed with simplicity at low temperatures, ensuring uniformity and yielding high-quality results. While it is a cost-effective method, it tends to produce weak bonding and exhibits high permeability. It is particularly valuable for producing metal oxide nanoparticles, requiring the presence of monomers in a colloidal solution to form a gel network. This procedure is commonly referred to as the wet chemical technique, involving subsequent hydrolysis and polycondensation reactions (Kamat et al., 2002; Danks et al., 2016).

4.2.10 Chemical and Physical Vapour Deposition Technique: The process involves the evaporation of substances to form a vapour phase, followed by saturation in an inert atmosphere to aid the condensation of metal nanoparticles. Subsequently, thermal treatment is administered in an inert atmosphere (Choy et al., 2003). Achieving uniform dispersion and homogeneous mixing of metal nanoparticles is facilitated by high-energy ball milling (He et al., 2023). These methods incorporate techniques such as sputtering and evaporation.

5. Sustainable Approaches to Synthesis and Functionalization

5.1 *Green Synthesis Methods for Metal Nanoparticles Using Biopolymers*

Green synthesis methods for metal nanoparticles using biopolymers involve the utilization of natural, biodegradable polymers derived from biological sources such as plants, bacteria, fungi, and algae. These methods offer several advantages over conventional chemical synthesis routes, including environmental friendliness, cost-effectiveness, and biocompatibility. Here are some common biopolymers and their applications in the green synthesis of metal nanoparticles:

5.1.1 Chitosan: Chitosan is a biopolymer derived from chitin, found in the exoskeletons of crustaceans such as shrimp and crabs. It has amino and hydroxyl groups that can reduce metal ions and stabilize the resulting nanoparticles. Chitosan-coated metal nanoparticles have been synthesized for various applications, including catalysis, sensing, and drug delivery (Verma et al., 2021; Mohammad et al., 2017).

5.1.2 Starch: Starch is a polysaccharide found in plants and serves as a reducing and stabilizing agent in the synthesis of metal nanoparticles. Starch-coated nanoparticles have been employed in fields such as food packaging, biomedicine, and environmental remediation (Mylkie et al., 2021).

5.1.3 Alginate: Alginate is a polysaccharide extracted from brown algae, consisting of guluronic and mannuronic acid residues. It acts as a reducing and stabilizing agent for the synthesis of metal nanoparticles, with applications in drug delivery, tissue engineering, and wastewater treatment (Yugay et al., 2020).

5.1.4 Cellulose: Cellulose is the most abundant biopolymer on Earth, derived from plant cell walls. It can be used as a template or stabilizer in the synthesis of metal nanoparticles. Cellulose-based nanocomposites have found applications in areas such as water purification, sensing, and biomedical engineering (Bethke et al., 2018).

5.1.5 Proteins: Proteins derived from various biological sources, such as soybeans, milk, and egg whites, can serve as reducing and stabilizing agents for the synthesis of metal nanoparticles. Examples include Bovine Serum Albumin (BSA), soy protein, and lysozyme. Protein-stabilized nanoparticles have been utilized in biomedical imaging, drug delivery, and catalysis (Oliveira et al., 2015; Turcsányi et al., 2023).

5.1.6 DNA: Deoxyribonucleic acid (DNA) can be utilized as a template or scaffold for the synthesis of metal nanoparticles through specific binding interactions. DNA-based nanostructures have shown promise in applications such as biosensing, imaging, and therapeutics (Zhang et al., 2018).

These biopolymer-based green synthesis methods typically involve the reduction of metal ions in the presence of the biopolymer under mild reaction conditions, such as room temperature and atmospheric pressure. The resulting metal nanoparticles exhibit well-defined size, shape, and surface properties, making them suitable for a wide range of applications in various fields.

5.2 Possible Sustainable Approaches to Functionalize Biopolymer-Metal Nanoparticle Composites

Functionalizing biopolymer-metal nanoparticle composites sustainably involves approaches that minimize environmental impact, optimize resource utilization, and enhance the biodegradability of the final product. Here are several sustainable approaches:

5.2.1 Green Synthesis Methods: Utilize environmentally friendly methods for synthesizing metal nanoparticles, such as green chemistry principles or biological methods using plant extracts or microorganisms (Mylkie et al., 2021). These methods often eliminate or reduce the need for harmful chemicals or high energy consumption (Oliveira et al., 2015; Turcsányi et al., 2023).

5.2.2 Biopolymer Sourcing: Choose biopolymers sourced from renewable and sustainable biomass, such as cellulose, chitosan, alginate, or starch. Avoid biopolymers derived from non-renewable resources or those associated with deforestation or habitat destruction (Ling et al., 2018).

5.2.3 Waste Utilization: Utilize waste biomass or by-products from agricultural, forestry, or food industries as sources for biopolymers. This reduces the environmental impact associated with waste disposal and adds value to underutilized resources (Ranganathan et al., 2020).

5.2.4 Bio-based Reducing Agents: Substitute traditional chemical reducing agents with bio-based alternatives derived from natural sources, such as plant extracts, sugars, or polyphenols. These agents can facilitate the synthesis of metal nanoparticles while minimizing environmental pollution (McGauran et al., 2021).

5.2.5 Surface Modification: Employ surface modification techniques to enhance the compatibility between biopolymers and metal nanoparticles, improving stability and performance without relying on environmentally harmful chemicals (Pavlovic et al., 2022).

5.2.6 Biodegradability Enhancement: Modify the structure of biopolymers or incorporate biodegradable additives to improve the overall biodegradability of the composite material. This ensures that the end product can be easily disposed of without causing long-term environmental harm (Sharma et al., 2018).

5.2.7 Life Cycle Assessment (LCA): Conduct a comprehensive LCA to evaluate the environmental impact of the entire lifecycle of biopolymer-metal nanoparticle composites, from raw material extraction to disposal. Use the findings to identify areas for improvement and optimize the sustainability of the manufacturing process (Yates and Barlow, 2013).

5.2.8 Circular Economy Principles: Design products with end-of-life considerations in mind, aiming for recyclability or compostability where possible. Implement closed-loop systems to recover and reuse materials, reducing waste and resource consumption (Dziuba et al., 2021).

5.2.9 Regulatory Compliance: Ensure compliance with relevant environmental regulations and standards throughout the production process. This includes proper waste management, pollution prevention measures, and adherence to safety protocols (Tirado-Kulieva et al., 2022).

6. Economic Considerations for Large-Scale Production and Commercialization

The large-scale production and commercialization of biopolymer-coated metal nanoparticles involve several economic considerations such as:

6.1 Production Cost: The cost of producing biopolymer-coated metal nanoparticles is a significant factor. It includes the cost of raw materials, such as metals and biopolymers, as well as the cost of synthesis processes, equipment, and labour. Optimizing production processes to reduce costs while maintaining quality is essential for commercial viability.

6.2 Scale of Production: Economies of scale play a crucial role in reducing production costs. Larger production volumes typically lead to lower per-unit costs due to spreading fixed costs over a greater number of units. However, achieving economies of scale may require significant upfront investment in production facilities and equipment.

6.3 Raw Material Sourcing: The availability and cost of raw materials, such as metals and biopolymers, can impact production economics. Access to reliable and cost-effective sources of these materials is essential for maintaining competitiveness.

6.4 Synthesis Method: The method used for synthesizing biopolymer-coated metal nanoparticles can influence production costs. Some synthesis methods may be more efficient, cost-effective, or scalable than others. Optimizing the synthesis process to minimize energy consumption, waste generation, and production time can help reduce costs.

6.5 Quality Control: Ensuring product quality is crucial for commercial success. Implementing robust quality control measures throughout the production process can help minimize defects and product variability, reducing the likelihood of costly recalls or rejections.

6.6 Regulatory Compliance: Compliance with regulatory requirements adds another layer of cost and complexity to the commercialization process. Understanding and adhering to relevant regulations regarding product safety, labelling, and environmental impact are necessary for market entry and ongoing operation.

6.7 Market Demand and Pricing: Assessing market demand and setting appropriate pricing strategies are essential for achieving profitability. Conducting market research to understand customer needs, preferences, and willingness to pay can help inform pricing decisions. Additionally, identifying niche markets or applications where biopolymer-coated metal nanoparticles offer unique value propositions can help differentiate products and command premium pricing.

6.8 Competitive Landscape: Analyzing the competitive landscape is critical for identifying potential competitors, assessing their strengths and weaknesses, and positioning products effectively. Differentiating biopolymer-coated metal nanoparticles based on factors such as performance, cost, and sustainability can help capture market share and maintain competitiveness.

7. Commercialization Strategies

7.1 Challenges and Opportunities in Commercializing Sustainable Materials

Commercializing sustainable materials presents both challenges and opportunities:

Challenges:

1. **Cost**: Sustainable materials often require innovative production processes or sourcing methods, which can initially be more expensive than conventional materials. This can deter adoption, especially if consumers or businesses prioritize lower costs over sustainability.
2. **Scale-up**: Transitioning from laboratory-scale production to large-scale manufacturing can be challenging for sustainable materials. Scaling up production while maintaining quality and consistency may require significant investment in infrastructure, technology, and expertise.
3. **Market Acceptance**: Convincing consumers or businesses to switch to sustainable materials may require overcoming entrenched preferences or perceptions. Educating the market about the benefits of sustainable materials, such as reduced environmental impact or improved performance, is essential but can be time-consuming and costly.
4. **Regulatory Hurdles**: Compliance with regulations and standards for sustainable materials can be complex and vary across regions. Navigating regulatory requirements adds time and costs to the commercialization process, particularly for materials with novel properties or applications.
5. **Supply Chain Complexity**: Sustainable materials often involve complex supply chains, incorporating various raw materials, production processes, and distribution channels. Ensuring transparency, traceability, and ethical sourcing throughout the supply chain is critical but can be challenging, particularly for global supply chains.

Opportunities:

1. **Growing Demand**: Increasing awareness of environmental issues and sustainability concerns is a driving demand for sustainable materials across industries. Companies that can provide environmentally friendly alternatives to conventional materials stand to capitalize on this growing market opportunity.
2. **Innovation Potential**: Sustainable materials present opportunities for innovation in material science, manufacturing processes, and product design. Developing new materials or improving existing ones to meet the sustainability criteria can lead to competitive advantages and market differentiation.
3. **Brand Value**: Adopting sustainable materials can enhance brand reputation and appeal to environmentally conscious consumers. Companies that demonstrate a commitment to sustainability through their products and practices can build stronger customer loyalty and attract new market segments.
4. **Risk Mitigation**: Using sustainable materials can help companies mitigate risks associated with environmental regulations, resource scarcity, and reputational damage. Diversifying

supply chains and reducing reliance on finite or environmentally harmful resources can improve resilience and long-term viability.

5. **Collaboration Opportunities**: Collaborating with suppliers, partners, and industry stakeholders can accelerate the development and adoption of sustainable materials. Partnerships across the value chain can facilitate knowledge sharing, resource pooling, and collective action to address common challenges and drive innovation.

7.2 Case studies of successful commercialization efforts

7.2.1 Dextran, a polymer renowned for its properties in creating edible films and coatings, was discovered to serve as a medium for the *in situ* synthesis of silver nanoparticles, typically ranging in size from 10 to 60 nm. The study by Bankura et al. (2012) concluded that these silver nanoparticles were effectively capped by dextran molecules, thereby attaching to the hydroxyl groups of dextran molecules and preventing their aggregation. This formulation demonstrated stability for over a month and exhibited potent antibacterial activity against various strains including *B. subtilis, B. cereus, E. coli, S. aureus*, and *P. aeruginosa.*

7.2.2 Several studies have explored the use of coating materials on fresh and minimally processed fruits and vegetables to mitigate spoilage and extend shelf life during post-storage. Sharanya Devadiga et al. (2016) found that apples and sapota covered with the tested coating extended their storage time to 25 days, while tomatoes, chili peppers, and eggplants showed extensions to 21, 23, and 30 days, respectively.

7.2.3 In recent years, bio-based and bio-coated materials have emerged as promising alternatives to conventional leather, addressing inherent issues in the leather industry. Meyer et al. (2021) highlight the rising demand for these materials among automakers seeking lower-CO2 emission alternatives to animal leather or PVC synthetic-coated fabric. These materials are increasingly utilized for complementary trim solutions on vehicle interiors, such as instrument panels, door panels, and centre consoles, with several nature-based alternatives and animal-free fibrous materials entering the market successfully.

7.2.4 Deserttex™ (Desserto®, Guadalajara, Mexico) represents a notable bio-coated fabric composed of a biopolymer material containing cactus fibres, aiming to develop a fully plant-based material. This fabric has garnered attention in recent years, with prestigious car brands like Mercedes-Benz and BMW incorporating it into their vehicle interiors (Mendanha, 2024).

7.2.5 Pilapong et al. (2014) introduced a smart magnetic nanoparticle-aptamer probe designed for targeted imaging and as a drug carrier for hepatocellular carcinoma treatment. This nanoprobe combined the cancer-specific capabilities of a DNA-based epithelial Cell Ddhesion Molecule (EpCAM) aptamer with the imaging capabilities of magnetic iron oxide nanoparticles (IONP).

8. Conclusion

The sustainable utilization and commercialization of biopolymer-coated metal nanoparticles hold immense promise across various sectors, ranging from healthcare to environmental conservation and beyond. As evidenced by the diverse applications highlighted in research and industry, these nanoparticles offer multiple benefits including enhanced biocompatibility, targeted drug delivery, antimicrobial properties, and eco-friendly alternatives to conventional materials. The successful commercialization of biopolymer-coated metal nanoparticles hinges on addressing key challenges such as scalability of production, regulatory compliance, cost-effectiveness, and understanding their interactions with biological systems. Collaborative efforts between interdisciplinary teams, biotechnologists, engineers, and business experts, are essential to navigate these challenges and realize the full potential of this technology.

Moreover, continued research and innovation are paramount to unlock new opportunities and address emerging needs in fields such as healthcare, agriculture, and sustainable manufacturing.

By harnessing the inherent properties of biopolymers and metal nanoparticles in a responsible and sustainable manner, we can pave the way for a future where advanced nanomaterials contribute to societal well-being while minimizing environmental impact. Hence, the sustainable use and commercialization of biopolymer-coated metal nanoparticles represent a paradigm shift towards more efficient, eco-friendly, and socially responsible technologies, shaping a brighter and more resilient future for generations to come.

References

Abbasi, H., Antunes, M. and Velasco, J. I. (2019). Recent advances in carbon-based polymer nanocomposites for electromagnetic interference shielding. Progress in Materials Science, 103: 319–373.

Abdul Rani, R., Zoolfakar, A. S., Mohamad Ryeeshyam, M. F., Ismail, A. S., Mamat, M. H., Alrokayan, S. et al. (2019). High Surface Area to Volume Ratio 3D Nanoporous Nb 2 O 5 for Enhanced Humidity Sensing. Journal of Electronic Materials, 48: 3805–3815.

Adschiri, T. and Yoko, A. (2018). Supercritical fluids for nanotechnology. The Journal of Supercritical Fluids, 134: 167–175.

Ahmad, A., Mukherjee, P., Senapati, S., Mandal, D., Khan, M. I., Kumar, R. et al. (2003). Extracellular biosynthesis of silver nanoparticles using the fungus Fusarium oxysporum. Colloids and surfaces B: Biointerfaces, 28(4): 313–318.

Ahmed, M. K., Menazea, A. A. and Abdelghany, A. M. (2020). Blend biopolymeric nanofibrous scaffolds of cellulose acetate/ε-polycaprolactone containing metallic nanoparticles prepared by laser ablation for wound disinfection applications. International Journal of Biological Macromolecules, 155: 636–644.

Aliotta, L., Cinelli, P., Coltelli, M. B., Righetti, M. C., Gazzano, M. and Lazzeri, A. (2017). Effect of nucleating agents on crystallinity and properties of poly (lactic acid)(PLA). European Polymer Journal, 93: 822–832.

Arif, Z. U., Khalid, M. Y., Ahmed, W. and Arshad, H. (2022). A review on four-dimensional (4D) bioprinting in pursuit of advanced tissue engineering applications. Bioprinting, 27: e00203.

Asgari, S., Saberi, A. H., McClements, D. J. and Lin, M. (2019). Microemulsions as nanoreactors for synthesis of biopolymer nanoparticles. Trends in Food Science & Technology, 86: 118–130.

Astruc, D. (2020). Introduction: nanoparticles in catalysis. Chemical Reviews, 120(2): 461–463.

Auffan, M., Rose, J., Wiesner, M. R. and Bottero, J. Y. (2009). Chemical stability of metallic nanoparticles: A parameter controlling their potential cellular toxicity *in vitro*. Environmental Pollution, 157(4): 1127–1133.

Augustine, R. and Hasan, A. (2020). Emerging applications of biocompatible phytosynthesized metal/metal oxide nanoparticles in healthcare. Journal of Drug Delivery Science and Technology, 56: 101516.

Aymonier, C., Loppinet-Serani, A., Reverón, H., Garrabos, Y. and Cansell, F. (2006). Review of supercritical fluids in inorganic materials science. The Journal of Supercritical Fluids, 38(2): 242–251.

Bang, J. H. and Suslick, K. S. (2010). Applications of ultrasound to the synthesis of nanostructured materials. Advanced Materials, 22(10): 1039–1059.

Bankura, K. P., Maity, D., Mollick, M. M., Mondal, D., Bhowmick, B., Bain, M. K. et al. (2012). Synthesis, characterization and antimicrobial activity of dextran stabilized silver nanoparticles in aqueous medium. Carbohydrate Polymers, 89(4): 1159–1165.

Barsan, M. M. and Brett, C. M. (2016). Recent advances in layer-by-layer strategies for biosensors incorporating metal nanoparticles. TrAC Trends in Analytical Chemistry, 79: 286–296.

Battegazzore, D., Sattin, A., Maspoch, M. L. and Frache, A. (2019). Mechanical and barrier properties enhancement in film extruded bio-polyamides with modified nanoclay. Polymer Composites, 40(7): 2617–2628.

Bethke, K., Palantöken, S., Andrei, V., Roß, M., Raghuwanshi, V. S., Kettemann, F. et al. (2018). Functionalized cellulose for water purification, antimicrobial applications, and sensors. Advanced Functional Materials, 28(23): 1800409.

Bhosale, R. R., Kulkarni, A. S., Gilda, S. S., Aloorkar, N. H., Osmani, R. A. and Harkare, B. R. (2014). Innovative eco-friendly approaches for green synthesis of silver nanoparticles. International Journal of Pharmaceutical Sciences and Nanotechnology (IJPSN), 7(1): 2328–2337.

Chattopadhyay, D. P. and Patel, B. H. (2014). Nano metal particles: synthesis, characterization and application to textiles. Manufacturing Nanostructures, 184–215.

Chiulan, I., Heggset, E. B., Voicu, S. I. and Chinga-Carrasco, G. (2021). Photopolymerization of bio-based polymers in a biomedical engineering perspective. Biomacromolecules, 22(5): 1795–1814.

Chmolowska, D., Hamda, N. and Laskowski, R. (2017). Cellulose decomposed faster in fallow soil than in meadow soil due to a shorter lag time. Journal of Soils and Sediments, 17: 299–305.

Choa, Y. H., Yang, J. K., Kim, B. H., Jeong, Y. K., Lee, J. S., Nakayama, T. et al. (2003). Preparation and characterization of metal/ceramic nanoporous nanocomposite powders. Journal of Magnetism and Magnetic Materials, 266(1-2): 12–19.

Choy, K. L. (2003). Chemical vapour deposition of coatings. Progress in Materials Science, 48(2): 57–170.

Danks, A. E., Hall, S. R. and Schnepp, Z. J. M. H. (2016). The evolution of 'sol–gel'chemistry as a technique for materials synthesis. Materials Horizons, 3(2): 91–112.

Daria, M., Krzysztof, L. and Jakub, M. (2020). Characteristics of biodegradable textiles used in environmental engineering: A comprehensive review. Journal of Cleaner Production, 268: 122129.

Das, A., Adhikari, J. and Saha, P. (2019). Electrospun polymeric nanocarbon nanomats for tissue engineering. In Nanocarbon and its Composites (pp. 91–122). Woodhead Publishing.

Deepa, K. G., Lekha, P. and Sindhu, S. (2012). Efficiency enhancement in DSSC using metal nanoparticles: A size dependent study. Solar Energy, 86(1): 326–330.

Delamarche, E., Mattlet, A., Livi, S., Gérard, J. F., Bayard, R. and Massardier, V. (2020). Tailoring biodegradability of poly (butylene succinate)/poly (lactic acid) blends with a deep eutectic solvent. Frontiers in Materials, 7: 7.

Devadiga, S. B., Shetty, D., Lakhsmi, G. and Packiyam, J. E. (2016). Synthesis of chitosan silver nanoparticles from chitin of crustacean shell and its applications. Int. J. Curr. Res. Chem. Pharmac. Sci., 3(05): 1–5.

Díaz, E., Aresti, J. and León, J. (2021). Evaluation of physicochemical and mechanical properties with the *in vitro* degradation of PCL/nHA/MWCNT composite scaffolds. Journal of Reinforced Plastics and Composites, 40(3-4): 134–142.

Dikusar, A. I., Globa, P. G., Belevskii, S. S. and Sidel'nikova, S. P. (2009). On limiting rate of dimensional electrodeposition at meso-and nanomaterial manufacturing by template synthesis. Surface Engineering and Applied Electrochemistry, 45: 171–179.

Dominici, F., Garcia Garcia, D., Fombuena, V., Luzi, F., Puglia, D., Torre, L. et al. (2019). Bio-polyethylene-based composites reinforced with alkali and palmitoyl chloride-treated coffee silverskin. Molecules, 24(17): 3113.

Dziuba, R., Kucharska, M., Madej-Kiełbik, L., Sulak, K. and Wiśniewska-Wrona, M. (2021). Biopolymers and biomaterials for special applications within the context of the circular economy. Materials, 14(24): 7704.

Eisa, W. H., Abdelgawad, A. M. and Rojas, O. J. (2018). Solid-state synthesis of metal nanoparticles supported on cellulose nanocrystals and their catalytic activity. ACS Sustainable Chemistry & Engineering, 6(3): 3974–3983.

Fan, C., Liang, Y., Dong, H., Yang, J., Tang, G., Zhang, W. et al. (2018). Guanidinium ionic liquid-controlled synthesis of zeolitic imidazolate framework for improving its adsorption property. Science of the Total Environment, 640: 163–173.

Fathi, F., Rashidi, M. R. and Omidi, Y. (2019). Ultra-sensitive detection by metal nanoparticles-mediated enhanced SPR biosensors. Talanta, 192: 118–127.

Feba Mohan, M. and Praseetha, P. N. (2023). Prospects of biopolymers based nanocomposites for the slow and controlled release of agrochemicals formulations. Journal of Inorganic and Organometallic Polymers and Materials, 33(12): 3845–3860.

Ferlic, P. W., Nogler, M., Weinberg, A. M., Kühn, K. D., Liebensteiner, M. and Coraça-Huber, D. C. (2020). Material modifications enhancing the antibacterial properties of two biodegradable poly (3-hydroxybutyrate) implants. Biomedical Materials, 16(1): 015030.

Furneaux, R. C., Rigby, W. R. and Davidson, A. P. (1989). The formation of controlled-porosity membranes from anodically oxidized aluminium. Nature, 337(6203): 147–149.

Gao, R., Pan, H., Kai, L., Han, K. and Lian, J. (2022). Microbial degradation and valorization of poly (ethylene terephthalate)(PET) monomers. World Journal of Microbiology and Biotechnology, 38(5): 89.

Gautam, A., Komal, P., Gautam, P., Sharma, A., Kumar, N. and Jung, J. P. (2021). Recent trends in noble metal nanoparticles for colorimetric chemical sensing and micro-electronic packaging applications. Metals, 11(2): 329.

George, A., Sanjay, M. R., Srisuk, R., Parameswaranpillai, J. and Siengchin, S. (2020). A comprehensive review on chemical properties and applications of biopolymers and their composites. International Journal of Biological Macromolecules, 154: 329–338.

George, S. M. (2010). Atomic layer deposition: an overview. Chemical Reviews, 110(1): 111–131.

Gréget, R., Nealon, G. L., Vileno, B., Turek, P., Mény, C., Ott, F. et al. (2012). Magnetic properties of gold nanoparticles: a room-temperature quantum effect. ChemPhysChem, 13(13): 3092–3097.

Grozdanov, A., Avella, M., Buzarovska, A., Gentile, G. and Errico, M. E. (2010). Reuse of natural fiber reinforced eco-composites in polymer mortars. Polymer Engineering & Science, 50(4): 762–766.

Guzmán, M. G., Dille, J. and Godet, S. (2009). Synthesis of silver nanoparticles by chemical reduction method and their antibacterial activity. Int. J. Chem. Biomol. Eng., 2(3): 104–111.

Hamrayev, H. and Shameli, K. (2021). Biopolymer-based green synthesis of zinc oxide (ZnO) nanoparticles. In IOP Conference Series: Materials Science and Engineering (Vol. 1051, No. 1, p. 012088). IOP Publishing.

Hassan, T. A., Rangari, V. K. and Jeelani, S. (2014). Value-added biopolymer nanocomposites from waste eggshell-based CaCO3 nanoparticles as fillers. ACS Sustainable Chemistry & Engineering, 2(4): 706–717.

He, H., Fan, G., Saba, F., Tan, Z., Su, Z., Xiong, D. et al. (2023). Enhanced distribution and mechanical properties of high content nanoparticles reinforced metal matrix composite prepared by flake dispersion. Composites Part B: Engineering, 252: 110514.

He, S., Guo, Z., Zhang, Y., Zhang, S., Wang, J. and Gu, N. (2007). Biosynthesis of gold nanoparticles using the bacteria Rhodopseudomonas capsulata. Materials Letters, 61(18): 3984–3987.

Heiligtag, F. J. and Niederberger, M. (2013). The fascinating world of nanoparticle research. Materials Today, 16(7-8), 262–271.

Ho, T. T. T., Dang, C. H., Huynh, T. K. C., Hoang, T. K. D. and Nguyen, T. D. (2021). In situ synthesis of gold nanoparticles on novel nanocomposite lactose/alginate: Recyclable catalysis and colorimetric detection of Fe (III). Carbohydrate Polymers, 251: 116998.

Hoque, M., Alam, M., Wang, S., Zaman, J. U., Rahman, M. S., Johir, M. A. H. et al. (2023). Interaction chemistry of functional groups for natural biopolymer-based hydrogel design. Materials Science and Engineering: R: Reports, 156: 100758.

Horue, M., Berti, I. R., Cacicedo, M. L. and Castro, G. R. (2021). Microbial production and recovery of hybrid biopolymers from wastes for industrial applications-a review. Bioresource Technology, 340: 125671.

Hu, P., Chen, L., Kang, X. and Chen, S. (2016). Surface functionalization of metal nanoparticles by conjugated metal–ligand interfacial bonds: impacts on intraparticle charge transfer. Accounts of Chemical Research, 49(10): 2251–2260.

Ilyas, R. A., Sapuan, S. M., Ibrahim, R., Abral, H., Ishak, M. R., Zainudin, E. S. et al. (2019). Effect of sugar palm nanofibrillated cellulose concentrations on morphological, mechanical and physical properties of biodegradable films based on agro-waste sugar palm (Arenga pinnata (Wurmb.) Merr) starch. Journal of Materials Research and Technology, 8(5): 4819–4830.

Iravani, S. (2011). Green synthesis of metal nanoparticles using plants. Green Chemistry, 13(10): 2638–2650.

Islam, M. A. and Islam, M. S. (2013). Electro-deposition method for platinum nano-particles synthesis. Engineering International, 1(2).

J Joye, I. and Julian McClements, D. (2016). Biopolymer-based delivery systems: challenges and opportunities. Current Topics in Medicinal Chemistry, 16(9): 1026–1039.

Jain, N., Singh, V. K. and Chauhan, S. (2017). A review on mechanical and water absorption properties of polyvinyl alcohol based composites/films. Journal of the Mechanical Behavior of Materials, 26(5-6): 213–222.

Kamat, P. V., Flumiani, M. and Dawson, A. (2002). Metal–metal and metal–semiconductor composite nanoclusters. Colloids and Surfaces A: Physicochemical and Engineering Aspects, 202(2-3): 269–279.

Karimi-Avargani, M., Bazooyar, F., Biria, D., Zamani, A. and Skrifvars, M. (2020). The special effect of the Aspergillus flavus and its enzymes on biological degradation of the intact polylactic acid (PLA) and PLA-Jute composite. Polymer Degradation and Stability, 179: 109295.

Khalid, M. Y., Al Rashid, A., Arif, Z. U., Ahmed, W. and Arshad, H. (2021). Recent advances in nanocellulose-based different biomaterials: Types, properties, and emerging applications. Journal of Materials Research and Technology, 14: 2601–2623.

Khalid, M. Y., Arif, Z. U., Ahmed, W. and Arshad, H. (2022). Evaluation of tensile properties of fiber metal laminates under different strain rates. Proceedings of the Institution of Mechanical Engineers, Part E: Journal of Process Mechanical Engineering, 236(2): 556–564.

Khalid, M. Y., Imran, R., Arif, Z. U., Akram, N., Arshad, H., Al Rashid, A. et al. (2021). Developments in chemical treatments, manufacturing techniques and potential applications of natural-fibers-based biodegradable composites. Coatings, 11(3): 293.

Kuppusamy, P., Yusoff, M. M., Maniam, G. P. and Govindan, N. (2016). Biosynthesis of metallic nanoparticles using plant derivatives and their new avenues in pharmacological applications–An updated report. Saudi Pharmaceutical Journal, 24(4): 473–484.

Landage, S. M., Wasif, A. I. and Dhuppe, P. (2014). Synthesis of nanosilver using chemical reduction methods. International Journal of Advanced Research in Engineering and Applied Sciences, 3(5): 14–22.

Larmier, K., Tada, S., Comas-Vives, A. and Copéret, C. (2016). Surface sites in Cu-nanoparticles: Chemical reactivity or microscopy? The Journal of Physical Chemistry Letters, 7(16): 3259–3263.

Li, C., Wu, J., Shi, H., Xia, Z., Sahoo, J. K., Yeo, J. et al. (2022). Fiber-based biopolymer processing as a route toward sustainability. Advanced Materials, 34(1): 2105196.

Li, D., He, Q., Cui, Y. and Li, J. (2007). Fabrication of pH-responsive nanocomposites of gold nanoparticles/poly (4-vinylpyridine). Chemistry of Materials, 19(3): 412–417.

Li, J., Yu, X., Martinez, E. E., Zhu, J., Wang, T., Shi, S. et al. (2022). Emerging biopolymer-based bioadhesives. Macromolecular Bioscience, 22(2): 2100340.

Lim, B. K. H. and San Thian, E. (2022). Biodegradation of polymers in managing plastic waste—A review. Science of The Total Environment, 813: 151880.

Ling, S., Chen, W., Fan, Y., Zheng, K., Jin, K., Yu, H. et al. (2018). Biopolymer nanofibrils: Structure, modeling, preparation, and applications. Progress in Polymer Science, 85: 1–56.

Liu, M., Ma, Y., Wu, H. and Wang, R. Y. (2015). Metal matrix–metal nanoparticle composites with tunable melting temperature and high thermal conductivity for phase-change thermal storage. ACS nano, 9(2): 1341–1351.

Liu, Y., Ahmed, S., Sameen, D. E., Wang, Y., Lu, R., Dai, J. et al. (2021). A review of cellulose and its derivatives in biopolymer-based for food packaging application. Trends in Food Science & Technology, 112: 532–546.

Mahmudin, L., Suharyadi, E., Utomo, A. B. S. and Abraha, K. (2015). Optical properties of silver nanoparticles for surface plasmon resonance (SPR)-based biosensor applications. Journal of Modern Physics, 6(08): 1071.

Malik, N., Shrivastava, S. and Ghosh, S. B. (2018). Moisture absorption behaviour of biopolymer polycapralactone (PCL)/organo modified montmorillonite clay (OMMT) biocomposite films. In IOP Conference Series: Materials Science and Engineering (Vol. 346, p. 012027). IOP Publishing.

Mazzanti, V., Malagutti, L. and Mollica, F. (2019). FDM 3D printing of polymers containing natural fillers: A review of their mechanical properties. Polymers, 11(7): 1094.

McGauran, T., Dunne, N., Smyth, B. M., Cunningham, E. and Harris, M. (2021). Poultry feather disulphide bond breakdown to enable bio-based polymer production. Polymers from Renewable Resources, 12(3-4): 92–110.

McNamara, K. and Tofail, S. A. (2017). Nanoparticles in biomedical applications. Advances in Physics: X, 2(1): 54–88.

Mendanha, D. (2024). Driving sustainability in the automotive industry: Bio-coated materials and modern strategies. Academia Materials Science, 1(2).

Meyer, M., Dietrich, S., Schulz, H. and Mondschein, A. (2021). Comparison of the technical performance of leather, artificial leather, and trendy alternatives. Coatings, 11(2): 226.

Miikkulainen, V., Leskelä, M., Ritala, M. and Puurunen, R. L. (2013). Crystallinity of inorganic films grown by atomic layer deposition: Overview and general trends. Journal of Applied Physics, 113(2).

Mochane, M. J., Sefadi, J. S., Motsoeneng, T. S., Mokoena, T. E., Mofokeng, T. G. and Mokhena, T. C. (2020). The effect of filler localization on the properties of biopolymer blends, recent advances: A review. Polymer Composites, 41(7): 2958–2979.

Mohammad, F., A Al-Lohedan, H. and N Al-Haque, H. (2017). Chitosan-mediated fabrication of metal nanocomposites for enhanced biomedical applications. Advanced Materials Letters, 8(2): 89–100.

Mondal, S., Nguyen, T. P., Hoang, G., Manivasagan, P., Kim, M. H., Nam, S. Y. et al. (2020). Hydroxyapatite nano bioceramics optimized 3D printed poly lactic acid scaffold for bone tissue engineering application. Ceramics International, 46(3): 3443–3455.

Morin-Crini, N., Lichtfouse, E., Torri, G. and Crini, G. (2019). Applications of chitosan in food, pharmaceuticals, medicine, cosmetics, agriculture, textiles, pulp and paper, biotechnology, and environmental chemistry. Environmental Chemistry Letters, 17(4): 1667–1692.

Muir, V. G. and Burdick, J. A. (2020). Chemically modified biopolymers for the formation of biomedical hydrogels. Chemical Reviews, 121(18): 10908–10949.

Mukherjee, P., Ahmad, A., Mandal, D., Senapati, S., Sainkar, S. R., Khan, M. I. et al. (2001). Fungus-mediated synthesis of silver nanoparticles and their immobilization in the mycelial matrix: A novel biological approach to nanoparticle synthesis. Nano Letters, 1(10): 515–519.

Mylkie, K., Nowak, P., Rybczynski, P. and Ziegler-Borowska, M. (2021). Polymer-coated magnetite nanoparticles for protein immobilization. Materials, 14(2): 248.

Narayan, N., Meiyazhagan, A. and Vajtai, R. (2019). Metal nanoparticles as green catalysts. Materials, 12(21): 3602.

Niculescu, A. G., Chircov, C., Bîrcă, A. C. and Grumezescu, A. M. (2021). Nanomaterials synthesis through microfluidic methods: An updated overview. Nanomaterials, 11(4): 864.

Oliveira, S. F., Bisker, G., Bakh, N. A., Gibbs, S. L., Landry, M. P. and Strano, M. S. (2015). Protein functionalized carbon nanomaterials for biomedical applications. Carbon, 95: 767–779.

Pan, J., Zhang, Z., Zhan, Z., Xiong, Y., Wang, Y., Cao, K. et al. (2020). In situ generation of silver nanoparticles and nanocomposite films based on electrodeposition of carboxylated chitosan. Carbohydrate Polymers, 242: 116391.

Pantidos, N. and Horsfall, L. E. (2014). Biological synthesis of metallic nanoparticles by bacteria, fungi and plants. Journal of Nanomedicine & Nanotechnology, 5(5): 1.

Pantidos, N. and Horsfall, L. E. (2014). Biological synthesis of metallic nanoparticles by bacteria, fungi and plants. Journal of Nanomedicine & Nanotechnology, 5(5): 1.

Pantidos, N. and Horsfall, L. E. (2014). Biological synthesis of metallic nanoparticles by bacteria, fungi and plants. Journal of Nanomedicine & Nanotechnology, 5(5): 1.

Parvizifard, M. and Karbasi, S. (2020). Physical, mechanical and biological performance of PHB-Chitosan/MWCNTs nanocomposite coating deposited on bioglass based scaffold: Potential application in bone tissue engineering. International Journal of Biological Macromolecules, 152: 645–662.

Pavlovic, M., Szerlauth, A., Muráth, S., Varga, G. and Szilagyi, I. (2022). Surface modification of two-dimensional layered double hydroxide nanoparticles with biopolymers for biomedical applications. Advanced Drug Delivery Reviews, 191: 114590.

Pedersen, H. and Elliott, S. D. (2014). Studying chemical vapor deposition processes with theoretical chemistry. Theoretical Chemistry Accounts, 133: 1–10.

Pedersen, H. and Elliott, S. D. (2014). Studying chemical vapor deposition processes with theoretical chemistry. Theoretical Chemistry Accounts, 133: 1–10.

Pfau, M. R., McKinzey, K. G., Roth, A. A. and Grunlan, M. A. (2020). PCL-based shape memory polymer semi-IPNs: The role of miscibility in tuning the degradation rate. Biomacromolecules, 21(6): 2493–2501.

Pilapong, C., Sitthichai, S., Thongtem, S. and Thongtem, T. (2014). Smart magnetic nanoparticle-aptamer probe for targeted imaging and treatment of hepatocellular carcinoma. International Journal of Pharmaceutics, 473(1-2): 469–474.

Polman, E. M., Gruter, G. J. M., Parsons, J. R. and Tietema, A. (2021). Comparison of the aerobic biodegradation of biopolymers and the corresponding bioplastics: A review. Science of the Total Environment, 753: 141953.

Prathna, T. C., Chandrasekaran, N., Raichur, A. M. and Mukherjee, A. (2011). Biomimetic synthesis of silver nanoparticles by Citrus limon (lemon) aqueous extract and theoretical prediction of particle size. Colloids and Surfaces B: Biointerfaces, 82(1): 152–159.

Rahmani, P., Shojaei, A., Sakorikar, T., Wang, M., Mendoza-Apodaca, Y. and Dickey, M. D. (2024). Liquid Metal Nanoparticles Physically Hybridized with Cellulose Nanocrystals Initiate and Toughen Hydrogels with Piezoionic Properties. ACS Nano.

Ranganathan, S., Dutta, S., Moses, J. A. and Anandharamakrishnan, C. (2020). Utilization of food waste streams for the production of biopolymers. Heliyon, 6(9).

Rangappa, S. M. and Siengchin, S. (2022). Moving towards biofiber-based composites: Knowledge gaps and insights. Express Polymer Letters, 16(5): 451–452.

Rangappa, S. M. and Siengchin, S. (2022). Moving towards biofiber-based composites: Knowledge gaps and insights. Express Polymer Letters, 16(5): 451–452.

Rastogi, P. and Kandasubramanian, B. (2019). Breakthrough in the printing tactics for stimuli-responsive materials: 4D printing. Chemical Engineering Journal, 366: 264–304.

Samantaray, P. K., Little, A., Haddleton, D. M., McNally, T., Tan, B., Sun, Z. et al. (2020). Poly (glycolic acid) (PGA): A versatile building block expanding high performance and sustainable bioplastic applications. Green Chemistry, 22(13): 4055–4081.

Saravanan, A., Kumar, P. S., Karishma, S., Vo, D. V. N., Jeevanantham, S., Yaashikaa, P. R. et al. (2021). A review on biosynthesis of metal nanoparticles and its environmental applications. Chemosphere, 264: 128580.

Sari, P. S., Thomas, S., Spatenka, P., Ghanam, Z. and Jenikova, Z. (2019). Effect of plasma modification of polyethylene on natural fibre composites prepared via rotational moulding. Composites Part B: Engineering, 177: 107344.

Sekhon, B. S. (2010). Supercritical fluid technology: an overview of pharmaceutical applications. International Journal of PharmTech Research, 2(1): 810–826.

Segala, K., Nista, S. V. G., Cordi, L., Bizarria, M. T. M., Ávila Júnior, J. D., Kleinubing, S. A. et al. (2015). Silver nanoparticles incorporated into nanostructured biopolymer membranes produced by electrospinning: A study of antimicrobial activity. Brazilian Journal of Pharmaceutical Sciences, 51: 911–921.

Sekhon, B. S. (2010). Supercritical fluid technology: An overview of pharmaceutical applications. International Journal of PharmTech Research, 2(1): 810–826.

Sharanya Devadiga, B., Shetty, D., Lakshmi, G. M. S., Packiyam, J. E. and Bhat, R. P. (2016). Synthesis of chitosan silver nanoparticles from chitin of crustacean shells and its applications. Int. J. Curr. Res. Chem. Pharm. Sci., 3: 1–5.

Sharma, B., Malik, P. and Jain, P. (2018). Biopolymer reinforced nanocomposites: A comprehensive review. Materials Today Communications, 16: 353–363.

Singh, B. N., Veeresh, V., Mallick, S. P., Jain, Y., Sinha, S., Rastogi, A. et al. (2019). Design and evaluation of chitosan/chondroitin sulfate/nano-bioglass based composite scaffold for bone tissue engineering. International Journal of Biological Macromolecules, 133: 817–830.

Stephen, D. P. and Palanisamy, S. B. (2023). Advances in biopolymer composites and biomaterials for the removal of emerging contaminants. Physical Sciences Reviews, 8(8): 1789–1809.

Su, W., Zhang, H., Xing, Y., Li, X., Wang, J. and Cai, C. (2021). A bibliometric analysis and review of supercritical fluids for the synthesis of nanomaterials. Nanomaterials, 11(2): 336.

Taketa, T. B., Bataglioli, R. A., Neto, J. B. M. R. and Beppu, M. M. (2021). Probing axial metal distribution on biopolymer-based layer-by-layer films for antimicrobial use. Colloids and Surfaces B: Biointerfaces, 199: 111505.

Taniguchi, I., Yoshida, S., Hiraga, K., Miyamoto, K., Kimura, Y. and Oda, K. (2019). Biodegradation of PET: Current status and application aspects. Acs Catalysis, 9(5): 4089–4105.

Tarique, J., Sapuan, S. M., Khalina, A., Sherwani, S. F. K., Yusuf, J. and Ilyas, R. A. (2021). Recent developments in sustainable arrowroot (Maranta arundinacea Linn) starch biopolymers, fibres, biopolymer composites and their potential industrial applications: A review. Journal of Materials Research and Technology, 13: 1191–1219.

Tirado-Kulieva, V. A., Sánchez-Chero, M., Jimenez, D. P. P., Sánchez-Chero, J., Santa Cruz, A. G. Y. and Velayarce, H. H. M. (2022). A critical review on the integration of metal nanoparticles in biopolymers: An alternative for active and sustainable food packaging. Current Research in Nutrition and Food Science Journal, 10(1): 01–18.

Tsang, M. P., Philippot, G., Aymonier, C. and Sonnemann, G. (2018). Supercritical fluid flow synthesis to support sustainable production of engineered nanomaterials: case study of titanium dioxide. ACS sustainable Chemistry & Engineering, 6(4): 5142–5151.

Turcsányi, Á., Ungor, D., Wojnicki, M. and Csapó, E. (2023). Protein-stabilized bimetallic Au/Ag nanoclusters as fluorescent reporters: Synthesis, characterization and their interactions with biocolloids. Journal of Molecular Liquids, 370: 121002.

Verma, D. K., Malik, R., Meena, J. and Rameshwari, R. (2021). Synthesis, characterization and applications of chitosan based metallic nanoparticles: A review. Journal of Applied and Natural Science, 13(2): 544–551.

Vinod, A., Sanjay, M. R., Suchart, S. and Jyotishkumar, P. (2020). Renewable and sustainable biobased materials: An assessment on biofibers, biofilms, biopolymers and biocomposites. Journal of Cleaner Production, 258: 120978.

Wang, B., Zhuang, X., Deng, W. and Cheng, B. (2010). Microwave-assisted synthesis of silver nanoparticles in alkalic carboxymethyl chitosan solution. Engineering, 2(5): 387.

Wang, L., Hasanzadeh Kafshgari, M. and Meunier, M. (2020). Optical properties and applications of plasmonic-metal nanoparticles. Advanced Functional Materials, 30(51): 2005400.

Xiong, S., Qi, W., Cheng, Y., Huang, B., Wang, M. and Li, Y. (2011). Universal relation for size dependent thermodynamic properties of metallic nanoparticles. Physical Chemistry Chemical Physics, 13(22): 10652–10660.

Yang, Y., Zan, J., Yang, W., Qi, F., He, C., Huang, S. et al. (2020). Metal organic frameworks as a compatible reinforcement in a biopolymer bone scaffold. Materials Chemistry Frontiers, 4(3): 973–984.

Yao, F., Zhu, P., Chen, J., Li, S., Sun, B., Li, Y. et al. (2023). Synthesis of nanoparticles via microfluidic devices and integrated applications. Microchimica Acta, 190(7): 256.

Yates, M. R. and Barlow, C. Y. (2013). Life cycle assessments of biodegradable, commercial biopolymers—A critical review. Resources, Conservation and Recycling, 78: 54-66.

You, H., Yang, S., Ding, B. and Yang, H. (2013). Synthesis of colloidal metal and metal alloy nanoparticles for electrochemical energy applications. Chemical Society Reviews, 42(7): 2880–2904.

Yugay, Y. A., Usoltseva, R. V., Silant'Ev, V. E., Egorova, A. E., Karabtsov, A. A., Kumeiko, V. V. et al. (2020). Synthesis of bioactive silver nanoparticles using alginate, fucoidan and laminaran from brown algae as a reducing and stabilizing agent. Carbohydrate Polymers, 245: 116547.

Zare, Y., Rhim, S. S. and Rhee, K. Y. (2021). The interphase degradation in a nanobiosensor including biopolymers and carbon nanotubes. Sensors and Actuators A: Physical, 331: 112967.

Zhang, Y. X. and Wang, Y. H. (2017). Nonlinear optical properties of metal nanoparticles: A review. RSC Advances, 7(71): 45129–45144.

Zhang, Y., Tu, J., Wang, D., Zhu, H., Maity, S. K., Qu, X. et al. (2018). Programmable and multifunctional DNA-based materials for biomedical applications. Advanced Materials, 30(24): 1703658.

Zhu, J., Li, Q., Che, Y., Liu, X., Dong, C., Chen, X. et al. (2020). Effect of Na2CO3 on the microstructure and macroscopic properties and mechanism analysis of PVA/CMC composite film. Polymers, 12(2): 453.

CHAPTER-16

Application of Silver Nanoparticles in Biopolymer-based Food Packaging or Edible Coatings

Snehlata Kujur and *Dinesh Prasad**

1. Introduction

Contamination of foods by microbes poses a significant challenge for the food industry, leading to wasted products and causing risks to public health through foodborne illnesses (Carbone et al., 2016). Packaging materials made from natural biopolymers are gaining increasing popularity due to their environmentally friendly, biodegradable, compostable, non-toxic, and biocompatible characteristics (Ali and Ahmed, 2018). Common biopolymers used in the production of these packaging materials encompass proteins, lipids, and polysaccharides (Zubair et al., 2021; Díaz-Montes et al., 2022). Biopolymers exhibit strong film and coating-forming abilities, allowing them to create thin barriers on food surfaces. These films and coatings are essential for preserving food quality, enhancing stability, and extending the shelf life by regulating the movement of gases, moisture and lipids between the food and its environment. They also provide protection against microbial contamination and help retain desirable compounds, such as flavor volatiles. Additionally, biopolymer-based materials can act as carriers for antimicrobial agents, antioxidants, colorants, flavor enhancers, vitamins, and other nutrients, improving both the sensory qualities and nutritional value of the packaged product (Xu et al., 2020; Salama et al., 2021; Pal et al., 2021). Recent studies have focused on the potential of edible packaging as carriers for probiotic microorganisms. Integrating living microorganisms into edible films and coatings provides two key benefits: it enhances consumer health by providing probiotics and helps to maintain food quality and safety by preventing the growth of spoilage or harmful microorganisms through the generation of antimicrobial compounds (Pop et al., 2019).

Although biopolymer packaging materials are gaining recognition as a substitute for synthetic alternatives, they are mostly used alongside traditional packaging systems rather than replacing them. This is primarily due to the limitations of biopolymers, such as their limited mechanical strength and vulnerability to moisture. To address these issues, many studies have investigated reinforcing biopolymer matrices with nanoparticles, which can greatly improve the functional characteristics of packaging materials (Kraśniewska et al., 2019; Plaza, 2015; Zambrano-Zaragoza et al., 2018).

Silver nanoparticles (AgNPs) are particularly notable in this context. They are being integrated into polymer matrices to develop cutting-edge nanocomposite materials for food packaging. AgNPs

Department of Bioengineering and Biotechnology, Birla Institute of Technology, Mesra, Ranchi, India-835215.
* Corresponding author: dinesh@bitmesra.ac.in
DOI: 10.1201/9781003558330-16

are recognized for their robust antimicrobial capabilities, effective against a broad spectrum of microorganisms, including yeast, bacteria, and mold. Integrating these nanoparticles into food packaging may prolong product shelf life by preventing the growth of spoilage and harmful microorganisms (Kraśniewska et al., 2019; Liao et al., 2019). Furthermore, AgNPs provide an advantage over other natural substances used in edible packaging, like essential oils or plant extracts, which can introduce strong smells and flavors that may affect the sensory qualities of the food. Silver nanoparticles have minimal impact on the sensory characteristics of food, making the packaged products more palatable and acceptable to consumers. This overview examines the current advancements in integrating silver nanoparticles into biopolymers to develop biodegradable films and coatings for food packaging.

2. Nanoparticles in Food Packaging and Coatings

Nanoparticles are increasingly recognized as valuable additives in food packaging and coatings because of their unique characteristics, such as a high surface area-to-volume ratio, increased mechanical strength, improved barrier properties, and antimicrobial effects. Recent research has explored various nanoparticles, such as titanium dioxide, silver, silica, and zinc oxide, for their potential roles in food packaging and coatings. These nanoparticles can be integrated into polymer matrices or used as surface coatings to provide benefits such as antimicrobial properties, UV protection, gas barrier enhancement, and preservation of sensory qualities.

A major advantage of nanoparticle-based food packaging and coatings is their ability to prevent the growth of pathogenic microorganisms and spoilage bacteria, thus prolonging the shelf life of perishable foods. They also help prevent oxidation reactions that can degrade food quality and nutritional value. However, there are still concerns regarding the security and regulatory aspects of using nanoparticles in food packaging and coatings. It is crucial to ensure the stability of nanoparticles within food matrices, assess their potential migration into food, and evaluate their prolonged effects on human health and the environment.

Packaging is crucial for maintaining food quality as well as minimizing waste. Despite the growing use of packaging, insufficient packaging still contributes significantly to food loss and waste. Nanotechnology offers a promising solution by enabling the development of materials with innovative properties. By incorporating nanoparticles into packaging materials for food, various characteristics can be enhanced, including ultraviolet, oxygen, carbon dioxide, moisture, and volatile barrier characteristics, along with mechanical attributes such as tensile strength, elongation, and yield strength. These enhancements help extend the shelf life of food products. Even small quantities of nanoparticles can significantly improve the properties of traditional packaging materials (Cushen et al., 2014; Ayhan et al., 2015).

However, the effectiveness of these improvements relies on multiple aspects, such as the kind of polymer selected, as well as the amount and characteristics of the nanoparticles (such as shape, size, and production method), the manufacturing process, and specific process conditions. Plastic-based packaging materials, which are not easily decomposed, pose recycling challenges. The growing concern over plastic pollution's environmental effects has sparked greater interest in creating biodegradable packaging materials from renewable resources. Global efforts are focused on creating renewable and biodegradable alternatives to synthetic polymer-based packaging. These sustainable solutions aim to reduce the environmental burden of plastic waste by offering materials that naturally degrade over time.

2.1 *Types of Nanoparticles Utilized in the Food Industry*

A diverse array of organic, inorganic, and hybrid nanoparticles plays a key role in advancing food packaging technologies. Their role is instrumental in addressing both quantitative and qualitative losses in food products. By bolstering mechanical strength and barrier properties, these nanoparticles make substantial contributions to the creation of efficient packaging solutions, ultimately prolonging the shelf life of diverse food items.

Calcium oxide nanoparticles

Calcium oxide nanoparticles, which are derived from natural sources such as scallop shells, have inherent antimicrobial properties. These nanoparticles are effective antimicrobial agents in various applications. Tang et al. (2013) discovered that CaO nanoparticles, when present at a concentration of 100 ppm, exhibit strong antibacterial properties important for food safety.

The antibacterial action of these nanoparticles is believed to be related to the concentration of oxygen (O_2). As the particle size reduces, the number of CaO particles per unit volume rises, resulting in an accelerated reaction rate. Tang et al. also explored the effectiveness of CaO nanoparticles against different strains of *E. coli* (such as *E. coli* O157, EPEC, and EHEC) at temperatures of 10°C and 18°C, highlighting their potential as a valuable antimicrobial agent for food preservation and safety applications.

Silicon dioxide (SiO_2) nanoparticles

Silicon dioxide (SiO_2), more commonly referred to as silica, exhibits strong antimicrobial properties at the nanoscale due to its high surface area, which effectively prevents the attachment and the proliferation of microbial cells on surfaces (Dhapte et al., 2014). Silica nanoparticles are often integrated into nanocomposites with other nanomaterials to enhance their antimicrobial effects. For example, Egger et al., (2009) discovered that a silver-silica nanocomposite demonstrated superior antimicrobial performance compared to traditional materials such as silver nitrate and silver zeolite. Similarly, Hetrick et al., (2009) used silica nanoparticles that release nitric oxide to target biofilms of various microorganisms, achieving an impressive 99% destruction rate, and also observing their potential for wound healing.

Aluminum oxide nanoparticlesTop of form

Alumina nanoparticles have diverse applications across various industries and are being thoroughly studied for their antimicrobial properties. These nanoparticles demonstrate antimicrobial activity observed at concentrations between 10 and 1000 μg/mL. In the case of *E. coli*, alumina nanoparticles have been observed to possess mild deactivation properties at high concentrations because of their capacity to scavenge free radicals, they prevent the degradation of the cell wall (Bai and Rai, 2011).

The antimicrobial properties and deactivation mechanism of alumina nanoparticles, particularly when embedded in materials, are associated with potential toxicity linked to electrostatic interactions near neutral pH. These nanoparticles, with their positively charged surfaces, tend to attach to bacterial surfaces, leading to cell death via interactions between the positively charged nanoparticles and the negatively charged bacterial cells (Mukherjee et al., 2011). Illustrated that alumina inhibits the growth of microalgae species such as Chlorella and Scenedesmus. They found that both bulk and nanoscale forms of alumina were toxic, causing a decrease in chlorophyll content in the exposed cells, with nanoparticles showing a more pronounced effect.

MgO nanoparticles

MgO nanoparticles offer several benefits, including high thermal stability, low cost, and strong antibacterial properties. Magnesium (Mg) is crucial for human health, and its deficiency can lead to issues like sleep disorders and low blood pressure. Nanoparticles made from metal oxides, such as MgO nanoparticles under 100 nm in size, show high activity due to their tiny particle size and extensive surface area, which are not present in their larger micro/macro counterparts (Jin and He, 2011).

The antimicrobial mechanism of MgO nanoparticles differs depending on whether the bacteria are Gram-positive or Gram-negative and concentration. For example, Krishnamoorthy et al. (2012) tested the antibacterial activity of MgO nanoparticles targeting *E. coli* and *P. aeruginosa*

(Gram-negative bacteria), and *S. aureus* (Gram-positive bacteria). They found that even at low concentrations, MgO nanoparticles demonstrated biocidal activity against *E. coli* (500 μg/mL), *S. aureus*, and *P. aeruginosa* (1000 μg/mL). The antimicrobial activity of these nanoparticles is associated with surface deficiencies or oxygen vacancies, which lead to lipid peroxidation and the production of Reactive Oxygen Species (ROS) (Al-Hazmi et al., 2012).

Gold nanoparticles

Gold has been valued in Western culture for its supposed medicinal properties since ancient times. Known as "nervine," it was believed to have therapeutic effects on neurological disorders. In the 16th century, gold was recommended for used to treat epilepsy, and by the 19th century, it was used in managing syphilis. However, it was not until Robert Koch identified the bacteriostatic effects of gold cyanide on Mycobacterium tuberculosis, which demonstrated the gold's potential for treating tuberculosis. This recognition led to its consideration for therapeutic use in the 1920s (Bai and Rai, 2011).

Selenium nanoparticles

Selenium (Se) is a crucial micronutrient for both humans and animals, with typical levels in adults ranging from 10 to 20 mg. The recommended daily consumption is around 40 μg, while the maximum permissible intake is about 300 μg per day. Excessive selenium intake, between 3200 and 6700 μg per day, can cause toxicity, with excess selenium being excreted through urine (Tran and Webster, 2013). Selenium has attracted interest in nanotechnology because of its significant biological characteristics, including antiviral, antibacterial, and antioxidant properties (Shakibaie et al., 2015). Nanotechnology allows for the modulation of chemo preventive substances like elemental selenium, potentially enhancing their efficacy or reducing toxicity. The immediate toxicity (LD50) of selenium nanoparticles in mice is notably lower compared to other selenium forms such as selenomethionine, selenite, and Se methylselenocysteine (Khiralla and El-Deeb, 2015).

Copper oxide nanoparticles

Copper provides several benefits as a biocidal agent due to its affordability and properties akin to noble metals like silver and gold. With advancements in nanotechnology, copper oxide nanoparticles have gained significant attention for their effectiveness against a broad spectrum of microorganisms (Xiong et al., 2015; Kalatehjari et al., 2015). These nanoparticles hold promising potential for various applications, including antimicrobial coatings, water purification, and medical devices, owing to their potent biocidal properties and compatibility with nanotechnology.

ZnO nanoparticles

ZnO nanoparticles have substantial potential to transform both food and agricultural systems. Their antimicrobial activity is especially noteworthy in view of the increasing need for alternative approaches to tackle diseases and prevent spoilage (Emami, 2011). ZnO, a zinc compound, can be utilized either on its own or in conjunction with other substances such as gelatin to aid in food preservation and prevent spoilage/Umamaheswari et al., 2015). Among metal oxides, ZnO exhibits strong inhibitory effects against a variety of microorganisms, including *S. aureus, Salmonella spp., E. coli, Listeria monocytogenes, and Campylobacter jejuni* (Emami, 2011).

Silver nanoparticles

Silver nanoparticles represent a cutting-edge advancement in biocidal agents, with the potential to lower rates of infection. Recognized for their excellent thermal resistance, safety profile, and broad-spectrum antimicrobial effectiveness, silver nanoparticles are utilized in a wide range of commercial products. Companies like Duretham and Kodak incorporate them into packaging materials to extend the shelf life of perishable goods and inhibit microbial growth, especially in food products (Jaiswal, 2010). The European Commission (2011) regulates the maximum release of silver nanoparticles into the food media to approximately 0.01 mg/kg. The antimicrobial effectiveness of silver nanoparticles are affected by factors such as their size, concentration, the species of target microorganisms, composition, and shape, with triangular and truncated forms under 10 nm showing particularly high reactivity (Pal, 2007).

Titanium dioxide (TiO_2)

Titanium dioxide, also referred to as titania or titanium (IV) oxide, naturally exists in three polymorphic forms: rutile, anatase and brookite. The United States Food and Drug Administration (FDA/USFDA) has authorized the use of TiO2 nanoparticles in a range of applications, including food, cosmetics, drugs, and surfaces in contact with food, such as cutting boards (Maneerat and Hayata, 2006). Despite their widespread use, durability, and cost-effectiveness, TiO2 nanoparticles have not been classified as Generally Recognized As Safe (GRAS) by the FDA or other regulatory bodies due to concerns about potential genotoxicity and cytotoxicity to human cells, as indicated by toxicology studies (Asare, 2012).

2.2 Silver Nanoparticles (AgNPs): Properties and Characteristics

Presently, silver nanoparticles (AgNPs) have gained significant attention in academic research because of their unique and favorable antibacterial characteristics against various types of bacteria. Silver (Ag) is known for its functional, antimicrobial, and non-toxic properties, making it a valuable metallic-based cation. Silver nanoparticles, which are derived from silver, are increasingly utilized to optimize the effectiveness of packaging materials (Carbone et al., 2016). These nanoparticles are produced by extracting silver ions from silver-containing salts using physical, chemical, or hybrid methods, with silver nitrate being the most common approach (Gajbhiye et al., 2009). Their key role in the packaging industry arises from their biocidal properties making them highly effective antimicrobial agents compared to other metallic nanoparticles used in packaging. Additionally, silver nanoparticles can act as nanofillers, enhancing various polymer matrices such as cellulose, polyethylene, pullulan, HPMC, agar hydrogel, starch, calcium alginate, sodium alginate, PLA, and chitosan (Abreu, 2015; de Moura et al., 2012). They improve the antimicrobial activity of these polymers due to their low volatility and high thermal resistance, making them superior antimicrobial agents among metallic nanoparticles (Youssef and Abdel-Aziz, 2013).

Specifically, Ag nanoparticles exhibit the highest antimicrobial activity in EVOH film matrices (Biniaś et al., 2020). They are effective against a broad spectrum of gram-positive and gram-negative bacteria, as well as fungi (Ansari et al., 2021). Their antimicrobial properties can be further enhanced by combining them with oxygen. However, they show reduced effectiveness against eukaryotic cells, such as certain strains of Staphylococcus aureus (Krychowiak et al., 2018).

The potential of silver (Ag) nanofillers to improve the mechanical, physical, and thermal properties of various polymers has been extensively researched. Ag nanoparticles are known to enhance properties of a barrier and mechanics (Jafarzadeh et al., 2021; Liao et al., 2019). However, their tendency to agglomerate and high surface functionality can limit their impact on these properties on their own. Ag nanoparticles can be used in conjunction with other nanofillers to address this. For example, combining Ag nanoparticles with cellulose nanocrystals (CNCs) has led to improved polymers' mechanical, barrier, and antimicrobial qualities (Xu et al., 2020; Pal et al., 2021). Additionally, incorporating silver-doped carriers such as zeolite, kaolinite, or halloysite into the polymer matrix can further enhance these properties, with zeolite significantly improving the performance of Ag nanoparticles (Burridge et al., 2011). This enhancement is attributed to better dispersion and stronger crosslinking within the polymer matrix.

Nevertheless, there is limited information on Ag nanoparticles' possible toxicity to the environment and human health, which requires cautious use in packaging materials. Ensuring the stability of Ag nanoparticles within the polymer matrix is crucial for both environmental and health safety (Yang et al., 2011).

2.3 Advantages and Challenges of Using Silver Nanoparticles

Silver nanoparticles are among the most promising nanomaterials for various commercial applications. Their versatility has led to widespread usage across industries, including electronics, healthcare, food storage, textile coatings, and environmental applications. In the health sector, they serve as effective antibacterial agents, finding applications in disinfecting medical devices, household appliances, and water treatment processes. The diverse range of applications underscores the extensive potential of silver nanoparticles to tackle various challenges and enhance numerous products and processes (Li et al., 2008). The ability of silver nanoparticles (AgNPs) to eradicate a wide range of harmful microorganisms, such as viruses, bacteria, yeasts, and fungi, has been demonstrated (Martinez, 2012; Verma and Maheshwari, 2019). AgNPs offer superior antimicrobial properties compared to bulk silver due to their exceptionally large surface area, which increases their contact with microorganisms (Toker et al., 2013). Moreover, they exhibit low volatility and remain stable at high temperatures (Youssef and Abdel-Aziz, 2013). AgNPs can be added to a variety of matrices, such as polymers and long-chain alcohols and citrates, which act as stabilizing agents (Toker et al., 2013). This integration can be achieved through several methods: by coating, absorbing, or directly incorporating the nanoparticles during the synthesis process (Martinez-Abad et al., 2012).

3. Biopolymers

Biopolymers refer to polymers sourced from plants, animals, and microbes. These materials, readily accessible and renewable, are commonly utilized in the production of environmentally friendly bioplastics (Azammi et al., 2020). Most biopolymers are biodegradable, breaking down into carbon dioxide (CO_2), methane (CH_4), water (H_2O), and other inorganic compounds through microbial action, though some exceptions exist (Christian, 2016). The degradation of biopolymers is influenced by factors such as the type of polymer, its chemical composition, and environmental conditions (Vinod et al., 2020). Biopolymers are utilized in various fields based on considerations such as price, availability, mechanical qualities, moisture absorption, thermal stability, resistance to degradation, and biocompatibility (Christian, 2016). The properties of a biocomposite, including its chemical composition, molecular weight, morphology, mechanical strength, and processing method, are determined by the biopolymer used (George et al., 2020). However, biopolymers do have some limitations, including their hydrophilic nature, limited mechanical strength, and relatively low durability in moist conditions (Mohamed et al., 2018).

3.1 Types of Biopolymers

Biopolymers can be broadly categorized into two types based on their source: both artificial and natural proteins, bacteria, and polysaccharides are the sources of natural biopolymers; synthetic biopolymers are produced via microbial fermentation, biotechnological processes, or petrochemical methods (Calori et al., 2020). Regarding their thermal properties, biopolymers are classified as elastomers, thermosets, or thermoplastics (George et al., 2020). Among these, PLA and PHA are among the most widely produced and applied biopolymers (Christian, 2016).

Polysaccharides, which consist of monosaccharides connected by glycosidic bonds, form the basis of various biopolymers, including chitosan, cellulose, starch, chitin, Hyaluronic Acid (HA), alginate, dextrin, carrageenan, and different gums. Cellulose, in particular, is the most prevalent carbohydrate and a highly promising biodegradable polymer present in nearly all plant materials (Bharathi et al., 2021). It is environmentally friendly, requiring minimal energy for production, and features excellent film-forming capabilities. Additionally, cellulose can be easily recycled through combustion (DeAzeredo, 2009). Structurally, cellulose is composed of microfibrils that assemble into cellulose fibers (Arrieta et al., 2014).

One type of polysaccharide that can be neutral or acidic is chitin known for its hydrophobic properties and its insolubility in water and many organic solvents (Park et al., 2017). Chitosan, derived from the deacetylation of chitin found in shrimp and other crustaceans, is the second most abundant natural amino polysaccharide (Arrieta et al., 2014). Chitosan is a good material for scaffolding because it has a number of advantageous characteristics, including oxygen permeability, water absorption, cytokine induction, and biodegradability (Bin et al., 2017).

Thermoplastic starches (TPS) are a category of starches that have been modified through thermal and/or mechanical processes to replace synthetic polymers (Peelman et al., 2013). Because of its broad availability, recyclability, renewable nature, and thermoplastic qualities, TPS is very promising in biopolymer composites (Nevoralová et al., 2019).

Proteins are complex biological substances composed of lengthy chains of amino acids. They can be classified based on their source into plant proteins (such as gluten, pea protein, zein, and soy protein) and animal proteins (such as collagen, keratin, gelatin, silk, casein, and whey).

Collagen, the most prevalent protein on Earth, is found in cartilage, tendons, bones, skin, and connective tissues (Calori et al., 2020). Its properties, such as biodegradability, water absorption, biocompatibility, cross-linking ability, high mechanical strength, and low antigenicity, make it suitable for bioactive molecule delivery and tissue engineering applications.

Gelatin is a water-soluble protein derived from the partial or controlled hydrolysis of collagen. It is valued for its abundance, biocompatibility, biodegradability, film-forming ability, gelling capacity, excellent gas barrier qualities, as well as affordability (Benjakul et al., 2016; Luo et al., 2022).

Soy Protein Isolate (SPI), a by-product of soybean oil extraction containing over 90% protein, presents a viable alternative to petroleum-based polymers. Its superior mechanical properties result from the polar and functional groups of amino acids that enable strong molecular interactions (Tian et al., 2018).

Wheat gluten, comprising about 50% gliadin and 50% glutenin proteins, is obtained from food processing by-products. Techniques such as gluten-based bioplastics are created using compression, extrusion, or injection molding (Jiménez-Rosado et al., 2019).

PLA (polylactide) is a noteworthy synthetic biopolymer with unique features making it a promising substitute for traditional polymers. PLA is biodegradable, renewable, and recyclable, providing an environmentally friendly alternative (Raneshkumar et al., 2021; Taib, 2023).

Polybutylene succinate (PBS) is a semi-crystalline thermoplastic polyester with properties similar to polypropylene (PP) (Amaya-Pinos, 2022). It is frequently used in the manufacture of bottles and films due to its favorable thermo-mechanical characteristics. PBS is produced from renewable resources, specifically 1,4-butanediol and succinic acid (Tamizhdurai et al., 2024).

Poly (Butylene Adipate-co-Terephthalate) (PBAT) is another notable polyester, recognized as a butylene adipate and terephthalate copolymer that degrades biologically (Pan et al., 2024). PBAT is synthesized through a melt polycondensation process and is widely used in various applications, particularly in packaging.

Polyvinyl alcohol (PVA) has hydrophilic properties and a semi-crystalline structure, which contributes to its mechanical strength and biocompatibility (Agrawal et al., 2023). These qualities make PVA suitable for both biological and industrial uses.

3.2 Types of Biopolymers Used in Food Packaging and Coatings

In recent years, various companies have launched a range of biopolymers for food packaging and coatings. According to Babu et al., (2013), these newly developed biopolymers can be categorized based on their raw materials into several types: starch-based polymers (like Mater-Bi), cellulosic polymers (like Cellophane), bio-based polyethylene (Bio-PE), aliphatic polyesters (like PLA), and polyhydroxyalkanoates produced by microbes (like poly-3-hydroxybutyrate (PHB)).

Polylactic acid (PLA)

Polylactic acid (PLA), often referred to as polylactide, stands out among biopolymers as one of the most promising and popular materials with a bright future in development (Balla et al., 2021). Its popularity stems from several key factors, including its biocompatibility, thermoplastic nature, and potential for diverse applications. The recent expansion of PLA usage can be attributed to advancements in producing high molecular weight PLA through economical methods. Three primary methods are used to generate a large molecular weight PLA: azeotropic dehydrative condensation and direct condensation polymerization (Desai et al., 2024), and polymerization via lactide formation followed by ring-opening polymerization (Tong and Cheng, 2009). Among these, the third method—lactide formation combined with ring-opening polymerization—is particularly crucial for generating commercially viable high molecular weight PLA (Balla et al., 2021). This technique has facilitated efficient production processes, leading to the broad adoption and expansion of PLA across various industries.

Starch

Starch-based materials are the most widely used among cost-effective biomaterials. Composed primarily of amylose and amylopectin, these materials have drawn interest in food packaging due to their abundance, biodegradability, and affordability. They can be utilized to create edible and bioresorbable films and coatings for packaging. However, starch-based polymers have some drawbacks, including strong hydrophilic properties that result in poor moisture barriers and weaker mechanical properties compared to traditional food packaging uses polymer films that are not biodegradable (Jahangiri et al., 2024; Kaur et al., 2021). To overcome these challenges, the properties and morphology of starch-based polymers can be improved by blending them with synthetic polymers. These blends are often employed for packaging films and sheets (Bastioli and Bettarini, 2020). High-starch-content blends (over 85%) are particularly useful for making foams that can replace polystyrene (PS) in loose fill packaging (Gupta et al., 2011). This blending strategy enables the creation of starch-based materials with improved properties, expanding their applicability in various packaging applications while addressing their inherent limitations

Cellulose

Indeed, the most common biopolymer on Earth is cellulose, serving as the main structural component of cell walls in all plants. Due to its abundance and unique properties, cellulose serves as a valuable resource for various industrial applications, including paper and textile manufacturing, as well as emerging fields such as bioplastics and biomedical materials (Gregory et al., 2021). The majority of cellulose produced is utilized in the manufacturing of paper and cardboard, where its fibrous structure provides strength and durability to these materials (Barhoum et al., 2020). Additionally, cellulose is processed into fibers or films and synthesized into various derivatives such as cellulose esters and ethers. Cellulose derivatives find extensive industrial applications, including the production of coatings, laminates, and optical films (Tajeddin, 2014). These cellulose derivatives exhibit a broad spectrum of properties and functionalities, making them highly valuable across multiple sectors, including packaging, textiles, pharmaceuticals, and cosmetics (Orasugh et al., 2018). The versatility of cellulose and its derivatives plays a crucial role in their extensive application in various industries. Research is actively focused on advancing cellulose derivatives for packaging uses, with cellulose esters like cellulose acetate showing particularly promising potential.

Chitosan

Chitosan, a cationic polymer derived from the exoskeletons' chitin of crustaceans like shrimp and crab, is known for its inherent antimicrobial properties. This makes it valuable for various applications, including films and coatings (Kumari and Kishore, 2020). The antimicrobial action of chitosan primarily stems from its ability to facilitate cell adhesion. The positively charged amine groups in chitosan interact with the negatively charged cell membranes of microorganisms, leading to leakage of intracellular contents and subsequent cell death (Ardean et al., 2021; Rashki et al., 2021). In food preservation, chitosan is used as a coating to protect fresh fruits and vegetables from fungal spoilage. It acts both by exerting direct antifungal effects and by physically blocking contaminants from reaching the produce. Research by Rhoades and Roller (2000) examined the antimicrobial efficacy of both degraded and native chitosan. They found that mild hydrolysis improved chitosan's inhibitory activity against spoilage yeasts, whereas highly degraded chitosan showed minimal antimicrobial effects. Overall, chitosan's antimicrobial properties, combined with its biocompatibility and biodegradability, make it a promising material for food packaging and preservation, providing effective protection against microbial spoilage while being safe for the environment and human health (Rodrigues et al., 2020).

3.3 Advantages of Biopolymer-Based Materials

Packaging is the process of assembling supplies for transportation, retail, distribution, storage, and final use. It ensures that products reach the retail market in a specified condition and at a reasonable cost. Traditional polymers used in food packaging are categorized based on their usefulness, characteristics, and the types of contents they hold. However, conventional plastics often face issues with contaminants, which can impact both the security and caliber of the packaged goods, along with the risk of these contaminants migrating into food, is a significant reason for the shift to biopolymers (Ncube et al., 2020). Another important factor is the environmental concern and the lack of biodegradability associated with plastics derived from petroleum.

Conventional plastics take an exceptionally long time to degrade, resulting in substantial environmental impact and contamination (Liu et al., 2022). Although these plastics were once deemed viable, their drawbacks have overshadowed the advantages of sustainable alternatives like biopolymers. Conventional plastics often contain toxic additives, limiting their secure application in the food packaging sector. In contrast, biopolymers provide significant benefits, including a 30–80% reduction in greenhouse gas emissions and a longer shelf life compared to traditional plastics (Kabir, 2010; Babu et al., 2013; Patti and Acierno 2022). Biopolymer-derived materials are generally biodegradable and can be composted, enriching the soil with nutrients and organic matter. This not only enhances retention of water and nutrients but also reduces the need for chemical inputs and helps prevent plant diseases.

Biopolymer mechanical properties

Various packaging applications demand biopolymers to have mechanical properties. For flexible packaging, such as films, flexibility is the most important characteristic. In contrast, the emphasis switches to characteristics like Young's modulus, which gauges stiffness and resistance to deformation, for rigid packaging, such as containers.

Polylactic acid (PLA) typically exhibits a higher modulus and lower elongation, making it suitable for rigid packaging needs (Swetha et al., 2023). On the other hand, starch-based biopolymers provide greater elongation and lower modulus, making them well-suited for flexible packaging applications where high flexibility is essential. PHAs are produced by bacterial fermentation. They are biodegradable, with properties that can be adjusted by altering their monomer composition. PHB (Polyhydroxybutyrate), a type of PHA, is brittle, while PHBV (Polyhydroxybutyrate-co-valerate) is more flexible (Dalton et al., 2022). Derived from chitin found in crustacean shells, chitosan is biodegradable and biocompatible (Hamed et al., 2016). It has good tensile strength and moderate elasticity. One of the main structural proteins in connective tissues' extracellular matrix is collagen (Halper, et al., 2014; Sorushanova et al., 2019). Gelatin, derived from collagen, demonstrates notable tensile strength and elasticity, which can be further improved through crosslinking. It finds applications in food and pharmaceuticals, with its mechanical properties varying significantly depending on its moisture content.

Thermal stability of biopolymers

Biopolymers encounter elevated temperatures during traditional polymer melting processes. To maintain their molecular weight and properties, it is crucial for biopolymers to possess adequate thermal stability and resist degradation. Thermal stability is a key consideration.

Polylactic acid (PLA) shares mechanical properties similar to traditional polymers, but its thermal properties are hindered by a low glass transition temperature (Tg) of around 60°C (Zhao et al., 2022). According to Khankrua et al., (2014), PLA melts at 190 to 250°C, but processing temperatures are often 20 to 100°C higher. Thermal degradation is a frequent challenge during PLA processing, but it can be alleviated by blending PLA with other polymers and modifying its stereochemistry (Nampoothiri et al., 2010).

Polyhydroxybutyrate (PHB) has a glass transition temperature (Tg) near room temperature. Specifically, Biopol, a copolymer of PHB with 90% PHB content, has a Tg ranging from 18 to 22°C (Fitzgerald, 2017; Bertini et al., 2012). However, PHB exhibits poor thermal stability above its melting point of approximately 175°C, though its stability can be improved with certain polymeric additives (Hong et al., 2011).

In contrast, starch and cellulose have higher thermal decomposition temperatures, around 300°C, compared to PLA and PHB (Arial, 2014). The thermal stability and decomposition of these materials are affected by their microstructures, modifications, and processing conditions (Liu et al., 2013).

Composting and recycling of biopolymers

Biopolymers differ from polymers derived from petroleum and are typically incompatible with traditional recycling methods. However, they are well-suited for composting. Biopolymers are typically not ideal for traditional recycling methods because they undergo biodegradation during their service life or disposed of streams and possess less chemical and thermomechanical resistance than traditional plastics in general. Composting is often a more suitable option for biopolymers due to its environmental benefits and lower energy costs compared to recycling. During composting, biopolymer packaging materials break down physiologically into carbon dioxide, water, and inorganic substances (Rydz et al., 2018).

For example, Polylactic acid (PLA) can be recycled, incinerated, or composted. It can undergo hydrolysis with steam or boiling water to produce lactic acid, which can then be recycled into monomers, facilitating molecular recycling. Similarly, Mater-Bi, a biopolymer, is both recyclable and biodegradable, much like pure cellulose (Hu, 2014; Bracciale et al., 2024). International standards like ASTM D-6400 certify that Mater-Bi materials are compostable (Goel et al., 2021).

4. Silver Nanoparticles in Biopolymer-Based Food Packaging

4.1 Incorporation Methods of Silver Nanoparticles into Biopolymers

Polymeric platforms are widely considered excellent carriers for silver nanoparticles (Ag NPs) because of their organized structure, side chains, branching, and various functional groups that effectively capture the nanoparticles (Azizi-Lalabadi et al., 2021). Rather than simply mixing Ag NPs into composite materials, different techniques are employed to incorporate them into nanocomposite matrices. The most common approach involves introducing silver cations into polymer or biopolymer chains, followed by their reduction using reducing agents. This method forms a polymeric matrix for Ag NPs, preventing aggregation and ensuring a consistent morphology and nanoscale size distribution. These attributes enhance the performance and effectiveness of the nanocomposite (Dallas et al., 2011).

Biopolymers, fabricated polymers, and dendrimers are frequently utilized to stabilize silver nanoparticles (Ag NPs) for antimicrobial applications. Among synthetic polymers, polyurethanes, poly (methyl methacrylate), nylons, and polyvinyl alcohol (PVA) are particularly notable for their use in producing silver nanocomposites across various industries. In biopolymeric matrices, materials such as starches, cellulose derivatives like carboxymethyl cellulose, carbohydrates such as agar, dextrose, and poly (γ-glutamic acid) and gums like gum acacia are employed to encapsulate Ag NPs (Dallas et al., 2011; Kyriakoudi et al., 2021; Azizi-Lalabadi et al., 2021). Dendrimers, known for their narrow particle size distribution, are especially suited for biological applications due to their compatibility. These dendrimers, including poly (ethylene imines) (PEI), poly(ethylenediamine), PAMAM (poly amido amine), and ethylenediamine, form organized and symmetrical assemblies that result in uniform, monodisperse Ag NPs, making them ideal for use in biological systems (Dallas et al., 2011).

In addition to chemical reduction and polymeric matrices for Ag ion reduction, several green methods have been developed. One such approach involves the photo-reduction of AgNO3 using titania nanoparticles or nanotubes exposed to UV-vis light at 365 nm, which creates thin films with smaller Ag NPs and improved antimicrobial properties (Dallas et al., 2011). Another method uses layer-by-layer platforms to create well-structured Ag NP assemblies. Magnetically oriented layer-by-layer structures, made from composites like montmorillonite, phosphazene, polyacrylamide, and polyacrylic acid, have shown enhanced antibacterial efficacy against bacteria that are both Gram-positive and Gram-negative (Cao et al., 2022).

Moreover, silver nanocomposites can be assembled through low-temperature plasma processes, which are considered an environmentally friendly technique as they avoid the need for chemical reagents. Magnetron sputtering at low pressures allows the synthesis of Ag NPs with minimal aggregation. Zhang et al. (2010) also developed highly conductive silver nanocomposites through low-temperature sintering of silver nanoparticles. These green fabrication techniques result in well-organized Ag nanocomposites with extended antimicrobial properties, making them well-suited for biological applications.

4.2 Effects of Silver Nanoparticles on Biopolymer Properties

Biopolymers have recently attracted significant interest as promising platforms for silver-based antibacterial systems because of their biocompatibility, non-toxicity, biodegradability, and eco-friendly properties (Dhiman et al., 2019; El Allaoui et al., 2024). They provide excellent matrices for incorporating metal compounds, facilitating the stabilization and dispersion of nanoparticles. For example, embedding silver in biopolymers such as polylactides, polyhydroxyalkanoates, cellulose, or chitosan results in biodegradable materials well-suited for controlled antimicrobial release in applications like active food packaging, food contact surfaces, and medical devices (Kuperkar et al., 2024).

Antimicrobial properties

Enhanced Antimicrobial Activity: Silver nanoparticles (AgNPs) endow biopolymers with potent bactericidal and fungicidal properties, significantly inhibiting the growth of various microorganisms (Liao et al., 2019). Raza et al. (2016) observed that spherical AgNPs, functioning as a weak acid, likely interact with and disrupt sulfur- and phosphorus-containing compounds, such as DNA (soft bases), within bacterial cells.

Additionally, these nanoparticles disrupt the membrane structure, ultimately leading to cell death. This is particularly beneficial for food packaging, where preventing microbial contamination is crucial.

Mechanical properties

Improved Strength and Durability: The incorporation of AgNPs can enhance the mechanical strength, elasticity, and durability of biopolymers by interacting with the polymer matrix and providing reinforcement and stability (Jamróz et al., 2019).

Increased Tensile Strength: AgNPs can increase the tensile strength of biopolymers, making them more resistant to physical stress and deformation (Yuan et al., 2021).

Barrier properties

Enhanced Barrier Performance: Silver nanoparticles (AgNPs) can significantly boost the barrier capabilities of biopolymers against gases like oxygen and moisture (Istiqola and Syafiuddin, 2020). This improvement is vital for food packaging, as it aids in maintaining freshness and prolonging the products' shelf life.

Reduced Permeability: The inclusion of AgNPs can diminish the biopolymers' ability to withstand gases and water vapor providing better protection for the contents (Jamróz et al., 2019).

Thermal stability

Enhanced Thermal Stability: Incorporating AgNPs can boost the thermal resistance of biopolymers, increasing their ability to withstand higher temperatures without degradation .This is important for applications requiring thermal processing or exposure to varying temperatures (Dash et al., 2021).

Optical properties

Altered Transparency and Color: AgNPs can affect the optical properties of biopolymers, potentially altering their transparency and color (Jamróz et al., 2019). The impact of AgNPs can be either beneficial or detrimental based on their specific use.

UV Protection: AgNPs can provide protection against UV light by absorbing and scattering it, which helps maintain the quality of packaged products (Mathew et al., 2019).

Cytotoxicity and safety

Potential Toxicity Reduction: When incorporated into biopolymers like chitosan, AgNPs can exhibit reduced cytotoxicity, making them safer for applications involving food or human tissues (Liao et al., 2019).

Health and Safety Concerns: Despite their benefits, the potential health risks associated with AgNPs, such as cytotoxicity, genotoxicity, and inflammatory responses, necessitate thorough safety evaluations and regulated usage levels (Ferdous and Nemmar, 2020; Nie et al., 2023). Silver nanoparticles greatly improve the characteristics of biopolymers, making them better suited for uses such as active food packaging. However, careful consideration of their potential health impacts and regulatory requirements is essential to ensure safe and effective use.

4.3 Antimicrobial Properties of Silver Nanoparticles in Food Packaging

Since ancient times, silver has been used as an antimicrobial agent for food storage, preservation, and medical procedures. Historically, It has been applied as a disinfectant and for the treatment of and burns (Nqakala et al., 2021; Kaiser et al., 2023). Recent studies have shown that silver wounds nanoparticles (AgNPs) are effective against bacteria, viruses, and fungi. Their antimicrobial activity is generally superior to Gram-positive bacteria in its efficacy against Gram-negative bacteria. This difference is Gram-positive bacteria have thick peptidoglycan layers in their cell walls, which create a rigid structure that limits the penetration of AgNPs (El-Seedi et al., 2019).

4.3.1 Parameters affecting antimicrobial activity

Various factors contribute to the antimicrobial efficacy of silver nanoparticles (AgNPs), either independently or in combination. These factors encompass the shape, size, morphology, surface charge(s), concentration of nanoparticles, and the particular microorganisms targeted (Varier et al., 2019; Hamouda et al., 2019). Here, we delve into some of the most critical factors in detail.

4.3.1.1 Size of Silver Nanoparticles (AgNPs)

The size and shape of nanoparticles are critical factors affecting their antimicrobial effectiveness. Studies show that smaller nanoparticles typically show greater antibacterial activity compared to larger ones (Tang and Zheng 2018; Dong et al., 2019). This is due to the greater ratio of smaller nanoparticles' surface to volume, which results in a greater concentration of ions or silver atoms. This increased concentration enhances their interaction with microbial cell walls or membranes and triggers the generation of reactive oxygen species (ROS) (Quinteros et al., 2016). Furthermore, smaller nanoparticles are more apt to enter microbial cells, enabling more effective interaction with intracellular components. Oliver et al. (2018) noted that the antimicrobial properties of silver nanoparticles larger than 20 nm are largely due to Ag+ ion release, whereas nanoparticles smaller than 10 nm are more effective because they can penetrate microbial cells and directly engage with cellular components and enzymes (Oliver et al., 2018). Similarly, Mei et al. (2017) demonstrated that ultra-small silver nanoclusters (AgNCs) show superior antimicrobial activity compared to larger AgNPs, due to their improved mobility, a high surface-to-volume ratio, and an increase in local surface concentration (Mei Zhang et al., 2017).

4.3.1.2 Shape of Silver Nanoparticles (AgNPs)

Besides the size, the shape of silver nanoparticles (AgNPs) significantly affects their antimicrobial activity by influencing how they interact with microbial cell walls or membranes. AgNPs have been studied for their antimicrobial properties in a variety of shapes, including polyhedral, spherical, oval, rectangular, and rod-shaped particles. Research indicates that spherical AgNPs often demonstrate more effective antimicrobial activity compared to rod-shaped ones (Bharti et al., 2021; Karade et al., 2021).

Additionally, triangular-shaped AgNPs have been shown to exhibit greater antimicrobial activity against *E. coli* than both spherical and rod-shaped nanoparticles (Acharya et al., 2018). However, dos Santos et al. (2017) found triangular AgNPs, spherical AgNPs demonstrated comparable antimicrobial efficacy against Streptococcus mutans. While some studies suggest that triangular AgNPs may be superior antimicrobial agents, there is no definitive evidence regarding the impact of nanoparticle shape on antimicrobial efficacy. More research is required to clarify the relationship between AgNP shape and their antimicrobial activity.

4.3.1.3 Concentration of Silver Nanoparticles (AgNPs)

The concentration of silver nanoparticles (AgNPs) is closely linked to their antimicrobial effectiveness. Research has shown that higher concentrations of AgNPs generally produce stronger antimicrobial effects, whereas lower concentrations result in diminished efficacy (Kumar et al., 2014, 2016).

Recent studies have indicated that the MinimumInhibitory Concentrations (MICs) of AgNPs for human pathogens like *S. mutans* and *S. oralis* range from 45 to 60 μg/mL. In contrast, MICs for the concentration of food spoilage bacteria can range from 15 to 90 μg/mL, including *Lactobacillus acidophilus* and *Lactobacillus fermentum* (Panpaliya et al., 2019). Additionally, the MIC for AgNPs against biofilms formed by foodborne pathogens, including *Staphylococcus aureus* and *Pseudomonas aeruginosa,* is approximately 62.5 μg/mL. Biofilm inhibition levels of 85 and 90% were achieved with AgNP quantities of 125 and 250 μg/mL, in that order (Gupta et al., 2019).

These findings underscore the importance of nanoparticle concentration in determining their antimicrobial efficacy and suggest potential applications for controlling microbial growth in various contexts, including healthcare and food safety.

4.4 Potential Health and Safety Concerns

The rise of commercialization of nanotechnology has sparked public concerns about the safety and migration of nanoparticles (NPs). A significant issue is the use of NPs in nanocomposite materials for use in food packaging and biomedical applications, with concerns about NPs potentially migrating from packaging into food and posing health risks to consumers (Addo Ntim et al., 2018). It is important to recognize that the toxicity of NPs can be affected by their size, shape, and surface properties (He and Hwang, 2016). Additionally, the application of nanoparticles in various fields is often restricted by regulatory approval processes that must be navigated before their implementation is authorized. Concerns are particularly high regarding nanoparticles due to their small size (Ummi et al., 2019). Many nanomaterials previously classified as "generally recognized as safe (GRAS)" may not have undergone nanoscale-specific safety assessments. A major challenge in evaluating the safety of nanomaterials is the insufficient scientific evidence on their bioavailability and toxicokinetics. Addressing these issues requires ongoing research into the effects of nanomaterials on human cells. Currently, there are no universally accepted standards for evaluating nanomaterials (King et al., 2018).

Researchers have found that silver nanoparticles (AgNPs) can cause inflammatory reactions, cytotoxicity, and genotoxicity in human cells (Cameron et al., 2018). However, studies by Regiel-Futura et al. (2017) and Jena et al. (2012) have proved that the chitosan is an effective polymeric matrix for stabilizing AgNPs. This combination not only offers excellent bactericidal and fungicidal properties but also mitigates or eliminates cytotoxic effects.

5. Edible Coatings with Silver Nanoparticles

5.1 Introduction to Edible Coatings

Extra-thin layers called edible coatings, which are typically thinner than edible films, are directly applied to the surface of food or other substances. While they can enhance the physical properties and quality of food, they are less rigid and more flexible compared to films. Edible coatings can also be used to enhance the appearance of products by adding color and gloss, making them more attractive to consumers (Arnon-Rips and Poverenov, 2018; Khalid et al., 2022).

In the food industry, particularly coatings, edible packaging, is commonly used to prolong the shelf life of food and protect it. Techniques like spray fluidization, dipping, brushing, falling and pan coatings, and spraying are examples of traditional methods for applying edible films. After application, moist products must be dried, while lipid-based coatings require cooling (Chettri et al., 2023; Grujic et al., 2017). Most edible films form as the solvent evaporates from a diluted biopolymer solution applied in thin layers. Safe solvents such as ethanol and water are frequently utilized in the food industry. Drying can be achieved using various methods, infrared energy, including ambient conditions, hot air, or microwave energy. The drying process impacts the film's properties, including its appearance, mechanical characteristics, and moisture and gas permeability (Thakhiew et al., 2013; Garcia-Tejeda et al., 2015; Al-Harrasi et al., 2022).

5.2 Release of Silver Nanoparticles from Edible Films and Coatings into Food

Active packaging using biopolymers infused with silver nanoparticles shows great promise for improving microbial food safety in packaged goods. The effectiveness of these packaging materials depends on how well antimicrobial agents diffuse from either directly on the food surface or into the headspace of the package using nanocomposite film (Khaneghah et al., 2018; Bahrami, et al., 2020). Research has mainly focused on nanoparticles moving away from commercially available synthetic food packaging materials such as polypropylene (PP), polyvinyl chloride (PVC), low-density polyethylene (LDPE), and polyethylene (PE) (Khajavi et al., 2019; Morais et al., 2020). However, data on the migration of silver nanoparticles from biopolymer-based packaging are scarce. For example, Nair et al. (2017) investigated the release of chitosan, cassava starch, and konjac glucomannan composite film containing silver nanoparticles into bread samples that were kept for a period of 14 days. Their results showed that the nano silver amount released from the film to the food was approximately 4.3–4.5 μg/L.

Marchiore et al. (2017) investigated the migration of silver nanoparticles in starch-based edible coatings applied to sausages. They measured the concentration of silver nanoparticles in the coating solution as 23.47 µg AgNPs /mL using ICP-MS. After coating, the amount of silver nanoparticles retained on the sausages was 5.3 ng AgNPs/g. The study also noted that common practices like washing and cooking could reduce the silver nanoparticle levels on the sausage surface to around 4 ng AgNPs/ g.

Both investigations demonstrated that the movement of active ingredients remained below the regulatory migration limit of 0.01 mg/kg of food, in accordance with Commission Regulation (EC) No. 450/2009 for "non-authorized substances." Since nano-sized silver is classified as a "non-authorized substance" for packaging materials, it requires regulation. However, new technologies that produce substances with significantly different nanoscale properties, such as nanoparticles, must be individually assessed for risk. These substances are not yet addressed by the functional barrier concept until more data is available on the new technology (European Commission, 2009; European Commission, 2011).

Regulations call for clarification on the total and specific migration of silver nanoparticles into food products. Key factors include the concentration of migrating nanoparticles, the type of polymer matrices used, and the migration mechanism influenced by diffusion rates. Additionally, nanoparticle size and morphology, along with physical and chemical factors like time and temperature, are crucial for assessing risks to human health and ensuring the microbial safety of packaged food (Zhang and Chen 2019; Kumar et al., 2020; Onyeaka et al., 2022).

5.3 Antimicrobial and Physical Characteristics of Edible Films and Coatings with Silver Nanoparticles

Antimicrobial properties

Biopolymers have proven to be effective in synthesizing and stabilizing silver nanoparticles (Alavi and Ravi, 2019; Basta et al., 2020). This polymer-assisted synthesis approach provides a uniform distribution of nanoparticles within the polymer matrix, thereby enhancing the structural stability and consistency of the resulting nanocomposite films. As a result, these films retain strong antimicrobial properties (Mlalila et al., 2018; Venkatesan et al., 2018; Shafey, 2020).

The process of creating nanocomposite films is influenced by the location where the nanoparticles are synthesized, resulting in two primary methods. The first method, known as *in situ*, uses the polymeric matrix as both the reaction medium for the formation of silver nanoparticles and their stabilizing agent. The second method, referred to as *ex situ*, involves using the polymeric matrix to disperse and stabilize silver nanoparticles that have been synthesized separately (Plazza, 2015).

Research by Bankura et al. (2012) demonstrated that dextran, a polymer advantageous for edible films and coatings, can serve as a medium for the *in situ* synthesis of silver nanoparticles, producing particles approximately 10–60 nm in size. The study revealed that dextran molecules effectively cap the silver nanoparticles by binding them to their hydroxyl groups, which prevents aggregation. These formulations remained stable for over a month and showed strong antibacterial properties against various bacteria, including *Bacillus cereus*, *Bacillus subtilis*, *Staphylococcus aureus*, *Escherichia coli*, and *Pseudomonas aeruginosa*. Further research has also investigated the *in-situ* formation of highly uniform silver nanoparticles within chitosan films (Kumar and Kaur, 2020; Ramadan et al., 2020; Zhao et al., 2022). These studies found that silver nanoparticles in chitosan matrices stayed well-dispersed for nearly three months without clumping, attributed to the robust interactions between the polymeric matrix and the metal particles due to chitosan's amine and hydroxyl groups (Lin et al., 2015).

Additionally, the biological characteristics of nanocomposite films have been demonstrated by incorporating silver nanoparticles into polymer matrices using the *ex situ* method. For example, HPMC, or hydroxypropyl methylcellulose polymers with silver nanofillers were successfully produced via this approach, where the polymer matrices act as stabilizers, preventing nanoparticle clumping and showing strong antimicrobial effectiveness against different bacterial strains (Jiménez et al., 2015; Qahir et al., 2021). Similarly, another study evaluated the antimicrobial efficacy of a trinary biocomposite film made from tragacanth, HPMC, and beeswax, incorporating pre-synthesized silver nanoparticles. This nanocomposite effectively suppressed the expansion of pathogens including *Salmonella Typhimurium*, *Bacillus cereus*, *Listeria monocytogenes*, *Pseudomonas aeruginosa*, *Streptococcus pneumoniae*, *Escherichia coli*, and *Klebsiella pneumonia* (Bahrami and colleagues, 2019). For improved antimicrobial performance, higher concentrations of silver nanoparticles are often required (Hamad et al., 2020; Bruna et al., 2021). Studies by Bahrami et al. (2019) and Youssef et al. (2014) revealed that increasing the amount of silver nanoparticles in biopolymers resulted in larger inhibition zones against tested microorganisms. Table 1 shows the antimicrobial impact of various biopolymer-based films embedded with silver nanoparticles. This antimicrobial effect is linked to the release of nanoparticles from the nanocomposites, enabling direct interaction with microbes (Sharma et al., 2020; Omerović et al., 2021).

Moreover, storage temperature is critical for preserving the antimicrobial properties of nanocomposite materials. Khalaf et al. (2013) examined pullulan-based films with silver nanoparticles and their ability to inhibit *S. aureus* and *L. monocytogenes* over 7 weeks at various temperatures of 4°C, 25°C, 37°C, and 55°C. The findings suggested that storing nanomaterials below 25°C is essential to retain their bactericidal activity, as higher temperatures substantially diminished their antimicrobial effectiveness.

Physical properties

The key benefits of nanomaterials lie in their ability to boost food protection and preservation by enhancing the structural and functional characteristics of film. The antimicrobial potential of silver nanoparticle-integrated biopolymers is shown in Figure 1. It is widely recognized that the qualities of biopolymer-based packaging materials are influenced by various factors, such as the origin of the biopolymer, the polymer chain's structural arrangement, processing techniques, and the level of crystallinity or cross-linking. Hence, when using edible films and coatings for food, it is essential to consider both the physical properties and functional features of biopolymer materials.

Polysaccharides and proteins, the main film-forming materials, are hydrophilic, making it challenging to develop coatings for perishable foods. To enhance their low moisture resistance, adding lipids to a hydrocolloid matrix is an effective approach (Galus and Kadzińska, 2015). Hydrocolloid-based materials provide excellent oxygen permeability in dry conditions (Fabra et al., 2013), but their effectiveness diminishes significantly in humid environments. Water vapor absorption leads to plasticization and swelling, weakening their oxygen barrier by disrupting the material's network (Kurek et al., 2014). Although proteins are typically ineffective moisture barriers, they provide good gas permeability and possess strong structural stability, which aids in maintaining an even layer on the surface of food. Different chemical, physical, and enzymatic treatments have been explored to enhance the performance of edible films and coatings (Mihalca et al., 2021). Polymer blending is particularly valuable, as composite materials leverage the strengths of different components while reducing their limitations. Enhancing barrier properties can help maintain food quality and extend shelf life while avoiding the use of chemical preservatives.

5.4 Applications of Silver Nanoparticles and Edible Coatings in Biopolymer-Based Food Packaging

Nanomaterials offer substantial benefits by improving food protection and preservation. Silver nanoparticle (AgNP)-based antimicrobial packaging provides a promising approach to the active food packaging solution that is the key to prolonging food shelf life and minimizing pathogen risks. The characteristics of biopolymer-based packaging materials are affected by several factors, such as their original source, the structural arrangement of the degree of cross-linking or crystallinity, the polymer chain, and the processing techniques. Consequently, when employing edible films and

Table 1. The antimicrobial properties of different biopolymer-based materials incorporating silver nanoparticles.

Polymeric frameworks	Method	Size of Silver Nanoparticles (nm)	Microorganisms Tested	Antimicrobial Impact of Nanocomposite Films/ Coatings	References
Agar/Banana-based	*In situ*	(a) 100 (b) 150–200	*L. monocytogenes* ATCC 15313, E. coli O157 ATCC 43895	*L. monocytogenes* (a) and (b) exhibit bacteriostatic activity; *E. coli* – (a) shows bacteriostatic activity, (b) exhibits bactericidal activity	Orsuwan et al., 2016
chitosan/ fucoidan	*In situ*	73 ± 9.54	*S. aureus E. coli*	Inhibition zone (mm): *S. aureus*: 2, *E. coli*: 3	Venkatesan et al., 2018
Chitosan	*In situ*	75–250	*E. coli*	Bactericidal effect on *E. coli*	Tankhiwale and Bajpai, 2010
Agar	*Ex situ*: Pre-synthesis of AgNPs through chemical methods	21.3–23.8	L. monocytogenes ATCC 19111 and *E. coli* O157 ATCC 11775	*L. monocytogenes* and *E. coli* (a) and (b) showed no inhibitory effect; (c) exhibited a bacteriostatic effect; (d) demonstrated a bactericidal effect.	Rhim et al., 2013
guar gum	*Ex situ*: AgNPs obtained from Sigma Aldrich, specifically Ag-Cu.	< 100	*L. monocytogenes* ATCC 19114 and *S. typhimurium* ATCC 14028	Reduction in the bacterial count of *L. monocytogenes*	Arfat et al., 2017
pectin	*Ex situ*: AgNPs pre-synthesis by biological methods	20–80	*L. monocytogenes* ATCC 15313 and *E. coli* O157 ATCC 43895	Inhibition zone (mm): *L. monocytogenes*: 3.9, *E. coli*: 8.4	Shankar et al., 2016
Gelatin	*Ex situ*: Commercial AgNPs product obtained from Sigma Aldrich, labeled as Ag-Cu.	< 100	*L. monocytogenes* ATCC 19114 and *S. typhimurium* ATCC 14028	Reduction in bacterial count: *L. monocytogenes* (a)-(d): 0.5-3.5 log CFU/mL; *S. Typhimurium* (a): no inhibitory effect, (b)-(d): reduction exceeding 7 log CFU/mL.	Arfat et al., 2017
sodium alginate/ chitosan	*Ex situ*: AgNPs pre-synthesis by biological methods	5–21	*B. cereus* MTCC 1305, *E. faecalis* MTCC 439, *E. coli*, *E. aerogenes* MTCC 2822 and *P. aeruginosa* MTCC 2488	Zone of Inhibition (mm): *B. cereus*: 6.0 *E. faecalis*: 5.1 *E. coli*: 1.9 *E. aerogenes*: 1.5 *P. aeroginosa*: 3.1	Sharma et al., 2012
sodium alginate	*Ex situ*: AgNPs pre-synthesis by biological methods	5–40	*S. aureus* ATCC 8739 *E. coli* ATCC 6538	inhibitory effect against both tested strains of bacteria was observed	Fayaz et al., 2009

pullulan/PVA	*Ex situ*	15–30	*S. aureus* ATCC 6538 *E. coli* ATCC 25922	Bactericidal effect observed across all concentrations against both tested bacterial strains.	Ul-Islam et al., 2013
Pullulan	*Ex situ*	100	*L. monocytogenes* ATCC 94229 *S. aureus* ATCC 11988	Inhibition zone (mm): *L. monocytogenes*: 25-30 *S. aureus* 15–23	Khalaf et al., 2013
HPMC	*Ex situ*: AgNPs pre-synthesis by chemical methods	(a) 41 (b) 100	*S. aureus* ATCC 25922 *E. coli* ATCC 25923	Inhibition zone (mm): *S. aureus*: (a) 3.11, (b) 1.35; *E. coli*: (a) 2.75, (b) 1.05	De Moura et al., 2012
cellulose	*in situ*	10–130	*E. coli*	inhibition zone for *E. coli* :20–25	Varsha et al., 2010
Gelatin	*Ex situ*	5–25	*L. monocytogenes*	Effective antimicrobial activity, prolonged shelf life	Tavassoli-Kafrani et al., 2016
PLA/PBAT	*In situ*	8–12	*S. aureus, E. coli*	Significant reduction in bacterial growth	Qin et al., 2015
Starch	*In situ*	30–50	*P. aeruginosa, E. coli*	Reduced bacterial colonization	Azarifar et al., 2020
Alginate/pectin	*Ex situ*	15–30	*B. cereus, E. coli*	Strong antimicrobial activity	Dammak et al., 2019
PVA/Starch	*In situ*	10–20	*S. aureus, E. coli*	Strong bactericidal activity	Shen et al., 2017
Gelatin/agar	*In situ*	25–35	*E. coli, S. aureus*	Effective antimicrobial activity, prolonged shelf life	Nidheesh et al., 2020

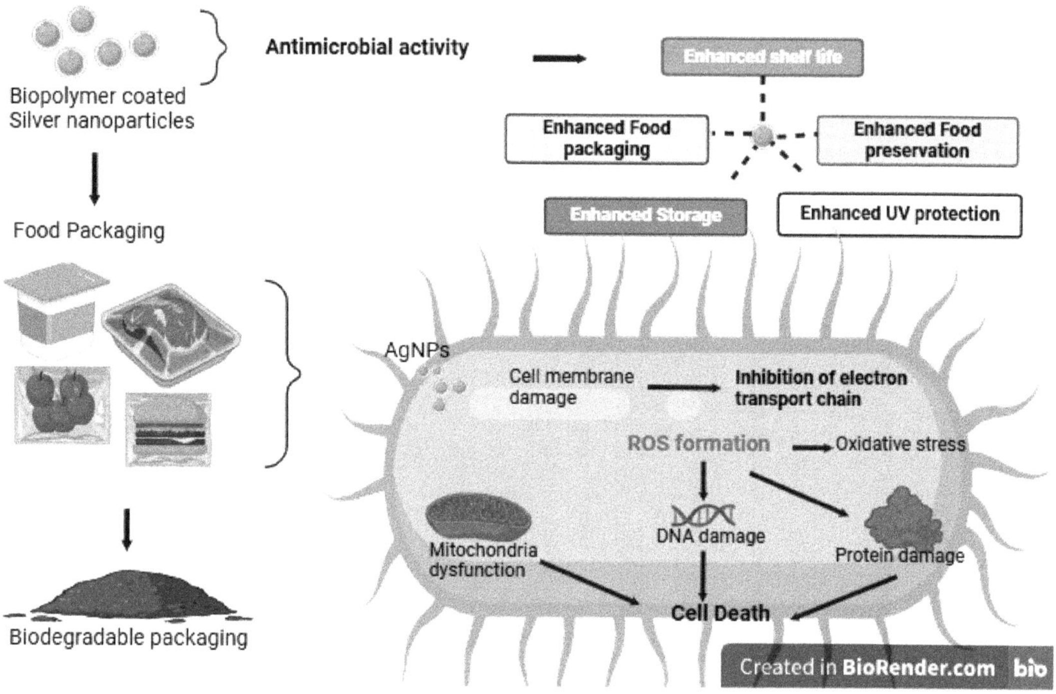

Figure 1. Antimicrobial properties of silver nanoparticles incorporated into biopolymer-based food packaging.

coatings in food applications, it is crucial to evaluate both the physical properties and functional characteristics of the biopolymer materials. Numerous studies have explored the effectiveness of various biopolymer matrices combined with silver nanoparticles in inhibiting foodborne pathogens (Table 2).

Recent research has concentrated on methods to reduce minimize microbial growth with antimicrobial food packaging, a form of active packaging (Campos-Requena et al., 2015; Fasihnia et al., 2018; Khaneghah et al., 2018; Kumar et al., 2020). This system integrates antimicrobial agents into packaging materials to create environments that can slow down, diminish, or prevent microbial growth (Huang et al., 2019; Kamarudin et al., 2022). Additionally, it helps minimize waste, protect food quality, retain appearance, and prolong the shelf life of packaged products (Ahmed and Kanchi, 2018; Czerwiński et al., 2021).

Biodegradable biopolymeric films present a promising alternative for food packaging due to their cost-effective production, renewable sources, and minimal environmental impact. These biodegradable polymers, already widely used in numerous applications (Song et al., 2018), are increasingly being considered as replacements for non-biodegradable plastics used in food containers. Commonly used biodegradable polymers include polysaccharides such as pullulan, cellulose, starch, agarose, and chitosan (Valencia-Chamorro et al., 2011; Dhall, 2013; Elsabee and Abdou, 2013). Cellulose, for example, interacts with electropositive transition metal atoms through electrostatic forces, leading to the adsorption of silver ions during immersion in silver nitrate. The porous structure of cellulose supports the synthesis and stabilization of silver nanoparticles (AgNPs). Absorbent pads made from cellulose are widely utilized in contemporary fresh food packaging strategies (Fernandez et al., 2012). Abundant and cost-effective, cellulose is biocompatible and can also be produced in an edible form.

Edible films and coatings have been employed for household food preservation for centuries, but their use in the industry of food processing is relatively recent and provides considerable potential for future advancements to cater to increasingly discerning consumers. Various applications of edible films and coatings have been explored with notable success (de Jesús Avena-Bustillos and

Table 2. Nanocomposite packaging utilizing AgNPs with biodegradable edible polymer matrices.

Polymer matrix	Food examined	Microorganisms tested	References
Cellulose with AgNPs	Meat of Beef	Overall aerobic bacteria, lactic acid bacteria, *Pseudomonas* species, *Enterobacteriaceae*	Fernandez et al. (2010a)
Cellulose with AgNPs	Pre-cut melon	Total mesophilic aerobic bacteria, psychrotrophic bacteria, yeasts, and molds	Fernandez et al. (2010b)
Pullulan with AgNPs	Turkey sliced deli meat	*L. monocytogenes* and *S. aureus*	Khalaf et al. (2013)
Pullulan with AgNPs	Beef and poultry	*S. aureus* and *L. monocytogenes*	Morsy et al. (2014)
Agar hydrogel with Ag-montmorillonite NPs	Fior di Latte cheese	*Pseudomonas* spp.	Incoronato et al. (2011)
Sodium alginate with AgNPs	Pears and carrots	*E. coli* and *S. aureus*	Fayaz et al. (2009
Sodium alginate + CaCl2 with AgNPs	Fior di Latte cheese	*Pseudomonas* spp., *Enterobacteriaceae* and *E. coli*	Mastromatteo et al. (2015)
Chitosan with AgNPs	Fresh-cut apple	*E. coli, S. aureus*	Kanmani and Rhim (2014)
Chitosan with AgNPs	Chicken breast	*Salmonella* spp., *L. monocytogenes*	Akhavan and Bakhshi (2015)
Gelatin with AgNPs	Fresh fish	*E. coli, S. aureus*	Roy and Rhim (2019)
Starch with AgNPs	Fresh strawberries	Total aerobic bacteria, yeasts, molds	Tawakkal et al. (2014)
Pectin with AgNPs	Fresh-cut melon	*E. coli, L. monocytogenes*	Jiang et al. (2011)
Soy protein isolate with AgNPs	Cheese	*L. monocytogenes, Pseudomonas* spp	Zhu et al. (2014)
Polylactic acid (PLA) with AgNPs	Fresh tomato	*E. coli, Salmonella* spp.	Shankar et al. (2015)
Polyvinyl alcohol (PVA) with AgNPs	Fresh vegetables	*E. coli, S. aureus*	Castro-Mayorga et al. (2016)
Polycaprolactone (PCL) with AgNPs	Fresh berries	*E. coli, Salmonella* spp.	Souza et al. (2019)
Polyhydroxyalkanoates (PHA) with AgNPs	Fresh spinach	*E. coli, L. monocytogenes*	Saifullah et al. (2016)
Polyethylene glycol (PEG) with AgNPs	Fresh broccoli	*E. coli, S. aureus*	López de Dicastillo et al. (2016)
Polyethylene (PE) + AgNPs	Fresh cucumber	*E. coli, L. monocytogenes*	Qian et al. (2012)
Polyvinyl chloride (PVC) + AgNPs	Fresh potatoes	*E. coli, S. aureus*	Fortunati (2012)

McHugh, 2012). The manufacturing of these coatings and films mainly involves two technological methods: the "Solvent-based method" and the "thermoplastic method." The solvent process is more widely used, as the thermoplastic process is limited to specific proteins like gluten and soy proteins.

Several techniques exist for applying edible films and coatings, selected based on the preferred properties of the film and the materials used. Here, only the most pertinent methods are outlined. The solvent removal technique involves dissolving a biopolymer in a film-forming solution with appropriate solvents and additives, casting the solution into a thin layer, and then drying it. Enrobing involves immersing the product in a coating solution, like chocolate, to form a thick layer. Techniques like drum coating, pan coating, and screw coating are similar, with pan coating being a batch process that can vary the coating thickness by adjusting the process duration, like when candy is covered in sugar and cocoa. Drum coating, another batch process, is suitable for applying thin or thick layers continuously to hard or solid foods. Screw coating applies a thin layer to foods that are firm and solid continuously. Coating for fluidized beds utilizes powder fluidization in an air stream while spraying a liquid coating, making it effective for applying thin layers to micronized particles (Palamanit et al., 2016). Spray coating, the most used method, applies liquid, suspension, or melted lipid layers of various thicknesses (Suhag et al., 2020). Individual wrapping, often used for meat products, involves encasing each item in an edible film.

Nowadays, edible films and coatings have diverse applications, frequently without consumers even realizing their presence. For instance, shoppers might question whether the glossy appearance of an apple in a store is natural or due to some other treatment. The use of edible films and coatings dates to the nineteenth century, when a patent was granted for coating meat with gelatin helps avoid dehydration and microbial deterioration (Dhiman et al., 2022). This innovation marked the inception of edible films and coatings in industrial food preparation. The summary of their uses are as follows.

Meat products

Enhancing a product's appearance on store shelves can provide a competitive advantage. Edible coatings not only help maintain visual appeal but also serve as carriers for antimicrobial and antioxidant agents. Coatings made from lipids, polysaccharides, proteins, and their composites have demonstrated impressive results (Hassan et al., 2018). The use of edible proteins for food packaging began with collagen casings and gelatin films for meat. Today film casings made of regenerate collagen are commonly used in the meat industry to wrap hams and form sausages (Song et al., 2021). Advances in film production now include extruded collagen fibers and additives such as cellulose and starch derivatives to improve mechanical properties. Adding antimicrobials to edible coatings further enhances food safety (Cutter, 2006). Developing new protein-based edible coatings remains crucial in the meat industry, as they help reduce moisture loss, limit lipid oxidation, prevent discoloration, and minimize water drippage (Shahidi and Hossain, 2022; Mohamed et al., 2020; Behbahani et al., 2017). The composition of edible casings affects meat products, making careful development of composite casings essential. For instance, Casings made from pectin or a blend of gelatin and sodium alginate with vegetable oil produce stable sausages, whereas gelatin-sodium casings alone may cause shrinkage. This highlights the importance of biopolymer type and vegetable oil content on sensory qualities, water loss, and lipid oxidation (Lu et al., 2007).

Bakery, cereal and confectionary coatings

In the twentieth century, dietary preferences evolved alongside changing lifestyles, with breakfast cereals illustrating this shift. Consumers generally favor sweet flavors, leading most cereal coatings to be sugar-based. Additional additives are used to enhance flavor, texture, and adhesion of the coating. Edible coatings are crucial for maintaining crunchiness and preventing dehydration, as cereals that lose these qualities in high humidity are often rejected by consumers (de Jesús Avena-Bustillos and McHugh, 2012).

Colorants are frequently added to make cereals more attractive. Products with nuts are prone to becoming soggy and experiencing lipid oxidation because of moisture absorption (Barden and Decker, 2016). Crunchy cereals combined with dried fruits can experience moisture migration, leading to undesirable physical and chemical alterations (Kong and Singh, 2016). Edible emulsion coatings derived from caseinates, or composite coatings made from chitosan help reduce water transfer, while protein-based coatings can prevent oxygen permeability and fat oxidation.

In the bakery industry, coatings are primarily used to enhance visual appeal and extend shelf life. Chocolate or sugar-based coatings improve the appearance of baked goods, while preservatives such as sodium benzoate can significantly prolong shelf life and ensure product safety. Confectionery items, more so than bakery products, use coatings to enhance visual appeal. For instance, shellac-based coatings provide a glossy finish to chocolate products and serve as oil barriers, preventing cocoa butter from migrating to the surface and staining packaging, which helps to avoid fat bloom.

Fresh and lightly processed produce

Although edible films hold great promise for fruits and vegetables, their commercial application is still limited. This limitation arises from difficulties in coating surfaces that are slippery, uneven, or irregular, and from the complexity of formulating ingredients to ensure uniform coverage. Edible films and coatings can assist in lowering transmission of water vapor and respiration rates by creating a protective outer layer, which also improves the appearance by maintaining cell vitality. These coatings are typically made from waxes or, more recently, from natural biopolymers and lipids. Recent research has investigated the impact of new edible coatings on the storage life and postharvest quality of fresh produce (Armghan Khalid et al., 2022; Yousuf et al., 2018). For example, according to Larotonda (2007), it was discovered that cherries coated with carrageenan lost 30% less water compared to uncoated cherries, and the coating helped preserve firmness and gloss during storage at 5°C. Likewise, antioxidants and antimicrobials can be found in edible coatings on fresh-cut fruit, extending its shelf life and improving food safety (Yousuf et al., 2021; Al-Tayyar et al., 2020).

Application of coating on fried foods

Using edible coatings on fried foods represent an innovative approach that caters to the growing demand for healthier eating options. These coatings form a protective barrier on the food's surface, reducing oil absorption during deep frying through the thermal gelation of hydrocolloids that occurs at temperatures exceeding 60°C. This technique leads to lower oil uptake and improved nutritional value. Edible films and coatings have already demonstrated success in enhancing meat products in this way. Figure 2 provides a visual overview of silver nanoparticle-based biopolymer applications in both packaging and edible coatings.

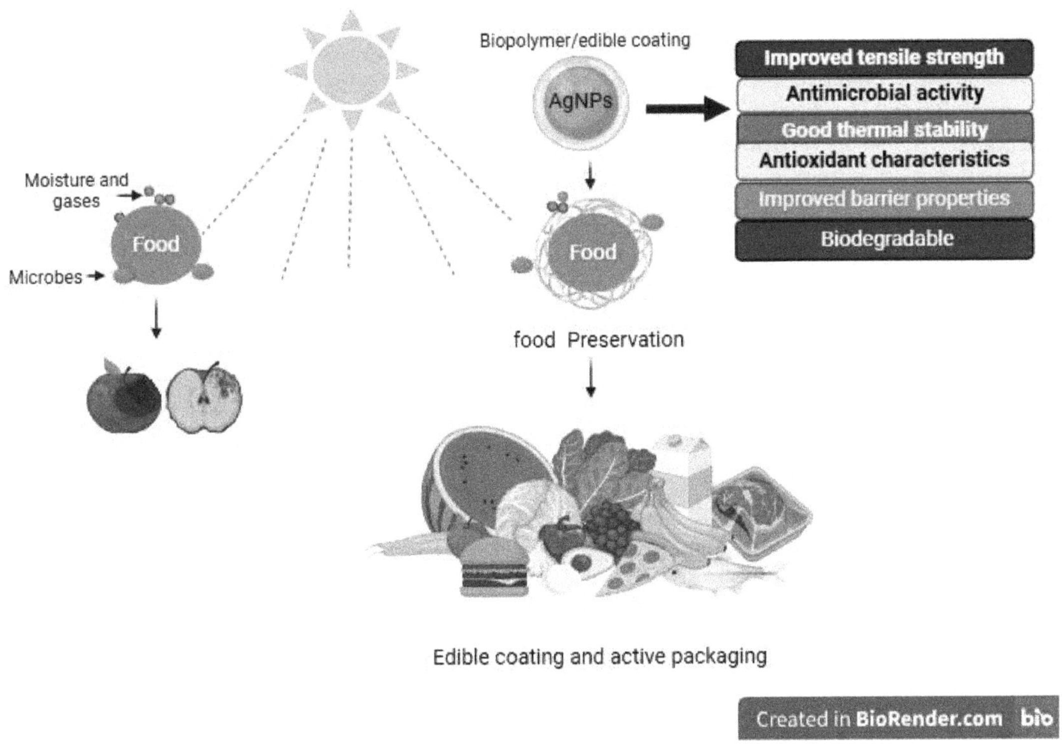

Figure 2. Application of biopolymer coated silver nanoparticles in food packaging and edible coatings.

6. Future Perspectives and Challenges

In the years ahead, silver nanoparticles may play a crucial part in advancing polymeric substances for active food packaging. However, a major issue with these materials is the insufficient understanding of the safety and potential risks of silver nanoparticles. Research has indicated that silver nanoparticles can be toxic to mammalian cells, causing issues such as cytotoxicity, genotoxicity, and inflammation in human cells (Cameron et al., 2018). Nonetheless, studies by Regiel-Futura et al., (2017) and Jena et al., (2012) have found that chitosan can effectively stabilize AgNPs, providing strong antibacterial and antifungal properties while reducing or eliminating cytotoxic effects. Exposure to nanomaterials can occur through skin contact, inhalation, or ingestion, with ingestion being the primary concern for nanocomposite materials used in food packaging due to the risk of nanoparticles migrating to the food surface. Thus, thorough toxicological evaluations are essential to assess the potential health risks associated with these materials.

Considering that the health and safety concerns related to nanomaterials are not yet fully understood, ensuring human safety is crucial when incorporating them into food packaging. To safeguard consumers from inadvertent exposure to potential risks, it is vital to establish maximum allowable concentrations of nanoparticles in food. Developing these regulatory standards poses a significant challenge for researchers today.

7. Conclusions

Biopolymers are an innovative type of packaging material that has substantial potential to decrease the use of synthetic polymers and thus reduce waste. By integrating silver nanoparticles into biopolymer matrices, these materials can be transformed into functional nanocomposites. The inclusion of silver nanoparticles changes the structure of the films and affects their physical and mechanical properties. Additionally, silver nanoparticles enhance the antimicrobial properties of edible films, rendering them ideal for food packaging applications to inhibiting microbial growth and extending product shelf life. Despite these advantages, it is important to investigate how silver nanoparticles transfer from bio-based packaging into the enclosed products and to evaluate their potential impact on public health and the ecosystem. Concerns about the safety of silver nanoparticles persist among consumers. Consequently, additional research is required to establish safe levels of silver nanoparticles in nanomaterials to avoid potential health risks.

References

Abreu, A. S., Oliveira, M., de Sá, A., Rodrigues, R. M., Cerqueira, M. A., Vicente, A. A. et al. (2015). Antimicrobial nanostructured starch based films for packaging. Carbohydrate Polymers, 129: 127–134.

Acharya, D., Singha, K. M., Pandey, P., Mohanta, B., Rajkumari, J. and Singha, L. P. (2018). Shape dependent physical mutilation and lethal effects of silver nanoparticles on bacteria. Scientific Reports, 8(1): 201.

Addo Ntim, S., Norris, S., Goodwin Jr, D. G., Breffke, J., Scott, K., Sung, L. et al. (2018). Effects of consumer use practices on nanosilver release from commercially available food contact materials. Food Additives & Contaminants: Part A, 35: 2279–2290.

Agrawal, P. K., Sharma, P., Singh, V. K. and Chauhan, S. (2023). A comprehensive review on the engineering of biocompatible polyvinyl alcohol composites with enhanced properties using carbonaceous fillers. Journal of Materials and Environmental Science, 14(5): 560–581.

Ahmed, S. and Kanchi, S. (2018). Handbook of Bionanocomposites. Jenny Stanford Publishing.

Akhavan, S. and Bakhshi, B. (2015). Antibacterial activity of chitosan-silver nanoparticles film on Listeria monocytogenes and Salmonella spp. in chicken breast. Food Control, 55: 147–153.

Alavi, M. and Rai, M. (2019). Recent progress in nanoformulations of silver nanoparticles with cellulose, chitosan, and alginic acid biopolymers for antibacterial applications. Applied Microbiology and Biotechnology, 103(21): 8669–8676.

Al-Harrasi, A., Bhatia, S., Al-Azri, M. S., Ullah, S., Najmi, A., Albratty, M. et al. (2022). Effect of drying temperature on physical, chemical, and antioxidant properties of ginger oil loaded gelatin-sodium alginate edible films. Membranes, 12(9): 862.

Al-Hazmi, F., Alnowaiser, F., Al-Ghamdi, A. A., Al-Ghamdi, A. A., Aly, M. M., Al-Tuwirqi, R. M. et al. (2012). A new large–scale synthesis of magnesium oxide nanowires: structural and antibacterial properties. Superlattices and Microstructures, 52(2): 200–209.

Ali, A. and Ahmed, S. (2018). Recent advances in edible polymer-based hydrogels as a sustainable alternative to conventional polymers. Journal of Agricultural and Food Chemistry, 66: 6940–6967.

Al-Tayyar, N. A., Youssef, A. M. and Al-Hindi, R. R. (2020). Edible coatings and antimicrobial nanoemulsions for enhancing shelf life and reducing foodborne pathogens of fruits and vegetables: A review. Sustainable Materials and Technologies, 26: e00215.

Amaya-Pinos, J. (2022). Thermo-mechanical study of the mixture of polylactic acid (PLA) obtained from potato starch with aliphatic copolyester PBSA (polybutylene succinate adipate). Dyna, 89(221): 142–150.

Ansari, M. A., Kalam, A., Al-Sehemi, A. G., Alomary, M. N., AlYahya, S., Aziz, M. K. et al. (2021). Counteraction of biofilm formation and antimicrobial potential of Terminalia catappa functionalized silver nanoparticles against Candida albicans and multidrug-resistant Gram-negative and Gram-positive bacteria. Antibiotics, 10(6): 725.

Ardean, C., Davidescu, C. M., Nemeş, N. S., Negrea, A., Ciopec, M., Duteanu, N. et al. (2021). Factors influencing the antibacterial activity of chitosan and chitosan modified by functionalization. International Journal of Molecular Sciences, 22(14): 7449.

Arfat, Y. A., Ejaz, M., Jacob, H. and Ahmed, J. (2017). Deciphering the potential of guar gum/Ag-Cu nanocomposite films as an active food packaging material. Carbohydrate Polymers, 157: 65–71.

Armghan Khalid, M., Niaz, B., Saeed, F., Afzaal, M., Islam, F., Hussain, M. et al. (2022). Edible coatings for enhancing safety and quality attributes of fresh produce: A comprehensive review. International Journal of Food Properties, 25(1): 1817–1847.

Arnon-Rips, H. and Poverenov, E. (2018). Improving food products' quality and storability by using Layer by Layer edible coatings. Trends in Food Science & Technology, 75: 81–92.

Arrieta, M. P., López, J., Rayón, E. and Jiménez, A. (2014). Disintegrability under composting conditions of plasticized PLA–PHB blends. Polymer Degradation and Stability, 108: 307–318.

Asare, N., Instanes, C., Sandberg, W. J., Refsnes, M., Schwarze, P., Kruszewski, M. et al. (2012). Cytotoxic and genotoxic effects of silver nanoparticles in testicular cells. Toxicology, 291(1-3): 65–72.

Asrarkulova, A. S. and Bulushova, N. V. (2018). Wheat gluten and its hydrolysates: Possible fields of practical use. Applied Biochemistry and Microbiology, 54: 825–833.

Ayhan, Z., Cimmino, S., Esturk, O., Duraccio, D., Pezzuto, M. and Silvestre, C. (2015). Development of films of novel polypropylene based nanomaterials for food packaging application. Packaging Technology and Science, 28(7): 589–602.

Azammi, A. N., Ilyas, R. A., Sapuan, S. M., Ibrahim, R., Atikah, M. S. N., Asrofi, M. et al. (2020). Characterization studies of biopolymeric matrix and cellulose fibres-based composites related to functionalized fibre-matrix interface. In Interfaces in Particle and Fibre Reinforced Composites (pp. 29–93). Woodhead Publishing.

Azarifar, M., Mirzaei, H. and Esmaiili, M. (2020). Biodegradable film production from starch and chitosan reinforced with silver nanoparticles. International Journal of Biological Macromolecules, 146: 1180–1188.

Azizi-Lalabadi, M., Garavand, F. and Jafari, S. M. (2021). Incorporation of silver nanoparticles into active antimicrobial nanocomposites: Release behavior, analyzing techniques, applications and safety issues. Advances in Colloid and Interface Science, 293: 102440.

Babu, R. P., O'Connor, K. and Seeram, R. (2013). Progress in biomaterials. Progress in Biomaterials, 2: 1–16.

Bahrami, A., Delshadi, R., Assadpour, E., Jafari, S. M. and Williams, L. (2020). Antimicrobial-loaded nanocarriers for food packaging applications. Advances in Colloid and Interface Science, 278: 102140.

Bahrami, A., Mokarram, R. R., Khiabani, M. S., Ghanbarzadeh, B. and Salehi, R. (2019). Physico-mechanical and antimicrobial properties of tragacanth/hydroxypropyl methylcellulose/beeswax edible films reinforced with silver nanoparticles. International Journal of Biological Macromolecules, 129: 1103–1112.

Bahrami, A., Mokarram, R. R., Khiabani, M. S., Ghanbarzadeh, B. and Salehi, R. (2019). Physico-mechanical and antimicrobial properties of tragacanth/hydroxypropyl methylcellulose/beeswax edible films reinforced with silver nanoparticles. International Journal of Biological Macromolecules, 129: 1103–1112.

Bai, A. J. and Rai, V. R. (2011). Bacterial quorum sensing and food industry. Comprehensive Reviews in Food Science and Food Safety, 10(3): 183–193.

Balla, E., Daniilidis, V., Karlioti, G., Kalamas, T., Stefanidou, M., Bikiaris, N. D. and Bikiaris, D. N. (2021). Poly(lactic acid): A versatile biobased polymer for the future with multifunctional properties—From monomer synthesis, polymerization techniques, and molecular weight increase to PLA applications. Polymers, 13(11): 1822.

Bankura, K. P., Maity, D., Mollick, M. M., Mondal, D., Bhowmick, B., Bain, M. K. et al. (2012). Synthesis, characterization and antimicrobial activity of dextran stabilized silver nanoparticles in aqueous medium. Carbohydrate Polymers, 89(4): 1159–1165.

Barden, L. and Decker, E. A. (2016). Lipid oxidation in low-moisture food: A review. Critical Reviews in Food Science and Nutrition, 56(15): 2467–2482.
Barhoum, A., Jeevanandam, J., Rastogi, A., Samyn, P., Boluk, Y., Dufresne, A. and Bechelany, M. (2020). Plant celluloses, hemicelluloses, lignins, and volatile oils for the synthesis of nanoparticles and nanostructured materials. Nanoscale, 12(45): 22845–22890.
Basta, A. H., Lotfy, V. F., Mahmoud, K. and Abdelwahed, N. A. (2020). Synthesis and evaluation of protein-based biopolymer in production of silver nanoparticles as bioactive compound versus carbohydrates-based biopolymers. Royal Society Open Science, 7(10): 200928.
Bastioli, C. and Bettarini, F. (2020). General characteristics, processability, industrial applications, and market evolution of biodegradable polymers. In Handbook of Biodegradable Polymers, 147–182.
Behbahani, B. A., Yazdi, F. T., Shahidi, F., Mortazavi, S. A. and Mohebbi, M. (2017). Principle component analysis (PCA) for investigation of relationship between population dynamics of microbial pathogenesis, chemical and sensory characteristics in beef slices containing Tarragon essential oil. Microbial Pathogenesis, 105: 37–50.
Benjakul, S., Nagarajan, M. and Prodpran, T. (2016). Films and coatings from collagen and gelatin. In Edible Films and Coatings, 121–142.
Bertini, F., Canetti, M., Cacciamani, A., Elegir, G., Orlandi, M. and Zoia, L. (2012). Effect of ligno-derivatives on thermal properties and degradation behavior of poly(3-hydroxybutyrate)-based biocomposites. Polymer Degradation and Stability, 97(10): 1979–1987.
Bharathi, S. V., Murugesan, P., Moses, J. A. and Anandharamakrishnan, C. (2021). Recent trends in nanocomposite packaging materials. Progress in Polymer Science, 68: 77–105.
Bharti, S., Mukherji, S. and Mukherji, S. (2021). Enhanced antibacterial activity of decahedral silver nanoparticles. Journal of Nanoparticle Research, 23: 1–18.
Biniaś, D., Biniaś, W., Machnicka, A. and Hanus, M. (2020). Preparation of antimicrobial fibres from the EVOH/EPC blend containing silver nanoparticles. Polymers, 12(8): 1827.
Bosetti, M., Masse, A., Tobin, E. and Cannas, M. (2002). Silver-coated materials for external fixation devices: In vitro biocompatibility and genotoxicity. Biomaterials, 23: 887–892.
Bracciale, M. P., De Gioannis, G., Falzarano, M., Muntoni, A., Polettini, A., Pomi, R. et al. (2024). Disposable Mater-Bi® bioplastic tableware: Characterization and assessment of anaerobic biodegradability. Fuel, 355: 129361.
Bruna, T., Maldonado-Bravo, F., Jara, P. and Caro, N. (2021). Silver nanoparticles and their antibacterial applications. International Journal of Molecular Sciences, 22(13): 7202.
Burridge, K., Johnston, J. and Borrmann, T. (2011). Silver nanoparticle–clay composites. Journal of Materials Chemistry, 21(3): 734–742.
Calori, I. R., Braga, G., de Jesus, P. D. C. C., Bi, H. and Tedesco, A. C. (2020). Polymer scaffolds as drug delivery systems. European Polymer Journal, 129: 109621.
Cameron, S. J., Hosseinian, F. and Willmore, W. G. (2018). A current overview of the biological and cellular effects of nanosilver. International Journal of Molecular Sciences, 19: 2030.
Campos-Requena, V. H., Rivas, B. L., Pérez, M. A., Garrido-Miranda, K. A. and Pereira, E. D. (2015). Polymer/clay nanocomposite films as active packaging material: Modeling of antimicrobial release. European Polymer Journal, 71: 461–475.
Cao, L., Huang, Y., Parakhonskiy, B. and Skirtach, A. G. (2022). Nanoarchitectonics beyond perfect order–Not quite perfect but quite useful. Nanoscale, 14(43): 15964–16002.
Carbone, M., Franco, B. D. G. M. and Conte-Junior, C. A. (2016). Microbial contamination of foods: Mechanisms, implications, and prevention. Journal of Food Quality, 39(6): 510–525.
Castro-Mayorga, J. L., Fabra, M. J., Pourrahimi, A. M., Olsson, R. T. and Lagaron, J. M. (2016). The impact of zinc oxide and silver nanoparticles as nanofillers in poly(lactic acid) films for food packaging applications: A comparative study. Journal of Food Engineering, 181: 31–39.
Chen, Y., Lu, K., Liu, H. and Yu, L. (2021). Starch-based packaging materials. In Bio-Based Packaging: Material, Environmental and Economic Aspects 1–26.
Chettri, S., Sharma, N. and Mohite, A. M. (2023). Edible coatings and films for shelf-life extension of fruit and vegetables. Biomaterials Advances, 213632.
Cho, M., Chung, H., Choi, W. and Yoon, J. (2005). Different inactivation behaviors of MS-2 phage and *Escherichia coli* in TiO2 photocatalytic disinfection. Applied and Environmental Microbiology, 71(1): 270–275.
Christian, S. J. (2016). Natural fibre-reinforced noncementitious composites (biocomposites). In Nonconventional and Vernacular Construction Materials, 111–126.
Cushen, M., Kerry, J., Morris, M., Cruz-Romero, M. and Cummins, E. (2014). Evaluation and simulation of silver and copper nanoparticle migration from polyethylene nanocomposites to food and an associated exposure assessment. Journal of Agricultural and Food Chemistry, 62: 1403–1411.

Cutter, C. N. (2006). Opportunities for bio-based packaging technologies to improve the quality and safety of fresh and further processed muscle foods. Meat Science, 74(1): 131–142.

Czerwiński, K., Rydzkowski, T., Wróblewska-Krepsztul, J. and Thakur, V. K. (2021). Towards impact of modified atmosphere packaging (MAP) on shelf-life of polymer-film-packed food products: Challenges and sustainable developments. Coatings, 11(12): 1504.

Dakal, T. C., Kumar, A., Majumdar, R. S. and Yadav, V. (2016). Mechanistic basis of antimicrobial actions of silver nanoparticles. Frontiers in Microbiology, 7(1831).

Dallas, P., Sharma, V. K. and Zboril, R. (2011). Silver polymeric nanocomposites as advanced antimicrobial agents: Classification, synthetic paths, applications, and perspectives. Advances in Colloid and Interface Science, 166: 119–135.

Dalton, B., Bhagabati, P., De Micco, J., Padamati, R. B. and O'Connor, K. (2022). A review on biological synthesis of the biodegradable polymers polyhydroxyalkanoates and the development of multiple applications. Catalysts, 12(3): 319.

Dammak, I., Sobral, P. J. A. and Aquino, A. (2019). Innovative materials based on the combination of alginate and pectin for food packaging applications. Journal of Food Engineering, 245: 105–115.

Dash, K. K., Kumar, A., Kumari, S. and Malik, M. A. (2021). Silver nanoparticle incorporated flaxseed protein-alginate composite films: Effect on physicochemical, mechanical, and thermal properties. Journal of Polymers and the Environment, 29: 3649–3659.

De Azeredo, H. M. C. (2009). Nanocomposites for food packaging applications. Food Research International, 42(9): 1240–1253.

de Jesús Avena-Bustillos, R. and McHugh, T. H. (2012). Role of edible film and coating additives. Edible coatings and films to improve food quality, 158–159.

De Moura, M. R., Mattoso, L. H. C. and Zucolotto, V. (2012). Development of cellulose-based bactericidal nanocomposites containing silver nanoparticles and their use as active food packaging. Journal of Food Engineering, 109: 520–524.

Desai, H., Mehta, T., Dwivedi, A., Shah, N., Parikh, D. and Saiyad, M. (2024). Novel catalyst-solvent system for high molecular weight polylactic acid synthesis via azeotropic solution polycondensation method. Materials Research Innovations, 1–8.

Dhall, R. K. (2013). Advances in edible coating for fresh fruits and vegetables: A review. Critical Reviews in Food Science and Nutrition, 53: 435–450.

Dhapte, V., Kadam, S., Moghe, A. and Pokharkar, V. (2014). Probing the wound healing potential of biogenic silver nanoparticles. Journal of wound care, 23(9): 431–441.

Dhiman, A., Chopra, R. and Garg, M. (2022). Edible film and coating for food packaging. In Biodegradable composites for packaging applications, 149–167.

Dhiman, N. K., Agnihotri, S. and Shukla, R. (2019). Silver-based polymeric nanocomposites as antimicrobial coatings for biomedical applications. In Nanotechnology in Modern Animal Biotechnology: Recent Trends and Future Perspectives 115–171.

Díaz-Montes, E. (2022). Polysaccharides: Sources, characteristics, properties, and their application in biodegradable films. Polysaccharides, 3(3): 480–501.

Dong, Y., Zhu, H., Shen, Y., Zhang, W. and Zhang, L. (2019). Antibacterial activity of silver nanoparticles of different particle size against Vibrio Natriegens. PloS One, 14(9): e0222322.

dos Santos Junior, V. E., Targino, A. G. R., Flores, M. A. P., Rodríguez-Díaz, J. M., Teixeira, J. A., Heimer, M. V. et al. (2017). Antimicrobial activity of silver nanoparticle colloids of different sizes and shapes against Streptococcus mutans. Research on Chemical Intermediates, 43: 5889–5899.

Echegoyen, Y. and Nerín, C. (2013). Nanoparticle release from nano-silver antimicrobial food containers. Food and Chemical Toxicology, 62: 16–22.

Egger, S., Lehmann, R. P., Height, M. J., Loessner, M. J. and Schuppler, M. (2009). Antimicrobial properties of a novel silver-silica nanocomposite material. Applied and Environmental Microbiology, 75(9): 2973–2976.

El Allaoui, B., Chakhtouna, H., Ouhssain, A., Kadmiri, I. M., Benzeid, H., Zari, N. and Bouhfid, R. (2024). Silver nanoparticle-decorated cellulose beads: Eco-friendly catalysts for efficient 4-nitrophenol reduction and antibacterial performance. International Journal of Biological Macromolecules, 273: 133078.

Elsabee, M. Z and Abdou, E. S. (2013). Chitosan based edible films and coatings: A review. Materials Science and Engineering, 33(4): 1819–1841.

El-Seedi, H. R., El-Shabasy, R. M., Khalifa, S. A. M., Saeed, A., Shah, A., Shah, A. T. et al. (2019). Anti-bacterial efficacy of silver nanoparticles against Gram-negative and Gram-positive bacteria. Journal of Genetic Engineering and Biotechnology, 17(1): 10.

Emami-Karvani, Z. and Chehrazi, P. (2011). Antibacterial activity of ZnO nanoparticle on gram-positive and gram-negative bacteria. Afr. J. Microbiol. Res., 5(12): 1368–73.

European Commission. (2009). Regulation (EC) No. 450/2009. Official Journal of the European Union, 135: 3–11.
European Commission. (2011). Commission regulation (EU) No 10/2011. Official Journal of the European Union, 12: 1–89.
Fabra, M. J., Busolo, M. A., Lopez-Rubio, A. and Lagaron, J. M. (2013). Nanostructured biolayers in food packaging. Trends in Food Science & Technology, 31: 79–87.
Fard, N. N., Noorbazargan, H., Mirzaie, A., Hedayati Ch, M., Moghimiyan, Z. and Rahimi, A. (2018). Biogenic synthesis of AgNPs using Artemisia oliveriana extract and their biological activities for an effective treatment of lung cancer. Artificial Cells, Nanomedicine, and Biotechnology, 46(sup3): S1047–S1058.
Fasihnia, S. H., Peighambardoust, S. H. and Peighambardoust, S. J. (2018). Nanocomposite films containing organoclay nanoparticles as an antimicrobial (active) packaging for potential food application. Journal of Food Processing and Preservation, 42(2): e13488.
Fayaz, A. M., Balaji, K., Girilal, M., Kalaichelvan, P. T. and Venkatesan, R. (2009). Myco-based synthesis of silver nanoparticles and their incorporation into sodium alginate films for vegetable and fruit preservation. Journal of Agricultural and Food Chemistry, 57(14): 6246–6252.
Fayaz, A. M., Balaji, K., Girilal, M., Yadav, R., Kalaichelvan, P. T. and Venketesan, R. (2009). Biogenic synthesis of silver nanoparticles and their synergistic effect with antibiotics: A study against Gram-positive and Gram-negative bacteria. Nanomedicine: Nanotechnology, Biology, and Medicine, 6(1): 103–109.
Fernández-Pan, I., Royo, M. and Ignacio Mate, J. (2012). Antimicrobial activity of whey protein isolate edible films with essential oils against food spoilers and foodborne pathogens. Journal of Food Science, 77(7): M383–M390.
Ferdous, Z. and Nemmar, A. (2020). Health impact of silver nanoparticles: A review of the biodistribution and toxicity following various routes of exposure. International Journal of Molecular Sciences, 21(7): 2375.
Fernandez, A., Soriano, E., Lopez-Carballo, G., Picouet, P., Lloret, E., Gavara, R. and Hernandez-Munoz, P. (2010a). Preservation of aseptic conditions in beef muscle packaged in a chitosan-coated film with ZnO nanostructures. Food Microbiology, 27(7): 882–889.
Fernandez, A., Soriano, E., Lopez-Carballo, G., Picouet, P., Lloret, E., Gavara, R. et al. (2010b). Active chitosan coatings for extended shelf life of fresh-cut melon. International Journal of Food Microbiology, 150(3): 210–216.
Fitzgerald, A. V. L. (2017). Preparation, characterisation and secondary crystallisation of PHB based copolymers and carbohydrate blends (Doctoral dissertation, University of Birmingham).
Fortunati, E., Armentano, I., Zhou, Q., Iannoni, A., Saino, E., Visai, L. et al. (2012). Multifunctional bionanocomposite films of poly (lactic acid), cellulose nanocrystals and silver nanoparticles. Carbohydrate Polymers, 87(2): 1596–1605.
Gajbhiye, M., Kesharwani, J., Ingle, A., Gade, A. and Rai, M. (2009). Fungus-mediated synthesis of silver nanoparticles and their activity against pathogenic fungi in combination with fluconazole. Nanomedicine: Nanotechnology, Biology and Medicine, 5(4): 382–386.
Galus, S. and Kadzińska, J. (2015). Food applications of emulsion-based edible films and coatings. Trends in Food Science & Technology, 45: 273–283.
García-Tejeda, Y. V., Salinas-Moreno, Y. and Martínez-Bustos, F. (2015). Acetylation of normal and waxy maize starches as encapsulating agents for maize anthocyanins microencapsulation. Food and Bioproducts Processing, 94: 717–726.
George, A., Sanjay, M. R., Srisuk, R., Parameswaranpillai, J. and Siengchin, S. (2020). A comprehensive review on chemical properties and applications of biopolymers and their composites. International Journal of Biological Macromolecules, 154: 329–338.
Goel, V., Luthra, P., Kapur, G. S. and Ramakumar, S. S. V. (2021). Biodegradable/bio-plastics: Myths and realities. Journal of Polymers and the Environment, 29: 3079–3104.
Goswami, S. R., Sahareen, T., Singh, M. and Kumar, S. (2015). Role of biogenic silver nanoparticles in disruption of cell–cell adhesion in *Staphylococcus aureus* and *Escherichia coli* biofilm. Journal of Industrial and Engineering Chemistry, 26: 73–80.
Gregory, D. A., Tripathi, L., Fricker, A. T., Asare, E., Orlando, I., Raghavendran, V. et al. (2021). Bacterial cellulose: A smart biomaterial with diverse applications. Materials Science and Engineering: R: Reports, 145: 100623.
Grujic, R., Vukic, M. and Gojkovic, V. (2017). Application of biopolymers in the food industry. In Advances in Applications of Industrial Biomaterials, 103–119.
Gupta, K. M. (2011). Starch-based composites for packaging applications. Handbook of Bioplastics and Biocomposites Engineering Applications, 189–215.
Gupta, K., Singh, S. P., Manhar, A. K., Saikia, D., Namsa, N. D. and Konwar, B. K. (2019). Inhibition of Staphylococcus aureus and Pseudomonas aeruginosa biofilm and virulence by active fraction of *Syzygium cumini* (L.) Skeels leaf extract: *In-vitro* and *in silico* studies. Indian Journal of Microbiology, 59(1): 13–21.
Halper, J. and Kjaer, M. (2014). Basic components of connective tissues and extracellular matrix: elastin, fibrillin, fibulins, fibrinogen, fibronectin, laminin, tenascins and thrombospondins. In Progress in Heritable Soft Connective Tissue Diseases, 31–47.

Hamad, A., Khashan, K. S. and Hadi, A. (2020). Silver nanoparticles and silver ions as potential antibacterial agents. Journal of Inorganic and Organometallic Polymers and Materials, 30(12): 4811–4828.

Hamed, I., Özogul, F. and Regenstein, J. M. (2016). Industrial applications of crustacean by-products (chitin, chitosan, and chitooligosaccharides): A review. Trends in Food Science & Technology, 48: 40–50.

Hamouda, R. A., Hussein, M. H., Abo-Elmagd, R. A. and Bawazir, S. S. (2019). Synthesis and biological characterization of silver nanoparticles derived from the cyanobacterium Oscillatoria limnetica. Scientific Reports, 9(1): 13071.

Hassan, B., Chatha, S. A. S., Hussain, A. I., Zia, K. M. and Akhtar, N. (2018). Recent advances on polysaccharides, lipids and protein-based edible films and coatings: A review. International Journal of Biological Macromolecules, 109: 1095–1107.

He, X. and Hwang, H.-M. (2016). Nanotechnology in food science: Functionality, applicability, and safety assessment. Journal of Food and Drug Analysis, 24: 671–681.

Hetrick, E. M., Shin, J. H., Paul, H. S. and Schoenfisch, M. H. (2009). Anti-biofilm efficacy of nitric oxide-releasing silica nanoparticles. Biomaterials, 30(14): 2782–2789

Hettiarachchi, M. A. and Wickramarachchi, P. A. S. R. (2011). Synthesis of chitosan stabilized silver nanoparticles using gamma ray irradiation and characterization. Journal of Science, University of Kelaniya, 6: 65–75.

Hong, S.-G., Gau, T.-K. and Huang, S.-C. (2011). Journal of Thermal Analysis and Calorimetry, 103: 967–975.

Hu, B. (2014). Biopolymer-based lightweight materials for packaging applications. Lightweight Materials from Biopolymers and Biofibers, 239–255.

Huang, Q., Qian, X., Jiang, T. andZheng, X. (2019). Effect of chitosan and guar gum based composite edible coating on quality of mushroom (Lentinus edodes) during postharvest storage. Scientia Horticulturae, 253: 382–389.

Ibrahim, H. M. M. (2015). Green synthesis and characterization of silver nanoparticles using banana peel extract and their antimicrobial activity against representative microorganisms. Journal of Radiation Research and Applied Sciences, 8(3): 265–275.

Idler, C., Venus, J. and Kamm, B. (2015). Microorganisms for the production of lactic acid and organic lactates. In Microorganisms in Biorefineries (pp. 225–273). Springer.

Incoronato, A. L., Conte, A., Buonocore, G. G. and Del Nobile, M. A. (2011). Use of hydrogel and silver nanoparticles packaging to prolong the shelf life of Fior di Latte cheese. Journal of Dairy Science, 94(4): 1697–1704.

International Life Science Institute (ILSI). (2015). NanoRelease Consumer Products. Retrieved from http://www.ilsi.org/ResearchFoundation/RSIA/Pages/NanoRelease1.aspx.

Istiqola, A. and Syafiuddin, A. (2020). A review of silver nanoparticles in food packaging technologies: Regulation, methods, properties, migration, and future challenges. Journal of the Chinese Chemical Society, 67(11): 1942–1956.

Jafarzadeh, S. and Jafari, S. M. (2021). Impact of metal nanoparticles on the mechanical, barrier, optical and thermal properties of biodegradable food packaging materials. Critical Reviews in Food Science and Nutrition, 61(16): 2640–2658.

Jahangiri, F., Mohanty, A. K. and Misra, M. (2024). Sustainable biodegradable coatings for food packaging: Challenges and opportunities. Green Chemistry.

Jaiswal, S., Duffy, B., Jaiswal, A. K., Stobie, N. and McHale, P. (2010). Enhancement of the antibacterial properties of silver nanoparticles using β-cyclodextrin as a capping agent. International Journal of Antimicrobial Agents, 36(3): 280–283.

Jamróz, E., Khachatryan, G., Kopel, P., Juszczak, L., Kawecka, A., Krzyściak, P. et al. (2020). Furcellaran nanocomposite films: The effect of nanofillers on the structural, thermal, mechanical and antimicrobial properties of biopolymer films. Carbohydrate Polymers, 240: 116244.

Jain, P. and Pradeep, T. (2005). Potential of silver nanoparticle-coated polyurethane foam as an antibacterial water filter. Biotechnology and Bioengineering, 90(1): 59–63.

Jamróz, E., Kulawik, P. and Kopel, P. (2019). The effect of nanofillers on the functional properties of biopolymer-based films: A review. Polymers, 11(4): 675.

Jena, P., Mohanty, S., Mallick, R., Jacob, B. and Sonawane, A. (2012). Toxicity and antibacterial assessment of chitosan-coated silver nanoparticles on human pathogens and macrophage cells. International Journal of Nanomedicine, 7: 1805–1818.

Jiang, Y., Wang, X., Zeng, Q. and Yan, X. (2011). Preparation and characterization of Ag-chitosan/PVA antibacterial nanofibers by electrospinning. Journal of Applied Polymer Science, 119(2): 1049–1060.

Jin, T. and He, Y. (2011). Antibacterial activities of magnesium oxide (MgO) nanoparticles against foodborne pathogens. Journal of Nanoparticle Research, 13: 6877–6885.

Jiménez, A., Fabra, M. J., Talens, P. and Chiralt, A. (2015). Polysaccharides as valuable materials in food packaging. In Functional Polymers in Food Science: From Technology to Biology, 211–251.

Jiménez-Rosado, M., Zarate-Ramírez, L. S., Romero, A., Bengoechea, C., Partal, P. and Guerrero, A. (2019). Bioplastics based on wheat gluten processed by extrusion. Journal of Cleaner Production, 239: 117994.

John, R. P., Nampoothiri, K. M. and Pandey, A. (2007). Fermentative production of lactic acid from biomass: An overview on process developments and future perspectives. Applied Microbiology and Biotechnology, 74: 524–534.

Jouki, M., Yazdi, F. T., Mortazavi, S. A. and Koocheki, A. (2014). Effect of glycerol concentration on edible film production from cress seed carbohydrate gum. Carbohydrate Polymers, 108: 162–168.

Kabir, A. (2010). Vibration damping property and flexural fatigue behavior of glass/epoxy/nanoclay composites (Doctoral dissertation, Concordia University).

Kabir, E., Kaur, R., Lee, J., Kim, K. H. and Kwon, E. E. (2020). Prospects of biopolymer technology as an alternative option for non-degradable plastics and sustainable management of plastic wastes. Journal of Cleaner Production, 258: 120536.

Kaiser, K. G., Delattre, V., Frost, V. J., Buck, G. W., Phu, J. V., Fernandez, T. G. et al. (2023). Nanosilver: An old antibacterial agent with great promise in the fight against antibiotic resistance. Antibiotics, 12(8): 1264.

Kalatehjari, P., Yousefian, M. and Khalilzadeh, M. A. (2015). Assessment of antifungal effects of copper nanoparticles on the growth of the fungus *Saprolegnia* sp. on white fish (*Rutilus frisii kutum*) eggs. The Egyptian Journal of Aquatic Research, 41(4): 303–306.

Kamarudin, S. H., Rayung, M., Abu, F., Ahmad, S. B., Fadil, F., Karim, A. A. et al. (2022). A review on antimicrobial packaging from biodegradable polymer composites. Polymers, 14(1): 174.

Kanmani, P. and Rhim, J. W. (2014). Antimicrobial and physical-mechanical properties of agar-based films incorporated with grapefruit seed extract. Carbohydrate Polymers, 102: 708–716.

Karade, V. C., Patil, R. B., Parit, S. B., Kim, J. H., Chougale, A. D. and Dawkar, V. V. (2021). Insights into shape-based silver nanoparticles: A weapon to cope with pathogenic attacks. ACS Sustainable Chemistry & Engineering, 9(37): 12476–12507.

Kaur, G., Sharma, S., Mir, S. A. and Dar, B. N. (2021). Nanobiocomposite films: A "greener alternate" for food packaging. Food and Bioprocess Technology, 14: 1013–1027.

Khiralla, G. M. and El-Deeb, B. A. (2015). Antimicrobial and antibiofilm effects of selenium nanoparticles on some foodborne pathogens. LWT-Food Science and Technology, 63(2): 1001–1007.

Khajavi, M. Z., Mohammadi, R., Ahmadi, S., Farhoodi, M. and Yousefi, M. (2019). Strategies for controlling release of plastic compounds into foodstuffs based on application of nanoparticles and its potential health issues. Trends in Food Science & Technology, 90: 1–12.

Khalaf, H. H., Zahran, M. K. and Easa, S. M. (2013). Incorporation of chitosan and silver nanoparticles into pullulan films for the extended shelf life of turkey deli meat. Carbohydrate Polymers, 96(2): 521–528.

Khalid, M. A., Niaz, B., Saeed, F., Afzaal, M., Islam, F., Hussain, M. et al. (2022). Edible coatings for enhancing safety and quality attributes of fresh produce: A comprehensive review. International Journal of Food Properties, 25(1): 1817–1847.

Khalil, H. A., Davoudpour, Y., Saurabh, C. K., Hossain, M. S., Adnan, A. S., Dungani, R. et al. (2016). A review on nanocellulosic fibres as new material for sustainable packaging: Process and applications. Renewable and Sustainable Energy Reviews, 64: 823–836.

Khaneghah, A. M., Hashemi, S. M. B. and Limbo, S. (2018). Antimicrobial agents and packaging systems in antimicrobial active food packaging: An overview of approaches and interactions. Food and Bioproducts Processing, 111: 1–19.

Khankrua, R., Pivsa-Art, S., Hiroyuki, H. and Suttiruengwong, S. (2014). Effect of chain extenders on thermal and mechanical properties of poly(lactic acid) at high processing temperatures: Potential application in PLA/Polyamide 6 blend. Polymer Degradation and Stability, 108: 232–240.

King, T., Osmond-McLeod, M. J. and Duffy, L. L. (2018). Nanotechnology in the food sector and potential applications for the poultry industry. Trends in Food Science & Technology, 72: 62–73.

Kong, F. and Singh, R. P. (2016). Chemical deterioration and physical instability of foods and beverages. In The Stability and Shelf Life of Food (pp. 43–76). Woodhead Publishing.

König, H. and Fröhlich, J. (2017). Lactic acid bacteria. In Biology of Microorganisms on Grapes, in Must and in Wine (pp. 3–41).

Kraśniewska, K. and Gniewosz, M. (2012). Substances with antimicrobial activity in edible films—a review. Polish Journal of Food and Nutrition Sciences, 62: 199–206.

Kraśniewska, K., Pobiega, K. and Gniewosz, M. (2019). Pullulan—Biopolymer with potential for use as food packaging. International Journal of Food Engineering, 15.

Krishnamoorthy, K., Manivannan, G., Kim, S. J., Jeyasubramanian, K. and Premanathan, M. (2012). Antibacterial activity of MgO nanoparticles based on lipid peroxidation by oxygen vacancy. Journal of Nanoparticle Research, 14(9): 1063.

Krychowiak, M., Kawiak, A., Narajczyk, M., Borowik, A. and Królicka, A. (2018). Silver nanoparticles combined with naphthoquinones as an effective synergistic strategy against *Staphylococcus aureus*. Frontiers in Pharmacology, 9: 816.

Kumar, A. and Kaur, H. (2020). Sprayed in-situ synthesis of polyvinyl alcohol/chitosan loaded silver nanocomposite hydrogel for improved antibacterial effects. International Journal of Biological Macromolecules, 145: 950–964.

Kumar, P., Mahajan, P., Kaur, R. and Gautam, S. (2020). Nanotechnology and its challenges in the food sector: A review. Materials Today Chemistry, 17: 100332.

Kumar, S., Singh, M., Halder, D. and Mitra, A. (2014). Mechanistic study of antibacterial activity of biologically synthesized silver nanocolloids. Colloids and Surfaces A: Physicochemical and Engineering Aspects, 449: 82–86.

Kumar, S., Singh, M., Halder, D. and Mitra, A. (2016). Lippia javanica: A cheap natural source for the synthesis of antibacterial silver nanocolloid. Applied Nanoscience, 6(7): 1001–1007.

Kumari, S. and Kishor, R. (2020). Chitin and chitosan: Origin, properties, and applications. In Handbook of Chitin and Chitosan (pp. 1–33). Elsevier.

Kuperkar, K., Atanase, L. I., Bahadur, A., Crivei, I. C. and Bahadur, P. (2024). Degradable polymeric bio(nano) materials and their biomedical applications: A comprehensive overview and recent updates. Polymers, 16(2): 206.

Kurek, M., Galus, S. and Debeaufort, F. (2014). Surface, mechanical, and barrier properties of bio-based composite films based on chitosan and whey protein. Food Packaging and Shelf Life, 1: 56–67.

Kyriakoudi, A., Spanidi, E., Mourtzinos, I. and Gardikis, K. (2021). Innovative delivery systems loaded with plant bioactive ingredients: Formulation approaches and applications. Plants, 10(6): 1238.

Langstraat, T. D., Jansens, K. J., Delcour, J. A., Van Puyvelde, P. and Goderis, B. (2015). Controlling wheat gluten cross-linking for high-temperature processing. Industrial Crops and Products, 72: 119–124.

Larotonda, F. D. (2007). Biodegradable films and coatings obtained from carrageenan from *Mastocarpus stellatus* and starch from Quercus suber (Doctoral dissertation, Universidade do Porto (Portugal)).

Lee, D., Cohen, R. E. and Rubner, M. F. (2005). Antibacterial properties of Ag nanoparticle loaded multilayers and formation of magnetically directed antibacterial microparticles. Langmuir, 21: 9651–9659.

Li, Q., Mahendra, S., Lyon, D. Y., Brunet, L., Liga, M. V., Li, D. et al. (2008). Antimicrobial nanomaterials for water disinfection and microbial control: Potential applications and implications. Water Research, 42(18): 4591–4602.

Liao, C., Li, Y. and Tjong, S. C. (2019). Bactericidal and cytotoxic properties of silver nanoparticles. International Journal of Molecular Sciences, 20(2): 449.

Lin, B., Luo, Y., Teng, Z., Zhang, B., Zhou, B. and Wang, Q. (2015). Development of silver/titanium dioxide/chitosan adipate nanocomposite as an antibacterial coating for fruit storage. LWT - Food Science and Technology, 63: 1206–1213.

Liu, L., Xu, M., Ye, Y. and Zhang, B. (2022). On the degradation of (micro) plastics: Degradation methods, influencing factors, environmental impacts. Science of the Total Environment, 806: 151312.

Liu, X., Wang, Y., Yu, L., Tong, Z., Chen, L., Liu, H. et al. (2013). Starch degradation. Starch, 65: 48–60.

López de Dicastillo, C., Guarda, A., Rodríguez, F. and Galotto, M. J. (2016). Antimicrobial effectiveness of novel nanocomposite films based on chitosan and ZnO nanoparticles in food packaging. Journal of Food Engineering, 174: 146–155.

Lu, L., Luo, X. G., Ji, C., Liu, B. and Yu, S. X. (2007). Effect of manganese supplementation and source on carcass traits, meat quality, and lipid oxidation in broilers. Journal of Animal Science, 85(3): 812–822.

Luo, Q., Hossen, M. A., Zeng, Y., Dai, J., Li, S., Qin, W. et al. (2022). Gelatin-based composite films and their application in food packaging: A review. Journal of Food Engineering, 313: 110762.

Maneerat, C. and Hayata, Y. (2006). Efficiency of TiO2 photocatalytic reaction on delay of fruit ripening and removal of off-flavors from the fruit storage atmosphere. Transactions of the ASABE, 49(3): 833–837.

Marchiore, N. G., Manso, I. J., Kaufmann, K. C., Lemes, G. F., de Oliveira Pizolli, A. P., Droval, A. A. et al. (2017). Migration evaluation of silver nanoparticles from antimicrobial edible coating to sausages. LWT - Food Science and Technology, 76: 203–208.

Martínez-Abad, A., Sánchez, G., Lagarón, J. M. and Ocio, M. J. (2012). On the different growth conditions affecting silver antimicrobial efficacy on *Listeria monocytogenes* and *Salmonella enterica*. International Journal of Food Microbiology, 158(2): 147–154.

Mastromatteo, M., Conte, A., Lucera, A., Saccotelli, M. A., Buonocore, G. G., Zambrini, A. V. et al. (2015). Packaging solutions to prolong the shelf life of Fiordilatte cheese: Bio-based nanocomposite coating and modified atmosphere packaging. LWT-Food Science and Technology, 60(1): 230–237.

Mathew, S., Mathew, J. and Radhakrishnan, E. K. (2019). Polyvinyl alcohol/silver nanocomposite films fabricated under the influence of solar radiation as effective antimicrobial food packaging material. Journal of Polymer Research, 26(9): 223.

Mazzuca, C., Micheli, L., Carbone, M., Basoli, F., Cervelli, E., Iannuccelli, S. et al. (2014). Gellan hydrogel as a powerful tool in paper cleaning process: A detailed study. Journal of Colloid and Interface Science, 416(15): 205–211.

Mei, L., Teng, Z., Zhu, G., Liu, Y., Zhang, F., Zhang, J. et al. (2017). Silver nanocluster-embedded zein films as antimicrobial coating materials for food packaging. ACS Applied Materials & Interfaces, 9(40): 35297–35304.

Mihalca, V., Kerezsi, A. D., Weber, A., Gruber-Traub, C., Schmucker, J., Vodnar, D. C. et al. (2021). Protein-based films and coatings for food industry applications. Polymers, 13(5): 769.

Mlalila, N., Hilonga, A., Swai, H., Devlieghere, F. and Ragaert, P. (2018). Antimicrobial packaging based on starch, poly (3-hydroxybutyrate) and poly (lactic-co-glycolide) materials and application challenges. Trends in Food Science & Technology, 74: 1–11.

Mohamed, S. A. N., Zainudin, E. S., Sapuan, S. M., Azaman, M. D. and Arifin, A. M. T. (2018). Introduction to natural fiber reinforced vinyl ester and vinyl polymer composites. In Natural fibre reinforced vinyl ester and vinyl polymer composites (pp. 1–25). Woodhead Publishing.

Mohamed, S. A., El-Sakhawy, M. and El-Sakhawy, M. A. M. (2020). Polysaccharides, protein and lipid-based natural edible films in food packaging: A review. Carbohydrate Polymers, 238: 116178.

Morais, L. D. O., Macedo, E. V., Granjeiro, J. M. and Delgado, I. F. (2020). Critical evaluation of migration studies of silver nanoparticles present in food packaging: A systematic review. Critical Reviews in Food Science and Nutrition, 60(18): 3083–3102.

Morsy, M. K., Khalaf, H. H., Sharoba, A. M. and El-Tanahy, H. H. (2014). Incorporation of antimicrobial agents into biodegradable packaging materials for application in food preservation. Food Science and Technology International, 20(6): 453–463.

Mukherjee, A., Sadiq, I. M., Prathna, T. C. and Chandrasekaran, N. (2011). Antimicrobial activity of aluminium oxide nanoparticles for potential clinical applications. Science Against Microbial Pathogens: Communicating Current Research and Technological Advances, 1: 245–251.

Murariu, M. and Dubois, P. (2016). PLA composites: From production to properties. Advanced Drug Delivery Reviews, 107: 17–46.

Nair, S. B., Alummoottil, N. and Moothandasserry, S. S. (2017). Chitosan-konjac glucomannan-cassava starch-nanosilver composite films with moisture resistant and antimicrobial properties for food-packaging applications. Starch, 69(1-2): 1600210.

Nampoothiri, K. M., Nair, N. R. and John, R. P. (2010). An overview of lactic acid production from biomass. Bioresource Technology, 101: 8493–8501.

Ncube, L. K., Ude, A. U., Ogunmuyiwa, E. N., Zulkifli, R. and Beas, I. N. (2020). Environmental impact of food packaging materials: A review of contemporary development from conventional plastics to polylactic acid based materials. Materials, 13(21): 4994.

Nevoralová, M., Koutný, M., Ujčić, A., Horák, P., Kredatusová, J., Šerá, J. et al. (2019). Controlled biodegradability of functionalized thermoplastic starch based materials. Polymer Degradation and Stability, 170: 108995.

Nidheesh, T., Swaroop, C., Sivaraj, S. and Natarajan, R. (2020). Development of gelatin/agar based edible packaging materials. Food Hydrocolloids, 108: 106066.

Nie, P., Zhao, Y. and Xu, H. (2023). Synthesis, applications, toxicity and toxicity mechanisms of silver nanoparticles: A review. Ecotoxicology and Environmental Safety, 253: 114636.

Nqakala, Z. B., Sibuyi, N. R., Fadaka, A. O., Meyer, M., Onani, M. O. and Madiehe, A. M. (2021). Advances in nanotechnology towards development of silver nanoparticle-based wound-healing agents. International Journal of Molecular Sciences, 22(20): 11272.

Oliver, S., Wagh, H., Liang, Y., Yang, S. and Boyer, C. (2018). Enhancing the antimicrobial and antibiofilm effectiveness of silver nanoparticles prepared by green synthesis. Journal of Materials Chemistry B, 6(24): 4124–4138.

Omerović, N., Djisalov, M., Živojević, K., Mladenović, M., Vunduk, J., Milenković, I. et al. (2021). Antimicrobial nanoparticles and biodegradable polymer composites for active food packaging applications. Comprehensive Reviews in Food Science and Food Safety, 20(3): 2428–2454.

Onyeaka, H., Passaretti, P., Miri, T. and Al-Sharify, Z. T. (2022). The safety of nanomaterials in food production and packaging. Current Research in Food Science, 5: 763–774.

Orasugh, J. T., Saha, N. R., Rana, D., Sarkar, G., Mollick, M. M. R., Chattoapadhyay, A. et al. (2018). Jute cellulose nano-fibrils/hydroxypropylmethylcellulose nanocomposite: A novel material with potential for application in packaging and transdermal drug delivery system. Industrial Crops and Products, 112: 633–643.

Orsuwan, A., Shankar, S., Wang, L.-F., Sothornvit, R. and Rhim, J.-W. (2016). Preparation of antimicrobial agar/banana powder blend films reinforced with silver nanoparticles. Food Hydrocolloids, 60: 476–485.

Otoni, C. G., Avena-Bustillos, R. J., Azeredo, H. M. C., Lorevice, M. V., Moura, M. R., Mattoso, L. H. C. et al. (2017). Recent advances on edible films based on fruits and vegetables—A review. Comprehensive Reviews in Food Science and Food Safety, 16(5): 1151–1169.

Pal, N., Banerjee, S., Roy, P. and Pal, K. (2021). Cellulose nanocrystalssilver nanoparticles-reduced graphene oxide based hybrid PVA nanocomposites and its antimicrobial properties. International Journal of Biological Macromolecules, 191: 445–456.

Pal, S., Tak, Y. K. and Song, J. M. (2007). Does the antibacterial activity of silver nanoparticles depend on the shape of the nanoparticle? A study of the gram-negative bacterium *Escherichia coli*. Applied and Environmental Microbiology, 73(6): 1712–1720.

Palamanit, A., Prachayawarakorn, S., Tungtrakul, P. and Soponronnarit, S. (2016). Performance evaluation of top-spray fluidized bed coating for healthy coated rice production. Food and Bioprocess Technology, 9(8): 1317–1326.

Pan, H., Yu, T., Zheng, Y., Ma, H., Shan, J., Yi, X. et al. (2024). Isolation, characteristics, and poly (butylene adipate-co-terephthalate)(PBAT) degradation mechanism of a marine bacteria Roseibium aggregatum ZY-1. Marine Pollution Bulletin, 201: 116261.

Panpaliya, N. P., Dahake, P. T., Kale, Y. J., Dadpe, M. V., Kendre, S. B., Siddiqi, A. G. et al. (2019). In vitro evaluation of antimicrobial property of silver nanoparticles and chlorhexidine against five different oral pathogenic bacteria. The Saudi Dental Journal, 31(1): 76–83.

Park, S. B., Lih, E., Park, K. S., Joung, Y. K. and Han, D. K. (2017). Biopolymer-based functional composites for medical applications. Progress in Polymer Science, 68: 77–105.

Patti, A. and Acierno, D. (2022). Towards the sustainability of the plastic industry through biopolymers: properties and potential applications to the textiles world. Polymers, 14(4): 692.

Peelman, N., Ragaert, P., De Meulenaer, B., Adons, D., Peeters, R., Cardon, L. et al. (2013). Application of bioplastics for food packaging. Trends in Food Science & Technology, 32(2): 128–141.

Plaza, H. (2015). Antimicrobial polymers with metal nanoparticles. International Journal of Molecular Sciences, 16(9): 2099–2116.

Pop, O. L., Pop, C. R., Dufrechou, M., Vodnar, D. C., Socaci, S. A., Dulf, F. V. et al. (2019). Edible films and coatings functionalization by probiotic incorporation: A review. Polymers, 12(12): 12.

Qahir, A., Khan, N., Hakeem, A. and Kamal, R. (2021). The antioxidant, antimicrobial, and clinical effects with elemental contents of pomegranate (*Punica granatum*) peel extracts: A review. Baghdad Journal of Biochemistry and Applied Biological Sciences, 2(01): 21–28.

Qian, L., Sun, Z. and Huang, Y. (2012). Antimicrobial active packaging film for food preservation. Packaging Technology and Science, 25(6): 293–302.

Qin, Y., Li, W. and Xiong, L. (2015). Development and characterization of novel starch/PBAT blend films for sustainable food packaging. Carbohydrate Polymers, 115: 144–151.

Quinteros, M. A., Cano Aristizabal, V., Dalmasso, P. R., Paraje, M. G. and Paez, P. L. (2016). Oxidative stress generation of silver nanoparticles in three bacterial genera and its relationship with the antimicrobial activity. Toxicology *in vitro*, 36: 216–223.

Rajeshkumar, G., Seshadri, S. A., Devnani, G. L., Sanjay, M. R., Siengchin, S., Maran, J. P. et al. (2021). Environment friendly, renewable and sustainable poly lactic acid (PLA) based natural fiber reinforced composites–A comprehensive review. Journal of Cleaner Production, 310: 127483.

Ramadan, M. A., Sharawy, S., Elbisi, M. K. and Ghosal, K. (2020). Eco-friendly packaging composite fabrics based on in situ synthesized silver nanoparticles (AgNPs) & treatment with chitosan and/or date seed extract. Nano-Structures & Nano-Objects, 22: 100425.

Ramalingam, B., Parandhaman, T. and Das, S. K. (2016). Antibacterial effects of biosynthesized silver nanoparticles on surface ultrastructure and nanomechanical properties of gram-negative bacteria viz. *Escherichia coli* and *Pseudomonas aeruginosa*. ACS Applied Materials & Interfaces, 8(7): 4963–4976.

Rashki, S., Asgarpour, K., Tarrahimofrad, H., Hashemipour, M., Ebrahimi, M. S., Fathizadeh, H. et al. (2021). Chitosan-based nanoparticles against bacterial infections. Carbohydrate Polymers, 251: 117108.

Raza, M. A., Kanwal, Z., Rauf, A., Sabri, A. N., Riaz, S. and Naseem, S. (2016). Size-and shape-dependent antibacterial studies of silver nanoparticles synthesized by wet chemical routes. Nanomaterials, 6(4): 74.

Regiel-Futyra, A., Kus-Liśkiewicz, M., Sebastian, V., Irusta, S., Arruebo, M., Kyzioł, A. et al. (2017). Development of noncytotoxic silver–chitosan nanocomposites for efficient control of biofilm forming microbes. RSC Advances, 7(86): 52398–52413.

Rhim, J. W., Wang, L. F. and Hong, S. I. (2013). Preparation and characterization of agar/silver nanoparticles composite films with antimicrobial activity. Food Hydrocolloids, 33(2): 327–335.

Rhoades, J. and Roller, S. (2000). Antimicrobial actions of degraded and native chitosan against spoilage organisms in laboratory media and foods. Applied and Environmental Microbiology, 66(1): 80-86.

Rodrigues, C., de Mello, J. M. M., Dalcanton, F., Macuvele, D. L. P., Padoin, N., Fiori, M. A. et al. (2020). Mechanical, thermal and antimicrobial properties of chitosan-based-nanocomposite with potential applications for food packaging. Journal of Polymers and the Environment, 28: 1216–1236.

Roy, S. and Rhim, J. W. (2019). Preparation of agar-based functional films incorporated with silver nanoparticles using a one-step green synthesis. Food Hydrocolloids, 87: 887–896.

Rydz, J., Musioł, M., Zawidlak-Węgrzyńska, B. and Sikorska, W. (2018). Present and future of biodegradable polymers for food packaging applications. Biopolymers for Food Design, 431–467.

Saifullah, M., Majeed, M. H., Shabbir, M. A. and Anwar, F. (2016). Green synthesis of silver nanoparticles and their incorporation into chitosan-based nanocomposite films for the extended shelf life of fresh produce. Journal of Food Science, 81(2): N348–N355.

Salama, A., Abouzeid, R. E., Owda, M. E., Cruz-Maya, I. and Guarino, V. (2021). Cellulose–silver composites materials: Preparation and applications. Biomolecules, 11(11): 1684.

Santos Junior, V. E., Targino, A. G. R., Flores, M. A. P., Rodríguez-Díaz, J. M., Teixeira, J. A., Heimer, M. V. et al. (2017). Antimicrobial activity of silver nanoparticle colloids of different sizes and shapes against *Streptococcus mutans*. Research on Chemical Intermediates, 43(10): 5889–5899.

Sarkar, P., Bosneaga, E. and Auer, M. (2009). Plant cell walls throughout evolution: Towards a molecular understanding of their design principles. Journal of Experimental Botany, 60(13): 3615–3635.

Shakibaie, M., Forootanfar, H., Golkari, Y., Mohammadi-Khorsand, T. and Shakibaie, M. R. (2015). Anti-biofilm activity of biogenic selenium nanoparticles and selenium dioxide against clinical isolates of *Staphylococcus aureus*, *Pseudomonas aeruginosa*, and *Proteus mirabilis*. Journal of Trace Elements in Medicine and Biology, 29: 235–241.

Shafey, A. M. E. (2020). Green synthesis of metal and metal oxide nanoparticles from plant leaf extracts and their applications: A review. Green Processing and Synthesis, 9(1): 304–339.

Shafizadeh, F. and Bradbury, A. G. W. (1979). Chemical composition of plant cell walls. Journal of Applied Polymer Science, 23(8): 1431–1442.

Shafizadeh, F. and Bradbury, A. G. W. J. (1979). Appl. Polym. Sci., 23: 1431–1442.

Shahidi, F. and Hossain, A. (2022). Preservation of aquatic food using edible films and coatings containing essential oils: A review. Critical Reviews in Food Science and Nutrition, 62(1): 66–105.

Shankar, S., Tanomrod, N., Rawdkuen, S. and Rhim, J. W. (2016). Preparation of pectin/silver nanoparticles composite films with UV-light barrier and properties. International Journal of Biological Macromolecules, 92: 842–849.

Shankar, S., Wang, L. F. and Rhim, J. W. (2015). Preparation and characterization of agar-based nanocomposite films reinforced with silver nanoparticles and clove essential oil for antimicrobial packaging applications. International Journal of Biological Macromolecules, 80: 3–12.

Sharma, R., Jafari, S. M. and Sharma, S. (2020). Antimicrobial bio-nanocomposites and their potential applications in food packaging. Food Control, 112: 107086.

Sharma, S., Sanpui, P., Chattopadhyay, A. and Ghosh, S. S. (2012). Fabrication of antimicrobial silver nanoparticle-sodium alginate-chitosan composite films. RSC Advances, 2(13): 5837–5843.

Shen, Y., Yang, D., Ye, R., Zhao, L. and Wang, C. (2017). Antimicrobial activity of silver nanoparticle-coated biodegradable films. Food Packaging and Shelf Life, 13: 1–9.

Silva, A. S., Costa, E. C., Reis, S., Spencer, C., Calhelha, R. C., Miguel, S. P. et al. (2022). Silk sericin: A promising sustainable biomaterial for biomedical and pharmaceutical applications. Polymers, 14(22): 4931.

Siracusa, V., Rocculi, P., Romani, S. and Rosa, M. D. (2008). Biodegradable polymers for food packaging: A review. Trends in Food Science & Technology, 19(12): 634–643.

Slavin, Y. N., Asnis, J., Hafeli, U. O. and Bach, H. (2017). Metal nanoparticles: Understanding the mechanisms behind antibacterial activity. Journal of Nanobiotechnology, 15(1): 65.

Song, D. H., Hoa, V. B., Kim, H. W., Khang, S. M., Cho, S. H., Ham, J. S. et al. (2021). Edible films on meat and meat products. Coatings, 11(11): 1344.

Song, R., Murphy, M., Li, C., Ting, K., Soo, C. and Zheng, Z. (2018). Current development of biodegradable polymeric materials for biomedical applications. Drug Design, Development and Therapy, 3117–3145.

Sorushanova, A., Delgado, L. M., Wu, Z., Shologu, N., Kshirsagar, A., Raghunath, R. et al. (2019). The collagen suprafamily: From biosynthesis to advanced biomaterial development. Advanced Materials, 31(1): 1801651.

Sotiriou, G. A. and Pratsinis, S. E. (2010). Antibacterial activity of nanosilver ions and particles. Environmental Science and Technology, 44(14): 5649–5654.

Souza, V. G. L., Fernando, A. L. and Pires, J. R. A. (2019). Biodegradable films based on biopolymers for food packaging applications: Challenges and future trends. Trends in Food Science & Technology, 91: 319–328.

Suhag, R., Kumar, N., Petkoska, A. T. and Upadhyay, A. (2020). Film formation and deposition methods of edible coating on food products: A review. Food Research International, 136: 109582.

Swetha, T. A., Bora, A., Mohanrasu, K., Balaji, P., Raja, R., Ponnuchamy, K. et al. (2023). A comprehensive review on polylactic acid (PLA)—Synthesis, processing, and application in food packaging. International Journal of Biological Macromolecules, 234: 123715.

Taib, N. A. A. B., Rahman, M. R., Huda, D., Kuok, K. K., Hamdan, S., Bakri, M. K. B. et al. (2023). A review on poly lactic acid (PLA) as a biodegradable polymer. Polymer Bulletin, 80(2): 1179–1213.

Tajeddin, B. (2014). Cellulose-based polymers for packaging applications. In Lignocellulosic Polymer Composites: Processing, Characterization, and Properties (pp. 477–498).

Tamizhdurai, P., Mangesh, V. L., Santhosh, S., Vedavalli, R., Kavitha, C., Bhutto, J. K. et al. (2024). A state-of-the-art review of multilayer packaging recycling: Challenges, alternatives, and outlook. Journal of Cleaner Production, 141403.

Tan, J., Abdel-Rahman, M. A. and Sonomoto, K. (2018). Biorefinery-based lactic acid fermentation: Microbial production of pure monomer product. Synthesis, Structure and Properties of Poly (lactic acid): 27–66.

Tang, S. and Zheng, J. (2018). Antibacterial activity of silver nanoparticles: structural effects. Advanced Healthcare Materials, 7(13): 1701503.

Tang, Z. X., Yu, Z., Zhang, Z. L., Zhang, X. Y., Pan, Q. Q. and Shi, L. E. (2013). Sonication-assisted preparation of CaO nanoparticles for antibacterial agents. Química Nova, 36: 933–936.

Tankhiwale, R. and Bajpai, S. K. (2010). Silver-nanoparticle-loaded chitosan lactate films with fair antibacterial properties. Journal of Applied Polymer Science, 115(4): 1894–1900.

Tavassoli-Kafrani, E., Shekarchizadeh, H. and Masoudpour-Behabadi, M. (2016). Development of edible films and coatings from alginates and carrageenans. Carbohydrate Polymers, 137: 360–374.

Tawakkal, I. S., Cran, M. J. and Bigger, S. W. (2014). Release kinetics of antimicrobial agents from agar-based films into a food simulant. International Journal of Biological Macromolecules, 69: 167–173.

Thakhiew, W., Devahastin, S. and Soponronnarit, S. (2013). Physical and mechanical properties of chitosan films as affected by drying methods and addition of antimicrobial agent. Journal of Food Engineering, 119(1): 140–149.

Toker, R. D., Kayaman-Apohan, N. L. H. A. N. and Kahraman, M. V. (2013). UV-curable nano-silver containing polyurethane based organic–inorganic hybrid coatings. Progress in Organic Coatings, 76(9): 1243–1250.

Thomas, V., Yallapu, M. M., Sreedhar, B. and Bajpai, S. (2007). A versatile strategy to fabricate hydrogel–silver nanocomposites and investigation of their antimicrobial activity. Journal of Colloid and Interface Science, 315(1): 389–395.

Tian, H., Guo, G., Fu, X., Yao, Y., Yuan, L. and Xiang, A. (2018). Fabrication, properties and applications of soy-protein-based materials: A review. International Journal of Biological Macromolecules, 120: 475–490.

Tong, R. and Cheng, J. (2009). Ring-opening polymerization-mediated controlled formulation of polylactide– drug nanoparticles. Journal of the American Chemical Society, 131(13): 4744–4754.

Tran, P. A. and Webster, T. J. (2013). Antimicrobial selenium nanoparticle coatings on polymeric medical devices. Nanotechnology, 24(15): 155101.

Travan, A., Pelillo, C., Donati, I., Marsich, E., Benincasa, M., Scarpa, T. et al. (2009). Non-cytotoxic silver nanoparticle-polysaccharide nanocomposites with antimicrobial activity. Biomacromolecules, 10(6): 1429–1435.

Ul-Islam, M., Khan, T., Khattak, W. A. and Park, J. K. (2013). Bacterial cellulose-MMTs nanoreinforced composite films: novel wound dressing material with antibacterial properties. Cellulose, 20: 589–596.

Ummi, A. S. and Siddiquee, S. (2019). Nanotechnology applications in food: Opportunities and challenges in food industry. In Nanotechnology Applications: Energy, Drug, and Food, 295–308.

Valencia-Chamorro, S. A., Palou, L., Del Rio, M. A. and Pérez-Gago, M. B. (2011). Antimicrobial edible films and coatings for fresh and minimally processed fruits and vegetables: A review. Critical Reviews in Food Science and Nutrition, 51(9): 872–900.

Varier, K. M., Gudeppu, M., Chinnasamy, A., Thangarajan, S., Balasubramanian, J., Li, Y. et al. (2019). Nanoparticles: antimicrobial applications and its prospects. Advanced nanostructured materials for environmental remediation, 321–355.

Varsha, C., Bajpai, S. K. and Navin, C. (2010). Investigation of water vapour permeation and antibacterial properties of nano silver loaded cellulose acetate film. International Food Research Journal, 17(2): 623–639.

Venkatesan, J., Singh, S. K., Anil, S., Kim, S.-K. and Shim, M. S. (2018). Preparation, characterization and biological applications of biosynthesized silver nanoparticles with chitosan-fucoidan coating. Molecules, 23(6): 1429.

Verma, P. and Maheshwari, S. K. (2019). Applications of Silver nanoparticles in diverse sectors. International Journal of Nano Dimension, 10(1): 18–36.

Vinod, A., Sanjay, M. R., Suchart, S. and Jyotishkumar, P. (2020). Renewable and sustainable biobased materials: An assessment on biofibers, biofilms, biopolymers and biocomposites. Journal of Cleaner Production, 120978.

Xiong, T., Dumat, C., Dappe, V., Vezin, H., Schreck, E., Shahid, M. et al. (2017). Copper oxide nanoparticle foliar uptake, phytotoxicity, and consequences for sustainable urban agriculture. Environmental Science & Technology, 51(9): 5242–5251.

Xu, C., Chen, W., Gao, H., Xie, X. and Chen, Y. (2020). Cellulose nanocrystal/silver (CNC/Ag) thin-film nanocomposite nanofiltration membranes with multifunctional properties. Environmental Science: Nano, 7(3): 803-816.

Yan, X., He, B., Liu, L., Qu, G., Shi, J., Hu, L. et al. (2018). Antibacterial mechanism of silver nanoparticles in *Pseudomonas aeruginosa*: Proteomics approach. Metallomics, 10(4): 557–564.

Yang, Y. K., He, C. E., He, W. J., Yu, L. J., Peng, R. G., Xie, X. L. et al. (2011). Reduction of silver nanoparticles onto graphene oxide nanosheets with N, N-dimethylformamide and SERS activities of GO/Ag composites. Journal of Nanoparticle Research, 13: 5571–5581.

Yazid, N. S. M., Abdullah, N., Muhammad, N. and Matias-Peralta, H. M. (2018). Application of starch and starch-based products in food industry. Journal of Science and Technology, 10(2).

Youssef, A. M. and Abdel-Aziz, M. S. (2013). Preparation of polystyrene nanocomposites based on silver nanoparticles using marine bacterium for packaging. Polymer-Plastics Technology and Engineering, 52(6): 607–613.

Youssef, A. M., Abdel-Aziz, M. S. and El-Sayed, S. M. (2014). Chitosan nanocomposite films based on Ag-NP and Au-NP biosynthesis by Bacillus subtilis as packaging materials. International Journal of Biological Macromolecules, 69: 185–191.

Yousuf, B., Qadri, O. S. and Srivastava, A. K. (2018). Recent developments in shelf-life extension of fresh-cut fruits and vegetables by application of different edible coatings: A review. Lwt, 89: 198–209.

Yousuf, B., Wu, S. and Siddiqui, M. W. (2021). Incorporating essential oils or compounds derived thereof into edible coatings: Effect on quality and shelf life of fresh/fresh-cut produce. Trends in Food Science & Technology, 108: 245–257.

Yuan, T., Zeng, J., Wang, B., Cheng, Z., Gao, W., Xu, J. et al. (2021). Silver nanoparticles immobilized on cellulose nanofibrils for starch-based nanocomposites with high antibacterial, biocompatible, and mechanical properties. Cellulose, 28: 855–869.

Zambrano-Zaragoza, M. L., González-Reza, R., Mendoza-Muñoz, N., Miranda-Linares, V., Bernal-Couoh, T. F., Mendoza-Elvira, S. et al. (2018). Nanosystem in edible coating: A novel strategy for food preservation. International Journal of Molecular Sciences, 19(3): 705.

Zhang, R., Moon, K.-S., Lin, W. and Wong, C. (2010). Preparation of highly conductive polymer nanocomposites by low temperature sintering of silver nanoparticles. J. Mater Chem., 20: 2018–23.

Zhang, H. and Chen, S. (2019). Nanoparticle-based methods for food safety evaluation. In Evaluation Technologies for Food Quality, 817–835.

Zhang, R., Moon, K.-S., Lin, W. and Wong, C. (2010). Preparation of highly conductive polymer nanocomposites by low temperature sintering of silver nanoparticles. Journal of Materials Chemistry, 20: 2018–2023.

Zhao, X., Liu, J., Li, J., Liang, X., Zhou, W. and Peng, S. (2022). Strategies and techniques for improving heat resistance and mechanical performances of poly (lactic acid)(PLA) biodegradable materials. International Journal of Biological Macromolecules, 218: 115–134.

Zhu, Y., Romain, C. and Williams, C. K. (2014). Sustainable polymers from renewable resources. Nature, 514(7520): 90–97.

Zou, L., Wang, J., Gao, Y., Ren, X., Rottenberg, M. E., Lu, J. et al. (2018). Synergistic antibacterial activity of silver with antibiotics correlating with the upregulation of the ROS production. Scientific Reports, 8(1): 11131.

Zubair, M., Pradhan, R. A., Arshad, M. and Ullah, A. (2021). Recent advances in lipid derived bio-based materials for food packaging applications. Macromolecular Materials and Engineering, 306(7): 2000799.

CHAPTER-17

Biopolymeric Nanoparticles for the Sustainable Removal of Pharmaceutical and Dye Wastes

Dhanam Jayam Gunaseelan,[1] *Mani Sivakumar,*[2] *Ganeshraja Ayyakannu Sundaram,*[3] *Vaishali C.V.,*[4] *Arun Alagarsamy*[1] and *Muneeswaran Thillaichidambaram*[5,*]

1. Introduction

Rapid globalization and industrialization have increased the rate of climate change which requires immediate attention. Both pollution and climate change have a deep impact on the Earth's ecosystem (Fletcher et al., 2024). The Earth is polluted with various newer chemicals coming from industries; most of these chemical wastes from industries were dumped into water sources with or without treatment that contaminate the water resources resulted in pure water scarcity. Water pollution is one of the most impactful environmental problems of industrialization and economic growth that create a threat to humans, living organisms reside and ecosystem (Hait et al., 2024; Khan et al., 2023b; Rehman et al., 2023; Sharma et al., 2024). Water is contaminated by several types of pollutants such as detergents, fertilizers, pesticides, dyes, pharmaceuticals, and heavy metals. However, industries such as pharmaceutical, textile, and cosmetics have been significant contributors of water pollution (Islam et al., 2023b; Jan et al., 2023).

The global annual production of pharmaceutical was over several hundred tons (Abdallah et al., 2024; Ciriminna et al., 2024). The increased production and use have resulted in a significant rise in pharmaceutical effluents. Annually, large amounts of pharmaceutical wastes which contains various contaminants such as solvent remains are discharged to the environment (Abdallah et al., 2024; Ahammad et al., 2023; Samal et al., 2022). Wastewater from the medicine manufacturing processes continues to discharge new Active Pharmaceutical Ingredients (APIs) into groundwater sources

[1] Department of Microbiology, Alagappa University Karaikudi – 630 003 Tamil Nadu, India.
[2] Department of Chemical Engineering, Faculty of Engineering, Mahidol University, Nakhon Pathom 73170, Thailand.
[3] Department of Research Analytics, Saveetha Dental College and Hospitals, Saveetha Institute of Medical and Technical Sciences, Poonamallee High Road, Chennai 600 077, Tamil Nadu, India.
[4] JSS Academy of Higher Education and Research, JSS Medical College, Sri Shivarathreeswara Nagara, Bannimantap, Mysuru, Karnataka 570015, India
[5] Lab of Toxicology, Department of Health Sciences, The Graduate School of Dong-A University, Busan 49315, Republic Korea.
* Corresponding author: tmuneeswaran@hotmail.com
DOI: 10.1201/9781003558330-17

such as freshwater bodies and urban water systems (Mazhandu and Mashifana, 2024; Sanusi et al., 2023; Sultana et al., 2023). This unregulated dumping of pharmaceutical wastes affects aquatic ecosystem and is harmful to living things (Benny et al., 2024). Non-steroidal anti-inflammatory drugs, antibiotics, hormones, antiepileptic drugs, antihistamines, blood lipid-lowering drugs, beta-blockers and steroidal antidepressants have been detected in water sources (Jayalakshmi et al., 2024). The overuse and improper disposal of these chemicals into the environment pose severe threats due to their highly persistent nature and toxicity (Singh et al., 2024).

Synthetic dyes are used in several industries including, textiles, tanning, food processing and pharmaceuticals (Islam et al., 2023b). They are acidic, basic, reactive, dispersive, azo-based, diazo-based, anthraquinone-based and complexed with metal. It was estimated that more than 10,000 dyes are used in dye industries that contribute to an approximate production of 700,000 tons annually (Rajendran et al., 2024; Tanaya et al., 2024). Moreover, during manufacturing or applications nearly 10–15% of these dyes were released into water bodies causing water pollution; most of these dyes are toxic, persistent and carcinogenic in nature (Dutta et al., 2024). It was found that 200,000 tons of such dyes were discharged from the textile industry alone every year and almost making its effluent polluted (Islam et al., 2023b). Further, these compounds are mutagenic, carcinogenic, and toxic; due their high colour intensity they can prevent light entering into water resulting in reduction of photosynthesis thus affects the aquatic life and human health (Dutta et al., 2024).

Removal of these pollutants from the environment is the need of the hour. Several water treatment methods such as ion exchange, ultrafiltration, reverse osmosis, oxidation, flocculation, and adsorption have been employed for the removal of these harmful pollutants (Khan et al., 2023a). Among them adsorption is commonly used to remove organic, anionic, or cationic pollutants (Ganiyu et al., 2023). The available methods are found to be inefficient due to increasing pollution which necessitates the development of new treatment techniques. More research is focused on sustainable development of biogenic materials as replacements for wastewater treatment. Traditional physical and chemical processes for treating dye wastewater are labourious, expensive, and less effective, making nanoparticles a promising solution due to their excellent surface properties and chemical reactivity.

Nanotechnology provides a variety of applications to treat polluted water that includes nanofibre filters, and carbon nanotubes. The smaller size and higher surface area of nanoparticles (NP) enhances adsorption and desorption processes (Zahmatkesh et al., 2023). Nanofiltration membranes, made of thin polymer films are used to reduce hardness, odour, colour, and heavy metal ions present in groundwater. In wastewater treatment, NP incorporated nanocomposite membranes are one of the innovative applications of nanotechnology (Rehman et al., 2023). However, the combination of NPs and materials like resin, bentonite, biopolymer, kaolin, rectorite, montmorillonite, biochar, and clinoptilolite have led to improved pollution degradation/removal efficiency (Anjum et al., 2024). Further, biopolymer nanocomposites are biocompatible with adjustable characteristics offer a promising, sustainable solution for the removal of dye/pharmaceutical wastes. Research of developing nanocomposites from environmentally friendly, renewable resources is essential for effective remediation and environmental protection (Nawaz et al., 2023; Rehman et al., 2023; Subash et al., 2023). This chapter reviews recent literature on using biopolymers, particularly cellulose, chitin, and chitosan-based products, for dye and pharmaceutical waste removal. The detailed information about the types of polymers used in different industries, characteristics features, advantages and their mode of removal are displayed in Figure 1.

2. Biopolymers

Biopolymers are extracted from living organisms, such as plants and microbes, rather than petroleum-based sources. They are classified as synthetic and natural polymers. Synthetic biopolymers are synthesized chemically or modified natural polymers; polylactic acid (PLA), polycaprolactone (PCL), and polyvinyl alcohol (PVA) are examples of synthetic polymers (Das et al.,

2023; Satchanska et al., 2024). Natural biopolymers are directly obtained from natural sources and cellulose, starch, chitosan, collagen, and alginate are some of the examples (Das et al., 2023; Elgarahy et al., 2023; Nawaz et al., 2023). Chitosan and cellulose are the most available natural biopolymers followed by starch. Chitin and chitosan are found in the exoskeleton of most organisms and extracted widely from the crustacean cell wastes, insects, cell wall of fungi and algae (Hisham et al., 2024). Cellulose is found in plants, fungi, bacteria, and algae (Kausar et al., 2023) and commonly extracted from agricultural wastes like pomelo peels, banana peels, corn stalks, and rice husks (Kausar et al., 2023; Ungureanu et al., 2023). While starch is largely stored as a reserved polymer in plant parts including leaves, fruits, stems, roots, and flowers and mostly extracted from stalks, roots, and crop seeds (Rehman et al., 2023). Biopolymers possess hydrophilic properties which helps to improve membrane resistance and water permeability in water treatment (Das et al., 2023; Elgarahy et al., 2023; Rehman et al., 2023). Hydrophilic functional groups form a thin water layer on the membrane's surface which prevents the deposition of foulants on the membrane. Also, the hydroxyl group of cellulose is responsible for the high degree of hydrophilicity, forms a hydrogen bond with water, thereby enhances water permeability (Kausar et al., 2023; Rehman et al., 2023). In addition, biopolymers are biodegradable and during decomposition often produce non-harmful by-products such as humus, carbon dioxide, biomass, and methane. Due to their biocompatibility and environmental benefits nanocomposites are considered for biomedical and water treatment applications (Al-Hazmi et al., 2024; Rehman et al., 2023; Sahu et al., 2023; Ungureanu et al., 2023). Natural polymers are effective in removing pollutants and cost-effective. Polysaccharides like chitin, chitosan, starch, cellulose are extensively studied for their ability to adsorb hydrocarbons, metals,

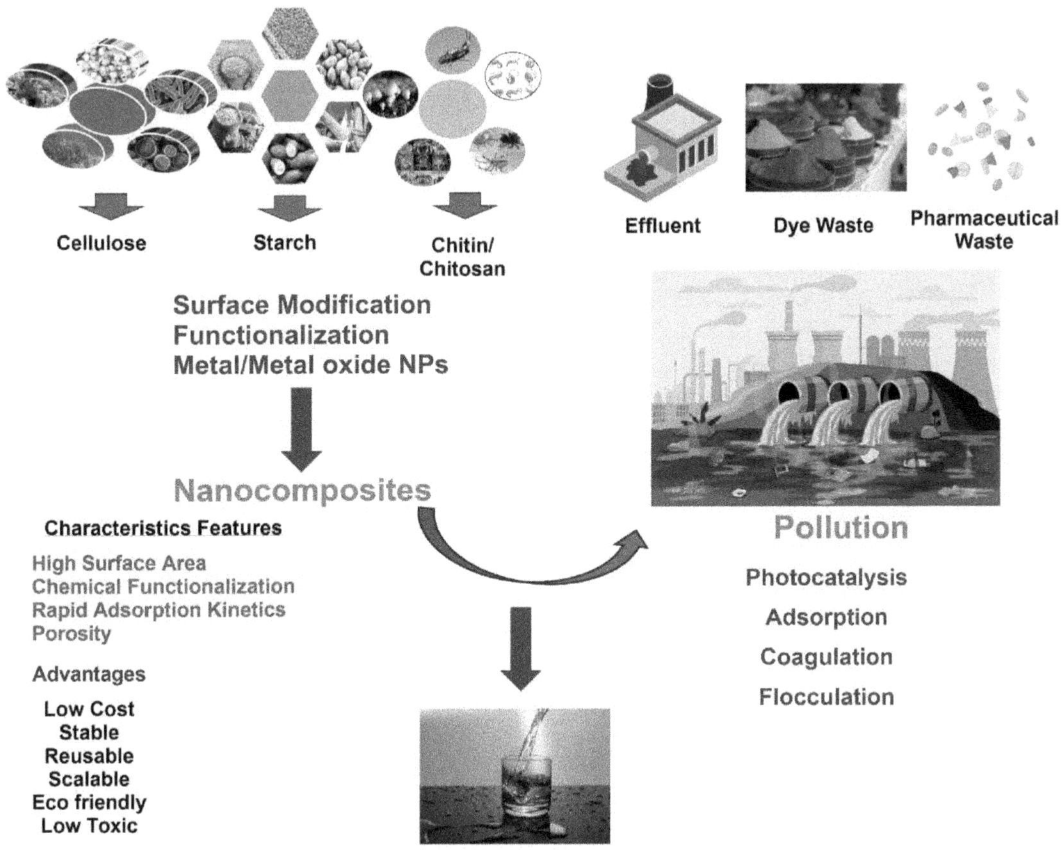

Figure 1. Schematic representation of the adsorptive removal of dyes, and pharmaceuticals using biopolymers such as starch, chitosan, and cellulose.

and organic pollutants (Ahmed and Mohamed, 2023). These biopolymers are also recognized for their eco-friendliness and high reactivity, which make them suitable for creating bio adsorbents.

3. Use of Biopolymers in Removal of Textile Wastes

Recently, the use of dyes in industries have increased rapidly and several researches were carried to find various methods to remove these dyes from water. Adsorption, Advanced Oxidation Processes (AOPs), coagulation-flocculation, electrochemical treatment, membrane filtration, and biological treatment are some of the recently developed technologies (Solayman et al., 2023). However, these technologies have major pitfalls such as labourious, costly, produce/utilize toxic materials/end products and complex (Radwan et al., 2023). Biopolymers such as chitosan, cellulose, and starch can be used as alternatives due to their tuneable properties (surface area, pore size) availability, cost-effective, biodegradability, and eco- friendly nature. Further, surface functionalization with hydroxyl (–OH), amine ($-NH_2$), carboxyl (–COO–), and amide ($-NHCOCH_3$) groups can enhance their compatibility, rigidity, flexibility, thermal and chemical stability (Khan et al., 2023a; Nawaz et al., 2023; Rehman et al., 2023). Figure 2 provides an overview of various types of dyes, their practical uses, and the biopolymer-based techniques used to remove them, as outlined by Al-Tohamy et al. (2022), Shao et al. (2021), Alfuhaid et al. (2023), and Rukhsar et al. (2024).

Adsorption is a surface-based process in which substance (adsorbate) accumulated on the solid surface (adsorbent) from its gaseous or aqueous environment (Rehman et al., 2023). Based on the interaction between the adsorbate and adsorbent adsorption can be divided into physisorption

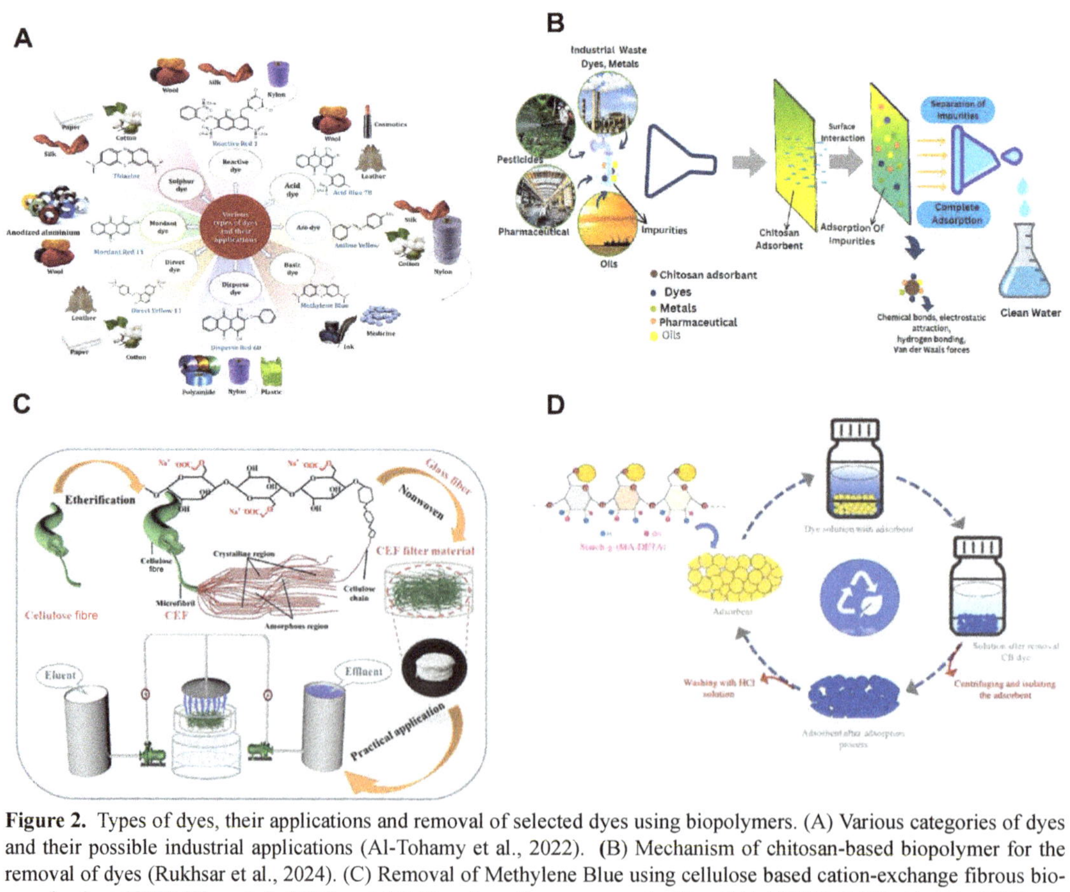

Figure 2. Types of dyes, their applications and removal of selected dyes using biopolymers. (A) Various categories of dyes and their possible industrial applications (Al-Tohamy et al., 2022). (B) Mechanism of chitosan-based biopolymer for the removal of dyes (Rukhsar et al., 2024). (C) Removal of Methylene Blue using cellulose based cation-exchange fibrous bio-adsorbent (CEF) (Shao et al., 2021) and (D) The adsorption process using St-g-(MA-DETA) (Alfuhaid et al., 2023).

and chemisorption (Agarwala and Mulky, 2023). Physisorption is reversible and weak; where weak intermolecular forces such as electrostatic, π-π, hydrophobic, dipole–dipole, Van der Waals interactions and hydrogen bonding is involved (Majeed et al., 2024; Salahshoori et al., 2024). Whereas, chemisorption is an irreversible process that involves strong chemical bonds such as covalent or ionic bonds (Agarwala and Mulky, 2023; Bensalah, 2024) between the adsorbate and adsorbent. Adsorption isotherms are used to describe adsorption efficiency of the adsorbent which relate the amount of substance adsorbed to its equilibrium concentration in solution (Mussa et al., 2023). Langmuir and Freundlich models are commonly used isotherms and other isotherms such as Tempkin, Dubinin–Radushkevich, and Brunauer–Emmett–Teller can also be used (Bensalah, 2024; Mussa et al., 2023). Adsorption technologies do not produce sludge, toxic substances, also they are simple, safe, efficient, and cost-effective which have favoured them for dye removal (Al-Gethami et al., 2024). Several factors such as the initial concentration of the adsorbate, pH, temperature, adsorbent properties, and contact time are found to influence adsorption efficiency (Khamis et al., 2024; Solayman et al., 2023). Cellulose, chitosan, and starch biopolymer-based adsorbents show promising results and some of the recent studies have highlighted their effectiveness in dye removal from aqueous solutions.

4. Cellulose-based Adsorbents

Cellulose/clay/sodium alginate (Na-Alg) composites prepared from paper mill waste removed up to 90% of Methylene Blue (MB) dye from water under optimal conditions (0.05 g dose, 30°C, pH 7 for cellulose-Na-Alg, and pH 11 for cellulose-Na-Alg-clay). The adsorption reaction was exothermic and spontaneous and followed pseudo-first order kinetics (Kausar et al., 2022). PEG-modified hydrogels (CS2/CMC2-PEG1.25) prepared from chitosan (CS) and carboxymethyl cellulose (CMC), crosslinked with epichlorohydrin and enhanced with polyethylene glycol (PEG) showed improved dye adsorption capacities that increased from 117.83 to 159.12 mg/g for Congo Red (CR) and from 110.2 to 136 mg/g for Methylene Blue (MB). Adsorption of CR was endothermic and followed pseudo-second-order while adsorption of MB was exothermic and followed pseudo-first-order. Both fitted in the Freundlich and Langmuir models, with CR showing a maximum capacity of 1053.88 mg/g and MB 331.72 mg/g (Zhu et al., 2023).

Inclusion of metal-organic framework with Cellulose Acetate (CA) membranes removed 98.2% of MB with 15 wt% PEG and 2 wt% MOFDPC NPs (Tahazadeh et al., 2021). A CO_2-responsive cellulose nanofibril aerogel created using poly (methacrylic acid-co-2-(dimethylamino) ethyl methacrylate) and carboxylated cellulose nanofibrils adsorbs anionic dyes such as MB, Naphthol Green B (NGB), and Methyl Orange (MO) via electrostatic interaction with maximum capacities of 598.8, 621.1 and 892.9 mg/g respectively, with equilibrium reached in 7–12 minutes (Yang et al., 2021). 3D network made of cellulose-based aerogel adsorbent (CPB) showed a high adsorption capacity of 1013.11 mg/g for MO at 25°C. and effectively removed anionic dyes from mixed systems due to its positive surface charge (Qiu et al., 2023).

pH-sensitive smart cellulose-derived nanofibrillar based hydrogel demonstrated excellent structural stability with rapid changes in swelling and surface charge in response to pH with adsorption capacities of 372.0 mg/g and 140.5 mg/g for Acridine Orange (AO) and MB (Jung et al., 2023). The addition of GO enhanced the dye adsorption capacity of composite material (GO/HEC/PGDE/Fe_3O_4) that showed maximum capacities of 277.78 mg/g, 233.33 mg/g, and 200.00 mg/g for MB, crystal violet (CV) and acid blue 25 respectively (Xu et al., 2023).

Poornachandhra et al. (2023) developed a nanocellulose-chitosan hydrogel bead (NCC–CH) that efficiently removed 94.75% crystal violet and 95.88% methylene blue due to its high surface area (25.77 m^2/g) under the optimal conditions of 0.5 g NCC–CH, 60 min, pH 9. Addition of polyethylene glycol to chitosan/carboxymethyl cellulose hydrogels increased the porosity and enhanced the adsorption of Congo Red (CR) about 159.12 mg/g (up from 117.83 mg/g) and MB about 136 mg/g (up from 110.2 mg/g). The adsorption was spontaneous and CR was endothermic, while MB was exothermic. CR fitted to both Freundlich and Langmuir models (max capacity 1053.88 mg/g) and MB fitted to Langmuir (max capacity 331.72 mg/g) (Zhu et al., 2023). Cellulose nanofibres incorporated with activated carbon and functionalized with EDTA showed a maximum of 99.14% MB removal under optimal conditions of pH 7, 2 g/L adsorbent, and 20 mg/L dye concentration for 180 minutes. The adsorption reaction followed a pseudo-second-order kinetic model and fitted best with the Langmuir isotherm (Qm = 60.61 mg/g) (Elmaghraby et al., 2023).

Cellulose nanofibre-based biocomposites (BCs) made with Poly 2-Acrylamide-2-Methyl-1-Propanesulfonic acid (PAMPS)-CNF and poly sodium 4-styrene sulphonate (PSSNa)-CNF had a high methylene blue (MB) adsorption capacities of 4536 mg/g and 11,930 mg/g respectively with electrostatic interactions (Oyarce et al., 2023). Cellulose nanocrystals (CNCs) achieved a maximum adsorption capacity of 746.24 mg/g in 100 minutes at pH 4. The reaction is chemisorption that followed the Langmuir model with pseudo-second-order kinetics with an adsorption energy of 17 kJ/mol and is endothermic as indicated by $\Delta H°$ (86.55 kJ/mol) and $\Delta S°$ (303.69 J/mol/K). The chemisorption involved π-π, hydrogen bonding, and electrostatic forces to form monolayer of dye (El-Fattah et al., 2024).

Biodegradable cellulose sheets from water hyacinths demonstrated high efficiency in adsorbing reactive red with a maximumcapacity of 71.43 mg/g (Sankar Santhosh et al., 2024). Magnetic Fe_3O_4 cellulose/graphene oxide nanocomposite demonstrated a removal efficiency of 97% for MB dye at pH 7, 7-h, and 0.0375 g of adsorbent in a 30 mg/L dye solution (Hegazy et al., 2024). Nanocellulose-based composed of 60% cellulose nanocrystals (CNCs), 15% carboxymethyl cellulose (CMC), zeolite, and citric acid removed 79.9% of MB dye with adsorption capacity of 38.3 mg/g at pH 7 via chemisorption. While at alkaline pH, 94.5% dye removed with 76.5 mg/g adsorption capacity (Ibrahim et al., 2024). CNC extracted from banana rachis fibres modified with charcoal (charcoal-CNC) composites effectively removed methyl orange and five heavy metals (Ni (II), Pb (II), Cr (III), Cd (II), Co (II)) from wastewater (Rahman et al., 2024).

Porous and flexible polyurethane (PU) mixed with Cellulose Nano-Whisker (CNW) activated carbon (AC) effectively adsorbed MB and basic violet 16 (BV16) dyes through chemical and physical mechanisms. Factors such as contact time, pH, adsorbent dosage, and temperature affected the removal efficiency with maximum adsorption capacities of 71.4 mg/g for BV16 and 67.1 mg/g for MB (Eyni Gavabari et al., 2024). Kaolin clay functionalized with cellulose (extracted from peanut shells) (KA/CEL) showed optimal conditions of 25% cellulose, 0.06 g adsorbent dose, pH 10 for MB, pH 7 for CR, 45°C, and contact times of 12.5 minutes for MB and 20 minutes for CR with a monolayer capacity of 291.5 mg/g for MB and 130.7 mg/g for CR (Reghioua et al., 2024). Bionanocomposite hydrogels made from sodium carboxymethyl cellulose-dextran sulphate and silver nanoparticle-modified zeolite (CMC-DS-AgZ) achieved high adsorption capacities of 344.82 mg/g and 454.55 mg/g for basic red 46 (BR46) and MB respectively at pH 7 with a 1 g/L of adsorbent and 6 h of contact time. The adsorption followed pseudo-second-order kinetics and the Langmuir isotherm model ($R^2 > 0.99$) (Benhalima et al., 2024).

MXene, ZnS, and chitosan-cellulose composite material (MX/ZnS/CC) removed anionic dyes through both adsorption and photocatalytic degradation with high adsorption capacity (1.29 g/g) and synergistic removal capacity (5.63 g/g). Chitosan-cellulose acts as both an adsorbent and stabilizer, also assists in the generation of free radical for dye degradation. Whereas, MXene enhances the photocatalytic performance by forming a Schottky heterostructure with ZnS thus improving the separation of photocarriers. The composite removes nearly 100% of 50 mg/L dye, shows an effective electrostatic adsorption and photocatalytic degradation through $•O^{2-}$, h^+, and •OH radicals (Lin et al., 2024). 3D bead made from bacterial cellulose microfilaments reinforced with chitosan and melamine 2D plates had a maximum adsorption capacity of 3168 mg/g for CR at pH 5.5. The adsorption kinetics well-described by pseudo-second-order, pseudo-first-order, and Avrami models and the isotherm fitted with Redlich-Peterson, Koble-Carrigan, and Hill models (Mosaffa et al., 2024). ZIF-8/UiO-66/PDA@CA nanofibres made of polydopamine (PDA)-modified binary ZIF-8/UiO-66 (MOFs) were added with Cellulose Acetate (CA) had an adsorption capacity of 82 mg/g for methylene blue (MB) dye (Eze et al., 2024).

5. Chitin and Chitosan-based Adsorbents

Recently published reviews of (Tarique et al., 2023) and (Hevira et al., 2024) highlighted the importance of chitin and chitosan-based composites in the removal of dye. For instance, chitosan-treated nanocomposite achieved a maximum adsorption capacity of 172.17 mg/g at pH 7.0 for MO (Waliullah et al., 2023) and removed 87% of MB within 2 hours (Amor et al., 2024) and the addition of pectin increased to 98.67% (Mohrazi and Ghasemi-Fasaei, 2023). Chitin extracted from insect–ealworm (*Tenebrio molitor*) achieved maximum adsorption of 121.15 mg/g and 138.55 mg/g for anionic dyes RB5 and RY84 and 3.22 mg/g and 59.56 mg/g for cationic dyes BV10 and BR46 respectively (Jóźwiak et al., 2023) while decoration with dopamine achieved 196.6 mg/g of MB (Li et al., 2023). Chitosan-zinc oxide (CS/ZnO) fibres showed a higher adsorption capacity for cationic dyes MB over MO, and Rhodamine B under light conditions (Li et al., 2021). Chitosan nanocomposites embedded with TiO_2 and ZnO showed photocatalytic activity of methylene blue and methyl orange dyes with respective degradation rates of 91.22% for and 88.9% after 5 hours under solar light (Rahman et al., 2023). Chitosan cross-linked TiO_2 nanocomposites (CS-TiO_2) showed increased removal efficiency with increasing pH, and a maximum at 6 after 120 minutes of contact time at an optimal dosage of CS-TiO_2 was between 0.10 mg/L and 0.20 mg/L. The adsorption process fit pseudo-first order, pseudo-second order, and intraparticle diffusion models (Meena et al., 2025).

3D macro-meso-microporous structured high-strength aerogel made from chitin and chitosan effectively adsorbs Cu (II) and Congo Red (CR) from water with an adsorption capacity of 59.21 mg/g and 2074 mg/g, respectively. The electrostatic interactions and chelation for Cu (II) and a mix of interactions for CR involved in adsorption (Kuang et al., 2023). Modified heterocyclic-magnetite chitosan nanocomposite (AOC@MC), removed 98.9% of Alizarin Red (AR) at a pH of 3.0, adsorbent dosage of 0.03 g/L, and contact time of 50 minutes, with the adsorption capacity of ~ 162 mg/g. The reaction followed pseudo-second-order kinetic model and Langmuir isotherm was found to effectively represent the adsorption process (Abuessawy et al., 2024). Magnetic chitosan-graphene oxide nanocomposite decorated with copper ferrite (MCSGO), was found to be effective in the removal of Safranin O (SAF) and Indigo Carmine (IC) dyes from wastewater with adsorption capacities of 112.6 mg/g and 66.15 mg/g, respectively. Notably, the adsorption process was both endothermic and spontaneous, and indicating favorable thermodynamic behavior (Ahmed et al., 2023). Magnetic nanohybrid (MGO@TETA@MACS) crafted by modifying magnetite nanoparticles decorated with graphene oxide using triethylenetetramine (TETA) and maleated chitosan (MACS) also effectively removes methylene blue (MB) by following the pseudo-first-order model. The Langmuir isotherm indicated a maximum capacity of 247.37 mg/g and the adsorption process is exothermic and spontaneous (Alsohaimi et al., 2023).

Addition of bentonite with chitosan was found to improve the adsorption of an anionic dye (Reactive Red 195, RR) and a cationic dye (Crystal Violet, CV) with removal rates of 99.4 and 99.8% within 20 minutes at pH values of 3 and 5 respectively. The adsorption was spontaneous and endothermic and followed pseudo-second-order kinetics and fitted with the Langmuir model (Elashery et al., 2023). Chitin nanofibre-polydopamine composite materials (CNDA) demonstrated a maximum MB adsorption capacity of 196.6 mg/g at an initial concentration of 50 mg/L. The enhancement in the adsorption is due to the presence of polydopamine. Additionally, CNDA showed high removal capacities for metal ions like Fe^{3+}, Mn^{2+}, Cu^{2+}, and Ni^{2+} (Li et al., 2023). Chitin-based magnesium oxide biocomposite (CHt@MgO) was developed with high efficiency in removing anionic MO and the optimal conditions were pH of 6, an adsorbent dose of 2 g/L, and a contact time of 120 minutes. The adsorption followed Langmuir model and pseudo-first-order kinetics with maximum adsorption capacity of 252 mg/g at 60°C (Majdoubi et al., 2023).

A pH-sensitive smart hydrogel created by Carboxymethyl cellulose nanofibrils (CMCNFs) and cationic Chitin nanofibrils (ChNFs) using green citric acid as a cross-linking agent showed high dye removal capacities of 372.0 mg/g and 140.5 mg/g for anionic AO and cationic MB respectively. The changes in surface charge with pH facilitated desorption of contaminants facilitates the reusability of the hydrogel (Jung et al., 2023). Graphene oxide dots loaded chitin (GO-chitin) flakes showed 80% removal for MO and Remazol Brilliant Blue at acidic pH levels within 30 minutes. The adsorption capacity was ranged between 75 and 95% following pseudo-second-order kinetics and the Freundlich isotherm model (Mishra et al., 2024). Transparent and water-redispersible α-chitin nanocrystals (ChNCs) extracted using Electron-Beam Irradiation (EBI) for oxidation and degradation of chitin in its dried state was found to be an effective adsorbent, interacting with cationic heavy metals (Cu^{2+} and Fe^{3+}) and organic blue dye through electrostatic attraction (Lee et al., 2023).

Chitin, derived from shrimp shells using glycerol/citric acid showed maximum adsorption capacity of 29.69 ± 0.2 mg/g for the removal of the anionic dye Congo Red (CR) at pH 7, contact time of 130 minutes, and temperature of 50°C. The process was fitted with the Langmuir isotherm model and suggested a monolayer adsorption of CR on the surface of the chitin (Gharbi et al., 2023). Chitin derived from the molts of the mealworm (Tenebrio molitor) removed both anionic (Reactive Black 5 (RB5), Reactive Yellow 84 (RY84)) and cationic (Brilliant Violet 10 (BV10), Brilliant Red 46 (BR46)) dyes from aqueous solutions. The maximum sorption capacities achieved at pH 2-3 for the anionic dyes were 121.15 mg/g for RB5 and 138.55 mg/g for RY84 and at pH 6 for cationic dyes were 3.22 mg/g for BV10 and 59.56 mg/g for BR46. The equilibrium time for the sorption of anionic dyes were 240-300 minutes while 180-240 minutes for cationic dyes (Jóźwiak et al., 2023).

6. Starch

Starch based polymers hold a promising future as an adsorbent in removing environmental pollutions particularly in dye removal and the status, challenges, and future perspectives of starch-based adsorbents were reviewed elsewhere (Khoo et al., 2023a). For instance, eco-friendly superabsorbent hydrogel (SAH) films from Cassava Starch (CSt) and polyacrylic acid (PAA) effectively adsorbed MB with a maximum capacity of 1,044 mg/g with a removal efficiency of 95%. The adsorption process was spontaneous and endothermic, followed a pseudo-second-order and Langmuir model, interacted strongly with dye via electrostatic interactions and hydrogen bonding (Srikaew et al., 2023). Phosphorylated Starch (PS) showed maximum adsorption capacities of 1036 mg/g for MB, 14994 mg/g for Cu, and 1535 mg/g for safranin (SF) and the adsorption equilibrium was reached within 10 minutes. The adsorption followed Pseudo-Second Order (PSO) kinetic model and the isotherm was fitted best by the Freundlich and Redlich-Peterson models (Azeroual et al., 2023). The amino-modified starch showed high removal efficiency in removing CV (65.31–80.46%) and Pb^{2+} (67.44–80.33%) under optimal conditions (pH 5, 10 mg/dm^3, 25°C, 180 min). The adsorption of CV was exothermic followed Langmuir and Freundlich isotherms and pseudo-second order model (Karić et al., 2023). While developing nanocomposite (Starch/MIL88A/GO) with different starch (0.05, 0.1, and 0.2 g), metal-organic framework (MIL88A), and graphene oxide showed photocatalytic degradation of Acid Blue 92 (AB92) dye. The composite with the highest starch concentration showed highest decolourization efficiency of 98.5% at 10 mg of the composite with a zero-order kinetic model. The bio-nanocomposite was largely reusable with 95% of efficiency after five cycles (Mohammad Samianifard et al., 2024).

Graphene Oxide (GO)-based starch/Poly (Vinyl Alcohol) (PVA) nanocomposite contained 3 g of GO effectively removed MB and MO; exhibited maximum adsorption capacities of 382 mg/g and 293.3 mg/g respectively and followed pseudo-second-order kinetics and Langmuir isotherm model. The adsorption processes were exothermic, and spontaneous with ΔH° values of −16.37 kJ/mol for MB and −13.72 kJ/mol for MO (Priya et al., 2023). Copolymerization of starch with maleic anhydride-diethylenetriamine (st-g-(MA-DETA)) showed a significantly better celestine dye removal properties than native starch indicated that the addition of maleic anhydride-diethylenetriamine improved the removal of starch (Alfuhaid et al., 2023). Acylhydrazone-functionalized with low-cost starch had a porous structure with high functional groups that contributed high adsorption efficiency for Safranin O (SO), Malachite Green (MG), Cu^{2+}, and sulphides and their respective maximum adsorption capacities were 2237.50 mg/g, 2101.66 mg/g, 410.70 mg/g, and 483.02 mg/g at 298.15 K. The contaminants were chemically adsorbed on the surface forming monolayer and the adsorption was spontaneous and exothermic (Wen et al., 2024).

Magnetic Starch-Based Adsorbent (MSBA) made using starch-grafted acrylic acid and Fe_3O_4@SiO_2 particles had a removal efficiency of 90.31% Crystal Violet (CV) dye at 308 K temperature, 10 mg/L CV concentration, pH 9.0, 0.10 g adsorbent dosage, and 50 minutes of contact time. The maximum adsorption capacity was 49.10 mg/g and followed first-order kinetics according to the Langmuir isotherm model. The adsorption mechanism involved electrostatic interactions and hydrogen bonding (Tan et al., 2023). Whereas decoration of natural starch (Starch/MIL100/$CoFe_2O_4$) with metal-organic framework (MIL100) and cobalt ferrite ($CoFe_2O_4$) nanoparticles effectively removed tetracycline (TCN) and methylene blue (MB). The photocatalytic degradation of TCN was 91.2% for Starch/MIL100/$CoFe_2O_4$ and 18% for starch alone while degradation of MB was about 96.6% for composite and 12% for the starch. The presence of salts such as NaCl, $NaNO_3$, and Na_2SO_4 was found to decrease the removal efficiency of TCN by 20%. The hydroxyl radicals were identified as the primary active species and the process followed first-order kinetics (Rabeie and Mahmoodi, 2024).

Composite made from Porous Starch Aerogel (PSA) and organically modified Ca-montmorillonite (OMMT) effectively removed MO dye. The amount of organoclay was found to enhance the removal efficiency, with maximum adsorption capacity of > 344.7 mg/g. The MO formed a monolayer on the surface via chemisorption (Jemai et al., 2023). Starch-modified NiFe-Layered Double Hydroxide (LDH) composite showed maximum adsorption capacities (qmax) for piroxicam-20, reactive blue 19, and reactive orange 16 were 2840 mg/g, 149.47 mg/g, and 182.4 mg/g, respectively. Adsorption kinetics followed the Elovich model, while the photocatalytic degradation of reactive red 120 followed a pseudo-first-order kinetic model with 90% removal efficiency within 3 hours under visible light. The electrons and holes were found to be involved in the photocatalytic degradation (Bansal and Pal, 2023).

7. Use of Biopolymers in Removal of Pharmaceutical Wastes

7.1 Chitosan

Chitosan (Cs) is a biopolymer derived from chitin through the deacetylation found in crustaceans (such as lobsters, shrimp, and crabs) and fungi. Chitosan is of interest in waste removal due to its low cost, biodegradability, hydrophobicity, non-toxicity, and excellent biocompatibility (Ahmed and Mohamed, 2023; Christina et al., 2024). However, low solubility of chitosan limits its applications in several fields and the improvement of its solubility and modification of its properties can be done via grafting, cross-linking, and composites with various materials has gained significant interest among researchers (Abd El-Monaem et al., 2022a; Jiang et al., 2024). The modified chitosan has improved solubility, gelling properties, and functionality, making it suitable for various uses as biocatalysts, antioxidants, antibacterial agents, wound dressings, and food preservatives (Picos-Corrales et al., 2023; Wan Yusof et al., 2023). Further, chitosan-based composites have shown great potential for efficiently adsorbing pharmaceutical pollutants, including antibiotics due to their diverse functional groups and eco-friendly properties (Abd El-Monaem et al., 2022a; Kaczorowska and Bożejewicz, 2024). Recently (Gkika et al., 2023; Kaczorowska and Bożejewicz, 2024), have reviewed the advances of chitosan-based adsorbents for the effective removal of pharmaceutical contaminants.

Several researches have shown that chitosan and its nano derivatives are highly effective at removing water contaminants (Benettayeb et al., 2023; El-Araby et al., 2024). Chitosan contains hydroxyl and amino groups (Benettayeb et al., 2023) that makes them suitable for the adsorption of pollutants especially when combined with magnetic composites (Sharifi et al., 2023; Wang and You, 2023). It was shown that the effectiveness of chitosan depends on dosage, pH, temperature, and contact time (Benettayeb et al., 2023). Chitosan, extracted from mud-crab shells showed high adsorption capacities of 66.91 mg/g for rifampicin (RIF), 11.00 mg/g for streptomycin (STM), and 24.21 mg/g for ibuprofen (IBU) from water (Shahrin et al., 2023). Chitosan (CS)-biochar composite synthesised using agricultural waste (orange peel (OB) and spent coffee grounds (SCB)) maximum sulphamethoxazole (SMX) adsorption capacities of 3.49 mg/g (OB), 7.65 mg/g (SCB), 7.24 mg/g (CS-OB), and 14.73 mg/g (CS-SCB). The adsorption efficiency was depended on pH, dosage, time, temperature, and initial SMX concentration. Ion exchange, π bonding, hydrogen bonding, and pore-filling are the mechanisms involved in the removal of SMX from water (Son Tran et al., 2023). Similarly, aspirin was efficiently removed using non-toxic chitosan modified with Spent Tea Waste Extract (STWE) with a maximum adsorption capacity of 61.7 mg/g. The chitosan-STWE adsorbent showed stability and low crystallinity. The removal efficiency was found to increase with increased bed height and aspirin concentration, but decreased with higher flow rates (Nordin et al., 2023).

Chitosan combined with activated carbon showed high adsorption capacity in removing diclofenac from wastewater. Diclofenac was removed through chemisorption, formed strong interactions with chitosan/activated carbon beads. The adsorbents remained effective up to ten regeneration cycles (Dago-Serry et al., 2024). While, chitosan/polyvinyl alcohol dopped with 0.2% graphene oxide and reduced graphene oxide effectively removed valsartan and diclofenac at pH 10 and 55°C within 3 hours (Malesic-Eleftheriadou et al., 2023). Similarly, chitosan-capped ZnS quantum dot and nickel ferrite nanoparticle hybrid (CZS-xNF) with varying nickel ferrite contents (5–20 wt%) photocatalytically degrade tetracycline. The hybrids can be reusable for up to four cycles (Kaur et al., 2024). Combination of UiO-66 metal-organic frameworks and chitosan on nano-silica excels in adsorption and photodegradation of tetracycline at a pH of 4.3, hybrid dose of 4.0 mg/mL and 10 minutes of radiation due to its enhanced conductivity and reduced the band gap, which improve charge transport and light absorption (Deymeh et al., 2024).

Ternary composites of hydroxyapatite (HAP), chitosan (CT), and magnetite (MNP) at a ratio of 2:5:3 (HAP:CT:MNP) achieved high adsorption capacities 535.6 mg/g for norfloxacin, 833 mg/g for ibuprofen, 625 mg/g for methylprednisolone, and 714 mg/g for deflazacort (Nayak et al., 2023). Chitosan/agar and chitosan/agar/SiO_2 nanocomposites showed maximum removal rates of 97.85% for amoxicillin and 99% for naproxen at initial concentration of 20 mg/L, adsorbent dose of 0.05 g, contact time of 10 mins, and pH 6 (Moradi et al., 2024). Immobilization of $CuFe_2O_4$ on chitosan ($CuFe_2O_4$@Chitosan) efficiently photocatalyically degrade ciprofloxacin (94.6%), ampicillin (92%), and erythromycin (90.3%) from water (Mohammed et al., 2023).

The combination of microalgae biomass and chitosan were used to form Magnetite-Chitosan-*Chlorella vulgaris* (MCC) and Magnetite-Chitosan-*Arthrospira platensis* (MCA) that effectively remove tetracycline, ciprofloxacin, and amoxicillin from wastewater (Mirizadeh et al., 2024). The reusable magnetic zinc ferrite/chitosan-curdlan (ZNF/CHT-CRD) composite efficiently removes tetracycline from water via adsorption, with a maximum adsorption capacity of 371.42 mg/g at a pH of 6 and a dosage of 0.65 g/L (Valizadeh et al., 2023). H_2TiO_7 nanotubes (TiNTs) functionalized with 5%, chitosan showed 80.79% adsorption/photocatalytic removal of clonazepam in 4 hours which is oxidized through OH^- ions (Castillo et al., 2023). Composite chitosan beads with molecularly imprinted with iron (III) hydroxide had a maximum adsorption capacity of 103.93 mg/g for ibuprofen at pH 5 (Stachowiak et al., 2023). The Chitosan–Walnut shells (Ch–W) composite successfully adsorbed levofloxacin from water, with a maximum uptake of 7.43 mg/g at a dose of 0.5 g/L, 45 minutes contact time, and pH 7 (Mahmoud and Fawzy, 2023). An Activated Carbon-Chitosan

Bead (ACCB) efficiently adsorbs diclofenac (DIF) from water with optimal conditions of pH 6.0, dosage 1.5 mg/mL, and temperature 40°C (Kumar et al., 2022).

7.2 Cellulose

Cellulose is a biopolymer and has a structure of β-D-glucopyranose units linked by β-1,4-glucosidic bonds which is found in lignocellulosic biomass with the degree of polymerization ranging between 100 and > 10,000 (Islam et al., 2023a). The rigidity crystallinity of cellulose is due to the hydroxyl groups which form hydrogen bonds (Nordin et al., 2024) and the abundance of hydroxyl groups makes cellulose a key focus in research (Etale et al., 2023). To effectively adsorb wastes, cellulose should be small in dimension, porous, with many active sites (Kausar et al., 2023; Nordin et al., 2024). Cellulose can be modified to enhance the ability to remove pollutants that can be made directly via attachment of functional groups (e.g., sulphation, esterification) or indirectly through polymerization of monomers onto its chains (Emenike et al., 2023; Magalhães et al., 2023; T. K et al., 2023).

Eco-friendly nano-adsorbents made of modified Graphene Oxide (GO) functionalized with cellulose nanofibres (CNFs) effectively remove ciprofloxacin and ofloxacin from water. The optimal adsorption performance was observed with 20 wt% carboxylated GO in esterified CNFs (Komal et al., 2022). Cellulose-based nanocomposite functionalized with β-cyclodextrin (β-CD), cucurbit (CUR), and calixarene (CAL) showed improved removal of propranolol (PRO), levofloxacin (LEV), and metformin (MET). Among them, β-CD/Cellulose nanocomposites showed the highest adsorption capacity for both and forms Van der Waals and electrostatic forces with them (Taherpoor et al., 2023). Sonolysis degradation by using recoverable palladium/gold nanocomposites supported on microcrystalline cellulose (MCC). PdAuNPs@MCC was highly efficient in the degradation of paracetamol (99%) and sulphamethazine (98%) at 90 minutes (Al-Ahmed et al., 2023). Whereas, activated hydro char derived from bamboo was immobilized in sodium carboxymethyl cellulose showed maximum adsorption capacity of 48.12 mg/g to paracetamol. The beads can also reduce COD levels in paracetamol-spiked sewage water (Senthil Kumar et al., 2024).

Novel adsorbent (MGO-MCell nanocomposite) made from cellulose and graphene oxide nanosheets effectively removes 91% of ibuprofen and 82% of cephalexin under optimal conditions of pH 7, at 90 minutes, and ambient temperature with maximum capacities of 99.01 mg/g for cephalexin and 30.21 mg/g for ibuprofen (Nekooei et al., 2023). In another study, hybrid material (GOMCC) contains microcrystalline cellulose (MCC) and a graphene oxide (GO) nanoparticle showed maximum metformin (MFM) adsorption capacity of 132.10 mg/g. The main mechanism of the removal of MFM is attributed to π–π interactions, lone pair interactions, sulphur interactions, and hydrogen bonding (Balasubramani et al., 2023). Cellulose biopolymers combined with hydro char/magnetite had a significantly larger surface area, showed > 90% removal of Fluoxetine HCl in pH 7 (Farghal et al., 2023).

Naeimi et al. (2022) developed a novel eco-friendly bentonite/cellulose@lead oxide (BT-CL-PbO) bio-nanocomposite by bonding cellulose extracted from barley waste with natural bentonite through covalent bonds followed by immobilization with lead oxide nanoparticles. The BT-CL-PbO showed 2.3 times higher catalytic degradation of ciprofloxacin (CIP) with a rate constant of 0.1309 minutes^{-1} compared to bare PbO nanoparticles. The degradation of CIP depended on several factors such as initial pH, CIP concentration, nanocomposite dosage, and anions. The degradation followed a pseudo-first-order kinetic model, and levels of photoexcited holes (h^+) were important for degradation of CIP (Naeimi et al., 2022).

Floating beads developed with zwitterionic UiO-66/ZIF-8 binary MOF/polydopamine@cellulose acetate (UiO-66/ZIF-8/PDA@CA) with high specific surface area removed tetracycline (TC) with equilibrium time of 1 hour and a maximum capacity of 290.69 mg/g at 25°C. Thermodynamic studies showed that the adsorption was endothermic and spontaneous (Abd El-Monaem et al., 2022b). Similarly, eco-friendly iron oxide nanoparticles and their cellulose-based magnetic nanocomposite synthesised using the aqueous extract of *Verbascum thapsus* showed maximum removal efficiency of (92.01%) Iat pH of 7, CIP concentration of 15 mg/L, nanocomposite dosage of 20 mg/L, and a contact time of 40 minutes. Adsorption reaction followed a pseudo-first-order kinetic model at rate constant of 1.21/minutes, fits Langmuir isotherm, with a maximum adsorption capacity of 168.03 mg/g (Azizi, 2021). Magnetic cellulose adsorbents made of cellulose extracted from *Pennisetum sinese* Roxb was highly efficient in removing tetracycline (TC) from water. The adsorption of TC was found to involve several mechanisms, such as intra-particle diffusion, chemical ion exchange, hydrogen bonding, and electrostatic interactions at higher temperature and pH 7. Also, humic acid and ionic strength of the medium affected the efficiency of adsorption (Sun et al., 2021).

Cellulose nanofibres (CNFs) from bamboo loaded with iron (III) nanoparticles (Fe (III)@ CNFs) showed a high adsorption capacities of 294.12 mg/g, 232.56 mg/g and 500.00 mg/g across a wide pH range for TC, chlortetracycline (CTC), and oxytetracycline (OTC) respectively. The presence of oxygen-containing functional groups and Fe (III) active metal sites in the nanofibres were found to be responsible for the adsorption of these antibiotics. Surface complexation, hydrogen bonding, electrostatic interactions, and van der Waals forces, with surface complexation between Fe (III) are a plausible mechanism (Lu et al., 2021). Carboxymethyl cellulose (CMC) and genipin-crosslinked carboxyalkyl-chitosan are mixed with (GCC) with Sulphonated Graphene Oxide (SGO) to synthesis novel sponge-like CMC/SGO–GCC material with multiple active sites. The adsorption of sulphamethoxazole (SMX) and sulphapyridine (SPD) is spontaneous and dependent on pH. The increasing temperature from 298 to 318 K decreased the maximum adsorption capacities of SMX and SPD from 312.28 to 272.83 mg/g and from 161.89 to 146.56 mg/g, respectively (Liu et al., 2021). Soft foam-like UiO-66/polydopamine/bacterial cellulose (UiO-66/PDA/BC) adsorbent with three-dimensional network structure and high hydrophilicity showed outstanding adsorption capacities for aspirin (149 mg/g) and TC (184 mg/g). Combination of chemical adsorption and π-π interactions was found to be responsible for its high efficiency (Cui et al., 2020).

7.3 Starch

Starch is a low-cost polysaccharide made up of glucopyranose units with the formula of $(C_6H_{10}O_5)$ n that consists of two types of polymer chains such as 20–30% amylose (linear chain with α-1,4-glycosidic bonds) and 70–80% amylopectin (branched chain with α-1,4-glycosidic and α-1,6-glycosidic bonds). The crystalline structure of starch is due to amylopectin. Common sources of starch are corn, rice, potato, and cassava (Abd El-Ghany et al., 2023). Starch (S) modified NiFe-Layered Double Hydroxide (LDH) composites (starch/NiFe-LDH) with a high porous structure showed improved performance in removing piroxicam-20, with enhanced photocatalytic activity and adsorption (Abd El-Ghany et al., 2023; Bansal and Pal, 2023; Khoo et al., 2023b).

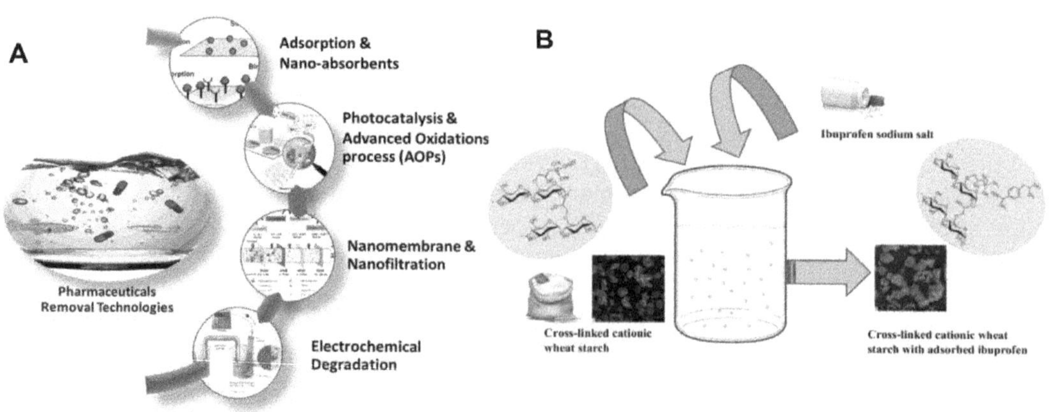

Figure 3. (A) Various technologies used for the removal of pharmaceutical wastes (Kumar et al., 2023) (B) synthesis of Cross-linked cationic wheat starches (CLCS) with quaternary ammonium groups for ibuprofen (IB) removal (Almonaityte et al., 2023).

Eco-friendly bio-nanocomposite (MMT/α-Fe_2O_3/ST) made from montmorillonite (MMT), α-Fe_2O_3, and starch (ST) achieved complete degradation of acetaminophen (ACT) (10 mg/L) in ~ 80 minutes under UVA-LED irradiation. The functional groups containing hydroxyl and superoxide radicals involved in the degradation and the presence of carbonate ions significantly inhibited ACT degradation (Pourmadadi et al., 2024). Cross-Linking of Cationic wheat Starches (CLCS) with varying Degrees of Substitution (DS) of quaternary ammonium groups, with and without the addition of CaO showed a maximum adsorption capacity of 550 mg/g for ibuprofen (IB). The electrostatic interactions between the quaternary ammonium groups and the carboxylate groups of IB were the primary driving force for adsorption. The addition of CaO improved the adsorptive properties of the composite thus improved the adsorption of IB (Almonaityte et al., 2023).

Whereas, crosslinked potato starch ester showed a maximum CIP adsorption capacity of 243.90 mg/g, that followed pseudo-first-order model and fitted with the Langmuir isotherm (Al-Jubory et al., 2023). Starch/zeolitic imidazole framework-67 (ZIF-67) modified magnetic montmorillonite nanocomposite effectively removed TC from aqueous solutions by following a pseudo-second-order model and fitted well with Langmuir isotherm. The maximum adsorption process is exothermic and spontaneous and showed maximum capacities of 40.24 mg/g for MMT, 66.1 mg/g for St/MMT-$MnFe_2O_4$, and 135.2 mg/g for the St/MMT-$MnFe_2O_4$-ZIF-67 composite (Sakhaei Niroumand et al., 2024). Corn starch nanoparticles (CSNP) effectively removed ibuprofen (IBU) and sulphamethoxazole (SUL) with respective removal efficiencies of 86.33% and 85.80%. The biosorption of IBU fitted well with Temkin isotherm model, while SUL fitted with the Langmuir isotherm model. Toxicological assessments using *Vigna mungo, Vigna radiata,* and zebrafish showed a significant toxicity reduction with increased the LC_{50} values for IBU and SUL after degradation process (Priyan and Narayanasamy, 2022).

8. Conclusion

In developing countries like India there is an alarming rate of pollution as well as demand for clean water the availability of sources is scarce which necessitates to advance technologies for effective elimination of emerging pollutants from water. The recent progress in eco-friendly approaches has led to the rise of development of "green" materials from natural sources. Usage of biopolymers have several pitfalls. However, incorporation of several adsorbents, functionalization, and surface modification of biopolymer and their composites has enhanced adsorption capacities, improved membrane performance, and efficiently remove pollutants. Further, theses composites have excellent mechanical properties, biocompatibility, and biodegradability, making them suitable candidates for wastewater treatment. These eco-friendly materials play a crucial role in the removal of various pollutants from wastewater, which emphasize the importance of developing innovative multifunctional biopolymer composites as an exciting opportunity of future research. However, these materials still create challenges such as scalability, cost-effectiveness, stability, and the regeneration of nanocomposites, as well as the disposal of used adsorbents. Further research and innovation in biopolymer-based nanocomposites might help in addressing these issues, pave the way for sustainable and effective water treatment systems. In addition, integration of biopolymer-based nanocomposites with advanced water treatment technologies, such as responsive materials and hybrid systems, provides new opportunities for energy-efficient and ecofriendly water purification.

References

Abd El-Ghany, N. A., Elella, M. H. A., Abdallah, H. M., Mostafa, M. S. and Samy, M. (2023). Recent advances in various starch formulation for wastewater purification via adsorption technique: A review. Journal of Polymers and the Environment, 31: 2792–2825.

Abd El-Monaem, E. M., Eltaweil, A. S., Elshishini, H. M., Hosny, M., Abou Alsoaud, M. M., Attia, N. F. et al. (2022a). Sustainable adsorptive removal of antibiotic residues by chitosan composites: An insight into current developments and future recommendations. Arabian Journal of Chemistry, 15: 103743.

Abd El-Monaem, E. M., Omer, A. M., Khalifa, R. E. and Eltaweil, A. S. (2022b). Floatable cellulose acetate beads embedded with flower-like zwitterionic binary MOF/PDA for efficient removal of tetracycline. Journal of Colloid and Interface Science, 620: 333–345.

Abdallah, M., Bethäuser, J., Tettenborn, F., Hein, A. and Hamann, M. (2024). Pharmaceutical consumption in human and veterinary medicine in Germany: Potential environmental challenges, 12.

Abuessawy, A., Fouda, A., Abdel-Rahman, A. A. H., Hawata, M. A. and Hamad, N. A. (2024). A new modified heterocyclic-magnetite chitosan nanocomposite for efficient alizarin red dye removal: Adsorption analysis and antibacterial activity. Journal of Polymers and the Environment, 32: 826–841.

Agarwala, R. and Mulky, L. (2023). Adsorption of Dyes from Wastewater: A Comprehensive Review, 10: 326–335.

Ahammad, N. A., Ahmad, M. A., Hameed, B. H. and Mohd Din, A. T. (2023). A mini review of recent progress in the removal of emerging contaminants from pharmaceutical waste using various adsorbents. Environmental Science and Pollution Research, 30: 124459–124473.

Ahmed, M. A., Ahmed, M. A. and Mohamed, A. A. (2023). Synthesis, characterization and application of chitosan/graphene oxide/copper ferrite nanocomposite for the adsorptive removal of anionic and cationic dyes from wastewater. RSC Advances, 13: 5337–5352.

Ahmed, M. A. and Mohamed, A. A. (2023). The use of chitosan-based composites for environmental remediation: A review. International Journal of Biological Macromolecules, 242: 124787.

Al-Ahmed, Z. A., Al-Hazmi, G. A. A., Munshi, A. M., Alamrani, N. A., Al-Qahtani, S. D., Habeebullah, T. M. et al. (2023). Recoverable palladium–gold nanocomposite based on microcrystalline cellulose for sono-catalytic degradation of pharmaceutical pollutants. Materials Chemistry and Physics, 296: 127219.

Al-Gethami, W., Qamar, M. A., Shariq, M., Alaghaz, A.-N. M. A., Farhan, A., Areshi, A. A. et al. 2024. Emerging environmentally friendly bio-based nanocomposites for the efficient removal of dyes and micropollutants from wastewater by adsorption: A comprehensive review. RSC Advances, 14: 2804–2834.

Al-Hazmi, H. E., Łuczak, J., Habibzadeh, S., Hasanin, M. S., Mohammadi, A., Esmaeili, A. et al. (2024). Polysaccharide nanocomposites in wastewater treatment: A review. Chemosphere, 347: 140578.

Al-Jubory, F. K., Abbas, A. S. and Mujtaba, I. M. (2023). Adsorptive removal of ciprofloxacin from simulated wastewater using crosslinked starch ester: Isotherms, kinetics, thermodynamics, modeling, and simulation for continuous operation. Chemical Engineering Research and Design, 200: 332–343.

Al-Tohamy, R., Ali, S. S., Li, F., Okasha, K. M., Mahmoud, Y. A. G., Elsamahy, T. et al. (2022). A critical review on the treatment of dye-containing wastewater: Ecotoxicological and health concerns of textile dyes and possible remediation approaches for environmental safety. Ecotoxicology and Environmental Safety, 231: 113160.

Alfuhaid, L., Al-Abbad, E., Alshammari, S., Alotaibi, A., Malek, N. and Al-Ghamdi, A. (2023). Preparation and Characterization of a Renewable Starch-g-(MA-DETA) Copolymer and Its Adjustment for Dye Removal Applications, 15: 1197.

Almonaityte, K., Bendoraitiene, J., Navikaite-Snipaitiene, V. and Rutkaite, R. (2023). Modified wheat starch granules for ibuprofen adsorption in aqueous solution. Water, Air, & Soil Pollution, 234: 447.

Alsohaimi, I. H., Alhumaimess, M. S., Alqadami, A. A., Hassan, H. M. A., Chen, Q., Alamri, M. S. et al. (2023). Chitosan-carboxylic acid grafted multifunctional magnetic nanocomposite as a novel adsorbent for effective removal of methylene blue dye from aqueous environment. Chemical Engineering Science, 280: 119017.

Amor, I. B., Hemmami, H., Laouini, S. E., Abdelaziz, A. G. and Barhoum, A. (2024). Influence of chitosan source and degree of deacetylation on antibacterial activity and adsorption of AZO dye from water. Biomass Conversion and Biorefinery, 14: 16245–16255.

Anjum, A., Gupta, D., Singh, B., Garg, R., Pani, B., Kashif, M. et al. 2024. Clay–polymer nanocomposites for effective water treatment: opportunities, challenges, and future prospects. Environmental Monitoring and Assessment 196: 666.

Azeroual, S., Wattati, H., Belfkira, A., Taourirte, M. and Jalal, R. (2023). Preparation and application of phosphorylated starch as a flocculant for cationic dyes and heavy metal. Colloids and Surfaces C: Environmental Aspects, 1: 100019.

Azizi, A. (2021). Green synthesis of iron oxide/cellulose magnetic recyclable nanocomposite and its evaluation in ciprofloxacin removal from aqueous solutions. Journal of the Iranian Chemical Society, 18: 331–341.

Balasubramani, K., Sivarajasekar, N., Sarojini, G. and Naushad, M. (2023). Removal of antidiabetic pharmaceutical (Metformin) using graphene oxide microcrystalline cellulose (GOMCC): Insights to process optimization, equilibrium, kinetics, and machine learning. Industrial & Engineering Chemistry Research, 62: 4713–4728.

Bansal, M. and Pal, B. (2023). Starch modified NiFe layered double hydroxide composites for better adsorption and photocatalytic removal of reactive dye and piroxicam-20 drug. Environmental Science and Pollution Research, 30: 73825–73848.

Benettayeb, A., Seihoub, F. Z., Pal, P., Ghosh, S., Usman, M., Chia, C. H. et al. (2023). Chitosan Nanoparticles as Potential Nano-Sorbent for Removal of Toxic Environmental Pollutants, 13: 447.

Benhalima, T., Sadi, A., Dairi, N. and Ferfera-Harrar, H. (2024). Multifunctional carboxymethyl cellulose-dextran sulfate/AgNPs@zeolite hydrogel beads for basic red 46 and methylene blue dyes removal and water disinfection control. Separation and Purification Technology, 342: 127001.

Benny, S. M., Gupta, S. D., Ismail, S. P. and Francis, D. (2024). Pharmaceutical Pollution of Water Bodies, Handbook of Water Pollution, pp. 371–416.

Bensalah, J. (2024). Removal of the textile dyes by a resin adsorbent polymeric: Insight into optimization, kinetics and isotherms adsorption phenomenally. Inorganic Chemistry Communications, 161: 111975.

Castillo, P. C. H.-D., Castro-Velázquez, V. and Rodríguez-González, V. (2023). Adsorption and photocatalytic-conjugated activity of a chitosan-functionalized titanate coating for the removal of the drug clonazepam from drinking water. Environmental Science and Pollution Research.

Christina, K., Subbiah, K., Arulraj, P., Krishnan, S. K. and Sathishkumar, P. (2024). A sustainable and eco-friendly approach for environmental and energy management using biopolymers chitosan, lignin and cellulose—A review. International Journal of Biological Macromolecules, 257: 128550.

Ciriminna, R., Della Pina, C., Luque, R. and Pagliaro, M. (2024). Reshoring fine chemical and pharmaceutical productions. Organic Process Research & Development, 28: 3026–3034.

Cui, J., Xu, X., Yang, L., Chen, C., Qian, J., Chen, X. et al. (2020). Soft foam-like UiO-66/Polydopamine/Bacterial cellulose composite for the removal of aspirin and tetracycline hydrochloride. Chemical Engineering Journal, 395: 125174.

Dago-Serry, Y., Maroulas, K. N., Tolkou, A. K., AbdelAll, N., Alodhayb, A. N., Khouqeer, G. A. et al. (2024). Composite super-adsorbents of chitosan/activated carbon for the removal of nonsteroidal anti-inflammatory drug from wastewaters. Journal of Molecular Structure, 1298: 137044.

Das, A., Ringu, T., Ghosh, S. and Pramanik, N. (2023). A comprehensive review on recent advances in preparation, physicochemical characterization, and bioengineering applications of biopolymers. Polymer Bulletin, 80: 7247–7312.

Deymeh, F., Ahmadpour, A., Allahresani, A. and Arami-Niya, A. (2024). Collaborative adsorption and photocatalytic degradation of high concentration pharmaceutical pollutants in water using a novel dendritic fibrous nano-silica modified with chitosan and UiO-66. International Journal of Biological Macromolecules, 275: 133534.

Dutta, S., Adhikary, S., Bhattacharya, S., Roy, D., Chatterjee, S., Chakraborty, A. et al. (2024). Contamination of textile dyes in aquatic environment: Adverse impacts on aquatic ecosystem and human health, and its management using bioremediation. Journal of Environmental Management, 353: 120103.

El-Araby, A., Janati, W., Ullah, R., Ercisli, S. and Errachidi, F. (2024). Chitosan, chitosan derivatives, and chitosan-based nanocomposites: Eco-Friendly Materials for Advanced Applications (a review). 11.

El-Fattah, W. A., Guesmi, A., Ben Hamadi, N., El-Desouky, M. G. and Shahat, A. (2024). A green synthesis of cellulose nanocrystals biosorbent for remediation of wastewater containing industrial dye. Colloids and Surfaces A: Physicochemical and Engineering Aspects 681: 132729.

Elashery, S. E. A., El-Bouraie, M. M., Abdelgawad, E. A., Attia, N. F. and Mohamed, G. G. (2023). Adsorptive performance of bentonite-chitosan nanocomposite as a dual antibacterial and reusable adsorbent for Reactive Red 195 and crystal violet removal: kinetic and thermodynamic studies. Biomass Conversion and Biorefinery.

Elgarahy, A. M., Eloffy, M. G., Guibal, E., Alghamdi, H. M. and Elwakeel, K. Z. (2023). Use of biopolymers in wastewater treatment: A brief review of current trends and prospects. Chinese Journal of Chemical Engineering, 64: 292–320.

Elmaghraby, N. A., Omer, A. M., Kenawy, E.-R., Gaber, M., Hassaan, M. A., Ragab, S. et al. (2023). Electrospun cellulose acetate/activated carbon composite modified by EDTA (rC/AC-EDTA) for efficient methylene blue dye removal. Scientific Reports 13: 9919.

Emenike, E. C., Iwuozor, K. O., Saliu, O. D., Ramontja, J. and Adeniyi, A. G. (2023). Advances in the extraction, classification, modification, emerging and advanced applications of crystalline cellulose: A review. Carbohydrate Polymer Technologies and Applications, 6: 100337.

Etale, A., Onyianta, A. J., Turner, S. R. and Eichhorn, S. J. (2023). Cellulose: A review of water interactions, applications in composites, and water treatment. Chemical Reviews, 123: 2016–2048.

Eyni Gavabari, S., Goudarzi, A., Shahrousvand, M. and Asfaram, A. (2024). Preparation of novel polyurethane/activated carbon/cellulose nano-whisker nanocomposite film as an efficient adsorbent for the removal of methylene blue and basic violet 16 dyes from wastewater. Separation and Purification Technology, 330: 125285.

Eze, E., Omer, A. M., Hassanin, A. H., Eltaweil, A. S. and El-Khouly, M. E. (2024). Cellulose acetate nanofiber modified with polydopamine polymerized MOFs for efficient removal of noxious organic dyes. Environmental Science and Pollution Research, 31: 29992–30008.

Farghal, H. H., Nebsen, M. and El-Sayed, M. M. H. (2023). Exploitation of expired cellulose biopolymers as hydrochars for capturing emerging contaminants from water. RSC Advances, 13: 19757–19769.

Fletcher, C., Ripple, W. J., Newsome, T., Barnard, P., Beamer, K., Behl, A. et al. (2024). Earth at risk: An urgent call to end the age of destruction and forge a just and sustainable future. PNAS Nexus 3.

Ganiyu, S. A., Suleiman, M. A., Al-Amrani, W. A., Usman, A. K. and Onaizi, S. A. (2023). Adsorptive removal of organic pollutants from contaminated waters using zeolitic imidazolate framework Composites: A comprehensive and Up-to-date review. Separation and Purification Technology, 318: 123765.

Gharbi, F. Z., Bougdah, N., Belhocine, Y., Sbei, N., Rahali, S., Damous, M. et al. (2023). Green and Fast Extraction of Chitin from Waste Shrimp Shells: Characterization and Application in the Removal of Congo Red Dye. 10: 599.

Gkika, D. A., Mitropoulos, A. C., Kokkinos, P., Lambropoulou, D. A., Kalavrouziotis, I. K., Bikiaris, D. N. et al. (2023). Modified chitosan adsorbents in pharmaceutical simulated wastewaters: A review of the last updates. Carbohydrate Polymer Technologies and Applications, 5: 100313.

Hait, M., Kashyap, N. K. and Bhardwaj, A. K. (2024). Emerging pollutants from urbanization and industrialization in the global south. pp. 39–87. *In*: Izah, S. C., Ogwu, M. C. and Hamidifar, H. (Eds.). Biomonitoring of Pollutants in the Global South. Springer Nature Singapore, Singapore.

Hegazy, S., Abdelwahab, N. A., Ramadan, A. M. and Mohamed, S. K. (2024). Magnetic Fe3O4-grafted cellulose/graphene oxide nanocomposite for methylene blue removal from aqueous solutions: Synthesis and characterization. Next Materials, 3: 100064.

Hevira, L., Ighalo, J. O. and Sondari, D. (2024). Chitosan-based polysaccharides for effective synthetic dye adsorption. Journal of Molecular Liquids, 393: 123604.

Hisham, F., Maziati Akmal, M. H., Ahmad, F., Ahmad, K. and Samat, N. (2024). Biopolymer chitosan: Potential sources, extraction methods, and emerging applications. Ain Shams Engineering Journal, 15: 102424.

Ibrahim, M. A., Salama, A., Zahran, F., Abdelfattah, M. S., Alsalme, A., Bechelany, M. et al. (2024). Fabrication of cellulose nanocrystals/carboxymethyl cellulose/zeolite Membranes for Methylene Blue Dye Removal: Understanding Factors, Adsorption Kinetics, and Thermodynamic Isotherms. 12.

Islam, M. D., Uddin, F. J., Rashid, T. U. and Shahruzzaman, M. (2023a). Cellulose acetate-based membrane for wastewater treatment—A state-of-the-art review. Materials Advances, 4: 4054–4102.

Islam, T., Repon, M. R., Islam, T., Sarwar, Z. and Rahman, M. M. (2023b). Impact of textile dyes on health and ecosystem: A review of structure, causes, and potential solutions. Environmental Science and Pollution Research, 30: 9207–9242.

Jan, S., Mishra, A. K., Bhat, M. A., Bhat, M. A. and Jan, A. T. (2023). Pollutants in aquatic system: A frontier perspective of emerging threat and strategies to solve the crisis for safe drinking water. Environmental Science and Pollution Research, 30: 113242–113279.

Jayalakshmi, K., Sangeetha, A. B., Sasikala, M., Selvi, C. and Paramasivam, M. (2024). Impact of pharmaceutical contaminants in the environment, environmental contaminants. Apple Academic Press, pp. 167–202.

Jemai, R., Djebbi, M. A., Boubakri, S., Ben Rhaiem, H. and Ben Haj Amara, A. (2023). Effective Removal of Methyl Orange Dyes Using an Adsorbent Prepared from Porous Starch Aerogel and Organoclay, 2: 209-229.

Jiang, R., Zhu, H.-Y., Zang, X., Fu, Y.-Q., Jiang, S.-T., Li, J.-B. et al. (2024). A review on chitosan/metal oxide nanocomposites for applications in environmental remediation. International Journal of Biological Macromolecules, 254: 127887.

Jóźwiak, T., Filipkowska, U., Bakuła, T., Bralewska-Piotrowicz, B., Karczmarczyk, K., Gierszewska, M. et al. (2023). The Use of Chitin from the Molts of Mealworm (*Tenebrio molitor*) for the Removal of Anionic and Cationic Dyes from Aqueous Solutions. 16: 545.

Jung, S., Kim, J., Bang, J., Jung, M., Park, S., Yun, H. and Kwak, H. W. (2023). pH-sensitive cellulose/chitin nanofibrillar hydrogel for dye pollutant removal. Carbohydrate Polymers, 317: 121090.

Kaczorowska, M. A. and Bożejewicz, D. (2024). The Application of Chitosan-Based Adsorbents for the Removal of Hazardous Pollutants from Aqueous Solutions—A Review, Sustainability.

Karić, N., Vukčević, M., Maletić, M., Dimitrijević, S., Ristić, M., Grujić, A. P. et al. (2023). Physico-chemical, structural, and adsorption properties of amino-modified starch derivatives for the removal of (in)organic pollutants from aqueous solutions. International Journal of Biological Macromolecules, 241: 124527.

Kaur, J., Renu, Kaur, M., Aggarwal, D., Kumar, V., Tikoo, K., Kaushik, A. et al. (2024). Unveiling the multifaceted applications of magnetically responsive chitosan capped ZnS QDs for sensing and annihilation of pharmaceutical drugs. Talanta, 266: 125084.

Kausar, A., Rehman, S. U., Khalid, F., Bonilla-Petriciolet, A., Mendoza-Castillo, D. I., Bhatti, H. N. et al. (2022). Cellulose, clay and sodium alginate composites for the removal of methylene blue dye: Experimental and DFT studies. International Journal of Biological Macromolecules, 209: 576–585.

Kausar, A., Zohra, S. T., Ijaz, S., Iqbal, M., Iqbal, J., Bibi, I. et al. (2023). Cellulose-based materials and their adsorptive removal efficiency for dyes: A review. International Journal of Biological Macromolecules, 224: 1337–1355.

Khamis, F., Hegab, H. M., Banat, F., Arafat, H. A. and Hasan, S. W. (2024). Comprehensive review on pH and temperature-responsive polymeric adsorbents: Mechanisms, equilibrium, kinetics, and thermodynamics of adsorption processes for heavy metals and organic dyes. Chemosphere, 349: 140801.

Khan, M. D., Singh, A., Khan, M. Z., Tabraiz, S. and Sheikh, J. (2023a). Current perspectives, recent advancements, and efficiencies of various dye-containing wastewater treatment technologies. Journal of Water Process Engineering, 53: 103579.

Khan, W. U., Ahmed, S., Dhoble, Y. and Madhav, S. (2023b). A critical review of hazardous waste generation from textile industries and associated ecological impacts. Journal of the Indian Chemical Society, 100: 100829.

Khoo, P. S., Ilyas, R. A., Uda, M. N. A., Hassan, S. A., Nordin, A. H., Norfarhana, A. S. et al. (2023a). Starch-Based Polymer Materials as Advanced Adsorbents for Sustainable Water Treatment: Current Status, Challenges, and Future Perspectives. 15: 3114.

Khoo, P. S., Ilyas, R. A., Uda, M. N. A., Hassan, S. A., Nordin, A. H., Norfarhana, A. S. et al. (2023b). Starch-Based Polymer Materials as Advanced Adsorbents for Sustainable Water Treatment: Current Status, Challenges, and Future Perspectives, Polymers.

Komal, Gupta, K., Nidhi, Kaushik, A. and Singhal, S. (2022). Amelioration of adsorptive efficacy by synergistic assemblage of functionalized graphene oxide with esterified cellulose nanofibers for mitigation of pharmaceutical waste. Journal of Hazardous Materials, 424: 127541.

Kuang, J., Cai, T., Dai, J., Yao, L., Liu, F., Liu, Y. et al. (2023). High strength chitin/chitosan-based aerogel with 3D hierarchically macro-meso-microporous structure for high-efficiency adsorption of Cu(II) ions and Congo red. International Journal of Biological Macromolecules, 230: 123238.

Kumar, A., Patra, C., Rajendran, H. K. and Narayanasamy, S. (2022). Activated carbon-chitosan based adsorbent for the efficient removal of the emerging contaminant diclofenac: Synthesis, Characterization and Phytotoxicity studies. Chemosphere, 307: 135806.

Kumar, S., Yadav, S., Kataria, N., Chauhan, A. K., Joshi, S., Gupta, R., Kumar, P. et al. (2023). Recent advancement in nanotechnology for the treatment of pharmaceutical wastewater: sources, toxicity, and remediation technology. Current Pollution Reports, 9: 110–142.

Lee, H., Heo, M. H., Jeong, H., Kim, S. Y., Yuk, J. S., Park, S. H. et al. (2023). Water-redispersible and high-yield α-chitin nanocrystals isolated using electron-beam irradiation as adsorbents to remove heavy metals and dye. Nanoscale, 15: 10990–11004.

Li, H., Hao, M.-X., Kang, H.-R. and Chu, L.-Q. (2021). Facile production of three-dimensional chitosan fiber embedded with zinc oxide as recoverable photocatalyst for organic dye degradation. International Journal of Biological Macromolecules, 181: 150–159.

Li, X., Xu, L., Gao, J., Yan, M., Bi, H. and Wang, Q. (2023). Surface modification of chitin nanofibers with dopamine as efficient nanosorbents for enhanced removal of dye pollution and metal ions. International Journal of Biological Macromolecules, 253: 127113.

Lin, J., Gao, D., Zeng, J., Li, Z., Wen, Z., Ke, F. et al. (2024). MXene/ZnS/chitosan-cellulose composite with Schottky heterostructure for efficient removal of anionic dyes by synergistic effect of adsorption and photocatalytic degradation. International Journal of Biological Macromolecules, 269: 131994.

Liu, Y., Nie, P. and Yu, F. (2021). Enhanced adsorption of sulfonamides by a novel carboxymethyl cellulose and chitosan-based composite with sulfonated graphene oxide. Bioresource Technology, 320: 124373.

Lu, L., Liu, M., Chen, Y. and Luo, Y. (2021). Effective removal of tetracycline antibiotics from wastewater using practically applicable iron(III)-loaded Cellulose Nanofibres, 8: 210336.

Magalhães, S., Fernandes, C., Pedrosa, J. F. S., Alves, L., Medronho, B., Ferreira, P. J. T. et al. (2023). Eco-Friendly Methods for Extraction and Modification of Cellulose: An Overview, 15: 3138.

Mahmoud, A. E. D. and Fawzy, M. (2023). Decontamination of levofloxacin from water using a novel chitosan–walnut shells composite: linear, nonlinear, and optimization modeling. Applied Water Science, 13: 244.

Majdoubi, H., Alqadami, A. A., Billah, R. E. K., Otero, M., Jeon, B.-H., Hannache, H. et al. (2023). Chitin-Based Magnesium Oxide Biocomposite for the Removal of Methyl Orange from Water. 20: 831.

Majeed, F., Razzaq, A., Rehmat, S., Azhar, I., Mohyuddin, A. and Rizvi, N. B. (2024). Enhanced dye sequestration with natural polysaccharides-based hydrogels: A review. Carbohydrate Polymers, 330: 121820.

Malesic-Eleftheriadou, N., Trikkaliotis, D. G., Evgenidou, E., Kyzas, G. Z., Bikiaris, D. N. and Lambropoulou, D. A. (2023). New biobased chitosan/polyvinyl alcohol/graphene oxide derivatives for the removal of pharmaceutical compounds from aqueous mixtures. Journal of Molecular Liquids, 387: 122673.

Mazhandu, Z. and Mashifana, T. (2024). Active pharmaceutical contaminants in drinking water: myth or fact? DARU Journal of Pharmaceutical Sciences.

Meena, H. M., Kukreti, S. and Jassal, P. S. (2025). Synthesis of a novel chitosan-TiO2 nanocomposite as an efficient adsorbent for the removal of methylene blue cationic dye from wastewater. Journal of Molecular Structure 1319: 139420.

Mirizadeh, S., Solisio, C., Converti, A. and Casazza, A. A. (2024). Efficient removal of tetracycline, ciprofloxacin, and amoxicillin by novel magnetic chitosan/microalgae biocomposites. Separation and Purification Technology 329: 125115.

Mishra, P. P., Das, P., Paul, S., Das, P., Manna, S., Basak, P. et al. (2024). Graphene oxide dots-loaded chitin flask: A Sustainable Adsorbent for Separating Multiple Dyes from Water. 34: e22249.

Mohammad Samianifard, S., Kalaee, M., Moradi, O., Mahmoodi, N. M. and Zaarei, D. (2024). Novel biocomposite (Starch/metal–organic framework/Graphene oxide): Synthesis, characterization and visible light assisted dye degradation. Journal of Photochemistry and Photobiology A: Chemistry, 450: 115417.

Mohammed, N. A. H., Shamma, R. N., Elagroudy, S. and Adewuyi, A. (2023). Copper ferrite immobilized on chitosan: A suitable photocatalyst for the removal of ciprofloxacin, ampicillin and erythromycin in aqueous solution. Catalysis Communications, 182: 106745.

Mohrazi, A. and Ghasemi-Fasaei, R. (2023). Removal of methylene blue dye from aqueous solution using an efficient chitosan-pectin bio-adsorbent: kinetics and isotherm studies. Environmental Monitoring and Assessment, 195: 339.

Moradi, O., Mhdavi, S. and Sedaghat, S. (2024). Synthesis of Nanocomposite Based Polysaccharide and Performance Evaluation for the Removal of Various Drug from Pharmaceuticals Wastewater. 9: e202305155.

Mosaffa, E., Patel, D., Ramsheh, N. A., Patel, R. I., Banerjee, A. and Ghafuri, H. (2024). Bacterial cellulose microfiber reinforced hollow chitosan beads decorated with cross-linked melamine plates for the removal of the Congo red. International Journal of Biological Macromolecules, 254: 127794.

Mussa, Z. H., Al-Ameer, L. R., Al-Qaim, F. F., Deyab, I. F., Kamyab, H. and Chelliapan, S. (2023). A comprehensive review on adsorption of methylene blue dye using leaf waste as a bio-sorbent: Isotherm adsorption, kinetics, and thermodynamics studies. Environmental Monitoring and Assessment, 195: 940.

Naeimi, A., Honarmand, M., Ali Chaji, M. and Khosravi, S. (2022). Green synthesis of bentonite/cellulose@lead oxide bio-nanocomposite with assistance of Pistacia Atlantica extract for efficient photocatalytic degradation of ciprofloxacin. Advanced Powder Technology, 33: 103441.

Nawaz, S., Tabassum, A., Muslim, S., Nasreen, T., Baradoke, A., Kim, T. H. et al. (2023). Effective assessment of biopolymer-based multifunctional sorbents for the remediation of environmentally hazardous contaminants from aqueous solutions. Chemosphere, 329: 138552.

Nayak, A., Bhushan, B. and Kotnala, S. (2023). Evaluation of hydroxyapatite-chitosan-magnetite nanocomposites for separation of pharmaceuticals from water: A mechanistic and comparative approach. Journal of Hazardous Materials Advances, 10: 100308.

Nekooei, A., Miroliaei, M. R., Shahabi-Nejad, M., Sheibani, G. and Sheibani, H. (2023). Cellulose-wrapped graphene oxide as efficient adsorbents for pharmaceutical contaminants. Inorganic Chemistry Communications, 154: 110997.

Nordin, A. H., Ngadi, N., Nordin, M. L., Noralidin, N. A., Nabgan, W., Osman, A. Y. et al. (2023). Spent tea waste extract as a green modifying agent of chitosan for aspirin adsorption: Fixed-bed column, modeling and toxicity studies. International Journal of Biological Macromolecules, 253: 126501.

Nordin, A. H., Yusoff, A. H., Husna, S. M. N., Noor, S. F. M., Norfarhana, A. S., Paiman, S. H. et al. (2024). Recent advances in nanocellulose-based adsorbent for sustainable removal of pharmaceutical contaminants from water bodies: A review. International Journal of Biological Macromolecules, 280: 135799.

Oyarce, E., Cantero-López, P., Roa, K., Boulett, A., Yáñez, O., Santander, P. et al. (2023). Removal of highly concentrated methylene blue dye by cellulose nanofiber biocomposites. International Journal of Biological Macromolecules, 238: 124045.

Picos-Corrales, L. A., Morales-Burgos, A. M., Ruelas-Leyva, J. P., Crini, G., García-Armenta, E., Jimenez-Lam, S. A. et al. (2023). Chitosan as an Outstanding Polysaccharide Improving Health-Commodities of Humans and Environmental Protection, Polymers.

Poornachandhra, C., Jayabalakrishnan, R. M., Prasanthrajan, M., Balasubramanian, G., Lakshmanan, A., Selvakumar, S. et al. (2023). Cellulose-based hydrogel for adsorptive removal of cationic dyes from aqueous solution: isotherms and kinetics. RSC Advances, 13: 4757and4774.

Pourmadadi, M., Aghababaei, N. and Abdouss, M. (2024). Performance and reaction mechanism of montmorillonite/α-Fe2O3/starch bio-nanocomposite as high-efficiency photocatalytic degradation of acetaminophen: Characterization, feasibility, and pathway. International Journal of Biological Macromolecules, 279: 135363.

Priya, V. S., Basha, S. K. and Kumari, V. S. (2023). Kinetics and adsorption performance of biosorbent starch/poly(vinyl alcohol)/graphene oxide nanocomposite for the removal of dyes. Journal of Umm Al-Qura University for Applied Sciences, 9: 529–547.

Priyan V, V. and Narayanasamy, S. (2022). Effective removal of pharmaceutical contaminants ibuprofen and sulfamethoxazole from water by Corn starch nanoparticles: An ecotoxicological assessment. Environmental Toxicology and Pharmacology, 94: 103930.

Qiu, C., Li, Y., Liu, H., Wang, X., Hu, S. and Qi, H. (2023). A novel crosslinking strategy on functional cellulose-based aerogel for effective and selective removal of dye. Chemical Engineering Journal, 463: 142404.

Rabeie, B. and Mahmoodi, N. M. (2024). Green and environmentally friendly architecture of starch-based ternary magnetic biocomposite (Starch/MIL100/CoFe2O4): Synthesis and photocatalytic degradation of tetracycline and dye. International Journal of Biological Macromolecules, 274: 133318.

Radwan, E. K., Abdel Ghafar, H. H., Ibrahim, M. B. M. and Moursy, A. S. (2023). Recent Trends in Treatment Technologies of Emerging Contaminants, 32: 7–25.
Rahman, O., Rahman, M. M. and Maniruzzaman, M. (2024). Removal of dye and heavy metals from industrial wastewater by activated charcoal-banana rachis cellulose nanocrystal composites filter. International Journal of Environmental Analytical Chemistry, 104: 1478–1496.
Rahman, T. U., Roy, H., Shoronika, A. Z., Fariha, A., Hasan, M., Islam, M. S. et al. (2023). Sustainable toxic dye removal and degradation from wastewater using novel chitosan-modified TiO2 and ZnO nanocomposites. Journal of Molecular Liquids, 388: 122764.
Rajendran, S., Kalairaj, A. and Senthilvelan, T. (2024). A comprehensive review on enzymatic decolorization of various azo dyes using laccase for the abatement of industrial pollution. Biomass Conversion and Biorefinery.
Reghioua, A., Atia, D., Hamidi, A., Jawad, A. H., Abdulhameed, A. S. and Mbuvi, H. M. (2024). Production of eco-friendly adsorbent of kaolin clay and cellulose extracted from peanut shells for removal of methylene blue and congo red removal dyes. International Journal of Biological Macromolecules, 263: 130304.
Rehman, M. U., Taj, M. B. and Carabineiro, S. A. C. (2023). Biogenic adsorbents for removal of drugs and dyes: A comprehensive review on properties, modification and applications. Chemosphere, 338: 139477.
Rukhsar, A., Iqbal, Z. F., Khan, M. S., Zainab, S. A., Nawaz, S., Kim, T. H. et al. (2024). Chitosan-Based Adsorbents and Catalysts for Removal of Toxic Pollutants from Water and Wastewater. Topics in Catalysis.
Sahu, A., Dosi, R., Kwiatkowski, C., Schmal, S. and Poler, J. C. (2023). Advanced Polymeric Nanocomposite Membranes for Water and Wastewater Treatment: A Comprehensive Review. 15: 540.
Sakhaei Niroumand, J., Peighambardoust, S. J. and Mohammadi, R. (2024). Tetracycline decontamination from aqueous media using nanocomposite adsorbent based on starch-containing magnetic montmorillonite modified by ZIF-67. International Journal of Biological Macromolecules, 259: 129263.
Salahshoori, I., Wang, Q., Nobre, M. A. L., Mohammadi, A. H., Dawi, E. A. and Khonakdar, H. A. (2024). Molecular simulation-based insights into dye pollutant adsorption: A perspective review. Advances in Colloid and Interface Science, 333: 103281.
Samal, K., Mahapatra, S. and Hibzur Ali, M. (2022). Pharmaceutical wastewater as Emerging Contaminants (EC): Treatment technologies, impact on environment and human health. Energy Nexus, 6: 100076.
Sankar Santhosh, A., Umesh, M., Kariyadan, S., Suresh, S., Salmen, S. H., Ali Alharb, S. et al. (2024). Fabrication of biopolymeric sheets using cellulose extracted from water hyacinth and its application studies for reactive red dye removal. Environmental Research, 240: 117466.
Sanusi, I. O., Olutona, G. O., Wawata, I. G. and Onohuean, H. (2023). Occurrence, environmental impact and fate of pharmaceuticals in groundwater and surface water: A critical review. Environmental Science and Pollution Research, 30: 90595–90614.
Satchanska, G., Davidova, S. and Petrov, P. D. (2024). Natural and Synthetic Polymers for Biomedical and Environmental Applications. 16: 1159.
Senthil Kumar, P., Shanmugapriya, M., Prasannamedha, G. and Rangasamy, G. (2024). Immobilization of hydrochar in cellulose beads for eradicating paracetamol from synthetic and sewage water. Environmental Pollution, 342: 123035.
Shahrin, E. W. E. S., Narudin, N. A. H., Shahri, N. N. M., Nur, M., Lim, J.-W., Bilad, M. R. et al. (2023). A comparative study of adsorption behavior of rifampicin, streptomycin, and ibuprofen contaminants from aqueous solutions onto chitosan: Dynamic interactions, kinetics, diffusions, and mechanisms. Emerging Contaminants, 9L 100199.
Shao, X., Wang, J., Liu, Z., Hu, N., Liu, M., Duan, C. et al. (2021). Cellulose based cation-exchange fiber as filtration material for the rapid removal of methylene blue from wastewater. Cellulose, 28: 9355–9367.
Sharifi, M. J., Nouralishahi, A. and Hallajisani, A. (2023). Fe3O4-chitosan nanocomposite as a magnetic biosorbent for removal of nickel and cobalt heavy metals from polluted water. International Journal of Biological Macromolecules, 248: 125984.
Sharma, K., Rajan, S. and Nayak, S. K. (2024). Chapter 1 - Water pollution: Primary sources and associated human health hazards with special emphasis on rural areas. pp. 3–14. *In*: Madhav, S., Srivastav, A. L., Chibueze Izah, S. and Hullebusch, E.v. (Eds.). Water Resources Management for Rural Development. Elsevier.
Singh, S., Sharma, P., Pal, N., Sarma, D. K. and Kumar, M. (2024). Antibiotic disposal challenges in India: investigating causes and effects. Environmental Monitoring and Assessment, 196: 325.
Solayman, H. M., Hossen, M. A., Abd Aziz, A., Yahya, N. Y., Leong, K. H., Sim, L. C. et al. (2023). Performance evaluation of dye wastewater treatment technologies: A review. Journal of Environmental Chemical Engineering, 11: 109610.
Son Tran, V., Hao Ngo, H., Guo, W., Ha Nguyen, T., Mai Ly Luong, T., Huan Nguyen, X. et al. (2023). New chitosan-biochar composite derived from agricultural waste for removing sulfamethoxazole antibiotics in water. Bioresource Technology, 385: 129384.
Srikaew, M., Jumpapaeng, P., Suwanakood, P., Kaiyasuan, C., Promarak, V. and Saengsuwan, S. (2023). Rapid synthesis and optimization of UV-photopolymerized cassava starch-based superabsorbent hydrogels as a

biodegradable, low-cost, and effective adsorbent for MB removal. Journal of Industrial and Engineering Chemistry, 118: 53–69.

Stachowiak, M., Cegłowski, M. and Kurczewska, J. (2023). Hybrid chitosan/molecularly imprinted polymer hydrogel beads doped with iron for selective ibuprofen adsorption. International Journal of Biological Macromolecules, 251: 126356.

Subash, A., Naebe, M., Wang, X. and Kandasubramanian, B. (2023). Biopolymer – A sustainable and efficacious material system for effluent removal. Journal of Hazardous Materials, 443: 130168.

Sultana, S., Sabir, M., Ullah, S., Ahmad, H. R. and Murtaza, G. (2023). Contamination of sewage water with active pharmaceutical ingredients: An emerging threat to food products and human health. pp. 193–231. *In*: Aftab, T. (Ed.). Emerging Contaminants and Plants: Interactions, Adaptations and Remediation Technologies. Springer International Publishing, Cham.

Sun, J., Cui, L., Gao, Y., He, Y., Liu, H. and Huang, Z., 2021. Environmental application of magnetic cellulose derived from *Pennisetum sinese* Roxb for efficient tetracycline removal. Carbohydrate Polymers, 251: 117004.

T. K, B.S., C. S, J.C., Sasi, S. and Arundhathi, C. K. (2023). Modification of cellulose. pp. 1–37. *In*: Thomas, S., Hosur, M., Pasquini, D. and Jose Chirayil, C. (Eds.). Handbook of Biomass. Springer Nature Singapore, Singapore.

Tahazadeh, S., Mohammadi, T., Tofighy, M. A., Khanlari, S., Karimi, H. and Motejadded Emrooz, H. B. (2021). Development of cellulose acetate/metal-organic framework derived porous carbon adsorptive membrane for dye removal applications. Journal of Membrane Science, 638: 119692.

Taherpoor, P., Farzad, F. and Zaboli, A. (2023). Performance evaluation and efficiency of functionalized cellulose nanocomposites for the removal of pharmaceutical contaminants from the aqueous environment. Journal of Molecular Liquids, 392: 123414.

Tan, Q., Jia, X., Dai, R., Chang, H., Woo, M. W. and Chen, H. (2023). Synthesis of a novel magnetically recyclable starch-based adsorbent for efficient adsorption of crystal violet dye. Separation and Purification Technology, 320: 124157.

Tanaya, K., Kumari, A., Singh, A. K. and Singh, D. (2024). Bioremediation: An Economical Approach for Treatment of Textile Dye Effluents. Water, Air, & Soil Pollution, 235: 516.

Tarique, J., Sapuan, S. M., Aqil, N. F., Farhan, A., Faiz, J. I. and Shahrizan, S. (2023). A comprehensive review based on chitin and chitosan composites. pp. 15–66. *In*: S. M, S., Ahmad, I. (Eds.). Composites from the Aquatic Environment. Springer Nature Singapore, Singapore.

Ungureanu, E. L., Mocanu, A. L., Stroe, C. A., Panciu, C. M., Berca, L., Sionel, R. M. et al. (2023). Agricultural Byproducts Used as Low-Cost Adsorbents for Removal of Potentially Toxic Elements from Wastewater: A Comprehensive Review. 15: 5999.

Valizadeh, K., Bateni, A., Sojoodi, N., Rafiei, R., Behroozi, A. H. and Maleki, A. (2023). Preparation and characterization of chitosan-curdlan composite magnetized by zinc ferrite for efficient adsorption of tetracycline antibiotics in water. International Journal of Biological Macromolecules, 235: 123826.

Waliullah, R. M., Rehan, A. I., Awual, M. E., Rasee, A. I., Sheikh, M. C., Salman, M. S. et al. (2023). Optimization of toxic dye removal from contaminated water using chitosan-grafted novel nanocomposite adsorbent. Journal of Molecular Liquids, 388: 122763.

Wan Yusof, W. R., Awang, N. Y. F., Azhar Laile, M. A., Azizi, J., Awang Husaini, A. A. S., Seeni, A. et al. (2023). Chemically modified water-soluble chitosan derivatives: Modification strategies, biological activities, and applications. Polymer-Plastics Technology and Materials, 62: 2182–2220.

Wang, M. and You, X.-y. (2023). Efficient adsorption of antibiotics and heavy metals from aqueous solution by structural designed PSSMA-functionalized-chitosan magnetic composite. Chemical Engineering Journal, 454: 140417.

Wen, Y., Xie, Z., Xue, S., Zhao, M., Liu, T. and Shi, W. (2024). Acylhydrazone-functionalized starch for efficient removal of hazardous dyes, heavy metal ion, and sulfides from wastewater: Adsorption behavior and mechanism analysis. International Journal of Biological Macromolecules, 279: 135461.

Xu, A., Gong, Y., Sun, Q., Li, L., Wang, F., Xiao, Z. et al. (2023). Recoverable cellulose composite adsorbents for anionic/cationic dyes removal. International Journal of Biological Macromolecules, 238: 124022.

Yang, L., Zhan, Y., Gong, Y., Ren, E., Lan, J., Guo, R. et al. (2021). Development of eco-friendly CO2-responsive cellulose nanofibril aerogels as "green" adsorbents for anionic dyes removal. Journal of Hazardous Materials, 405: 124194.

Zahmatkesh, S., Hajiaghaei-Keshteli, M., Bokhari, A., Sundaramurthy, S., Panneerselvam, B. and Rezakhani, Y. (2023). Wastewater treatment with nanomaterials for the future: A state-of-the-art review. Environmental Research, 216: 114652.

Zhu, H., Chen, S., Duan, H., He, J. and Luo, Y. (2023). Removal of anionic and cationic dyes using porous chitosan/carboxymethyl cellulose-PEG hydrogels: Optimization, adsorption kinetics, isotherm and thermodynamics studies. International Journal of Biological Macromolecules, 231: 123213faf.

Index

N

O

P

R

S

T

V

W

About the Editors

Ayat F. Hashim, Ph.D.

Dr. Ayat F. Hashim is currently working as a Researcher at the Fats and Oils-National Research Centre, Food Industries and Nutrition Institute, National Research Centre. She has been working on Nano Lipid Carriers and Polymer Nanocomposites as Effective Nutraceuticals Delivery Vehicles. She has published 36 articles and 12 book chapters in Elsevier and Springer Nature. Also, she has one Book Project entitled: Biopolymeric Nanoparticles for Agricultural Applications that has been formally accepted for publication and may be published in Springer's "Nanotechnology in the Life Sciences". She traveled to Londrina the State University of Londrina, Brazil, Department of Food Science & Technology where she engaged in a short training project in 2019 relevant to the preparation of edible coating film using natural biopolymers. Two years before that, she also traveled to the University of Wroclaw, Poland for another short-term training program in the field of nanoencapsulation of nutraceutical oils for functional food applications. She is a principal investigator for the project funded by the National Research Centre (NRC), Egypt (2023) entitled "Novel SMART Nanofibers as Visual Color Sensors for Edible Oil Deterioration". Also, she was the principal investigator for project number F/5853 funded by the International Foundation for Science (IFS), Sweden (2016). She is interested in the synthesis of nanomaterials, especially functional nanoemulsion, encapsulation, characterization, green synthesis, characterization, and application in pharmaceuticals, food, and agriculture.

Rabab M.A. El-Sherif

Prof. Dr. Rabab El-Sherif Dean of Faculty of Postgraduate Studies for Nanotechnology, Cairo University, Egypt. Professor of Physical Chemistry, Faculty of Science, Cairo University, Ex-Vice Dean of the faculty of Navigation Science and space technology- Beni Suif University. Fellow of National Defense College - Nasser Higher Military Academy. Ambassador of Egypt in NEF (Next Einstein Forum) in Africa. Member in the specialized council of cultural and knowledge, Academy of scientific research and technology, Higher ministry of education, Egypt. Member of the Royal Society of Chemistry, and Ex-president of EYAS (Egyptian Young Academy of Sciences). Expert in Electrochemistry, Fuel cells, Solar cells, Corrosion protection, Smart Coating, bio-implants, nano-composites, environmental studies, bio nanotechnology.

Farah K. Ahmed

Graduated from Biotechnology Program, Faculty of Agriculture, Cairo University and currently a Master student at Department of Biotechnology, Faculty of Nanotechnology Postgraduate, Cairo University.

Master Internship Faculty of Nanotechnology Postgraduate, Cairo University and Agricultural Genetic Engineering Research Institute. Working on a scientific project, abiotic stress (salinity) on different strains of rice by using molecular methods at the Agricultural Genetic Engineering Research Institute

[23/05/2023 – Current]. She has published 4 review articles and 15 book chapters in Elsevier and Springer Nature. Also, she has one Book project entitled: Biopolymeric Nanoparticles for Agricultural Applications has been formally accepted for publication and may be published in Springer's "Nanotechnology in the Life Sciences."

Sabu Thomas, PhD.

Prof. Sabu Thomas, former Vice Chancellor of Mahatma Gandhi University, is now a full-time professor of Polymer Science and Engineering at the School of Chemical Sciences of Mahatma Gandhi University. He is the Founder Director and Professor of the International and Interuniversity Centre for Nanoscience and Nanotechnology. Thomas is a leading figure in the field of nanoscience, polymer science, and engineering, with a focus on polymer nanocomposites, elastomers, polymer blends, interpenetrating polymer networks, polymer membranes, green composites, nanomedicine, and green nanotechnology. His groundbreaking inventions have significantly impacted the development of new materials for automotive, space, housing, and biomedical fields. Thomas has received numerous international awards, including the Fellowship of the Royal Society of Chemistry, Distinguished Professorship from Jozef Stefan Institute, and the National Education Leadership Award for Excellence in Education. He has published over 1400 peer-reviewed research papers, co-edited 220 books, and is the inventor of 15 patents.

For Product Safety Concerns and Information please contact our EU
representative GPSR@taylorandfrancis.com
Taylor & Francis Verlag GmbH, Kaufingerstraße 24, 80331 München, Germany

www.ingramcontent.com/pod-product-compliance
Ingram Content Group UK Ltd.
Pitfield, Milton Keynes, MK11 3LW, UK
UKHW052215190926
475702UK00006B/1034
9781032874883